ENGINEERING LIBRARY

T4-ABK-639

Understanding UMTS Radio Network Modelling, Planning and Automated Optimisation

Understanding UMTS Radio Network Modelling, Planning and Automated Optimisation

Theory and Practice

Edited by

Maciej J. Nawrocki
Wrocław University of Technology, Poland

Mischa Dohler
France Télécom R&D, France

A. Hamid Aghvami
King's College London, UK

John Wiley & Sons, Ltd

Copyright © 2006   John Wiley & Sons Ltd, The Atrium, Southern Gate, Chichester,
                        West Sussex PO19 8SQ, England

                        Telephone   (+44) 1243 779777

Email (for orders and customer service enquiries): cs-books@wiley.co.uk
Visit our Home Page on www.wiley.com

All Rights Reserved. No part of this publication may be reproduced, stored in a retrieval system or transmitted in any form or by any means, electronic, mechanical, photocopying, recording, scanning or otherwise, except under the terms of the Copyright, Designs and Patents Act 1988 or under the terms of a licence issued by the Copyright Licensing Agency Ltd, 90 Tottenham Court Road, London W1T 4LP, UK, without the permission in writing of the Publisher. Requests to the Publisher should be addressed to the Permissions Department, John Wiley & Sons Ltd, The Atrium, Southern Gate, Chichester, West Sussex PO19 8SQ, England, or emailed to permreq@wiley.co.uk, or faxed to (+44) 1243 770620.

Designations used by companies to distinguish their products are often claimed as trademarks. All brand names and product names used in this book are trade names, service marks, trademarks or registered trademarks of their respective owners. The Publisher is not associated with any product or vendor mentioned in this book.

This publication is designed to provide accurate and authoritative information in regard to the subject matter covered. It is sold on the understanding that the Publisher is not engaged in rendering professional services. If professional advice or other expert assistance is required, the services of a competent professional should be sought.

Other Wiley Editorial Offices

John Wiley & Sons Inc., 111 River Street, Hoboken, NJ 07030, USA

Jossey-Bass, 989 Market Street, San Francisco, CA 94103-1741, USA

Wiley-VCH Verlag GmbH, Boschstr. 12, D-69469 Weinheim, Germany

John Wiley & Sons Australia Ltd, 42 McDougall Street, Milton, Queensland 4064, Australia

John Wiley & Sons (Asia) Pte Ltd, 2 Clementi Loop #02-01, Jin Xing Distripark, Singapore 129809

John Wiley & Sons Canada Ltd, 22 Worcester Road, Etobicoke, Ontario, Canada M9W 1L1

Wiley also publishes its books in a variety of electronic formats. Some content that appears in print may not be available in electronic books.

British Library Cataloguing in Publication Data

A catalogue record for this book is available from the British Library

ISBN-13 978-0-470-01567-4 (HB)
ISBN-10 0-470-01567-5 (HB)

Typeset in 9/11pt Times by Integra Software Services Pvt. Ltd, Pondicherry, India.
Printed and bound in Great Britain by Antony Rowe Ltd, Chippenham, England.
This book is printed on acid-free paper responsibly manufactured from sustainable forestry in which at least two trees are planted for each one used for paper production.

Contents

Preface	xiii
Acknowledgments	xvii
List of Acronyms	xix
Notes on Editors and Contributors	xxix
PART I INTRODUCTION	1
1 Modern Approaches to Radio Network Modelling and Planning	3
Maciej J. Nawrocki, Mischa Dohler and A. Hamid Aghvami	
1.1 Historical aspects of radio network planning	3
1.2 Importance and limitations of modelling approaches	5
1.3 Manual versus automated planning	7
References	9
2 Introduction to the UTRA FDD Radio Interface	11
Peter Gould	
2.1 Introduction to CDMA-based networks	11
2.2 The UTRA FDD air interface	15
2.2.1 Spreading codes	15
2.2.2 Common physical channels	20
2.2.3 Dedicated physical channels	27
2.3 UTRA FDD key mechanisms	29
2.3.1 Cell breathing and soft capacity	29
2.3.2 Interference and power control	31
2.3.3 Soft handover and compressed mode	32
2.4 Parameters that require planning	34
2.4.1 Signal path parameters	34
2.4.2 Power allocation	35
2.4.3 System settings	35
References	35

3 Spectrum and Service Aspects 37
Maciej J. Grzybkowski, Ziemowit Neyman and Marcin Ney

3.1 Spectrum aspects 37
   3.1.1 Spectrum requirements for UMTS 38
   3.1.2 Spectrum identified for UMTS 39
   3.1.3 Frequency arrangements for the UMTS terrestrial component 39
   3.1.4 Operator spectrum demands 45
3.2 Service features and characteristics 46
References 52

4 Trends for the Near Future 55
Maciej J. Nawrocki, Mischa Dohler and A. Hamid Aghvami

4.1 Introduction 55
4.2 Systems yet to be deployed 56
   4.2.1 UTRA TDD 56
   4.2.2 TD-SCDMA 57
   4.2.3 Satellite segment 58
4.3 Enhanced coverage 60
   4.3.1 Ultra High Sites (UHS) 61
   4.3.2 High Altitude Platform System (HAPS) 61
4.4 Enhanced capacity 61
   4.4.1 Hierarchical Cell Structures (HCS) 61
   4.4.2 High Speed Downlink Packet Access (HSDPA) 62
   4.4.3 High Speed Uplink Packet Access (HSUPA) 63
   4.4.4 Orthogonal Frequency Division Modulation (OFDM) 64
4.5 Heterogeneous approaches 64
   4.5.1 Wireless LANs 64
   4.5.2 Wireless MANs (WiMAX) 65
4.6 Concluding Remarks 65
References 65

PART II MODELLING 67

5 Propagation Modelling 69
Kamil Staniec, Maciej J. Grzybkowski and Karsten Erlebach

5.1 Radio channels in wideband CDMA systems 69
   5.1.1 Electromagnetic wave propagation 69
   5.1.2 Wideband radio channel characterisation 73
   5.1.3 Introduction to deterministic methods in modelling WCDMA systems 75
   5.1.4 Deterministic methods: comparison of performance 79
5.2 Application of empirical and deterministic models in picocell planning 80
   5.2.1 Techniques for indoor modelling 80
   5.2.2 Techniques for outdoor-to-indoor modelling 82
5.3 Application of empirical and deterministic models in microcell planning 84
   5.3.1 COST 231 Walfisch–Ikegami model 85
   5.3.2 Manhattan model 87
   5.3.3 Other microcellular propagation models 88

5.4	Application of empirical and deterministic models in macrocell planning	90
	5.4.1 Modified Hata	90
	5.4.2 Other models	91
5.5	Propagation models of interfering signals	94
	5.5.1 ITU-R 1546 model	94
	5.5.2 ITU-R 452 model	100
	5.5.3 Statistics in the Modified Hata model	104
5.6	Radio propagation model calibration	105
	5.6.1 Tuning algorithms	106
	5.6.2 Single and multiple slope approaches	108
Appendix: Calculation of inverse complementary cumulative normal distribution function		110
References		111

6 Theoretical Models for UMTS Radio Networks 115
Hans-Florian Geerdes, Andreas Eisenblätter, Piotr M. Słobodzian, Mikio Iwamura, Mischa Dohler, Rafał Zdunek, Peter Gould and Maciej J. Nawrocki

6.1	Antenna modelling	115
	6.1.1 Mobile terminal antenna modelling	117
	6.1.2 Base station antenna modelling	118
6.2	Link level model	122
	6.2.1 Relation to other models	123
	6.2.2 Link level simulation chain	124
	6.2.3 Link level receiver components	126
	6.2.4 Link level receiver detectors	128
6.3	Capacity considerations	134
	6.3.1 Capacity of a single cell system	134
	6.3.2 Downlink power-limited capacity	134
	6.3.3 Uplink power-limited capacity	137
6.4	Static system level model	139
	6.4.1 Link level aspects	140
	6.4.2 Propagation data	141
	6.4.3 Equipment modelling	142
	6.4.4 Transmit powers and power control	144
	6.4.5 Services and user-specific properties	146
	6.4.6 Soft handover	147
	6.4.7 Complete model	148
	6.4.8 Applications of a static system-level network model	149
	6.4.9 Power control at cell level	152
	6.4.10 Equation system solving	157
6.5	Dynamic system level model	161
	6.5.1 Similarities and differences between static and dynamic models	161
	6.5.2 Generic system model	162
	6.5.3 Input/output parameters	164
	6.5.4 Mobility models	164
	6.5.5 Traffic models	165
	6.5.6 Path loss models	167
	6.5.7 Shadowing models	168

	6.5.8 Modelling of small scale fading	169
	6.5.9 SIR calculation	170
	References	172
7	**Business Modelling Goals and Methods** *Marcin Ney*	**177**
	7.1 Business modelling goals	177
	7.1.1 New business planning	177
	7.1.2 Infrastructure development	178
	7.1.3 Budgeting	179
	7.2 Business modelling methods	179
	7.2.1 Trends and statistical approach	180
	7.2.2 Benchmarking and drivers	181
	7.2.3 Detailed quantitative models	181
	7.2.4 Other non-quantitative methods	182
	References	183
	PART III PLANNING	**185**
8	**Fundamentals of Business Planning for Mobile Networks** *Marcin Ney*	**187**
	8.1 Process description	187
	8.1.1 Market analysis and forecasting	187
	8.1.2 Modelling the system	189
	8.1.3 Financial issues	190
	8.1.4 Recommendations	190
	8.2 Technical investment calculation	191
	8.2.1 CAPEX calculation methods	191
	8.2.2 OPEX calculation methods	196
	8.2.3 The role of drivers: Sanity checking	197
	8.3 Revenue and non-technical related investment calculation	198
	8.3.1 Input parameters and assumptions	198
	8.3.2 Revenue calculation methods	199
	8.3.3 Non-technical related investments	199
	8.4 Business planning results	199
	8.4.1 Business plan output parameters	200
	8.4.2 Business plan assessment methods	200
	References	201
9	**Fundamentals of Network Characteristics** *Maciej J. Nawrocki*	**203**
	9.1 Power characteristics estimation	203
	9.1.1 Distance to home base station dependency	203
	9.1.2 Traffic load dependency	207
	9.2 Network capacity considerations	210
	9.2.1 Irregular base station distribution grid	210
	9.2.2 Improper antenna azimuth arrangement	212
	9.3 Required minimum network size for calculations	214
	References	218

10 Fundamentals of Practical Radio Access Network Design — 219
Ziemowit Neyman and Mischa Dohler

- 10.1 Introduction — 219
- 10.2 Input parameters — 222
  - *10.2.1 Base station classification* — 222
  - *10.2.2 Hardware parameters* — 222
  - *10.2.3 Environmental specifics* — 229
  - *10.2.4 Technology essentials* — 231
- 10.3 Network dimensioning — 238
  - *10.3.1 Coverage versus capacity* — 238
  - *10.3.2 Cell coverage* — 239
  - *10.3.3 Cell Erlang capacity* — 249
- 10.4 Detailed network planning — 251
  - *10.4.1 Site-to-site distance and antenna height* — 252
  - *10.4.2 Site location* — 254
  - *10.4.3 Sectorisation* — 256
  - *10.4.4 Antenna and sector direction* — 259
  - *10.4.5 Electrical and mechanical tilt* — 260
  - *10.4.6 Temporal aspects in HCS* — 263
- References — 268

11 Compatibility of UMTS Systems — 271
Maciej J. Grzybkowski

- 11.1 Scenarios of interference — 272
  - *11.1.1 Interference between UMTS and other systems* — 272
  - *11.1.2 Intra-system interference* — 274
- 11.2 Approaches to compatibility calculations — 275
  - *11.2.1 Principles of compatibility calculations* — 275
  - *11.2.2 Minimum Coupling Loss (MCL) method* — 280
  - *11.2.3 Monte Carlo (MC) method* — 283
  - *11.2.4 Propagation models for compatibility calculations* — 284
  - *11.2.5 Characteristics of UTRA stations for the compatibility calculations* — 286
- 11.3 Internal electromagnetic compatibility — 286
- 11.4 External electromagnetic compatibility — 292
  - *11.4.1 UMTS TDD versus DECT WLL* — 292
  - *11.4.2 Compatibility between UMTS and Radio Astronomy Service* — 294
  - *11.4.3 Compatibility between UMTS and MMDS* — 295
- 11.5 International cross-border coordination — 296
  - *11.5.1 Principles of coordination* — 296
  - *11.5.2 Propagation models for coordination calculations* — 297
  - *11.5.3 Application of preferential frequencies* — 298
  - *11.5.4 Use of preferential codes* — 300
  - *11.5.5 Examples of coordination agreements* — 301
- References — 305

12 Network Design – Specialised Aspects — 309
Marcin Ney, Peter Gould and Karsten Erlebach

- 12.1 Network infrastructure sharing — 309
  - *12.1.1 Network sharing methods* — 309

	12.1.2 Legal aspects	313
	12.1.3 Drivers for sharing	314
12.2	Adjacent channel interference control	315
12.3	Fundamentals of Ultra High Site deployment	318
	References	320

PART IV   OPTIMISATION   321

13 Introduction to Optimisation of the UMTS Radio Network   323
Roni Abiri and Maciej J. Nawrocki

13.1	Automation of radio network optimisation	324
13.2	What should be optimised and why?	325
13.3	How do we benchmark the optimisation results?	326
	13.3.1 Location based information	327
	13.3.2 Sectors and network statistical data	328
	13.3.3 Cost and optimisation efforts	330
	References	331

14 Theory of Automated Network Optimisation   333
Alexander Gerdenitsch, Andreas Eisenblätter, Hans-Florian Geerdes, Roni Abiri, Michael Livschitz, Ziemowit Neyman and Maciej J. Nawrocki

14.1	Introduction	333
	14.1.1 From practice to optimisation models	334
	14.1.2 Optimisation techniques	335
14.2	Optimisation parameters for static models	339
	14.2.1 Site location and configuration	340
	14.2.2 Antenna related parameter	340
	14.2.3 CPICH power	344
14.3	Optimisation targets and objective function	345
	14.3.1 Coverage	345
	14.3.2 Capacity	346
	14.3.3 Soft handover areas and pilot pollution	347
	14.3.4 Cost of implementation	348
	14.3.5 Combination and further possibilities	348
	14.3.6 Additional practical and technical constraints	348
	14.3.7 Example of objective function properties	349
14.4	Network optimisation with evolutionary algorithms	354
	14.4.1 Genetic algorithms	355
	14.4.2 Evolution strategies	357
	14.4.3 Practical implementation of GA for tilt and CPICH	361
14.5	Optimisation without simulation	366
	14.5.1 Geometry-based configuration methods	366
	14.5.2 Coverage-driven approaches	368
	14.5.3 Advanced models	369
	14.5.4 Expected coupling matrices	372
14.6	Comparison and suitability of algorithms	373
	14.6.1 General strategies	374
	14.6.2 Discussion of methods	374
	14.6.3 Combination of methods	375
	References	375

15 Automatic Network Design — 379
Roni Abiri, Ziemowit Neyman, Andreas Eisenblätter and Hans-Florian Geerdes

- 15.1 The key challenges in UMTS network optimisation — 379
  - 15.1.1 Problem definition — 379
  - 15.1.2 Matching UMTS coverage to GSM — 380
  - 15.1.3 Supporting high bit rate data services — 381
  - 15.1.4 Handling dual technology networks — 382
- 15.2 Engineering case studies for network optimisation — 382
  - 15.2.1 Example network description — 383
  - 15.2.2 Pre-launched (unloaded) network optimisation — 383
  - 15.2.3 Loaded network optimisation — 389
- 15.3 Case study: optimising base station location and parameters — 395
  - 15.3.1 Data setting — 396
  - 15.3.2 Optimisation approach — 397
  - 15.3.3 Results — 399
  - 15.3.4 Conclusions — 402
- References — 403

16 Auto-tuning of RRM Parameters in UMTS Networks — 405
Zwi Altman, Hervé Dubreil, Ridha Nasri, Ouassim Ben Amor, Jean-Marc Picard, Vincent Diascorn and Maurice Clerc

- 16.1 Introduction — 405
- 16.2 Radio resource management for controlling network quality — 406
- 16.3 Auto-tuning of RRM parameters — 408
  - 16.3.1 Parameter selection for auto-tuning — 408
  - 16.3.2 Target selection for auto-tuning — 410
  - 16.3.3 Fuzzy logic controllers (FLC) — 410
  - 16.3.4 Case study: Auto-tuning of macrodiversity — 412
- 16.4 Optimisation strategies of the auto-tuning process — 415
  - 16.4.1 Off-line optimisation using Particle Swarm approach — 416
  - 16.4.2 On-line optimisation using reinforcement learning — 421
- 16.5 Conclusions — 425
- Acknowledgement — 425
- References — 425

17 UTRAN Transmission Infrastructure Planning and Optimisation — 427
Karsten Erlebach, Zbigniew Jóskiewicz and Marcin Ney

- 17.1 Introduction — 427
  - 17.1.1 Short UTRAN overview — 428
  - 17.1.2 Requirements for UTRAN transmission infrastructure — 428
- 17.2 Protocol solutions for UTRAN transmission infrastructure — 430
  - 17.2.1 Main considerations for ATM layer protocols in current 3G networks — 430
  - 17.2.2 MPLS-architecture for future 3G transmissions — 443
  - 17.2.3 The path to direct IP transmission networking — 444
- 17.3 End-to-end transmission dimensioning approach — 446
  - 17.3.1 Dimensioning of Node B throughput — 446
  - 17.3.2 Traffic dimensioning of the ATM network — 451
  - 17.3.3 Traffic dimensioning of the IP-Network — 452

17.4	Network solutions for UTRAN transmission infrastructure	456
	17.4.1 Leased lines	456
	17.4.2 Point-to-point systems	457
	17.4.3 Point-to-multipoint systems – LMDS	460
	17.4.4 WiMAX as a potential UTRAN backhaul solution	468
17.5	Efficient use of WiMAX in UTRAN	472
	17.5.1 Dimensioning of WiMAX for UTRAN infrastructure	472
	17.5.2 Current WiMAX limitations	473
17.6	Cost-effective radio solution for UTRAN infrastructure	474
	17.6.1 RF planning aspects	474
	17.6.2 Throughput dimensioning	475
	17.6.3 Methods of finding optimal LMDS network configurations	476
	17.6.4 Costs evaluation of UTRAN infrastructure – software example	485
	17.6.5 Example calculations and comparison of results	487
References		493

Concluding Remarks   **497**

Index   **501**

Preface

Yet another book on UMTS? Not quite!

Our prime goal is to encourage the readership to understand why certain things happen in the UMTS Radio Access Network and others do not, which parameters are strongly coupled and which are not and what the **analytical dependencies** are between them. Thus, we try to minimise explaining system performance only on a case-by-case basis, which is the general case for many related books on the market, but rather equip the readership with fairly generic mathematical tools which allow complex system performances and dependencies to be understood, analysed and, above all, optimised.

Also – 'automated' – a small additional word in the title of this book which makes the big difference: a difference to the scope of this book, a difference to the life of thousands of network optimisation engineers, a difference to everybody making use of wireless voice or data services in one way or another.

While the 3rd generation (3G) UMTS standard may seem an 'old hat' to the euphoric academic research community, the number of people trying to understand, deploy and hone this very sophisticated wireless communication system increases on a daily basis. They can only begin to grasp that, unlike the 2nd generation (2G) GSM standard, UMTS is indeed very flexible across all communication layers in providing a whole raft of services. They soon come to realise, however, that this flexibility comes at the non-negligible price of increased complexity, a prolonged system learning curve and much higher risks in return for investment.

Did you know that in a wrongly dimensioned UMTS network a faulty 3G terminal in London may influence a communication link in Edinburgh? Did you know that a 3dB planning error in pilot transmission power, which determines the size of each cell, may potentially cost an operator millions of pounds? Or, put it in other words, why the salary of a 3G-contract's sales man in Edinburgh is dependent on the transmission power levels in London? If you did not know, this book will give you a clue as to why all the parameters in UMTS are so highly dynamic and interdependent. If you did know, you will appreciate that optimising such systems is both vital and inescapable.

Optimisation has been known to civilisation from its very beginnings – the wheel being a prominent example which, by trial and error, fortunately emerged to be round. In contrast, given the vast number of its interdependent UMTS network parameters, optimisation by means of trial and error is clearly not an option. Only the early UMTS test trials and preliminary network rollouts were conducted manually, mainly using the experience of 2G network optimisation engineers. The currently deployed, operational UMTS networks have been partially optimised by means of software programs which yield satisfactory solutions for given input conditions.

And here lies the trick! The input conditions may vary on an hourly basis, an example of which is the temporarily varying terminal density in central London that results from the rush hour. Given the highly dynamic nature of UMTS, the optimal radio design would require many parameters to be reconfigured frequently and continuously, something clearly not viable given the large network size, limited processing power and long convergence times of numerical optimisation routines. To introduce automated optimisation routines embedded into UMTS base stations and the network backbone is the natural direction to take.

A successful radio network optimisation, be it automated or manual, can only be accomplished by appropriate prior network planning, which in turn must rely on precise network modelling. The book will discuss these three complementary subjects related to the UMTS radio network, i.e. *modelling*, *planning* and *optimisation*. They are dealt with in great theoretical depth facilitating an understanding of the UMTS network behaviour and, importantly, an abstraction of the presented theory to other beyond-3G networks that rely, in one form or another, on CDMA technology. The theoretical analysis is underpinned by professional field experience from the first commercially successful UMTS network implementations, thereby enriching the understanding of a 3G network design.

Modelling is examined theoretically and practically at various levels and covers a wide range of aspects that have significant importance on the overall 3G network planning process: simplified as well as very detailed models of the UMTS radio network and its behaviour, modelling of geographical data as well as propagation with a special attention to the wideband character of the radio channel, all in terms of the actual UMTS radio network elements deployed. The important issue of investment business modelling is included as well. The models serve as a basis for development of network planning methods and sophisticated automatic network design procedures.

Planning considers various planning stages, starting with business planning and including the following technical requirements: network dimensioning including coverage/capacity considerations, influence of traffic on the required number of both radio and non-radio network elements, detailed network planning with computer aided design and comprehensive aspects that need to be taken into account, such as infrastructure sharing, cross-border co-ordination etc.

Optimisation means achieving the highest profit by an operator with the lowest possible expenses and is characterised by good investment business planning as well as tuning the network parameters and infrastructure for optimal performance. This covers the challenges and goals of an *automated optimisation processes*, the selection of appropriate cost functions and optimisation algorithms as well as the computational complexity of an implementation. Automated network tuning of RRM parameters, as the highest level of optimisation activities, becomes increasingly important for correct network operation.

The subject of planning and optimisation in the book also relates to the *UTRAN transmission infrastructure*, where significant amounts of money are spent by network operators. This part of the network needs to be planned efficiently but is usually somehow neglected and hence requires careful attention. To this end, Chapter 17 concentrates solely on the issue of UTRAN transmission infrastructure planning and optimisation.

The theoretical approach, coupled with practical examples, makes this book a *complete and systematic compendium*, serving a wide spectrum of readership ranging from college students to professional network engineers. The healthy mix of academics, ex-academics, industrial members of both small and large telecom companies having written this compendium guarantees that the important issue of UMTS radio network tuning is reflected in a fair, comprehensive and knowledgeable manner. Ideally, this book ought to be read from the beginning to the end; however, each chapter can be read stand-alone, which is why some natural overlap between the chapters occurs.

The reader is also invited to visit the *book's website*, where complete lists of acronyms, abbreviations and variables are available, as well as figures and some optimisation examples (http://www.zrt.pwr.wroc.pl/umts-optimisation). This website will also include a dynamic forum, allowing modelling, planning and optimisation experts around the globe to share thoughts and experiences.

We dedicate this book to the student who, we trust, will understand the problems associated with current system design and inject new knowledge into future wireless communication system designs; to the network designer and optimiser who, we hope, will comprehend the parametric interdependencies and use this to implement automated solutions; and to managers and CEOs who will come to believe that there is hope of effectively running these networks, acquired, not so long ago, for such substantial sums.

Enjoy reading.

Dr Maciej J. Nawrocki
Dr Mischa Dohler
Prof A. Hamid Aghvami

Acknowledgments

As the editors of this book, we would first of all like to express our sincere gratitude to our knowledgeable co-authors, without whom this book never would have been accomplished. It is their incredible expertise combined with their timely contributions that have facilitated this high quality book to be completed and published on time. We have endeavoured to acknowledge their respective contributions within each chapter.

We would like to thank Sarah Hinton at Wiley, who initiated this book and believed in its success, as well as Olivia Underhill and Mark Hammond, also at Wiley, for their continuous support, trust and patience in and during the preparation of this manuscript. While Sarah and Mark have inspired us, it was Olivia who ran the daily business in getting this project finished.

We are also very grateful to the reviewers, both anonymous and eponymous, who have helped considerably in improving the contents of this book. We are grateful for the comments received from Michael Livschitz from Schema Ltd, Israel, Jose Gil from Motorola, UK, Zbigniew Górski from Polska Telefonia Cyfrowa sp. z o.o., Poland, Krystian Sroka form DataX sp. z o.o., Poland, Maciej Zengel from Telekomunikacja Polska S.A., Poland, Amir Dan, Independent Consultant, Israel, Prof. Thomas Kürner and Andreas Hecker from Braunschweig Technical University, Germany, Theodora Karveli, King's College London, and Dr Seyed Ali Ghorashi, King's College London. We wish to thank them that they have supported and pushed for the publication of the manuscript.

We owe special thanks to our numerous colleagues, with whom we had lengthy discussions related to the topic of automated UMTS optimisation; they are mainly academic colleagues from King's College London and Wroclaw University of Technology and industrial colleagues from France Télécom R&D and the UK Mobile Virtual Centre of Excellence.

Certainly, our employers, Wroclaw University of Technology, France Télécom R&D and King's College London, have to be thanked for generously allocating us time and resources to complete this manuscript. Maciej Nawrocki would like to thank Prof. Tadeusz Więckowski and Prof. Daniel J. Bem for being his unequalled masters in both research and academia, as well as Piotr Kocyan for his vital help in the early stages of the design of the book's outline. Mischa Dohler is infinitely in debt to Gemma, his wife; he would also like to thank his colleagues at Tech/Idea, France Télécom R&D, Grenoble, for creating such a fantastic working environment, and in particular Marylin Arndt and Dominique Barthel, both at France Télécom R&D, Grenoble, in giving sufficient freedom for this work to be completed.

As for the 'technical' support, we would like to thank Schema Ltd, Israel, for kindly making their optimisation tool available for the analysis of our case studies. A special thanks goes to Tomasz 'Yankes' Pławski for his knowledgeable translations and language corrections, as well as to Prof. Ian Groves for correcting, proof-reading and commenting on some of the book chapters.

Finally, we are infinitely grateful to our families for their understanding and support during the time we devoted to writing and editing this book.

List of Acronyms

2D	2 Dimensional
2G	2nd Generation
3D	3 Dimensional
3G	3rd Generation
3GPP	3rd Generation Partnership Project
3GPP2	3rd Generation Partnership Project 2
3GPPiP	3rd Generation Partnership Project For Internet Protocol
AAL	Atm Adaption Layer
ABR	Available Bit Rate
AC	Admission Control
ACF	Auto-Correlation Function
ACIR	Adjacent Channel Interference Ratio
ACLR	Adjacent Channel Leakage Ratio
ACP	Adjacent Channel Protection
ACS	Adjacent Channel Selectivity
ADC	Analog-to-Digital Converter
ADSL	Asymetric Digital Subscriber Line
AES	Advance Encryption Standard
AFP	Automatic Frequency Planning
AICH	Acquisition Indictor Channel
AIS	Alarm Indication Signal
AM	Amplitude Modulation
AMC	Adaptive Modulation And Coding
AMPS	American Mobile Phone System
ANN	Artificial Neural Networks
ANSI	American National Standards Institute
AoA	Angle of Arrival
ARPU	Average Revenue Per User
ARQ	Automatic Repeat Request
ASP	Application Service Profider
ATL	Above The Line
ATM	Asynchronous Transfer Mode
AWGN	Additive White Gaussian Noise
AXC	ATM Cross Connect

BCH	Broadcast Channel
BER	Bit Error Rate
BGAN	Broadband Global Area Network
BGP-4	Border Gateway Protocol-4
BH	Busy Hour
BiCG	Bi-Conjugate Gradient
BiCGSTAB	Bi-Conjugate Gradient Stabilised
BIM	Broadcast Interface Module
BLER	Block Error Rate
BS	Base Station
BSC	Base Station Controller
BTL	Below The Line
BTS	Base Transceiver Station
CAC	Call Admission Control
CAD	Computer Aided Design
CAPEX	Capital Expenditure
CBC	Cipher Block Chaining
CBR	Constant Bit Rate
CC	Cross Connect
CC	Continuity Check
CCCHs	Common Control Channels
CCIR	International Radio Consultative Committee
CCPCH	Common Control Physical Channels
CCS	Central Controller Station
CCS7	Common Channel Signalling System 7
CDMA	Code Division Multiple Access
CDV	Cell Delay Variation
CE	Channel Element
CEPT	European Conference Of Postal And Telecommunications
CER	Cell Error Rate
CES	Circuit Emulation Services
CGS	Conjugate Gradient Square
CIR (C/I)	Carrier-To-Interference
CL	Cone Launching
CLP	Cell Loss Priority
CLR	Cell Loss Rate
CMR	Cell Misinsertion Rate
CN	Core Network
C-NBAP	Common Node B Application Protocol
CPE	Customer Premises Equipment
CPICH	Common Pilot Channel
CPM	Conference Preparatory Meeting
CPS	Common Part Sublayer
CRC	Cyclic Redundancy Check
CRS	Central Radio Station
CS	Convergence Sublayer
CS	Central Station
CS	Circuit Switched
CSU	Channel Service Unit
CTA	Cordless Terminal Adapter

CTD	Cell Transfer Delay
CWTS	China Wireless Telecommunications Standard
DAMA	Demand Assigned Multiple Access
DBS	Direct Broadcast Satellite
DDP	Delivery Duty Paid
DDU	Delivery Duty Unpaid
DECT	Digital Enhanced Cordless Telecommunications
DEM	Digital Elevation Models
DiffServ	Differentiated Services
DL	Downlink
DLCI	Data Link Connection Identifier
D-NBAP	Dedicated Nodeb Application Protocol
DPCCH	Dedicated Physical Control Channel
DPCH	Dedicated Physical Channel
DPDCH	Dedicated Physical Data Channel
DSS	Digital Data Service
DTX	Discontinuous Transmission
DVB	Digital Video Broadcasting
DVMRP	Distance Vector Multicast Routing Protocol
EA	Evolutionary Algorithm
EBIDTA	Earnings Before Interest, Taxes, Depreciation And Amortisation
ECC	Electronic Communications Committee
ECC PT1	ECC Project Team 1
ECPs	European Common Proposals
ECTRA	European Committee For Telecommunications Regulatory Affairs
E-DCH	Enhanced Dedicated Channel
EIRP	Equivalent Isotropic Radiated Power
EM	Electro-Magnetic
EMC	Electro-Magnetic Compatability
ERC	European Radiocommunication Committee
ERC TG1	Erc Task Group 1
ERO	European Radiocommunications Office
ERP	Effective Radiated Power
ES	Evolution Strategies
ESA	European Space Agency
ETSI	European Telecommunication Standard Institute
FACH	Forward Access Channel
FCS	Fast Cell Selection
FDD	Frequency Division Duplex
FDMA	Frequency Division Multiple Access
FEC	Forward Error Coding
FER	Frame Erasure Rate
FH	Frequency Hopping
FIR	Finite Impulse Response
FIS	Fuzzy Inference Systems
FLC	Fuzzy Logic Controllers
FPLMTS	Future Public Land Mobile Telecommunications Systems
FS	Fixed Service
FWA	Fixed Wireless Access
GA	Genetic Algorithm

GDP	Gross Domestic Product
GEO	Geosynchronous Orbit
GFC	Generic Flow Control
GGSN	Gateway Gprs Serving Node
GIS	Geographical Information System
GMLC	Gateway Mobile Location Center
GMRES	Generalised Minimum Residual
GO	Geometrical Optics
GoS	Grade of Service
GPRS	General Packet Radio Service
GSM	Global Standard For Mobiles
GTD	Geometrical Theory Of Diffraction
HAP	High Altitude Platform
HARQ	Hybrid Automatic Repeat Request
HCR	High Chip Rate
HCS	Hierarchical Cell Structure
HEC	Head Error Control
HEO	High Earth Orbit
HF	High Frequency
HLR	Home Location Register
HSDPA	High Speed Downlink Packet Access
HS-DPCCH	High-Speed Dedicated Physical Control Channel
HS-DSCH	High-Speed Downlink Shared Channel
HS-PDSCH	High-Speed Physical Downlink Shared Channel
HS-SCCH	High-Speed Shared Control Channel
HSUPA	High Speed Uplink Packet Access
HT	Hilly Terrain
I	In-Phase
I-4	Inmarsat-4 (Satellite)
IB	In Band
ICMP	Internet Control Message Protocol
iDCS	Instant Dynamic Channel Selection
IDU	Indoor Unit
IEC	International Electrotechnical Commission
IF	Intermediate Frequency
IIM	Interactive Interface Module
IM	Image Method
IMA	Inverse Multiplexing For ATM
IMS	Intelligent Multimedia Systems
IMSI	International Mobile Subscriber Identity
IMT	International Mobile Telecommunication Group
IMT-2000	International Mobile Telecommunications – 2000
IMT-DS	IMT Direct Spread
IMT-FT	IMT Frequency Time
IMT-MC	IMT Multi Carrier
IMT-TC	IMT Time Code
IN	Intelligent Network
INA	Interactive Network Adapter
IntServ	Integrated Services
IP	Internet Protocol

IPR	Intellectual Property Rights
IRC	Interference Rejection Combining
IRR	Internal Rate Of Return
IS-95, -136	Interim Standard-95, -136
ISDN	Integrated Services Digital Network
ISI	Intersymbol Interference
IS-IS	Intermediate System To Intermediate System
ISP	Internet Service Provider
IT	Information Technology
ITU	International Telecommunication Union
ITU WP8F	ITU Working Party 8f
ITU-R	ITU Radiocommunication Sector
ITU-T	ITU Telecommunication Standardisation Sector
IWF	Inter Working Function
J-RRM	Joint Radio Resource Management
KPI	Key Performance Indicator
LA	Location Area
LAN	Local Area Network
LANE	LAN Emulation
LCR	Low Chip Rate
LEO	Low Earth Orbit
LI	Length Indicator
LL	Leased Lines
LMDS	Local Multipoint Distribution System
LMMSE	Linear Minimum Mean-Square Error
LNA	Low Noise Amplifier
LOS	Line of Sight
LRD	Long-Range Dependence
LSP	Label Switched Path
LSR	Label Switched Router
LSS	Loss of Synchronisation Signal
LTP	Long-Term Plan
LTP	Long-Term Perspective
MAC	Medium Access Control
MAI	Multiple Access Interference
MBP	Measurement Based Prediction
MC	Monte-Carlo
MC-CDMA	Multi-Carrier CDMA
MCL	Minimum Coupling Loss
MCN	Mobile Network Code
Mcps	Mega Chips Per Second
MCR	Minimum Cell Rate
MD	Macrodiversity
MEO	Medium Earth Orbit
MHA	Mast Head Amplifier
MID	Message Identifier
MIMO	Multiple-Input Multiple-Output
MIS	Management Information Systems
ML	Maximum Likelihood
MMDS	Multipoint Multimedia Distribution System

MMS	Multimedia Message Service
MMSC	Multimedia Message Service Center
MOP	Multi-Objective Optimisation
MOSPF	Multicast OSPF
MoU	Minutes of Usage
MP	Measurement Point
MPLS	Multi Protocol Layer Switching
MPM	Multi Path Propagation Model
MP-MP	Multipoint-Multipoint
MRC	Maximal Ratio Combining
MS	Mobile Station
MSC	Main Switch Controller
M-SCLR	Maximum Sector Capacity Limited Range
MSE	Medium/Small Enterprise
MSS	Mobile Satellite Services
MT	Moble Terminal
MTP	Mid-Term Plan
MTU	Maximum Transmission Unit
MTU	Maximum Transfer Unit
MVNO	Mobile Virtual Network Operator
MW	Microwave
MWM	Multi-Wall Model
NIU	Network Interface Unit
NLOS	Non Line Of Sight
NMS	Network Management System
NOC	Network Operations Centre
NP	Non-Polynomial
NPV	Net Present Value
NRT	Non-Real Time
NRT-VBR	Non-Real Time Variable Bit Rate
NTP	Network Time Protocol
OAM, O&M	Operations And Maintenance
OC-3/12	Optical Container 3/12
ODU	Outdoor Unit
OFDM	Orthogonal Frequency Division Modulation
OOB	Out Of Band
OPEX	Operational Expenditure
OSI	Open Systems Interconnection
OSPF	Open Shortest Path First
OSVF	Orthogonal Spreading Vector Format
P2P	Point To Point
PA	Power Amplifier
PAMA	Pre-Assigned Multiple Access
PBX	Private Branch Exchange
PC	Power Control
PCCPCH	Primary-CCPCH
PCH	Paging Channel
PCM	Pulse Code Modulation
PCMCIA	Personal Computer Memory Card International Association
PCR	Peak Cell Rate

PCS	Personal Communication Systems
PCU	Packet Control Unit
PC-UTD	Perfectly Conducting-UTD
pdf	Probability Density Function
PDH	Plesiochronous Digital Hierarchy
PDP	Power Delay Profile
PDP	Policy Decision Point
PDSCH	Physical Downlink Shared Channel
PHS	Personal Handyphone System
PHY	Physical Layer
PICH	Paging Indictor Channel
PIFA	Patch Inverted F Antenna
PIM	Protocol Independent Multicast
PIR	Peak Information Rate
PLMN	Public Land Mobile Network
PMP	Point-To-Multipoint
PN	Pseudo-Noise
PNNI	Private Network-To-Network Interface
POI	Point Of Interconnection
POTS	Plain Old Telephony System
PP	Portable Profile
PPP	Point to Point Protocol
PRACH	Physical Random Access Channel
PS	Particle Swarm
PS	Packet Switched
PSD	Power Spectral Density
PSK	Phase Shift Keying
PSTN	Public Switched Telephone Network
PTI	Payload Type Identifier
PTP	Point-To-Point
PVC	Permanent Virtual Circuit
Q	Quadrature-Phase
QAM	Quadrature Amplitude Modulation
QF	Quality Factor
QMR	Quasi-Minimal Residual
QoS	Quality of Service
QPSK	Quadrature Phase Shift Keying
RA	Rural Area
RAB	Radio Access Bearer
RACH	Random Access Channel
RAN	Radio Access Network
RAS	Radio Astronomy Service
RAT	Radio Access Technology
RB	Radio Bearer
RBF	Radial Basis Function
RCT	Remote Controlled Tilt
RET	Remote Electrical Tilt
RF	Radio Frequency
RFP	Radio Fixed Profile
RIP	Routing Information Protocol

RL	Ray Launching
RL	Reinforcement Learning
RLC	Radio Link Control
RMS	Root-Mean Square
RNC	Radio Network Controller
ROI	Return Of Investment
RR	Radio Regulations
RRC	Radio Resource Control
RRC	Root-Raised Cosine
RRM	Radio Resource Management
RS	Repeater Station
RSCP	Received Signal Code Power
RSSI	Received Signal Strength Indicator
RT	Real Time
RT-VBR	Real Time Variable Bit Rate
Rv	Victim Receiver
Rx	Receiver
SA	Smart Antenna
SAC	Subscriber Acquisition Cost
SAG	Spectrum Aspect Group
SAR	Segmentation And Reassembly Sublayer
SCCPCH	Secondary-CCPCH
SCH	Synchronisation Channel
SCLR	Sector Capacity Limited Range
SCR	Sustainable Cell Rate
SDH	Synchronous Digital Hierarchy
SDL	Simplified Data Link
S-DMB	Satellite Digital Multimedia Broadcasting
SDU	Service Data Unit
SE	Spectrum Engineering
SEAMCAT	Spectrum Engineering Advanced Monte-Carlo Analysis Tool
SF	Spreading Factor
SFH	Synthesised Frequency Hopping
SGSN	Service GPRS Serving Node
SHO	Soft Hand Over
SINR	Signal-to-Noise and Interference Ratio
SIR	Signal-to-Interference Ratio
SISO	Single Input Single Output
SLA	Service Level Agreement
SLG	Service Level Guarantee
SME	Small and Medium Enterprises
SMS	Short Message Service
SMSC	Short Message Service Center
SNP	Sequence Number Pointer
SNR	Signal-to-Noise Ratio
SOHO	Small Office, Home Office
SONET	Synchronous Optical Network Technologies
SOR	Successive Overrelaxation Method
SPVC	Semi-Permanent Virtual Circuit
SRB	Signalling Radio Bearer

SRC	Subscriber Retention Cost
SRD	Short-Range Dependence
SRI-E	Satellite Radio Interface – E
SSCS	Service Specific Convergence Sublayer
STB	Set Top Box
STD	Standard Deviation
STDCC	Swept Time Delay Cross Correlation
STU	Set Top Unit
SW-CDMA	Satellite WCDMA
SWOT	Strengths, Weaknesses, Opportunities And Threats
TACS	Total Access Communication System
TCH	Traffic Channel
TCP	Transmission Control Protocol
TCU, TC	Transcoder Unit
TDD	Time Division Duplex
TDM	Time Division Multiplexing
TDMA	Time Division Multiple Access
TD-SCDMA	Time Division-Synchronous CDMA
TE	Terminal Equipments
Ti	Interfering Transmitter
TIS	Technical Information Systems
TL	Tabu List
TPC	Transmit Power Control
TPM	Transversal Propagation Plane Model
TS	Terminal Station
TTA	Telecommunications Technology Association (South Korea)
TTI	Transmit Time Interval
TU	Typical Urban
TV	Television
Tx	Transmitter
UBR	Unspecified Bit Rate
UDP	User Datagram Protocol
UE	User Equipment
UHS	Ultra High Site
UL	Uplink
U-MSC	Utran Main Switched Controller
UMTS	Universal Mobile Telecommunications System
UNI	User Network Interface
UTD	Uniform Theory of Diffraction
UTRA	UMTS Terrestrial Radio Access
UTRAN	UMTS Terrestrial Radio Access Network
UUI	User-to-User Indication
UWGW	UMTS Wireless Gateway
VAS	Value Added Service
VBR	Variable Bit Rate
VC	Virtual Channel
VCI	Virtual Channel Identifier
VLR	Visitor Location Register
VoIP	Voice over IP
VP	Virtual Path

VPI	Virtual Path Identifier
VPM	Vertical Propagation Plane Model
WACC	Weighted Average Cost of Capital
WAN	Wide Area Network
WAP	Wireless Access Protocol
WARC	World Administrative Radio Conference
WI	Walfisch-Ikegami
WiMAX	Worldwide Interoperability For Microwave Access
WIS	Weighted Independent Set
WLAN	Wireless Local Area Network
WLL	Wireless Local Loop
WRC	World Radiocommunication Conference
WSI	Weighted Independent Set Problem
WWW	World Wide Web

Notes on Editors and Contributors

Maciej J. Nawrocki obtained his MSc and PhD degrees in Telecommunications from Wrocław University of Technology, Poland, in 1997 and 2002 respectively where he currently holds an Assistant Professor position. From 2004 to 2005, he also worked as a research fellow in the Centre for Telecommunications Research, King's College London, under the prestigious EU FP6 Marie Curie Intra European Fellowship focusing on UMTS radio network optimisation algorithms. In his research, he specialised in CDMA network planning and optimisation, intra/inter-system EMC as well as in software design for efficient simulation in telecommunications. Prior to his telecom research, Maciej has been for four years part of the R&D team of Microtech International Ltd, Poland, designing specialised hardware and software. In 2000, he played an important role in the consulting team working for the Polish Ministry of Telecommunications during UMTS license bidding. He gives consultancy services to a large number of companies including operators, vendors and governmental institutions in the area of radio network planning, optimisation and coordination. This includes consultation activities in the area of radio network planning and optimisation software where he has been responsible for software and product development. Maciej has participated in a number of research projects in leading positions, and is the author of a number of scientific papers. He is a member of the IEEE.

Mischa Dohler obtained his MSc degree in Telecommunications from King's College London, in 1999, his Diploma in Electrical Engineering from Dresden University of Technology, Germany, in 2000, and his PhD from King's College London, in 2003. He was a lecturer at the Centre for Telecommunications Research, King's College London, until June 2005. He is now in the R&D department of France Télécom working on embedded and future communication systems. Prior to Telecommunications, he studied Physics in Moscow. He has won various competitions in Mathematics and Physics, and participated in the third round of the International Physics Olympics for Germany. He is a member of the IEEE and he has been the Student Representative of the IEEE UKRI Section, member of the Student Activity Committee of IEEE Region 8 and the London Technology Network Business Fellow for King's College London. He has published over 50 technical journals and conference papers, holds several patents, co-edited and contributed to several books, and has given numerous international short-courses. He has been a TPC member and co-chair of various conferences and is a member of the editorial board of the EURASIP journal.

A. Hamid Aghvami is presently the Director of the Centre for Telecommunications Research at King's College London. He has published over 300 technical papers and given talks on invitation all over the world on various aspects of Personal and Mobile Radio Communications, as well as giving courses on the subject worldwide. He was Visiting Professor at NTT Radio Communication Systems Laboratories in 1990 and Senior Research Fellow at BT Laboratories from 1998 to 1999. He is currently Executive Advisor to Wireless Facilities Inc., USA, and Managing Director of Wireless Multimedia Communications Ltd. He leads an active research team working on numerous mobile and personal communications projects for 3G and 4G systems, these projects are supported both by the government and industry. He is a distinguished lecturer and a member of the Board of Governors of the IEEE Communications Society. He has been member, chairman, vice-chairman of the technical programme and organising committees of a large number of international conferences. He is the founder of PIMRC and ICT. He is a Fellow Member of the Royal Academy of Engineering, the IEEE and the IEE.

Roni Abiri obtained his BSc and MSc degrees in Electrical Engineering from Tel-Aviv University in 1980 and 1991, both with honours. During 1980–1995 he worked in R&D labs on various elements of communication systems. In 1995, he started working at Pelephone, Israel's first cellular network, in charge of the Radio part of the planned CDMA network. As this network was one of the first large scale CDMA networks in the world, he was involved in planning and optimisation aspects, whilst simultaneously addressing the needs of a commercial network. Roni joined Schema – a leading cellular optimisation company – in 2000 as CTO. In this role, he defined and helped to develop software products for cellular network planning and optimisation, for all major radio technologies. Roni obtained global recognition for his contribution to this discipline. In 2005, he moved to Intel – Mobility Group where he is currently responsible for the development of UMTS supporting chip-sets.

Zwi Altman received the BSc and MSc degrees in electrical engineering from the Technion-Israel Institute of Technology, in 1986 and 1989, and the PhD degree in electronics from the Institut National Polytechnique de Toulouse, France, in 1994. He was a Laureate of the Lavoisier scholarship of the French Foreign Ministry in 1994, and from 1994 to 1996 he was a Post-Doctoral Research Fellow in University of Illinois at Urbana Champaign. In 1996 he joined France Télécom R&D, where he has been involved in mobile network engineering and optimisation. He is currently the project coordinator of the European CELTIC Gandalf project. Dr Altman was in the winning team of the 2003 Innovation Prize of France Télécom. He has published over 80 journals and conference papers and holds four patents. His domains of interest include mobile communications, autonomic networking and automatic cell planning.

Ouassim Ben Amor received the diploma of the École Nationale Supérieure des Télécommunications (ENST) and the DEA degree of Computer Science and Networks from the University of Paris 6 (Pierre & Marie Curie), Paris, France in 2005. He is currently working in France Télécom on Internet and multi-media applications. His research interests include mobile communications, optimisation and Internet applications.

Maurice Clerc received his MS degree in Mathematics (algebra and complex functions) from the Université de Villeneuve, France, and the Eng. degree in computer science from the Institut Industriel du Nord, Villeneuve d'Asq, France, in 1972. His current research interests include cognitive science, non-classical logics and swarm intelligence. He has written the first book devoted entirely to Particle Swarm Optimization and has received the 2005 IEEE Transactions on Evolutionary Computation award for a paper on the same topic.

Vincent Diascorn received the diploma of the Institut National des Télécommunications, Evry, France, in 2004. Since then, he has been working in the Research and Development centre of France Télécom. His research interests include mobile communications, real time IP based services, network design and optimisation.

Hervé Dubreil graduated from the École Polytechnique in 1998, from the École Nationale Supérieure des Télécommunications (ENST) in 2000 and received the DEA degree in digital telecommunication systems and the PhD from the ENST in 2001 and 2005 respectively. Since 1998, he has been an engineer of the French Telecommunication Corps. In 2000 he joined France Télécom R&D as a R&D engineer in the radio interface and engineering for mobile networks. He has specialised in the design strategy, parameter setting and capacity estimation of UMTS networks. His recent studies concern dynamic parameter settings of multi-system mobile networks (GSM/GPRS/EDGE/UMTS).

Andreas Eisenblätter studied Computer Engineering, Computer Science, Mathematics, and Philosophy in Stuttgart, Hagen, Heidelberg (Germany), and Urbana-Champaign (Illinois, USA). He has degrees in Computer Engineering (from the Berufsakademie Stuttgart) and Mathematics (from the University of Heidelberg). He received his PhD in Mathematics from the Technische Universität Berlin, Germany, in 2001. His thesis on 'Frequency Assignment in GSM Networks' was awarded two international prizes. He holds a researcher position at the Zuse Institute Berlin since 1995 and heads a project in the DFG Research Center Matheon, 'Mathematics for key technologies'. He is a co-founder and managing director of atesio GmbH, a company specialising in telecommunication network optimisation. His research interests and professional activities include the optimisation of WLL/PMP radio communication systems, GSM/GPRS/UMTS RANs, WLANs and SS7 signalling networks. He has been active in several international projects, authored more than 20 scientific publications, co-edited and contributed to books, and is on the editorial board of the International Journal on Mobile Network Design and Innovation.

Karsten Erlebach obtained his Diploma in Telecommunication Engineering from Kassel University in 1994, where he participated in the development of CDMA – Spread Spectrum systems. From there he gathered in-depth knowledge leading teams in roll outs for several mobile operators and vendors in Asia, Africa and Europe. In 2003 he returned to Germany. He then became a senior specialist in the Access System Engineering department of o2 Germany, which became recently a part of Telefónica. He is now working on the current and future transmission strategy and performance enhancement.

Hans-Florian Geerdes studied Engineering Mathematics with a minor in Telecommunications in Berlin and Barcelona. He received his Master in 2003 from the Berlin University of Technology, Germany. Since 2003, he is a member of the DFG Research Center Matheon and researcher at Zuse Institute Berlin, currently working towards his PhD. His research interests are integer programming, combinatorial optimisation and their application to problems arising in wireless communication networks. His research focus currently is planning, dimensioning and optimisation of UMTS radio networks. His Master's thesis has won prizes from the German Operations Research Society and the German National Mathematical Society.

Alexander Gerdenitsch received the Dipl.-Ing. degree (MS) in Mechatronics Engineering from the Johannes Kepler Universität Linz, Austria, and the Dr. techn. degree from Technische Universität Wien, Vienna, Austria (TU Wien). During his diploma studies he investigated the implementation of a digital predistorter for linearisation of UMTS power amplifiers in the uplink. In his doctoral thesis, he studied the influence of various base station parameters on network capacity, and developed algorithms for automatic tuning of those parameters. From 2002 to 2004, he was working at the Institut für Nachrichtentechnik und Hochfrequenztechnik of Technische Universität Wien, where he was a member of the mobile communications group. The focus of his work was on UMTS network planning and optimisation. Since October 2004 he has worked for Motorola GmbH Austria as Technical Account Manager.

Peter Gould obtained his BEng Degree in Electronics from the University of Southampton in 1991. He joined Multiple Access Communications Limited (MAC Ltd) as an engineer shortly after graduating and he is now the Technical Director of the company. Since joining MAC Ltd, he has worked on a wide range of different projects including the development of a 32 Mbps quadrature amplitude modulation (QAM) modem, capacity and link budget analyses of the GSM and cdmaOne (IS-95) technologies, the detailed analysis of new network architectures and numerous radio propagation studies. He has also given training courses covering a number of different subjects including radio propagation, teletraffic modelling and radio network optimisation, as well as courses on specific technologies such as GSM, cdmaOne and UMTS. He has presented papers at a number of conferences and he is the co-author of a book entitled *GSM, cdmaOne and 3G Systems* (Wiley 2001). He has acted as an evaluator for the European Commission's Information Society Technologies research programme. He is a member of the Institution of Electrical Engineers (IEE) and a chartered engineer.

Maciej J. Grzybkowski obtained his MSEE degree from the Military University of Technology (MUT), Warsaw, Poland, in 1971 and PhD degree from MUT in 1989. He has been a lecturer at the Military College of Signal Corps, Zegrze, Poland, from 1971 to 1990. In 1990 he joined the National Institute of Telecommunications (NIT), Wroclaw Branch, Poland as a Senior Expert. From 1998 to 2002 he had been Assistant Professor at the Institute of Telecommunication and Acoustics of Wroclaw University of Technology. In 2002, he joined the Electromagnetic Compatibility Dept. of NIT again. He is now working as an adjunct professor in the radio-communication field, i.e. mobile radio-communication, frequency management, compatibility of radio-communication systems, cross-border coordination and computer systems of frequency coordination. He is part of the teams preparing the Polish position to ITU WRCs and RRCs and the Polish National Table of Frequency Allocations, as well as a participant to the CEPT Working Groups. He is a member of the IEEE and Association of Polish Electrical Engineers. He is author and co-author of many publications and papers in radio-communication, radio-wave propagation and compatibility of mobile systems.

Mikio Iwamura received his BSc and MSc degrees in Electrical Engineering from the Science University of Tokyo in 1996 and 1998, respectively. In 1998 he joined the R&D division of NTT Mobile Communications Network, Inc. (now NTT DoCoMo, Inc.) and worked on various issues regarding standardisation and development of 3G radio access, especially on cell planning optimisation. Joining the Centre for Telecommunications Research, King's College London, in 2002, he obtained his PhD degree in Telecommunications from King's College London in 2006. He has now returned to the R&D division of NTT DoCoMo and is now a 3GPP delegate for standardisation of radio network protocol aspects of the long term evolution of 3G. He has published over 20 technical journals and conference papers, and holds several patents internationally.

Zbigniew Jóskiewicz received his MSc and PhD degrees in Telecommunications from Wroclaw University of Technology, Poland, in 1994 and 2002 respectively. He joined the Institute of Telecommunication and Acoustics of Wroclaw University of Technology in 1994. Since 1998, he has been working as a lecturer in mobile communication systems. His research interests concern modern mobile and wireless communication systems aspects as well as electromagnetic compatibility of devices and systems, i.e. spectrum management, methods of emission measurement, EMC of ITE and EMC in radiocommunication systems. He has published 28 technical journals and conference papers. He is the Organising Chairman of Wroclaw International Symposium and Exhibition on Electromagnetic Compatibility and EMC Section Secretary of Electronic and Telecommunication Committee of Polish Academy of Science.

Michael Livschitz obtained his MSc degree in Mathematics from Moscow Electronic Engineering Institute, MSc degree in Automation from Moscow Chemical Machine Building Institute and PhD study in Hybrid Expert Systems at the Moscow Chemical-Technology Institute. He had worked on optimisation in different fields such as Gas Transportation, Aircraft Landing System and Cargo Ship Loading. The last seven years he had been developing optimisation and simulation algorithms for telecommunication systems. He is now in Schema Ltd. (Israel), working on simulation and optimisation of 3G and 4G cellular networks. He has published about 30 technical journals and conference papers in simulation, optimisation and artificial intelligence, and holds several patents.

Ridha Nasri received his Engineering Diploma and his MSc degree in Telecommunications with distinction from the higher school of communications of Tunis (SupCom), Tunisia, in 2002 and 2004 respectively. He is currently pursuing his PhD in telecommunication networking at the University of Pierre and Marie Curie, France. His PhD is a CIFRE convention between the University and France Télécom R&D. From 2002 to 2004, he was a radio engineer in Tunisia Telecom where he focused on radio network planning and optimisation. During the summer of 2004, he had been awarded the informatisation expert grade from Korean Agency for Digital Opportunity and Promotion (KADO), South Korea, under the programme of World Summit on the Information Society (WSIS). His research interests are in the area of wireless communication systems including mobile network planning and automatic parameterisation of multi-system networks (GSM, UMTS, WLAN). He has published over 10 technical papers and has been involved in some research projects related to autonomic mobile networking. In addition to his active research activities, he has served as a session chair for ISCCSP2004 and as a reviewer for IEEE Transactions on Vehicular Technology.

Marcin Ney obtained his MSc degree with honours in Telecommunications from Warsaw University of Technology in 1998. He has taken various positions in PTK Centertel (Orange Poland) technical division from 1996 until now, where his responsibility covers radio network planning and optimisation, new technologies, systems, platforms and services introduction, project management, license bidding (GSM, UMTS, WiMAX), GIS systems development, network dimensioning, business planning, and others. He is now leading the department responsible for technical strategy development, business planning and modelling, new product development process and technical programmes management. He is a member of the IEEE and SIT (Polish Telecommunications Engineers Association). He has published a number of technical journals and conference papers and has spoken at a number of radiocommunication conferences and internal France Télécom Group summits.

Ziemowit Neyman obtained his MSc degree in Electrical Engineering from Dresden University of Technology, Germany, in 1996. He is a member of the IEEE. From August 1996 until December 1997, he participated at the Siemens turn key project for Era GSM in Poland for RF network related activities: planning, deployment and optimisation. From January 1998 until July 2000, he had been with Viag Interkom (currently Telefónica o2) in Germany, where he dealt with RF network evolution aspects, especially the introduction of GPRS and UMTS technologies. From August 2000 until July 2002, he joined Telecom Network Consultant Ltd, UK, as a principal consultant. At that time, he developed the strategic roll-out plan for the UMTS RF network of viag Interkom in one of the four markets, provided GSM/GPRS/EDGE/UMTS RF training courses to corporate clients and designed the WLAN network for Invisible Networks Ltd, UK. After that, he supported Schema Ltd, Israel, as an independent consultant in the evaluation of Schema's optimisation solution for the NTT DoCoMo FOMA RF network, Japan, as well as in the UTRAN vendor evaluation process for Partner (Orange), Israel. From January until October 2003, he joined Siemens in Poland. He was the leader of GSM/GPRS/EDGE group dealing with RF engineering and RF tool development and programming. From October 2003 until December 2004, he had been with Schema Ltd as a senior consultant for UTRAN audit and optimisation for Vodafone KK, Tokyo, Japan, RFI/RFQ process for Schema RF optimisation solutions and as a project manager of EVDO RF network optimisation for Pelephone, Israel. From January

2005 onwards, he has been providing consultancy services to Telefónica o2, Berlin, Germany, in the area of UTRAN optimisation, troubleshooting and parameter planning, FOA processing and evaluation.

Jean-Marc Picard received the diploma of the École Polytechnique, Palaiseau, France, in 1998, and the diploma in electrical engineering of the Technical University of Aachen Germany, in 2001. Since then, he has been working in FTR&D, the R&D centre of France Télécom. His research interests include mobile communications, optimisation and digital signal processing.

Piotr M. Słobodzian received the MSc and PhD degree from the Worcław University of Technology, Wrocław, Poland, in 1993 and 1998 respectively. Since 1998, he has been with the Radio Department, Institute of Telecommunications, Teleinformatics and Acoustics, Wrocław University of Technology, where he is currently an assistant professor. In 1999, he obtained the Swiss Fellowship and joined, for nine months, the Laboratory of Electromagnetics and Acoustics at the Swiss Federal Institute of Technology (LEMA-EPFL), Lausanne, Switzerland, where he started his work concerning application of the Integral Equations – Method of Moments (IE-MoM) approach in analysis of shielded microstrip circuits. His research interests focus on computational electromagnetics, antenna theory and technology, and antenna measurement techniques. He has published over 30 technical journals and conference papers, and holds one patent.

Kamil Staniec obtained his MSc degree in Telecommunications and Computer Science from the International Faculty of Engineering at the Technical University of Lodz, Poland, in 2001. Prior to graduation, he also studied Telecommunications at the Technical University of Denmark, in 1999. Currently, he is pursuing his PhD studies at Wroclaw University of Technology. His chief field of expertise is focused on the sensitivity analysis of deterministic modelling techniques of the indoor radiowave propagation. He is also the main scientist in the project sponsored by the Polish Ministry of Science on the radiowave propagation modelling for the needs of the EMC analysis of modern indoor wireless systems. He has contributed to COST 286 activities and is the author of several conference papers devoted to both radiowave propagation modelling and EMC in Wireless LAN systems.

Rafał Zdunek received the MSc and PhD degrees in telecommunications from Wroclaw University of Technology, Poland, in 1997 and 2002 respectively. Since 2002, he has been a lecturer in the Institute of Telecommunications, Teleinformatics and Acoustics, Wroclaw University of Technology, Poland. In 2004, he was a visiting associate professor in the Institute of Statistical Mathematics, Tokyo, Japan. Since 2005, he is working as a research scientist in the Brain Science Institute, RIKEN, Saitama, Japan. His areas of interest include numerical methods and inverse problems, especially in image reconstruction. He has published over 30 journals and conference papers.

Part I

Introduction

1

Modern Approaches to Radio Network Modelling and Planning

Maciej J. Nawrocki, Mischa Dohler and A. Hamid Aghvami

The 3rd Generation (3G) Universal Mobile Telecommunications System (UMTS) radio access network relies upon novel, more flexible and efficient communication methods, a consequence of which is that novel modelling and planning approaches become of prime importance to the network's roll-out success. In this chapter, we will briefly consider the historical developments of radio network modelling and planning, thereby highlighting the need for a more modern approach to the subject. Equally importantly, we alert the reader to the limitations of modelling tools. The chapter concludes with a discussion of the advantages, disadvantages and limitations of both manual and automated optimisation processes.

1.1 HISTORICAL ASPECTS OF RADIO NETWORK PLANNING

One of the co-editors was working as a radio network planning consultant for one of the emerging UMTS networks in the late 1990s. It soon emerged that his point of view on network planning for 3G was surprisingly different from that of the operator's engineers. A likely explanation for this would be that neither party, at that time, had had any practical experience of UMTS network planning. On the one hand was an academic UMTS background and on the other was the operators' extensive GSM network planning experience. Once discussion started among both parties, it turned out that both sides had very different points of view on virtually all network planning aspects, because... they simply viewed the network using very different planning parameters. While the more academic approach recognised the multitude of parameters influencing a UMTS radio network, the concern of the operators was more the appropriate selection of 3G base site locations and their static configuration. The operators' view was largely driven by the suggestion of some consultants that the main issue in 3G network planning was capacity and coverage, which, because no real planning tools were then available, were derived either analytically or using manual measurement regimes.

Understanding UMTS Radio Network Modelling, Planning and Automated Optimisation Edited by Maciej J. Nawrocki, Mischa Dohler and A. Hamid Aghvami © 2006 John Wiley & Sons, Ltd

So what, really, is network planning? And, is network optimisation part of network planning or is this included in network optimisation? This book is going to give the answers to the above questions and, hopefully, to many more. As will also become apparent, neither the academic nor the operators' initial approach was right or wrong. Without going into too much detail, let us dwell upon a couple of issues related to network planning and optimisation.

Let us start with the choice of an appropriate base site location. Traditionally, the choice has been to locate base sites at the intersects of a triangular grid, thereby covering hexagonally shaped cells. There are many reasons why this approach has been successfully applied to network planning problems, the main reason being its analytical simplicity. Indeed, such a cell-layout uniformly covers any flat surface and is also well approximated by omnidirectional base site antenna elements. The choice of a hexagonal cell shape is further justified when three-sector sites are being used; or was it the hexagonal shape which stipulated the use of three sectors? Either answer certainly contains elements of the truth; however, it is worth pondering on the usefulness of the hexagonal assumption. Indeed other cell layouts are equally possible and are used, as e.g. rectangular grids in Point-to-Multipoint (PMP) systems such as Local Multipoint Distribution Systems (LMDS) (see Chapter 17). Perhaps surprisingly, it turns out that when applied to cellular system, square-shaped cells may also constitute a good modelling approach, which is facilitated by the horizontal plane antenna characteristics (assuming 90° sectors).

Regular shapes, be they hexagonal or square, largely result from the prior assumption of an homogeneous propagation environment. However, once one looks into the real propagation behaviour, the beauty of regular cell shapes has, sadly, to be abandoned. In the radio propagation community, it is widely known that propagation conditions in urban areas can destroy any regular cell and/or frequency plan, as street canyons become strong ducts for interference which far exceeds the levels anticipated using regular planning approaches. Hexagonality is also lost when the traffic becomes severely non-uniform, something observable not only in CDMA networks but also in any kind of cellular network, albeit in different forms. So, does the choice of medium access technique really impact the cell shapes, site locations and the network planning process as a whole? It definitely does.

This brings us to the next related issue, that of access technique, which is the primary driver for the choice of network planning and optimisation techniques; generally, we distinguish among three mainstream techniques in mobile systems: Frequency Division Multiple Access (FDMA), Time Division Multiple Access (TDMA) and Code Division Multiple Access (CDMA). They guarantee that competing users access the wireless medium in an ideally non-interfering manner, where access is granted in time, frequency or code. For instance, FDMA enables the simultaneous and interference-free use of a given frequency band by many users, where each user is assigned a sub-part of the entire band. In TDMA, users are assigned orthogonal time slots and in CDMA distinct spreading codes. In summary, access techniques facilitate an optimal use of available resources and their flexible management. To increase flexibility, hybrids of these techniques are desirable and, indeed, widely used in currently deployed communication systems.

First generation networks were using pure FDMA as a fairly inflexible access technique. The basics of network planning consisted of the aggregation of cells into clusters with each cell in the cluster having a unique frequency allowing the proper allocation of frequencies to the whole network. The system performance was limited by the level of inter-cell interference and interference sources were located far from the victim cell. Transmission over a particular link was possible if the signal-to-interference ratio (SIR) *was greater* than a given threshold, but this lead to an overhead in transmission power resulting in a non-optimal use of spectrum and power resources. The first automated solutions for network planning were developed to address this, most notably the Automatic Frequency Planning (AFP) approach [1] which assured good network quality, given appropriate site location selection and antenna configuration.

The planning and optimisation situation did not change greatly with the introduction of second generation (2G) networks. They were, of course digital, thereby easing the stringent requirements on signal-to-interference ratio when compared with the analogue predecessor. Although they were still heavily reliant on FDMA, advances in technology enabled hybrid extensions to the time domain, thereby leading to more flexible FDMA/TDMA networks (e.g. GSM, IS-136); some systems also started to utilise the time domain more intensively (e.g. DECT, PHS). For these 2G systems, new concerns in network planning emerged: synchronisation proved to be an important issue, as well as dynamic channel allocation in conjunction with interference avoidance in both the time and frequency domains; however, frequency planning which takes into account frequency hopping (FH) remained the main issues for GSM-like systems [1].

The exception to 2G TDMA/FDMA developments was the North-American IS-95 standard, which was the first commercial cellular system based on CDMA technology. Although it was also a 2G system, network planning here drifted into new and unknown areas, not all of which were identified or appreciated, despite 3G being solely dependent on CDMA technology. CDMA, conceptually an excellent idea in which all mobiles use the same frequency band at all times differentiated only by code and power level, requires numerous practical problems to be solved before the network can be referred to as efficient. This process may take years to evolve which only automated solutions can accelerate.

So, what are the problems from a network planning perspective with CDMA in UMTS? To begin with, CDMA based networks are very sensitive to traffic distributions and traffic volumes. Such traffic dependency does not exist in this form in FDMA and/or TDMA based systems and the network for these systems can be simply characterised, e.g., by the capacity (here referred to as hard capacity) regardless of the traffic in the network. In contrast, the 3G system performance depends heavily on the traffic conditions, making many performance measures *soft* (e.g. soft capacity, soft handoff). Consequently, the network planning engineer cannot plan the network a priori to find a generally optimum solution, because any solution is valid only for a given traffic condition; when the traffic changes, the optimal plan also changes. This demands, at least from a theoretical point of view, the use of automated procedures which attempt dynamically to adapt network parameters to the (near) instantaneous traffic distribution (see Chapter 16).

All of the above effects are amplified by the re-use of the same frequency in every cell, i.e. in all neighbouring cells. Therefore, the coordination distance based planning, which hexagonal and other grid design relies upon, cannot be used for CDMA networks in the same sense as for FDMA based networks; most established methods of cellular network design thus prove useless. Furthermore, a characteristic of CDMA based networks is that the SIR needs to be *equal* to a given threshold; this is in contrast to FDMA requirements where the SIR should be *greater* then a threshold and where energy and spectrum use are less than optimal. The strong influence of transmission powers and the resulting interference in CDMA systems make the system performance dependent not only on the traffic level in a given cell, but also on that in neighbouring and further cells; the system is said to be *interference limited*. Since interference is the prime concern in CDMA networks, the process of network planning attempts to minimise this occurring across all cells simultaneously. Such a planning process is the main topic of this book and it is not an easy task since dynamically interdependent interference occurs virtually everywhere in the network. Indeed, as is now well recognised, it is relatively easy to make a CDMA network operate, but to make it operate well is extremely difficult.

1.2 IMPORTANCE AND LIMITATIONS OF MODELLING APPROACHES

The planning of any modern radiocommunication network is carried out by means of computer algorithms, which numerically calculate the desired network characteristics and thereby assist the network planning engineer during the design process. To evaluate and compute any of these characteristics, a model of the physical surroundings and user behaviour must be made available. Such a model ought to be as precise as possible, where the modelling quality determines the consistency and applicability of the obtained results to the real world network deployment; an imprecise modelling renders any planning and optimisation approach, no matter how precise, useless.

Because of its importance, Part II of this book is dedicated to those modelling aspects which are crucial in UMTS network planning and optimisation. Each network operates within a given environment which must be reflected as precisely as possible in a proper model. Terrain topography as well as building layout in urban areas constitutes an important part of many network planning solutions. Digital Elevation Models (DEM) of the terrain within which the network operates need to be provided, not only for planning purposes but also to fulfil international procedures, such as cross-border coordination. Models are needed for a plethora of radio communication systems, ranging from TV broadcast systems to cellular and microwave point-to-point systems. In addition to the DEM, modelling of urban environments is also crucial for network planning; this is usually accomplished by means of clutter maps or, in more detailed approaches, through 3D building maps. All these models are a vital prerequisite for another part of the virtual modelling environment used by modern network planners – *propagation modelling*.

The propagation modelling process might appear to be the easiest of the modelling components, since all phenomena of electromagnetism are well quantified and appropriate approximations to Maxwell's equations readily applicable. In reality, however, propagation models are likely to be the most inaccurate part of this virtual modelling world. This may be for many reasons, the inaccuracy of terrain and building models being the important one. Building models can have errors in vertices in both the horizontal and vertical directions and, even when a 3D map has a high resolution, these errors can be significant compared with reality. Resolution and accuracy do not mean the same thing. Unfortunately, today's scanning technology facilitates only the modelling of general building geometries without precisely taking into account windows, roof shapes, gutters, vegetation, surface roughness etc. Even if a higher precision were possible, the electrical parameters of all the surroundings are generally unknown. Furthermore, statistical propagation modelling, whilst useful for evaluating performance tendencies, is of little use in given real-world network environments.

For a moment, let us consider that we have managed to obtain very high resolution maps of geometrical and electrical parameters describing the urban environment of our network roll-out area. A non-negligible problem soon arises, apart from the cost of this kind of map, how are we to process this huge amount of data? Advances in modern computing technology are breathtaking, but it is still not enough for *efficient* and *accurate* propagation computations. Operators are happy today when the standard deviation of an error in propagation prediction is about 8 dB; but, from a link budget perspective, 8 dB can easily mean a doubling of the communication range! Of course large supercomputers can be employed for this kind of modelling work, but the associated costs are not justified. Paradoxically, although computing power increases dramatically every year, the systems we want to model and simulate also become more complicated. It seems that solutions for propagation modelling will rely on reasonable simplifications for a long time to come, where the degree of simplification will slowly be reduced over the forthcoming years.

Temporal changes to the propagation environment are another important factor which make propagation models inaccurate. Such temporal variations are very difficult to account for; just imagine, e.g., the changes of vegetation throughout the seasons or changes in electrical parameters of the ground and walls after the rain. We have known of many situations where the network was well planned and optimally functioning in the summer, but when winter came and leaves disappeared from the trees, the call drop probability increased significantly because interfering sites became visible in hilly terrain (so-called 'boomer' cells). In addition to this, digital maps should be regularly updated to ensure that all newly constructed buildings are included.

An important part of the entire modelling process relates to users' behaviour. Users can have, e.g., various preferences in service usage or the financial budget allocated for mobile phone calls. These effects are difficult to measure by the operator and are more appropriate to sociological or psychological research than engineering studies. Even if mathematical models of users' behaviour can be made (of any aspect), obtaining realistic values for such a model parameters might be impossible. User density and mobility as well as their temporal changes during the day or week are also hard to capture. All these issues create uncertainty in UMTS planning, where the network performance heavily depends on the traffic.

The modelling of the system and network behaviour relies in one way or another on all the constituent models. There are a number of components and phenomena which must be modelled with the highest possible precision, such as antennas, receivers, transmitters, access techniques, codecs, actions taken in the whole Radio Resource Management (RRM) domain and many more. The modelling can precisely reflect transceiver structures and channel variations, making direct use of link-level modelling approaches, or be more general and use more system-level modelling approaches; dynamic effects can be simplified or passed over into static or dynamic system-level simulators. An ideal solution would clearly be to have a full dynamic simulator which includes complete link level models; this is the dream of many researchers and engineers, but some time still must pass before computing solutions enable simulations of every detail for every user within a network with thousands of users within reasonable computing times.

From this brief overview, the question arises of how far we should go into the modelling of the surrounding world for cellular network planning? Interesting considerations related to this concern are presented in [2]. There is also one particular trend that has become visible in the past few years: if reality cannot be modelled precisely enough so that errors and simplifications do not impair the planning results, use the reality instead – this makes *measurements* increasingly important, not only for network planning validation but also as an input to the network planning and optimisation processes.

In summary, this section, albeit lacking in technical depth, serves to emphasise the various modelling issues that are important to the radio network planning process and hence to the level of service satisfaction experienced by the network users.

1.3 MANUAL VERSUS AUTOMATED PLANNING

The models from Section 1.2 form the basis for the simulation software used by network planning engineers to dimension and plan the network. Such radio communication system design requires highly sophisticated computer aided design (CAD) tools. These tools facilitate refined network planning, the complexity of which is generally beyond the scope of any manual approach. Planning tools and optimisation tools must be distinguished. A planning tool consists of a database, a Geographical Information System (GIS), propagation modules etc., and is used to predict the network performance for a given set of system (*analysis*). An optimisation tool takes this data and tries to find such system parameters as will achieve the best network performance (*synthesis*, aided by analysis; see Chapter 13).

Radio network planning can be defined as *designing a network structure and its configuration to meet certain quality requirements*. There may be a number of criteria defined by the operator, and usually these differ from operator to operator. The criteria can take into account such topics as coverage, Quality of Service (QoS), equipment and other costs, revenues from network operation (see Chapters 7 and 8) etc. and can be used to verify the network quality either by the engineer (manual decision) or by the software (automated decision). A schematic presentation of the modelling, planning and optimisation tasks in the network design process is shown in Figure 1.1.

Having a new generation network to deploy, operators start from some simple planning rules (manual planning) to get acquainted with the system behaviour. The first networks are usually rather badly planned since the transition to any new technology requires some time to understand the relationships between any change of configuration and the response in the network; such a testing approach is typically applied to trials prior to the official network launch. Under these simplified operational conditions, the operator is usually unable to learn and recognise all the aspects of running the network since there is usually insufficient traffic demand to experience extreme load conditions. When the

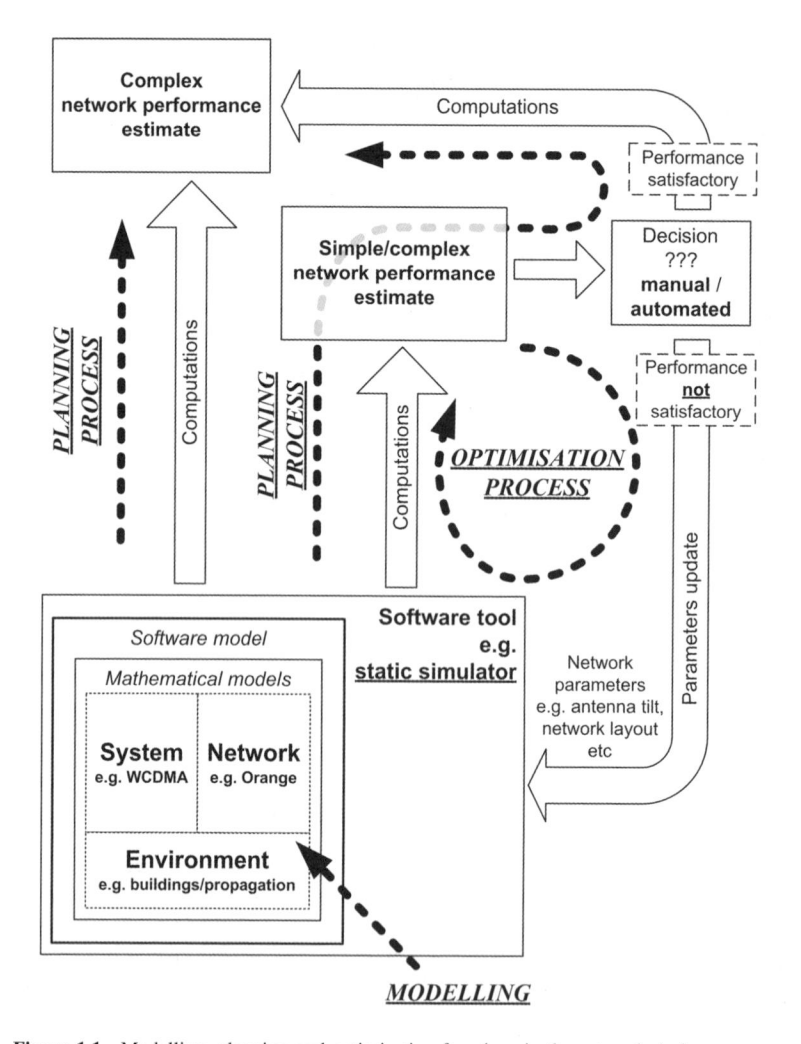

Figure 1.1 Modelling, planning and optimisation functions in the network design process.

network starts its commercial operation, however, the traffic increases and the operator faces new situations in which the system cannot serve the increased traffic; this requires some changes to the network configuration and is usually accomplished by means of optimisation. It must be noted that the term *optimisation* is sometimes misunderstood in the context of *optimisation being part of the pre-launch network planning*. From an operational network point of view, and the definition proposed at the beginning of this section, the optimisation process should be treated as part of the ongoing network planning and all these activities should be performed over the whole network operation period, e.g. perhaps for 10 years or more. Historically, some operators have had two separate departments, one for planning and another for optimisation; however, this type of operation has many times proved inefficient, thus partially answering the above dilemma.

With an ever increasing complexity of real networks, planning and optimisation requires the use of sophisticated software tools. Their sole task is to assist the planning engineer and help him or her process the enormous amount of relevant information into a comprehendible form. Such a process is usually accomplished by means of a number of modules within a given planning tool, e.g. a database, the GIS, propagation and CDMA Monte-Carlo simulation modules. Optimisation tools are even more important in this than planning tools since they use automated search processes on good initial network designs, something beyond human ability. Although modern software solutions are able to deliver a full network design by means of a single mouse click if this is accepted unchecked, there are risks. The uncritical acceptance of information received from planning or optimisation tools is indeed very dangerous; we have known of cases where inferior tools were preferred simply because they required less involvement of the radio design engineer. It should be recognised that CAD software is used to help people, but not as a substitute for them. Tools should be user friendly, accurate, fast and flexible (ease of connecting third-party modules) and allow group working on a common project.

In summary, the ideal situation would be to encourage experienced planning engineers to use automated software in the initial network planning and optimisation stages, but to rely on their own intuition and experience in the final network verification stage.

REFERENCES

[1] http://www.datax.pl.
[2] Trevor Gill, 'Radio planning and optimisation – the challenge ahead', 4th International Conference on 3G Mobile Communication Technologies, London, 25–27 June 2003, pp. 28–30.

2

Introduction to the UTRA FDD Radio Interface

Peter Gould

In this chapter, we will give a simple – but vital – introduction to the UTRA FDD radio interface, so the reader can easily follow and comprehend later sections of the book. We will start with some general CDMA-based principles, which are then followed by a description of the UTRA physical channels and a discussion of some key mechanisms that are used in the UTRA FDD system. The chapter concludes with a brief description of those parameters that we consider to be of most importance in the network optimisation process.

2.1 INTRODUCTION TO CDMA-BASED NETWORKS

As already mentioned in the previous chapter, in first generation cellular systems (e.g. TACS and AMPS) frequency division multiple access (FDMA) was used to allow several users to communicate simultaneously with a network base station. In FDMA each user is given its own radio carrier frequency for the duration of a call and this frequency will only be used by this user within the local area. The carrier frequency can be used by another user when the distance between the two users is sufficient to prevent any interference between them and this is the concept of cellular frequency reuse. In second generation systems (e.g. Digital AMPS and GSM), different users were also separated in the time domain to produce time division multiple access (TDMA). In TDMA systems, different users can share the same carrier frequency in the same local area, but only one user can transmit or receive at any point in time.

In third generation networks (e.g. UMTS), the different users in the same local area are distinguished by means of a 'spreading code' and this means that all users can access the same carrier frequency at the same time. This is termed code division multiple access (CDMA). The spreading code will be unique to a particular user within a local area and it is used to spread the narrowband user data signal up to a wideband signal for transmission. At the receiver, the user's unique spreading code is used again to de-spread the user's radio signal, before the user data is recovered from the transmitted signal.

Understanding UMTS Radio Network Modelling, Planning and Automated Optimisation   Edited by Maciej J. Nawrocki, Mischa Dohler and A. Hamid Aghvami   © 2006 John Wiley & Sons, Ltd

To explain the spreading and de-spreading processes in more detail we shall consider them first in the time domain and then in the frequency domain. In the time domain, spreading simply consists of multiplying a lower rate data signal by a higher rate spreading code as shown in Figure 2.1. The upper trace in Figure 2.1 shows the user data, represented as a square wave with logical levels of $+1$ and -1. In the centre trace we show the user's unique spreading code and in the lower trace we show the multiplication of the user data and the spreading code, which will form the transmitted signal for this particular user. The transmitted signal will be used to modulate a radio frequency (RF) carrier, which will then be transmitted via an appropriate antenna. Clearly, the bandwidth of the low rate user data has been spread to a significantly higher bandwidth as a result of this multiplication process.

At the CDMA receiver, the user's data signal is recovered by de-spreading the received signal. The de-spreading process consists of multiplying the received signal by the user's spreading code and this process is shown in Figure 2.2. The upper trace is the baseband signal received at the input to the CDMA receiver, which results from the demodulation of the RF carrier. This is identical to the lower trace in Figure 2.1. The central trace is the user's spreading code and this is identical to the central trace shown in Figure 2.1. Finally, the lower trace is the multiplication of the received signal and the user's spreading code and it represents the recovered user data. There are two important points to note from the simple example shown in Figure 2.2. First, multiplying the received signal by the user's spreading code has the effect of decreasing the bandwidth of the user's signal, i.e. it is de-spread. Secondly, this de-spreading process occurs only if the user's spreading code that is generated at the receiver is synchronised with the spreading code in the received signal. Synchronisation is a key factor in CDMA technologies and we will see that several radio channels are dedicated to the process of synchronisation in the UTRA FDD system.

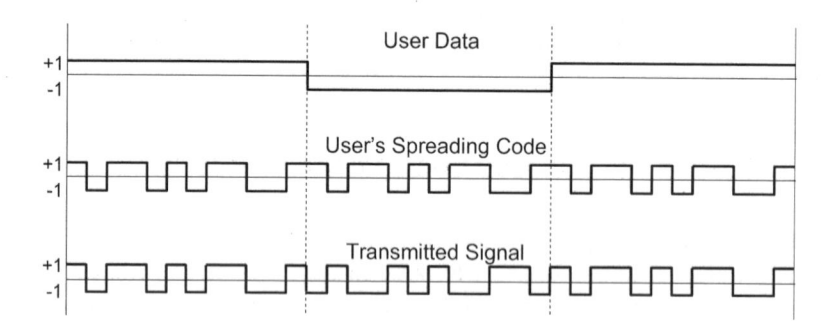

Figure 2.1 The spreading process.

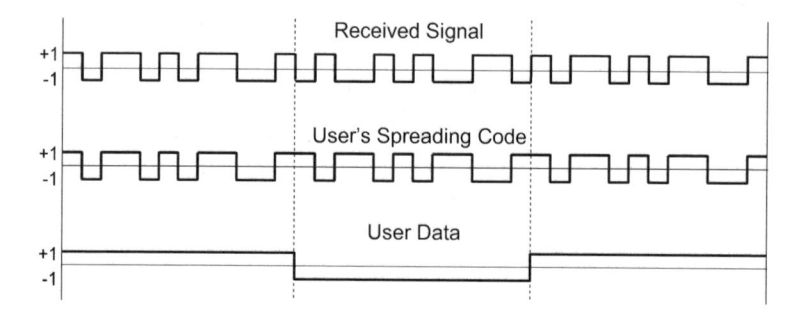

Figure 2.2 The de-spreading process.

Introduction to the UTRA FDD Radio Interface

Having examined the manner in which a user's data signal can be spread and de-spread in a CDMA system, we now consider how this technique can be used to provide multiple access in a cellular network. Figure 2.3 shows what happens if we use a different spreading code in the de-spreading process at the CDMA receiver. The upper trace shows the received signal, generated using the user's spreading code shown in Figure 2.1 and Figure 2.2. The central trace in Figure 2.3 shows a different spreading code to that shown in Figure 2.1 and Figure 2.2, i.e. it belongs to a different user. Finally, the lower trace in Figure 2.3 shows the multiplication of the upper and central trace and it represents the output of the de-spreading process.

The key point to note from Figure 2.3 is that the received signal is not de-spread and the user data is not regenerated at the output of the de-spreading process. In fact, the signal at the output of the de-spreader has a similar bandwidth to the transmitted signal.

To fully understand how we can separate a number of CDMA signals, each using different spreading codes, we now consider the spreading and de-spreading process in the frequency domain. In Figure 2.4 we show a power spectral density (PSD) frequency domain representation of the signals arriving at the CDMA receiver. The signals will be centred on the RF carrier frequency, f_c, and the first nulls in the spectrum will appear at $\pm 1/T_c$ either side of this frequency, where T_c is the period of a bit in the spreading code. We note that, in order to distinguish between the information-carrying bits in the user data and the bits in the user spreading codes, we tend to use the term 'chips' to refer to the bits in the spreading code. Therefore, T_c is the chip period and $1/T_c$ is the 'chip rate' of the spreading codes.

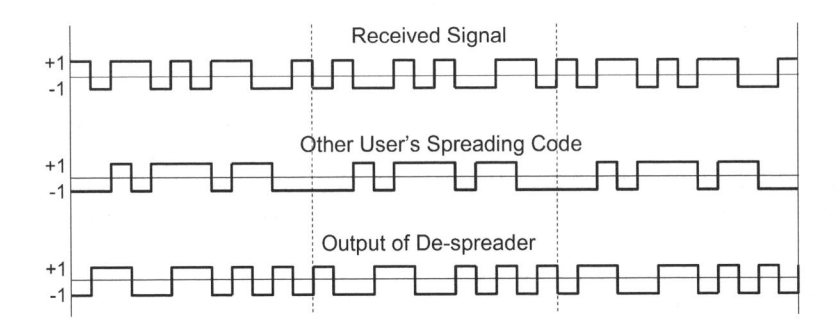

Figure 2.3 The effects of other users.

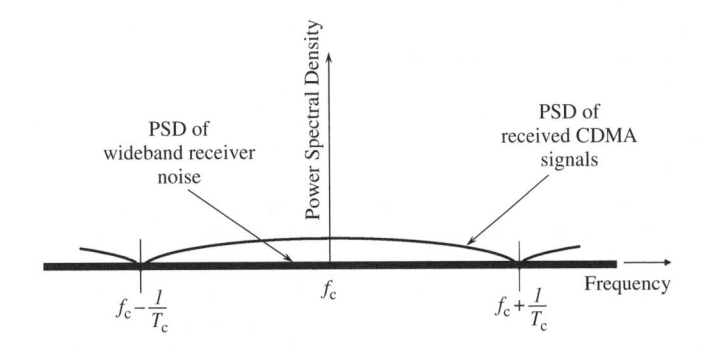

Figure 2.4 The signals at the input to a CDMA receiver in the frequency domain.

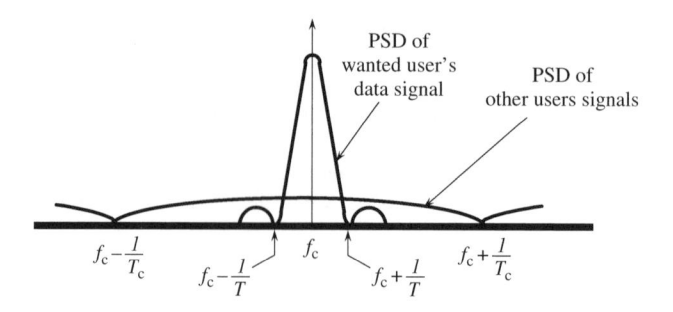

Figure 2.5 The signals at the output of the de-spreading process in the frequency domain.

Figure 2.5 shows what happens to the signals as they pass through the de-spreading process. The signal from the wanted user (i.e. the user whose code was used in the de-spreading process) has its bandwidth decreased such that the first nulls in the spectrum occur at $\pm 1/T$ either side of the carrier frequency, where T is the bit period of the user data. The spectrum characteristics of the signals from the other CDMA users (i.e. those users with different codes from that used in the de-spreading process) remain unchanged and these signals are not de-spread.

By applying a bandpass filter to the signal at the output of the de-spreading process we can reject most of the power in the signals of the other users, whilst allowing most of the power in the wanted signal to pass. This allows us to significantly increase the signal-to-interference ratio (SIR) of the wanted signal, which will allow us to recover the user data in the face of interference from other users.

One of the key parameters in a CDMA system is the processing gain, which is essentially the ratio of the user data rate to the CDMA chip rate (or the chip rate of the spreading code). The processing gain tells us by how much we can improve the SIR of the wanted signal as it passes through the de-spreading process. For example, if we have a user data rate of 9.6 kbps and a chip rate of 3.84 Mchips/s, then the processing gain is 400 or 26 dB. In other words, the de-spreading process will improve the SIR of the wanted signal by 26 dB. If we assume we require an SIR of, say, 10 dB to decode the user data at the output of the de-spreading process, then we can tolerate an input SIR of -16 dB. If each user is received at the same power, then this means that we can receive signals from 41 users and still be able to recover the data from each user.

The fact that CDMA systems can work with a negative SIR means that they can operate with single cell frequency reuse, i.e. the same carrier frequencies can be used on every cell within a network. This is in contrast to other cellular systems such as GSM and TACS, where carrier frequencies can only be reused within the network at cells that are far enough apart to ensure a sufficiently large positive SIR.

Let us now consider what happens if we increase the data rate of our users from 9.6 kbps to, say, 96 kbps. In our example our processing gain will decrease to 40 or 16 dB and, assuming we still require an SIR of 10 dB at the output of our de-spreader, we can tolerate an SIR of -6 dB at the receiver input. This means that we can support only five users in our CDMA system. In other words, by increasing the data rate of the users by 10 times, we decrease the number of interferers (i.e. other users) that we can tolerate by 10 times, from 40 to just 4.

In this section we have attempted to provide a very simple introduction to the concept of CDMA for those readers who are unfamiliar with the technology. We have aimed to provide a high level overview of the spreading and de-spreading process and we have introduced the concept of processing gain. As you read through this book, other important CDMA features and concepts will be introduced, particularly those associated with the UTRA FDD implementation of the CDMA technology. Those readers who would like to obtain a more thorough grounding in the CDMA technology are referred to references [1] and [2].

2.2 THE UTRA FDD AIR INTERFACE

In this section we examine the fundamentals of the UTRA FDD air interface and the mechanisms that are used to support user mobility and manage the system interference. Before we continue, let us introduce two important terms that are commonly used in conjunction with the UTRA FDD technology. First, we have the *Node B*, which is the term used to describe a base station in UTRA FDD. Secondly, we have *user equipment*, or UE, which is the term used to describe the mobiles or terminals.

2.2.1 SPREADING CODES

We will start our look at the UTRA FDD air interface by examining the different types of codes that are used. There are two broad categories of code used, namely channelisation codes and scrambling codes. As their name suggests, channelisation codes are used to distinguish between the different physical channels on the uplink and downlink paths. Scrambling codes, on the other hand, are used to distinguish between different Node Bs on the downlink and different UEs on the uplink.

The channelisation codes used in UTRA FDD are known as orthogonal variable spreading factor (OVSF) codes and they are essentially a set of Walsh codes of different lengths. Walsh codes have the attractive property that, when they are synchronised in time, they are orthogonal. In other words, if you take two different Walsh codes of equal length and multiply them together and sum over the length of the Walsh code, the result will be zero, provided the two codes are time aligned. To demonstrate this property, let us take two Walsh codes each of eight chips in length and multiply them together chip-by-chip, as shown in Figure 2.6. If we sum the resulting chip sequence we get a value of zero (i.e. there are always an equal number of $+1$'s and -1's).

If the two Walsh codes shown in Figure 2.6 are time shifted with respect to each other by, say, one chip period, as shown in Figure 2.7, then the result of the multiplication and summing process (or the correlation process) is no longer zero. This shows that in order to achieve orthogonality, the Walsh codes received on the different channels in a CDMA system must be time aligned.

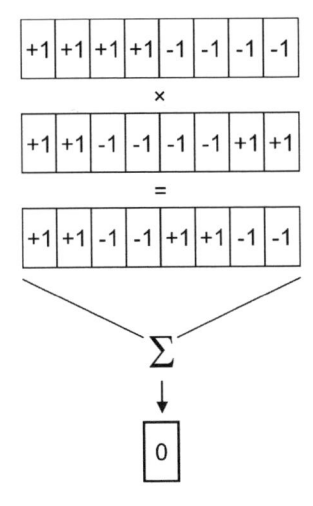

Figure 2.6 The orthogonality property of Walsh codes.

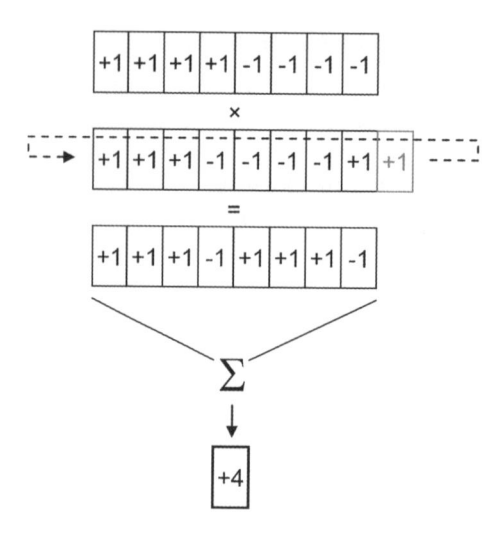

Figure 2.7 The cross correlation of two different Walsh codes with a non-zero offset.

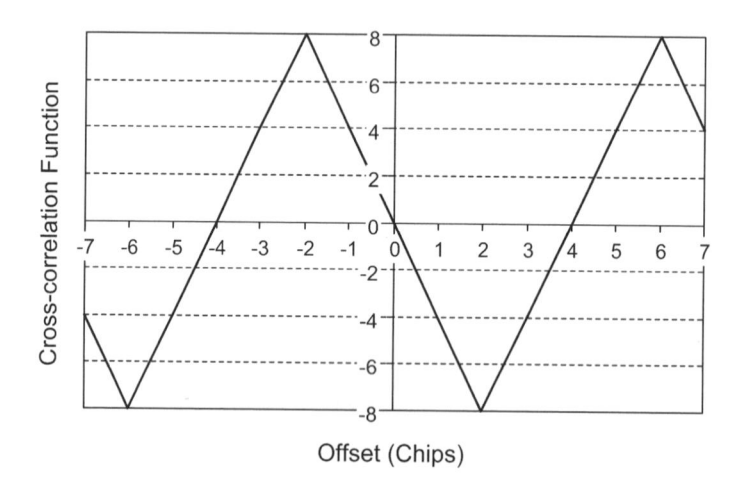

Figure 2.8 An example cross correlation function of two eight-chip Walsh codes.

In Figure 2.8 we show the result of the correlation between the two Walsh codes shown in Figure 2.7 for different chip offsets. This is known as the cross correlation function of these two codes and it further demonstrates the requirement for code synchronisation to maintain orthogonality. We can see that if we have an offset of ±2 or ±6 chips, then we have large values for the cross correlation function and this would manifest itself as strong interference between the different channels.

The implication of this orthogonality property is that if we assign two users different Walsh codes as their spreading codes, then we can completely remove the signal (or interference) from one user as we de-spread the signal from the other user. This is the reason why Walsh codes are used to differentiate between different channels in the UTRA FDD system. In the UTRA FDD system, the different user data rates are supported by using Walsh codes of different lengths and these are formed based on a simple tree structure whereby each code of length N can be used to form two further codes of

Introduction to the UTRA FDD Radio Interface

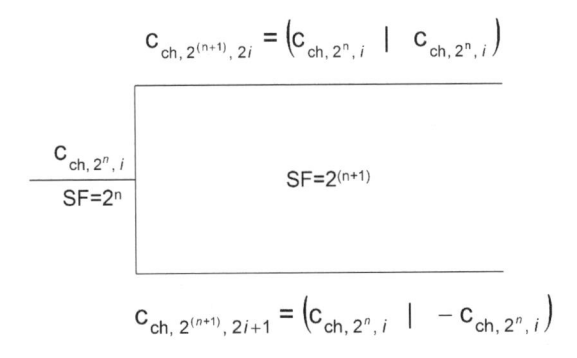

$$C_{ch, 2^{(n+1)}, 2i} = \left(C_{ch, 2^n, i} \mid C_{ch, 2^n, i} \right)$$

$$\dfrac{C_{ch, 2^n, i}}{SF=2^n} \quad SF=2^{(n+1)}$$

$$C_{ch, 2^{(n+1)}, 2i+1} = \left(C_{ch, 2^n, i} \mid -C_{ch, 2^n, i} \right)$$

Figure 2.9 Channelisation code generation process.

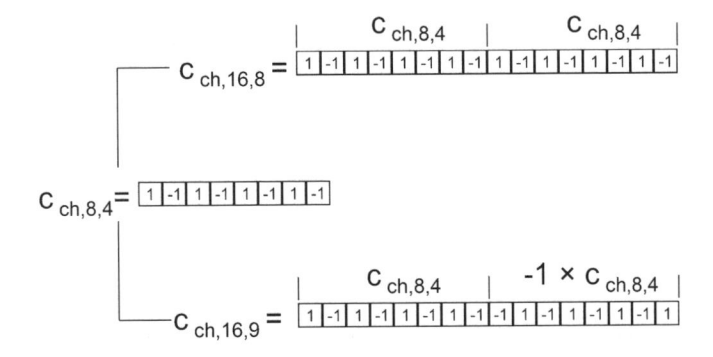

$$C_{ch,16,8} = \underbrace{[1|\text{-}1|1|\text{-}1|1|\text{-}1|1|\text{-}1|}_{C_{ch,8,4}} \underbrace{1|\text{-}1|1|\text{-}1|1|\text{-}1|1|\text{-}1]}_{C_{ch,8,4}}$$

$$C_{ch,8,4} = [1|\text{-}1|1|\text{-}1|1|\text{-}1|1|\text{-}1]$$

$$C_{ch,16,9} = \underbrace{[1|\text{-}1|1|\text{-}1|1|\text{-}1|1|\text{-}1|}_{C_{ch,8,4}} \underbrace{\text{-}1|1|\text{-}1|1|\text{-}1|1|\text{-}1|1]}_{-1 \times C_{ch,8,4}}$$

Figure 2.10 An example of the generation of two channelisation codes.

length $2N$. This process is shown in Figure 2.9, where $c_{ch,x,y}$ represents the y-th channelisation code of length x chips, or spreading factor (SF) of x. The process essentially consists of taking the base channelisation code and copying it twice to form one new channelisation code of twice the length and performing the same process, but inverting the second copy of the code to get the second new channelisation code.

Figure 2.10 provides an example of the channelisation code generation process in action. We start with $c_{ch,8,4}$ (i.e. the fourth channelisation code that is eight chips in length). This is copied twice to produce $c_{ch,16,8}$ (i.e. the eighth channelisation code that is 16 chips in length). The code is copied once, inverted and copied again to produce $c_{ch,16,9}$.

In the UTRA FDD system, spreading codes with lengths of four chips up to 512 chips can be used in powers of two and, therefore, the code 'tree' has eight sets of branches with the first set of branches consisting of four four-chip codes, the second set of branches consisting of eight eight-chip codes and so on until we reach the eighth set of branches with 512 512-chip codes. In Figure 2.11 we show the section of the code tree containing the eight-chip and 16-chip codes.

When formed in this way, the codes have the property that two codes from different branches of the tree will be orthogonal regardless of their length. This is demonstrated in Figure 2.12 where we have shown the cross correlation of the $c_{ch,16,8}$ code with the $c_{ch,8,3}$ code. The diagram shows that regardless of whether we sum the result over the length of the 16-chip code or the eight-chip code, the result is zero.

In Figure 2.13 we show what happens if we use two codes from the same branch of the tree. In this case the cross correlation of the two codes is not zero and we cannot use these two codes at the same time to support different channels.

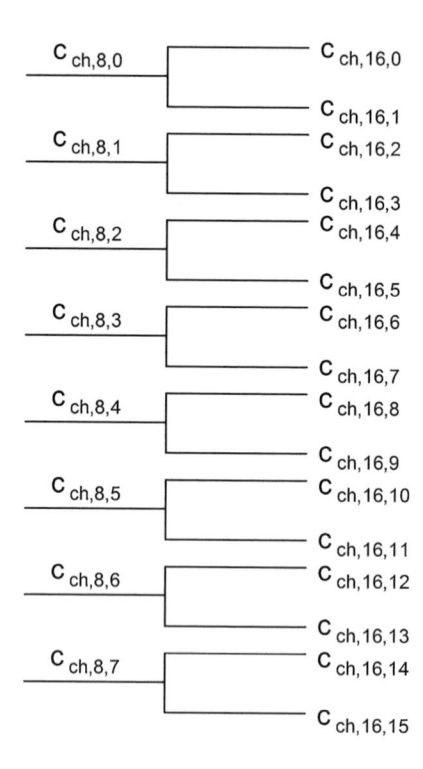

Figure 2.11 The 8-chip and 16-chip codes within the code tree.

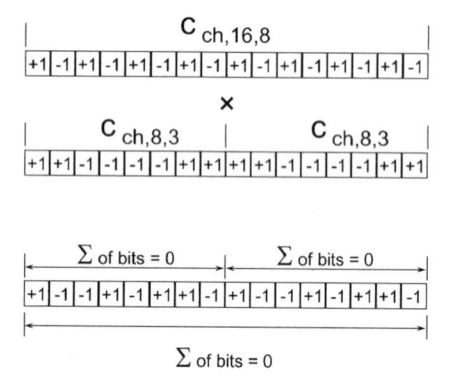

$C_{ch,16,8}$
| +1 | -1 | +1 | -1 | +1 | -1 | +1 | -1 | +1 | -1 | +1 | -1 | +1 | -1 | +1 | -1 |

×

$C_{ch,8,3}$ | $C_{ch,8,3}$
| +1 | +1 | -1 | -1 | -1 | -1 | +1 | +1 | +1 | +1 | -1 | -1 | -1 | -1 | +1 | +1 |

Σ of bits = 0 Σ of bits = 0
| +1 | -1 | -1 | +1 | -1 | +1 | +1 | -1 | +1 | -1 | -1 | +1 | -1 | +1 | +1 | -1 |

Σ of bits = 0

Figure 2.12 The orthogonality property of different length OVSF codes.

This means that we cannot use every code in the code tree at the same time and we have a set of rules that govern which codes can be used simultaneously. A code can only be used if none of the codes in the path from the code to the root of the code tree are already in use *and* none of the codes in the sub-tree below the code are in use. This is shown in diagrammatic form in Figure 2.14.

As we have seen, synchronisation is a key requirement with the use of Walsh codes. However, Walsh codes do not lend themselves to being used for synchronisation. We have already seen that the

Introduction to the UTRA FDD Radio Interface

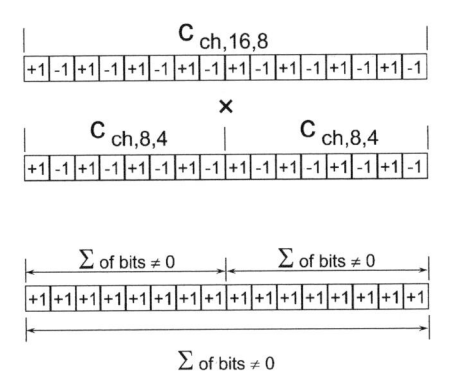

```
                     C ch,16,8
        |+1|-1|+1|-1|+1|-1|+1|-1|+1|-1|+1|-1|+1|-1|+1|-1|
                        ×
         C ch,8,4                 C ch,8,4
        |+1|-1|+1|-1|+1|-1|+1|-1|+1|-1|+1|-1|+1|-1|+1|-1|
```

|← Σ of bits ≠ 0 →|← Σ of bits ≠ 0 →|
|+1|+1|+1|+1|+1|+1|+1|+1|+1|+1|+1|+1|+1|+1|+1|+1|
|←   Σ of bits ≠ 0   →|

Figure 2.13 The cross correlation of codes from the same branch of the code tree.

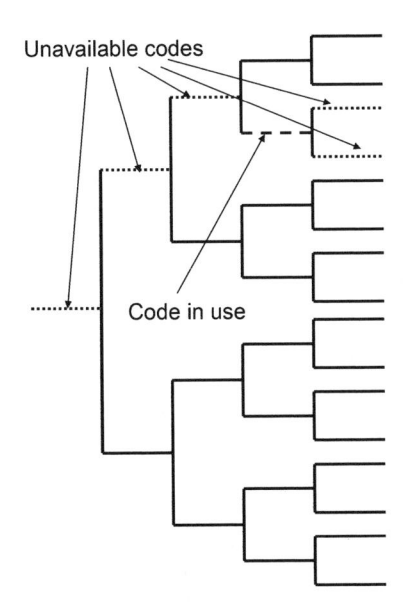

Figure 2.14 An example of the code allocation rules.

cross correlation function of Walsh codes has strong responses at non-zero offsets (see Figure 2.8). In Figure 2.15 we show an example autocorrelation function of an eight-chip Walsh code (i.e. the correlation of a Walsh code with itself at different offsets). This shows that the autocorrelation function not only has a strong peak at zero offset, but also has peaks of similar magnitude at offsets of ±4 chips. Since synchronisation is usually achieved by finding peaks in the autocorrelation function of a code, these multiple peaks mean that Walsh codes are not suited to the synchronisation function.

Therefore, in the UTRA FDD system, the OVSF channelisation codes are used in conjunction with a set of scrambling codes with a different set of autocorrelation and cross correlation properties. The scrambling codes are used to distinguish between the different Node Bs on the downlink and they are used to distinguish between the different UEs on the uplink. The scrambling codes are known as Gold codes and there are a total of 8192 of them available in the UTRA FDD system. Gold codes

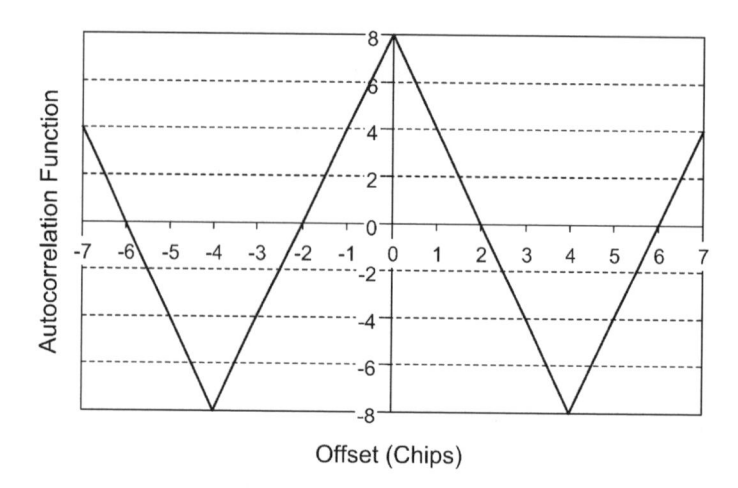

Figure 2.15 An example autocorrelation function of a Walsh code.

have very good autocorrelation and cross correlation properties, as shown in Figure 2.16. Figure 2.16a shows the autocorrelation function of one of the scrambling codes and it is clear the code has a strong autocorrelation peak at zero offset and very low values of autocorrelation at non-zero offsets. Figure 2.16b shows the cross correlation between two different scrambling codes and this demonstrates that the cross correlation between two different codes results in very small values, regardless of the relative offset between the two codes. This means that we do not need to synchronise the scrambling codes received from different sources.

As their name would suggest, the scrambling codes are used to scramble rather than spread the transmitted signals at the Node B and the UEs. The bandwidth of the data signal is spread to the CDMA chip rate using the OVSF codes and the signal is then scrambled by the scrambling code. This two-stage approach allows the properties of both the OVSF codes and the Gold codes to be exploited. The OVSF codes provide orthogonality, provided time synchronisation can be maintained. Therefore, they are used to distinguish between different channels transmitted from the same source, i.e. either a Node B or a UE. Provided the different OVSF codes are aligned when they are transmitted, then they will still be aligned when they arrive at the receiver. We note that multipath propagation has the effect of decreasing the degree of orthogonality to some extent and this effect is accounted for within the network planning process by including an orthogonality factor. In situations where signals arrive at a receiver from a number of different sources (e.g. the signals that arrive at a Node B from the UEs that it is serving), the synchronisation of the codes cannot be guaranteed. Therefore, in the UTRA FDD system Gold codes are used to scramble the signals transmitted by different sources to decrease the interference from these different sources.

2.2.2 COMMON PHYSICAL CHANNELS

Having examined the various CDMA codes that are used in the UTRA FDD system, we now move on to examining the different physical layer channels that are used on the UTRA FDD radio interface. We will start by examining the downlink physical channels (i.e. those transmitted by the Node B) and then we will move on to consider the uplink channels (i.e. those transmitted by the UEs). We note that this is designed to provide a brief overview of the different channel types used in UTRA FDD and

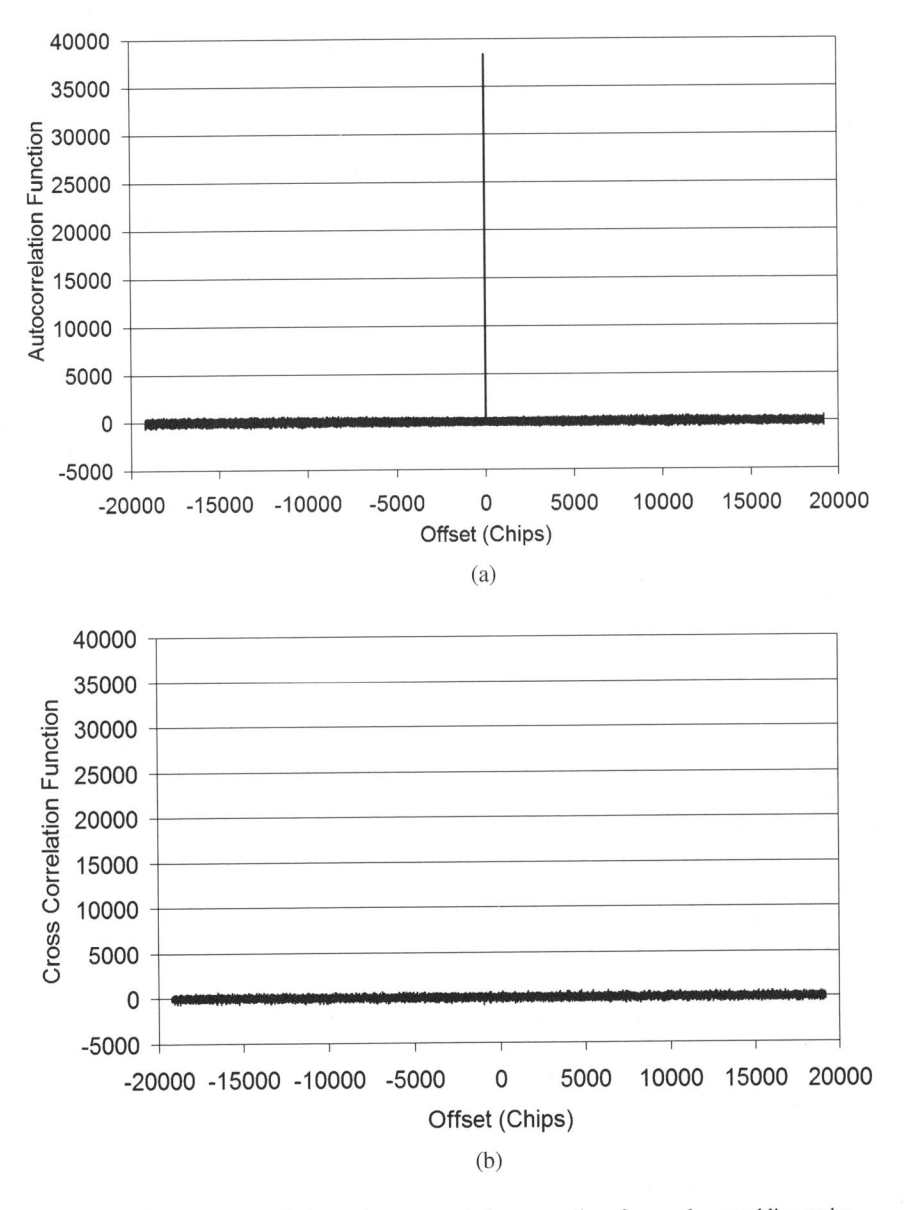

Figure 2.16 The autocorrelation and cross correlation properties of example scrambling codes.

some channels have been omitted from our description. The reader is referred to references [2] and [3] for a more in-depth and complete discussion of the different physical channels.

2.2.2.1 Synchronisation Channels

The first task that a UE must perform after it has been switched on is to 'find' the UTRA FDD network, i.e. it must locate and decode signals coming from at least one Node B. The first radio channel that we will consider is the one used by the UE to initially detect the presence of a Node B and this is called the synchronisation channel (SCH). The SCH consists of a 256-chip sequence, known as the primary synchronisation code, which is transmitted at 3.84 Mchip/s. This is the CDMA chip rate used by the UTRA FDD system. This sequence lasts for 66.67 μs and it is transmitted once every 666.67 μs, i.e. it is transmitted with a 10 % duty cycle. The primary synchronisation code is the same for all Node Bs.

When a UE is first switched on, it uses a technique known as swept time delay cross correlation (STDCC) to look for the primary synchronisation code on the chosen operating frequency. The correlation part of this technique essentially consists of multiplying the received radio signal by a locally generated version of the chip sequence and integrating (or summing) the result over the period of the sequence. We note that this is very similar to the de-spreading process described in the earlier section. However, in this case the local version of the chip sequence is generated at a slightly slower or faster chip rate than the sequence in the received signal so that the two sequences gradually 'slip' past each other in time, i.e. the correlation process 'sweeps' in time. In this way, the UE performs a correlation of the two identical chip sequences, one in the received signal and one locally generated in the UE, for a range of different time offsets. When the two sequences align in time, all of the '+1' and '−1' levels will align and the result will be a constant '+1' level for the duration of the code. If the two sequences are not aligned, then there will be '+1' and '−1' levels in the resulting signal. This process is shown very simply in Figure 2.17. In Figure 2.17a we have shown a chip sequence multiplied by a time synchronised copy of itself and the result is an all +1's sequence which, when summed, produces a value of +7. In Figure 2.17b the two chip sequences are no longer aligned and they are offset by one chip period, i.e. the two sequences have slipped in time by one chip period. In this case the sum of the output is now −2.

In Figure 2.18, we use the process shown in Figure 2.17 to derive the autocorrelation function (ACF) of the primary synchronisation code. We can see that the function has a strong peak at an offset of zero chips, i.e. when the two sequences are aligned. We can also see that when the sequences are not aligned, the output of the correlation process is very much lower than the peak level. The UE uses this autocorrelation characteristic of the primary synchronisation code to determine two pieces of important information. First, if the UE detects a high value at the output of the autocorrelation process, it knows that it has found a nearby Node B that could potentially offer service. Secondly, it knows that the time offset associated with the peak in the autocorrelation function represents the start of a 2560-chip timeslot on the transmissions from the Node B.

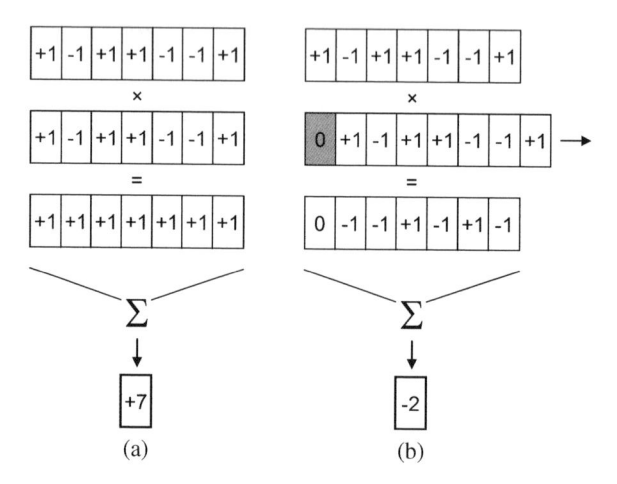

Figure 2.17 The autocorrelation process.

Introduction to the UTRA FDD Radio Interface

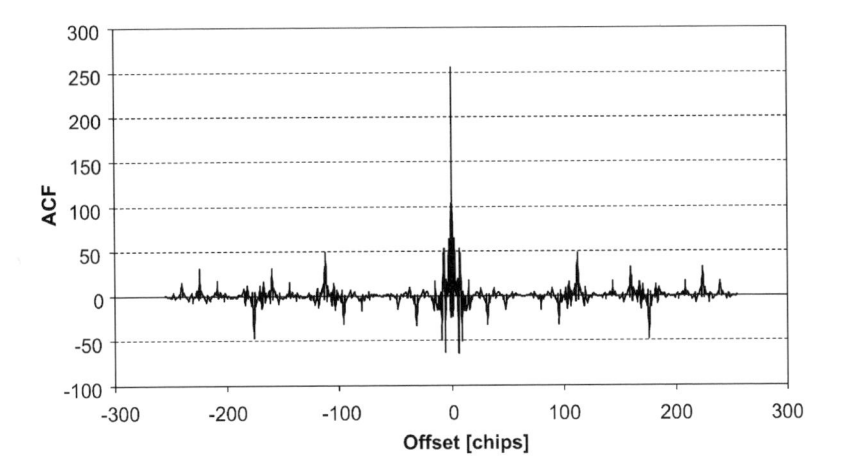

Figure 2.18 The autocorrelation function of the primary synchronisation code.

Having detected a Node B and identified the start of a timeslot, the next information the UE must learn is the identity of the other CDMA codes used at the Node B and also further information about the timeslot and frame structure used on the Node B. The SCH also consists of a second CDMA code known as the secondary synchronisation code and this is also 256 chips in length and it is transmitted at the same time as the primary synchronisation code. There are in fact 16 different secondary synchronisation codes that can be transmitted and the sequence in which these codes are transmitted is used to indicate to the UE the group of primary scrambling codes that are used at the Node B. The primary scrambling code is used to scramble all of the other channels transmitted by the Node B. The sequence of the secondary synchronisation codes also indicates the start of a 10 ms frame, which consists of 15 timeslots and this allows the UE to gain full time synchronisation with the Node B transmissions.

2.2.2.2 Pilot Channels

The next type of downlink channel we shall consider is the pilot channels. These channels are used to provide a coherent reference for the decoding of some of the other downlink channels that do not carry their own reference. In the UTRA FDD system, the main pilot channel is called the common pilot channel (CPICH) and it can consist of a single primary CPICH and a number of secondary CPICHs. Every cell in the network will transmit a single primary CPICH and this essentially consists of an unmodulated version of the cell's primary scrambling code, which is 38400 chips in length and it is transmitted at 3.84 Mchips/s, i.e. it is repeated once every 10 ms or once every frame. Since the UE already knows when the frame starts from the secondary synchronisation code, it will know when the primary scrambling code starts since this is aligned with the frame boundaries.

If we compare Figure 2.16a and Figure 2.18, we can see that the scrambling codes have much smaller values for the autocorrelation function when the offset is non-zero than the primary synchronisation code and, therefore, the CPICH can be used to provide a better estimate of the propagation channel than the SCH. We also note that all Node Bs will use the same primary synchronisation code and, as a result, the UE will not necessarily be able to differentiate between signal components arriving from different Node Bs. There are 512 different primary scrambling codes available within UTRA FDD and, therefore, it should be possible to ensure that a single primary scrambling code is used only once in a particular area. This allows the UE to associate the signal components received over different paths with the correct Node Bs.

In practice, the UE will need to update its estimate of the propagation channel more rapidly than once every 10 ms, particularly if the UE is moving. Therefore, the UE may not be able to perform a correlation over the entire length of the primary scrambling code and it will perform a partial correlation over only part of the code. In Figure 2.19 we show the partial autocorrelation and cross correlation functions for one of the scrambling codes. If we compare this with Figure 2.16 we can

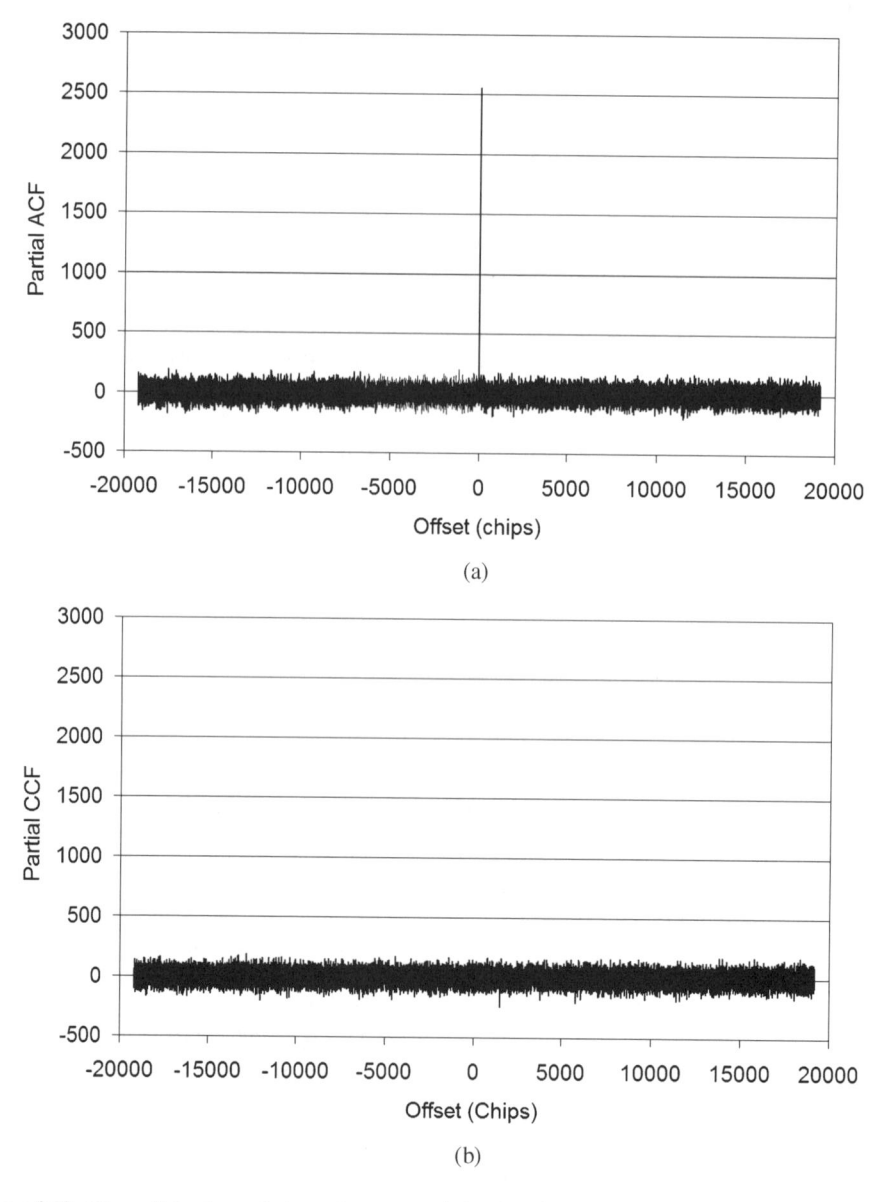

Figure 2.19 The partial autocorrelation and cross correlation functions of the scrambling code for 2560 chips.

see that the value of the cross correlation function and the value of the autocorrelation function for non-zero offsets is greater relative to the autocorrelation peak at a zero offset. However, the difference is still large and the CPICH is still effective for providing an accurate estimate of the radio channel even if only a partial correlation is performed.

The secondary CPICH is very similar to the primary CPICH in that it is used to provide a coherent reference for demodulation of other channels. However, whereas the primary CPICH must be transmitted over an entire cell, the secondary CPICH is designed to be transmitted over only parts of a cell and its main function is to support coverage 'hot spots' created by separate directional antennas at the Node B. Where a pilot channel is to be used as a coherent reference for another channel, it is important that the pilot signal is transmitted over an identical radio channel to the other physical channels. In this context the radio channel includes both the transmitting and receiving antennas, as well as the physical propagation channel that exists between the Node B and the UE. Therefore, if we want to set up a small coverage hot spot within a particular area by using a directional antenna to transmit a set of traffic channels, we cannot use the primary CPICH as a coherent reference for these channels because it will not be using the same transmitting antenna. Therefore, the secondary CPICH is used to provide the common coherent phase reference in this hot spot and it is transmitted using the same directional antenna as the other physical channels that are associated with the hot spot. The second CPICH will be scrambled using one of the 15 secondary scrambling codes that are associated with the primary scrambling code that has been assigned to the Node B. The secondary CIPCH can also use any of the 256-chip channelisation codes.

2.2.2.3 Common Control Channels

In addition to the synchronisation and pilot channels, the Node B transmits a number of common control channels that are available to all the UEs within the coverage area of a Node B. In this section we will examine each of these channels in turn and explain their function and general format. If the reader would like a more detailed explanation of the construction of each of these channels then this can be found in references [2] and [3] or the system specifications themselves, which can be accessed via the Third Generation Partnership Project website [4].

Common control physical channels
Having gained full synchronisation with the Node B, the UE must now learn more information about the cell and this is achieved by decoding the common control physical channels (CCPCH). There are two CCPCHs, namely the primary-CCPCH (PCCPCH) and the secondary-CCPCH (SCCPCH). The PCCPCH is used to carry a higher layer 'transport' channel known as the broadcast channel (BCH). The BCH carries a range of information that is required by all of the UEs within the coverage area of a Node B including the identity of the network operator, the identity of the Node B (or cell) and information on the other channels transmitted from the Node B. The BCH also carries information that allows a UE to decide how to switch between Node Bs as it moves around the network either in idle mode (i.e. when it is not actively engaged in a call) or in connected mode (i.e. when it is actively engaged in a call). The PCCPCH carries data at a fixed rate of 30 kbps, it is always transmitted on channelisation code $c_{ch,256,1}$ and it is always scrambled using the cell's primary scrambling code.

The SCCPCH carries two transport channels, namely the forward access channel (FACH) and the paging channel (PCH). The FACH is used to communicate with UEs within a particular cell without the need to allocate them a dedicated channel. It can be used to transmit channel assignment messages to a UE and it can also be used to transmit limited amounts of user packet data. Since the SCCPCH is a common channel, information that is transmitted on the FACH must be individually addressed to the particular UE for which it is intended. As its name would suggest, the PCH is used to transmit paging messages to the UEs within a cell, i.e. messages alerting the UE to the presence of an incoming call. Again, the paging messages must be individually addressed to the particular UEs. The SCCPCH can be transmitted at any data rate ranging from 30 to 1920 kbps and it can use any channelisation code. The SCCPCH can also be scrambled using either a primary scrambling code or a secondary scrambling code and each cell will have one or a number of SCCPCHs.

Indicator channels
The UTRA FDD system has two additional downlink common control channels known as 'indicator' channels because they are used to rapidly indicate an important piece of information to the UE. The first indicator channel is the paging indicator channel (PICH) and this is used by the Node B to indicate to the UEs in a cell whether or not they should expect to find a paging message addressed to them on the PCH. The PICH is used to remove the need for a UE to continually listen to the PCH for incoming paging messages and this, in turn, helps to improve the battery life of the UE by decreasing the power consumption in idle mode. When a UE registers on the network it will be assigned to one of a number of paging groups. If the network would like to page a UE belonging to a particular paging group it transmits a flag belonging to that group on the PICH. If a UE sees a flag belonging to its paging group on the PICH it will then decode the next frame that is transmitted on the PCH to check whether it contains a paging message with its specific address. The flags associated with a particular paging group are transmitted at regular intervals on the PICH and the UE is able to enter 'sleep mode' and conserve its battery power in the periods between the transmission of PICH flags.

The second indicator channel is the acquisition indicator channel (AICH) and this is used to assist the UE to access the network. As we will see later in this chapter, it is important that all UEs use just enough transmitted power on the uplink to maintain a satisfactory connection with the Node B. Using too little power will result in a poor link quality and a bad user experience and using too much power will result in excessive levels of interference within the system, which will decrease the system capacity and degrade the link quality for other users. When the UE is engaged in a call, a fast uplink closed loop power control mechanism is used to regulate the UE's transmitted power. However, when the UE first accesses the network this feedback loop does not exist and an alternative means must be found to establish the correct transmit power for the UE. The UE can gain a rough estimate of the amount of uplink power required to reach a Node B by measuring the received power from the Node B and by using information received on the BCH regarding the transmit power of the Node B. However, this open loop estimate of the transmit power will have a large margin of error because the uplink and downlink transmissions occur at different frequencies and, as a result, the multipath fading on each link can be quite different. Therefore, the approach that is adopted in UTRA FDD is that the UE estimates the uplink power required to reach the Node B based on the downlink received power and then subtracts a margin to account for the potential difference between the uplink and downlink fading. It then transmits a message on the uplink physical random access channel (PRACH) at this power and listens to the AICH for a response. The PRACH message contains a random 'preamble signature' that the UE selects from a number of signatures that the Node B indicates are available. If the Node B hears the UE's transmissions, then it will respond by setting a flag on the AICH to indicate that it has heard a particular preamble signature. If the UE sees that the flag corresponding to its selected preamble is not set, then it assumes that it was not heard by the Node B and it will transmit another randomly selected preamble signature at a slightly increased power. Eventually, the Node B will hear the UE and respond by setting the relevant flag on the AICH, at which point the UE will then transmit the remainder of its access message. In this way, the amount of transmit power used by the UE to access the network can be kept to a minimum at the expense of a small delay in the UE accessing the network.

Random access channel
So far in this chapter we have considered only downlink channels, i.e. those channels that are transmitted by the Node B. As far as common signalling channels are concerned, we must also consider the uplink

physical random access channel (PRACH), which is used by the UE to initially access the network either to register with the network following power on, to inform the network that it has moved into a new location area, to establish a user-initiated call or to respond to a page. Since the network does not know when UEs will want to establish a link with a Node B, it cannot share the PRACH resources between the different UEs using rigid code or timing constraints. Instead, the PRACH uses a number of randomisation algorithms to provide the UEs with access to the shared resource.

We have already explained how the power control mechanism works on the PRACH in association with the downlink AICH. The PRACH also includes a number of other mechanisms to decrease the probability that two UEs will interfere with each other, or collide, as they try to access the PRACH resources. First, the PRACH preamble signature can be scrambled with one of a number of different codes and the UE randomly selects one of the codes from the ones that are available. The PRACH is also organised into a timeslot structure, or access slots, and the UE randomly chooses the access slot in which it transmits its preamble signature.

Packet data control channel
In Release 5 of the UMTS specifications a new feature, known as high-speed downlink packet access (HSDPA), was added to the system and this involved the introduction of two new downlink physical channels, namely the high-speed physical downlink shared channel (HS-PDSCH) and the high-speed shared control channel (HS-SCCH). In this section we consider the HS-SCCH and in the next section we consider the HS-PDSCH.

The HS-SCCH is used to control the use of the associated shared HS-PDSCH channels and it always uses a 128-chip channelisation code and the data rate is 60 kbps. The data that are transmitted on this channel are arranged in such a way that the information required by the UE to decode the data transmitted on the HS-PDSCH is transmitted first, to allow the UE to start decoding the HS-PDSCH before the HS-SCCH has been fully received. This initial information consists of details of the channelisation codes used on the HS-PDSCH and also the modulation scheme used. The second part of the HS-SCCH includes a check sum to allow the UE to identify errors in the transmitted control information and also a number of parameters relating to the hybrid automatic repeat request (HARQ) mechanism used on the HS-PDSCH.

2.2.3 DEDICATED PHYSICAL CHANNELS

Having examined the various common control channels, we now turn our attention to the 'dedicated' channels, i.e. those channels that are dedicated to point-to-point communications between a particular Node B and a UE. We will consider the downlink and uplink dedicated channels separately in the following sections.

2.2.3.1 Downlink Dedicated Channels

There are two types of downlink dedicated channels, the dedicated physical channel (DPCH) and the physical downlink shared channel (PDSCH). The downlink DPCH consists of two components, the dedicated physical data channel (DPDCH) and the dedicated physical control channel (DPCCH) and these are time-multiplexed together. The DPDCH is used to carry user data from the Node B to the UE and the DPCCH is used to carry signalling information to the UE consisting of pilot bits to assist in the decoding of the DPDCH, power control bits to rapidly adjust the transmit power of the UE and transport format bits to describe the structure of the transmitted information. The DPCH can use channelisation codes with lengths from 4 to 512 chips and the channel bit rate can range from 15 to 1920 kbps. This provides an available raw throughput on the DPDCH of between 3 and 1872 kbps before channel coding. It is also possible to allocate multiple downlink DPCHs to the same

UE, each with a different channelisation code, to increase the data rate beyond that possible with a single channel. The downlink DPCH can be scrambled with either the cell's primary or secondary scrambling code.

As its name suggests, the PDSCH is a channel that can be shared by a number of UEs. The main purpose of the PDSCH is to carry 'bursty' high rate data between the Node B and the UE where the transmissions are infrequent. It provides an alternative to establishing a DPCH for each transmission, which could unnecessarily consume a large proportion of the shorter (i.e. higher data rate) channelisation codes. The Node B still requires a dedicated signalling path to a UE and, therefore, any UE that is assigned to a PDSCH will also establish a downlink DPCH. However, the data rate of the DPCH can be very low and, hence, it will consume only a small portion of the overall channelisation code tree resources. The spreading factor used on the PDSCH can be varied from one 10ms frame to the next and more than one UE can share the PDSCH resources simultaneously.

As mentioned in the previous section, in Release 5 of the UMTS specifications a new feature, known as HSDPA was added to the system and this involved the introduction of the HS-SCCH and HS-PDSCH. The HS-PDSCH carries the transport channel known as the high-speed downlink shared channel (HS-DSCH), which is used to transfer user data from the Node B to the UE. The HS-PDSCH has a number of features that distinguish it from channels defined in earlier versions of the UMTS specifications. First, 16-level quadrature amplitude modulation (QAM) is supported in addition to quadrature phase shift keying (QPSK), which is supported in earlier versions of the specification. The introduction of this higher level modulation scheme provides a doubling of the throughput on the HS-PDSCH compared with the QPSK modulation scheme. The second unique characteristic of the HS-PDSCH is its use of a shorter interleaving period or transmit time interval (TTI) of 2 ms compared with a minimum of 10 ms in Release '99. This allows the UE to decode the received data and request retransmissions more quickly, hence decreasing the overall latency in the data channel. Finally, the packet scheduling is handled within the Node B itself, whereas in Release '99 the packet scheduling was handled at the Radio Network Controller (RNC). This means that, when packet retransmissions are required, these can be carried out much more rapidly and, again, the overall latency in the data channel is further decreased.

The HS-PDSCH always uses a spreading factor of 16 and this means that a total of 15 channelisation codes are available for this channel, since some of the channelisation code resources must be reserved for common control channels. The channel can support multicode transmission which means that, depending on the capabilities of the UE, all 15 codes could be assigned to one UE, giving a maximum theoretical data rate of 14.4 Mb/s. The codes can also be shared between different UEs allowing the channel resources to be multiplexed between different users. For example, three UEs could each be assigned five codes allowing a data rate of 3.6 Mb/s to be delivered to each user.

2.2.3.2 Uplink Dedicated Channels

In common with the downlink, the uplink has two dedicated channels, namely the DPDCH and the DPCCH. These channels are similar to those used on the downlink; however, the main difference is that they are code multiplexed rather than time multiplexed. The DPCCH carries control information, including pilot bits, channel format information, feedback information (when closed loop transmit diversity is used) and power control bits and it always uses a 256-chip spreading code ($c_{ch,256,0}$). The DPDCH is used to carry user data and higher layer control information and it can use channelisation codes with lengths of 256 down to 4 chips. Both channels are scrambled using the UE-specific scrambling code.

In order to increase the uplink data rate beyond that supported by a single channel, a UE can transmit up to six DPDCHs. If more than one DPDCH is transmitted, then they will all use four-chip channelisation codes.

In Release 5 of the UMTS system specifications, a further uplink channel called the high-speed dedicated physical control channel (HS-DPCCH) was added to support HSDPA. This channel is used to signal whether or not a downlink packet has been received correctly as part of the HARQ scheme and it is also used to indicate the quality of the downlink channel to assist the Node B in choosing suitable channel configurations in the downlink direction. The HS-DPCCH contains 30 bits of information per 2 ms sub frame and, therefore, it has a throughput data rate of 15 kbps and it always uses a 256-chip channelisation code.

2.3 UTRA FDD KEY MECHANISMS

Having described the various channels that go to make up the UTRA FDD physical layer, we now turn our attention to some of the characteristics and features of the UMTS system that are important to the overall performance of a network.

2.3.1 CELL BREATHING AND SOFT CAPACITY

The first characteristic to examine is 'cell breathing', since it is vital to understand this concept before we can hope to understand the performance of a CDMA network. Cell breathing is used to describe the way in which the coverage of a Node B changes (or breathes) in response to changes in the network load. To show this effect let us start with a very simple example of a UE with a fixed transmit power communicating with a single nearby Node B. The signal from the UE is received and decoded at the Node B receiver in the presence of thermal noise only. If the UE now moves away from the Node B, a point will be reached where the strength of the received signal at the Node B is just sufficient to overcome the thermal noise and maintain an adequate link quality. At this point we can consider the UE to be located at the edge of the coverage cell. If we now add a second active UE to the network, this UE will cause interference to the signal from the first UE at the Node B receiver. In order to overcome this additional interference, the first UE must deliver more power to the Node B receiver and, since the UE has a fixed transmit power, this can only be achieved by the first UE moving closer to the Node B. As the first UE moves closer to the Node B, a point will be reached where its received signal strength at the Node B is just sufficient to overcome the effects of the thermal noise and the interference introduced by the second UE. This point effectively becomes the new cell boundary. As more and more UEs are added to the system, and hence the first UE suffers more and more interference, the cell boundary moves closer to the Node B, i.e. the cell shrinks. Conversely, as UEs leave the system and the interference decreases, the cell will expand. In a practical network, the network load (i.e. the number of users accessing the network) will vary considerably throughout the day and this will cause the cells in the network to 'breathe' in response to this load.

To examine this cell breathing effect in more detail, let us consider the classic uplink CDMA equation for the ratio of energy per bit to interference and noise power spectral density:

$$\frac{E_b}{I_0 + N_0} = \frac{\frac{C}{R_b}}{N_0 + \frac{\alpha(1+i)(N-1)C}{W}} = \frac{G_p}{\frac{N_0 W}{C} + \alpha(1+i)(N-1)} \quad (2.1)$$

where C is the required received power, R_b is the user bit rate, N_0 is the power spectral density of the noise, W is the CDMA chip rate, N is the number of active users, α is the transmitter duty cycle under discontinuous transmission, G_p is the processing gain (i.e. W/R_b) and i is the ratio of the interference generated from neighbouring cells (intercell interference) to interference generated from within the serving cell (intracell interference). Rearranging Equation (2.1) to solve for C, we have:

$$C = \frac{N_0 W \left(\frac{E_b}{I_0 + N_0}\right)}{G_p - \left(\frac{E_b}{I_0 + N_0}\right)\alpha(1+i)(N-1)}. \qquad (2.2)$$

Figure 2.20 shows a plot of the required received power, C, against the number of users N based on Equation (2.2) and assuming a processing gain of 256, a value for $E_b/(I_0+N_0)$ of 7 dB, a value for α of 50%, a value for i of 55% and a receiver noise figure of 5 dB. This shows the manner in which the required received power increases as the number of users increases. The increase in received power is gradual at first, but then it starts to increase more rapidly as more users are added to the network. At some point we reach a value for N that causes the denominator in Equation (2.2) to become zero and, hence, C goes to infinity. Since no practical transmitter can generate an infinite amount of power, this value of N can never be reached in a practical system and it is termed the 'pole capacity' of the network. If a practical network starts to approach its pole capacity then it can become unstable, with the transmit power requirements of the UEs varying dramatically for very small changes in the network load. Therefore, practical networks are usually designed to operate at a certain fraction of their pole capacity and new calls are rejected once this limit is reached. For a more detailed description of these effects please refer to Chapters 9 and 10.

The system load in the uplink direction can be measured in terms of equivalent 'noise rise' at the Node B, which is defined as the additional power that must be delivered by a UE at the Node B to overcome the interference generated by other UEs. Returning to Figure 2.20, we can see that with a single user on the network, this UE must be received with a power of −120.5 dBm. However, if the network load increases to 10 users, then each UE must deliver a power of −119.9 dBm at the Node B receiver, i.e. an increase or noise rise of 0.6 dB. In a practical network, an operator may choose to limit the network load to 75% of the pole capacity and this equates to a noise rise of 6 dB. Once the Node B detects that the total received noise and interference power at its receiver is 6 dB greater than the thermal noise alone, it will reject any new calls.

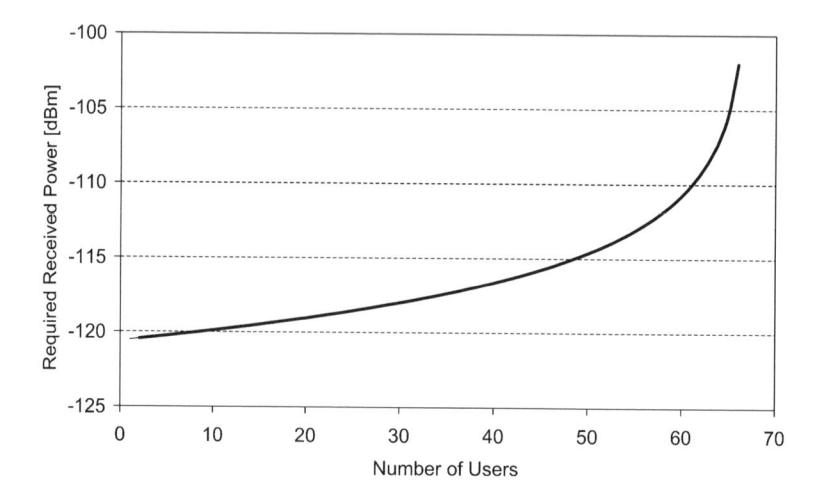

Figure 2.20 The relationship between the received power and the number of users.

This technique of using noise rise to control the load on a Node B is the simplest form of call admission control (CAC) and it can be used to limit the degree of cell shrinkage within the network. If we assume a pathloss slope of 36 dB per decade increase in distance, a 6 dB rise in required received power represents a 32 % decrease in cell range. In other words, if a network operator is going to allow its network to reach 75 % of its pole capacity, then it must allow for a cell shrinkage of 32 % in range (or roughly 47 % in area), compared with an unloaded network, to ensure that coverage holes do not appear in the network at times of peak load.

In an urban area, a cell shrinkage of 32 % may be acceptable since the Node Bs are likely to be closely packed to support the high levels of offered traffic. However, in a rural area, where the Node Bs will be less densely packed and maximising the coverage of individual Node Bs is more critical, an operator may choose to operate at a lower maximum load and thereby decrease the maximum degree of cell shrinkage. For example, by limiting the maximum noise rise to 3 dB, the load will be limited to 50 % of the pole capacity and the cell shrinkage will be limited to 17 % of the cell radius.

The above discussion clearly demonstrates the concept of soft capacity. If we consider a FDMA or TDMA system, then the capacity of the system is limited by the number of frequency channels or timeslots available at a base station. Once all the channels or timeslots are filled, then the system is full and no more new calls can be accepted. The networks are designed to operate right up to this 'hard' capacity limit. In CDMA networks, on the other hand, the capacity limit is soft and network capacity and coverage can be traded off against each other. By increasing the maximum permitted noise rise at the Node B receiver, the network capacity can be increased, but this increase is at the expense of decreasing the coverage of the Node B.

2.3.2 INTERFERENCE AND POWER CONTROL

The capacity and coverage of a CDMA system is directly linked to the amount of interference generated by the Node Bs and UEs within the system. Interference is an inherent feature of CDMA systems and it cannot be avoided, but mechanisms must be put in place to ensure that it is kept to the minimum level required to serve the needs of the network users. Power control is the main technique used in a CDMA network to control interference and an effective power control mechanism is critical to the performance of a CDMA network.

The performance of a CDMA system will be optimised if each transmitter uses just sufficient power to support an adequate link quality between itself and the receiver. If too little transmit power is used, then the link quality will be poor and this will lead to a bad user perception of the system quality. If the transmit power is too high, then the transmitter will cause excessive interference to other users and this will have the impact of degrading the link quality of these other users. Power control is particularly challenging in mobile radio networks because the effects of mobile movement and multipath fading mean that the attenuation imposed by the radio channel can vary dramatically in a very short space of time. The dynamic range of the radio channel attenuation is also very large as the mobile moves around the network and its distance to the Node B changes.

The most challenging link as far as power control is concerned is the uplink. Each UE will transmit from a different location and the uplink power control mechanism must ensure that they are all received at the Node B at the correct power to meet the SIR target value and any differences imposed by the attenuation of radio channels between the Node B and the UEs are removed. On the downlink the situation is less challenging because all of the channels are transmitted from the same point (i.e. the Node B transmitter) and, since they will all pass through the same radio channel on their way to a UE's receiver, the power of each channel will remain the same relative to the other channels. Therefore, on the downlink, power control is used to adjust the relative power between the different channels to assist UEs that are suffering from increased interference from neighbouring Node Bs and it is not used to overcome the effects of different propagation channels.

On the uplink, the power control mechanism uses both an open loop and a closed loop technique. The open loop technique is used when the UE initially accesses a Node B and there is no feedback path to implement the closed loop mechanism. In the open loop mechanism, the UE measures the downlink power received from the Node B and uses this to determine its own transmit power. This open loop technique allows the UE to adapt its transmit power to slow changes in the radio link attenuation caused by changes in the distance between the UE and the Node B and the effects of objects causing shadow fading on this link. However, the uplink and downlink use different frequency bands and, as a result, the multipath fading on each link will be different. Therefore, the open loop mechanism cannot be used to compensate for the effects of multipath fading and a closed loop mechanism is used once a bi-directional link between the Node B and the UE has been established to provide a more accurate means of controlling the UE transmitted power.

The closed loop mechanism consists of the Node B measuring the SIR on the received uplink dedicated physical channel from the UE. This SIR value is compared with a target value and a power control command is sent by the Node B to the UE based on this comparison. If the measured SIR is below the target SIR, then the power control command will instruct the UE to increase its transmit power. Conversely, if the measured SIR is above the target value, the power control command will instruct the UE to decrease its power. In its normal mode of operation, the UE will respond to the power control commands by either increasing or decreasing its transmit power by the step size, which can be either 1 or 2 dB. More complex power control command processing algorithms are also available that allow smaller step sizes to be emulated and also allow power control to be turned off. The power control commands are sent to the UE once every timeslot (i.e. once every 667 μs), which means that the UE can alter its transmit power 1500 times per second.

The SIR target at the Node B is set by an outer control loop that operates more slowly than the fast, inner power control loop. In the outer loop, the bit error rate (BER), block error rate (BLER) or frame erasure rate (FER) experienced on the uplink dedicated channel is compared against a target BER/BLER/FER for a given service and, based on this comparison, the target SIR at the Node B is adjusted up or down to bring the measured BER/BLER/FER in line with the target value.

Closed loop power control is also supported on the downlink. In this case the UE measures the SIR of the downlink dedicated channel and compares this against a target SIR derived from a higher layer outer loop. If the SIR is below its target, the UE will send a power control command to the Node B requesting an increase in the power allocated to its dedicated channels. Conversely, if the measured SIR is above the target, the UE will issue a power control command asking the Node B to decrease the power allocated to its dedicated channels.

2.3.3 SOFT HANDOVER AND COMPRESSED MODE

As we have discussed earlier, CDMA networks can use the same frequency in every cell and users are distinguished by means of codes. This means that it is a relatively simple task for a UE to decode the signals from more than one Node B simultaneously by despreading the single received radio signal using a number of different scrambling and channelisation codes. This technique is exploited in CDMA to support a feature known as soft handover, whereby a UE can communicate with more than one Node B simultaneously as it moves between cells in the network. The soft handover has a number of advantages compared with the hard handover used in FDMA and TDMA systems such as TACS and GSM, in which simultaneous communication with more than one base station is not allowed. First, the soft handover between neighbouring cells should be more reliable, since it allows the UE to establish a communication with the target Node B before communications with the original serving cell are relinquished, i.e. it is a 'make-before-break' handover. If any problems occur with the establishment of a link with the target cell, the UE can still maintain communications with the network via the original cell. In systems, such as GSM, where the handovers are 'break-before-make', there is always a chance that the mobile is unable to establish a link with the target cell and, since the link to the original cell has been broken, it is more difficult to re-establish that link and the call may drop.

Another advantage of soft handover is the ability to exploit the macrodiversity gains between different Node Bs. Most cellular systems use multi-antenna diversity at the base station receiver and more recent cellular systems can also support transmit antenna diversity at the base station transmitter. However, these 'microdiversity' techniques are mainly aimed at mitigating the effects of multipath fading and they have very little impact on the larger scale effects of shadow fading. If a mobile can establish simultaneous links to different Node Bs, then, given the spatial separation of the Node Bs, each link is likely to experience a different shadow fading characteristic. By combining the information received on each individual link, the effects of shadow fading can be mitigated to some degree and this is termed macrodiversity.

In the UTRA FDD system, soft handover is controlled by means of an active set, which contains all of the Node Bs with which the UE is currently communicating. If the active set contains more than one Node B, then the UE is deemed to be in soft handover. On average we might expect around 30 % of the UEs to be in soft handover in a typical network. On the downlink, all of the Node Bs in a UE's active set will transmit the same user data to the UE. The UE will receive the signals from each Node B and use combining techniques (e.g. maximal ratio combining) to determine the most likely transmitted data pattern. The macrodiversity gains mean that the UE will be able to demand less power from each Node B compared to the situation where it was communicating with a single Node B.

On the uplink, the UE will continue to transmit a single uplink signal, but this will be received by more than one Node B when the UE is in soft handover. Each Node B will individually decode the user data and send this to the RNC with an indication of the quality of the recovered data. The RNC will select the data from the Node B with the best quality on a frame-by-frame basis. This technique is known as switched diversity and the associated gains can be used to decrease the overall transmit power of the UE.

If a UE is in soft handover between two sectors of the same Node B, this is known as softer handover and the Node B can perform maximal ratio combining with the uplink signals received on each sector.

One complication that arises with soft handover involves the interpretation of the uplink power control commands received by the UE. The Node Bs involved in the soft handover process may provide the UE with conflicting power control commands. However, the UMTS specifications define a mechanism for combining different power control commands received at the UE into a single command that is implemented by the UE. The combining mechanism takes into account the relative reliability of the different received commands and it also tends to be biased towards the UE decreasing its power. In other words, if the UE receives equal numbers of 'power up' and 'power down' commands, it will tend to decrease its power.

Although the UTRA FDD system can support soft handovers, it must also have the ability to support hard handovers. This type of handover is required when the UE moves between different CDMA radio carriers or between systems (e.g. moving from UTRA FDD to GSM). This presents a problem to the UE since it must make measurements on a different frequency to assess the suitability of the new radio carrier to support the ongoing call, but there are no natural breaks in the downlink transmissions that would allow the UE to do this. There are two possible solutions to this problem. One solution would be to include two separate radio frequency front ends in every UE. This would allow the UE to make measurements of one radio carrier whilst still continuing to decode the downlink transmissions from its serving Node B on a different radio carrier. Whilst this approach is relatively straightforward from the network point of view, it adds a level of complexity to the terminal that, in many cases, would be unacceptable.

Therefore, a second option exists whereby gaps are opened up in the downlink transmissions to give the UE an opportunity to retune to another radio channel and make a measurement. Unfortunately, the amount of data that must flow between the Node B and the UE in the downlink direction does not necessarily decrease during the periods when these measurements are required and this means that the Node B must transmit at a higher data rate on either side of the measurement gaps to ensure that the same amount of data can be transferred. This mode of operation is referred to as compressed mode because of the manner in which the data is compressed into the transmission periods on either side of the measurement gap.

The simplest way to increase the data transmission rate is to decrease the channel spreading factor. For example, the spreading factor could be decreased from 64 to 32, thereby doubling the channel throughput and allowing the Node B to operate with a 50 % transmission duty cycle. Since the decrease in spreading factor leads to a reduction in processing gain on the channel, the Node B will need to increase the transmit power for a particular UE to compensate for this effect. Code puncturing may also be used as a means of increasing the user data rate without changing the spreading code. In this case, some of the coded data bits are not transmitted and the channel decoding process at the receiver is relied upon to recover these 'lost' bits. This is similar to the situation that would occur if errors were imposed on these bits during transmission, but the receiver has the added advantage that it will know which data bits have been removed, whereas it does not usually know which bits contain errors. This puncturing process has the effect of decreasing the error correcting capabilities of the channel coding (i.e. the power of the code) and this has a similar effect to decreasing the processing gain, i.e. the Node B needs to allocate more power to the UE during the puncturing periods.

In some cases, it may also be possible to decrease the amount of data that is transmitted between the Node B and the UE during periods when the UE must make inter-frequency measurements. This requires the schedule for compressed mode operation to be communicated to the higher layers in the protocol stack and these higher layers then restrict the amount of information presented to the physical layer during compressed mode operation.

2.4 PARAMETERS THAT REQUIRE PLANNING

When we consider the parameters that require planning and optimisation within a UTRA FDD network, these broadly break down into three main categories, namely signal path, power allocation and system settings. We will briefly examine each of these categories in the following sections.

2.4.1 SIGNAL PATH PARAMETERS

In this category, we place all those parameters that have an impact on the signal path between the Node B and the UE. These include such parameters as:

a) the placement of the Node B sites,
b) the degree of sectorisation used at a site,
c) the number of transmitting and receiving antennas used at the Node B,
d) the height of the Node B antennas,
e) the beamwidth of the Node B antennas,
f) the direction (azimuth) of the Node B antennas,
g) the downtilt of the Node B antennas and
h) the use of tower top amplifiers at the Node B receiver.

In general these parameters are chosen when a Node B site is initially introduced into the network and they remain fixed for the lifetime of the site. However, adjustments to parameters such as antenna beamwidth and downtilt can be used to address network capacity and coverage problems as part of an optimisation procedure.

2.4.2 POWER ALLOCATION

In this category, we place all those parameters that are associated with the allocation of power to the different channels within the UTRA FDD system. This includes the fraction of the overall Node B transmitted power that is allocated to the pilot channels, the synchronisation channels and the common control channels. These channels represent a system overhead and any resources consumed by these channels will not be available for the revenue-generating traffic channels. However, the common channels also provide the means by which the UEs detect and access the network and, hence, they must be transmitted with sufficient power to ensure that they can be successfully decoded by all UEs within the network. In addition, the UEs use the received pilot channel power from the Node Bs to make handover decisions and changes in the pilot power transmitted by different Node Bs can be used to alter the coverage area of individual Node Bs.

2.4.3 SYSTEM SETTINGS

In this category, we place all of the parameters that are used to control the behaviour of the UTRA FDD network and the associated UEs. There is a wide range of different parameters in this category and they are too great in number to list here. We also note that, whilst many of the parameters will be common across all UTRA FDD networks, other parameters will be specific to particular implementations of the system, i.e. they will be vendor-specific. This category of parameters will relate to all aspects of the system behaviour, including

a) network acquisition and access by the UE,
b) call admission control and radio resource allocation within the network,
c) uplink and downlink power control,
d) handover control and
e) radio link failure control.

In general many of the system settings that are available to the network operator will not be adjusted and will remain at their default settings. Other parameters may be set based on the type of site (e.g. rural or urban, high capacity or low capacity) and they may not change for the lifetime of the site, whilst others will be adjusted periodically as part of regular optimisation processes, as introduced in the later parts of this book.

Before proceeding to more detailed planning and optimisation issues, however, we will briefly dwell on the service and spectrum aspects of 3G communication systems in the next chapter.

REFERENCES

[1] Viterbi, A.J., *CDMA Principles of Spread Spectrum Communication*, Addison-Wesley, Reading, MA, 1995.
[2] Steele, R., Lee, C.-C. and Gould, P., *GSM, cdmaOne and 3G Systems*, John Wiley & Sons, Ltd/Inc., Chichester, 2001.
[3] Holma, H. and Toskala, A. (ed.), *UTRA FDD for UMTS – Radio Access For Third Generation Communications*, Third Edition, John Wiley & Sons Ltd/Inc., Chichester, 2004.
[4] 3rd Generation Partnership Project (3GPP) website, http://www.3gpp.org.

3

Spectrum and Service Aspects

Maciej J. Grzybkowski, Ziemowit Neyman and Marcin Ney

As part of the introductory part of the book, we will now briefly dwell on two loosely related yet very important issues, i.e. the 3G spectrum allocation and service provision. We will elaborate on the historical development behind today's choice of frequency bands which, in one way or another, influences solutions provided by optimisation techniques. To complete the overview of 3G, we have also included a section on the way 3G handles service provision by means of bearer and associated quality of service (QoS) parameters. The later also have a profound influence on planning and optimisation techniques.

3.1 SPECTRUM ASPECTS

Already in the early 1980s of the past century, the International Telecommunication Union (ITU) was conducting research in the area of universal telecommunication systems, aimed at multimedia transmissions – called at that time 'Future Public Land Mobile Telecommunications Systems' (FPLMTS) and nowadays known as 'International Mobile Telecommunications – 2000' (IMT-2000). Research was conducted by both the ITU Radiocommunication sector (ITU-R) and the ITU Telecommunication Standardisation Sector (ITU-T). This included analysis of services, technical solutions and operating systems, as well as the electromagnetic spectrum requirements.

In the 1990s, when Europe decided to implement a Universal Mobile Telecommunications System (UMTS) compatible with the IMT-2000 standards family, the spectrum requirements were analysed by the European Radiocommunication Committee (ERC), established by the European Conference of Postal and Telecommunications Administrations (CEPT). Currently, spectrum related aspects are simultaneously considered by two organisations: Project Team PT1 established by the ECC (Electronic Communications Committee is a successor of the ERC after being merged with European Committee for Telecommunications Regulatory Affairs, ECTRA) and the European Telecommunication Standard Institute (ETSI) – under the universal 3rd Generation Partnership Project (3GPP). Independent research is also being conducted by the UMTS Forum – an association of operators, manufacturers and regulatory

---

Understanding UMTS Radio Network Modelling, Planning and Automated Optimisation Edited by Maciej J. Nawrocki, Mischa Dohler and A. Hamid Aghvami © 2006 John Wiley & Sons, Ltd

bodies. Results of this research are presented at the ITU-R Forum by Working Party WP8F (former TG 8/1) responsible on IMT-2000 related issues.

The current 21st century research is aimed at presenting at the international forum realistic spectrum requirements for further system development and to identify availability of frequency bands for UMTS/IMT-2000.

3.1.1 SPECTRUM REQUIREMENTS FOR UMTS

Even early analysis conducted by the ITU-R clearly concluded that the introduction of 3G systems will require vast technological changes, whilst the new quality of services (e.g. multimedia) will lead to increased spectrum requirements. In the 1980s, the first estimates were that for all operators a frequency spectrum amount of about 130 to 180 MHz [1] should be sufficient. Considering this, it was assumed that each operator will be granted at least 2×20 MHz of spectrum, and that at each geographical location there will be at least two operators. It was later determined that in the early stage of 3G systems a spectrum of 230 MHz shall be sufficient, where ITU-R has recommended approximately 60 MHz for use by personal stations and approximately 170 MHz for use by mobile stations [2]. The spectrum requirements were discussed during the World Administrative Radio Conference WARC-92, where it was agreed that terrestrial FPLMTS (UMTS/IMT-2000) component will be allocated a total of 155 MHz of frequency spectrum, whilst the satellite component will have 2×30 MHz.

Evolution of mobile communication systems lead to increased spectrum requirements and the early assumptions stated above were quickly verified. Under workgroup WP8 ITU (with significant involvement of ECC PT1 – former ERC TG1 – as well as UMTS Forum Spectrum Access Group SAG), a methodology for spectrum assessment was developed as a response to the requirement for fulfilment of various services provided by UMTS/IMT-2000. This methodology was presented individually for both the terrestrial and the satellite components in ITU-R Recommendations [3,4].

Calculations performed before the World Radiocommunication Conference – 2000 (WRC-2000) in regards to future frequency band usage requirements [5] concluded that the bands defined by WARC-92 (called initial) would not be sufficient to fully and effectively satisfy the radio resource requirements for both the terrestrial and satellite components, and would hence allow for implementation of only the base system – without room for future development and without the ability to use its full technical potential. Considering this, additional frequency bands for UMTS usage were sought in Europe within the available radiocommunications spectrum, which served as the basis for submitting European Common Proposals (ECPs) for WRC-2000. Similar proposals were also submitted by other countries from all over the world.

The Conference Preparatory Meeting (CPM) report and WRC-2000 ECPs indicated that in order to satisfy future frequency band requirements for UMTS/IMT-2000 at the year 2010 assumed traffic, in addition to the earlier WARC-92 specifications, it will be necessary to allocate:

- 160 MHz for the terrestrial component;
- 2×30 MHz for the satellite component.

It was also concluded that in order to fulfil the requirement for additional 160 MHz of the spectrum for the terrestrial component, WRC-2000 should identify a single, universal set of frequencies to be used by all countries. Unified universal frequencies should allow for worldwide roaming and would also reduce costs and complexity of implementing IMT-2000. The WRC-2000 was also supposed to identify the part of the spectrum currently used by 2G systems, which could be available to UMTS/IMT-2000 in the future. The above spectrum requirements are now fulfilled, with WRC-2000 identifying sufficient amounts of additional spectrum.

Research pertaining to future spectrum requirements as per further UMTS/IMT-2000 development is continuously being carried out by ITU-R WP 8F. Currently, new ITU documents are being created towards a preliminary draft new report on spectrum requirements for the future development of UMTS/IMT-2000 and systems beyond IMT-2000 and towards methodology for calculation of spectrum requirements for the future development of UMTS/IMT-2000 and systems beyond IMT-2000 from the year 2010 onwards. These documents should be finished before WRC-2007 in year 2007. An interesting proposal for UMTS spectrum requirements identification in Europe was developed under the WINNER project, Work Package 6 Task 2 [6].

3.1.2 SPECTRUM IDENTIFIED FOR UMTS

Third generation systems forced the introduction of advanced technologies. Considering the planned full integration of the terrestrial and satellite components, it was necessary to allocate for them a relatively close frequency bands. Considering this, WARC-92 designated the frequency bands 1885–2025 MHz and 2110–2200 MHz for use by UMTS/IMT-2000, where the satellite component shall comprise of 1980–2010 MHz and 2170–2200 MHz sub-bands. These bands are known in Europe as the *core band*, where the size of this core band was defined in No. 5.388 (Article 5 of ITU Radio Regulations) [7]. The band should be made available for IMT-2000 in accordance with Resolution 212 (Rev.WRC-95, Rev.WRC-97) [2]. In Europe, because of current DECT radiocommunication systems in use, the frequency band for UMTS was limited to 1900–2025 MHz and 2110–2200 MHz [8,9].

A diagram illustrating the frequency bands designated for UMTS/IMT-2000 by the WARC-1992 conference at various parts of the world is presented in Figure 3.1.

In addition to the frequency bands for IMT-2000 (1885–2025 MHz and 2110–2200 MHz), the WRC-2000 has identified the following additional (extended) frequency bands for the use by IMT-2000, as indicated in RR Nos. 5.384A and 5.317A and in Resolutions 223 (WRC-2000), 224 (WRC-2000) and 225 (WRC-2000) [7]:

- For the terrestrial component: 1710–1885 MHz, 2500–2690 MHz and parts of the band 806–960 MHz used or planned to be used for mobile systems.
- For the satellite component: 1525–1544 MHz, 1545–1559 MHz, 1610–1626.5 MHz, 1626.5–1645.5 MHz, 1646.5–1660.5 MHz, 2483.5–2500 MHz, 2500–2520 MHz and 2670–2690 MHz.

with the remarks that some countries may deploy IMT-2000 systems in bands other than identified in the Radio Regulations and that identification of these bands does not establish priority in the RR and does not preclude use of the bands for any other services to which these bands are allocated. The additional (extended) frequency bands identified for future UMTS/IMT-2000 development are also presented in Figure 3.1.

Considering the above, the frequency bands 2500–2520 MHz and 2670–2690 MHz were allocated to both segments, whilst their detailed selection was left up to the individual countries. In Europe, the terrestrial component will use the entire 2500–2690 MHz band [10] which should be made available for use by UMTS/IMT-2000 systems by 1 January 2008, subject to market demand and national licensing schemes, whilst the frequency bands used by current 2G systems should be available to UMTS/IMT-2000 only in the far future. The satellite component was satisfied by being designated the frequency bands adequate for Mobile Satellite Services (MSS).

3.1.3 FREQUENCY ARRANGEMENTS FOR THE UMTS TERRESTRIAL COMPONENT

A frequency arrangement is recommended for more effective spectrum usage. The harmonised worldwide frequency arrangements should facilitate international roaming, simplify assurance of

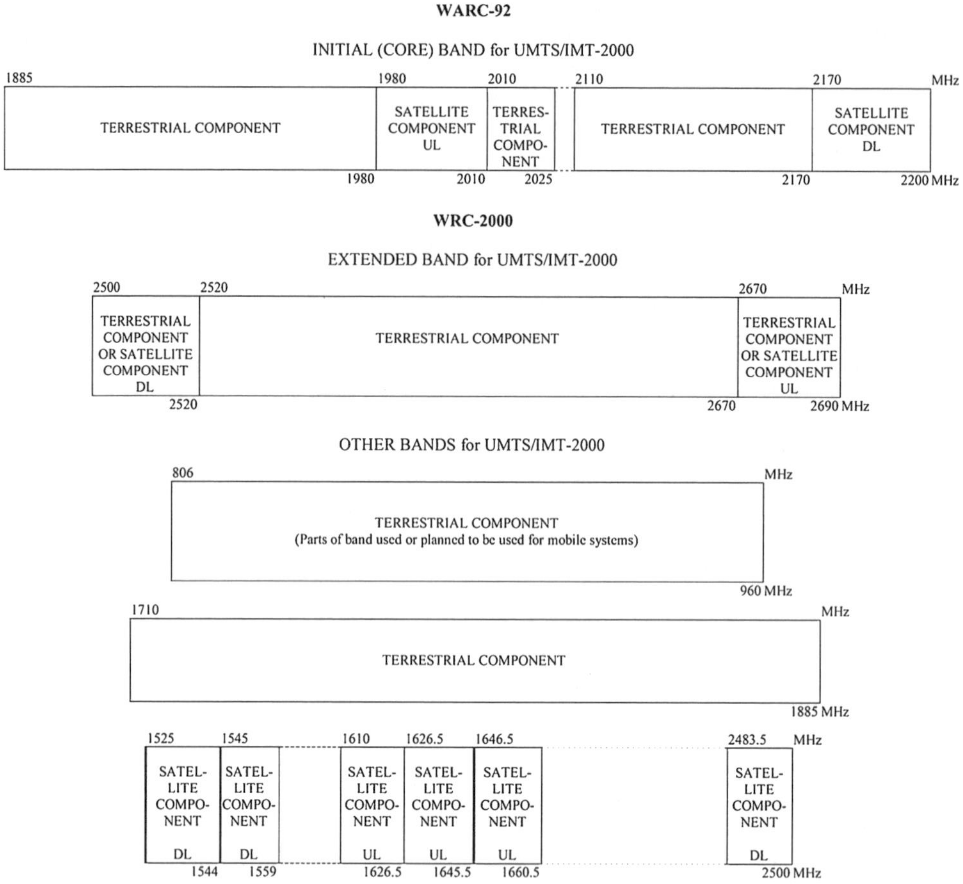

Figure 3.1 Frequency bands identified for UMTS/IMT-2000 by WARC-92 and WRC-2000.

internal system compatibility, especially at the borderline regions, and ultimately lead to reduction of UMTS/IMT-2000 network implementation as well as terminal unit costs.

3.1.3.1 Core band: 1885–2170 MHz

In the earlier versions of the Recommendation [11], the ITU recommended to assume two modes of operation for the base frequency band of the terrestrial component – frequency division duplex (FDD) and time division duplex (TDD) – and to take into account that TDD and FDD system installation in adjacent blocks should be done with respect to interference and mitigation techniques. According to the ITU requirement, the recommended frequency arrangement of the core band should be as follows:

- unpaired frequency bands 1885–1920 MHz and 2010–2025 MHz are intended for TDD operation;
- paired bands 1920–1980 MHz and 2110–2170 MHz are intended for paired FDD operation, provided that duplex direction for FDD carriers in these bands is mobile transmit within the lower band and base transmit within the upper band. Duplex separation should be equal to 190 MHz.

Spectrum and Service Aspects 41

```
1885        1920                          2010       2110                          MHz
┌─────┬──────────────────────────┬──────┬───────────────────────────────┐
│ TDD │         FDD              │ TDD  │         FDD                   │
│OPERA-│      OPERATION          │OPERA-│       OPERATION               │
│TION │                          │ TION │                               │
│     │         UL               │      │         DL                    │
└─────┴──────────────────────────┴──────┴───────────────────────────────┘
  1920                         1980   2025                          2170 MHz
```

Figure 3.2 ITU-R frequency arrangement of the UMTS/IMT-2000 terrestrial component in the core band.

This arrangement is illustrated by Figure 3.2. In view of this base scheme, detailed solutions can slightly differ, as in some countries, within the lower part of the 1885–2025 MHz band there is Digital Enhanced Cordless Telecommunications (DECT) operating at 1880–1900 MHz, or Personal Handyphone System (PHS) operating at 1893.5–1919.6 MHz, and in some other countries there are Personal Communication Systems (PCS) based on North American standards, using 80 MHz duplex separation within the 1850–1990 MHz band.

In European countries, in view of the DECT systems indicated above, the CEPT ERC has recommended to limit the lower UMTS terrestrial component frequency band, i.e. to start at 1900 MHz, and decided to harmonise spectrum usage in this band [12]. The frequency arrangement within the limited core band does not vary from that recommended by the ITU; however, this specified in more detail as how it ought to be used. The Decision [12] specifies among other things that:

- the channel raster is 200 kHz and the carrier frequency is an integer multiple of 200 kHz;
- FDD carrier spacing between public operators is a minimum of 5.0 MHz. FDD carrier spacing within a public operators spectrum is variable, based on a 200 kHz raster, and may be less than 5.0 MHz;
- the frequency band 2010–2020 MHz is identified for self provided applications operating in self coordinating mode;
- TDD carrier spacing between public operators is a minimum of 5.0 MHz. TDD carrier spacing within a public operators spectrum is variable, based on a 200 kHz raster, and may be less than 5.0 MHz;
- carrier spacing between TDD and FDD carriers is a minimum of 5.0 MHz between public operators.

However, in the case of the duplex TDD operation, the CEPT ECC assumed the possibility of lower chip rates, below 3.84 Mcps (with a nominal channel spacing of 5 MHz) allowing in the Recommendation [13] for 1.28 Mcps (with a nominal channel spacing of 1.6 MHz). In this case, a 5 MHz frequency block assigned to one 3.84 Mcps UTRA TDD carrier may alternatively be arranged to contain up to three 1.28 Mcps UTRA TDD subcarriers. The European arrangement scheme is illustrated in Figure 3.3.

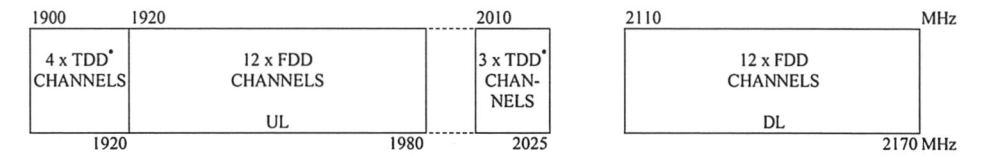

```
1900      1920                        2010      2110                          MHz
┌────────┬──────────────────────────┬────────┬───────────────────────────────┐
│4 x TDD*│      12 x FDD            │3 x TDD*│      12 x FDD                 │
│CHANNELS│      CHANNELS            │ CHAN-  │      CHANNELS                 │
│        │                          │ NELS   │                               │
│        │         UL               │        │         DL                    │
└────────┴──────────────────────────┴────────┴───────────────────────────────┘
       1920                       1980     2025                          2170 MHz
```

*) A 5 MHz channel may be split up to three UTRA TDD subcarriers with 1.6 MHz channel spacing.

Figure 3.3 CEPT frequency arrangement of the UMTS/IMT-2000 terrestrial component in the core band.

3.1.3.2 Extended band: 2500–2690 MHz

Possible arrangements of the extended frequency bands for the terrestrial component are described in [11]. Considering the foreseen (so far unknown) asymmetric radio traffic, it was assumed that the extended bands can be split into several sub-bands, where duplex access would be segmented. These sub-bands can be used for:

- FDD uplink and downlink operation (with centre gap which may optionally be used for TDD);
- independent FDD downlink operation (can be combined with any FDD pairing core bands);
- TDD operation.

The ITU did not specify in the documents the size of these sub-bands. The ITU-R assumed seven scenarios, which can be applied by various countries with regard to arrangement of the extended frequency bands, including three base arrangements. These scenarios are shown in Figure 3.4. It should also be noted that the Recommendation [11] does not define the centre gap, nor the FDD duplex separation. It is known that the size of the centre gap and the spacing of the concerned duplex separation mainly depend on the maximum transmitted power of downlink and uplink and on a foreseen isolation between transmitter and receiver (duplexer – design of receive and transmit filters with a low insertion loss).

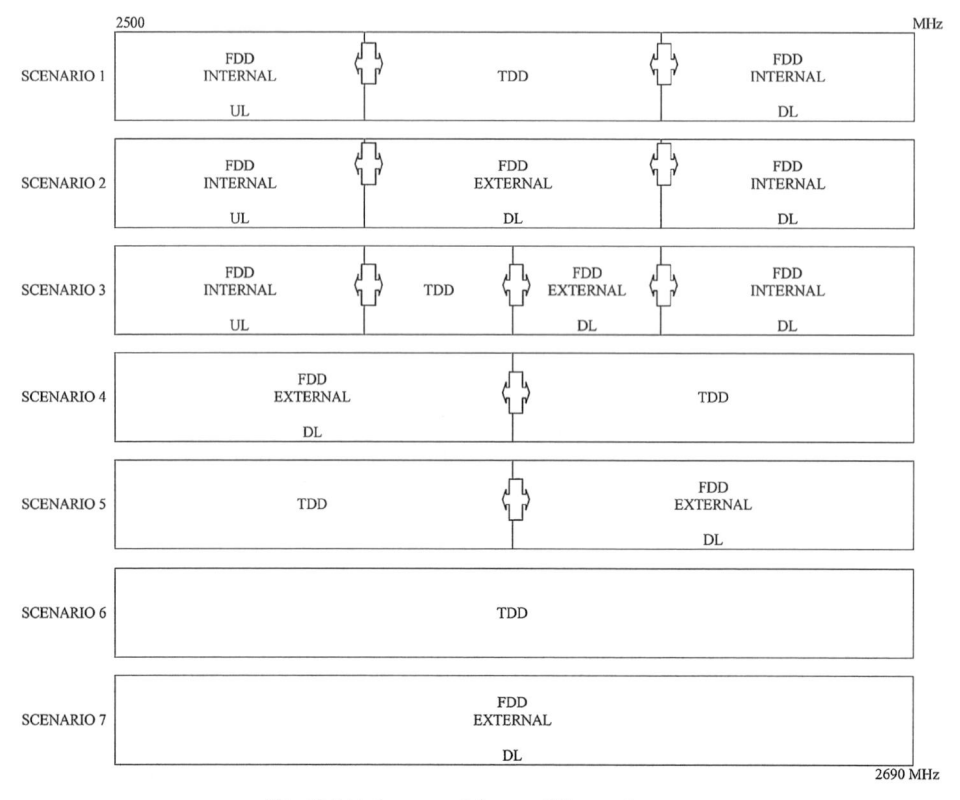

Note: Administrations can use whole or part of this arranged spectrum

Figure 3.4 ITU-R frequency arrangements of the UMTS/IMT-2000 terrestrial component in the extended band.

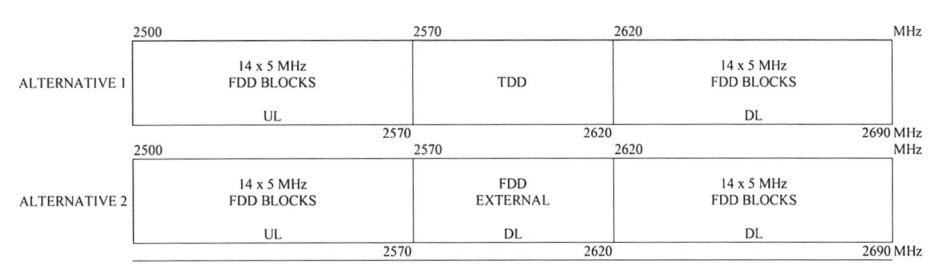

Spectrum and Service Aspects

	2500		2570		2620		MHz
ALTERNATIVE 1		14 x 5 MHz FDD BLOCKS UL		TDD		14 x 5 MHz FDD BLOCKS DL	
			2570		2620		2690 MHz
	2500		2570		2620		MHz
ALTERNATIVE 2		14 x 5 MHz FDD BLOCKS UL		FDD EXTERNAL DL		14 x 5 MHz FDD BLOCKS DL	
			2570		2620		2690 MHz

Note: Any guard bands taken from 2570 – 2620 MHz band should be specified by national administrations to ensure adjacent band compatibility at FDD/TDD and FDD Internal/External boundaries

Figure 3.5 CEPT frequency arrangements of the UMTS/IMT-2000 terrestrial component in the extended band.

The arrangement described in draft revision of Recommendation [11] worked out by ITU-R WP8F has three possible frequency arrangements, i.e. C1, C2 and C3. Arrangements C1 and C2 refer appropriately to scenarios 1 and 2 of Figure 3.4., whilst arrangement C3 provides for flexible use of either TDD or FDD throughout the band with no specific blocks.

Usage and size of individual frequency blocks was recently defined for Europe in a decision issued by ECC pertaining to harmonisation of extended frequency band for the terrestrial UMTS/IMT-2000 component [14]. It states that the frequencies for the FDD operation shall be available at the frequency band 2500–2570 MHz paired with 2620–2690 MHz with the mobile transmitting within the lower sub-band and base transmitting within the upper sub-band. This decision suggested two alternative solutions, where the 50 MHz centre gap can be used for duplex TDD or for an external FDD downlink; however, any guard bands required to ensure adjacent band compatibility at 2570 MHz and 2620 MHz boundaries should be decided on a national basis and included within this gap. Frequency blocks (channels) will be assigned as integer multiples of 5.0 MHz, noting that within the extended band the FDD duplex separation shall be 120 MHz. Both frequency channel arrangement variants are illustrated by Figure 3.5.

It should be noted that the European solution corresponds to scenarios 1 and 2 published in the ITU-R Recommendation [11]. The problems of utilising the 2500–2690 MHz band in Europe have been presented in the CEPT Reports to the European Commission [15,16].

3.1.3.3 Other bands

Taking into account other frequency bands foreseen for the terrestrial component of IMT-2000 by WRC-2000, the ITU-R recommends two arrangements within the band of 824–960 MHz (indicated as A1 and A2) and six arrangements within the band of 1710–2170 MHz (indicated as B1–B6, where B1, B2 and B3 serve as the base arrangements) [11]. Arrangement B1 is identical to the basic band with the addition of 5 MHz for duplex TDD in the unpaired sub-band. It is also assumed that individual countries can implement only selected parts of these frequency arrangements. Another assumption is that TDD used in the unpaired sub-bands under certain conditions can also be used in paired uplink sub-bands and/or at the centre gap (between uplink and downlink FDD sub-bands), but excluding the 2025–2110 MHz band. This recommendation also defines the size of the centre gaps, but does not state the width of the frequency blocks (channels). Arrangements A1 and A2 are presented in Table 3.1, whilst arrangements B1–B6 are given in Table 3.2. These are also illustrated by Figures 3.6 and 3.7.

Of course, it should be noted that some of the above frequency arrangements will be possible at a much later time, because they include frequency bands currently intensively used by 2G systems.

Table 3.1 Frequency arrangements in the 824–960 MHz band.

No. of frequency arrangements	FDD uplinks (MHz)	FDD downlinks (MHz)	FDD duplex separation (MHz)	Centre gap (MHz)
A1	824–849	869–894	45	20
A2	880–915	925–960	45	10

Table 3.2 Frequency arrangements in the 1710–2170 MHz band.

No. of frequency arrangements	FDD uplinks (MHz)	FDD downlinks (MHz)	FDD duplex separation (MHz)	Centre gap (MHz)	Unpaired spectrum (e.g. for TDD) (MHz)
B1	1920–1980	2110–2170	190	130	1880–1920, 2010–2025
B2	1710–1785	1805–1880	95	20	—
B3	1850–1910	1930–1990	80	20	1910–1930
B4 (harmonised with B1 and B2)	1710–1785 1920–1980	1805–1880 2110–2170	95 190	20 130	1900–1920, 2010–2025
B5 (harmonised with B3 and parts of B1 and B2)	1850–1910 1710–1755 1755–1805	1930–1990 1805–1850 2110–2160	80 95 355	20 50 305	1910–1930 — —
B6 (harmonised with B3 and parts of B1 and B2)	1850–1910 1710–1770	1930–1990 2110–2170	80 400	20 340	1910–1930 —

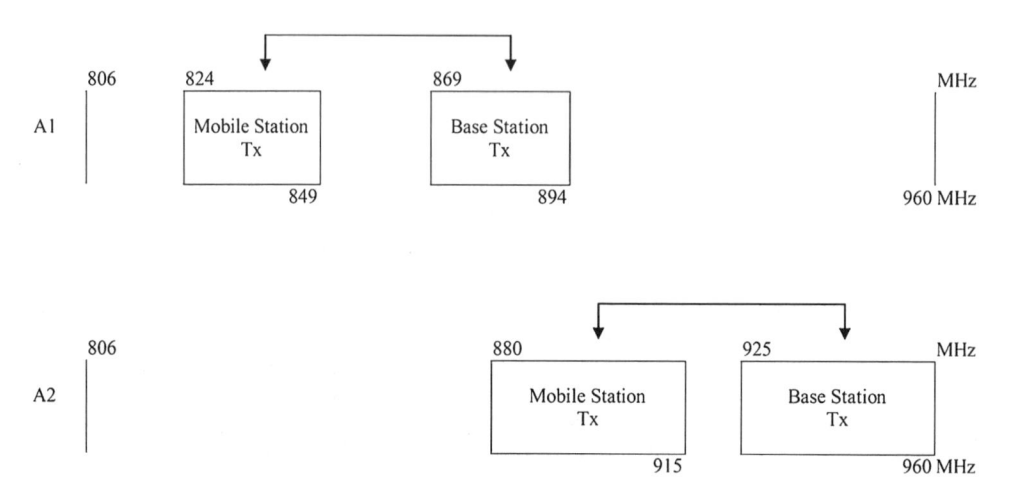

Figure 3.6 Variants of ITU-R frequency arrangements of the UMTS/IMT-2000 terrestrial component in the 824–960 MHz band.

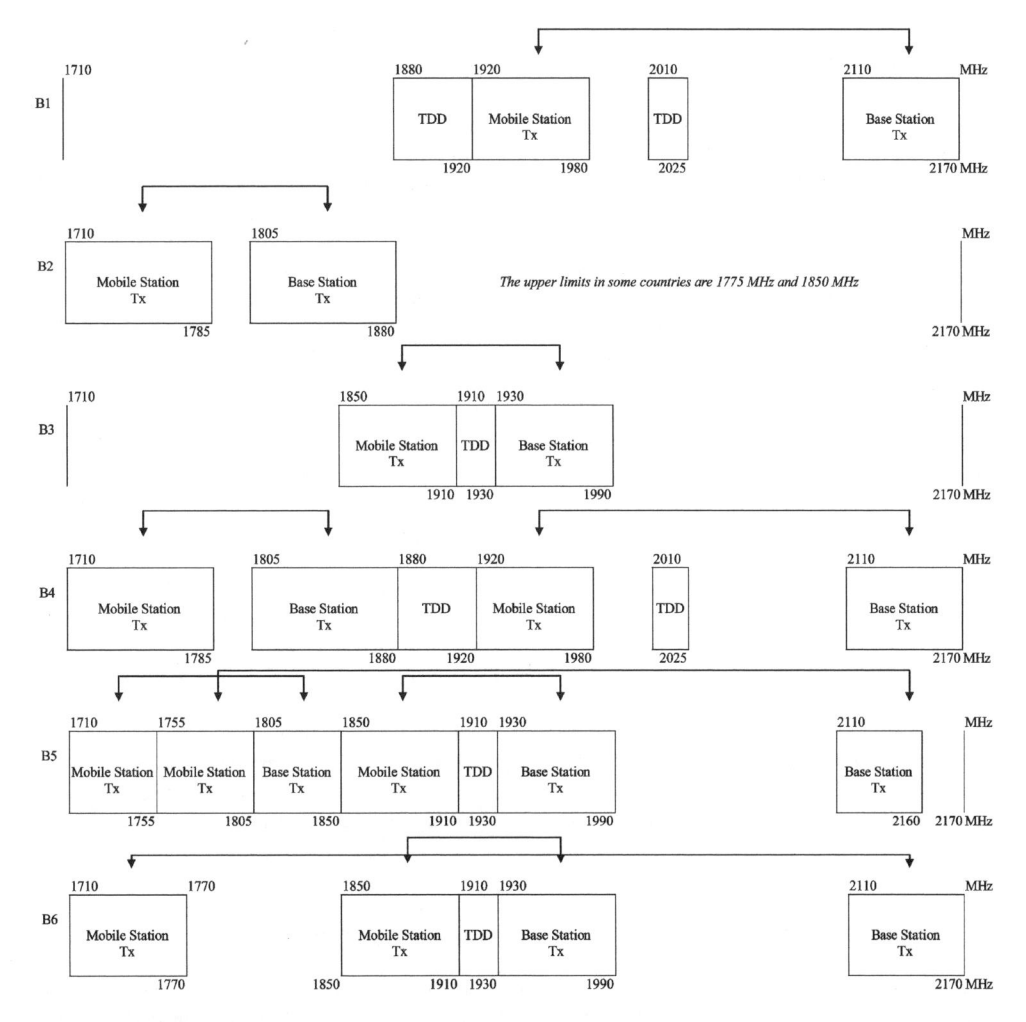

Figure 3.7 Variants of ITU-R frequency arrangements of the UMTS/IMT-2000 terrestrial component in the 1710–2170 MHz band.

3.1.4 OPERATOR SPECTRUM DEMANDS

Radiocommunication system operators were interested from the very beginning in works pertaining to 3G systems and proper spectrum allocation. They also carried out their own research in order to be able to specify their individual spectrum demands. A few years before implementation of the first UMTS/IMT-2000 networks, operators and hardware manufacturers under the UMTS Forum announced their research results concerning spectrum demands. This research took into account factors such as characteristics of potential market, service, traffic, density of future users, forecast of penetration and infrastructure and technological aspects. In Report [17], the UMTS Forum announced the spectrum demands for years 2005 and 2010. This estimation was built on the assumption that radio traffic generated from wideband and multimedia services as well as from narrowband services will be carried by UMTS. It was assessed that the general spectrum demand can reach 410 MHz for the terrestrial component and 50 MHz for the satellite component in 2005 and 580 MHz for the terrestrial component

(UMTS and evolved 2G systems) and 90 MHz for the satellite component in 2010; note that the moment the request was issued, in Europe the terrestrial component was assigned a total of 155 MHz and satellite component of 2×30 MHz of the spectrum. During the start-up phase in 2002 the spectrum demands of licensed UMTS operators were estimated at 2×20 MHz.

Another analysis provided by UMTS Forum [18] pertained to the minimum spectrum demand per public terrestrial UMTS operator in the initial phase of system development. Eight different scenarios were analysed assuming various parts of spectrum given to an operator, and various factors, such as maximum available data rates, offered bit quantity, asymmetry factors and spectral efficiencies. Based on this analysis, the UMTS Forum recommended 2×15 MHz (in paired sub-bands) $+5$ MHz (in unpaired sub-bands) as the preferred minimum spectrum requirement per public UMTS operator in the first phase; however, from a technical point of view, the minimum spectrum requirement was specified as 2×10 MHz (in paired sub-bands) $+5$ MHz (in unpaired sub-bands). In Report [19], the operators stated that their demands can be fulfilled only by an additional 187 MHz of spectrum for the terrestrial component as the other part of operators' demands will be provided by the currently used 2G systems spectrum.

The operator spectrum demands were fulfilled in the year 2000, when the WRC allotted UMTS-IMT-2000 with an additional frequency band (190 MHz spectrum located within 2500–2690 MHz). Taking into account better radio wave propagation at lower frequencies (longer range coverage, larger cells) operators announced lately that a new band of frequencies should be sought for UMTS/IMT-2000. This new band, referred to as 'New Coverage Extension Band', should be allocated within the 'digital dividend' (following from the future switch-off analogue and move to digital television broadcasting) – amount of spectrum taken from the bands designated for TV broadcasting and harmonised within 470–600 MHz. Operators, manufacturers and regulators associated with the UMTS Forum consider that the 2×30 MHz of paired spectrum would provide a viable minimum Coverage Extension Band for UMTS/IMT-2000 [20].

It is clear that the amount of available frequency spectrum will directly influence any optimisation solution; for example, having two paired FDD bands available for a given user and service distribution will yield a different optimisation solution, if compared to the case of three paired FDD bands. Also, the actual choice of services will be influential, where associated principles are described in the subsequent section.

3.2 SERVICE FEATURES AND CHARACTERISTICS

The driving principle behind the UMTS network design is the need for support of a variety of traffic sources. The design concept ought to be flexible so as to carry different application data over the same medium in a capacity efficient way; example application data include video conferencing and telemetry, video streaming, short message services, voice conversations, broadband Internet access, TV broadcast, FAX delivery services, etc. Those applications differ in a variety of QoS parameters, such as required bandwidth, data rate variation during transmission, sensitivity to latency, etc. However, they have in common that the service should be delivered from source to destination according to its given constraints, so as to satisfy the customer's needs. This is complicated by the fact that any telecommunication system has limited resources. Therefore, the concept of Quality of Service (QoS) and Bearer has been introduced in UMTS [21]. It caters for the end-to-end QoS and allows the following between sender and receiver:

- Negotiation of bearer parameters (bearer type, bit rate, delay, BER, up/down link symmetry, protection, etc.) for data transmissions; the parameters of bearer services may be renegotiated during a connection, triggered by, e.g., handover or worsening of radio conditions;
- Parallel bearer services (service mix);
- Real-time and non-real-time communication modes;
- Circuit switched and packet oriented bearers;

- Supports scheduling (and pre-emption) of bearers (including control bearers) according to a given priority;
- Adaptation to quality, traffic and network load, as well as radio conditions (in order to optimise the link in different environments).

Figure 3.8 presents the UMTS end-to-end QoS architecture. The *bearer* is a service providing QoS between two defined points in the network. It consists mainly of user plane, control signalling and QoS management functionality. The layered structure implicates that upper layers are using lower layers' services to provide its own QoS. Within the network, a connection is maintained between two Terminal Equipments (TEs) with certain mechanisms to secure the requested connection quality. This is important from the customer point of view, who expects a telecommunication system to be a network that supports services, independent of the used technology.

For any type of connection, a TE must communicate via a Mobile Terminal (MT) to get access to network resources; this is possible with the *TE/MT Local Bearer Service*. Throughout the network, the *UMTS Bearer Service* is used. In the case that the other TE belongs to a different network than the serving UMTS network, the *External Bearer Service* demands to preserve the data transmission between the serving UMTS and the external networks. *Radio Access Bearer* (RAB) and *Core Network (CN) Bearer Services* belong to the *UMTS Bearer Service*. The *CN Bearer Service* is responsible to maintain the connectivity with external networks. The *RAB Service* deals with user mobility within

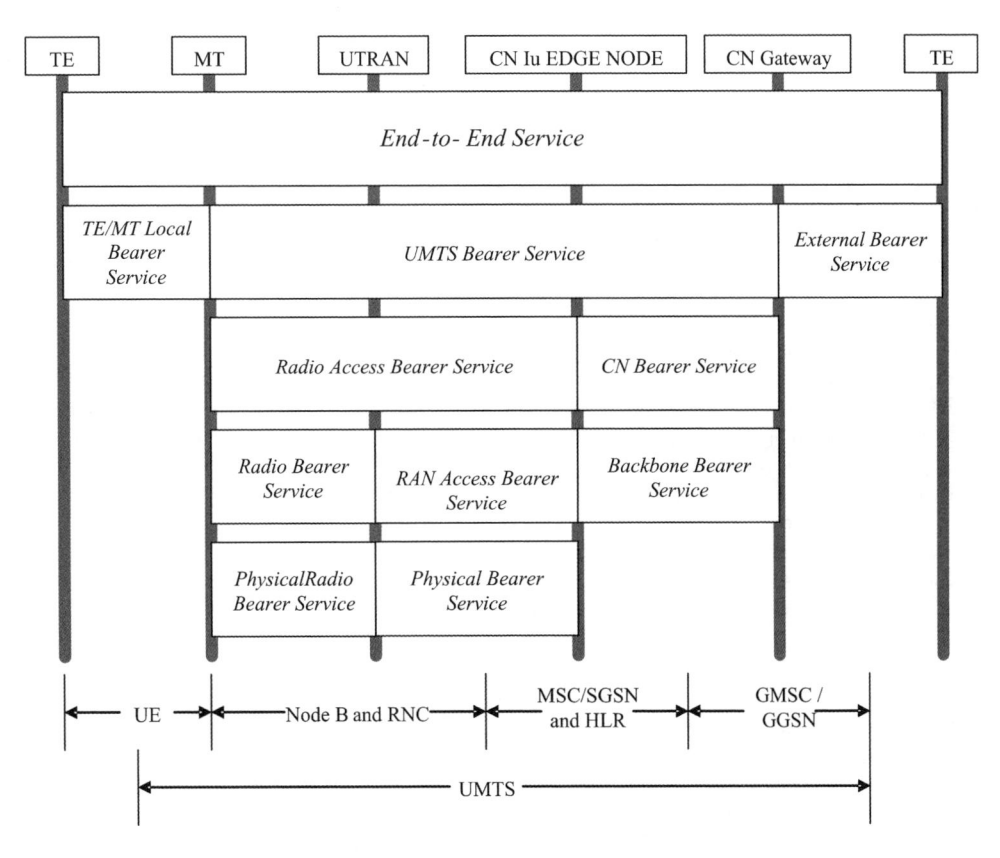

Figure 3.8 UMTS QoS and bearer architecture.

the network and UTRAN specific issues. It consists of the *Radio Bearer* (RB) and the *RAN Access Bearer Services*. The *RB Service* is responsible for the radio interface and the *RAN Access Iu Bearer Service* for data transmission between UTRAN and CN.

Any radio network planning and optimisation particularly need to take into account *RAB Service*, *RB Service* and *Physical Radio Bearer Service*.

The complex nature of the air interface requires special handling of particular bearers. On the other hand, there are varieties of traffic sources, which may require a different treatment. The combination of each different traffic source, together with the air interface behaviour for QoS support, would lead to a massive number of possible solutions within of the RB. Since the real implementation for the RB should be robust and provide QoS within practical possibilities, the traffic has been quantified into a few classes. The UMTS specifications hence define four QoS traffic classes:

1. *Conversational Class*
2. *Streaming Class*
3. *Interactive Class*
4. *Background Class*.

The main differentiation parameter among the classes is the sensitivity to time delay variations. The most delay sensitive class is the *Conversational Class*; it means that delay variation from a certain point will lead to significant service degradation and hence could lead to a termination of the connection. For instance, if a speech frame arrives at a user with a delay greater than 400 ms, it is perceived as significant service damage. In contrast, the *Background Class* is the most insensitive to delay variations; therefore, certain delays in the data delivery would not lead directly to a service deficiency. For instance, when using an e-mail service, there is a high user tolerance of messages arriving at different times at the destination, depending usually on the spare bandwidth to support it. Important, however, is that data arrives without any loss.

The *Conversational Class* represents *Real Time* (RT) applications. The transfer time ought to be low due to the conversational nature of the scheme. It is restricted by the human perception of audio or video conversations. Moreover, the time relation between two entities should be preserved in the same manner as for real time streams. Consequently, the delay must be limited and fixed and no buffering is allowed, the BER may vary depending of data stream bandwidth, but the bit rate should be guaranteed. The traffic, to some extent, is symmetric. For low data rate applications (e.g. 12 kbps) the BER should be less than 10^{-3}, for higher data rates (e.g. 128 kbps) the BER should be less than 10^{-6}.

Typical examples of the *Conversational Class* applications are speech and video conversations, which are connection oriented services and are supported by the *Circuit Switched* CN domain. For example, for speech applications, the delay should be less than 100 ms, but the hard limit is 400 ms and jitter should be less than 1 ms.

There are a number of *Packet Switched* applications, which require the support of *Conversational QoS Class*: voice over IP and video conferencing over IP. Real time conversation is always performed between peers (or groups) of live end users.

The *Streaming Class* represents *Real Time* (RT) applications. The important factor here is the time relation between information entities within the same data stream, which needs to be preserved. Generally, there are no restrictions on (low) time delay and the data streams are only one way from server to user. As a result, the delay may minimally vary, buffering of the data stream is allowed for jitter smoothing, the BER should be kept low (less than 10^{-6}) and a given bit rate should be guaranteed; the traffic is asymmetric.

Typical examples of the *Streaming Class* applications are audio and video streaming (e.g. radio or TV broadcast over Internet, usually the delay should be less than 10 s and jitters less than 1 ms), monitoring, ftp and data base access.

The *Interactive Class* represents *Non-Real Time* (NRT) *Best Effort* applications. It is characterised by a request–response pattern of the end user, meaning that a customer (human or machine) is requesting

Table 3.3 Main attributes of QoS traffic classes.

		QoS Traffic Class			
		Conversional RT	Streaming RT	Interactive NRT	Background NRT
Main Attributes	Transfer delay	Rigorous	Rigorous	Looser	Not constrained
	Jitter (transfer delay variation)	Constrained	Constrained	Not constrained	Not constrained
	Low BER	Not constrained	Not constrained	Needs to be supported	Needs to be supported
	Guaranteed bit rate	Needs to be supported	Needs to be supported	No constraints	No constraints

data from a remote equipment or server through an online session, e.g. web browsing. The response is expected to arrive within a certain time, although there are no strict requirements on a minimum delay or delay variations; they ought to be moderate though. Buffering of data is allowed for the sake of jitter smoothing; also, the bit rate should be guaranteed and the traffic is asymmetric. Another importer parameter is the preservation of payload content, which should arrive at the destination error free (i.e. a BER less than 10^{-6}).

Typical examples of the *Interactive Class* applications are WWW applications, data base retrieval, server access and automatic data base enquiries by tele-machines and pooling for measurements collection.

The *Background Class* represents *Non-Real Time* (NRT) *Best Effort* applications. The traffic class is optimised to support communication between machine-to-machine communication. The schema applies when a destination does not expect data within a certain time, e.g. when a computer (server) receives or sends data in the background. There are no restrictions on minimum delay or delay variations. Buffering of data is allowed for jitter smoothing. The bit rate does not have to be guaranteed and the traffic is asymmetric. The content should be preserved, thus it ought to arrive error free at the destination (i.e. a BER less than 10^{-8}).

Typical examples of the *Background Class* applications are e-mail download, calendar update or SMS delivery.

The main attributes of the above described QoS traffic classes are presented in Table 3.3.

The RAB Service is characterised by the following attributes [21]:

- *Traffic class*: Conversational, Streaming, Interactive and Background (as defined below).
- *Maximum bit rate* (kbps): Maximum number of bits delivered by the UTRAN and to the UMTS within a period of time; it specifies the upper limit of the bit rate.
- *Guaranteed bit rate* (kbps): Guaranteed number of bits delivered by the UMTS within a period of time; the QoS requirements described by other attributes are valid up to the *Guaranteed bit rate*.
- *Delivery order* (y/n): It indicates whatever bearer service delivery should support in-sequence *Service Data Unit* (SDU) delivery or not.
- *Max SDU size* (bytes): maximum permitted size of an SDU.
- *SDU format information* (bits): It contains all possible sizes of SDUs; this may be used by UTRAN in RLC transparent mode, which is used to minimise the transmission delays and hence the retransmission.
- *SDU error ratio*: It is described as a portion of erroneous or lost SDUs. In the case that resources are reserved for a bearer service, the SDU error ratio is not dependent on the current load.
- *Residual BER* (bit error ratio): It indicates the undetected bit errors (bit error ratio) in the delivered SDUs, when error detection has been demanded; else it shows the bit error ratio in the delivered SDUs.

- *Delivery of erroneous SDUs* (y/n/–): It points out how the erroneous SDUs should be treated; thus, it may support or not (–) the detection of erroneous SDUs, and when a SDU is detected as erroneous, it can be forwarded (y) or not (n).
- *Transfer Delay* (ms): It is defined as the maximum delay of the 95th percentile of the delay distribution for all delivered SDUs during lifetime of a bearer service;
- *Traffic handling priority*: It is utilised within the Interactive traffic class; it describes the relative importance of SDUs supported by one bearer service in contrast to SDUs supported by another bearer of the Interactive class; this attribute is used by the scheduling algorithm for proper order of different packet transmission.
- *Allocation/retention priority*: It is related to the subscription and cannot be negotiated or set by UE; it defines the relative importance of a bearer service in contrast to other bearer services; it may be used by admission control and resource allocation procedures.
- *Source Statistics Descriptor* ('speech'/'unknown'): It is specified within the traffic source; it may be used by relevant network elements (RAN, SGSN, GGSN) to calculate the statistical multiplexing gain, when admission control is performed.
- *Signalling Indication* (yes/no): It indicates the signalling nature of submitted SDUs and is only for Interactive Traffic Class; in a case the attribute is set to 'yes', UE should set the traffic handling priority to '1'.

The dependencies between QoS attributes and traffic classes have been depicted in Table 3.4.

The mapping between RAB and RB is internal to the UTRAN Radio Resource Management (RRM) and is not standardised; it is thus vendor specific. To each RAB, a *Signalling Radio Bearer* (SRB) is assigned to maintain the control information flow between UTRAN and UE, as depicted in Figure 3.9.

Table 3.4 QoS Attributes of UMTS bearer service.

		QoS Traffic Class			
		Conversational RT	Streaming, RT	Interactive, NRT	Background, NRT
QoS Attributes	Maximum bit rate	x	x	x	x
	Guaranteed bit rate	x	x		
	Delivery order	x	x	x	x
	Max SDU size	x	x	x	x
	SDU format information	x	x		
	SDU error ratio	x	x	x	x
	Residual BER	x	x	x	x
	Delivery of erroneous SDUs	x	x	x	x
	Transfer delay	x	x		
	Traffic handling priority			x	
	Allocation/retention priority	x	x	x	x
	Source statistics descriptor	x	x		
	Signalling indication			x	

Spectrum and Service Aspects 51

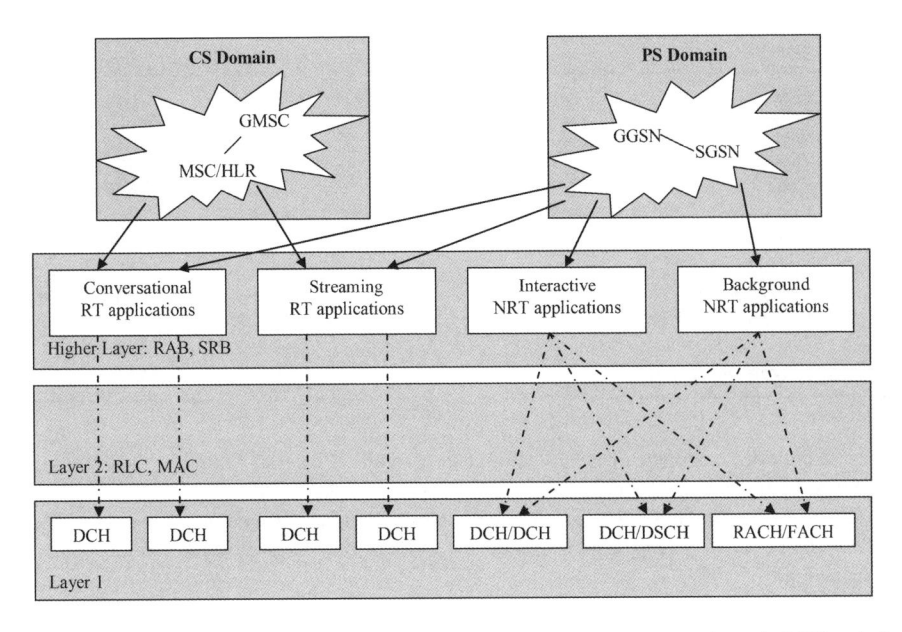

Figure 3.9 RAB mapping (RLC – Radio Link Control, MAC – Medium Access Control, DCH – dedicated channel, DSCH – downlink shared channel, RACH – random access channel, FACH – forward access channel).

The following RRM procedures are involved in the control of QoS:

- *Admission Control* estimates at user access and service request the potential load and interference rise in a case the requested service would be accepted, following it grants an access to the user with predefined QoS parameters, when the load and interference increase due to the new connection would be acceptable or not in other case.
- *Load/Congestion Control* monitors continuously the system interference to maintain it under acceptable, predefined thresholds; in the case that the interference would exceed a threshold, it applies a mechanism to reduce it.
- *Power Control* maintains the transmit power at a minimum required level; the power control (PC) procedures include fast PC (inner closed loop) to maintain the instantaneous E_b/N_0 of a service at a minimum target level, slow PC (outer closed loop) to adjust the target E_b/N_0 according to changing radio channel conditions, and open loop PC to limit the transmit power of the PRACH access frames.
- *Handover* sustains the connections between the radio network and moving UEs; the mobility of users is controlled by soft handover (intra-frequency) or hard handover (inter-frequency or inter-RAT) procedures.
- *Packet Scheduling* supports the *Interactive* and *Background Classes* applications by distributing the available radio resources between all packet sources; it controls the assigned bit rates and also the order of different sources packet transmissions.

There are many different RBs, which may fulfil the requirements of a RAB, depending on:

- current radio condition;
- load and resource utilisation;
- bearer attributes.

Table 3.5 Examples of RAB.

RAB	QoS Class	Domain	Max data rate (kbps) UL	Max data rate (kbps) DL	Application example
Single	Conversational, RT	CS	12.2	12.2	AMR speech
Single	Conversational, RT	CS	64	64	Video call
Single	Streaming, RT	PS	12.2	64	Radio station stream
Single	Streaming, RT	PS	12.2	384	TV station stream
Multi	Conversational, RT	CS	12.2	12.2	AMR speech
	Background, NRT	PS	64	64	SMS
Multi	Conversational, RT	CS	64	64	Video call
	Background, NRT	PS	64	256	WWW

Table 3.5 provides examples of RABs which can be realised by RBs [22]. The data rate given for each RAB is the maximum data rate that can be supported by that RAB.

The aim of this section has been to equip the reader with a background on the bearer concept and associated QoS parameter mappings. Further details on QoS, RAB, RB and SRB can be obtained from [21], [22] and [23].

REFERENCES

[1] CCIR, Report 1153, *Future Public Land Mobile Telecommunications Systems*, Geneva 1990.
[2] World Radiocommunication Conference WRC-97 (Geneva, 1997), Resolution 212 (Rev. 97), *Implementation of International Mobile Telecommunications-2000 (IMT-2000)*.
[3] ITU, ITU-R Recommendation M.1390, *Methodology for the calculation of IMT-2000 terrestrial spectrum requirements*.
[4] ITU, ITU-R Recommendation M.1391, *Methodology for the calculation of IMT-2000 satellite spectrum requirements*.
[5] ITU, ITU-R Report M.2023, *Spectrum requirements for International Mobile Telecommunications-2000 (IMT-2000)*.
[6] WINNER, IST-**2003-507581**, D 6.2, T. Irnich et al., *Methodology for estimating the spectrum requirements for 'further developments of IMT-2000 and systems beyond IMT-2000'*.
[7] ITU, ITU-R Radio Regulations, Geneva 2004.
[8] ERC, *ERC Decision of 30 June 1997 on the frequency bands for the introduction of the Universal Mobile Telecommunications System (UMTS)*, ERC/DEC/(97)07.
[9] ERC, *ERC Decision of 28 March 2000 extending ERC/DEC/(97)07 on the frequency bands for the introduction of terrestrial Universal Mobile Telecommunications System (UMTS)*, ERC/DEC/(00)01.
[10] ECC, *ECC Decision of 15 November 2002 on the designation of frequency band 2500–2690 MHz for UMTS/IMT-2000*, ECC/DEC/(02)06.
[11] ITU, ITU-R Recommendation M.1036, *Frequency arrangements for implementation of the terrestrial component of International Mobile Telecommunications-2000 (IMT-2000) in the bands 806–960 MHz, 1710–2025 MHz, 2110–2200 MHz and 2500–2690 MHz*.
[12] ERC, *ERC Decision of 29 November 1999 on the harmonised utilisation of spectrum for terrestrial Universal Mobile Telecommunications System (UMTS) operating within the bands 1900–1980 MHz, 2010–2025 MHz and 2110–2170 MHz*, ERC/DEC/(99)25.
[13] ECC, ECC Recommendation (02)10, *Harmonised utilisation of spectrum for 1.28 Mcps UTRA TDD option in connection with ERC/DEC/(99)25*.
[14] ECC, *ECC Decision of 18 March 2005 on harmonised utilisation of spectrum for IMT-2000/UMTS systems operating within the band 2500–2690 MHz*, ECC/DEC/(05)05.
[15] ECC, CEPT Report 001, Report from CEPT to the European Commission under Mandate 4, *Frequency usage to facilitate a co-ordinated implementation in the community of third generation mobile and wireless*

communication systems operating in additional frequency bands as identified by the WRC-2000 for IMT-2000 systems, 15 November 2002.
[16] ECC, CEPT Report 002, Report from CEPT to the European Commission on the 5th Mandate on IMT-2000/UMTS, *Harmonisation of the frequency usage within the additional frequency band of 2500–2690 MHz to be made available for IMT-2000/UMTS systems in europe*, 12 November 2004.
[17] UMTS Forum, Report No. 1, *A regulatory framework for UMTS*, 1997.
[18] UMTS Forum, Report No. 5, *Minimum spectrum demand per public terrestrial UMTS operator in the initial phase*, 1998.
[19] UMTS Forum, Report No. 6, *UMTS/IMT-2000 spectrum*, 1998.
[20] UMTS Forum, Report No. 38, *Coverage extension bands for UMTS/IMT-2000 in the bands between 470–600 MHz*, 2005.
[21] 3GPP, *Quality of Service (QoS) concept and architecture*, TS 23.107.
[22] 3GPP, *Typical examples of RABs and RBs supported by UTRA*, TR 25.993.
[23] Heikki Kaaranen, Ari Ahtiainen, Lauri Laitinen, Siamäk Naghian, Valtteri Niemi, *UMTS Networks, Architecture, Mobility and Services*, John Wiley & Sons Ltd/Inc., 2001.

4

Trends for the Near Future

Maciej J. Nawrocki, Mischa Dohler and A. Hamid Aghvami

4.1 INTRODUCTION

The aim of this book is to equip the reader with analytical tools that facilitate the modelling, planning and optimisation of the radio part of 3G communication systems. 3G, however, is a system that is constantly on the move. There are many worldwide meetings taking place each week to discuss modifications, improvements and further 3G standardisation [1]. While the 1990's saw the Release '99 crystallising, today we are already discussing Release 6. What are these trends all about? And most importantly, are the techniques exposed in this book applicable to those future radio access network evolvements?

It is often said that to understand the future, one has to understand the past. The roadmap of 3G has so far been a very complex concoction of politics, needs and technology; and this is unlikely to change in the near future. The grand vision, however, has always been to design a worldwide common wireless communication system that would provide highest possible data rates anywhere and anytime. A contending of technologies and associated intellectual property rights (IPR) thus began which resulted in the submission of one terrestrial indoors system, nine terrestrial systems and six satellite systems. The ITU eventually adopted not a single, but a family of five complementary standards for terrestrial radio interfaces as being IMT-2000 compatible at their 1999 meeting in Helsinki [2,3], of which four are still being supported and standardised:

1. IMT Direct Spread   (IMT-DS; UTRA FDD)
2. IMT Multi Carrier   (IMT-MC; CDMA2000)
3. IMT Time Code   (IMT-TC; UTRA TDD / TD SCDMA)
4. IMT Frequency Time   (IMT-FT; DECT).

This family of standards, applied to

- satellite systems, for providing global seamless coverage;
- cellular FDD systems, for providing a national and international network;
- cellular TDD systems, for providing high capacity voice and data in hot spots;
- licence exempt systems, for flexible and cheap voice and data networks,

Understanding UMTS Radio Network Modelling, Planning and Automated Optimisation   Edited by Maciej J. Nawrocki, Mischa Dohler and A. Hamid Aghvami   © 2006 John Wiley & Sons, Ltd

is expected to realise an effective and profitable 3G communication network, where multi-mode terminals are used to accomplish the much needed global roaming. Of the four standards, however, only three are deployed today.

The DECT system (standardised by ETSI) covers indoor areas and last-mile connections, i.e. cordless telephony and WLL [4]; it operates in the 1880–1900 MHz band which is licensed for WLL (last mile) systems. Just below this spectrum allocation is GSM1800 and immediately above it is UMTS TDD. Inter-system interference may well prove to be a serious problem between DECT and TDD mode, as further detailed in Section 11.4.1. Note that there are also planning issues for WLL systems when there is more then one network operating in the same area, since this is a TDMA/TDD system and inter-operator synchronisation will be needed to avoid any significant loss of capacity.

UTRA FDD (standardised by 3GPP [1]) and CDMA2000 (standardised by 3GPP2 [5]) are terrestrial cellular communication systems which are built on CDMA technology and expected primarily to operate outdoors. Both systems are now commercially deployed, mainly in urban hot spots throughout major cities on the planet.

In contrast to UTRA FDD and CDMA2000, the terrestrial wideband UTRA TDD (standardised by 3GPP) and its Chinese narrowband companion TD-SCDMA (also standardised by 3GPP) are yet to be commercially deployed; however, given the investment in development costs and background IPR, associated parties are likely to push for deployment sooner rather than later. Also yet to be deployed is the satellite part of 3G, which has witnessed neither final selection nor standardisation; however, as for the case of IMT-TC, this system is expected to reach deployment eventually. Since these systems are all CDMA based, we will briefly discuss their near-future developments in Section 4.2.

Furthermore, 3G is all about revenue from a large number of ideally high usage customers. The customer, however, already has a large range of different wireless data services available, some of which can already deliver much higher data rates at lower price, albeit with drastically reduced coverage; current examples are public or private WLANs [6]; upcoming WiMAX [7], etc.

To ensure it is competitive, 3G must deliver comparable data rates over a wide coverage; this is precisely what the 3GPP Release 5 intends and any future releases will encompass. We will thus witness a push for better coverage, higher data rates and, perhaps, unlikely marriages (i.e. convergence) between cellular and other wireless communication systems, as discussed in Sections 4.3–4.5.

4.2 SYSTEMS YET TO BE DEPLOYED

This section deals with systems yet to be standardised and/or deployed. We will highlight the reason of their delayed deployment and also its relation to the UTRA FDD optimisation presented in this book.

4.2.1 UTRA TDD

The UTRA TDD standard, as mentioned earlier, has been selected by the ITU as one of the IMT-2000 communication standards; it has been part of all 3GPP releases. As the name suggests, UTRA TDD operates in time division duplex, i.e. the same spectral band is utilised for uplink and downlink. The wireless medium is accessed in a hybrid TDMA and CDMA fashion, i.e. each frame is slotted and within each slot several CDMA codes can be utilised. Therefore, user data can be scheduled in time, i.e. each slot carries data for a different user using the same access code; or, data can be scheduled by means of a CDMA code, i.e. within the same slot, user data is spread by different codes; or, a hybrid of both access methods is feasible, which makes the approach quite flexible. Each time slot in a frame can be allocated to uplink or downlink.

The objective in pushing for UTRA TDD was to have a standard available fulfilling the following characteristics:

- provision of high data rates in hot spots;
- support of asymmetric uplink and downlink traffic;
- applicability of more sophisticated signal processing schemes.

TDD combined with the time slotted approach clearly facilitates the support of asymmetric uplink and downlink, as more time slots can be utilised towards the higher traffic direction. Assumptions to date have been that the downlink will dominate the traffic direction, which has been justified by an increase in multimedia download activities and Web browsing. However, these assumptions may well be reversed soon, because emerging peer-to-peer and interactive multimedia applications also require high data rates in the uplink.

Hot spot traffic implies that communication takes place over comparably short distances, resulting in short signal round trip times, hence shorter guard times between uplink and downlink time slots and an increased spectral efficiency; this has been another reason why the TDD option is particularly applicable to data hot spots.

Note finally that the shorter scrambling sequences of TDD, compared with its FDD counterpart, allows less sophisticated multiuser algorithms to be used and thus further boosts the TDD system capacity. Company internal studies have also revealed that the complexity of these multiuser schemes is not drastically higher when compared with more conventional systems [8]. In addition, the use of a single frequency band for uplink and downlink facilitates the application of capacity boosting techniques that require some form of channel state feedback, such as beamforming.

Although, one field trial has been reported in Japan, no UTRA TDD system has yet been deployed. So why is TDD not deployed yet? The most likely reason is because it lost its marketplace to another wireless system –Wireless Local Area Networks (WLAN, IEEE 802.11 family [6]). WLANs were not very sophisticated when first deployed, e.g. no QoS mechanism was catered for, but they were (and are) cheap and effective. Hot spot areas, such as airports, train stations, cafes, are now cluttered with WLANs, often in competition with each other and leaving no opportunity for TDD. In addition, the TDD data rates cannot really match current and anticipated WLAN data rates, making TDD even less attractive.

A clear advantage of 3GPP's standardisation efforts is the interoperability between FDD and TDD, which means that a single UMTS terminal could easily switch between both systems as the need arises. More than one chip manufacturer, however, has already announced the availability of a single baseband chip comprising GSM, GPRS, UMTS and WLAN. In addition, some operators have already gone so far as to provide data services through proprietary or co-sponsored hot spot WLAN networks. This complicates the market position of UTRA TDD, despite new business ideas emerging all the time (e.g. private wireless indoors PBX systems). Only time will tell whether it can find a commercially viable niche.

There are also drawbacks with the UTRA TDD system, some of which are summarised below:

- comparably low multipath diversity resulting from short indoor multipath delay profiles;
- an interference problem between adjacent TDD sites, where a time slot may be used for the uplink in one cell, but for the downlink in the adjacent one.

The interference problem will be the driving factor in network planning and optimisation, if UTRA TDD is to be deployed. Given its localised hot spot deployment, however, no sophisticated radio tuning algorithms would be needed; the emphasis would be more on suitable MAC and scheduling protocols that can handle the available time slots and codes, as well as the occurring interference.

4.2.2 TD-SCDMA

TD-SCDMA is the Chinese contribution to ITU's IMT-2000 specification for 3G wireless mobile services. It has been submitted by the China Wireless Telecommunications Standard (CWTS) group, who spent much effort, time and money to produce an independent ITU compliant standard. The main concern of the Chinese government was that the majority of the W-CDMA IPR lay outside its reach, namely with the US; furthermore, CDMA2000 was not seen as an option because it heavily relies on the Global Positioning System (GPS) of the US, thereby posing a security and reliability threat within China.

As 3G in China is anticipated to be a US$100 billion market, the decision on its inclusion into the IMT-2000 family was not a surprise given that it carried significant implications for national and international operators, manufacturers and vendors. In March 2001, all technical schemes of the TD-SCDMA standard were accepted by 3GPP and were included in its Release 4. TD-SCDMA has now been accepted not only by the ITU, but also by an industrial alliance of operators, manufacturers and vendors.

The main physical layer differences between TD-SCDMA and UTRA TDD are the chip-rate and the synchronous uplink. The chip-rate which, after Release 4, has been set to one third of UTRA TDD rate is 1.28 Mcps, occupying just one third of the 5 MHz 3G spectral band; this clearly allows three communication channels to be utilised simultaneously. Having three usable channels also facilitates simple cell planning because the re-use factor can now be a multiple of three and the synchronisation of uplink and downlink results in higher spectral efficiencies from the reduced loss in orthogonality, which can be further enhanced by means of smart antennas and multiuser detectors.

It has successfully been demonstrated that 2 Mbps can be delivered over this reduced bandwidth, which is a significant increase in capacity when compared with UTRA FDD or CDMA2000. This is critical to China's communication infrastructure, particularly for deployment in densely populated or metropolitan areas. Further information on the standard and its spectral enhancements using advanced equalisers, beamforming, adaptive multi-rate codecs, joint detection or adaptive multiuser detection algorithms, etc. can be gathered from the TD-SCDMA Forum [9].

As to the downside of TD-SCDMA, arguably the same issues as for UTRA TDD apply. One of the concerns with TDD is its unsuitability to rural and remote areas, where communication distances of several tens of kilometres have to be accomplished. Perhaps the actual weakness of TD-SCDMA, as is frequently argued, is its two-year delay in proposals and test runs when compared with UTRA FDD and CDMA2000. First tests were only conducted in 2001.

Given that by 2001, Siemens had invested US$1 billion into TD-SCDMA and set up joint laboratories with Datang in Beijing and across Europe, it is difficult to imagine that TD-SCDMA will not happen. It will, however, go through the same uncertain periods as any other 3G network deployment worldwide. Notably, the need for killer applications and the need for careful network planning and optimisation.

The modelling, planning and optimisation techniques offered in this book are applicable to the Chinese TD-SCDMA, albeit with some modifications. Most notably, the synchronous uplink drastically reduces multipath interference which needs to be reflected in the analysis using a reduced factor to describe the loss in orthogonality. Furthermore, the reduced spreading factor and the increased number of communication bands require appropriate modifications. Fortunately, however, these changes are easily implemented, thereby paving the way for automated optimisation solutions for TD-SCDMA networks.

4.2.3 SATELLITE SEGMENT

Satellites have been circling earth for many decades and using these as a means to communicate is not new to humanity. Although many different satellite communication services exist, they are today generally referred to as Mobile Satellite Services (MSS), which provide personalised unicast but mainly multicast or broadcast services to handheld terrestrial mobile terminals. There are currently numerous 2G satellite systems commercially available which provide voice and data service, however, which are not compatible between themselves or with terrestrial 2G systems; example systems include Globalstar,

Skybridge, Thuraya, Orbcomm, ICO, Iridium, Inmarsat, etc. They currently serve as a niche market for a few businessmen, remote and/or international companies and government with voice and medium rate data services. This niche market was initially hoped to be much larger, where the satellite service providers have not anticipated the rapid expansion of terrestrial 2G communication systems in the 1990s. Despite the large investment costs, the target market is unlikely to change, thereby requiring new business models to make MSS a success; only time will tell whether investments will be returned and the services proved profitable.

An MSS system enjoys some advantages, such as:

- robust coverage in remote, desolate and disaster areas, as well as over water and in the air;
- fast network roll-out when compared with terrestrial deployment;
- true global roaming;
- convenient for multicast or broadcast services;
- no small scale fading (rather on/off shadowing).

However, it also suffers from many disadvantages, such as:

- expensive launch and maintenance of satellite transceiver;
- comparably large terrestrial terminals resulting from the large communication distances involved;
- very low spectral efficiency per area;
- current lack of spectrum to provide medium to high data rates to a large customer base.

Above list is partially exclusive to terrestrial systems; however, the aim of the ITU was to establish the satellite segment as an integral and complementary part of a truly global 3G system, where the same user terminal is capable of communicating with an orbital satellite, a cellular basestation and an indoor DECT station. It was not to be seen as an alternative to the terrestrial 3G service and, accordingly, major satellite communication companies and organisations submitted a total of six satellite proposals as IMT-2000 compatible:

1. 'A Specification' by European Space Agency (ESA); called Satellite W-CDMA (*SW-CDMA*); based on UTRA FDD; maximum rate of 144 kbps; Low Earth Orbit (LEO), Medium Earth Orbit (MEO), Geosynchronous Orbit (GEO) and High Earth Orbit (HEO) are feasible.
2. 'B Specification' also by ESA; called *SW-C/TDMA*; hybrid code and time division multiple access; maximum rate of 144 kbps; LEO, MEO, GEO and HEO are feasible.
3. 'C Specification' by South Korea's Telecommunications Technology Association (TTA); called *SAT-CDMA*; based on CDMA technology; maximum rate of 144 kbps; 48 LEO satellites are stipulated.
4. 'D Specification' by ICO Global Communications; called *ICO RTT*; combination of FDMA & CDMA; maximum rate of 38.4 kbps; 12 MEO satellites are stipulated.
5. 'E Specification' by Inmarsat (Horizons); called *Horizons*; combination of TDM with TDMA & FDMA; maximum rate of 512 kbps; geostationary satellites are stipulated.
6. 'F Specification' by Satcom2000; called *Iridium*; uses both FDMA/TDMA and FDMA/CDMA; maximum rate of 144 kbps; 96 LEO satellites are stipulated.

As of today, only the frequency bands have been reserved to 1980–2010 MHz for uplink and 2170–2200 MHz for downlink (see Section 3.1.2), but none of the above satellite radio transmission technologies have explicitly been standardised. However, considerable efforts have been made between the European ESA and the South Korean TTA in terms of harmonising their CDMA based approaches with the UTRA FDD releases. In doing so, any future WCDMA terrestrial terminal with minimal ability to reconfigure will also be able to use the satellite segment of 3G.

And while business analysts and terrestrial operators view the satellite market with suspicion, Inmarsat have gone ahead independently and recently launched their vision of global data communication – a broadband global area network (BGAN). The launch of the first Inmarsat-4 (I-4) satellite took place in March 2005 and the service is designed to deliver 3G compatible broadband data and voice services to mobile terrestrial users. The I-4 is 60 times more powerful, and has 16 times more network capacity than its I-3 predecessors [10] making it one of the most powerful commercial satellites launched to date. The satellite is deployed in geostationary orbit 36 000 km above the Indian Ocean and footprints Europe, Africa, the Middle East, the Indian sub-continent, most of Asia Pacific, and Western Australia. The launch of a second I-4 is has been accomplished in November 2005 to cover the Americas. The two I-4 satellites are the backbone of Inmarsat's next generation BGAN satellite network. It is an IP and circuit-switched system that will offer voice and high-bandwidth services, including Internet access, videoconferencing and other services, at speeds of up to 492 kbps.

The cost of launch and maintenance of the I-4 BGAN satellites is huge and Inmarsat is hoping that terrestrial GPRS and 3G services will be the main drivers for the take-up of their services. Inmarsat well recognise that once 3G data rates are available to the business or leisure market, these will be required beyond the boundaries of the current 3G network deployment – giving them a head-start of a few years. Also, because BGAN is fully UMTS compliant, a 3G user can take out the SIM card from the terrestrial handset and put it into an Inmarsat terminal; this, however, requires that terminal technology develops and roaming agreements will have to be put in place.

This development by the satellite operators clearly indicates that satellite businesses are willing to share the wireless voice and data market and – together with the large customer base of terrestrial operators – create a truly global three-dimensional communication system. The challenge for satellite service providers lies less in the technology as such, but more in overcoming the perception that satellite communications are bulky, expensive and only for niche markets.

Once fully compliant 3G satellite systems are rolled-out (spaced-out is possibly a better phrase to use), planning and optimisation issues arise, as well as issues relating to their impact on terrestrial 3G systems; although with the availability of few satellites, the planning and optimisation phases are significantly less difficult than for terrestrial systems. Further, the frequency bands for the satellite segment of 3G both need to be and are different from their terrestrial allocations. Although this book certainly helps in understanding 3G WCDMA systems analytically, including the satellite parts, it is of marginal use to the satellite 3G radio designer.

4.3 ENHANCED COVERAGE

3G operators have agreed very stringent requirements for geographical service availability, for instance, UK operators have to provide 80% of the population with coverage by the end of 2007 [11]. This generally requires a large number of 3G base stations to be installed, with the associated costs of:

- physical base site
- expensive base station equipment
- installation of base stations
- upgrades and maintenance.

The number of base sites must be minimised; however, for a required level of coverage and limited spectrum, the capacity of each cell decreases with increasing coverage area per base station. Indeed splitting a larger cell into a number of smaller cells is known to increase the capacity. Therefore, trading capacity with base site deployment costs is a fundamental part of 3G optimisation. To make things even more complicated, a CDMA system suffers from the effect of 'cell breathing', where capacity can be traded against coverage. In this section, we review some promising techniques related to coverage extension that do not jeopardise but rather aid the 3G system capacity.

4.3.1 ULTRA HIGH SITES (UHS)

Ultra High Sites are formed by means of a multi-sector base station mounted at a high location, for instance a TV tower or a high building. The achieved coverage clearly exceeds currently deployed macro cells; however, the capacity per area is decreased as the serving area increases. The topic of UHS is discussed in Section 12.3 and not elaborated further here.

4.3.2 HIGH ALTITUDE PLATFORM SYSTEM (HAPS)

High Altitude Platform Systems (HAPS) are small aircrafts or balloons placed in the stratosphere at around 20 km altitude [12]. They are considered to be a hybrid between satellite and terrestrial systems, thereby enabling personalised unicast but mainly multicast or broadcast broadband services to be delivered. HAPS will not replace existing technologies, but rather complement these in an integrated fashion. HAPS are designed to support different cell sizes ranging from micro cells (up to 1 km), macro cells (up to 20 km) and regional cells (larger than 20 km). Frequency bands have been allocated to HAPS to deliver broadband services; they may also use the IMT-2000 bands to complement the 3G network infrastructure.

HAPS are clearly suited to a natural or man-made disaster emergency service application; they need only a sparse ground-based infrastructure and may be rapidly deployed to any location. The currently envisaged applications – other than 3G – are remote sensing, navigation and surveillance.

The problems and deployment challenges currently faced by HAPS include the still fairly high production costs; however, with a maturing industry, this cost ought to fall steeply. Also, depending where the HAP is deployed, weather and wind conditions may influence deployment and maintenance. If these drawbacks can be overcome, then HAPS will likely form part of an integrated 3G communication network.

In the case of 3G, HAPS can be deployed in a hierarchical manner (see Section 4.4.1); in that case, the frequency and power usage of HAPS have to be designed carefully in conjunction with terrestrial 3G systems. The tools and approaches outlined in this book are then well suited to such a design process.

4.4 ENHANCED CAPACITY

Once sufficient coverage for a given system load is established, the aim of the network designer is to maximise the system capacity of the 3G network even further at minimum cost. There are numerous ways of increasing the capacity, which mainly reduces to:

- the choice of a proper network topology;
- spectrally efficient physical layer (PHY) mechanisms;
- efficient medium access control (MAC) mechanisms.

Some novel trends related to the 3G network roll-out are discussed in the following sections.

4.4.1 HIERARCHICAL CELL STRUCTURES (HCS)

A good solution towards finding a suitable trade-off, aided by several factors as explained below, is the deployment of HCS [13,14,15]. As its name suggests, a hierarchy of overlaying cell layers is deployed, where satellite, macro, micro and pico cells may cover the same geographical area simultaneously. Interference between the layers is clearly not permitted, which necessitates a proper separation in at least one of the below:

- frequency
- location
- time.

Separation in frequency is the most common, albeit least efficient, approach among operators, because spectrum is scarce and the use of two bands in the same area clearly reduces system capacity. However, it has been shown that capacity gains can be achieved if slow moving users are confined to the underlying micro cell layer, whereas fast moving users are delegated to the overlaying macro cell layer. The argument being that, usually, slower moving users require higher data rates, whereas faster moving users require respectively lower rates. A further argument for such separation is the fact that fast moving users would require a large number of handovers during their movement, thereby generating significant signalling overhead and hence decrease the overall system capacity [16].

Separation in location allows, for instance, an outdoor macro cell to overlay an indoor micro or pico cell, enabled by the natural wave attenuation of building walls. This facilitates frequency reuse in the same geographical area and hence a drastic boost to the system capacity.

Separation in time is a fairly novel technique which has been introduced in [17]. The suggestion here was to embed a micro cell into a macro cell, both communicating within the same geographical area and within the same frequency band. Without coordination between micro and macro cell, interference would clearly occur which would drastically reduce the system capacity; however, it has been suggested that the micro cell user reports the experienced macro cell interference level to the micro cell base station which transmits only when the interference level is below a predefined threshold. Since the micro cell is fairly small, the likelihood that it generates significant interference to a macro cell user is small. The thus proposed HCS architecture stipulates real-time traffic to be served by the macro cell, but nonreal-time traffic by the micro cell, which schedules the data packets in an opportunistic manner. In [17], it has been indicated that significant performance gains can be achieved.

The interesting contributions in [18] have shown that cell throughput in an HCS can be maximised by co-locating the micro and macro cell site but using a different frequency; however, if fairness of QoS *and* cell throughput was to be improved, the micro-cell ought to be deployed at the edges of the macro cell.

To facilitate HCS deployment, proper handover mechanisms are needed and these are catered for in all IMT-2000 specifications; however, there has been some concern about the currently specified soft handover mechanism in UTRA FDD for use in contiguous micro cellular coverage areas. Micro cells may thus not be designed to perform optimally until equipment designed to a later release of the standards is available.

Hierarchical Cell Structures require careful modelling, planning and optimisation. The material presented in this book, although not explicitly dealing with the topic, allows similar techniques to be applied to HCSs. They have been excluded here because the number of different HCS configurations is large, leading to numerous case studies rather than a fundamental comprehension. We have, however, included a fairly fundamental section on temporal throughput and related planning analysis of HCSs using the same frequency band, which can be found in Section 10.4.6.

4.4.2 HIGH SPEED DOWNLINK PACKET ACCESS (HSDPA)

The 3GPP Release 5 specifications focused primarily on HSDPA [19,20], also referred to as 3.5G, which was designed to provide initial data rates of up to about 10 Mbps to support packet-based multimedia services. Release 6 further included MIMO transceiver techniques, which facilitate even higher data transmission rates of up to 20 Mbps. The latest HSDPA implementations includes adaptive modulation and coding (AMC), multiple-input multiple-output (MIMO), hybrid automatic repeat request (HARQ), fast scheduling, fast cell selection (FCS) and advanced transceiver design. An example of the later is intersymbol interference (ISI) mitigating linear minimum mean-square error (LMMSE) equalizers, which are very effective in HSDPA since, because of its TDM nature, individual users are given larger powers when compared with UTRA FDD.

Theoretical investigations into HSDPA have been conducted at great technical depth, whereas the commercial success has yet to be demonstrated. Currently, HSDPA is beginning to reach deployment status in North America. DoCoMo has announced that it will introduce HSDPA from the second half of 2006. Also, European operators will follow soon. HSDPA has another serious commercial contender, the 3G high speed packet data technology CDMA-2000 1x-EvDO (*e*volution, *d*ata-*o*ptimised) [21], which has already been commercially deployed in Japan, South Korea and the Americas.

The planning and optimisation process of a 3G network in general very much depends, as will become apparent in this book, on the anticipated mix of services, requiring particular attention in planning HSDPA roll-out. HSDPA planning initially requires the same fundamental planning as UTRA FDD, e.g. pilot planning. However, HSDPA planning also requires meticulous designing of the packet scheduler, which depends on the traffic mix, associated QoS, etc. Because of the time division functionalities of HSDPA, the scheduler design is actually a fundamental issue. As the aim of this book is not to deal with the optimisation of specific network topologies but rather to give some basic understanding and tools, HSDPA has not been included here. It is understood, though, that the majority of the described techniques are equally applicable to an HSDPA network. Furthermore, the interested reader is referred to the last edition of the excellent book by J. Laiho, A. Wacker and T. Novosad [22].

4.4.3 HIGH SPEED UPLINK PACKET ACCESS (HSUPA)

High Speed Uplink Packet Access is a data access technique for 3G networks with very high upload data rates of up to 5.8 Mbps. The specifications for HSUPA are part of Release 6, with Europe planning first deployments in late 2007. Clearly, the idea of HSUPA is to complement HSDPA for accomplishing a truly bi-directional, interactive wireless broadband experience; only together will they enable symmetrical data communications, thus supporting multimedia, Voice over IP (VoIP), etc. [20].

Technically, HSUPA will use an uplink enhanced dedicated channel (E-DCH) with dynamic link adaptation methods as already enabled in HSDPA, i.e. shorter transmission time intervals, thereby enabling faster link adaptation, and also a hybrid ARQ with incremental redundancy, thereby making retransmissions more effective.

A packet scheduling mechanism is envisaged to operate on a request–grant principle, where the terminals request data to be transmitted and the scheduler dictates when and how many terminals will be allowed to send data. In addition to scheduled transmissions, Release 6 also incorporates a terminal self-initiated transmission mode, which is expected to be useful for VoIP services, as they require extremely short delay times, low jitter and constant bandwidth.

Release 6 comprises further enhanced scheduling methods based on long term and short term grants. Long term grants are given to multiple terminals which are allowed to send their data simultaneously. Short term grants, however, allow terminals to be multiplexed in the time domain rather than the code domain (cf. long term scheduling). Importantly, in order to allow the multiplexing of the uplink transmissions of several terminals in *both* code and time domain, the channelisation and scrambling codes are not shared between terminals (as is currently the case on the shared downlink channel in HSDPA).

For the optimisation process it is important to note that, unlike previous specifications, the ratio between the power of DPDCH and DPCCH will be controlled by the Node B, so as to facilitate a better adaptation to link and cell load conditions. Also, unlike HSDPA, soft and softer handovers are supported for packet transmissions, where the serving Node B is allowed both power-up and down commands and the remaining Node Bs participating in a handover only power down commands.

Since HSUPA is still being standardised during the preparation of this book, it is excluded here. However, as already previously mentioned, the described techniques can be utilised to optimise any future HSUPA network.

4.4.4 ORTHOGONAL FREQUENCY DIVISION MODULATION (OFDM)

Orthogonal Frequency Division Modulation divides the signal into several narrow-band sub-carriers. These sub-carriers are designed to be narrower than the coherence bandwidth of the wireless propagation channel. This guarantees the fading on each sub-carrier to be flat and hence eliminates the need for complex equalizers in the receiver. This useful property may eliminate the need for complex WCDMA receiver architectures in realising high data rates.

Such high data rates are implemented by means of a low spreading factor, for instance in HSDPA with a spreading factor of 16. The use of a low spreading factor, however, makes the system less resilient to time-dispersive propagation channels; the channel becomes more prone to ISI. For a conventional WCDMA Rake receiver, this self-interference is known to be a performance limiting factor. To combat the ISI, more sophisticated WCDMA receivers are needed, an example being the chip-level based LMMSE equaliser; although such an equaliser is currently of prohibitively high complexity.

The 3GPP has, accordingly, sought to establish the potential benefits of introducing OFDM to the UMTS downlink [23], which is seen as an evolution of current 3G system developments. OFDM, however, has its own drawbacks, having a high peak-to-average power ratio, thereby requiring expensive linear amplifiers – a serious drawback for cost-effective implementation in mobile terminals.

In the context of this book, an OFDM approach would require considerable changes to the planning and optimisation philosophy of 3G networks. In fact, if not used in conjunction with some MC-CDMA spreading/scrambling techniques, GSM-like frequency planning patterns would be needed and, for that reason, OFDM is not considered further here in the planning and auto-tuning techniques.

4.5 HETEROGENEOUS APPROACHES

It has long been a vision to have a single 3G communication standard governing the entire wireless world; instead, for historical, political and economical reasons, 3G will have to be able to communicate with already existing or fast-emerging standards. Release 6, for instance, envisages the unlikely marriage between 3G and WLAN systems, something unthinkable only a couple of years ago.

A 3G network together with any other wireless communication system, both of which are capable of communicating with each other at various system levels, is often referred to as a *heterogeneous network*. The strength of intercommunication is typically referred to as the *degree of coupling*.

The provision of heterogeneous network topologies is conceptually a very attractive notion; however, it is certainly a challenge to the network designer. Here, coupling between the networks of possibly different characteristics can be provided, leading to open, loose, tight and very tight coupling. The stronger the coupling the more optimally resources will be utilised. However, this comes with an increased effort in the definition and implementation of required interfaces. A suitable trade-off for specific systems is thus a major part of ongoing research [24].

A few heterogeneous approaches are described below, some of which clearly aim to improve the capacity, coverage and maintenance costs of 3G systems.

4.5.1 WIRELESS LANs

IP-based wireless technologies have received a strong technological and economical boost recently. This has been fostered by various standards, e.g. IEEE 802.11x, 802.16x, 802.20x etc. These technologies are currently evolving towards higher broadband data rates and/or support of continuous mobility in wide service areas. Although at different stages of development and deployment, these standards are competing among themselves. In that context, the need to provide evolved 3G inter-working with these technologies and networks becomes mandatory; this has been reflected in 3GPP Release 6 with a specific work item for 3GPP-WLAN inter-working. The inter-working in Release 6 is thus defined in a very flexible way, enabling different multi-radio scenarios.

From a radio planning and optimisation point of view, although both networks are conjoined, they can be dealt with independently, though joint radio resource management (RRM) schemes are required, which is beyond the scope of this book.

4.5.2 WIRELESS MANs (WiMAX)

The capacity and capabilities of wireless MANs, such as WiMAX, are expected to be comparable with 3G HSDPA [7]. WiMax is a fixed wireless access IEEE standard (IEEE 802.16x). IEEE 802.16a and its enhanced standard IEEE 802.16-2004 do not support mobility; however, the new versions of these standards with mobility features are now being standardised (IEEE 802.16e). In its early roll-out stages HSDPA will be targeting mobile data and voice users, whereas WiMAX is expected to deliver broadband data to companies and suburban areas. In the later stages, serious competition is expected between Mobile WiMAX and HSDPA. Here, HSDPA has the clear advantage of an already established infra-structure, whereas WiMAX will require new infra-structure if it was to penetrate the coverage areas of HSDPA. Whether there will be a marriage between 3G and Mobile WiMAX will depend on the way WiMAX solves its mobility and coverage problem; if no solution can be found, a heterogeneous approach may benefit both sides; if a solution can be found, then both systems will be competing vehemently for the same customers.

WiMAX, however, has also been designed for and can hence very well be applied to point-to-multipoint broadband wireless access systems, a prominent example being the UTRAN backhaul network. The applicability of WiMAX to cater for the needs of the UTRAN transmission infrastructure is hence discussed in Sections 17.4.4 and 17.5.

4.6 CONCLUDING REMARKS

New releases of standards will emerge from 3GPP, no doubt, and all with the aim of improving capacity and flexibility to satisfy the ever increasing demands of the end user. What will happen in the further future with the landscape of wireless communications is not clear at all. A strengthening opinion of late is demanding an opening of the spectrum, with abilities of the competing wireless systems to bid for spectrum on demand. Also, in the advent of software defined radio, it may one day be possible that only transmission power levels and some other general requirements are standardised, so that theoretically anybody could run his/her wireless technology and service.

Such futuristic thoughts, however, are still far off and the subsequent chapters offer the reader a unique insight into the behaviour of CDMA based 3G systems. We hope that, whatever the past, present or future, the presented analytical approach to modelling, planning and optimisation of wireless access networks will be of great use to a wide audience of academics, industrials and policy decision makers for many years to come.

REFERENCES

[1] 3rd Generation Partnership Project (3GPP) website, http://www.3gpp.org.
[2] International Telecommunication Union (ITU) website, http://www.itu.int.
[3] ITU Press Release, 'IMT-2000 Radio Interface Specifications Approved in ITU Meeting in Helsinki,' 5, November 1999.

[4] Digital Enhanced Cordless Telecommunications (DECT) website, http://www.dect.org.
[5] 3rd Generation Partnership Project 2 (3GPP2) website, http://www.3gpp2.org.
[6] IEEE Working Group for Wireless Local Area Networks (WLANs) website, http://grouper.ieee.org/groups/802/11/.
[7] Worldwide Interoperability for Microwave Access (WiMAX) Forum website, http://www.wimaxforum.org.
[8] Marylin Arndt, *et al.*, 'Integration Cost Estimation of UMTS TDD Radio Interface into GSM/UMTS FDD Terminals,' *RADIUM PROJECT*, France Telecom R&D, internal, October 2002.
[9] TD-SCDMA Forum website, www.tdscdma-forum.org.
[10] Inmarsat I4 website, http://countdown.inmarsat.com/inside_I4.
[11] UK 3G press release, http://www.cellular-news.com/3G/uk.php.
[12] T.C. Tozer, D. Grace, 'High-altitude platforms for wireless communications,' *Electronics & Communication Engineering Journal*, vol. 13, issue 3, pp. 127–137, June 2001.
[13] S. Hamalainen, H. Lilja, J. Lokio and M. Leinonen, 'Performance of a CDMA based hierarchical cell structure network,' in *8th IEEE International Symposium on Personal, Indoor and Mobile Radio Communications (PIMRC 97)*, vol. 3, pp. 863–866, 1997.
[14] Lauro Ortigoza-Guerrero, A. Hamid Aghvami, *Resource Allocation in Hierarchical Cellular Systems*, Artech House Books, 1999.
[15] S.A. Ghorashi, F. Said and A.H. Aghvami, 'Forward link capacity of hierarchically structured cellular CDMA systems with isolated microcells,' *IEICE Trans. Commun.*, vol. E86-B, no. 5, pp. 1698–1701, May 2003.
[16] S.A. Ghorashi, F. Said and A.H. Aghvami, 'Handover Rate Control in Hierarchically Structured Cellular CDMA Systems,' *14th IEEE Int. Symp. Personal, Indoor and Mobile Radio Communications (PIMRC)*, vol. 3, pp. 2083–2087, September 2003.
[17] S.A. Ghorashi, F. Said and A.H. Aghvami, 'Beamforming and Intelligent Scheduling for Layer Separation in HCS', *U.K. Patent Application 0216-291.5*, filed on 15 July 2002.
[18] Sangbum Kim, Daehyoung Hong and Jaeweon Cho, 'Hierarchical cell deployment for high speed data CDMA systems,' *IEEE Wireless Communications and Networking Conference (WCNC2002)*, vol. 1, p. 7–10, 17–21 March 2002.
[19] 3GPP TS 25.855, 'High Speed Downlink Packet Access (HSDPA); Overall UTRAN description', http://www.3gpp.org.
[20] Harri Holma and Antti Toskala, *HSDPA/HSUPA for UMTS: High Speed Radio Access for Mobile Communications*, John Wiley & Sons, Ltd/Inc., 2006.
[21] CDMA2000 standard family, 'cdma2000 High Rate Packet Data Air Interface Specification,' C.S0024-A v2.0, http://www.3gpp2.org.
[22] Jaana Laiho, Achim Wacker and Tomas Novosad, *Radio Network Planning and Optimisation for UMTS*, John Wiley & Sons, Ltd/Inc., 2006.
[23] 3GPP TSG-RAN WG1,'TR25.892 Feasibility Study of OFDM for UTRAN Enhancement', v1.1.0, March 2004, available at ftp://ftp.3gpp.org/Specs/archive/25_series/25.892.
[24] IST EVEREST project, http://www.everest-ist.upc.es.

Part II
Modelling

5

Propagation Modelling

Kamil Staniec, Maciej J. Grzybkowski and Karsten Erlebach

Ever since Personal Communication Systems appeared there has been an increasing demand for effective tools to model radiowave propagation. The purpose of this chapter is to give an overview of existing models, with particular emphasis placed on deterministic (theoretical), site-specific approaches. Other techniques will be briefly mentioned as well, but with less attention, as the way in which they describe the radio channel does not always provide sufficient information for the planning of modern wideband wireless systems. To better understand the idea behind these models, some place will be devoted to the analysis of physical phenomena which assist the propagation of the electromagnetic (EM) waves and their interactions with the environment.

5.1 RADIO CHANNELS IN WIDEBAND CDMA SYSTEMS

This section is devoted to some fundamentals required to comprehend modern radio propagation modelling approaches. We shall start with reviewing some principles related to EM wave propagation, and use this as a basis for the understanding of how radio channels are modelled.

5.1.1 ELECTROMAGNETIC WAVE PROPAGATION

For a uniform (i.e. one in which planes of equal phase and amplitude are overlapping), flat electromagnetic wave (i.e. moving in straight lines with all wavefronts in parallel) a general solution to Maxwell's equations can be found in Equation (5.1) that describes a wave travelling in $\bar{z}$ direction, with the E-component polarised in $\bar{x}$ direction:

$$E(t, z) = E_0 e^{-\gamma z} e^{j(\omega t - \beta z)\hat{x}} \qquad (5.1)$$

where the amplitude of the disturbance is described by the factor $E_0 e^{-\gamma z}$ and experiences exponential attenuation (with exponent γ) as it propagates through a lossy medium. The term $e^{j(\omega t - \beta z)\hat{x}}$ represents the phase of the wave.

In the simplest, free-space case, the received power at any distance d from the source can be found from Equation (5.2), where $P_{Rx}(d)$ is the distance dependent received power, P_{Tx} the transmission power, G_T and G_R the gains of the transmit and receiving antennas respectively and λ the operating wavelength. Equation (5.2) is also sometimes referred to as Friis transmission formula. To hold true, it assumes constant gain antennas (in contrast to constant aperture antennas) and a perfect matching of transmitting and receiving antennas. In real life terrestrial communications, however, this model will be unsatisfactory.

$$P_{Rx}(d) = P_{Tx} G_T G_R \left(\frac{\lambda}{4\pi d}\right)^2 \tag{5.2}$$

More accurate models should obviously take into account all kinds of interactions of the travelling wave with the surrounding environment. This knowledge of effects of the intervening terrain on the propagation is a useful piece of information in deploying modern radio systems. Deterministic models do account for the interaction mechanisms by treating them as, so-called, propagation primitives, which is especially evident in models based on ray-tracing. This is a technique that exploits principles of geometrical optics (GO) by using the concept of rays that have direction and position, amplitude and phase corresponding to the total path travelled and all kinds of interactions with the propagation environment. Another assumption is that the size of the obstacles is greater than the wavelength. This way of modelling radio waves allows each of the propagation phenomena to be considered separately, and their net result determines the final value of the electric field at the input of the receiver; this will be described in some more detail later in this chapter. The first of the primitives is the *specular reflection* (Figure 5.1a)

By *specular*, it is understood that two parallel rays incident on the same plane will travel parallel after reflection. Maxwell's equations are satisfied if all three rays (i.e. incident, reflected and refracted) lie in the same plane and the portion of the incident field that will be reflected is determined by the Fresnel reflection coefficients, $\Gamma_{\|,\perp}$, $\|$ denoting parallel and $\perp$ perpendicular polarisation with respect

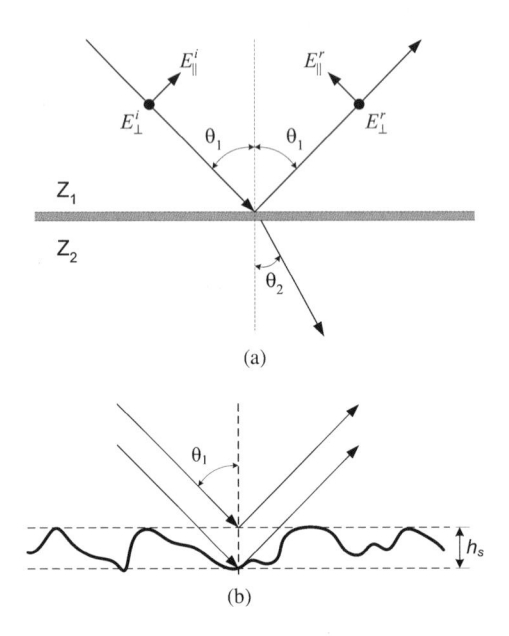

Figure 5.1 (a) Specular reflection; (b) Scattered reflection.

Propagation Modelling

to the plane of incidence, as quantified by Equations (5.3a) and (5.3b). Here, Z_1 and Z_2 are the wave impedances of the two respective mediums, and θ_1 and θ_2 the impinging and refracted angles respectively; the latter are related by means of Snell's law which, assuming lossless media with the same magnetic properties, is given in Equation (5.3c). This allows obtaining the reflected field as given in Equation (5.4).

$$\Gamma_{\parallel} = \frac{Z_1 \cos \theta_1 - Z_2 \cos \theta_2}{Z_1 \cos \theta_1 + Z_2 \cos \theta_2} \tag{5.3a}$$

$$\Gamma_{\perp} = \frac{Z_2 \cos \theta_1 - Z_1 \cos \theta_2}{Z_2 \cos \theta_1 + Z_1 \cos \theta_2} \tag{5.3b}$$

$$\frac{\cos \theta_1}{Z_1} = \frac{\cos \theta_2}{Z_2} \tag{5.3c}$$

$$E^R_{\parallel,\perp} = \Gamma_{\parallel,\perp} E^I_{\parallel,\perp} \tag{5.4}$$

Equation (5.4) refers to the case of an ideally smooth surface, which again in reality is never found. More probable is that a surface will be semispecular or diffuse and the incident energy will be scattered over a wider range of angles (see Figure 5.1b), thereby reducing the strength of the specular component [1,2]. It is common to use the Rayleigh criterion (Equation (5.5)) as a test of the surface roughness, which defines the critical height of surface protrusions h_{cr} (understood as the difference between the maximum and minimum protuberance). The surface height, h_s, being considered rough if $h_s > h_{cr}$.

$$h_{cr} = \frac{\lambda}{8 \cos \theta_1} \tag{5.5}$$

If, therefore, for a given incident angle θ_1 a surface is found rough, the amount of energy radiated in the specular direction diminishes by ρ_s, which is a function of the surface standard deviation σ_s about the mean, derived in [3] and reproduced in Equation (5.6).

$$\rho_s(\sigma_h) = \exp\left[-8\left(\frac{\pi \sigma_s \cos \theta_1}{\lambda}\right)^2\right] \tag{5.6}$$

The assumption in Equation (5.6) is that the surface protuberances are Gaussian distributed, in which case the surface is classified as a Gaussian rough scattering model and the modified Fresnel's coefficients will take on the form as in Equation (5.7).

$$\left(\Gamma_{\parallel,\perp}\right)_{Gauss} = \rho_s(\sigma_s) \Gamma_{\parallel,\perp} \tag{5.7}$$

In [1], a modification to Equation (5.7) has been proposed for the averaged Fresnel's coefficient (Equation 5.8) in situations when no knowledge on the surface roughness is available.

$$\left(\Gamma_{\parallel,\perp}\right)_{Gauss} = [1 + \rho_s(\sigma_s)] \frac{\Gamma_{\parallel,\perp}}{2} \tag{5.8}$$

While reflection and scattering can be nicely modelled by the laws of geometrical optics, diffraction (see Figure 5.2) should also be included in these procedures for the sake of accuracy.

First analytical solution to the problem of diffraction by an impedance wedge, a shape often encountered in the indoor or urban environment, with constant impedance on each face has been solved by Maliuzhinets [4] in the form of a set of homogenous equations. Due to its computational complexity, approximations of the Maliuzhinets function have been elaborated in [5]. The method will

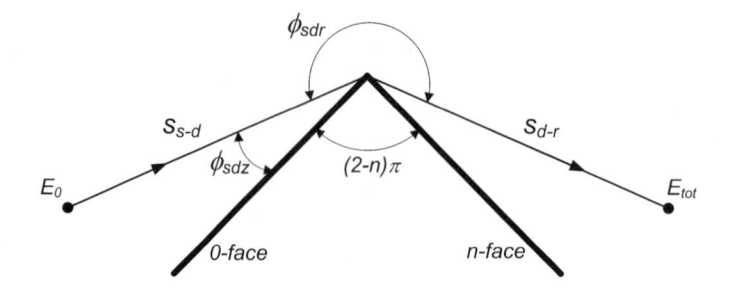

Figure 5.2 Ray geometry for diffraction by the edge.

serve as a reference for the analysis of the other diffraction approaches as discussed in this section. The Geometrical Theory of Diffraction (GTD) was proposed in [6], in which an extension to GO was provided in order to account for the effects of rays reflected from the side of the diffraction edge and their interference with direct and diffracted fields. The drawback of GTD was the fact that it failed in the vicinity of the shadow regions, i.e. on the line joining the source, diffracting edge and the receiving point. This drawback was eliminated in [7], introducing the Uniform Theory of Diffraction (UTD), where the use of Fresnel's integrals generate continuous diffracted fields at the shadow and reflection boundaries for a perfectly conducting wedge (often referred to as PC-UTD). Since in real situations it is more reasonable to assume that a diffracting wedge has finite conductivity, Luebbers [8] and Holm [9] have developed a heuristic formula for a diffraction coefficient for imperfectly conducting wedges for both horizontal (soft) and vertical (hard) polarisation; see Equation (5.9a). The net electric field E_{tot} at the reception point equals to Equation (5.10), where E_0 represents the source (initial) electric field [9].

$$D_{\|,\perp} = D(L_s, \phi_{sdr}, \phi_{sdz}, n) = D^{(1)} + D^{(2)} + \Gamma_{0,\|,\perp} D^{(3)} + \Gamma_{n,\|,\perp} D^{(4)} \tag{5.9a}$$

where Γ_0 and Γ_n are the reflection coefficients for the zero- and n-face, respectively (as in Figure 5.2). If either surface is perfectly conducting, $\Gamma_{0,n}$ equals to -1 for horizontal and $+1$ for vertical polarisation. The constituents $D^{(m)}$ are defined by Equation (5.9b):

$$D^{(m)} = D^{(m)}(L_s, \phi_{sdr}, \phi_{sdz}, n) = -\frac{e^{-j\pi/4}}{2n\sqrt{2\pi k}} \cot \gamma^{(m)} F\left(2kL_s n^2 \sin^2 \gamma^{(m)}\right) \tag{5.9b}$$

where
k – wave number,
$L_s - d_{s-d} \cdot d_{d-r}/(d_{s-d} + d_{d-r})$
$n\pi$ – wedge exterior angle.

$$F(x) = 2j\sqrt{x}e^{jx} \int_{\sqrt{x}}^{\infty} e^{-j\tau^2} d\tau \text{ – transition function (Fresnel integral)}$$

$$\gamma^{(1)} = [\pi - (\phi_{sdr} - \phi_{sdz})]/2n$$

$$\gamma^{(2)} = [\pi + (\phi_{sdr} - \phi_{sdz})]/2n$$

$$\gamma^{(3)} = [\pi - (\phi_{sdr} + \phi_{sdz})]/2n$$

$$\gamma^{(4)} = [\pi + (\phi_{sdr} + \phi_{sdz})]/2n$$

$$E_{\text{tot}} = E_i D_{\|,\perp} \sqrt{\frac{d_{s-d}}{d_{d-r}(d_{d-r}+d_{s-d})}} e^{-ikd_{d-r}} \qquad (5.10)$$

where E_i is the field incident on the wedge and equals Equation (5.11)

$$E_i = E_0 \frac{e^{-jkd_{s-d}}}{d_{s-d}} \qquad (5.11)$$

In Equation (5.10), the expression under the square root (often referred to as attenuation factor) represents the decrease of the diffracted wave amplitude with the distance from the source.

The above effects of free-space propagation, reflection, scattering and diffraction contribute in a linear fashion to the electrical field experienced at the receiving antenna. Once this signal is received and down-converted, it can be represented by a complex channel coefficient h, as will be evident from the next section.

5.1.2 WIDEBAND RADIO CHANNEL CHARACTERISATION

In practical situations, an emitted signal will propagate by interacting with the surrounding environments that involves reflections from objects, transmissions through obstacles, diffraction on edges and scattering from rough surface. Thus, the signal arriving at the receiver will not come in a single fringe, but as a pack of signals with random amplitudes, phases, angles of arrival, and short time delays, being delayed copies of the original signal. Once collected within a certain time span at a receiver, they sum up vectorially, accounting for their relative phase differences, which causes some copies to overlap constructively if both are in phase or cancel out otherwise. Such behaviour leads to *small scale fading*, which is a typical propagation effect, especially in the indoor and urban environment. Hence, a radio channel can be mathematically represented at any point in a three-dimensional space as a linear, time-invariant filter (Figure 5.3) of an impulse response given by Equation (5.12a):

$$h(t) = \sum_{i=0}^{N_{\text{multipath}}-1} a_i(t)\delta[t-\tau_i]e^{j\theta_i^{ph}(t)} \qquad (5.12a)$$

$$H(f) = \int_{-\infty}^{\infty} h(t) e^{-j2\pi ft} dt \qquad (5.12b)$$

Here, $N_{\text{multipath}}$ is the number of multipath components, $\theta_i(t)$ and $a_i(t)$ are respectively the time varying phase and amplitude of each component and τ_i its delay. The frequency response $H(f)$ can be easily obtained from the Fourier transform of $h(t)$ (Equation 5.12b). Therefore, since either $h(t)$ or $H(f)$ are needed for the exhaustive characterisation of the radio channel, only one of these should be measured (or accurately predicted), while the other one will be obtained by means of the Fourier transform or its

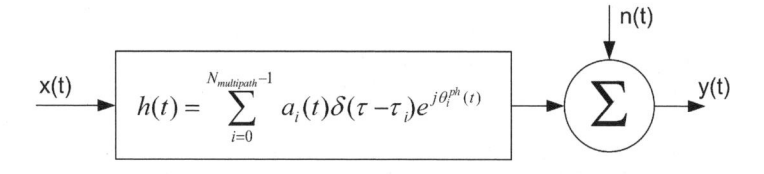

Figure 5.3 A schematic representation of a time-dispersive radio channel.

inverse. Now, assuming that the signal is transmitted over an Additive White Gaussian Noise (AWGN) radio channel, the output signal will be in the form of Equation (5.13).

$$y(t) = \int_{-\infty}^{\infty} x(\tau)h(\tau)d\tau + n(t) \qquad (5.13)$$

Crucial parameters with which to identify the characteristics of such a channel are [10,11] the mean excess delay τ_m, the root-mean square (rms) delay spread τ_{rms}, the delay window, the total energy, the coherence (correlation bandwidth) $B_{x\%}(f)$ and the Power Delay Profile (PDP). The parameters of particular importance to the PCS systems design are PDP, τ_{rms} and $B_{x\%}(f)$. The PDP [11] gives the time distribution of the received signal power from a transmitted impulse, and is defined by Equation (5.14). It is often used to quantify time dispersion in mobile channels, and hence characterises the channel's frequency selectivity. The mean excess delay, τ_m, is the averaged multipath delay and has the sense of the first moment of the PDP. Finally, τ_{rms}, having the sense of the second central moment of the PDP, is a measure of the channel time dispersiveness and determines the maximum symbol rate achievable by a communication system before intersymbol interference (ISI) occurs.

$$P_h(t) = h(t)h^*(t) = |h(t)|^2 = \sum_{i=0}^{N_{multipath}-1} a_i^2 \delta(t - \tau_i) \qquad (5.14)$$

$$\tau_m = \frac{\sum_i \tau_i P_h(\tau_i)}{\sum_i P_h(\tau_i)} \qquad (5.15)$$

$$\tau_{rms} = \sqrt{\frac{\sum_i (\tau_i - \tau_m)^2 P_h(\tau_i)}{\sum_i P_h(\tau_i)}} \qquad (5.16)$$

In [10], $B_{x\%}(\Delta f_{x\%})$ is defined by Equation (5.17) as the Fourier transform of the PDP, where $\Delta f_{x\%}$ denotes the frequency range, for which signal components are correlated at $x\%$, where x is usually chosen from between 50 and 90.

$$B_{x\%}(\Delta f_{x\%}) = \int_{t_0}^{\tau_{max}} \mathrm{PDP}(\tau) e^{-j2\pi \Delta f_{x\%} \tau} d\tau \qquad (5.17)$$

The coherence bandwidth in other words is a statistical measure of a frequency range over which the attenuation is constant and linear in phase. The coherence bandwidth has been related by inverse proportionality to τ_{rms} Equation (5.18), where values of α_B span between 5 and 50, which corresponds to the correlation values of respectively 50 and 90 %.

$$B_{x\%} = \frac{1}{\alpha_B \cdot \tau_{rms}} \qquad (5.18)$$

As reported in [11–14], where the dispersive parameters of various radio channels were both measured and simulated, the values of τ_{rms} lie between 5 and 300 ns in indoor environments. In microcells, τ_{rms} is usually found between 0.35 and 2 μs. In macrocells, values around 5 μs and more were measured for rural and hilly terrains. Regardless of the type of environment, the channels turned out to be less dispersive under Line Of Sight (LOS) than under Non Line of Sight (NLOS). In each type of environment, the lower bounds of τ_{rms} were reported under LOS conditions, whereas the upper limits where measured under NLOS conditions. This is an expected outcome since in LOS a strong direct signal dominates over other (reflected and/or diffracted) components. However, in NLOS situations, the reception occurs mainly due to multipath echoes, many of which possess similar amplitude, spread over a larger time.

As a rule of thumb (see e.g. [14,15]) one may assume that if a digital signal has a symbol duration exceeding ten times the rms delay spread, then an equaliser will not be required to achieve BER $\leq 10^{-3}$. Equation (5.19) relates this rule to the maximum achievable data rate.

$$\max(R_b) = 0.1/\tau_{\text{rms}} \qquad (5.19)$$

In narrowband systems, small scale fades affect the signal's entire spectrum in a uniform manner leading to the conclusion that the channel transfer function H is frequency-independent. In wideband systems, in contrast, only a part of the total signal spectrum is likely to be threatened by the deep fade, the rest will be left intact, an effect known as *selective fading*. In the narrowband case, $B_{x\%}(\Delta f_{x\%})$ forms only a fraction of the signal's spectral width, or in other words $H = H(f)$.

Multipath propagation has also a degrading effect on the received signal due to multiple, mutually delayed, interfering symbol components, referred to as ISI. In this way, ISI puts an upper bound on the system's maximum data rate without equalisation. Destructive in narrowband systems like GSM, multipath mechanism has proven to be beneficial in CDMA systems, owing to path diversity. Due to this feature the receiver (RAKE) is able to discriminate separate components of the received signal (see Chapter 2). This is only possible when the relative path differences between multipath components are not shorter than the length of a single pulse (chip) of the spreading sequence. Signals from individual paths can be separated, weighted and summed up to assure a maximum signal strength.

5.1.3 INTRODUCTION TO DETERMINISTIC METHODS IN MODELLING WCDMA SYSTEMS

Since empirical models have been widely described in the literature, the following subsection will be devoted to the description of deterministic (theoretical) models. This class of models is a relatively new approach and, because some of the deterministic models require the use of fast computation machines, their application was very limited until more or less the last decade. As will subsequently be demonstrated, encouraging results can be obtained with hybrid models, linking advantages of both deterministic (accurate) and empirical (fast) models.

By and large the most commonly used deterministic model is the one based on the technique referred to as *ray tracing* (with many derivative techniques). It has been successfully applied to modelling radio coverage in microcells [15–18] (although [19] and [20] report satisfactory results obtained with ray-tracing models at distances of up to 11 and 13 km, respectively) and, most suitably, to indoor environments [21–24]. The main idea behind this method is to discretise the wavefront into a number of individual rays, each of which will carry a part of the total transmitted power. The received field therefore is the superposition of all EM components reaching the receiver. Since the technique imitates the real radio wave propagation, it is possible to account for such propagation phenomena as reflection, transmission through obstacles and diffraction. A method that accounts for scattering in ray tracing is presented in [25]; a randomly oriented reflection plane is defined for every ray/wall interaction.

Image Method (IM) is an example of the ray tracing implementation [15,22,26]. Initially, reflection and transmission coefficients are calculated from the environment database containing the geometrical structure (walls coordinates) and electric parameters of construction materials. The core of this method is to find all images (virtual sources) of the real source with respect to all planes (walls) of the given environment, as depicted in Figure 5.4.

In this example, a propagation path with three reflections is found. To do so, the I_3 image of the transmitter Tx is determined relative to plane 3. I_3 is next reflected relative to plane 4, thus producing image $I_{3,4}$. Finally, $I_{3,4,2}$ is obtained by reflecting $I_{3,4}$ against plane 2. Once the procedure is finished, a receiver Rx is located at a desired place and the propagation path with three reflections is determined by tracing back through intersection points $r_1 \rightarrow r_2 \rightarrow r_3$. Similarly, propagation paths can be found

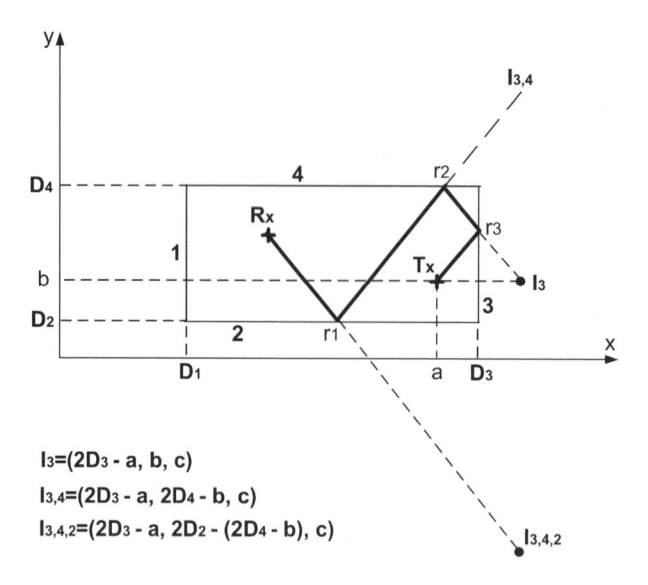

Figure 5.4 Determining the propagation path with the image method (IM).

for any number of reflections. A great advantage of this method is that for a given location of Rx, the path tracing procedure need be carried out only once, irrespective of the later Tx locations.

In the Ray Launching (RL) method, the transmitted wave is discretised into N rays with equal angular separations α_S. By doing so, each ray will carry a portion of the power proportional to the solid angle between any neighbouring rays. To assure constant α_S, the radiation sphere is first approximated with an icosahedron (with 12 vertices) and one ray is passed through each vertex with $\alpha_S = 63.4°$. Next, in order to reduce the separation, new vertices are generated by tessellating each of the 20 triangles of the solid (i.e. joining the midpoints of each triplet of neighbouring sides into smaller triangles) and setting as shown in Figure 5.5a. If the ray resolution is unsatisfactory, the tessellation process can be repeated S_T times, where S_T is the tessellation frequency and is incremented by one after each tessellation ($S_T = 1$ for an initial, untessellated icosahedron).

The total number of rays $N(S_T)$ that can be launched is therefore the number of vertices after S_T tessellations, and is given by Equation (5.20).

$$N(S_T) = 10 \cdot 4^{S_T - 1} + 2 \qquad (5.20)$$

Once S_T has been selected to achieve a desired resolution, the rays are launched through each vertex and traced separately as they traverse the environment. On its way, each ray interacts with the surrounding, so that at the reception point, the i-th ray amplitude is expressed by Equation (5.21).

$$E_i = \frac{\sqrt{Z_0 P_0}}{4\pi d L_{Di}(d)} G_{ti} G_{ri} \prod_m D(\theta_{mi}) \prod_j \Gamma(\theta_{ji}) \prod_k T(\theta_{ki}) e^{-j\frac{2\pi d}{\lambda}} \qquad (5.21)$$

where G_{ti}, G_{ri} are respectively the transmitter and receiver gains for the i-th ray, Z_0 the free space impedance (approximately 377 Ω), d the total path length of the i-th ray, $D(\theta_{mi})$, $\Gamma(\theta_{ji})$, $T_F(\theta_{ki})$, respectively, the diffraction, reflection and transmission coefficients, λ the wavelength, L_{Di} a spreading factor (see Section 5.1.1) and $\Gamma(\theta_{ji})$, $T_F(\theta_{ki})$ are, respectively, the reflection and transmission coefficient. At the reception end, a single point in space is replaced with a reception sphere to account for resolution loss as rays diverge from the source. The radius of the sphere will thus be a function

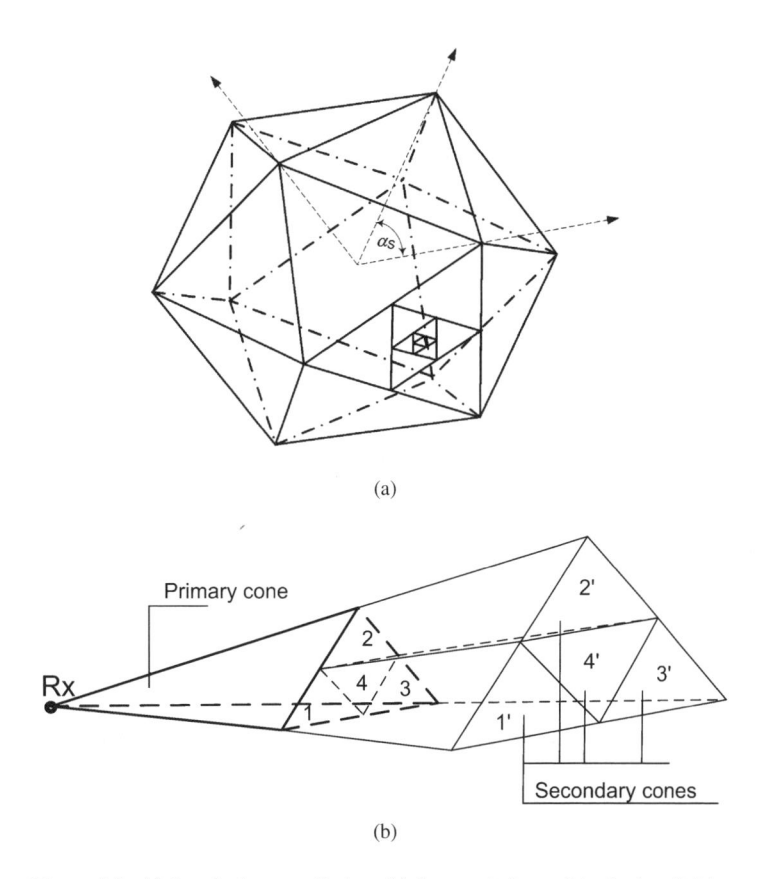

Figure 5.5 (a) Icosahedron tessellation; (b) Cone emission and its further division.

of the total distance d travelled by the ray and is equal to $2d\alpha_s/\sqrt{3}$. The main disadvantages of the RL method are the long computation time (as compared with IM) and a loss of resolution unless the tessellation frequency is very large.

A last modification of the ray tracing technique presented here is the Cone Launching (CL) technique. The wavefront is now split into separate cones, instead of rays (see Figure 5.5b). In this way, the loss of resolution is avoided. One of the problems with this technique is that, despite the spherical nature of the real wavefront, in CL the simulated plane spread between each triplet of vertices will always remain flat. Focusing on one such triplet one can notice that at its vertices both wavefronts (the real and the simulated one) will be identical, however closer to the midpoint of the triplet the difference between the actual (bulged) and the simulated (flat) wavefront will be more significant. To mitigate this effect, after having travelled some defined distance, each initial (primary) cone will undergo further divisions into secondary cones, as illustrated in Figure 5.6. The method requires sophisticated techniques to account for the fact that the cross section of the cone is very likely to change shape, for instance when only a part of it is reflected from a reflecting object.

The latest, fast growing, propagation prediction method utilises the concept of Artificial Neural Networks (ANNs). These can be thought of as data processing systems imitating the behaviour of a human brain. In general, irrespective of the type of ANN systems used for a particular problem, they are all characterised by the following features: *parallel structure* resulting in high computation speed

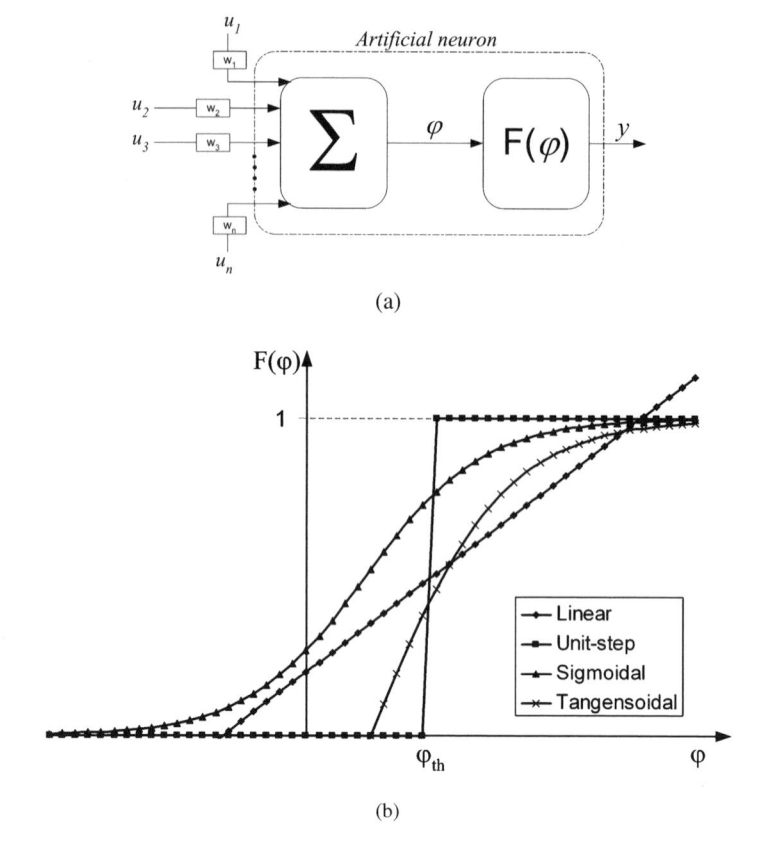

Figure 5.6 (a) Artificial neuron model; (b) Activation functions.

and *adaptation* to the environment in the process of learning. Much like in the case of biological neural networks, the basic elements in ANN are artificial neurons (Figure 5.6a), composed of a summation and activation block. The input signals, multiplied by respective weights (or a vector of weights $\vec{w}$), are summed up to create an output signal φ. This signal is then fed into an activation function (or transfer function) $F(\varphi)$. The simplest form of an activation function is the linear function $F(\varphi) = k\varphi$ or a unit-step function that assumes zero for all values of φ smaller than a threshold φ_{th} and takes on one for all φ greater than φ_{th}. Most sophisticated functions give a closer approach to the non-linear transfer function of the biological neuron; most commonly used are the sigmoidal and tangensoidal functions (Figure 5.6b).

As will be shown in subsequent sections concerning modelling, the ANN method has been applied to all types of environments, giving very satisfactory results, provided that appropriate input $u_i (i = 1 \ldots n)$ parameters have been defined. One possible method of selecting input parameters is to use those parameters, which exist in empirical models (e.g. streets width, buildings heights or the number of intersected walls and their attenuation). The great advantage of the ANN method is its ability to do fast computing of an output vector $\vec{v}$ given an input vector $\vec{u}$ without any explicit knowledge on the analytical transfer function $\vec{y} = f(\vec{u})$ [27]. After the selection of the input vector $\vec{u}$, containing parameters which have the strongest impact on the value of electric field distribution, extensive measurements need to be carried out to provide *training patterns*, which are then used to adjust the weights vector $\vec{w}$. If the ANN structure (perceptron) and the vector u_i have been properly defined, the procedure will converge after some training iterations to yield accurate results.

5.1.4 DETERMINISTIC METHODS: COMPARISON OF PERFORMANCE

A comparison of IM with RL leads to the following conclusions:

- In IM, the computation time grows exponentially with the predefined number of reflections, whereas for RL this relation is linear.
- The image search in IM assures that exactly all the propagation paths are found. As was mentioned above with the RL method, it may come to the loss of resolution due to the divergence of the traced rays and thus a possible omission of receiver.

Comparing RL to CT leads to the following:

- The possible loss of resolution (RL) does not exist in the CT method (no empty gaps between launched cones allows all points of the scanned space to be covered).
- The execution time of the CL methods becomes considerably greater than that of RL, especially for more than two reflections.

Comparing both ray tracing methods (RL, CT) with the ANN method, the following can be concluded:

- The greatest advantage of the ANN method is the fast computation time; the procedure of tracing the actual propagation path is now replaced with the training of the artificial neural network. Moreover, with ANN it is possible to process (by training) large quantities of data (measurement results).
- No physical radio propagation relations need to be defined prior to simulations performed with the ANN method – in the process of training, weights are adjusted according to presented patterns (in the form of measurement results).
- Since RL calculates the total received field as a sum of field components, it is suitable for the calculation of the time and angular dispersion of the received signal. This information may be crucial in systems utilising multiple-input multiple-output (MIMO) techniques. The ANN method, in turn, is applicable to cases when only the electric field level (or coverage maps) is desired.

Table 5.1 presents some summary of performance results obtained by different authors with ANN and RL/RT methods. At this point, it should be reminded that a significant spread in the values of $|\mu_{err}|$ and σ_{err} obtained with different models inside either group is attributed to the fact that the research performed therein concerned various propagation environments – ranging from microcellular (with large σ_{err}) to rural (with small σ_{err}). Therefore, the purpose of results presented in Table 5.1 was to demonstrate that, on average, the outcomes in terms of absolute mean error $|\mu_{err}|$ and standard error deviation σ_{err} attained with ANN and RL/RT do not allow to unequivocally grant either technique superiority over the other.

Furthermore, relatively large values of the standard deviations attained with the RL/RT method is due to the fact that the ray tracing technique bears intrinsic errors by simplifying any real environment (simplification is required in order to make ray tracing algorithms more tractable). The inclusion of a greater number of propagation details, of course, positively affects the accuracy of simulations. The ANN method has therefore the advantage of being resistant to propagation details; however, measurement campaigns are an unavoidable step in order to train a network to a particular environment.

Table 5.1 Simulation results compared with measurements for ANN and RL/RT techniques.

	Site	$|\mu_{err}|$	σ_{err}
Artificial neural networks	Munich [23]	0.9	6.0
	Valenzia [23]	3.5	6.5
	Belgrade [24]	0.6	8.2
	University of Vienna [28]	2.5	6.0
	University of Stuttgart [28]	3.5	4.6
	Inst. de Telecomunicações (Lisbon) [28]	8.7	3.2
	Munich [28]	1.2	7.4
	Helsinki (Finland) [28]	1.4	7.4
	Mannheim (Germany) [29]*	0.0	4.7
	Kavala (Greece) [29,30]*	3.7	3.9
	Oia village on Santorini Island (Greece) [29]*	2.7	2.9
	Average:	2.61	4.62
Ray launching/ray tracing	Defined by authors [19]	1.3	4.4
	Munich (irregular street layout) [26]	4.1	10.7
	Munich (regular street layout) [26]	2.4	17.1
	University of Stuttgart [31]	1.5	8.6
	Villa of Marconi (Italy) [31]	4.1	4.1
	Virginia Tech Campus, USA [17]	1.5	4.8
	Karlsruhe downtown [18]	1.6	1
	Worcester Polytechnic Institute (USA) [32]	<1	2.3
	Munich (Germany) [33]	1.5	6.9
	University of Stuttgart [34]	1.15	4.73
	Average:	2.22	6.66

* A hybrid model was used (Cost 231 Walfish-Ikegami + dominant path method).

5.2 APPLICATION OF EMPIRICAL AND DETERMINISTIC MODELS IN PICOCELL PLANNING

It is useful in the planning of wireless systems to distinguish between two basic propagation environments: indoor and outdoor, each exhibiting specific features to propagating EM waves. Since no unique cell size constraints for different environments have been defined, we will use those provided in [35]. In this section, attention will be paid to the picocell modelling, which is generally characterised by a small cell radius (<100m), low transmit powers, and where both the user and the base stations are placed indoors. Examples of environments belonging to this category include Small Office Home Office (SOHO), Medium/Small Enterprise (MSE), airports, shopping plazas, train stations, etc. The indoor space exhibits different features than outdoor areas. In particular, the coverage range is ultimately limited by the building's outer walls; also, LOS conditions are rarely met and the radio waves propagate by means of reflections, diffraction and transmission through obstacles.

5.2.1 TECHNIQUES FOR INDOOR MODELLING

The simplest model for predicting the path loss is given by the Motley–Keenan formula (Equation 5.22) [36].

$$L = 20\log\left(\frac{4\pi fd}{c}\right) + k_f \cdot 10\log A_{floor} + k_w \cdot 10\log L_{wall} \qquad (5.22)$$

where k_f is the number of floors traversed, A_{floor} is the floor attenuation factor, k_w is the number of walls traversed and L_{wall} is the wall attenuation factor. The model is an empirical and simplified representation of propagation loss; since it models the path loss only, it does not include effects like multipath fading. Besides, all intervening floors and walls are assumed to be of the same types with equal attenuations, F and W respectively.

The COST 231 Indoor [37], also known as the Multi-Wall Model (MWM), is a more refined path loss formula:

$$L = 20\log\left(\frac{4\pi f d}{c}\right) + \sum_{i=1}^{M} k_{wi} L_{wi} + k_f^{[(k_f+2)/(k_f+1)-0.46]} L_f \qquad (5.23)$$

where k_{wi} is the number of penetrated walls of type i, k_f is the number of penetrated floors, L_{wi} is the loss of wall type i, L_f is the loss between adjacent floors, M is the number of different wall types and finally i is the number of wall types. Above equation accounts for the observation that the total floor loss is a non-linear function of the number of penetrated floors. Since the exact knowledge of all wall attenuations is usually unavailable, a simplification can be proposed [37] by differentiating between only two types of walls, namely the light wall L_{w1} (e.g. a partition wall made of plasterboard or a light concrete wall, thinner than 10 cm) and the heavy wall L_{w2} (load-bearing, made of concrete or brick, thicker than 10 cm). It must be noticed that the loss factors in Equation (5.23) are not actually physical losses but rather model coefficients fitted to the measured path loss curve. Thus, propagation effects such as furniture shadowing or signal guiding in corridors are already implicitly included. Default values of L_{w1} and L_{w2} are given in [37] and equals 3.4 dB and 6.9 dB, respectively.

Since the interest in indoor channel modelling has been mainly inspired by the development of wideband short-range systems, a lot of efforts have been made to investigate the broadband characteristic of various indoor radio channels. In [38], it was demonstrated that, due to multipath, rays arrive at a receiver in clusters. Each of these clusters is composed of another cluster of closely spaced rays. As a result, a double-Poisson inter-arrival time process has been proposed to describe the behaviour of the rays' inter-arrival time. A similar behaviour has been theoretically confirmed in [39] using a ray tracing model, thereby extending the model by the observation that the distribution of relative angles of arrival is best approximated by a Laplacian distribution. The authors in [40] performed a number of measurements to reveal that significant multipath components were decreasing in number and power as the room size was increased. Multipath components were also analysed as a function of the distance between the transmitter and the receiver, showing that, as the distance between them increases, the number of multipath components tends to decrease.

Most deterministic models for the indoor propagation are based on ray tracing, mainly due to its applicability to model time dispersion of a given channel along with the path loss. The main disadvantage of these methods, however, lies in their high computational demands. An alternative method of calculating the electric field distribution indoors is described in [28]. The method consists of two steps: (1) determination of so-called *dominant paths* and (2) finding the path loss L, according to the Equation (5.24).

$$L = 20 \cdot p_L \cdot \log d + \sum_{j=1}^{n} L_i(\theta_{fc}, j) + \sum_{i=1}^{m} L_{wi} - \alpha_w \qquad (5.24)$$

where p_L is the indoor path loss exponent, $L_i(\theta_{fc}, j)$ is a loss function due to interaction (depends on the changing propagation direction of j-th path), θ_{fc} is the angle between the former and the current direction of propagation, i is the interaction number, t_j is the j-th wall transmission loss and α_w is a waveguiding factor. A great advantage of this model is that, unlike in standard ray launching methods, all those rays between the transmitter and the receiver that interact with the same set of walls are replaced with a single representative ray – a dominant path, as shown in Figure 5.7. The model has been applied to both single and multiple floor buildings with satisfactory result [23,24,26,28,29].

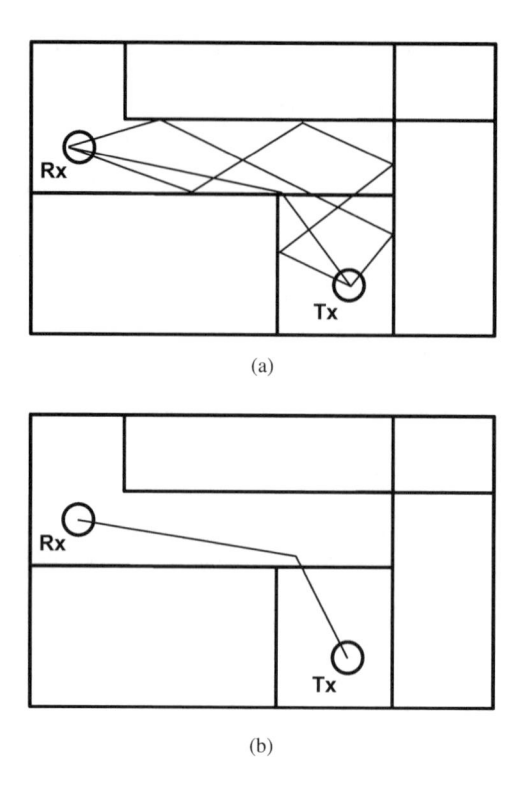

Figure 5.7 (a) Multiple-reflections path; (b) Dominant path (based on [28]).

5.2.2 TECHNIQUES FOR OUTDOOR-TO-INDOOR MODELLING

For many microcellular coverage predictions, the buildings are assumed to be either perfectly conducting (so that no building penetration is assumed) or to introduce a constant loss. In [41], for instance, the mean building penetration loss was found from measurements to oscillate around 20 dB, whereas in [42] this loss was determined to range from 20 to 40 dB. A more in-depth approach was presented in [43], where a simple building penetration formula (Equation 5.25) has been proposed, assuming that the external wall dielectric permittivity $\varepsilon_r = 5$:

$$L = \alpha_b \cdot d_{in} - 20 \log T_{F1} - 20 \log T_{F2} \qquad (5.25)$$

where α_b denotes the building specific attenuation coefficient, d_{in} is the length of the ray penetrating the building (from the entrance point r1 to the exit point r2), finally T_{F1} and T_{F1} represent the transmission coefficients of the walls through which a ray enters (at point r1) and through which it exits the building (at point r2). The value of α_b represents the sum of all losses due to wave penetration through internal walls, furniture and persons. Its value was measured for 22 different buildings [43] and its averaged value was 2.1 dB/m.

The authors in [44] have proposed a simple to use Equation (5.26a) for the median path loss, where L_{ex} is the attenuation of the external wall, analogous to L_{w2} in Equation (5.23). The formula describes situations where the receiver and transmitter are both located in the outermost rooms. When from either side this condition is not fulfilled, some extra loss should be added to account for attenuation introduced by external walls, for instance using Equation (5.23).

$$L_{50\%} = L_{\text{Modified\_Hata}} + 2 \cdot L_{\text{ex}} \qquad (5.26a)$$

Furthermore, it had been noted in [44] that the actual path loss L should consider the sum of both median path loss $L_{50\%}$ and some Gaussian variations $T(G(\sigma))$, see Equation (5.41), where σ is the standard deviation of the Gaussian process. Appropriate formulas for calculating the Gaussian distributed $T(G(\sigma))$, as well as uniform and Rayleigh distributed variations, are provided in [44].

In some outdoor-to-indoor measurements [42,45,46], it has been demonstrated that the building penetration loss decreases with the increasing floor number, the effect known as the *floor height gain*. The existence of this effect is attributed to the influence of surrounding buildings which diffract and reflect energy and thus contribute to the total power received inside a building. As could be expected, the effect ceases at higher (the seventh [46] or the fifth [42] and above) floors because, first, in such cases it is more probable that LOS conditions hold and, second, if the floor height exceeds the average building heights, less contributions from neighbouring buildings will occur. In [42], a formula, Equation (5.27), that expresses its dependence of frequency has been fitted to the height gain measurement:

$$G_n(N_f) = 2.9 \cdot N_f + 1.16 \qquad (5.27)$$

Above equation applies only up to the fifth floor (the floor number being represented by N_f), because for higher floors the measured height gain curve begins to level off; a floor height of 3 m was assumed.

The model presented in [45] considers parameters such as the angle of incidence and building properties which may affect the penetration loss. The penetration loss for the LOS case (Figure 5.8a) is given by Equation (5.28).

$$L = 32.45 + 20\log(f) + 20\log(d_{\text{ex}} + d_{\text{in}}) + L_{\text{ex}} + L_{\text{WGe}} \cdot \left(1 - \frac{d_\perp}{d_{\text{ex}}}\right)^2 + \max(L_{\text{in,LOS}}, L_{\text{in-ext,LOS}}) \qquad (5.28)$$

where $L_{\text{in,LOS}} = L_{wi} \cdot k_w$ and $L_{\text{in-ext,LOS}} = \alpha_b \cdot (d_{in} - 2)\left(1 - \frac{d_\perp}{d_{\text{ex}}}\right)^2$, L_{ex} is the loss in the external wall illuminated at angle $\theta_{\text{ex}} = 90°$, L_{WGe} is the additional loss in the external wall when illuminated at $\theta_{\text{ex}} = 0°$, k_w is the number of penetrated internal walls, L_{wi} is the loss in the i-th internal wall. The following parameter values are recommended in the model:

- L_{ex}: 4–10 dB (7 and 4 dB for concrete and wood, respectively);
- L_{wi}: 4–10 dB (7 and 4 dB for concrete and wood/plaster, respectively);
- L_{WGe}: about 20 dB
- α_b: 2.1 dB/m (according to [43,46]).

Figure 5.9 shows a NLOS case and a path loss formula related to an outside reference loss L_{out} (minimum of L1 and L2 in Figure 5.8b), which is given by Equation (5.29).

$$L = L_{\text{out}} + L_{\text{ex}} + W_{\text{Wge}} + \max\left(L_{\text{in,LOS}}, L_{\text{b,NLOS}}\right) - G_{\text{FH}} \qquad (5.29)$$

where $L_{\text{b,NLOS}} = \alpha_b \cdot d_{in}$ and $G_{\text{FH}} = \begin{cases} n \cdot G_n \\ h \cdot G_h \end{cases}$

G_n is the floor height gain (dB/floor), while G_h is the height gain (dB/m). The former applies to 3 m high storeys (with attenuation of 1.5–2 dB/floor or 4–7 dB/floor, since two groups of values have been reported), while the latter is suitable for taller (4–5 m) floors, with attenuation within 1.1–1.6 dB/m (at 1800 MHz). L_{Wge} is the extra angle-dependent penetration loss suffered by the wave when impinging

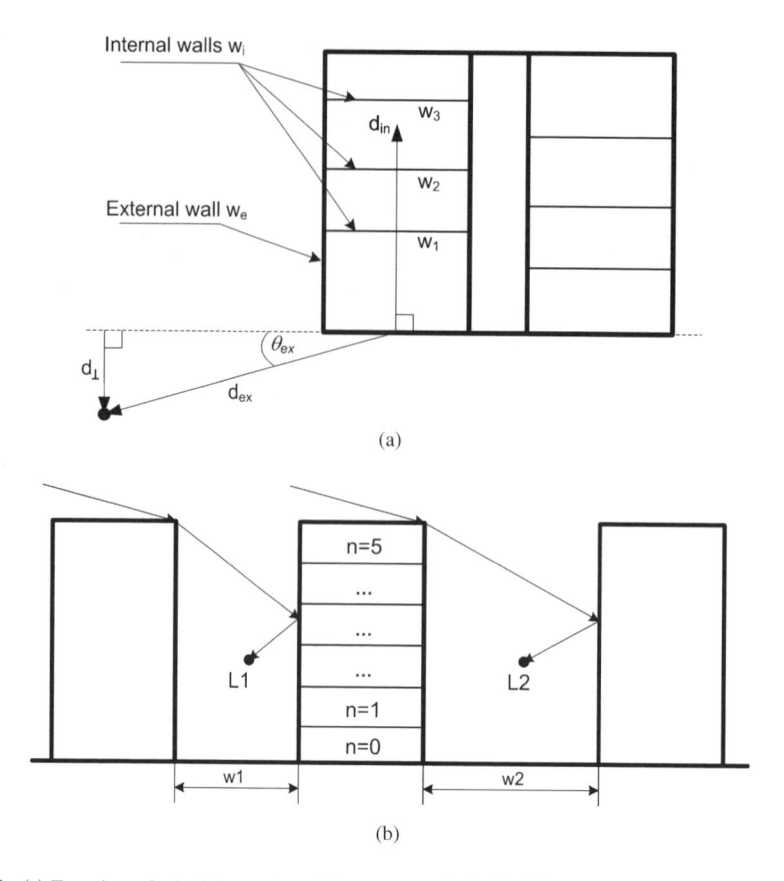

Figure 5.8 (a) Top view of a building and model parameters for LOS; (b) Side view for NLOS (based on [45]).

the external wall from a non-perpendicular direction. Its value is suggested to fall between 3 and 5 dB at 900 MHz adding 2 dB per octave increase in frequency [45]. L_{out} can be calculated, for instance, with Berg's recursive formula or with the COST 231 Walfisch–Ikegami (WI) model, both of which will be treated in greater detail in Section 5.3.

5.3 APPLICATION OF EMPIRICAL AND DETERMINISTIC MODELS IN MICROCELL PLANNING

A microcell is understood as an area ranging from 0.1 to 1 km from the base station [35]. In microcellular design, models should account for phenomena typical for urban environments, like reflection from building walls, wave tunnelling in street canyons, single or multiple diffraction on rooftops and corners or buildings penetration. Apart from deterministic models to be discussed herein, which do account for these effects, some of the empirical microcellular models will also be presented.

As will be analysed in greater detail in Section 5.6, a commonly used approach to describe propagation in microcellular environment has proved to reveal double-slope behaviour. Models that consider this phenomenon make distinction between two regions separated by a 'break point'.

Before this point, the received signal strength experiences regular deep fades due to the combination of constructive/destructive additions of the two received signals – direct and ground-reflected one. This is because at shorter distances both rays are still relatively strong which makes their vector summation very sensitive to their phase differences. The envelope of these summed rays follows more or less a free space loss exponent $n = 2$. At further distances, after the break point, the strengths of both signals have diminished to the degree where both in-phase or out-of-phase addition effect does not noticeably contribute to signal oscillations any more. The signal envelope now monotonically declines with a greater loss exponent (e.g. $n = 4$).

Referring to the Fresnel theory, the break point can be considered as the distance for which the ground begins to obstruct the first Fresnel zone and the distance from the transmitter to this point is given by Equation (5.30) (although the actual, measured value of d_{br} appears to be shorter than the theoretical one due to the influence of pedestrians, vehicles and other obstacles).

$$d_{br} = 4\frac{h_t h_r}{\lambda} \tag{5.30}$$

where h_t and h_r are the heights of transmitter and receiver antennas, respectively.

5.3.1 COST 231 WALFISCH–IKEGAMI MODEL

By and large, the COST 231 Walfisch–Ikegami (WI) model [47,48,51] is the most widely used empirical model today, being an extension of the models from J. Walfisch and F. Ikegami. It has been adopted as a standard model for 3G IMT 2000/UMTS systems [49]. It is valid within the following constraints:

- transmitter height h_t: 4–50 m
- receiver height h_r: 1–3 m
- transmitter to receiver separation d_{tot}: 0.02–5 km.

The model is based on the assumption that the transmitted signal propagates through multiple diffractions over rooftops (Figure 5.9a). Therefore, it considers the buildings in the vertical plane between the transmitter and the receiver, which are characterised as diffracting half-screens of equal heights, widths and separations (Figure 5.9a). At the terminal, the received field is composed of two rays – the direct multiple diffracted and diffracted-single-reflected one – which are vector summed to account for their phases. In order for the above vertical plane propagation assumption to hold true and thereby keeping the mean error and the standard deviation low (+3 dB and 4–8 dB, respectively [45]), it is advisable to ensure that $h_t \gg h_{Rf}$. This is because as h_t approaches the roof height, the model performance degrades considerably. Since the model was intended for microcellular field prediction and amendment to Ikegami's original model was made to account for the 'street canyon' effect by adding the orientation loss (L_{ori}) to the rooftop-to-street loss (L_{rts}), this is an empirical correction term attained from the calibration with measurements. Lastly, the model performs poorly if the height of the terrain varies in the calculation area or when the building's coverage is not uniform. With respect to the building heights, in [50] a thorough investigation has been carried out to conclude that non-uniform building heights result in an increased value of propagation loss by a shift of a few decibels, whereas the trend in the propagation loss curve does not change as compared to the case of uniform buildings heights.

The WI model considers data describing the urban environment such as the heights of buildings h_{Rf}, widths of roads w_s, building separation d_s, road orientation to the direct radio path φ_{ori} etc. (Figure 5.9b).

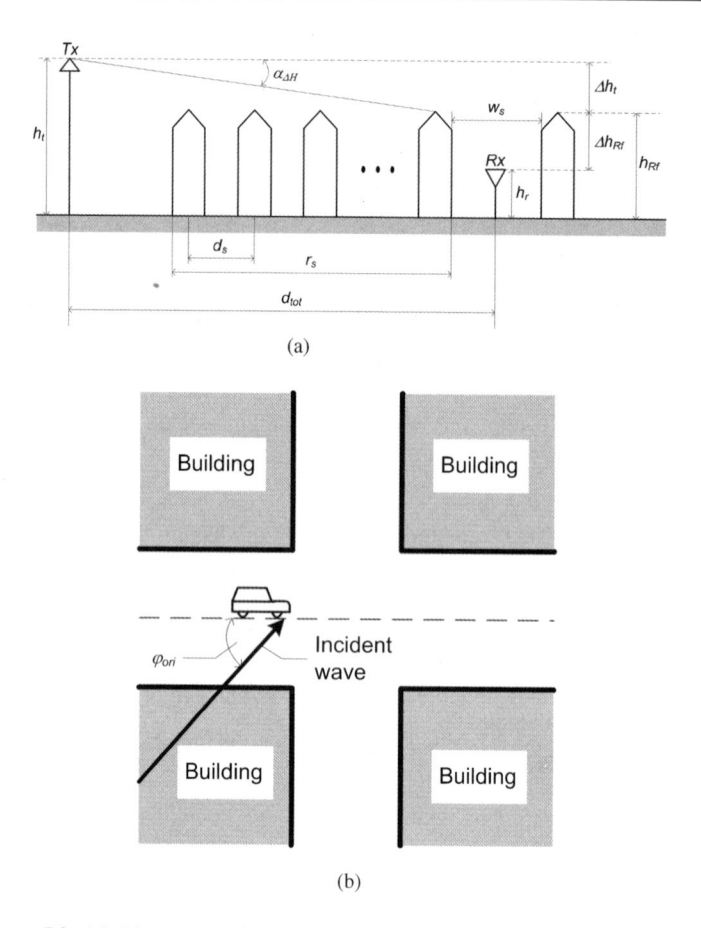

Figure 5.9 Model parameters in COST 231 Walfisch–Ikegami model (based on [51]).

The model distinguishes between LOS (Equation 5.31a) and NLOS (Equations 5.31b,c) situations:

$$L = 42.6 + 26\log(d_{tot}[km]) + 20\log(f[MHz]) \tag{5.31a}$$

$$L = L_{bf} + L_{rts} + L_{msd} \quad \text{for } L_{rts} + L_{msd} > 0 \tag{5.31b}$$

$$L = L_{bf} \quad \text{for } L_{rts} + L_{msd} \leq 0 \tag{5.31c}$$

where L_{rts} represents the coupling of an EM wave travelling along multiple screens into the street and is given by Equation (5.32)

$$L_{rts} = -16.9 - 10\log(w_s[m]) + 10\log(f[MHz]) + 20\log(\Delta h_{Rf}[m]) + L_{ori} \tag{5.32}$$

$$L_{ori} = \begin{vmatrix} -10 + 0.354\varphi_{ori}[deg] & \text{for } 0° \leq \varphi_{ori} < 35° \\ 2.5 + 0.075(\varphi_{ori}[deg] - 35) & \text{for } 35° \leq \varphi_{ori} < 55° \\ 4.0 - 0.114(\varphi_{ori}[deg] - 55) & \text{for } 55° \leq \varphi_{ori} < 90° \end{vmatrix} \tag{5.33}$$

$$\Delta h_{Rf} = h_{Rf} - h_r$$

$$\Delta h_t = h_t - h_{Rf}$$

Propagation Modelling

L_{msd} describes the diffraction loss on multiple buildings (represented by absorbing screens) and can be computed with Equation (5.34).

$$L_{\text{msd}} = L_{\text{bsh}} + k_a + k_d \log(d_{\text{tot}}[km]) + k_f \log(f[MHz]) - 9\log(d_s[m]) \qquad (5.34)$$

where

$$L_{\text{bsh}} = \begin{cases} -18\log(1+\Delta h_t[m]) & \text{for } h_t > h_{\text{Rf}} \\ 0 & \text{for } h_t \leq h_{\text{Rf}} \end{cases}$$

$$k_a = \begin{cases} 54 & \text{for } h_t > h_{\text{Rf}} \\ 54 - 0.8 \cdot \Delta h_t[m] & \text{for } d_{\text{tot}} \geq 0.5[km] \text{ and } h_t \leq h_{\text{Rf}} \\ 54 - 0.8 \cdot \Delta h_t[m] \cdot \dfrac{d_{\text{tot}}[km]}{0.5} & \text{for } d_{\text{tot}} < 0.5[km] \text{ and } h_t \leq h_{\text{Rf}} \end{cases}$$

$$k_d = \begin{cases} 18 & \text{for } h_t > h_{\text{Rf}} \\ 18 - 15\left(\dfrac{\Delta h_t}{h_{Rf}}\right) & \text{for } h_t \leq h_{\text{Rf}} \end{cases}$$

$$k_f = -4 + \begin{cases} 0.7\left(\dfrac{f[MHz]}{925} - 1\right) & \text{for medium sized city and suburban centres with medium tree density} \\ 1.5\left(\dfrac{f[MHz]}{925} - 1\right) & \text{for metropolitan centres} \end{cases}$$

The term k_a accounts the increase in the path loss for the case when the transmitter antennas are located below the rooftops of the adjacent buildings. The other two terms k_d and k_f, represent the dependence of the multi-screen diffraction loss versus distance and frequency, respectively. In case no precise data on the buildings heights and the widths of roads are available, the following values are recommended:

$$h_{Rf} = 3[m] \cdot \{number\_of\_floors\} + roof\_height$$

$$roof\_height = \begin{cases} 3\,m & \text{for pitched roofs} \\ 0\,m & \text{for flat roofs} \end{cases}$$

$$d_s = 20 \ldots 50\,m$$

$$w_s = d_s/2$$

$$\varphi_{\text{ori}} = 90°$$

The predicted path loss values stay in a reasonable agreement with measured data as long as $r_s > d_{\text{set}}$ (e.g. see [51]), where d_{set} is called the 'settled field'-distance (Equation 5.35). Otherwise, grazing incidence occurs where the COST 231 WI model performs poorly.

$$d_{\text{set}} = \frac{\lambda \cdot d_{\text{tot}}^2}{\Delta h_t^2} \qquad (5.35)$$

5.3.2 MANHATTAN MODEL

Since the COST 231 WI model is best suited for situations when h_t is much greater than the mean buildings height, the main emphasis is placed on propagation in the vertical plane, accounting multiple diffractions by successive edges as a dominant propagation mechanism. Manhattan model (also known as Berg's model) lends itself best for situations where propagation occurs in horizontal plane rather than vertical plane (as is assumed in WI model). This is the case when h_t is found below the mean buildings height (and in this respect Manhattan model can be considered complimentary to WI). When the base station is located below the average buildings height, it should be assumed that the signal reaching the receiver will traverse street canyons in the horizontal plane, rather than diffract on rooftops.

In [52], a recursive model was proposed (and widely adopted [35,53,54]), which accounts for arbitrary street crossing angles and curved streets (with linear sections) particularly suitable in Manhattan-type environments for microcell path loss calculations. The path loss is found from the modified free space formula, but the physical distance is replaced with an 'illusory' distance. The path loss formula makes use of the dual slope behaviour (thus considers contributions from both LOS and NLOS cases) by defining a parameter $D_{br}(x)$, which varies depending on whether the distance from the transmitter is before or after the break point:

$$L^{(N_{sss})} = 20\log\left[\frac{4\pi d_{N_{sss}}}{\lambda} D_{br}\left(\sum_{j=1}^{N_{sss}} s_{j-1}^{street}\right)\right] \text{ where } D_{br}(x) = \begin{cases} x/d_{br} & \text{for } x > d_{br} \\ 1 & \text{for } x \leq d_{br} \end{cases} \quad (5.36)$$

The parameter N_{sss} is the number of straight street sections between the transmitter and the receiver along the shortest path and may take on an arbitrary value, but not less than two. The final 'illusory' distance $d_{N_{sss}}$ at the N_{sss}-th section is obtained recursively by adding all the preceding illusory distances d_j according to Equation (5.37) with initial values $k_0 = 1$ and $d_0 = 0$:

$$k_j = k_{j-1} + d_{j-1} \cdot q_{j-1}^{angle}$$
$$d_j = k_j \cdot s_{j-1}^{street} + d_{j-1} \quad (5.37)$$

The above parameter s_j^{street} represents the physical distance in metres. The parameter q_j^{angle} stands for the angle dependence of the path loss and equals zero for perfectly aligned streets, i.e. $\theta_j^{street} = 0°$. In another extreme case, for perpendicular streets, a proper choice of q_j^{angle} should be within 0.5 and 1.0, continuously increasing with increasing angle, such as proposed in [52]:

$$q_j^{angle}(\theta_j^{street}) = \left(\theta_j^{street} \cdot \frac{q_{90}^{angle}}{90}\right)^v \quad (5.38)$$

where $q_{90}^{angle} = 0.5$ and $v = 1.5$ (for these values, the predictions yield results closest to those measured). Figure 5.10 illustrates a simple case of applying the Berg's model with all necessary parameters included. Since a conclusive criterion of selecting either of the above two methods is missing, [35] proposes to make calculations of the path attenuation with the method that accounts for the effects of propagation occurring above rooftops (COST 231 WI) and performs the same search with the Berg's recursive method that, in turn, accounts for the propagation through the streets. The final path loss value will be the minimum between the findings of both methods, i.e.

$$L = \min(\text{COST 231 WI; Manhattan model}) \quad (5.39)$$

5.3.3 OTHER MICROCELLULAR PROPAGATION MODELS

Discussing radiowave prediction in microcellular environments, one should also consider the usage of fully deterministic models, like the ones based on ray tracing. Successful implementation of these models has been numerously proven in the literature [15–19]. The use of ray tracing models is quite

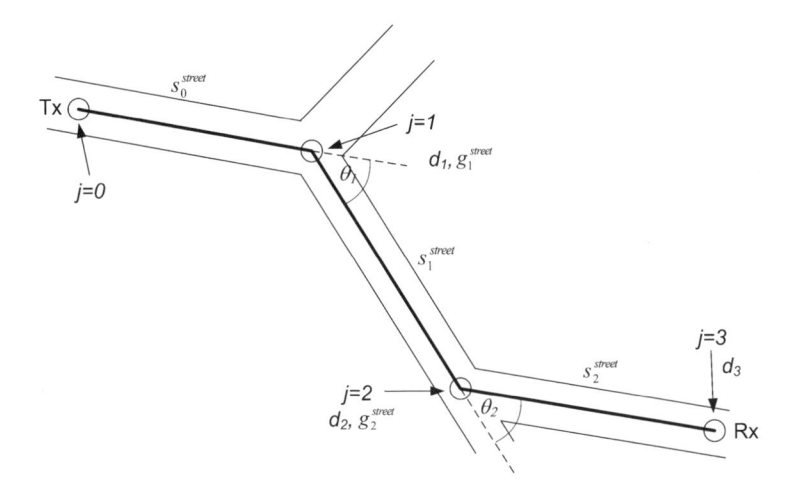

Figure 5.10 Illustration of the Berg's model configuration (based on [52]).

attractive for the new generation wideband systems in that, besides coverage predictions, they allow to extract all desired information on the multipath radio channel and display them in form of power delay profiles (PDP). This is possible because, in ray tracing techniques, the rays imitate the actual wave propagation and allow to store data for each traced ray individually, i.e. their amplitudes, phase, lengths, delays and angles of arrival (AoA). Although computationally far more demanding than empirical models, deterministic methods offer a great amount of precision. An example for this is [55], where signal power and time-dispersion channel parameters were predicted with ray tracing using a commercially available city map; despite limited accuracy of the map, simulated results were in close agreement with measurements.

In [56,57], the authors thoroughly investigated all microcellular propagation mechanisms and based on them defined three basic sub-models:

1. Vertical Propagation Plane Model (VPM)
2. Transversal Propagation Plane Model (TPM)
3. Multi Path Propagation Model (MPM).

Simulations were performed with each of these models separately, as well as in combination, and compared with measurements carried out in Berlin, Bologna and Leipzig. As expected, the greatest accuracy was obtained when all submodels were included in the simulations. Additionally, whenever vegetation was detected on the ray path, its influence was also incorporated as an extra loss due to penetration through and diffraction over it. An important conclusion was that multi path propagation is relevant only within a distance of up to approximately 500 m to the base station (BS). In other words, considering neighbouring scatterers when a mobile station is within the radius of 500 m from the BS improves significantly the simulation accuracy. Further away from the BS, scattering effects appear to be negligible.

As a final comment to microcellular propagation modelling, a relation developed in [47–53,55,58] between the path loss L and delay spread τ_{rms} is presented in Equation (5.40). It may serve as a direct source of information on the degree of channel time dispersion that can be expected in the given environment. It may also facilitate determination of the maximum cell size based on the delay spread characteristics.

$$\tau_{rms} = e^{0.065 \cdot L} \text{ns} \qquad (5.40)$$

5.4 APPLICATION OF EMPIRICAL AND DETERMINISTIC MODELS IN MACROCELL PLANNING

A macro cellular region according to [35] extends to distances beyond 1 km and is often subdivided into small (<2 km) and big macro cells. For the purpose of macrocellular design, the radio wave suffers attenuation predominantly due to the free space loss and losses due to interactions with the ground or larger obstacles (reflection, diffraction). Therefore, it is useful to know the terrain data such as the average height, clearance angle, size of major obstacles, vegetation height etc. The most popular empirical models (terrain models) that account for these macro-parameters and are applicable to macro cell planning at frequencies around 2 GHz include Modified Hata [44], COST 231 Hata [51] and ITU-R P.1546 [54]. Some diffraction models are given by Giovaneli [59], Deygout [60,61], Bullington [62] or knife-edge diffraction [63].

5.4.1 MODIFIED HATA

In the Modified Hata model, the general path loss formula L is given by Equation (5.41) as the sum of a median path loss $L_{50\%}$ and a term $T(G(\sigma))$ describing Gaussian variations [44]. In this section, only frequency ranges applicable to UMTS system will be covered, i.e. 1500–2000 and 2000–3000 MHz. The model is valid within the following bounds:

- frequency range f: 30 MHz–3 GHz
- distance range d: 0.02–100 km
- base station height h_{BS}: 1–200 m
- mobile station height h_{MS}: 1–200 m.

$$L(f, h_t, h_r, d, env) = L_{50\%} + T(G(\sigma)) \quad (5.41)$$

where
  h_t: transmitter antenna height
  h_r: receiver antenna height
  $h_{MS} : \min(h_t, h_r)$
  $h_{BS} : \max(h_t, h_r)$
  env: environment type (outdoor/indoor, rural/urban/suburban, propagation below or above roof)

If h_{MS} and/or h_{BS} are below 1 m, a value of 1 m should be used instead. Antenna heights above 200 m might lead to considerable errors. The model is broken up into several cases with different median path loss formulas:

- Case 1: d ≤ 0.04 km

$$L_{50\%} = 32.4 + 20\log(f) + 10\log(d^2 + (h_{BS} - h_{MS})^2/10^6)$$

- Case 2: 0.04 km < d < 0.1 km

$$L_{50\%} = L_{50\%}(0.04) + \frac{\log(d) - \log(0.04)}{\log(0.1) - \log(0.04)} (L_{50\%}(0.1) - L_{50\%}(0.04))$$

- Case 3: d ≥ 0.1 km

$b(h_{BS}) = \min\{0, 20\log(h_{BS}/30)\}$

$$\alpha = \begin{cases} 1 & for\ d \leq 20[km] \\ 1 + (0.14 + 1.87 \cdot 10^{-4} f + 1.07 \cdot 10^{-3} h_{BS})(\log(d/20))^{0.8} & for\ 20[km] < d \leq 100[km] \end{cases}$$

1. Sub-case 1: Urban

    $1500\,\text{MHz} < f \leq 2000\,\text{MHz}$

    $$L_{50\%} = 46.3 + 33.9\log(f) - 13.82\log(\max\{30, h_{\text{BS}}\})$$
    $$+ [44.9 - 6.55\log(\max\{30, h_{\text{BS}}\})]\log(d)^\alpha - a(h_{\text{MS}}) - b(h_{\text{BS}})$$

    $2000\,\text{MHz} < f \leq 3000\,\text{MHz}$

    $$L_{50\%} = 46.3 + 33.9\log(2000) + 10\log(f/2000) - 13.82\log(\max\{30, h_{\text{BS}}\})$$
    $$+ [44.9 - 6.55\log(\max\{30, h_{\text{BS}}\})]\log(d)^\alpha - a(h_{\text{MS}}) - b(h_{\text{BS}})$$

2. Sub-case 2: Suburban

    $$L_{50\%} = L_{50\%}(\text{urban}) - 2\left\{\log\left[(\min\{\max\{150, f\}, 2000\})/28\right]\right\}^2 - 5.4$$

3. Sub-case 3: Open areas

    $$L_{50\%} = L_{50\%}(\text{urban}) - 4.78\left\{\log\left[\min\{\max\{150, f\}, 2000\}\right]\right\}^2$$
    $$+ 18.33\log\left[\min\{\max\{150, f\}, 2000\}\right] - 40.94$$

For any of the above cases, if $L_{50\%}$ is found below the free space attenuation, the free space attenuation L_{bf}, given in Equation (5.42), should be used instead.

$$L_{\text{bf}} = 32.44 + 20\log d[km] + 20\log f\,[\text{MHz}] \tag{5.42}$$

5.4.2 OTHER MODELS

As was already mentioned in Section 5.1, a great deal of success has been achieved by applying neural network techniques to evaluating electric field distributions. Despite differences, all types of ANNs exhibit the following common features:

- parallel structure which allows to accelerate computation speed;
- environment adaptation – the ability to learn from a changing environment.

In [23], a Multilayer Perceptron is used, together with a backpropagation learning rule, to find an attenuation term α_{building} which accounts for the effect of buildings. It is then added to the free space propagation L_{bf} so that the final path loss equals $L = L_{\text{bf}} + \alpha_{\text{building}}$. The inputs to the perceptron are base station height, mean distance between consecutive buildings, mobile station height, angle of the last diffraction, etc. Since the propagation loss is dependent on the building heights relative to the line joining the transmitter and receiver, these relative values were regarded to be representative of building heights rather than their absolute heights above the ground level. The mean errors oscillated around zero, and standard deviations of predictions with respect to measurements were about 7 dB. A similar approach was presented in [29], where an ANN network was used as a correcting agent to the COST 231 WI model; this decreased the mean error to zero and the standard deviation to 4.7 dB. It also shortened the computation time by up to five times, when compared to the COST 231 models.

In [64], thirteen different realisations of a Radial Basis Function (RBF) neural network were constructed, trained and finally tested against measurements. The simplest realisations included the distance between transmitter and receiver, the width of the street, the building separation and the buildings height. Any *n*-th model RBF_n would contain an extra parameter as compared the preceding model RBF_{n-1}. For instance RBF_2 included, beside all RBF_1 input parameters, also the difference between H_b and H_m. RBF_3, in turn, would include the street orientation in addition to the parameters used for the training of RBF_2 and so forth. The experiment was broken up into two parts – in the first (submodels $RBF_1 - RBF_7$), the trained ANN was used for predicting the path loss directly, whereas in the second, the ANN was used to compensate for the errors obtained by applying the COST 231 WI model; this has been summarised in Figure 5.11.

As could be envisaged for both types of models and both types of environments (urban and suburban), the inclusion of more input factors resulted in a greater accuracy as compared to measurements. To emphasise the advantages of ANNs, all results were compared to those obtained with the Single Slope Model, Walfisch–Bertoni and COST 231 WI model. Both the mean error and the standard deviation were the least for the ANN (on the order of 2.7 dB and 2.55 dB, respectively).

In [56,57], measurement campaigns were carried out at distances spanning from the closest vicinity up to a few kilometres away from the transmitter. Three different models were applied: a full 3D

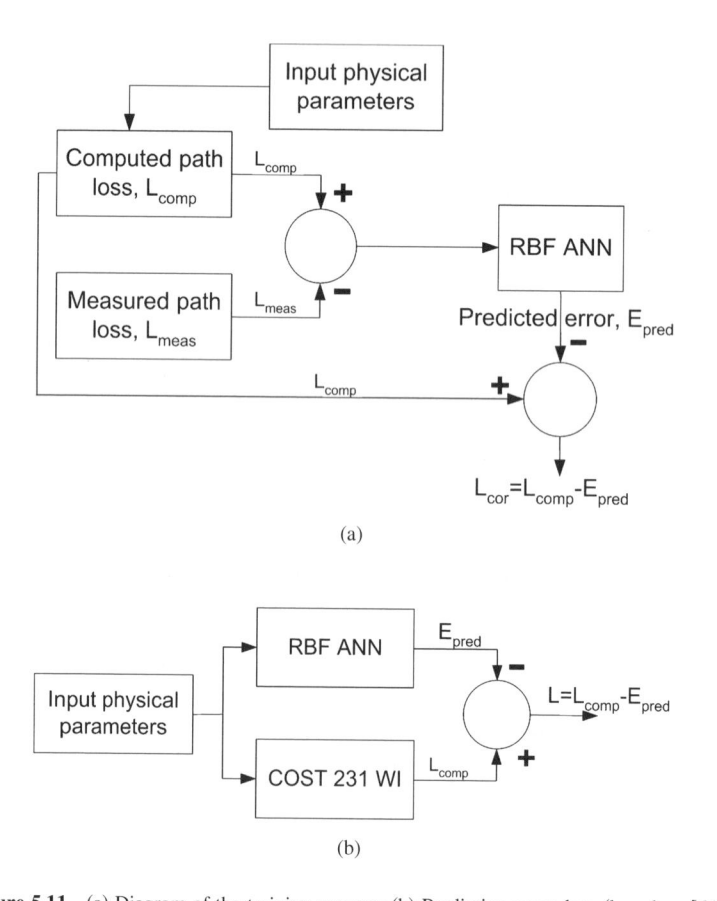

Figure 5.11 (a) Diagram of the training process; (b) Prediction procedure (based on [64]).

ray tracing (reference) model, a vertical plane and a horizontal plane model to discover the degree to which propagation occurs in each of these two planes. The conclusions are given below:

- in the proximity of the transmitter, the main contributor to the received field is the horizontal plane propagation (with multiple wall reflections and diffractions on vertical building edges);
- in the horizontal plane at a distance far away from the transmitter, multipath propagation can be neglected – propagation over rooftops (in vertical plane) predominates;
- simulation results are more accurate when building edges were modelled as dielectric wedges rather than in the classical knife-edge form.

Extensive research performed in [56] covered all main aspects of automated configuration and optimisation of large-scale UMTS networks. The part concerning adaptive propagation model selection is particularly interesting for this book. For this purpose, a general classification of models into five types was done based on the distinction between low- and high-resolution area databases required in each (Figure 5.12a):

1. Macro cell models using low-resolution data (M1).
2. Small macro cell models using low high resolution data (M2).
3. Microcell models using high-resolution data (M3).
4. Outdoor-to-indoor models using high-resolution data (M4).
5. Indoor and indoor-to-outdoor models using high-resolution data (M5).

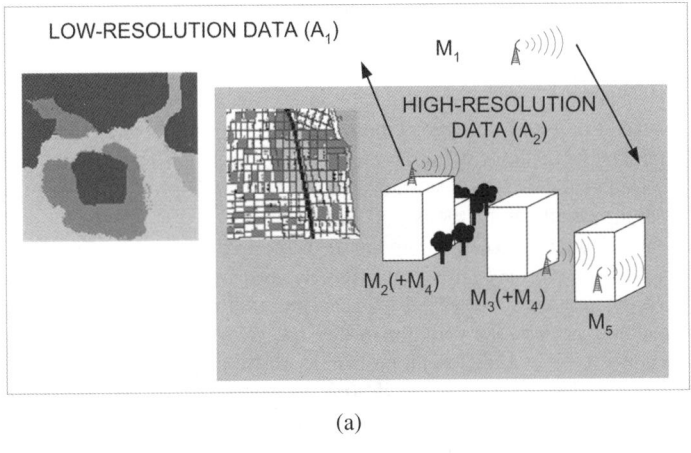

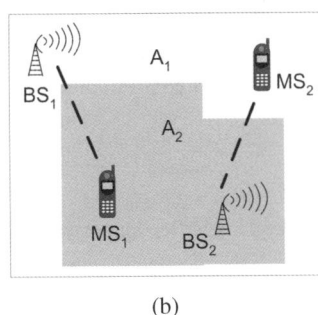

Figure 5.12 (a) Propagation areas and model types; (b) Case of mixed propagation area A_i and model type M_j, [56]. (Reproduced by permission of Thomas Kürner).

Next, 19 propagation models (e.g. COST 231 Hata or Walfisch–Ikegami, E-Plus ray tracing model, Manhattan (Berg's) recursive model, Motley–Keenan model) were assigned to one of those five generic types ($M_1 - M_5$). Typically, low resolution databases are available for all environments, while high resolution data can be found only for dense urban areas. The corresponding areas were defined as follows:

- A_1: area where only low-resolution data is available;
- A_2: area where also high-resolution data is available;
  - A_{2a}: outdoor areas;
  - A_{2b}: indoor areas.

Since in reality it occurs quite often that the Mobile Station (MS) and BS are located in different areas (Figure 5.12b), an appropriate selection of both resolution map, A_i, and propagation model type, M_j, has to done. In [56], however, precise rules have been elaborated that enable intelligent and smooth switching between areas and model types in an automated and time-of-execution optimum way.

5.5 PROPAGATION MODELS OF INTERFERING SIGNALS

In mobile radio communications, the interfering signals, affecting a victim receiver, are subject to path loss and random fluctuations that are dependent on location and time; these interfering signal level variations may be on a medium or small scale. Therefore distinguished are two kinds of signal fluctuations – the medium scale fading, called shadowing, and the small scale fading, called multipath fading. It is commonly known that the signal fluctuations occur due to changes of conditions in the atmosphere, terrain irregularities, clutter and mobility of mobile stations.

Non-fluctuating signals may be observed mainly near a transmitter in the LOS zones, where there are no changes of the atmosphere conditions, no signal reflections or scatterers and no movement of the mobiles. Therefore, fluctuations, being an effect of the multipath propagation, may occur where the transmitting as well as the receiving antennas are placed in a diverse environment (i.e. urban, hilly terrain or mountain), especially along the NLOS paths.

The interfering signals may affect the reception of the desired signals at different radio paths, even at large distances. Along these long radio paths, fluctuations of the interfering signal strength occur, as well as between the mobile stations and between the base (stable) stations. There is no typical shadowing or fast fading, because the variation of the signal level is related to both location and time.

Shadowing is known to obey a lognormal probability distribution, whereas the envelope of fading phenomena is most often described by Rayleigh, Rice or Nakagami distributions; some further information about these probability distributions are given in [65]. However, the propagation models used in practice, when describing the interference impact onto the terminal's reception, most often refer to the lognormal distribution. Below presented models have been elaborated by the ITU-R Study Group 3 and the Working Group SE (Spectrum Engineering) of the ECC CEPT. These models should be used for both internal and external IMT-2000/UMTS compatibility calculations (see Chapter 11).

5.5.1 ITU-R 1546 MODEL

The propagation model published in the ITU-R Recommendation P 1546 [54] is a statistical method applicable to field strength calculations for:

- terrestrial radio communication systems;
- frequency range 30 to 3000 MHz;
- pathways of the type outdoor-outdoor land, sea and mixed;
- pathways in diverse terrain configuration and diverse coverage;

- radio path lengths from 1 to 1000 km;
- calculations of the type point-to-point and point-to-area.

The model may hence be successfully used in calculations of interference power in the presently developed UMTS systems. The procedures applied in these calculations comprise the consideration of propagation mechanism of the ground-wave, and the field strength is calculated after reading the field strength values of the published statistical propagation curves. Those curves concern the propagation over the land, warm or cold sea, and they are drawn for nominal frequency values 100, 600, and 2000 MHz, effective heights of transmitter/base antennas, an effective radiated power (ERP) of 1 kW, and for a nominal time variability of 1, 10 or 50 %, as well as for a nominal location variability of 50 % in a representative 500 × 500 m² with assumed representative clutter height. Subsequently, interpolations or extrapolations are being made with respect to the parameters of the curves differing from the nominal curves, and the calculated field strength values are corrected according to prevailing system parameters of the transmitter, receiver and the radio pathway. The propagation curves for a nominal frequency of 2000 MHz, and field strength values exceeded at 50 % of the locations and 50 % of time (this is the median used for the determination of the useful field strength), are shown in Figure 5.13. The curves for the same frequency represent field strength values exceeded at 50 % of the locations and 1 % of time (used for determination of the interference field strength) are shown in Figure 5.14.

The field strength for 50 % locations and T % time may be calculated using the following formula:

$$E(50, T) = E_{\text{curve}}(50, T) + L(h_r) + L(\Theta_{\text{tca}}) + L(c) \; [\text{dB}(\mu\text{V/m})] \tag{5.43}$$

where

$E_{\text{curve}}(50, T)$ is field strength value in dBμV/m read directly for the required distance d[km] from the propagation curves (for 50 % locations and T % time) amended using the corrections for really radiated power and the effective height h_1 of the transmitter/base antenna;

$L(h_r)$ in dB is the correction for the receiving antenna elevation different of clutter height;

$L(\Theta_{\text{tca}})$ in dB is the correction for the radio path attenuation connected with the terrain clearance angle;

$L(c)$ in dB is the correction for building clutter for a (short) pathway in an urban or suburban terrain.

Additionally, if needed, one calculates:

the correction $L(h_t)$ in dB for negative effective height of transmitter antenna (below 0 m);
for the case of the point-to-area calculations, the correction $L(Q)$ in dB for a location variability other than 50 % (scaling from 1 to 99 %);
and the resultant field strength for mixed land-sea paths.

The calculated value of the field strength should not exceed the value obtained for free space – e.g. for 1 kW ERP, the free-space field strength is $E_{\text{fs}} = 106.9 - 20 \log(d_{\text{km}})$ in [dB(μV/m)].

For the UMTS system, the field strength value calculated for a frequency different from 2000 MHz is equal to:

$$E = E_{\text{inf}} + (E_{\text{sup}} - E_{\text{inf}}) \log(f/f_{\text{inf}}) / \log(f_{\text{sup}}/f_{\text{inf}}) \; [\text{dB}(\mu\text{V/m})] \tag{5.44}$$

where

f is the frequency for which the field strength is calculated (MHz);
f_{inf} is the lower nominal frequency (600 MHz);
f_{sup} is the upper nominal frequency (2000 MHz);
E_{inf} is the field strength value obtained using the curves for f_{inf};
E_{sup} is the field strength value obtained using the curves for f_{sup}.

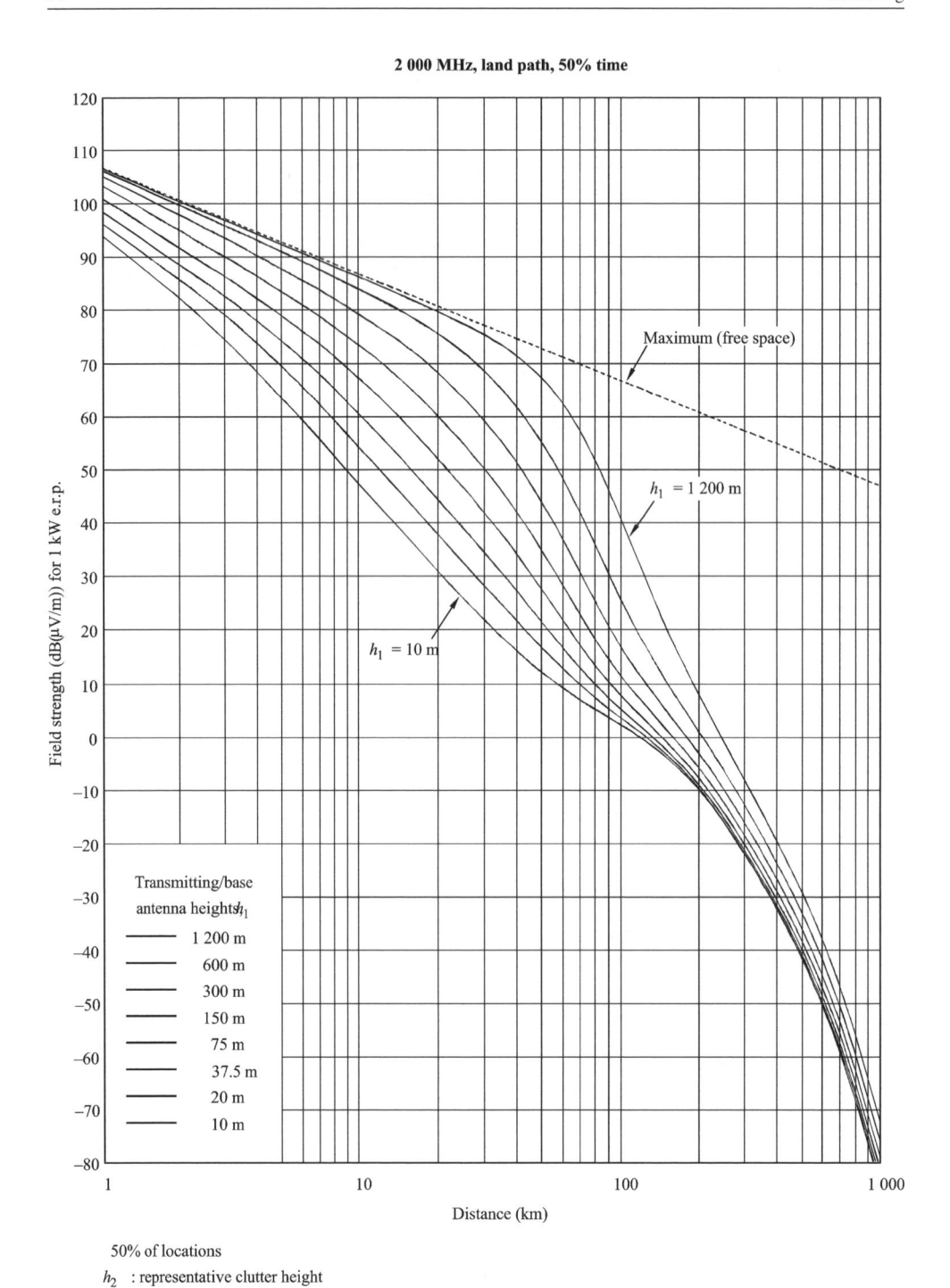

Figure 5.13 Family of propagation curves, 2000 MHz, land path, 50 % time [54] (Reproduced by permission of ITU).

Propagation Modelling

2 000 MHz, land path, 1% time

Field strength (dB(µV/m)) for 1 kW e.r.p. vs *Distance (km)*

Transmitting/base antenna heights h_1:
- 1 200 m
- 600 m
- 300 m
- 150 m
- 75 m
- 37.5 m
- 20 m
- 10 m

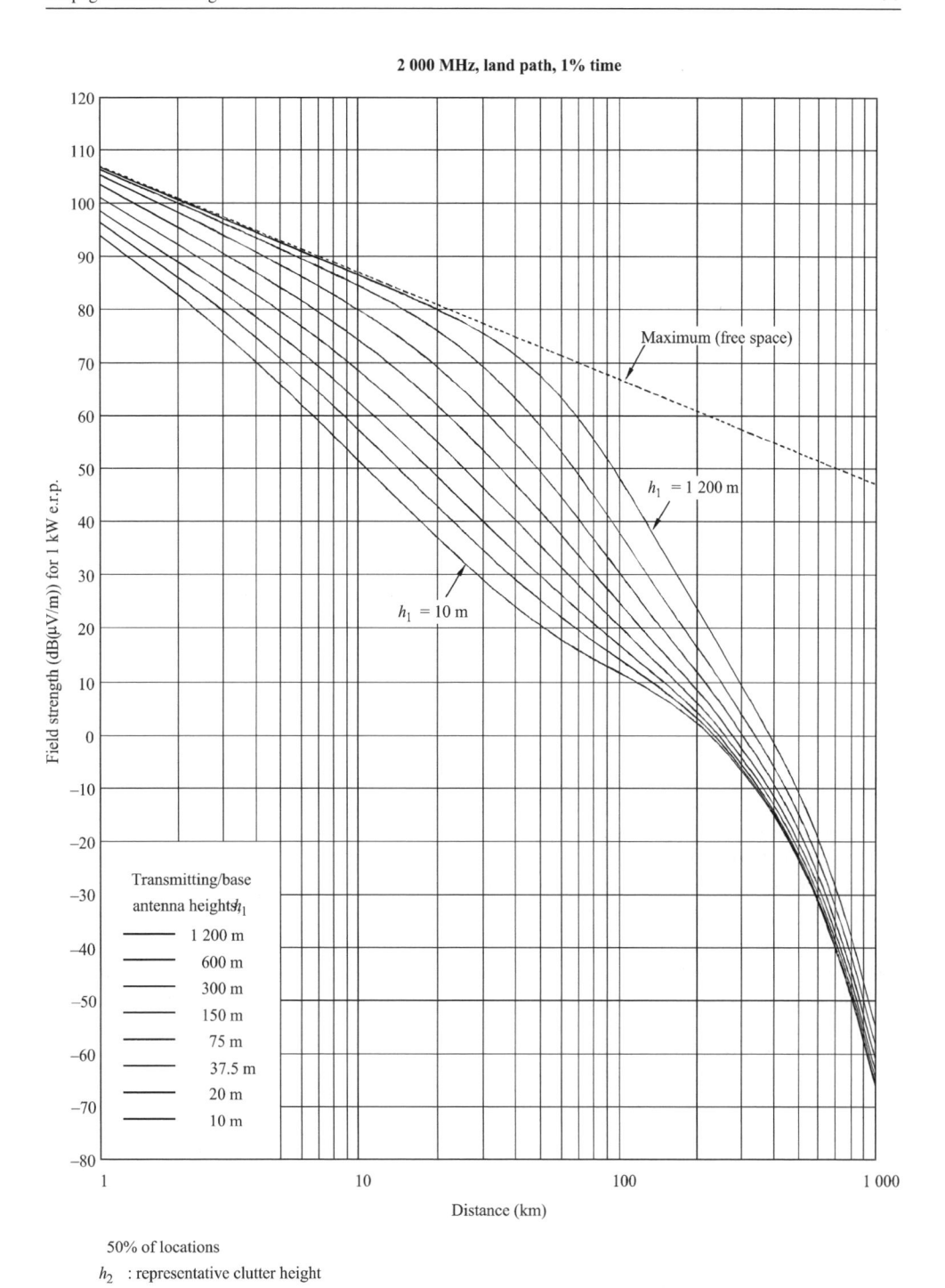

50% of locations
h_2 : representative clutter height

Figure 5.14 Family of propagation curves, 2000 MHz, land path, 1% time [54] (Reproduced by permission of ITU).

The quantity h_1 of the transmitter/base antenna height is related to the idea of an effective antenna height h_{eff}. The transmitter/base effective antenna height is a parameter which enables to take into account the shape of the terrain near the transmitter antenna. It helps to attain greater accuracy in the field strength assessment. It is defined as the height of the electrical centre of an antenna above the average level of the terrain determined along a certain piece of propagation path. The method of estimation of a transmitter effective antenna height depends on the type of propagation path (land, sea), the length for land paths (below or above 15 km) and on the availability of data concerning the terrain shapeup (for the paths shorter than 15 km).

For example, for sea paths, the effective height is equal to the physical height of the antenna. In the case of land paths longer than 15 km, $h = h_{\text{eff}}$, where the antenna height is understood as the antenna elevation above the average terrain level as measured in the distance range of 3 to 15 km from the transmitter antenna in the direction of the receiver. For land paths shorter than 15 km, if the terrain shapeup data are available, then:

$$h_1 = h_a + (h_b - h_a) \cdot (d_{\text{km}})/15 \text{[m]} \tag{5.45}$$

where h_b is the antenna height above the averaged terrain height between 0, $2d$ and d kilometre, and h_a is the height of the antenna above the ground. Recommendation [54] gives also the method of evaluating h_1 if there is no data available concerning the terrain shapeup near the transmitting antenna.

The propagation curves have been drawn for nominal values of the height h_1: 10, 20, 37.5, 70, 150, 300, 600 and 1200 m. The readout of field strength for those values is simple. For other than nominal (in the range 10–3000 m), the height h_1 and field strength E may be calculated using:

$$E = E_{\text{inf}} + (E_{\text{sup}} - E_{\text{inf}}) \log(h_1/h_{\text{inf}})/\log(h_{\text{sup}}/h_{\text{inf}}) \text{ [dB}(\mu\text{V/m)]} \tag{5.46}$$

where
h_{inf} is the proximate lower nominal height (600 m if $h_1 > 1200$ m);
h_{sup} is the proximate higher nominal height (1200 m if $h_1 > 1200$ m);
E_{inf} is the field strength value for h_{inf} for a distance d;
E_{sup} is the field strength value for h_{sup} for a distance d.

The Recommendation [54] gives also a method of field strength evaluation when $0 \le h_1 < 10$ and when h_1 takes on negative values, i.e. the antenna is below the surrounding terrain. Diffraction effects and tropospheric scattering must then be considered, and with respect to the field strength calculated for $h_1 = 0$, a correction $L(h_1)$ is introduced. This method cannot by applied if the antenna height h_1 is smaller than the surrounding clutter height.

If the calculations are made with the help of the propagation curves, a reference height of the receiving antenna h_{ref} is used; it represents the height of buildings surrounding the receiving antenna. Examples of the reference heights are 20 m for urban area, 30 m for dense building development urban area and 10 m for suburban and rural areas; for sea paths, $h_{\text{ref}} = 10$ m. If the height of the receiving antenna is different from h_{ref}, then the field strength value obtained by means of an appropriate curve needs to be corrected using $L(h_r)$, connected with the elevation angle of the arriving ray. The calculation method holds if a receiving mobile antenna height is greater than 1 m over ground and 3 m over sea.

The correction introduced into calculations connected with the terrain clearance angle Θ_{tca} amends the calculation accuracy; this is due to consideration of the terrain shapeup in the surrounding of the receiving antenna. The angle Θ_{tca} is being evaluated along a 16 km length, by leading a tangent line

Propagation Modelling

from the receiving (i.e. victim in the case of interference) antenna to the highest terrain obstruction along the considered path fragment in the direction of transmitting (interfering) antenna. Its value is determined as follows:

$$\Theta_{tca} = \Theta - \Theta_r \text{ [degrees]} \quad (5.47)$$

where:
Θ in degrees is the elevation angle of tangent line;
$\Theta_r = \arctan\left(\dfrac{h_{1s} - h_{2s}}{1000d}\right)$ in degrees is the reference angle;
$h_{1,2s}$ in meters is the transmitter/receiver antenna height above sea level, d is in kilometers.

When evaluating the angle Θ_{tca}, the earth's surface curvature should be neglected. The correction of the field strength value is evaluated using the diagram given in Figure 5.15.

The correction for short urban/suburban paths refers to paths below 15 km. It is related to additional field strength attenuation caused by building development (of the height h_{ref}) surrounding the transmitter antenna. The correction value is given as follows:

$$L(c) = -3.3(\log(f))(1 - 0.85\log(d))(1 - 0.46\log(1 + h_a - h_{ref})) \text{ [dB]} \quad (5.48)$$

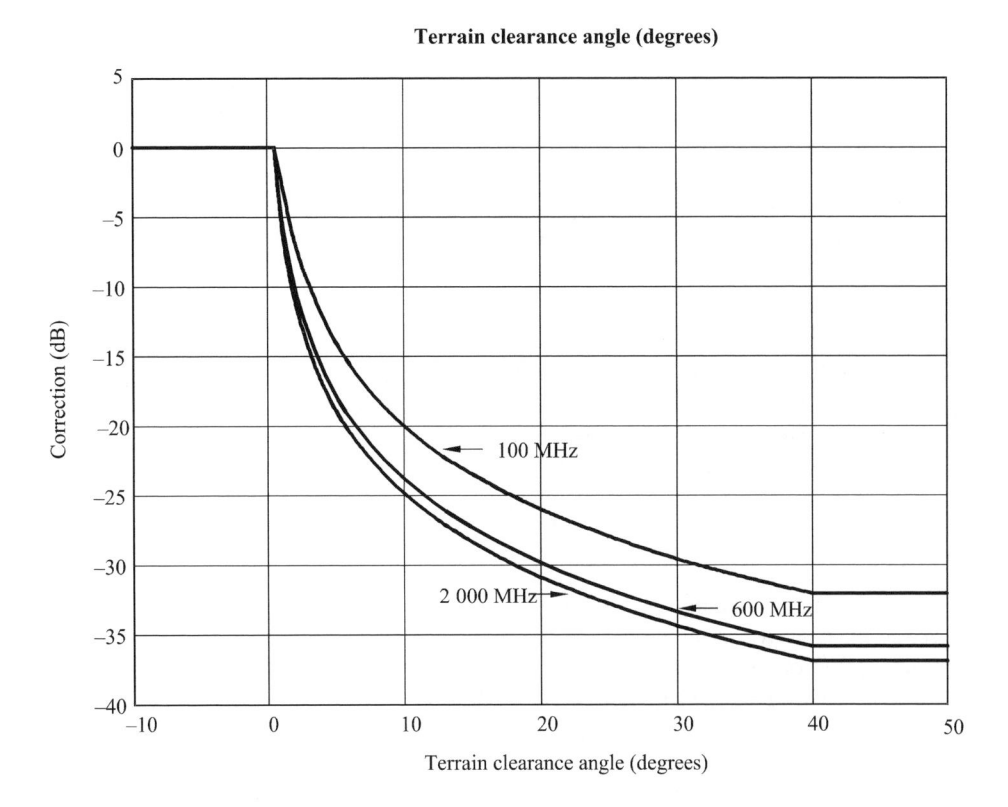

Figure 5.15 Terrain clearance angle correction $L(\Theta_{tca})$ for the nominal frequencies [54] (Reproduced by permission of ITU).

where f is in MHz, d in km, h_a in m and h_{ref} also in m. The correction may be used if $h_1 - h_{ref} < 150$ m. In the cases, when there is a need of evaluating the field strength values exceeded for the other time percentage than nominal, an interpolation procedure is to be used, i.e.

$$E = E_{sup}(Q_{inf} - Q_t)/(Q_{inf} - Q_{sup}) + E_{inf}(Q_t - Q_{sup})/(Q_{inf} - Q_{sup}) \; [\text{dB}(\mu\text{V/m})] \qquad (5.49)$$

where:
 t is the percentage time for which the prediction is required;
 t_{inf} is the lower nominal percentage time;
 t_{sup} is the upper nominal percentage time;
 $Q_t = Q_i(t/100)$;
 $Q_{inf} = Q_i(t_{inf}/100)$;
 $Q_{sup} = Q_i(t_{sup}/100)$;
 E_{inf} is the field strength value obtained for time percentage t_{inf};
 E_{sup} is the field strength value obtained for time percentage t_{sup};
 $Q_i(x)$ is the inverse complementary cumulative normal distribution function.

This method is valid only for interpolating field strengths exceeding the percentage times in the range from 1 to 50 %; any extrapolation outside this range is not valid. A method for the calculation of $Q_i(x)$ is given in the appendix at the end of this chapter.

If the statistics of the received signal due to a moving terminal in a point-to-area communications system has to be taken into account, then one clearly has to consider multipath fading, variations of the local terrain coverage and changes of the length and geometry of the radio paths (e.g. in the mountains). In the method described here, a correction has been determined for location variability $L(Q)$ due to above effects with respect to the median of the field strength, i.e.

$$L(Q) = Q_i(q/100)\sigma_L(f) \; [\text{dB}] \qquad (5.50)$$

where $Q_i(x)$ is the inverse complementary cumulative normal distribution function and σ_L is the standard deviation of the Gaussian distribution of the local mean in the receiving area. For the 500×500 m area in urban environments, we have

$$\sigma_L(f) = K + 1.3\log(f) \; [\text{dB}] \qquad (5.51)$$

where f is in MHz and K is equal to 1.2 when omnidirectional antennas at car-roof height are used. When areas other than 500×500 m, other environments or other antenna heights are considered, then the standard deviation will differ from the described above. This correction is also only valid for percentage locations in the range of 1 to 99 % and for land paths not adjacent to the sea.

5.5.2 ITU-R 452 MODEL

Some methods of evaluating the (microwave) interference in the frequency range from 0.7 to 50 GHz between land stations are presented in the Recommendation ITU-R P.452 [66]. The land stations here are assumed to operate point-to-point and be placed in the open terrain, up to 10 000 km apart. The extracted calculation procedures take into account the following mechanisms of interference propagation:

- long-term
  - above the line of direct visibility;
  - diffraction;
  - tropospheric scattering.

- short-term

  - abnormal propagation (tropospheric ducts, reflections/scattering in elevated tropospheric layers);
  - scattering in the hydrometeors.

Calculations conducted with the use of the ITU-R 452 model require the provision of:

- basic input data (working frequency, required time percentage not exceeded by basic transmission loss, latitude and longitude of interfering and affected stations, antenna centre heights above the ground, and above mean sea levels, and the gain of antennas in the direction of the horizon);
- requirements concerning the 'worst month' and the 'average year';
- radiometeorological data (i.e. tropospheric refraction indexes and determination of climatic regions within the continental zones, at the sea shores, on seas and other 'large area' waters with radiuses of a minimum of 100 km, as well as islands on these waters, and also with the refraction connected value of the effective earth radius);
- profile of a radio-path;
- choice of the propagation method.

The way in which interfering signals propagate and the appropriate propagation model used for minute calculations depend on the type of the interference path. To this end, the following interference paths are distinguished: LOS with first Fresnel zone clearance, LOS with sub-path diffraction (partly covered by first Fresnel zone) and trans-horizon path. For each of these path types, a basic transmission loss $L_b(p)$ not exceeded for $p\%$ of the time is determined as follows: For the LOS path:

$$L_b(p) = L_{b0}(p) + L_{ht} + L_{hr} \text{ [dB]} \qquad (5.52)$$

where

$L_{b0}(p)$ is the foreseen basic path attenuation not exceeded for $p\%$ of the time, and determined for a path with the LOS model, taking into account the multipath propagation effects, and tropospheric absorption, as well as the respective dry air and water vapour attenuations according to Recommendation ITU-R P.676 [67];

L_{ht}, L_{hr} are the respective additional attenuations resulting from the protection against local interference due to signal dispersing off objects in the surrounding of both interfering and interfered station antennas (existence of local clutter).

The analytical expressions are

$$L_{b0}(p) = 92.5 + 20\log f + 20\log d + E_s(p) + L_g \text{ [dB]} \qquad (5.53)$$

and

$$L_{ht,hr} = 10.25 \times e^{-d_k}\left(1 - \tanh\left[6\left(\frac{h_{t,r}}{h_a} - 0.625\right)\right]\right) - 0.33 \text{ [dB]} \qquad (5.54)$$

where
  f is expressed in GHz and d in km;
  d_k in km is the distance from nominal clutter point to the antenna;
  $h_{t,r}$ in m is the interferer/victim antenna height above local ground level;
  h_a in m is the nominal clutter height above local ground level;
  $E_s(p)$ is the correction due to multipath and focusing effects; and
  L_g in dB is the total gaseous absorption, see [67].

The analytical expression of $E_s(p)$ is given as

$$E_s(p) = 2.6(1 - e^{-d/10})\log(p/50) \text{ [dB]} \tag{5.55}$$

The parameter L_g characterising the gaseous attenuation on terrestrial paths may be omitted in the calculations in the case of the UMTS system, due to its operational frequencies $f < 3\,\text{GHz}$.

For the LOS path with sub-path diffraction, the basic transmission loss is given by:

$$L_b(p) = L_{b0}(p) + L_{ds}(p) + L_{ht} + L_{hr} \text{ [dB]} \tag{5.56}$$

where $L_{ds}(p)$ – attenuation in dB, for $p\%$ of time, is given by the diffraction loss calculated over LOS path with sub-path obstruction by application of the diffraction model described in Recommendation ITU-R P.526 [63].

The method of calculating the attenuation $L_{ds}(p)$ depends on whether $p\%$ of the time is smaller or greater from $\beta_0\%$, the latter being the probability of existence of the super-refractive layer in the low atmosphere. β_0 is the time percentage for which refractive index lapse-rates exceeding $100\,\text{N-units/km}$ (super-refractive layer) can be expected in the lowest 100 m of atmosphere. The method of evaluation of the parameter β_0 for the path centre location is determined in [66].

For $p = 50\%$, $L_{ds}(50\%)$ is computed using the method described in [63] for a median effective Earth radius $a_e = k \cdot a$ (where: a is the true Earth radius of 6371 km and for a standard atmosphere $k = 1.33$; $a_e = 8500\,\text{km}$).

For $p \le \beta_0$, $L_{ds}(\beta_0)$ is computed using the method described in [63], assuming an effective Earth radius of $a_e = 19\,100\,\text{km}$ and using the shapes of obstructions as the knife edges (path profile components) identified for the median case.

For $\beta_0 < p < 50\%$, $L_{ds}(p)$ is given by:

$$L_{ds}(p) = L_{ds}(50\%) - F_i(p)[L_{ds}(50\%) - L_{ds}(\beta_0)] \text{ [dB]} \tag{5.57}$$

where $F_i(p)$ is the interpolation factor given by $F_i(p) = Q(p/100)/Q(\beta_0/100)$, where $Q(x)$ is the inverse cumulative normal distribution function ($Q(x) = -Q_i(x)$).

If the analysis concerns a trans-horizon path (shown in Figure 5.16) then:

$$L_b(p) = -5\log(10^{-0.2L_{bs}} + 10^{-0.2L_{bd}} + 10^{-0.2L_{bam}}) + L_{ht} + L_{hr} \text{ [dB]} \tag{5.58}$$

where
- $L_{bs}(p)$ is the basic transmission loss predicted for $p\%$ of the time determined due to tropospheric scattering, described with greater details in Chapter 4.4 of Recommendation [66], depending on frequency, distance, gains of transmitting-/receiving antennas, path centre sea-level refractivity and gaseous absorption (which may be omitted for $f < 3\,\text{GHz}$);
- $L_{bd}(p)$ is the basic transmission loss predicted for $p\%$ of time, determined for the diffraction path, and dependent on frequency, distance and excess transmission loss $L_d(p)$ (calculated similar to $L_{ds}(p)$) and corrected for multipath effects (similar to $E_s(p)$ by changing the distance d for the sum of distances-to-horizon obstacles ($d_{lt} + d_{lr}$));
- $L_{bam}(p)$ is the basic transmission loss predicted for $p\%$ of time obtained by the use of ducting/layer reflection loss which depends on fixed coupling losses, site shielding loses, angular-distance depending losses with the special procedure for modifying of line-of-sight loss for the trans-horizon path.

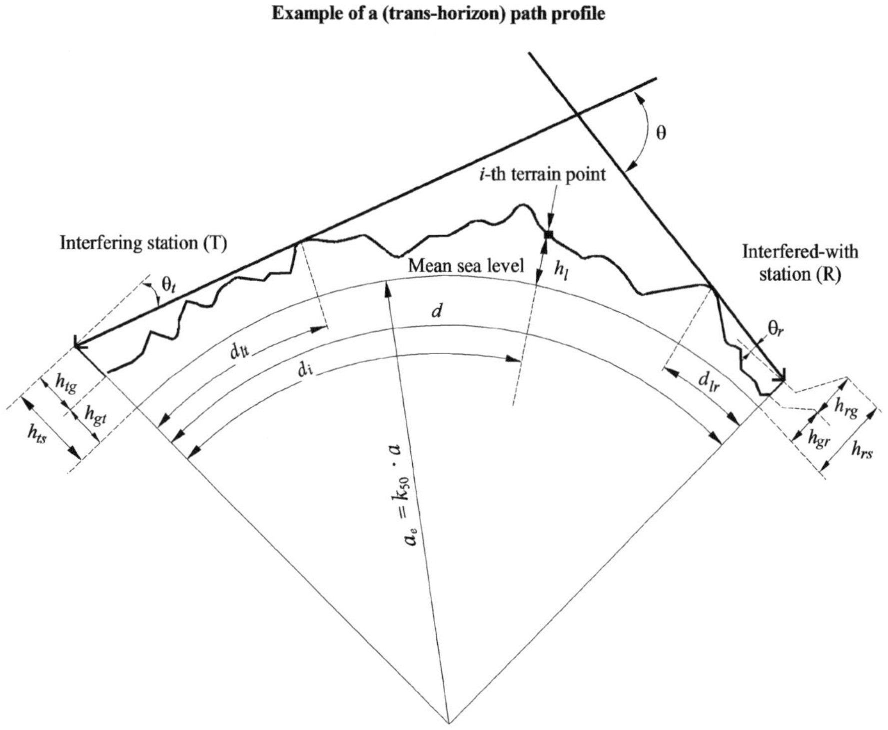

Example of a (trans-horizon) path profile

Note 1 – The value of θ_t as drawn will be negative.

Figure 5.16 Trans-horizon path profile [67] (Reproduced by permission of ITU).

The method of evaluating the basic transmission losses $L_{bs}(p)$, $L_{bd}(p)$ and $L_{bam}(p)$, making use of given radio path parameters shown in Figure 5.16, is described in detail in [66]. For this, the following trans-horizon radio path parameters, related to the actual terrain, should be taken into account for calculating above losses:

a_e	effective Earth radius, in km,
d	great-circle path distance, in km,
d_i	great-circle distance of the i-th terrain point from the interferer, in km,
d_{ii}	incremental distance for regular path profile data, in km,
d_{lt}	horizon distance, in km, measured from the interfering antenna
d_{lr}	horizon distance, in km, measured from the interfered-with antenna
h_{ts}	interferer antenna height, in m, above mean sea level
h_{rs}	interfered-with (victim) antenna height, in m, above mean sea level
θ_t	horizon elevation angle above local horizontal, in mrad, measured from the interfering antenna
θ_r	horizon elevation angle above local horizontal, in mrad, measured from the interfered-with antenna
θ	path angular distance, in mrad,
h_{gt}	height of the smooth-Earth surface above mean sea level at the interfering station location, in m,

h_{gr} height of the smooth-Earth surface above mean sea level at the interfered-with station location, in m,
h_i height of the i-th terrain point above mean sea level, in m,
h_{tg} ground height of interfering station, in m,
h_{rg} ground height of interfered-with station, in m.

In propagation analysis concerning the interference propagation to and from 3G systems, the application of this model is highly recommended for distances greater than 20 km. For smaller distances, the Modified Hata model may be used.

5.5.3 STATISTICS IN THE MODIFIED HATA MODEL

Modified Hata model generally should be used in outdoor–outdoor interference calculations at relative short distances, i.e. in principle below 20 km. An evaluation of the median radio path loss $L_{50\%}$ of the Modified Hata method by the CEPT WG SE has been described in Section 5.4 in great detail. It has been assumed that the distribution of the path loss variation is the well known shadowing lognormal distribution. The standard deviation for the distribution of the path loss calculated by Modified Hata has been assessed in [68] as well as in [44]. The outcome clearly depends on distance between transmitting and receiving antennas, where:

- when $d \leq 0.04$ km, then

  $\sigma = 3.5$ dB,

- when 0.040 km $< d \leq 0.1$ km, then:

  $\sigma = 3.5 + 141.7(d - 0.04)$ [dB] for propagation above the roofs, and
  $\sigma = 3.5 + 225(d - 0.04)$ [dB] for propagation below the roofs,

- when 0.1 km $< d \leq 0.2$ km, then

  $\sigma = 12$ dB for propagation above the roofs, and
  $\sigma = 17$ dB for propagation below the roofs

- when 0.2 km $< d \leq 0.6$ km

  $\sigma = 12 - 7.5(d - 0.2)$ [dB] for propagation above the roofs, and
  $\sigma = 17 - 20(d - 0.2)$ [dB] for propagation below the roofs, and finally,

- when $d > 0.6$ km, then

  $\sigma = 9$ dB.

In the case, when combined indoor–outdoor or indoor–indoor (different buildings) propagation models at interference calculations should be used, the median of the path loss is the sum of the Modified Hata model and the losses due to traversed walls and ceilings (see Section 5.2.1). The standard deviation of the slow shadow fading distribution depends on the location of transmitter and receiver antennas; when one is located indoors and other outdoors or both are indoors in different buildings, the standard deviation of the outdoor lognormal distribution increases due to the additional variations caused by heterogeneous building materials and changes of the relative location of antennas in the buildings with respect to walls, ceilings, corridors, staircases, windows, doors etc. The resultant standard deviation can be expressed as:

$$\sigma = \sqrt{\sigma_{Hata}^2 + n\sigma_{add}^2}$$

where:

σ_{Hata} is the distance dependent standard deviation explained above,

σ_{add} is the additional standard deviation of the signal (conventional value assumed in [44] and [68] is 5 dB), and

n is the number of the external walls crossed by the interfering signal.

For the indoor–indoor scenario, when both transmitter and receiver antennas are located in the same building, the Modified Hata model is not applicable. In this case, the Motley–Keenan formula or Multi-Wall Model may be applied. Variation distribution of path loss is lognormal, and the value of the standard deviation should be determined individually, mainly influenced by the trajectories of the radio paths, the number and design of floors, any room furniture, the form of rooms and corridors etc. The default value of the standard deviation assumed in indoor–indoor calculations of the interference path loss distribution, according to Monte Carlo simulation methodology shown in [44,68], is 10 dB.

5.6 RADIO PROPAGATION MODEL CALIBRATION

The accuracy of statistical models can be significantly enhanced for the specific market by a thorough calibration. The first calibration should be initially performed in the Greenfield phase prior to RF-design and site deployment. After this, a constant process to keep and enhance the quality of the model portfolio is required. Model recalibration is of a particular need, when:

- new markets were entered that contain new types of clutter;
- average cell sizes within markets change drastically due to densification, new frequencies or new technologies (e.g. HSDPA, HSUPA);
- clutters significantly changed through the years;
- terrain and clutter data in the RF-Planning tools have been updated.

Enhancements are usually implemented to diversify the set of models, e.g. to take into account:

- strong topological changes (e.g. hilly and flat models);
- seasonal changes, particularly on rural models at agricultural clutters;
- specific propagation environments on certain sites or clusters.

Most statistic macro cell models used in the industry are derived from Hata or Walfisch–Ikegami, as detailed in the previous sections. For these types of models several tuning algorithms have been developed which are introduced in the next sections.

Prior to surveying a cell, the following (minimum) steps should be taken:

1. Ensure adequate clearance of the transmitting antenna. Macrocell models assume that the transmitting antenna is located above the clutter.
2. Measure the antenna height and tilt accurately. In the case of an omnidirectional antenna, ensure that the antenna is truly vertical (0 tilt).
3. Measure the cable loss, antenna gain and calibrate the signal at the mobile unit (aim for an overall gain of 0 dB).
4. Verify the site coordinates with a GPS and a map.
5. Calibrate the transmitter in a minimum of 1 dB steps.

Drive tests should ideally meet the following criteria:

1. Data should cover all the areas surrounding the cell in order to cover all the clutter classes in the serving area.
2. If possible, line-of-sight routes should be avoided or measured separately, so that clutter and diffraction losses can be properly studied. Additionally, local clutter losses at the mobile can be

better studied if the survey routes encircled the base station. Driving radially puts greater emphasis on path losses.
3. A high number of data samples should be collected for statistical confidence. Important is not necessarily a high number of drive tests but a high number of bins within the distances from the base stations that you are interested in. Usually, all roads between 350 and 2000 m from the base station should be tested.
4. Another factor to take into account is the sampling rate. This can be distance/speed or time-dependent. In urban areas, distance-dependent sampling is preferable, since it avoids collecting a large number of samples when the vehicle is stationary.
5. Avoid collecting measurements in areas where the signal level is at or below the noise floor, typically −110 dBm. If this is not possible, many network planning tools provide a filtering facility that deals with this problem.

Additionally an optical filtering of the drive test data based on photos from the site or maps should be undertaken in a Geographical Information System (GIS) tool. The goal is the elimination of:

- samples that are not taken from the correct terrain height; this happens when the drive test vehicle moves on higher bridges or within tunnels and road canyons; these sample falsify the total result;
- samples that are taken from bins located out of the 3 dB points of antennas' horizontal main lobe; this issue does not appear if an omnidirectional antenna has been used, because the horizontal main lobe does not exist;
- samples that have been taken in areas where the clutter information is not up-to-date.
- samples taken in (or behind) constructions that generate a significant dielectric effect within the path; examples are some steel-made suspension bridges or several parallel railway tracks causing Faraday or resonance effects within the electromagnetic field.

With regard to the above, it is very beneficial to be on site and in the car when the measurements are performed. This gives a chance to crosscheck the actual situation regarding clutter, terrain and constructions.

5.6.1 TUNING ALGORITHMS

The most common tuning algorithms are based on iterative approaches intended to minimise the standard deviation between the Hata coefficients and the drive test data. Figure 5.17 shows the level-to-distance relationship of a suburban test drive in a major German city at 2.160 GHz.

The most important outputs of the statistical tools are the average error and the standard deviation of the entire path as well as the average error per distance. The average error $\bar{x}_e$ indicates whether the model is under-performing (positive value) or over-predicting (negative value) compared with measurement data; it is calculated as:

$$\bar{x}_e = \frac{1}{N_s} \sum_{i=1}^{N_s} \left(x_{\text{predicted}} - x_{\text{measured}}\right)_i \tag{5.59}$$

where N_s is the number of samples, $x_{\text{predicted}}$ is the power level of prediction and x_{measured} is the measured power level.

The average error value per distance is derived in the same way as the total average error, but for sections of bins having the same distance to the antenna location. The standard deviation (STD) estimates the variability of the prediction model (around average value).

In the case of single slope model, the path loss L for the general model is given by:

$$L = C_1 + C_2 \log(d) + C_3 L_d + C_4 \log(h_e) + C_5 \log(h_e) \log(d) + C_6 L_c + C_7 d \tag{5.60}$$

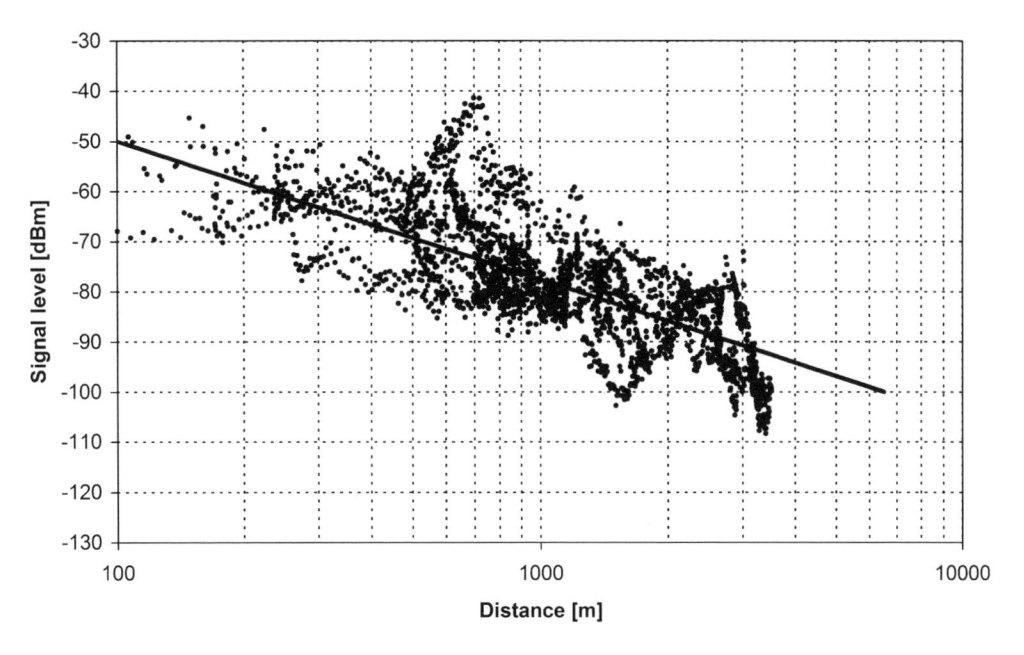

Figure 5.17 Typical drive test sample.

where C_1–C_7 are weighting factors (described below), d is the distance between transmitter and receiver, L_d is loss due to diffraction, h_e is effective mobile height above ground and L_c is loss of the clutter. The constants C_1–C_7 can be summarised as:

C_1 – constant describing the intercept
C_2 – slope factor
C_3 – diffraction weight
C_4 – effective height weight
C_5 – distance/height weight
C_6 – clutter weight
C_7 – distance weight.

The objective of the calibration exercise is to achieve a minimum STD and an average error between measured and predicted data for each clutter group. Usually, this minimisation is originated like the following example:

1) Filter the section to be calibrated (minimum dB, distance from site, number of knife edges, minimum bin size).
2) Initialise the model with parameter offset values.
3) Change C_1 and C_2 in a manner that a minimum in the standard deviation is reached. Then repeat the procedure for C_3. The diffraction weight should not increase above 1.0.
4) Once a new minimum is reached, start varying C_2 in smaller steps. After this, start varying C_3 in lower steps. Keeping the new values for C_2 and C_3 constant, repeat the procedure for C_4.
5) Now vary C_2, C_3 and C_4 in very small steps until a new minimum is reached. Usually, C_3 and C_4 are not correlated, so that a re-tuning of C_3 is usually not required. Keeping the new values for C_2, C_3 and C_4 constant, repeat the procedure for C_5.

6) Regarding C_6, it is a global clutter weight factor that is multiplied with the clutter correction factors. This allows to leave C_6 constant, while all clutter factors get optimised with respect to:

 - a statistical optimum (lowest standard deviation, average error of 0) for each clutter;
 - an engineering approach; this should consider that the clutter correction factors relative to each other are in plausible limits (e.g. 'dense urban' should not have a more optimistic correction factor than 'open' or 'suburban' etc.).

7) C_7 is usually assumed to be equal to zero.

A different and quicker approach compared to an iterative tuning is the determinant based method. The goal here is to find an absolute minimum of the deviation between the predicted and the surveyed samples.

In general, it has to be stated that neither the automated tuning algorithm, nor Measurement Based Prediction (MBP), nor the determinant approach will always deliver plausible results from an engineering point of view. For this reason, a validation of the coverage produced by the model is a prerequisite before implementing the model.

5.6.2 SINGLE AND MULTIPLE SLOPE APPROACHES

Many RF planning tools enable the addition of a second slope at a breakpoint to refine the statistical performance of the model. This is usually done by defining an additional case for the general model parameters C_1 and C_2 (intercept and slope). The idea is to define a breakpoint at a certain distance D from the antenna (Figure 5.18).

The difference to the single slope approach is presented in Equation (5.61).

$$L = C_1 + C_2 \log(d) + C_3 L_d + C_4 \log(h_e) + C_5 \log(h_e) \log(d) + C_6 L_c + C_7 d \tag{5.61}$$

$$C_1 = \begin{cases} C_{1a} & for\ d < D \\ C_{1b} & for\ d > D \end{cases} \tag{5.61a}$$

$$C_2 = \begin{cases} C_{2a} & for\ d < D \\ C_{2b} & for\ d > D \end{cases} \tag{5.61b}$$

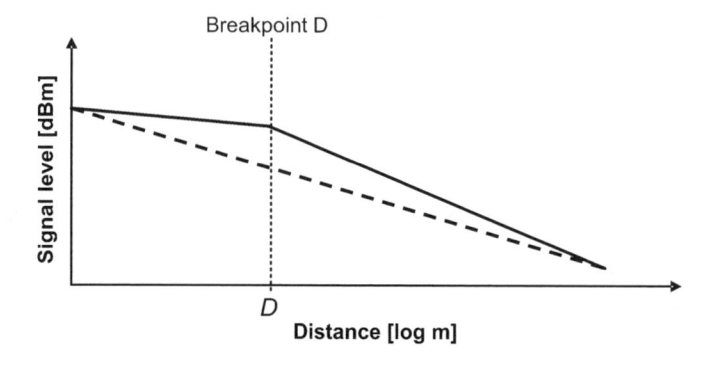

Figure 5.18 Principle of a dual slope approach.

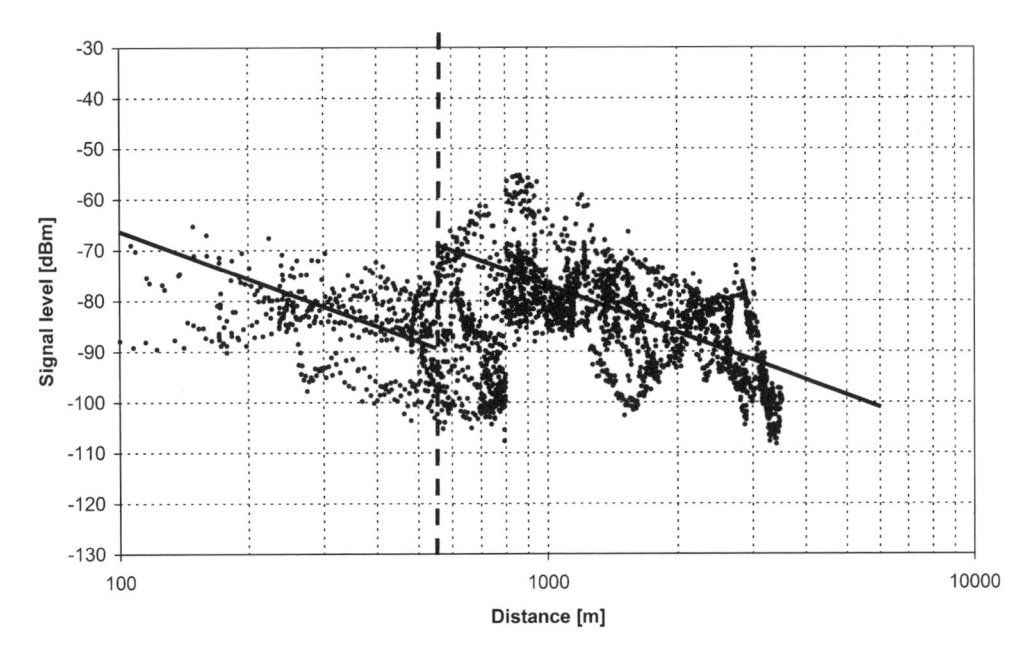

Figure 5.19 Discontinuity of the prediction at the breakpoint due to erroneous C_{1a}.

While this enhancement of statistical confidence is actually desirable, the tuning of a dual slope model bears the great risk of a discontinuity at the defined breakpoint if coverage gaps appear within an area of given cell service. Figure 5.19 shows an example for such a discontinuity, which is just caused by setting one parameter too low; in this case the false parameter is C_{1a}.

The next case shows a miscalibration by applying an erroneous slope (C_2) prior to the breakpoint (Figure 5.20). This can be detected by monitoring either the error between the prediction and the measurements in small distance intervals or the error vs distance graph.

The errors vs distance evaluation can also be used to evaluate whether a single slope model can be enhanced by a dual slope model, and where the breakpoint should be applied. Applying a dual slope can increase the statistical confidence in the case that a breakpoint exists. A sign for an existence of such a breakpoint is an 'arch' within the graph 'error vs prediction' and a high standard deviation in certain sections compared to the standard deviation in the entire path. The peak of the arch can then considered to be a breakpoint. As mentioned above, a dual slope approach increases the risk of discontinuities at the breakpoint, so after the tuning to the statistical optimum, the area around the breakpoint has to be checked and, if required, adjusted towards prediction continuity.

To conclude, the main emphasis of this chapter has been to discuss the use of various radio channel modelling approaches within the planning framework of 3G systems. We have dealt with generic propagation issues, the purpose of which has been to yield the foundations for understanding standardised 3G channel models. We have also touched upon the very important topic of radio propagation model calibration. This chapter hence constitutes an important foundation for the subsequent chapters, which will rely in one form or another on the radio channel models described above.

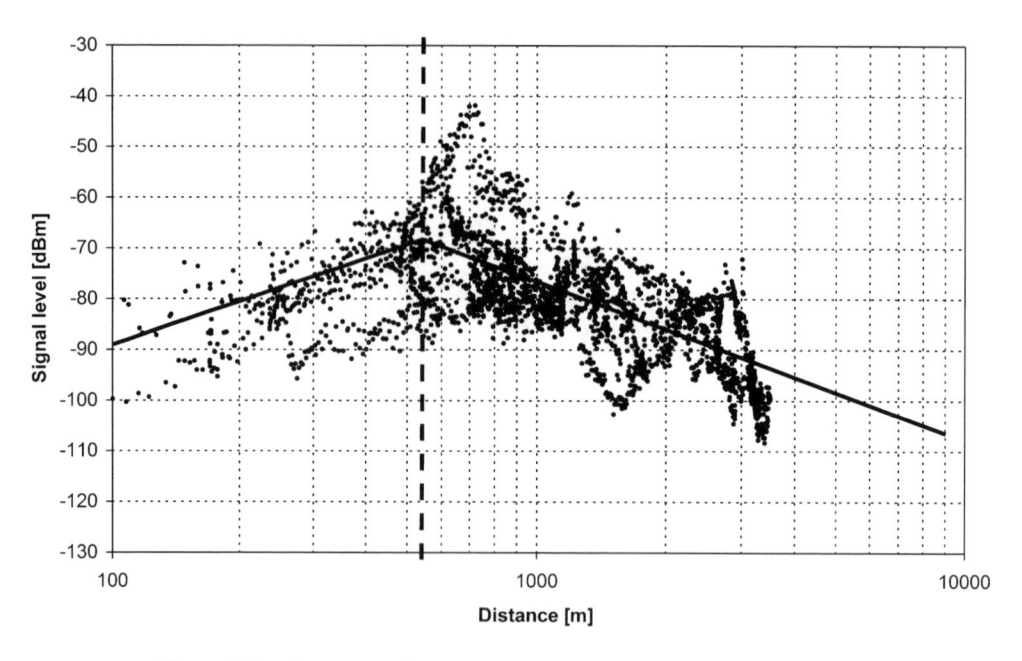

Figure 5.20 Discontinuity of the prediction at the breakpoint due to erroneous C_2.

APPENDIX: CALCULATION OF INVERSE COMPLEMENTARY CUMULATIVE NORMAL DISTRIBUTION FUNCTION

The brief analysis presented here is based on [54]. The inverse complementary cumulative normal distribution function, $Q_i(x)$, for $0.01 \leq x \leq 0.99$, may be approximated as follows:

$$Q_i(x) = T(x) - \xi(x) \qquad \text{if } x < 0.5$$
$$Q_i(x) = -\{T(1-x) - \xi(1-x)\} \qquad \text{if } x \geq 0.5$$

where

$$T(x) = \sqrt{[-2\ln(x)]}$$
$$\xi(x) = \frac{[(C_2 \cdot T(x) + C_1) \cdot T(x)] + C_0}{[(D_3 \cdot T(x) + D_2) \cdot T(x) + D_1] \cdot T(x) + 1}$$

and

$C_0 = 2.515517$
$C_1 = 0.802853$
$C_2 = 0.010328$
$D_1 = 1.432788$
$D_2 = 0.189269$
$D_3 = 0.001308$.

REFERENCES

[1] O. Landron, M.J. Feuerstein, T.S. Rappaport, 'A comparison of theoretical and empirical reflection coefficients for typical exterior wall surfaces in a mobile radio environment', *IEEE Transactions on Antennas and Propagation*, vol. 44. no. 3, pp. 341–351, March 1996.

[2] O. Landron, M.J. Feuerstein, T.S. Rappaport, 'In situ microwave reflection coefficient measurements for smooth and rough exterior wall surfaces', *IEEE Vehicular Technology Conference*, Secaucus, NJ, pp. 77–80, 18 May 1993.

[3] W.S. Ament, 'Toward a theory of reflection by a rough surface', *Proceedings of Institute of Radio Engineers IRE*, vol. 41, no.1, pp. 142–146, January 1953.

[4] G.D. Maliuzhinets, 'Excitation, reflection and emission of surface waves from a wedge with given wave impedances', *Soviet Physics: Doklady*, vol. 3, pp. 752–754, 1958.

[5] M. Aidi, J. Lavergnat, 'Approximation of the Maliuzhinets function', *Journal of Electromagnetic Waves and Applications*, vol. 10, pp. 1395–1411, 1996.

[6] H.B. Keller, 'Geometrical theory of diffraction', *Journal of the Optical Society of America*, vol. 52, no. 2, pp. 116–130, February 1962.

[7] R.G. Kouyoumjian, P.H. Pathak, 'A uniform geometrical theory of diffraction for an edge in a perfectly conducting surface', *Proceedings of the IEEE*, vol. 62, no. 11, pp. 1448–1461, November 1974.

[8] R.J. Luebbers, 'Finite conductivity uniform GTD versus knife edge diffraction in prediction of propagation path loss', *IEEE Transactions on Antennas and Propagation*, vol. AP32 no. 1, pp. 70–76, January 1984.

[9] P. Holm, 'A new heuristic UTD diffraction coefficient for nonperfectly conducting wedges', *IEEE Trans. on Antennas and Propagation*, vol. 48, no. 8, 1211–1219, August 2000.

[10] ITU, ITU-R Recommendation P.1145, 'Propagation data for the terrestrial land mobile service in the VHF and UHF bands'.

[11] N. Moraitis, A. Kanatas, G. Pantos, P. Constantinou, 'Delay spread measurements and characterization in a special propagation environment for PCS microcells', *13th IEEE International Symposium on Personal, Indoor and Mobile Communications*, Lisbon, Portugal, vol. 3, pp. 1190–1194, September 2002.

[12] J.B. Andersen, 'Radio Channel Characterisation', *COST 231 Final Report*, COST Office, European Commission, Brussels, Belgium, 1999.

[13] H. Hashemi, 'The indoor radio propagation channel', *Proceedings of IEEE*, vol. 81, no. 7, pp. 943–968, July 1993.

[14] J.C. Chuang, 'The effects of Time Delay Spread on Portable Radio Communications Channels with Digital Modulation', *IEEE Journal on Selected Areas in Communications*, vol. 5, no. 5, pp. 879–889, June 1987.

[15] C.W. Trueman, R. Paknys, J. Zhao, D.Davis, B. Segal, 'Ray tracing algorithm for indoor propagation', *Applied Computational Electromagnetics Society. Proc 16th Annual Review of Progress in Applied Computational Electromagnetics*, pp. 493–500, Monterey, CA., March 2000.

[16] M.E.C. Rodrigues, L.A.R. Ramirez, L.A.R. Silva Mello, F.J.V. Hasselmann, 'A ray tracing technique for coverage predictions in micro cellular environments', *Journal of Microwaves and Optoelectronics*, vol. 3, no. 5, pp. 1–17, July 2004.

[17] S. Seidel, T.S. Rappaport, 'Site-specific propagation prediction for wireless in-building personal communication system design', *IEEE Transactions on Vehicular Technology*, vol. 43, no 4. pp. 879–891, 1994.

[18] M. Dottling, T. Zwick, W. Wiesbeck, 'Ray tracing and imaging techniques in urban pico and micro cell wave propagation modelling', *IEE 10th International Conference on Antennas and Propagation*, no. 436, pp. 2.311–2.315, Edinburgh, April 1997.

[19] M. Feistel, A. Baier, 'Performance of a three-dimensional propagation model in urban environments', *IEEE Personal, Indoor and Mobile Radio Communications*, vol. 2, pp. 402–407, Toronto, Canada September 1995.

[20] J-P. Rossi, Y. Gabillet, 'A mixed ray launching/tracing method for full 3-D UHF propagation modelling and comparison with wide-band measurements', *IEEE Transactions on Antennas and Propagation*, vol. 50, no. 4, pp. 517–523, April 2002.

[21] Scott.Y. Seidel, Theodore S. Rappaport, 'A ray tracing technique to predict path loss and delay spread inside buildings', *Proc. IEEE GLOBECOM'92 Conference*, Orlando, FL, pp. 649–653, December 1992.

[22] R.A. Valenzuela, 'Ray tracing prediction of indoor radio propagation', *5th IEEE International Symposium on Personal, Indoor, Mobile Radio Communications*, pp. 140–144, September 1994.

[23] R. Fraile, N. Cardona, 'Macrocellular coverage prediction for all ranges of antenna height using neural networks', *IEEE 1998 International Conference on Universal Personal Communications*, vol. 1, Florence, Italy, pp. 21–25, October 1998.
[24] A. Neskovic, N. Neskovic, D. Paunovic, 'Macrocell electric field strength prediction model based upon artificial neural networks', *IEEE Journal on Selected Areas in Communications*, vol. 20, no. 6, pp. 1170–1177, August 2002.
[25] D. Didascalou, J. Maurer, W. Wiesbeck, 'A novel stochastic rough-surface scattering representation for ray-optical wave propagation modelling', *Proceedings of the International Conference on Electromagnetics In Advanced Applications ICEAA2001*, Torino, Italy, pp. 171–174, September 2001.
[26] M. Lott, B. Walke, 'On the performance of an advanced 3D ray tracing method', *Proceedings of European Wireless (EW'99)*, Munich, Germany, October 1999.
[27] T. Binzer, F.M. Landstorfer, 'Radio network planning with neural networks', *Proc. 52nd IEEE Vehicular Technology Conference – Fall*, vol. 2, Boston, MA, pp. 811–817, 2000.
[28] G. Wölfle, R. Wahl, P. Wildbolz, P. Wertz, 'Dominant Path Prediction Model for Indoor and Urban Scenarios', *AWE Communications GmbH*, University of Stuttgart, Germany http://www.awe-communications.com/Propagation/dpm/dpm.htm.
[29] B. Gschwendtner, F. Landstorfer, 'Adaptive propagation modelling based on neural network techniques', *IEEE 46th Vehicular Technologies Conference*, vol.2, pp. 623–626, 28 April–1 May, 1996.
[30] I. Popescu, A. Kanatas, P. Constantinou, I. Nafornită, 'Applications of general regression neural networks for path loss prediction', *University of Oradea (Romania)*, Technical University of Timisoara, National Technical University of Athens http://hermes.etc.utt.ro/docs/cercetare/articole/aprnnplp2002.pdf.
[31] G. Wöffle, P. Wertz, F.M. Landstorfer: 'Performance, Accuracy and Generalization Capability of Indoor Propagation Models in Different Types of buildings', *10th IEEE International Symposium on Personal, Indoor and Mobile Radio Communications*, September 1999.
[32] G. Yang, K. Pahlavan and J.F. Lee, 'A 3D propagation model with polarization characteristics in indoor radio channels', *Proceedings of the IEEE Globecomm*, Houston, TX, vol. 2, pp. 1252–1256, 29 November–2 December 1992.
[33] P. Wertz, R. Hoppe, D. Zimmermann, G. Wölfle, F.M. Landstorfer, 'Enhanced Localization Technique within Urban and Indoor Environments', 3rd *COST 273 MCM-Meeting* in Guildford, UK, COST 273, TD(02)033, January 2002.
[34] R. Hoppe, G. Wölfle, P. Wertz, F.M. Landstorfer, 'Advanced ray-optical wave propagation modelling for indoor environments including wideband properties', *European Transactions on Telecommunications*, vol. 14, pp. 61–69, 2003.
[35] ETSI TR 101 112 V3.2.0 (1998–04), Universal Mobile Telecommunications System (UMTS); Selection procedures for the choice of radio transmission technologies of the UMTS (UMTS 30.03 version 3.2.0).
[36] J.M. Keenan, A.J. Motley, 'Radio coverage in buildings', *British Telecom Technology Journal*, vol. 8, no. 1, January 1990.
[37] E. Damoso, L.M. Correia, 'Digital mobile radio towards future generations systems', *COST 231 Final Report*, COST Office, European Commission, Brussels, Belgium, 1999.
[38] A.A.M. Saleh, R.A. Valenzuela, 'A statistical model for indoor multipath propagation', *IEEE Journal on Selected Areas in Communications*, vol. SAC-5, no. 2, pp. 128–137, February 1987.
[39] G. German, Q. Spencer, L. Swindlehurst, R. Valenzuela, 'Wireless indoor channel modelling: statistical agreement of ray tracing simulations and channel sounding measurements', *Intl. Conf. Acoustics, Speech, Signal Processing (ICASSP 2001)*, vol. 4, Salt Lake City, UT, pp. 778–781, May 2001.
[40] J.G. Wang, A.S. Mohan, T.A Aubrey, 'Angles-of-arrival of multipath signals in indoor environments', *IEEE 46th Vehicular Technology Conference*, vol.1, pp. 155–159, 28 April–1 May 1996.
[41] R. Hoppe, G. Wölfle, G.G. Landstorfer, 'Measurement of building penetration loss and propagation models for radio transmission into buildings', *IEEE VTS 50th Vehicular Technology Conference – Fall*, vol.4, pp. 2298–2302, September 1999.
[42] T. Kürner, A. Meier, 'Prediction of outdoor-to-indoor coverage in urban areas at 1.8 GHz', *IEEE Journal on Selected Areas in Communications*, vol. 20, no. 3, pp. 496–506, April 2002.
[43] Y.L.C. de Jong, M.H.J.L. Koelen, M.H.A.J. Herben, 'A building-transmission model for improved propagation prediction in urban microcells', *IEEE Trans. on Veh. Techn*, vol. 53, no. 2, pp. 490–502, March 2004.
[44] ITU, ITU-R Report SM.2028-1, 'Monte Carlo simulation methodology for the use in sharing and compatibility studies between different radio services or systems', 2002.

[45] J.E. Berg, 'Building penetration', Digital Mobile Radio Toward Future Generation Systems (*COST 231 Final Report*). Brussels, Belgium, 1998.
[46] E.F.T. Martijn, M.H.A.J. Herben, 'Characterization of radio wave propagation into buildings at 1800 MHz', *IEEE Antennas and Wireless Propagation Letters*, vol. 2, no. 9, pp. 122–125, 2003.
[47] J. Walfisch, H. Bertoni, 'A theoretical model of UHF propagation in urban environments', *IEEE Transactions on Antennas and Propagation*, vol. 36, no. 12, pp. 1788–1796, December 1988.
[48] L. Maciel, H. Bertoni, H. Xia, 'Unified approach to prediction of propagation over buildings for all ranges of base station antenna height', *IEEE Vehicular Technology Conference*, vol. 42, no.1, pp. 41–45, 1993.
[49] P. Mege, 'Frequency assignment and licensing. Addendum 1: IMT 2000/ UMTS Radio Planning Procedures', ITU, Document 1-HNB-SM/50-E, ch. 3, 9. January 2003.
[50] D. Crosby, S. Greaves, A. Hopper, 'The effect of building height variation on the multiple diffraction loss component of the Walfisch-Bertoni model', *14th IEEE Proceedings on Personal, Indoor and Mobile Radio Communications Conference*, China, vol. 2, pp. 1805–1809, September 2003.
[51] T. Kürner, 'Propagation prediction models. Propagation models for macro-cells', *COST 231 Final Report*, COST Office, European Commission, Brussels, Belgium, 1999.
[52] J.E. Berg, 'A recursive method for street microcell path loss calculations', *6th IEEE International Symposium on Personal, Indoor and Mobile Radio Communications*, vol. 1, pp. 140–143, September 1995.
[53] H.W. Son, N.H. Myung, 'A deterministic ray tube method for microcellular wave propagation prediction model', *IEEE Transactions on Antennas and Propagation*, vol. 47, no. 8, pp. 1344–1350, August 1999.
[54] ITU, ITU-R P.1546, 'Method for point-to-area predictions for terrestrial services in the frequency range from 30 MHz to 3000 MHz'.
[55] K. Kimura, J. Horikoshi, 'Prediction of milimeter-wave multipath propagation characteristics in mobile radio environment', *IEICE Trans. Electron.*, vol. E82-C, no. 7, pp. 1253–1259, July 1999.
[56] T. Kürner et al., MOMENTUM: Models and simulations for network planning and control of UMTS. Final report on automatic planning and optimisation, IST-2000-28088, October 2003.
[57] K. Rizk, R. Valenzuela, S. Fortune, D. Chizhik, F. Gardiol, 'Lateral, full-3D and vertical plane propagation in microcells', *COST 259 TD (98) 47*, Bern, Switzerland, February 1998.
[58] M.J. Feuerstein, K.L. Blackard, T.S. Rappaport, S.Y. Seidel, H.H. Xia, 'Path loss, delay spread, and outage models as functions of antenna height for microcellular system design', *IEEE Transactions on Vehicular Technology*, vol. 43, no. 3, pp. 487–498, August 1994.
[59] C.L. Giovaneli, 'An analysis of simplified solution for multiple knife-edge diffraction', *IEEE Transactions on Antennas and Propagation*, vol. 32, no. 3, pp. 297–301, March 1984.
[60] J. Deygout, 'Multiple knife-edge diffraction of microwaves'. *IEEE Transactions on Antennas and Propagation*, vol. AP14, no. 4, pp. 480–489, 1966.
[61] J. Deygout, 'Correction factor for multiple knife-edge diffraction', *IEEE Transactions on Antennas and Propagation*, vol. 39, no. 8, pp. 1256–1258, 1991.
[62] K. Bullington, 'Radio propagation for vehicular communications', *IEEE Trans. on Vehicular Technology*, vol. VT-26, no. 4, pp. 295–308, November 1977.
[63] ITU, ITU-R Recommendation P.526, 'Propagation by diffraction'.
[64] I. Popescu, A. Kanatas, E. Angelou, I. Nafornită, P. Constantinou, 'Applications of generalized RBF-NN for path loss prediction', *13th IEEE International Symposium on Personal, Indoor and Mobile Radio Communications*, vol.1, pp. 484–488, September 2002.
[65] ITU, ITU-R Recommendation P.1057, 'Probability distributions relevant to radiowave propagation modelling'.
[66] ITU, ITU-R Recommendation P.452, 'Prediction procedure for the evaluation of microwave interference between stations on surface of the Earth at frequencies above about 0.7 GHz'.
[67] ITU, ITU-R Recommendation P. 676, 'Attenuation by atmospheric gases'.
[68] CEPT, ERC Report 68, 'Monte Carlo simulation methodology for the use in sharing and compatibility studies between different radio services or systems', Naples February 2000, rev. in Regensburg, May 2001 and Baden, June 2002.

6

Theoretical Models for UMTS Radio Networks

Hans-Florian Geerdes, Andreas Eisenblätter, Piotr M. Słobodzian,
Mikio Iwamura, Mischa Dohler, Rafał Zdunek, Peter Gould and
Maciej J. Nawrocki

The aim of this chapter is to provide an in-depth analysis of theoretical modelling approaches in UMTS radio network planning. It constitutes a key chapter in this part of the book related to modelling, and many herein discussed aspects will be utilised in later chapters. The exposure of the subject is structured such that we deal with modelling approaches in a layered manner, i.e. we will commence with theoretical antenna and link level modelling and proceed then with static and dynamic system level modelling. A set of parameters is exchanged between these models, where antenna patterns and link level results have a direct impact onto the performance of system level simulations, whereas system simulator settings are required to determine the operating conditions of the link level simulator.

6.1 ANTENNA MODELLING

Antennas constitute a very important part of wireless communication systems. An antenna is defined as a device for radiating and receiving radio waves [1]. We can hence distinguish between transmitting and receiving antennas; they are connected by a radio frequency (RF) channel, which is a medium for carrying signals from a transmitter to a receiver. In addition to this, modern wireless systems use antennas as spatial filters, i.e. exploit their directional properties to optimise the radiation of radio waves in some directions and suppress it in others. Antenna parameters may strongly influence the performance of a wireless system at both link and system levels, and therefore their adequate modelling is essential for a reliable performance prediction. Basic antenna parameters describe how and where the radio waves are transmitted (or from where received), and some of them may differ considerably depending on the role which antennas play in a given system. The differences concern mainly the directive characteristics of antennas. For example, when considering antennas used in cellular systems, antennas of mobile terminals receive and transmit radio waves almost equally in all directions. This is not the case for the base station antennas, which are designed to service only a specific spatial region.

All antenna parameters are strictly related to the size and geometrical structure (shape) of the antenna. Therefore, a detailed specification of the antenna geometry is sufficient for an accurate determination of most antenna parameters. Nevertheless, a precise and rigorous antenna modelling is extremely complex, since it is based on principles and methods of electromagnetism, and generally consists in solving partial differential equations (Maxwell's equations) under suitable boundary conditions [2,3]. On the other hand, the simultaneous rigorous antenna and system modelling is highly ineffective from the link and system level viewpoint due to relatively high computational effort, which needs to be devoted to antenna modelling itself. Consequently, the link and system level antenna modelling is based on the general results of the antenna theory and exploits very simple models, which describe only selected antenna parameters. Such models have proved to be sufficiently accurate and very often are specified in various recommendations; see, for example, recommendations of the International Radio Consultative Committee (CCIR) for parabolic reflector antennas used in the Direct Broadcast Satellite (DBS) system [4].

When modelling antennas for cellular communication systems, we need to consider two types of antennas, namely very small (electrically small) antennas, which are installed in mobile terminals, and the base station antennas, which are usually built as linear antenna arrays. Regardless of the type of antennas, their modelling involves a similar set of parameters, among which the most important are the following ones: antenna radiation pattern, directivity, gain, antenna polarisation and finally antenna bandwidth. Definitions of all these parameters, conforming to standards of the International Electrotechnical Commission (IEC), can be found in [1], and their extensive explanation is given, for example, in [5]. Nevertheless, before we start specific considerations concerning the antenna modelling we will briefly recall some of them.

An *antenna radiation pattern* is defined as a function representing radiation properties of the antenna in the 3D space. The most important radiation property is the spatial distribution of radiated energy (for the power pattern) or electric field intensity (for the amplitude field pattern) over a sphere surrounding the antenna in the far-field zone [5]. In practice, instead of the full 3D antenna pattern a set of two-dimensional patterns is usually determined (measured or calculated), to give the most needed information. For example, for the majority of terrestrial wireless systems, the antenna radiation properties are described by means of a couple of 2D radiation patterns (the principal patterns or principal cuts). Since such antennas are usually linearly polarised, the principal cuts are determined in two perpendicular planes, horizontal and vertical ones, crossed along the direction of maximum radiation (see $F_H(\varphi)$ and $F_V(\theta)$ in Figure 6.1 respectively). For linearly polarised antennas, these planes are also referred to as the H-plane and E-plane, depending on the antenna polarisation. The most important characteristics of the antenna radiation pattern are the half-power beamwidth (or -3 dB beamwidth) and the side lobe level (in dB).

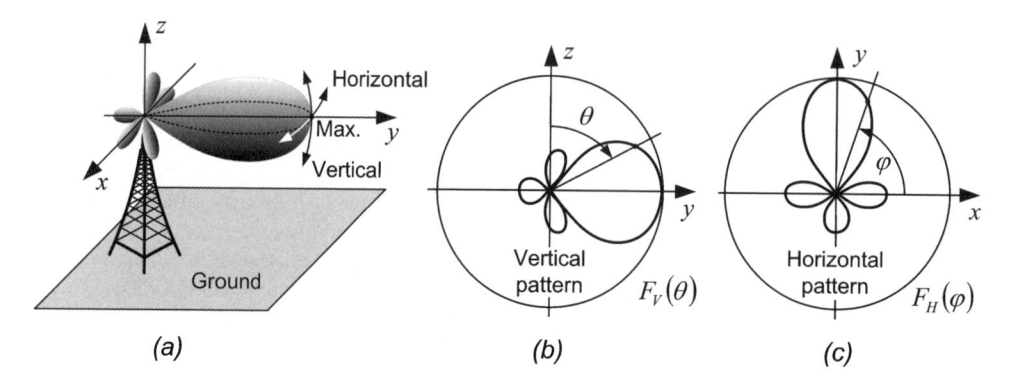

Figure 6.1 The three-dimensional antenna radiation pattern (a) and its respective vertical (b) and horizontal (c) cuts.

Polarisation of an antenna in a given direction (when not stated explicitly, the direction of maximum radiation is assumed) is defined with respect to the electric-field vector, and is classified according to the shape of the curve traced by the time varying *E*-field vector's end in space. Consequently, we may distinguish the linear, circular or elliptical polarisation. Additionally, for linearly polarised antennas, we usually define the horizontal, vertical or slant polarisation.

Directivity of an antenna is defined as the ratio of the radiation intensity in a given direction from the antenna to the radiation intensity averaged over all directions. If the direction is not specified, then the direction of maximum radiation is implied, and the antenna directivity is denoted as D_0. In practice, another parameter closely related to directivity is of greater importance, namely the antenna gain, which takes into account the antenna efficiency as well as its directional characteristics. Therefore, the antenna gain is of crucial importance for the link budged calculations of a communication system.

The *gain* $G(\theta, \varphi)$ (absolute gain) is defined as the ratio of the radiation intensity in a given direction (θ, φ), to the radiation intensity that would be obtained from the lossless isotropic source, fed by the same power. With respect to the link budget calculations, the relative maximum gain G_0 is more important and is defined as the ratio of the absolute gain in the direction of maximum radiation to maximum absolute gain of the reference antenna. In most practical cases, the reference antenna is a lossless isotropic source or a lossless half-wavelength dipole, and the relative gain is expressed in dBi (or simply dB) or dBd, respectively. It is worth mentioning that the gain definition does not take into account losses arising from impedance and polarisation mismatches, and we hence need to include them in the link budged separately. However, sometimes the impedance mismatch is included in the gain, usually in the measured one; in such a case, the gain is referred to as the *realised gain* [6].

Finally, the *bandwidth* of an antenna is defined as the range of frequencies within which a given antenna characteristic conforms to a specified standard. Usually, the characteristic is input impedance, radiation pattern shape, beamwidth or polarisation.

Having described all fundamental antenna parameters, we can begin dealing with issues involved in antenna modelling.

6.1.1 MOBILE TERMINAL ANTENNA MODELLING

Most of contemporary mobile terminals for 2G cellular systems are equipped with electrically small, compact, internal antennas of PIFA-like structure (Patch Inverted F Antenna) [7], and similar ones will also be used in 3G terminals. Such antennas are inherently non-directive and exhibit low gains, usually of 0–2 dBi. Their radiation patterns are almost omnidirectional in the horizontal plane (assuming the antenna is vertically polarised), and the vertical ones are similar to those of a half-wavelength dipole, but usually have strong irregularities and more than two deep zeros [7,8]. All the radiation patterns may also vary considerably in frequency, and the direction of maximum radiation does not always coincide with the horizontal plane. In addition to this, internal antennas exhibit low discrimination of orthogonal polarisation (the horizontal one), especially in the vertical plane. This quality, however, need not be a disadvantage, since a terminal may take all potential positions (including horizontal ones) during its operation.

Taking into account the aforementioned facts, we can draw a conclusion that the best way to model the mobile terminal antenna is to assume uniform radiation (receive) in all directions, and the antenna gain of 0 dBi. Hence, the antenna model may be described for all operating frequencies by the following radiation pattern (power pattern):

$$F^i(\theta, \varphi)_{[dBi]} = G_0 = 0 \qquad (6.1)$$

As we can see, the proposed model is extremely simple, but on average reflects the performance of terminal antennas reasonably well (in terms of time and location).

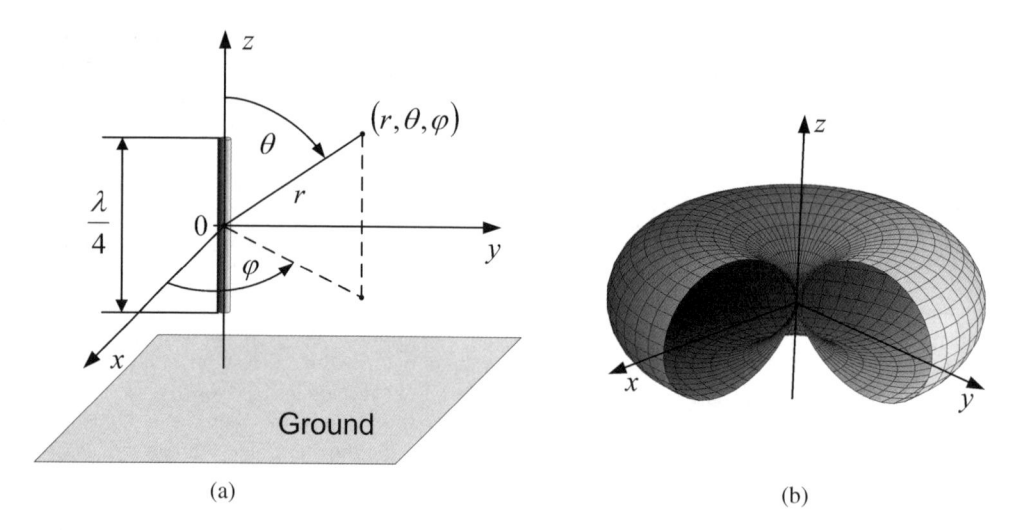

Figure 6.2 Dipole antenna arrangement in the spherical coordinate system (a) and its normalised power pattern (b).

In order to increase the model accuracy we may assume that the antenna, for example, has the form of a λ/4 dipole with the sinusoidal current distribution [5]. For such an antenna, the radiation properties are described by the following normalised power pattern:

$$F^d(\theta, \varphi)_{[\text{dBi}]} = 10 \cdot \log \left\{ \left[\frac{\cos\left[\frac{\pi}{4} \cdot \cos(\theta) - \frac{1}{\sqrt{2}}\right]}{\sin(\theta) \cdot \left[1 - \frac{1}{\sqrt{2}}\right]} \right]^2 \right\} \quad (6.2)$$

Equation (6.2) has been derived for a vertically polarised dipole aligned along the z-axis, as shown in Figure 6.2a (it has been assumed that the ground did not influence the antenna properties). The antenna pattern described in Equation (6.2) is depicted in Figure 6.2b.

6.1.2 BASE STATION ANTENNA MODELLING

Generally, base station antennas have the form of linear antenna arrays [5], which are intended to obtain suitably directive characteristics in order to increase the radiation towards the serviced area (usually a sectorised area) and suppress it towards other ones. Consequently, the radiation patterns of such an antenna have a relatively narrow main beam, giving rise to the gain ranging from 10 to 20 dB, and side and back lobe levels usually between −15 and −25 dB. Additionally, the antenna is linearly polarised and typically operates with vertical polarisation (slant polarisation, i.e. so-called X-pol, is also exploited, but only in the receiving mode). In practice, radiation properties of a base station antenna are characterised by means of two radiation patterns, the horizontal and vertical ones, and the gain. All these parameters are obtained by performing measurements on the antennas.

With respect to the wireless system performance, the base station antenna modelling is aimed to provide information concerning spatial distribution and strength of signals radiated from an antenna towards a serviced area as well as adjacent areas, potentially exposed to interference. Therefore, the antenna radiation capabilities should be known at least for all directions, which point to areas of interest. In general, however, most up-to-date CAD tools use a very simple antenna model, which is based on two principal patterns (two-dimensional ones) determined in two planes, i.e. the horizontal and vertical one. Unfortunately, such a model is justified only for a special case, which will be discussed in the following.

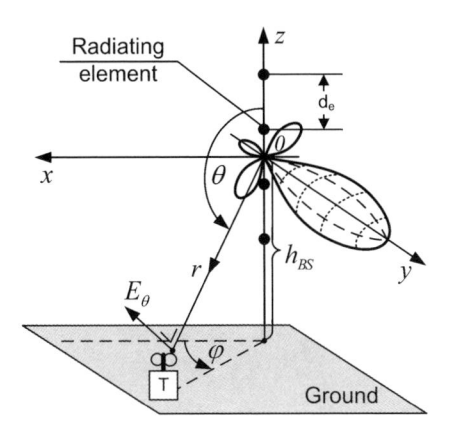

Figure 6.3 Linear antenna array arrangement in the spherical coordinate system and the relative position of a mobile terminal.

In order to begin the analysis, let us assume that the base station antenna is a linear array of N elements, positioned uniformly and symmetrically along the z-axis, with the antenna radiation maximum directed along the y-axis, as shown in Figure 6.3. Next, let $g(\theta, \varphi)$ and $F(\theta, \varphi)$ denote a three-dimensional far-field amplitude patterns of the element and antenna array, respectively. Since, it is assumed that the antenna operates in vertical polarisation, the patterns describe angular variations of only one component of the total radiated field, namely E_θ. Using a simplified description of the linear antenna array [5], the pattern $F(\theta, \varphi)$ may be expressed as (provided N is an even number):

$$F(\theta, \varphi) = g(\theta, \varphi) \cdot \sum_{n=1}^{N} a_n \cdot e^{j \cdot k \cdot \left(\frac{N+1}{2} - n\right) \cdot d_e \cdot \cos(\theta)} \tag{6.3}$$

where a_n are the excitation coefficients of the array elements, d_e denotes the distance between the elements and $k = 2\pi/\lambda$, where λ is the wavelength. The sum following $g(\theta, \varphi)$ is referred to as the array factor, which is responsible for the antenna pattern shaping.

Using Equation (6.3), the horizontal and vertical antenna patterns, defined as two-dimensional cuts of $F(\theta, \varphi)$, as illustrated respectively in Figure 6.4a and 6.4b, can be described by the following equations:

$$F_H(\varphi) = F\left(\theta = \frac{\pi}{2}, \varphi\right) = g\left(\frac{\pi}{2}, \varphi\right) \cdot \sum_{n=1}^{N} a_n \tag{6.4a}$$

$$F_V(\theta) = F\left(\theta, \varphi = \frac{\pi}{2}\right) = g\left(\theta, \frac{\pi}{2}\right) \cdot \sum_{n=1}^{N} a_n \cdot e^{j \cdot k \cdot \left(\frac{N+1}{2} - n\right) \cdot d_e \cdot \cos(\theta)} \tag{6.4b}$$

Both the equations contain only a small fraction of information required to describe $F(\theta, \varphi)$. Therefore, in general, they are insufficient for rigorous modelling of the three-dimensional radiation capabilities of the antenna. A special case occurs when $F(\theta, \varphi)$ is separable with respect to θ and φ. In such a case, Equation (6.3) can be rewritten as

$$F(\theta, \varphi) = F(\theta, \varphi = const) \cdot F(\theta = const, \varphi) = F_V(\theta) \cdot F_H(\varphi) \tag{6.5}$$

and as we can see, only two principal radiation patterns are sufficient for a full description of $F(\theta, \varphi)$. When modelling the base station antenna performance, we need to determine the relative gain of the

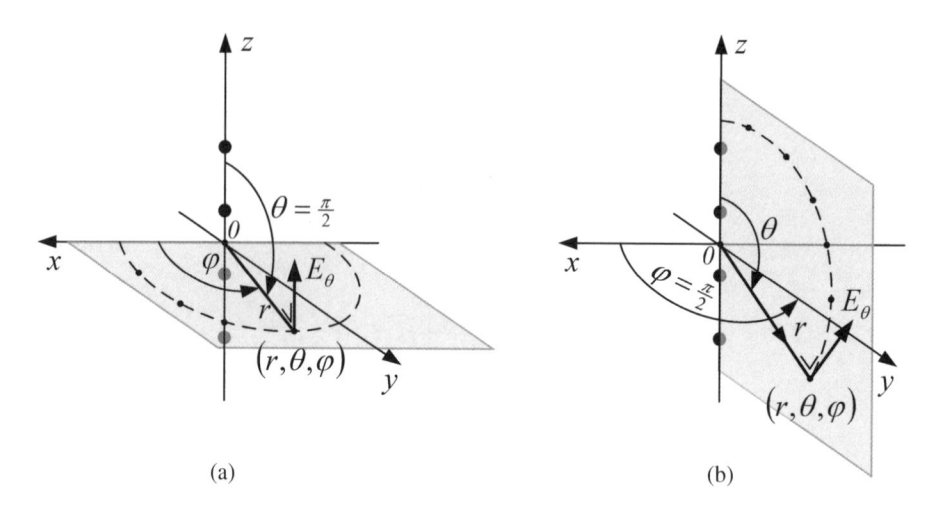

Figure 6.4 Principal planes for the horizontal (a) and vertical (b) pattern determination.

antenna (in dB) in the direction of a mobile terminal, the position of which can be uniquely described by a pair of angles (θ_0, φ_0). The value of the gain can be calculated using the following simple equation:

$$G(\theta_0, \varphi_0)_{[dB]} = G_{0_{[dB]}} + F_V(\theta_0)_{[dB]} + F_H(\varphi_0)_{[dB]} \quad (6.6)$$

where, $G_{0_{[dB]}}$ is the antenna gain, F_V and F_H are the vertical and horizontal relative power patterns of the antenna, respectively (note that all the quantities are given in dB). The antenna model described in Equation (6.6) can be easily implemented in practice, since all the contributing quantities form a set of standard antenna parameters, usually provided by antenna manufacturers.

Upon analysing Equation (6.3), we can draw the conclusion that the separation given in (6.5) is possible only when $g(\theta, \varphi)$ is separable with respect to θ and φ (the array factor does not influence the separation). Unfortunately, vast majority of radiating elements applied in practice have a radiation pattern which is not separable; hence, in such cases the antenna modelling requires the full 3D radiation pattern. A rigorous and efficient antenna modelling could be based on closed-form expressions for $F(\theta, \varphi)$ or $g(\theta, \varphi)$; however, for most practical antenna arrays their radiation patterns are so complex that such expressions are not available at all. Another possibility is to apply a full-wave numerical modelling of the antenna, but such approach is highly ineffective in terms of computational effort, and in addition it cannot always be accomplished. Consequently, the only way of antenna modelling is to use a set of 2D patterns measured at discrete values of φ, i.e $F(\theta, \varphi_i)$, $i = 1 \ldots K$, where K is chosen so that $F(\theta, \varphi)$ can be sufficiently well approximated. In practice, not the whole 3D radiation pattern is required to model antenna performance, and the most important information is contained in the following range of angles (referring to Figure 6.3): $0 \leq \varphi \leq 2\pi$ and $\pi/2 \leq \theta \leq \pi$. The range can be further restricted to values, which are related directly to the shape of the area serviced by the antenna. For example, if we assume that the area has the form of a triangular sector, as illustrated in Figure 6.5, then it suffices to determine the antenna radiation pattern for the following range of angles θ and φ:

$$\frac{\pi - \beta_{sector}}{2} = \varphi_{min} \leq \varphi \leq \varphi_{max} = \frac{\pi + \beta_{sector}}{2} \quad (6.7a)$$

$$\pi - arctg\left(\frac{d}{h_{BS}}\right) = \theta_{min} \leq \theta \leq \theta_{max} \cong \pi \quad (6.7b)$$

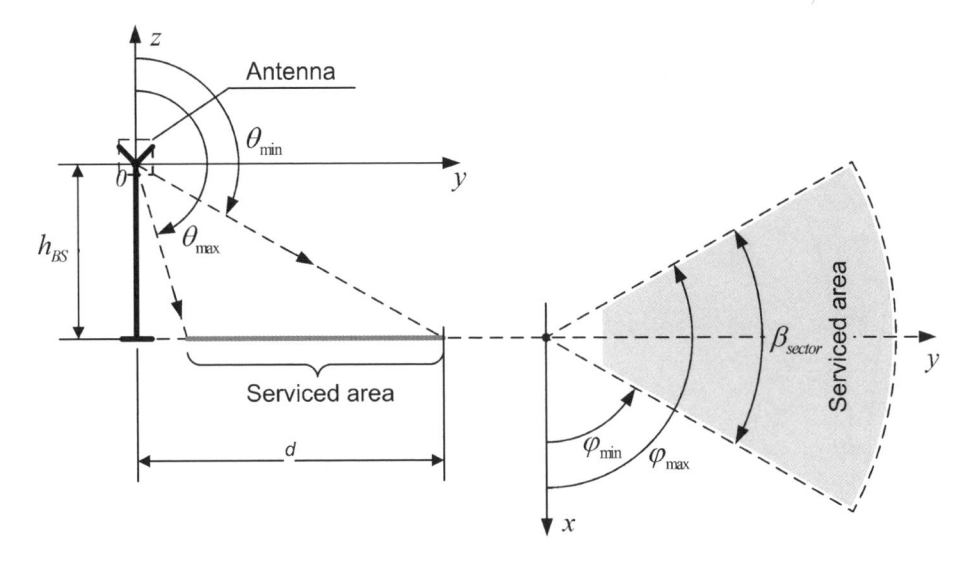

Theoretical Models for UMTS Radio Networks 121

Figure 6.5 Configuration of a sector area serviced by the base station antenna.

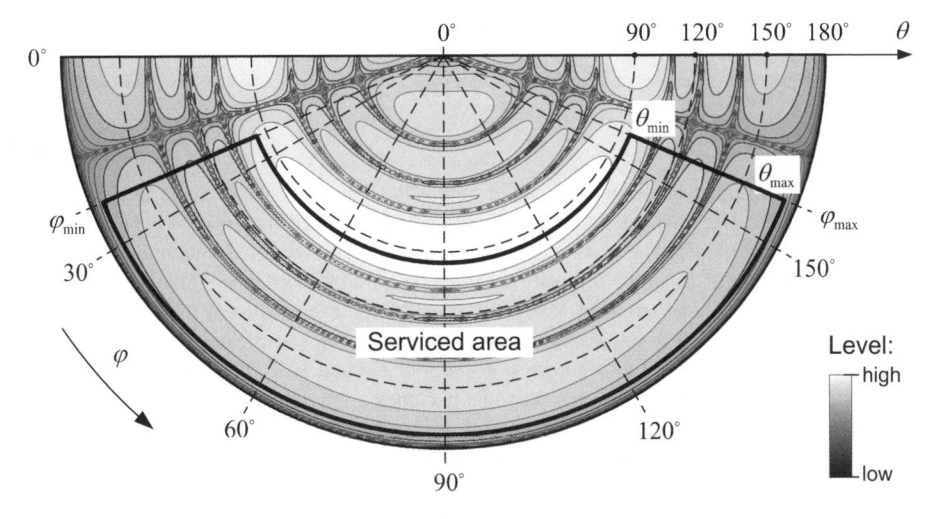

Figure 6.6 Contour plot of the 3D radiation pattern of the base station antenna and the associated serviced area.

The range of angles associated with the serviced area can be greatly illustrated on the contour plot of the 3D radiation pattern of the antenna, as shown in Figure 6.6. The plot has been obtained upon assuming that $F(\theta, \varphi)$ is separable and the principal plane patterns of the antenna are shown in Figure 6.7. The serviced area is bordered by the black thick line, and its shape has been obtained for $h_{BS} = 25\,m$, $d = 400\,m$ and $\beta_{sector} = 140°$.

The contour plot of the 3D radiation pattern is also convenient for considering the following two issues: determination of the base station antenna mechanical tilt as well as determination of antenna radiation towards undesired areas (the main reason for interference generation). The former issue requires some discussion since it may become quite involved. The value of the mechanical tilt can

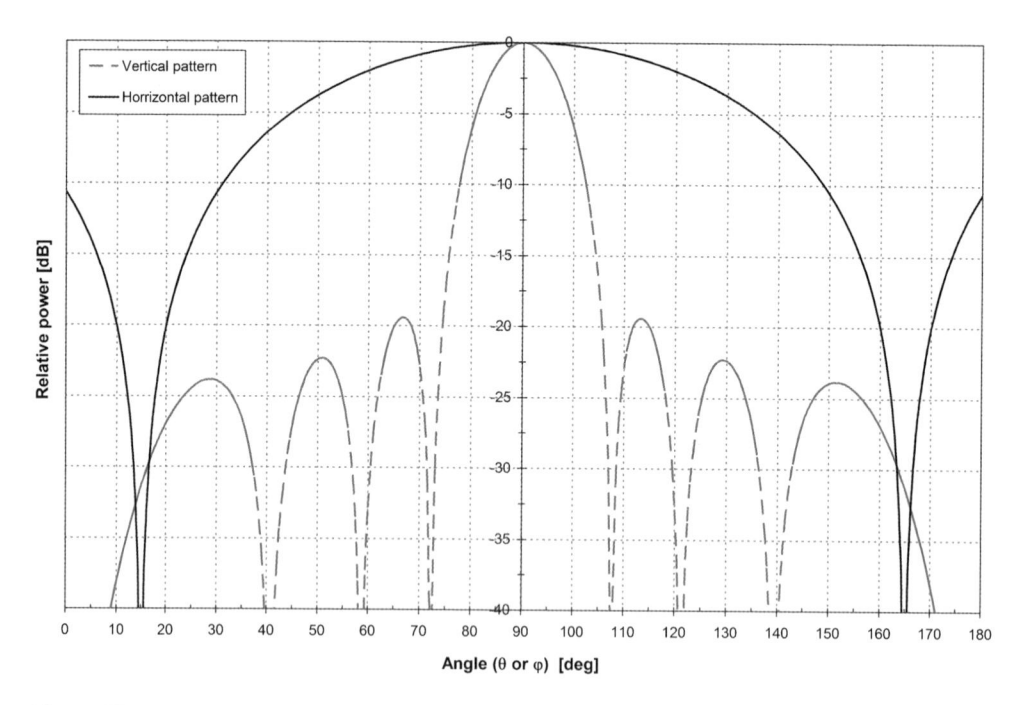

Figure 6.7 The principal plane radiation patterns of the base station antenna used to draw the contour plot in Figure 6.6.

be easily determined upon analysing the pattern along the line of constant angle φ (i.e. $\varphi = 90°$), and choosing the value so that the main lobe of the pattern is aimed towards the central part of the serviced area. This procedure is quite simple and does not require further explanation. The real problem concerns determination of E_θ in the original (not tilted) system of coordinates. The set of 2D antenna patterns, determined (e.g. measured) to model the 3D radiation capabilities of the antenna, describes the electric field in the tilted (rotated around the x-axis) system of coordinates (let us designate it as E_θ'). In order to get E_θ we need to calculate it using both E_θ' and E_φ'. The second component starts to contribute considerably to E_θ when the tilt angle becomes large (larger than 15–20°). For small tilt angles (up to 4–7°), we can recover the information about E_θ using only E_θ' or simply by shifting the whole contour plot down by the value of the tilt angle.

In summary, the base station antenna modelling can be easily accomplished when the radiation pattern of the antenna array is separable with respect to angles θ and φ. When this is not the case, the modelling requires full 3D information about the co-polar (E_θ') and cross-polar (E_φ') radiation pattern, which can be obtained either analytically or empirically.

Having completed antenna modelling, we will now proceed with modelling aspects at the link layer, where both together provide important performance parameters/metrics to any system level simulator.

6.2 LINK LEVEL MODEL

This section summarises the most important issues related to link level modelling in UMTS. Since it is a very well explored topic, we will confine ourselves to a brief summary without repeating generally available results.

6.2.1 RELATION TO OTHER MODELS

As previously mentioned, the exact way of modelling the UMTS network is to simulate it as a whole at every time moment. This would include a full simulation of every aspect related to the radio channel, RF front-end, transmitting and receiving baseband algorithms and associated link level performances, medium access and link control protocols and associated system level performances etc. Such a modelling approach is clearly far too complex and a layered modelling approach with given input/output relationships between the models is usually adopted. An example modelling approach is given below, which incidentally follows the structure of this book:

- radio channel modelling (Chapter 5);
- RF front-end modelling, notably the antennas (Section 6.1);
- link level modelling (Section 6.2);
- static and dynamic system level modelling (Sections 6.4 and 6.5).

The link level model is an important constituent in the entire UMTS system performance simulation process. It links the effects of radio channel and RF front-end to produce a gamut of results useful for the static and dynamic system level simulators. Below, we summarise in greater detail the input/output relationships between the respective models.

An important set of input parameters to the radio channel model is comprised of:

- antenna beamwidth (from antenna model);
- type of environment (from system level model).

All modern communication systems are optimised to the radio channel they operate in; for instance, GSM uses equalisers of some optimised equalisation depth, UMTS uses a Rake receiver with an optimised number of fingers etc. Since modern systems become increasingly dependent on the properties of the radio channel, the radio channel model ought to be as precise as possible.

As outlined in Chapter 5, with above parameters, the radio channel can be modelled to a varying degree of precision by means of deterministic, statistical and/or empirical modelling approaches. Such a modelling yields the following far from complete, but fairly essential, set of input parameters to the link level simulator:

- number of multipath components (MPCs) resolvable at the receiver;
- type of fading statistics of these MPCs (Rayleigh, Rice, Nakagami etc.);
- resulting average power per MPC;
- temporal characteristics of each MPC (fast/slow fading, coherence time etc.);
- statistics and temporal characteristics of shadowing.

In addition, the system level simulator informs the link level simulator about:

- type of radio bearer (TTI, choice of encoder, interleaver etc.);
- type of physical channels (i.e. type of control and data);
- average received signal and interference power (determining the operating SINR point).

With the above set of parameters from the channel and system level simulator, a chosen RF and baseband configuration produces the following set of link level output parameters per control and data channel, which is to be fed into the static and dynamic system level simulations:

- coded (and uncoded) block error rate (BLER) vs. E_b/N_0 and also vs channel PDP;
- the CIR at which a given mode operates at a given error rate;
- cyclic redundancy check (CRC) indicators.

The aim of the link level simulation is hence to quantify the performance of the RF equipment and radio channel in sufficient detail, whereas the role of the system level simulation is to quantify the performance between network elements. Since the link level simulator usually reflects a real RF and baseband system with great precision, it ought correctly be referred to as link level emulator; in contrast, no matter to which degree of sophistication, a system level simulator will always rely on some simplifications.

6.2.2 LINK LEVEL SIMULATION CHAIN

In Figure 6.8, we depict a typical link level simulation chain that is also applicable to systems other than UMTS. Generally, we distinguish four parts, namely:

1. outer modem, which comprises all processing on bit level;
2. inner modem, which comprises all processing on chip/symbol level;
3. RF components, which comprises effects such as sampling etc.;
4. radio channel, which comprises the effects of time-varying MPCs.

For the 3GPP UMTS standard, the four parts can be further detailed as:

1. outer modem:
   - (de-)interleaver;
   - (de-)multiplexing;
   - rate matching;
   - channel en/decoding (Viterbi or Turbo);
   - CRC check.

2. inner modem:
   - synchronisation;
   - root-raised cosine (RRC) matched filter;
   - acquisition (MPC delay estimation and finger assignment);
   - channel estimator;
   - detector (Rake/LMMSE/etc.).

3. RF components:
   - analog/digital converters;
   - nonlinearities of various components;
   - phased locked loop (PLL).

4. radio channel:
   - delayed multipath components according to 3GPP channel models;
   - statistics of each MPC;
   - Doppler spread of each MPC.

Note that the inner and outer modem settings differ for the different UMTS traffic classes, where the exact number procedures can be inferred from the 3GPP standards [9–15]; we will omit further details because they can be found in numerous books, e.g. [16].

We shall, however, briefly dwell on the complexity of above processes, where the most complex are clearly those performed at sample and chip level; it is hence an ultimate link level and PHY layer

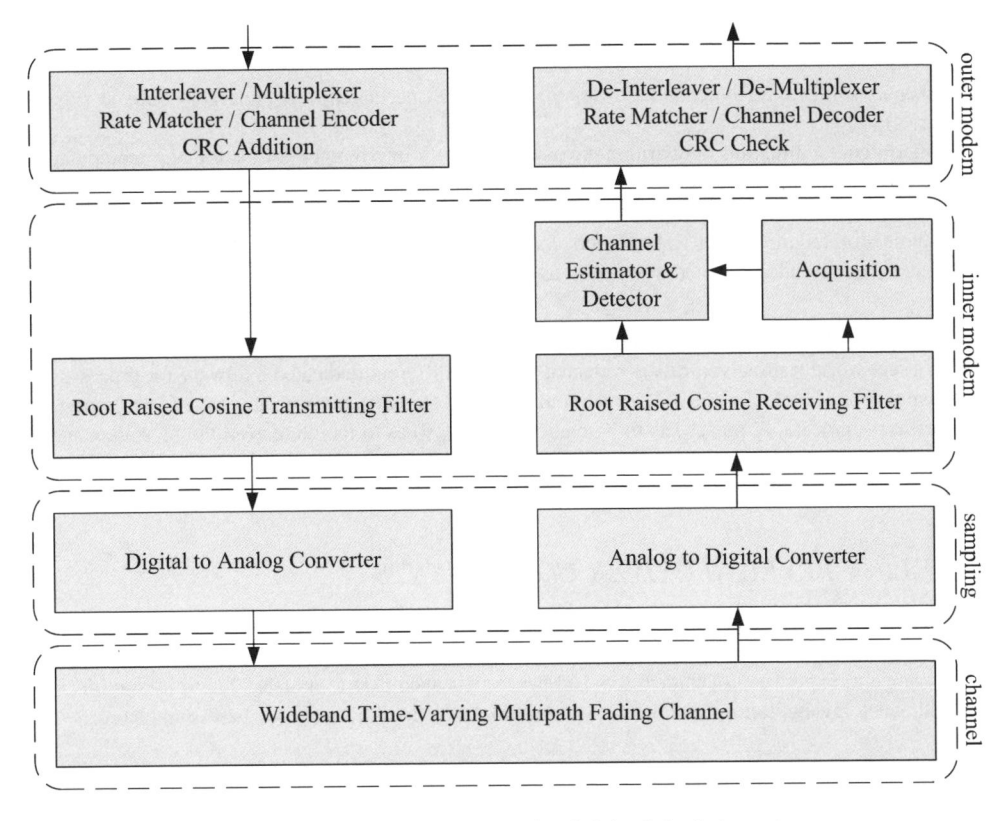

Figure 6.8 Typically simulated transceiver link level simulation structure.

design goal to minimise the amount of processes performed at these levels. With this in mind, it is important to understand which process happens at which sampling rates in a UMTS system:

- continuous:
  - received signal (consists of spread, scrambled and transmit filtered transmit DPCH, CPICH and other control channel signals which undergo a multipath fading channel);
  - RF Rx chain (receive antenna, IF stages, low noise amplifiers, analogue filters, input to ADC);
- samples:
  - filter (output samples of AD converter are fed into matched filter; sampling rate above chip rate for better time resolution at acquisition unit);
  - acquisition (correlates on CPICH to estimate multipath delays; channel estimation should be avoided here!);
  - hold/sample (input to Rake finger has to be delayed by appropriate number of samples and then sampled at chip rate);
- chips:
  - despread (the delayed finger inputs have to be despread and descrambled);
  - integrator (output of despreader has to be integrated/averaged over one symbol);
  - LMMSE (the preferred LMMSE operates at chip level; compared to LMMSE operating at symbol level, complexity is higher but performance drastically better);

- symbols/bits:

  - channel estimation (the channel estimation per finger per antenna is preferably done at symbol level);
  - interference estimation (any arithmetic associated with interference estimation, e.g. computation of cross-correlation matrix);
  - combining (channel compensation according to detection-dependent combining);
  - summation (addition over Rake fingers and receive antennas with adjusted delays);
  - outer modem (all outer modem operations).

Above parts of an UMTS transceiver vary in sampling, memory, load/store and processing complexity. In a real-world transceiver, this is reflected by using different dedicated hardware for processes of different complexity. As for the link level simulator, the different sampling rates have to be catered for and sufficient processing power has to be made available, so as to facilitate even the most demanding parts of the link level transceiver architecture to be simulated.

6.2.3 LINK LEVEL RECEIVER COMPONENTS

The most important part of a link level simulator is clearly the receiver chain, because significant gains, but also failures, can be achieved here. This part is also usually not standardised, allowing for a fair competition between manufacturers and hence improved user terminals. Due to its significance, we will hence review various components of the link level simulator receiver baseband chain.

6.2.3.1 RRC Matched Filter

The impulse response of the chip-matched root-raised cosine (RRC) filter is:

$$RC_0(t) = \frac{\sin\left(\pi \frac{t}{T_c}(1-\beta)\right) + 4\beta \frac{t}{T_c} \cos\left(\pi \frac{t}{T_c}(1+\beta)\right)}{\pi \frac{t}{T_c}\left(1 - \left(4\beta \frac{t}{T_c}\right)^2\right)} \qquad (6.8)$$

where T_c is the chip duration and the roll-off factor is $\beta = 0.22$. The RRC matched filter is traditionally simulated with a finite number of taps.

6.2.3.2 Acquisition

One of the most crucial entities in a WCDMA receiver is the unit which identifies the delays of the impinging MPCs and assigns the Rake fingers accordingly; this entity is henceforth referred to as the *searcher*.

In the downlink, the searcher traditionally obtains delays from the CPICH which is continuously transmitted at higher power level and in parallel to the data streams. For the simulator it is important to know how many pilot symbols are actually used as well as the sample rate, length of the cross-correlation window and other factors.

During power up, loss of session or hand-over, the searcher clearly has to perform a complete update on channel delays and Rake assignments, henceforth referred to as *full channel acquisition*. On the other hand, once the modem is operational, an incremental update with shorter correlation windows and hence lower complexity suffices, and is henceforth referred to as *partial channel acquisition*. The frequency at which either method is performed will depend on the communication scenario and eventually on the manufacturer.

There are numerous techniques to estimate the channel delays and hence influence the Rake receiver performance, where performance is usually traded against complexity. We have chosen to describe an iterative-type channel acquisition process without any further signal processing techniques, such as weighted multi-slot averaging, etc. Also, the delay estimation is only performed once per antenna since the delay profile will not differ among the receiving antenna elements.

As for the full channel acquisition, given the received signal that is supposed to contain the CPICH and the codes that were used to spread and scramble the pilot symbols, the channel estimator in the link level simulator must compute the following:

- number of valid paths;
- delays of the valid paths, with respect to the beginning of the slot;
- complex attenuation of the valid paths for MPC cancellation and estimation purposes, which is traditionally performed by means of several iterations of computation of the cross-correlation between the received signal and a local copy of the *ideal* received signal as if no multipath propagation had occurred;
- selection of the delay that yields the highest magnitude correlation peak;
- estimation of the identified path's complex attenuation;
- validation of the delay and complex attenuation;
- suppression of a weighted local copy of the ideal pilot signal to allow identification of the *next* strongest path with the same process.

This algorithm is considered to be of average performance compared to other available acquisition techniques.

As for partial channel acquisition, it is sometimes preferable to lower the overall link level simulator complexity to perform only a simple form of MPC estimation known as *tracking*, or *partial channel acquisition*. Given the channel model that was estimated for a previous slot, the tracking module in the simulator will only adjust the delays and complex attenuations for the current slot, without having to deal with a cross-correlation on the whole search window length. For each previously identified path, the tracking module adjusts the delay and the complex attenuation by computing the cross-correlation over a window that is about one chip long.

6.2.3.3 Channel Estimator

Channel estimation has to be performed per resolved multipath and per antenna; therefore, the estimation process itself has to be done at the lowest possible rate. There are two main options of implementing the channel estimating into the link level simulation platform, i.e. using:

1. CPICH channel (extra operations at chip level due to despreading operation, but much higher reliability); and/or
2. the pilot symbols embedded into the slot (no extra chip level operations but less reliability).

The natural choice is to use the CPICH channel since it is transmitted with higher power than the data channels and hence audible with a higher reliability. There are situations, however, when channel estimation from the slot pilots becomes an attractive option, which is when the terminal is at the cell edge; this is because the data channels are power controlled, whereas the CPICH is not.

6.2.3.4 Power Control

Power control ought to be implemented into the link level simulator, because different physical traffic and control channels are transmitted at different power levels and hence cause mutual multiple access interference of different magnitude. One has the choice:

- to assume ideal error-free power control;
- to assume ideal power control with an additive (Gaussian) error component or
- to simulate the erroneous power control in both up and downlinks, which requires some additional system level input, such as path loss, shadowing etc.

6.2.4 LINK LEVEL RECEIVER DETECTORS

The detector is probably the most important component of a link level simulator, which is why we have dedicated an entire subsection to it. The role of the detector is to obtain the symbol stream to be decoded from the chip stream with the aid of the side-information provided by the searcher. The theory behind detection and estimation processes is vast and well explored, with a gamut of available detectors trading complexity against performance [17].

Clearly, the optimum choice of any detection is the Maximum Likelihood (ML) detector which searches through all possible transmitted sequences and decides thereupon on the most likely transmitted sequence. The metrics which is to be maximised is $\eta = \|x - Fs\|_{R_x^{-1}}^2$, where x is the incoming data stream, F is a detection filter including the spreading and scrambling codes, s is the candidate signal stream with the most likely to be chosen and R is the temporal-spatial correlation matrix. No need to say that, with an exponential dependency of complexity on sequence length and number of bits per symbol, such detector is prohibitively complex. Less complex detectors with the caveat of inferior performance are hence used in analysis, simulations and practice.

6.2.4.1 SISO MRC Rake Combiner (Baseline)

The simplest of all realisations, and henceforth the baseline of our investigations, is the single input single output (SISO) antenna Rake receiver, based on the maximum ratio combining (MRC) principle. Such a receiver is classified as a linear receiver which ignores multiple access interference (MAI) in the detection process.

A classical Rake architecture is depicted in Figure 6.9, where the receiver RF chain ends with an analog-to-digital converter which provides the signal stream at sampling rate. The stream is then receiver matched filtered by means of a root-raised cosine filter. The filtered stream is then fed into the acquisition unit, which estimates the relative delays between the MPCs at sample resolution. The acquisition unit then informs each Rake finger about the delays, which then can delay the incoming stream appropriately and sample it at chip rate. Thereafter, the de-spreading and de-scrambling operation are performed at chip-level for the data stream and the pilot stream, both of which are then averaged over all chips per symbol. The CPICH symbols are utilised to estimate the complex channel coefficients, which are used to calculate the combining weight coefficients w according to the MRC algorithm, i.e. $w = h$, where h is the estimated channel coefficient on the channel path of interest. The complex conjugate of the combing weight w^* is then multiplied with the de-spread data stream at symbol rate. Finally, the outputs of the Rake fingers are added and fed into the symbol estimators and the outer modem.

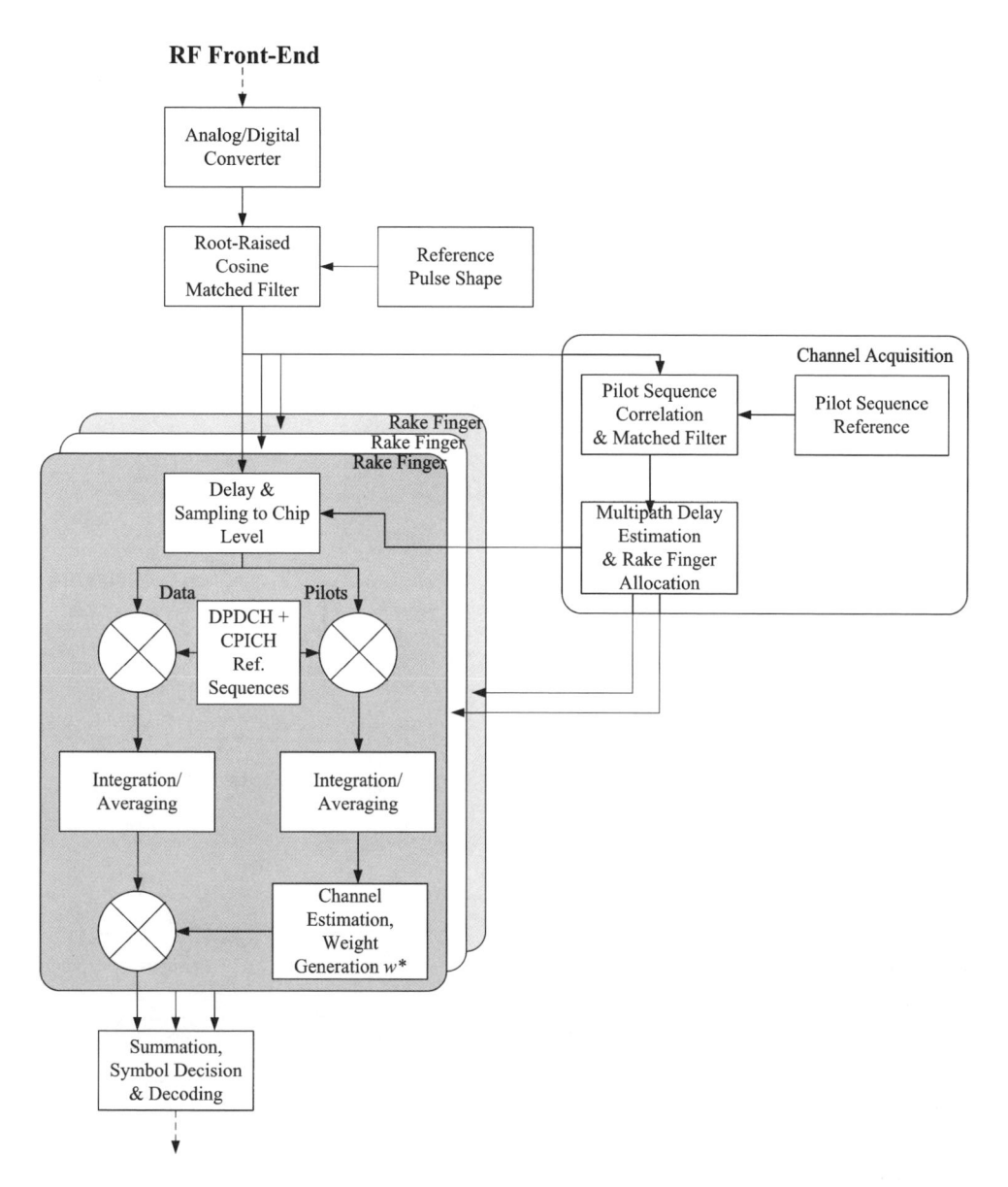

Figure 6.9 Link level simulator building blocks for SISO MRC Rake combiner.

6.2.4.2 SIMO MRC Rake Combiner

An extension of the above SISO MRC is the receiver with multiple antenna elements, as catered for by the 3GPP. Again, such a receiver is classified as a linear receiver which ignores MAI in the detection process; however, the detection process itself is enhanced by the provision of a spatial domain. Some typical building blocks for a link level simulator are shown in Figure 6.10.

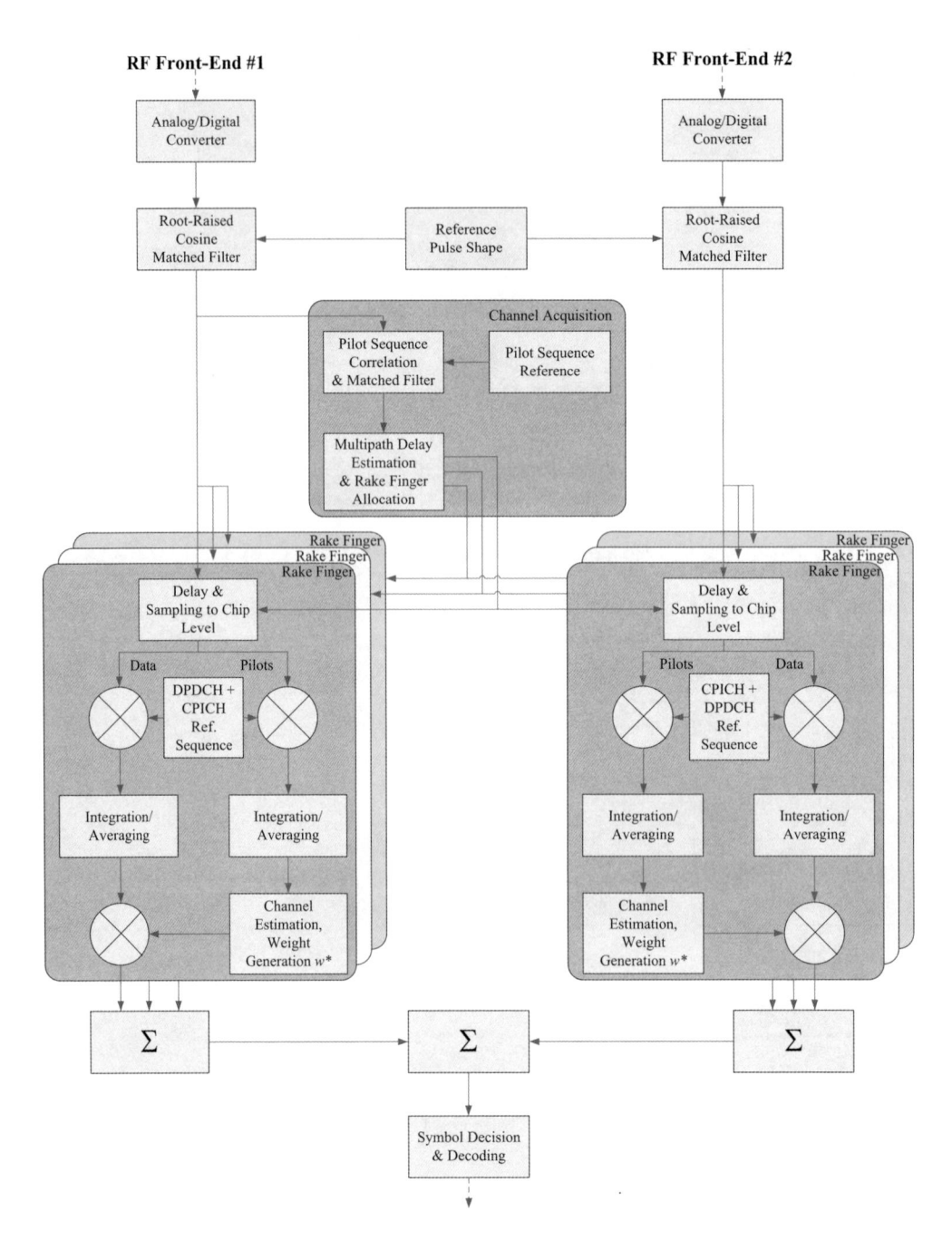

Figure 6.10 Link level simulator building blocks for SIMO MRC Rake combiner (example of 2 Rx antennas).

The changes with respect to the baseline SISO MRC receiver are as follows:

- The acquisition unit in the simulator can theoretically estimate the delays of the MPCs from one receive antenna only, as the power delay profile (PDP) can be assumed to be the same between the antennas; however, the estimates can be made more reliable using more than one receive antenna.
- Each finger of the Rake receiver has to de-spread the stream of each receive antenna separately, so as to facilitate the channel estimation and weight estimation for each receive antenna separately, where $w_n = h_n$ and h_n is the estimated channel coefficient of the n-th receive antenna; finally, the contributions of each antenna need to be added within each finger.

6.2.4.3 SIMO IRC Rake Combiner

An algorithmic extension of the above SIMO MRC is a receiver for SIMO Interference Rejection Combining (IRC), which requires the existence of multiple receive antennas. Again, such a receiver is classified as a linear receiver which now mitigates multiple access interference (MAI) in the detection process by means of additionally available spatial domains [18]. This generally requires some additional signal processing, such as calculation of correlation matrices and inverses thereof.

The general IRC receiver architecture is the same as depicted in Figure 6.11 with the following changes with respect to the baseline SISO MRC receiver:

- The acquisition unit in the simulator can theoretically estimate the delays of the MPCs from one receive antenna only, as the PDP can be assumed to be the same between the antennas; however, the estimates can be made more reliable using more than one receive antenna.
- Each finger of the Rake receiver has to de-spread the stream of each receive antenna separately, so as to facilitate the channel estimation and weight estimation for each receive antenna separately, where the weight vector for the channel path of interest is given by $\boldsymbol{w} = \mathbf{R}^{-1}\boldsymbol{h}$ with $\boldsymbol{h}$ the estimated channel coefficient vector on the channel path of interest and $\mathbf{R}$ is the spatial correlation matrix; finally, the contributions of each antenna need to be added within each finger.
- The spatial correlation matrix $\mathbf{R}$ is based on the incoming signal chip streams and ought to be calculated at chip level; it is the same (up to a multiplicative constant) for all the channel paths delayed by a multiple of the chip duration.

6.2.4.4 SISO LMMSE Combiner

A more complex receiver, albeit classified as linear, is the linear minimum mean square error (LMMSE) receiver which is a Bayesian detector modelling the symbols as zero mean, independent, identically distributed random variables. A typical way to implement the LMMSE detector into the link level simulator is to equalise the signal stream by means of a finite impulse response (FIR) filter prior to descrambling and despreading. In the SISO case, the LMMSE exploits the temporal properties of the received signal stream only.

The changes with respect to the baseline SISO MRC receiver are as follows:

- The channel estimation has to be performed at chip level.
- The covariance matrix of the received signal has to be calculated at chip level.
- An equalisation FIR filter has to be implemented instead of the Rake.

6.2.4.5 SIMO LMMSE Combiner

The temporal LMMSE receiver can be extended to the spatial domain, where now an estimation is jointly performed over the temporal and spatial domains. The simulation complexity clearly grows

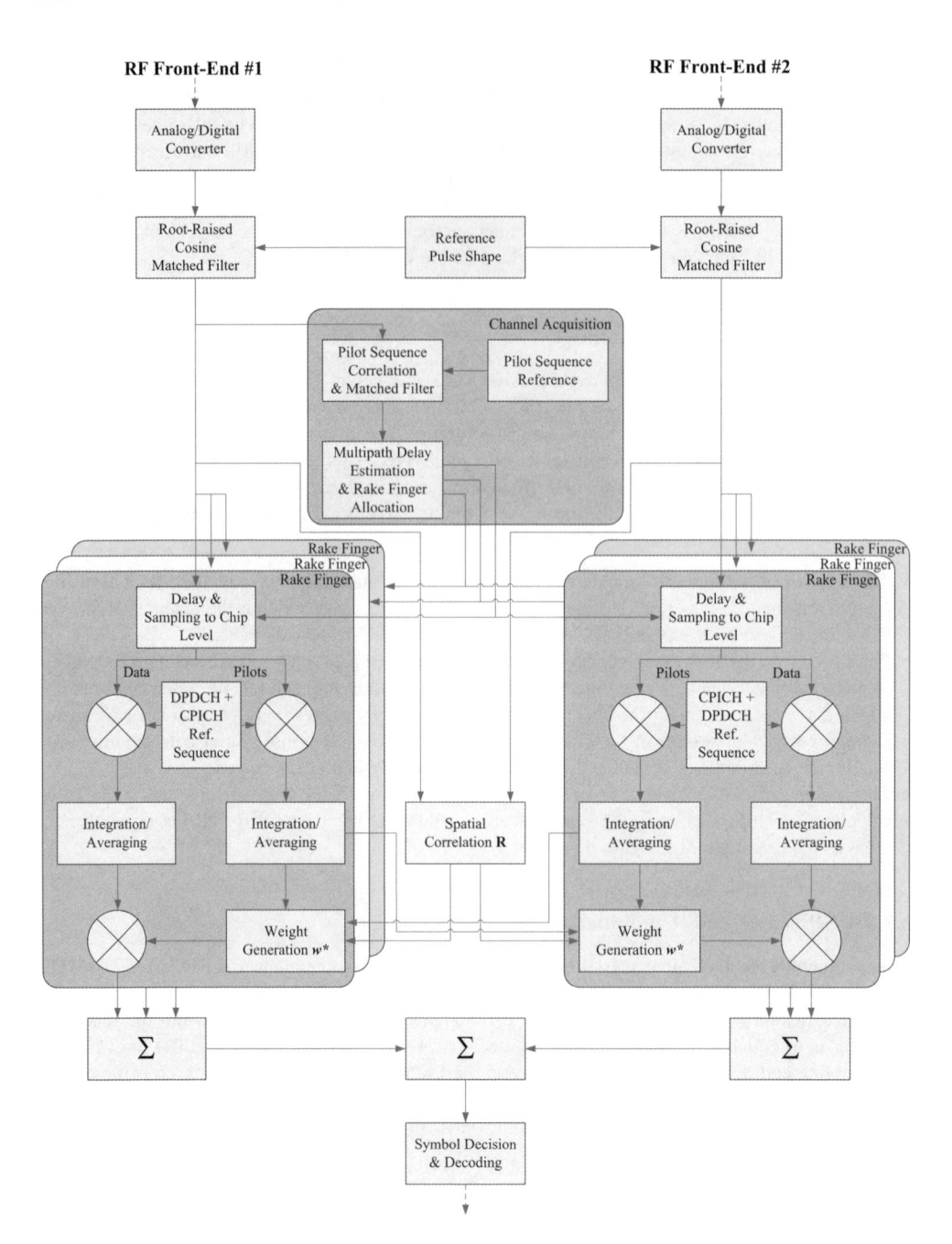

Figure 6.11 Link level simulator building blocks for SIMO IRC combiner (example of 2 Rx antennas).

exponentially with the additional degrees of freedom available, which is the reason why sub-optimal solutions have been introduced [19,20]. In order to minimise the simulation and implementation complexity, we depict in Figure 6.12 a simulation architecture which simply sums the outputs of the temporal chip-level equalisers applied independently on each antenna, each equaliser being simulated as a low-complexity method described in [21].

The changes in the link level simulation platform with respect to the SISO LMMSE receiver are as follows:

- One chip-level equaliser has to be computed and implemented for each antenna.
- The equalisers' outputs are summed before descrambling and despreading.

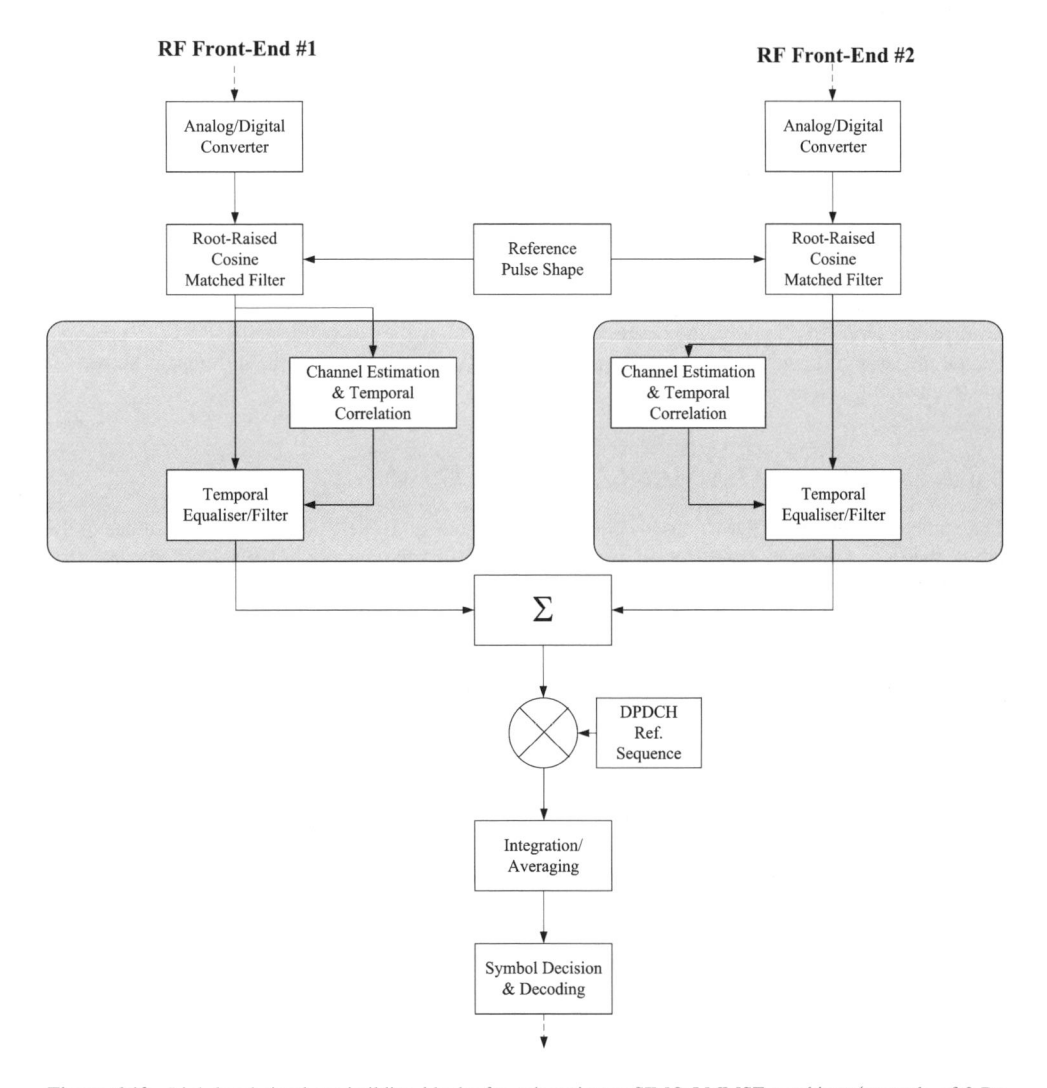

Figure 6.12 Link level simulator building blocks for sub-optimum SIMO LMMSE combiner (example of 2 Rx antennas).

6.2.4.6 SIMO MPIC Combiner

More complex signal detectors can be simulated, such as the multipath interference cancellation receiver which cancels multipath interference in an iterative manner [22]. However, we will omit such architectures here, because they require a great detail of description.

To conclude this section, with the above given link level simulator building blocks comprised of outer modem, inner modem, RF components and the radio channel, the required link level performance metrics, i.e. BLER, can be produced and passed on to the system level simulator discussed in Sections 6.4 and 6.5.

6.3 CAPACITY CONSIDERATIONS

This section is concerned with analytical considerations related to some issues involved in system level modelling of a CDMA based cellular system. We will start with the analysis of a single isolated cell in the static state. Such an approach enables a closed-form analytical solution to the problem of distributing optimal power among all users of the system. Based on the basic CDMA theory, a formula for the *capacity* of a single cell will be derived. The formula is based on limits imposed by the base and mobile station total available transmit power. The theory developed in this section allows us to investigate, in a relatively simple manner, boundary states of the system and to determine the influence of the system parameters on the actual capacity. Since capacity considerations are fundamental to CDMA system planning and operation, these simple closed-form expressions can be very useful, especially for CAD and planning tools to estimate, for example, initial values of system level parameters. Additionally, the single cell approach allows us to better understand the principle behind the issue of power distribution and power control, and to understand the differences between a single and multi-cell case.

6.3.1 CAPACITY OF A SINGLE CELL SYSTEM

The number of users of a CDMA based cellular system is inherently self-interference limited due to the use of the same frequency bands for all users and the loss in orthogonality of the spreading codes (see Section 2.2). However, the destructive influence of interference can be minimised by controlling the power level of each signal so that it arrives at the intended receiver, i.e. at base station (BS) or mobile station (MS), with at least the minimum required signal-to-interference ratio (SIR). Unfortunately, each admission of a new user to the system provokes an increase of the signal level, which starts to tend very fast to infinity beyond a certain number of users. The number of users cannot increase ad infinitum, since it is bounded by the so-called pole capacity of the system; see Chapters 2, 9 or [23]. The maximum supportable number of users can be restricted by the power-limited capacity, on the one hand, and the number of available spreading codes, on the other. In practice, however, the former restriction is more important than the latter one, and hence in this section we will consider only the power-limited system capacity.

A simple expression for the estimation of the CDMA downlink power-limited capacity may be found, for example, in [24]. In this section, an exact formula for uplink and downlink system capacity will be derived, and the derivation is based on the basic CDMA theory and uses, to a certain extent, the approach described in [25], [26] and [27], which has been successfully exploited and improved by many researchers.

6.3.2 DOWNLINK POWER-LIMITED CAPACITY

In order to determine the power-limited system capacity on the downlink, let us analyse a single isolated cell of a cellular CDMA system at a given instant of time (in a static state), as shown in

Theoretical Models for UMTS Radio Networks

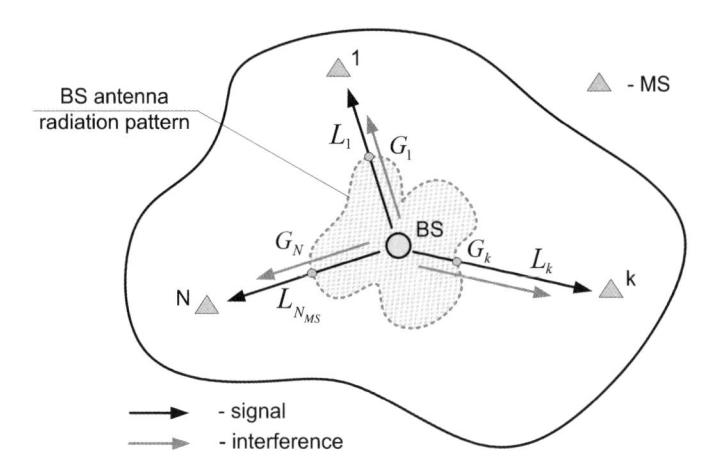

Figure 6.13 A single isolated cell with N_{MS} users (downlink case).

Figure 6.13. The base station allocates power P_k to the k-th user admitted to the cell. A signal of power P_k is also transmitted towards all the other users resulting in interference. Assuming that the antenna gain of each mobile station is equal to unity (0 dB), the power of the signal received at the k-th MS receiver can be expressed as:

$$C_k = P_k G_k L_k \tag{6.9}$$

where P_k is the transmitted power allocated on the BS, G_k stands for the BS antenna gain in the direction of the k-th user, and L_k is the path loss between the BS and the k-th user. On the other hand, all the 'desired' signals transmitted to the rest of the users are also transmitted to user k along the same path (i.e. $G_k L_k$) and must be treated, obviously, as interference. Thus, the total power of interfering signals can be expressed as:

$$I_k = G_k L_k \cdot \sum_{\substack{j=1 \\ j \neq k}}^{N_{MS}} P_j \tag{6.10}$$

where P_j is the transmitted power allocated to the BS for the j-th user.

The 'desired' signal that reaches user k must meet the required SIR or carrier-to-interference (CIR) ratio level. For a given value of CIR, the signal level at the k-th MS receiver must meet the following relationship:

$$\forall k \in \{1, \ldots, N_{MS}\} \quad C_k = \left(\frac{E_b}{N_0 + I_0}\right)_k \cdot R_{b_k} \cdot \left(N_0 + (1-\alpha) \cdot \frac{I_k}{B}\right) \tag{6.11}$$

where for each k, $E_b/(N_0 + I_0)$ is the required ratio of bit energy to noise and interference power spectral density, R_b is the data transmission rate, B stands for the communication channel bandwidth (the total spread bandwidth) and α is the average orthogonality factor of the spreading codes used. Upon substituting Equations (6.9) and (6.10) into Equation (6.11) and assuming that the transmission parameters are identical for all users, which generally is not a restrictive assumption, we arrive at:

$$\forall k \in \{1, \ldots, N_{MS}\} \quad w \cdot P_k G_k L_k - G_k L_k \cdot \sum_{\substack{j=1 \\ j \neq k}}^{N_{MS}} P_j = D_p \tag{6.12}$$

where

$$w = \frac{B}{\left(\frac{E_b}{N_0+I_0}\right) \cdot R_b \cdot (1-\alpha)} \quad (6.13a)$$

$$D_p = \frac{B \cdot N_0}{1-\alpha} \quad (6.13b)$$

Using Equation (6.12) we are able to determine the total power allocated to all users, as well as the power allocated to a given single user. The respective equations take the following form:

$$P_{tot} = \sum_{k=1}^{N_{MS}} P_k = \frac{D_p}{w+1-N_{MS}} \cdot \sum_{k=1}^{N_{MS}} \frac{1}{G_k L_k} \quad (6.14)$$

and

$$\forall k \in \{1, \ldots, N_{MS}\} \quad P_k = \left(P_{tot} + \frac{D_p}{G_k L_k}\right) \cdot \frac{1}{w+1} \quad (6.15)$$

where $w+1$ is equal to the pole capacity (N_{pole}) of the system [23].

An inspection of Equation (6.14) reveals that the total allocated power (P_{tot}) does not depend on the actual distribution of the users over the cell, but on an 'average' path loss (i.e. $G_a L_a$) resulting from their distribution. Hence, if we assume that

$$\sum_{k=1}^{N_{MS}} \frac{1}{G_k L_k} = N_{MS} \cdot \frac{1}{G_a L_a} \quad (6.16)$$

we can rewrite Equation (6.14) in the following form:

$$P_{tot} = \sum_{k=1}^{N_{MS}} P_k = \frac{D_p}{G_a L_a} \cdot \left(\frac{N_{MS}}{N_{pole} - N_{MS}}\right). \quad (6.17)$$

In order to derive an equation for the power-limited capacity of a single cell CDMA system, we may analyse two distinct criteria based on the available transmit power: the first one, concerning the total available transmit power that can be allocated for all users, and the second one, the available transmit power per single user. In the following, however, we shall confine ourselves only to the first criteria due its practical importance. Consequently, the power-limited capacity of the system can be determined using the following equation:

$$P_{tot} = \sum_{k=1}^{N_{MS}} P_k \leq P_{ta}^{BS} \quad (6.18)$$

where P_{ta}^{BS} is the total available transmit power on the BS. Substitution of Equation (6.17) into Equation (6.18) yields:

$$N_{max}^d \leq N_{pole} \cdot \frac{1}{1 + \left(\frac{D_p}{P_{ta}^{BS} \cdot G_a L_a}\right)} \quad (6.19)$$

where N_{max}^d is the maximum number of users on the downlink, which can be admitted to the cell serviced by the BS with a given P_{ta}^{BS}. Upon analysing Equation (6.19) we can clearly see that the upper bound to N_{max}^d is the pole capacity of the system (N_{pole}), and this fact results in a straightforward way from the above equation (i.e. when we allow $P_{ta}^{BS} \to \infty$).

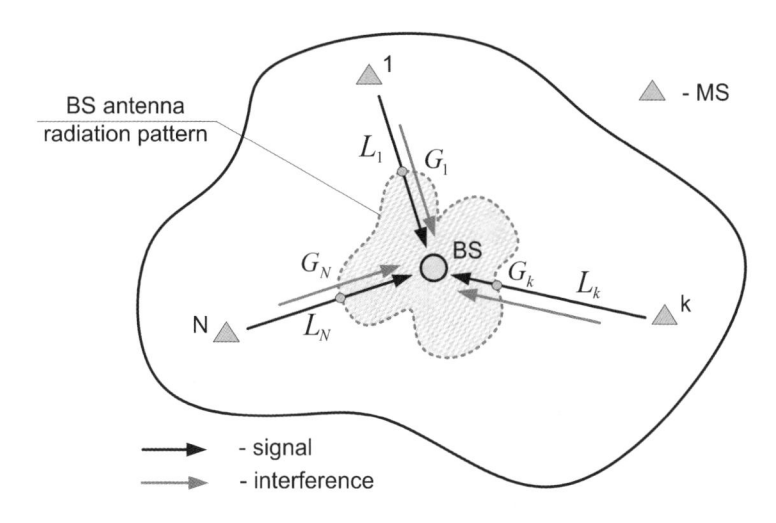

Figure 6.14 A single isolated cell with N_{MS} users (uplink case).

6.3.3 UPLINK POWER-LIMITED CAPACITY

To investigate the question of the uplink power-limited capacity, we shall follow the approach of the previous section. Consequently, let us analyse a single isolated cell of a cellular CDMA system at a given instant of time for the uplink case, as shown in Figure 6.14.

The power of the signal received at the k-th BS receiver is described by Equation (6.9). The signal transmitted by user k towards the BS and intended for its k-th receiver is also experienced by all the remaining BS receivers. This process results, obviously, in interference, but this time all interfering signals are transmitted along distinct paths (i.e. $G_j L_j$, $j \neq k$). Therefore, the total power of the interfering signals received at the k-th BS receiver can be expressed as:

$$I_k = \sum_{\substack{j=1 \\ j \neq k}}^{N_{MS}} P_j G_j L_j \qquad (6.20)$$

where P_j stands for power transmitted by the j-th user. Upon substituting Equations (6.9) and (6.20) into Equation (6.11), and assuming again that the transmission parameters are identical for all users, we get:

$$\forall k \in \{1, \ldots, N_{MS}\} \quad w \cdot P_k G_k L_k - \sum_{\substack{j=1 \\ j \neq k}}^{N_{MS}} P_j G_j L_j = D_p \qquad (6.21)$$

where w and D_p are described by Equations (6.13), provided the average orthogonality factor of the spreading codes is equal to zero ($\alpha = 0$).

Using Equation (6.21) we can derive the following relations:

$$\sum_{k=1}^{N_{MS}} P_k G_k L_k = \frac{N_{MS}}{w + 1 - N_{MS}} \cdot D_p \qquad (6.22)$$

$$\forall k \in \{1, \ldots, N_{MS}\} \quad P_k G_k L_k = \frac{1}{w + 1 - N_{MS}} \cdot D_p \qquad (6.23)$$

and

$$\forall k \in \{1, \ldots, N_{\text{MS}}\} \quad \sum_{\substack{j=1 \\ j \neq k}}^{N_{\text{MS}}} P_j G_j L_j = \frac{N_{\text{MS}} - 1}{w + 1 - N_{\text{MS}}} \cdot D_{\text{p}} \qquad (6.24)$$

The ratio of Equations (6.23) and (6.24) equals the CIR at the input of the k-th BS receiver, and depends only on the total number of users admitted to the cell:

$$\forall k \in \{1, \ldots, N_{\text{MS}}\} \quad \left(\frac{C}{I}\right)^{\text{BS}} = \frac{P_k G_k L_k}{\sum_{\substack{j=1 \\ j \neq k}}^{N_{\text{MS}}} P_j G_j L_j} = \frac{1}{N_{\text{MS}} - 1} \qquad (6.25)$$

Making use of Equations (6.16) and (6.21)–(6.23) we can write:

$$P_{\text{tot}} = \sum_{k=1}^{N_{\text{MS}}} P_k = \frac{D_{\text{p}}}{G_{\text{a}} L_{\text{a}}} \cdot \left(\frac{N_{\text{MS}}}{N_{\text{pole}} - N_{\text{MS}}}\right) \qquad (6.26)$$

and

$$\forall k \in \{1, \ldots, N_{\text{MS}}\} \quad P_k = \frac{1}{N_{\text{pole}} - N_{\text{MS}}} \cdot \frac{D_{\text{p}}}{G_k L_k} \qquad (6.27)$$

where N_{pole} is the pole capacity of the system in the uplink, which is, obviously, equal to $w + 1$. As we can see, Equation (6.26) has exactly the same form as (6.17) for P_{tot} in the downlink case. Nevertheless, each of them yields a different result since the uplink and downlink pole capacity are different, in general.

In the present case, the power-limited capacity of the system can be determined from the following equation:

$$\forall k \in \{1, \ldots, N_{\text{MS}}\} \quad P_k \leq P_{\text{a}}^{\text{MS}} \qquad (6.28)$$

where P_{a}^{MS} is the total available transmit power of the MS terminal. Introducing Equation (6.27) into Equation (6.28) yields:

$$\forall k \in \{1, \ldots, N_{\text{MS}}\} \quad N^{(k)} \leq N_{\text{pole}} - \frac{D_{\text{p}}}{P_{\text{a}}^{\text{MS}} \cdot G_k L_k} \qquad (6.29)$$

Equation (6.29) reveals that the power-limited capacity is limited, in fact, by the 'worst case' path loss. Therefore, the maximum number of users, which can be admitted to the cell on the uplink, can be determined as follows:

$$N_{\text{max}}^{\text{u}} = \min_{k=1, \ldots, N_{\text{MS}}} \{N^{(k)}\} \qquad (6.30)$$

Finally, it can be stated that the power-limited capacity of a single cell CDMA system on the uplink depends on the 'worst case' path loss, i.e. $\min\{G_k L_k\}$, so we can write:

$$N_{\text{max}}^{\text{u}} \leq N_{\text{pole}} - \frac{D_{\text{p}}}{P_{\text{a}}^{\text{MS}} \cdot \min_{k=1, \ldots, N_{\text{MS}}} \{G_k L_k\}} \qquad (6.31)$$

The upper limit to N_{max}^u is the pole capacity N_{pole}, which results in a straightforward way from the above equation, e.g. when we allow $P_a^{MS} \to \infty$.

The analytical expressions derived in this section describe all important characteristics of the single-cell CDMA-FDD system and allow us finally to determine its capacity. In general, basic characteristics of the multi-cell system are similar, to a certain degree, to those of the single-cell system, and some of them can be treated as a generalisation of characteristics described in this section. This fact will become more obvious after reading the subsequent sections of the book.

6.4 STATIC SYSTEM LEVEL MODEL

This section is focused on FDD UMTS radio network as a whole, i.e. as a complex system. Here, dynamic aspects are neglected and a *static* system level model is developed. In Sections 6.4.1–6.4.6, the various aspects of a UMTS system that are relevant to a static network model are treated. It is shown how they can be taken into account in the system model. In Section 6.4.7, the explicit system model is given. Section 6.4.8 then indicates how a static system model can be used – notably for *Monte-Carlo Simulation* – and discusses its limitations. For practical implementations, a more efficient formulation given in Section 6.4.9 is normally used, in which individual users are treated implicitly rather than explicitly. Methods for solving the model are provided in Section 6.4.10.

There is no standard notation for radio network modelling. The symbols used in this book are introduced throughout this section, Table 6.1 gives a complete overview. A UMTS radio network is composed of a set $\mathcal{N}$ of antennas. Each antenna is configured individually, the ensemble of the antennas constitutes the entire system. For assessing the performance of the network and potential issues, the *loaded* network has to be considered. In the following, the underlying concepts are introduced.

Table 6.1 Notation.

Symbol	Domain	Description
$\mathcal{N}$		Set of antenna installations (cells) in the network
$\mathcal{M}$		Set of mobiles
$\mathcal{M}_i$		Mobiles served by antenna installation i
m	$m \in \mathcal{M}$	Mobile
S		Set of services
A		Network area
$P_m^\uparrow$	$\in \mathbb{R}_+$	Uplink transmit power from mobile m
$P_{im}^\downarrow$	$\in \mathbb{R}_+$	Downlink transmit power from installation i to mobile m
P_i^{CPICH}	$\in \mathbb{R}_+$	(Downlink) pilot transmit power from installation i
$P_i^{otherCCH}$	$\in \mathbb{R}_+$	(Downlink) common channels transmit power from installation i
$\bar{P}_i^\downarrow$	$\in \mathbb{R}_+$	Total transmit power of installation i
$\bar{C}_i^\uparrow$	$\in \mathbb{R}_+$	Total received power at installation i
$P_{max,i}^{BS}$	$\in \mathbb{R}_+$	Maximum feasible output power for installation i
$L_{mi}^\uparrow$	[0,1]	Uplink attenuation factor between mobile m and installation i
$L_{im}^\downarrow$	[0,1]	Downlink attenuation factor between installation i and mobile m
N_i, N_m	≥ 0	Noise at installation i/mobile m
$\nu_m^\uparrow, \nu_m^\downarrow$	[0,1]	Uplink/downlink activity factor of mobile m
α_m	[0,1]	Orthogonality factor for mobile m, $\bar{\alpha} := 1 - \alpha$
$CIR_m^\uparrow, CIR_m^\downarrow$	≥ 0	Uplink/downlink CIR target for mobile m
CIR_m^{CPICH}	≥ 0	Pilot CIR requirement for mobile m
$\gamma^{(SHO)}$	$\geq 0\,dB$	SHO window

User Snapshot Under the static point of view, time 'freezes' at an arbitrary instance; the task is to analyse the network's state in the given situation. The concept of load for the static model is hence that of a *snapshot*. A snapshot consists of a set $\mathcal{M}$ of users having individual properties. These properties include user location, the service the user accesses and the type of mobile equipment. The static approach inherently ignores dynamic effects that influence the system. However, dynamic influences can be taken into account to a limited degree. Methods to do so are treated later in this section.

System of Linear (In-) Equalities For a signal to be successfully decoded at the receiver with WCDMA technology, the ratio of the received strength of the desired signal to all interfering signals – including noise interior and exterior to the system – must reach a specific threshold. This ratio, as already introduced in previous sections, is called carrier-to-interference ratio (CIR); the threshold is called *CIR target*. Transmission can take place if the CIR target is met (or exceeded). Formally, the following inequality must be satisfied:

$$\frac{\text{Strength of desired signal}}{\text{Noise} + \sum \text{strength of interfering signals}} \geq \text{CIR requirement} \qquad (6.32)$$

Because UMTS uses sophisticated power control technology, the inequality is usually assumed to hold with *equality* for power controlled channels, see Section 6.4.4.3. The CIR requirement is then also called CIR *target*. Spread spectrum technology allows the right-hand side of this inequality to be smaller than one, i.e. the desired signal strength may be (much) weaker than the interference.

The basic prerequisite for a user to be connected with the network is the ability to receive the pilot channel. Apart from this, CIR inequalities are considered for the dedicated channels in the uplink and downlink in today's static system level models. A user from a snapshot is served by the network if all applicable CIR inequalities hold (pilot signal, downlink and/or uplink). The static system level model can be used to determine whether a given combination of users can be served by the network and, if so, under what conditions. A static system level model of the UMTS network thus consists of a collection of CIR inequalities of the type shown in Equation (6.32).

6.4.1 LINK LEVEL ASPECTS

The WCDMA scheme of UMTS affects the system model in two principal aspects. The first one is related to the *CIR requirements*. The second is the *orthogonality* of downlink signals within the same cell.

6.4.1.1 CIR Requirements

The right-hand side in Equation (6.32) is specific to the situation. In general, in order to successfully decode a higher data rate, a higher CIR requirement must be fulfilled; coding also plays a role. The required CIR is determined with the help of link-level simulations, see Section 6.2. Simulations determine the relationship between block error probability and E_b/N_0 for the different radio bearers. The decision, which bearer and error probability apply to a specific user, is made on the basis of the user's *service*. The type of service a user requests is usually part of the information in a snapshot. For each service, a target error probability is assumed, leading to a minimum E_b/N_0 value; the value is usually interpolated from the link-level lookup table. The E_b/N_0 requirement also depends on the user's speed, because the influence of multipath variation varies with velocity; users who move faster are usually more demanding.

The E_b/N_0 requirement is converted into a CIR requirement using the specific processing gain. The CIR requirements for a user $m \in \mathcal{M}$ for uplink and downlink are denoted by $CIR_m^\uparrow$ and $CIR_m^\downarrow$, respectively. For example, data in [28] suggests an E_b/N_0 target value of 7.5 dB for a downlink speech user moving at 3 km/h. The processing gain is calculated as the ratio of the chipping rate (3.84 Mcps) over the data bit rate (12.2 kbps) and amounts to about 25 dB for a speech bearer. This results in a downlink CIR value of

$$CIR^\downarrow = 7.5\,\text{dB} - 25\,\text{dB} = -17.5\,\text{dB}$$

Other values derived from the same source are $CIR^\downarrow = -12.1\,\text{dB}$ (video telephony), $CIR^\uparrow = -19.5\,\text{dB}$ (speech telephony), $CIR^\uparrow = -14.6\,\text{dB}$ (video telephony).

6.4.1.2 Orthogonality

Each radio network controller (RNC) selects orthogonal downlink transmission codes for its mobiles (we neglect the use of secondary scrambling codes). In theory, transmissions with orthogonal codes do not mutually interfere. Due to multipath propagation, however, the signals partly lose this property. Interference from the same cell is typically modelled to be affected by an *orthogonality factor* $\alpha \in [0, 1]$, with $\alpha = 1$ meaning perfect orthogonality and $\alpha = 0$ no orthogonality. Interfering signals from the same cell are multiplied by $(1 - \alpha)$, we will use the notation $\bar{\alpha} := 1 - \alpha$.

The value of α depends on the specific propagation environment. In rural areas with few obstacles, orthogonality is preserved better (higher values for α), whereas in urban environments the impact of multipath fading are more severe. As the orthogonality loss depends on the individual channel profile (notably, the value of power delay spread), the value is specific to each user's location or even the link in more accurate models [29,30]. Example values from [28] are $\alpha = 0.327$ for urban environment or $\alpha = 0.938$ for rural environment.

6.4.2 PROPAGATION DATA

Radio signal propagation is one of the key factors in UMTS radio network assessment. Propagation models and signal strength prediction have extensively been treated in Chapter 5.

6.4.2.1 Attenuation

The main output of propagation modelling for the purpose of static system models are *attenuation factors* $L_{im}^\uparrow$ and $L_{mi}^\downarrow$ for each pair $i \in \mathcal{N}$, $m \in \mathcal{M}$ of antenna and mobile. The factors $L_{im}^\uparrow$ and $L_{mi}^\downarrow$ also include losses and gains from cabling, hardware and user equipment, see Section 6.4.3. It is essential to carry uplink and downlink as separate values, because the influence of equipment is in general different for the both directions. Propagation data sometimes also includes a *power delay profile* (PDP) of the channel. This data can be used to calculate a link's delay spread, which in turn can be used to increase the accuracy of orthogonality factor and CIR target estimations.

In the case of *smart antennas*, however, the attenuation factors cannot be modelled depending only on sender and receiver. In this case, the antenna adapts to the position of the mobile it maintains a link to. The attenuation of interfering signals depends on whether they are emitted close to the sending mobile. The attenuation factors are thus specific to a triple of sender, receiver and interferer.

6.4.2.2 Channel Variations

Channel variations are divided into *medium scale* fading (also *shadowing*) and *small scale* fading (also simply *fading*); the later is usually subdivided into *slow/fast* and *frequency-selective/non-selective* fading. For the influence of fast fading onto a static system level model, see Section 6.4.4.3. Shadow fading can be modelled as a random value applied to all attenuation values $L^\uparrow$ and $L^\downarrow$. The global effect of shadowing is then taken into account, e.g. in Monte-Carlo simulation, see Section 6.4.8.2.

The following are the typical assumptions on the stochastics of shadowing: The values are lognormally distributed, a common value for the distribution's standard deviation in cellular outdoors systems is 8 dB. The values for different users are independent if the distance between them exceeds a predefined *correlation length* depending on the environment. Shadow fades on the different links of a given user are correlated. This correlation is typically modelled based on the difference in the angles of arrival/departure in the horizontal plane.

6.4.3 EQUIPMENT MODELLING

The specific receiver and transmitter hardware influences the system from the static perspective in three main ways: First, there are limits on the powers that may be transmitted. Second, the equipment attenuates or amplifies (directional antenna, mast head amplifier) the signal travelling through it. Third, the equipment might alter the noise perceived at the receiver. Section 6.4.3.1 discusses sources of noise and its treatment in the system model; after that, the main impact of the equipment properties is outlined.

6.4.3.1 Interference and Noise

All signals received from other sources than the transmissions in the UMTS network in question are referred to as noise. Besides thermal noise, which is calculated using the system's bandwidth and the average spectral noise density, other sources of noise might play a role. These can be subsumed under a location specific *noise floor* value. As the latter may be frequency-dependent, the noise values of uplink and downlink should be handled separately in an FDD system. They will be denoted by $N^\uparrow$ and $N^\downarrow$. There is also equipment-specific noise generated in the equipment; see Section 6.4.3.2.

6.4.3.2 Mobile Terminal

The UMTS standard specification defines four power classes for user equipment [31]. A mobile has transmit power limits P_{max}^{MS} and P_{min}^{MS}. The maximum power value P_{max}^{MS} is the limiting factor for uplink coverage limited systems. Values for the maximum output power are in the range $P_{max}^{MS} = 21\,\text{dBm}$ (Power Class 4) to $P_{max}^{MS} = 33\,\text{dBm}$ (Power Class 1). The nominal value has to be decreased by a margin for allowing power control to equalise fades; see Section 6.4.4.3. The mobile itself adds to attenuation and noise.

Coverage

A coverage evaluation depends crucially on the assumptions on the mobile terminal's properties. The mobile's receiver's absolute *sensitivity* or E_c threshold – denoted by P_{limit}^{RSCP} – is the limit for E_c coverage. Typical values for P_{limit}^{RSCP} are about $-120.0\,\text{dBm}$. Because the pilot signal also has to be properly received in temporary fades, a *margin* to this nominal value has to be respected. The absolute value of this margin (in the range of 10–15 dB) depends on the fading characteristics. For determining E_c/I_0 coverage, a certain E_c/I_0 threshold CIR_m^{CPICH} is assumed above which the pilot signal can properly be decoded.

6.4.3.3 Base Station

One important parameter of the base station is the *maximum transmit power*, denoted by $P_{max,i}^{BS}$ for base station $i \in \mathcal{N}$. This power limit usually governs the capacity of downlink capacity limited systems. A typical value for a macro cell base station is $P_{max,i}^{BS} = 20\,\text{W}\,(43\,\text{dBm})$.

As a piece of technical equipment, the base station's receiver has a noise figure that is to be taken into account for noise calculations. For accurate calculations, it is important to include this information also for possible equipment between base station and antenna (e.g. mast head amplifiers). They have a noise figure and the signal is attenuated when traversing additional cables. In [28], for example, an additional loss of 3 (for $L^{\downarrow}$, downlink) and 1 dB (for $L^{\uparrow}$, uplink) is calculated for the connection between antenna and base station.

6.4.3.4 Rake Receiver Efficiency Factor

The most common form of receiver used in CDMA terminals and base stations is the Rake receiver. Here, we briefly consider how the performance of the Rake receiver can be modelled in a static system-level model. The reader is referred to Section 6.2.3 [32] and [33] for more detailed descriptions of the operation of the Rake receiver.

In simple terms, the Rake receiver consists of a number of separate receivers, or fingers, that can be locked onto the individual multipath components of the received signal. Each finger consists of a correlator and it can be locked to a particular multipath component by ensuring the locally generated spreading code used in the correlator is synchronised with the spreading code in the received multipath component. Also, each finger can use a different spreading code, so this means that individual fingers in the terminal's Rake receiver can be locked to multipath components from different base stations and this is the means by which the terminal receives signals from more than one base station in soft and softer handover.

The Rake receiver also contains a search finger that is used to find significant multipath components from each of the signals that the base station or terminal wishes to receive. The search finger uses the swept time delay cross correlation (STDCC) technique to identify and measure these multipath components and the Rake receiver ensures that its receiver fingers are always locked to the strongest multipath components in the wanted signal. The signals received on each finger are combined using maximal ratio combining (MRC) whereby their individual phase shifts are adjusted so that they are all co-phased and each one is weighted by its received power. The resulting combined signal is then used to extract the transmitted data.

The Rake receiver will only be able to resolve individual multipath components if they are separated by at least the chip period of the CDMA spreading code. In the UTRA FDD system, the chip rate is 3.84 Mchip/s and, therefore, the minimum resolvable multipath delay is 260 μs. If two or more paths arrive at the receiver with a relative delay of less than 260 μs, then the Rake receiver will not be able to resolve these individual paths and they will be treated as a single composite path.

Since a typical Rake receiver might contain, say, six receiver fingers, situations may arise where the received energy in some significant resolvable multipath components cannot be captured and utilised in the decoding process. This effect can be modelled using a *Rake efficiency factor*, which is defined as the ratio of the total power captured by the Rake receiver to the total available power in the received signal. The Rake efficiency factor will depend both on the number of fingers available within the receiver, which may be different for the terminals and the base stations, and also the number of significant resolvable multipath components within the radio channel. Since the number of multipath components may be different in different types of propagation environments (e.g. urban, suburban, rural), different rake efficiency factors can be used in each environment.

6.4.3.5 Adjacent Channel Interference

Although the nominal channel spacing in the UTRA FDD system is 5 MHz, the radio signal transmitted by the Node Bs and the UEs is not completely contained within this channel bandwidth. The adjacent channel leakage ratio (ACLR) is used to quantify the amount of power from a particular transmitter

Table 6.2 ACLR and ACS values for the UTRA FDD Node B and UE.

	First adjacent channel		Second adjacent channel	
	Node B	UE	Node B	UE
ACLR	45 dB	33 dB	50 dB	43 dB
ACS	45 dB	33 dB	—	—

that leaks into an adjacent channel and it is defined as the ratio of the transmitted power to the power measured after an appropriately applied receiver filter in the adjacent channel. In addition, practical receivers are not able to completely suppress the signal received in adjacent radio channels and the ability of a receiver to suppress adjacent signals is measured by its adjacent channel selectivity (ACS). The ACS is defined as the ratio of the receiver filter attenuation on the assigned frequency to the receiver filter attenuation on the adjacent channels.

The values of ACLR and ACS for the UTRA FDD Node B and UE are presented in Table 6.1. The values for ACLR and the ACS value for the UE are taken directly from the UTRA FDD specifications [31,34]. However, the value of ACS for the Node B is not explicitly provided in the specifications and this has been derived from test conditions that are detailed in the specifications [35]. The overall adjacent channel interference on a link is therefore a combination of the ACLR of the transmitter and the ACS of the receiver and this can be characterised by a third parameter known as the adjacent channel protection (ACP). The ACP essentially defines the level up to which an interfering adjacent channel signal is suppressed at the receiver and it is given by the following equation:

$$\text{ACP} = \frac{1}{\frac{1}{\text{ACLR}} + \frac{1}{\text{ACS}}} \quad (6.33)$$

where ACP, ACS and ACLR are expressed in linear terms. Using Equation (6.33) to calculate the ACP for the uplink and the downlink on the first adjacent channel, we find that the value is 33 dB in both cases since the UE ACLR dominates on the uplink and the UE ACS dominates on the downlink.

For an accurate evaluation, transmissions in adjacent channels can be considered in the same way as transmissions in the wanted channel, but the attenuation factors ($L^\downarrow$ and $L^\uparrow$) have to be increased by adding (in logarithmic terms) the ACP value. Since transmission powers are coupled, all available channels theoretically have to be considered. If this level of accuracy is not necessary or impossible *to achieve* (e.g. if the necessary data on other operators' networks are not available), the power emitted on neighbouring channels can be estimated or measured. The adjacent channel interference is then considered as an adjacent channel interference map of location-specific values that add to the noise N in the simulations of the victim network, as described in [35].

6.4.4 TRANSMIT POWERS AND POWER CONTROL

Transmission powers play a paramount role in analysing UMTS radio systems. The central goal of the static model at hand is to determine the (average) power levels that the power controlled *dedicated* channels assume. Apart from these channels, there are *control* channels, which are not subject to power control, and *shared* channels. This section first discusses how the latter two kinds of channels are reflected in a static model. After that, the static modelling of power control and channel variations are treated in detail.

6.4.4.1 Shared Channels

Modelling shared channels is beyond the scope of the static system level model presented here. One reason for this is that these channels adapt dynamically to traffic (and radio propagation) conditions. The additional interference caused by shared channels, however, has to be included. The simplest way of doing so is by considering a fixed amount of power emitted by the antenna (analogously to control channels, see Section 6.4.4.2). In the research literature, however, mostly HSDPA has been considered. In [36], a method for integrating a dedicated HSDPA simulation module with static snapshot analysis is reported.

6.4.4.2 Control Channels

There is no power control for control channels. From the system level point of view, the most important control channel is the pilot channel (CPICH). The power spent on the CPICH channel is denoted by P^{CPICH}. It is possible to have an individual value P_i^{CPICH} for the CPICH power for every cell $i \in \mathcal{N}$. The power spent on all remaining control channels is denoted by P_i^{otherCCH}. This is usually chosen depending on the CPICH level, P_i^{otherCCH} is then a multiple of P_i^{CPICH}. Occasionally, the power spent on all control channels will be denoted by $P^{CCH} := P^{CPICH} + P^{\text{otherCCH}}$.

6.4.4.3 Dedicated Channels

Each link on a dedicated channel is power controlled, see Section 2.2.3. Power control aims at preventing senders from causing excessive interference for the other links by using the least amount of power to maintain the connection.

Perfect power control
For static system level models, it is common to assume that the UMTS power control can equalise channel variations; the transmitter instantaneously adjusts its power exactly to the level required to exactly meet the CIR requirement. This assumption is called *perfect power control*; the inequality in Equation (6.32) can then be assumed to always hold with equality. The CIR requirement is in this case often called CIR *target*.

The assumption of perfect power control and the disregard of power control loops has been found to introduce a systematic underestimation of transmit powers. For increasing the static model's accuracy, link level simulations including inner and outer loop power control can be used. The error in average power values can then be reduced by adding a corresponding margin to the CIR targets $CIR^{\uparrow}$ and $CIR^{\downarrow}$. The margins should be specific to the user's speed and link state (uplink/downlink, SHO mode).

Effects of channel variation
For equalising fades, transmit power necessarily has to be increased above the average value. For doing so, a certain 'buffer' must be available in addition to the average transmit power. This means that the mobile's nominal maximum transmission power effectively has to be reduced by a certain *transmit power control (TPC) headroom*. The necessary headroom depends on the statistics of the fading [37], it can amount to up to 5 dB (for slowly moving users) [28].

Even if the assumption that all channel variations are equalised by power control were true, some effects of channel fading could still be noticed. The reason for this is that power control on a link applies only to the link's transmitter, not to the interferers. Interferers are power controlled by the TPC commands of their respective receivers. As fast fading on different links is mostly uncorrelated, the interferes are unlikely to be, e.g., in a deep fade simultaneously with the link's receiver. Although the links are individually power controlled, they are not power controlled as a whole.

For this reason, a higher average interference power than expected is observed [37]. In the system level model, this can be taken into account by an *interference raise* value. This value is to be added (in logarithmic scale) to the attenuation values $L_{mi}^\uparrow$ and $L_{im}^\downarrow$ whenever antenna i is *not* serving user m. Typical values for the interference raise are in the range of 1–2 dB [28].

6.4.5 SERVICES AND USER-SPECIFIC PROPERTIES

The service a user accesses and the user's mobility are usually also part of the information in a snapshot. Both service – see Section 3.2 – and mobility influence how much radio resources have to be spent for serving a user.

6.4.5.1 Bearer Association

The service provided by Layer 2 for the transfer of user data between user equipment and UTRAN is called a radio bearer. The bearer(s) a radio link has to support depends on the user's service. For some services (speech, video), there is a canonical bearer mapping. For packet switched services, there are usually several possibilities. What type of bearer a user link is mapped onto in different situations depends on the policy of the vendor and/or provider.

6.4.5.2 Transmit Activity and Data Source Models

Services differ in the characteristics of data flow on the link over time. For static simulation, this is relevant as (almost) no physical transmission takes place and no interference is generated if no data has to be transmitted.

Circuit switched services
In a speech conversation, each of the two users involved is assumed to speak roughly 50 % of the time. No user data is transmitted in silence periods (discontinuous transmission, DTX). Dedicated control channels keep transmitting, but they are usually neglected. There are services that cause significant traffic in only one direction, e.g. downlink data streaming (control traffic in the reverse direction is negligible).

This is taken into account in the form of *activity factors*. For each user $m \in \mathcal{M}$, there are two activity factors, $v_m^\uparrow$ for the uplink and $v_m^\downarrow$ for the downlink. For voice, a typical choice of activity factor is $v^\downarrow = v^\uparrow = 0.5$. To account for signalling frames etc. even in silence periods (e.g. with AMR codec), a higher value is sometimes chosen, e.g. 0.67 for voice.

The activity factor can be used to compute the *average* power that is sent over a period of time in order to support a link. This average value is used for determining the average interference contribution to other links in the snapshot. Without it, the interference power would be overestimated. It is hence only taken into account in the denominator of Equation (6.32) (not in the numerator).

Interactive packet switched services
A detailed modelling of an interactive service distinguishes between two levels of activity. During a session, a user is usually considered to request a number of *packet calls* separated by a much longer period of inactivity (*reading time*). Within a packet call, typically a large bandwidth is allocated to the user and data is transmitted with high activity. The *connection activity factor* denotes the probability that a data user is actually in an active packet call. For users in a packet call, a *transmit activity factor* $v_m^\uparrow / v_m^\downarrow$ is used to calculate the average emitted power for a particular bearer allocation.

6.4.5.3 Mobility

An important piece of information on the user is mobility, notably the user's *speed*, which largely determines the characteristics of fast fading. A faster moving user is less susceptible to fading, because the fades happen within one transmission time interval (TTI) and can hence be mitigated by the outer channel coder. This influences all link-level related aspects: SHO margins, CIR targets etc.

6.4.5.4 Body and Usage Loss

According to the situation of the user, additional losses might occur that have to be taken into account for radio link budget and attenuation calculation. A user holding a mobile device to his ear causes a *body loss*. A user in a car has additional attenuation. In [28], the value for body loss is 3 dB while the additional attenuation caused by a car is 8 dB.

6.4.6 SOFT HANDOVER

The influence of SHO on the dynamic handover process cannot be modelled in a static model. What can be considered, though, is the *soft/softer handover gain* on received/transmitted power due to diversity. Soft handover gains depend on user speed and the difference between the link attenuations to the base stations in the active set [37]. In general, a slowly moving terminal is most affected by fading (see explanation above); thus, SHO has a beneficial effect here. The gain is larger if the base stations i in the active set have similar attenuation factors to the mobile m. If the signals can be combined with maximum ratio combining (softer handover), the gain is substantially larger. Different gains apply to downlink and uplink.

To determine whether or not a mobile maintains connections to more than one cell, a static version of the active set can be used. To this end, a SHO margin $\gamma^{(SHO)}$ (corresponding to the add/remove window) and a maximum active set size $n^{(MAX)}$ are introduced. Let $i^* := \arg\max_{i \in \mathcal{N}} L_{im}^{\downarrow} \cdot P_i^{CPICH}$ be the cell with the strongest signal. The cells to which a mobile $m \in \mathcal{M}$ can potentially maintain connections are the cells whose pilot signal is at most $\gamma^{(SHO)}$ weaker:

$$\left\{ i \mid L_{i^*m}^{\downarrow} P_{i^*}^{CPICH} - L_{im}^{\downarrow} P_i^{CPICH} \leq \gamma^{(SHO)} \right\}$$

If this set contains more than $n^{(MAX)}$ elements, then the weakest links are removed. This method ignores all dynamic effects, especially hysteresis.

6.4.6.1 Uplink

In the uplink, SHO provides a transmit power gain. The gain can be accounted for in the system level model by decreasing the link's CIR target $CIR_m^{\uparrow}$ accordingly. In addition, channel element consumption in the base stations in the active set have to be considered.

6.4.6.2 Downlink

Also in the downlink, a transmit power gain compared to the single link case applies. However, altogether more radio resources might be needed than in the single link case because several cells transmit to the mobile. Power balancing algorithms ensure that all links transmit with the same power (otherwise, SHO gains vanish).

Under favourable conditions, soft handover power gains might amount to up to 5 dB (softer handover), but it diminishes quickly for link gain differences of more than a few dB. Specific values for SHO gains in different cases can be found in [28,37].

6.4.7 COMPLETE MODEL

The goal of the complete model is to evaluate the network's performance on a given set of mobile users. To this end, it has to be established to which antennas a user can be linked. The simplest assignment method is to assign user m to the cell with the highest received E_c power, also commonly denoted as the best server:

$$\arg\max_{i \in \mathcal{N}} L_{im}^{\downarrow} P_i^{\text{CPICH}}$$

In practice, cell assignment based on E_c/I_0 is equally important. For single-carrier systems, this is equivalent to E_c-based assignment as the spectral noise density I_0 at the user's position is invariant. It is also possible to consider cell assignment probabilities. In any case, the set $\mathcal{M}$ of all mobiles is subdivided into sets $\mathcal{M}_i$ of mobiles served by antenna i: $\mathcal{M} = \bigcup_i \mathcal{M}_i$. If soft handover is to be considered, a mobile can be assigned to several base stations.

6.4.7.1 Uplink

Denote the *uplink* transmission power of a mobile $m \in \mathcal{M}$ by $P_m^{\uparrow}$. If mobile m is connected to antenna $i \in \mathcal{N}$, the received signal strength at antenna i is $L_{mi}^{\uparrow} P_m^{\uparrow}$ during transmission. Recall that N_i stands for the received background noise at antenna $i \in \mathcal{N}$. The interfering signals are the transmission from all other mobiles $n \neq m$. Their influence is, however, only accounted for weighed with their average activity factor $\nu_n^{\uparrow}$. The uplink version of Equation (6.32) for the transmission from m to i with $CIR_m^{\uparrow}$ as CIR target thus reads

$$\frac{L_{mi}^{\uparrow} P_m^{\uparrow}}{N_i + \sum_{n \neq m} L_{ni}^{\uparrow} \nu_n^{\uparrow} P_n^{\uparrow}} = CIR_m^{\uparrow} \qquad (6.34)$$

Note that the power $P_m^{\uparrow}$ does not add to the interference on the link to mobile m. This can be condensed by writing

$$\bar{C}_i^{\uparrow} := N_i + \sum_{m \in \mathcal{M}} L_{mi}^{\uparrow} \nu_m^{\uparrow} P_m^{\uparrow} \qquad (6.35)$$

for the *average* total received power at antenna $i \in \mathcal{N}$. The equation then simplifies to

$$\frac{L_{mi}^{\uparrow} P_m^{\uparrow}}{\bar{C}_i^{\uparrow} - L_{mi}^{\uparrow} \nu_m^{\uparrow} P_m^{\uparrow}} = CIR_m^{\uparrow} \qquad (6.36)$$

6.4.7.2 Downlink

Let the total *average* output power of installation i be

$$\bar{P}_i^{\downarrow} := \sum_{m \in \mathcal{M}_i} \nu_m^{\downarrow} P_{im}^{\downarrow} + P_i^{\text{CPICH}} + P_i^{\text{otherCCH}} \qquad (6.37)$$

In the downlink version of Equation (6.32) related to transmission from i to m, there is another caveat: signals from the own cell are assumed to be orthogonal with an orthogonality factor of $\bar{\alpha}_m$:

$$\frac{L_{im}^{\downarrow} P_{im}^{\downarrow}}{L_{im}^{\downarrow} \bar{\alpha}_m \left(\bar{P}_i^{\downarrow} - \nu_m^{\downarrow} P_{im}^{\downarrow}\right) + \sum_{j \neq i} L_{jm}^{\downarrow} \bar{P}_j^{\downarrow} + N_m} = CIR_m^{\downarrow} \qquad (6.38)$$

6.4.7.3 CPICH

For the *pilot*, the situation is simpler. Technically speaking, the received chip energy of the pilot signal – called the pilot channel's E_c or CPICH E_c – relative to the total power spectral density I_0 has to lie above a threshold CIR_m^{CPICH}. When computing the spectral density I_0 as the denominator in the pilot version of Equation (6.32), no benefit due to orthogonality applies and even the pilot's own contribution appears as interference:

$$\frac{L_{im}^{\downarrow} P_i^{CPICH}}{N_m + \sum_{i \in \mathcal{N}} L_{im}^{\downarrow} \bar{P}_i^{\downarrow}} \geq CIR_m^{CPICH} \qquad (6.39)$$

Transmit powers are determined by solving the system of linear Equations (6.36) and/or (6.38) for all users, depending on whether they are active in the uplink and downlink. If Equation (6.39) fails to hold for a user, the user is put to E_c/I_0-outage.

6.4.7.4 Served Users

It is also part of the snapshot evaluation to determine which users cannot be served by the network. Evidently, users are put to outage if they lack E_c or E_c/I_0 coverage. Another reason that prohibits service for a given user m is if the resulting uplink power $P_m^{\uparrow}$ exceeds the mobile's output power limit. If any of these conditions applies, the user has to be removed and powers have to be reevaluated. However, if the system's *radio resources* are insufficient for serving all users, load control and call admission is required to determine how to spend the available resources on the users, see Section 6.4.10.2.

6.4.8 APPLICATIONS OF A STATIC SYSTEM-LEVEL NETWORK MODEL

For a given user snapshot, the network's performance can be measured in detail. For insights on the *average* or *expected* performance of a network, the Monte-Carlo method can be used.

6.4.8.1 Measurable Performance Figures

The direct use of the above static system level model is to determine link powers and thereby the cell's emitted and received powers for a given set of served users in a snapshot.

Radio resource consumption
A prominent performance figure is *cell load*, a percentage value. The load of a cell i is determined using the average output powers $\bar{P}_i^{\downarrow}$ and $\bar{C}_i^{\uparrow}$. In the downlink, a popular definition of a cell's load is the fraction of the maximum output power that is needed to serve the users:

$$\text{load}_i^{\downarrow} := \frac{\bar{P}_i^{\downarrow}}{P_{\max,i}^{BS}} \qquad (6.40)$$

In the uplink, the load of cell i can be defined via the *noise rise*, the ratio of total received power to the noise exterior to the system. As the noise is always included in the received power, this ratio is never smaller than one.

$$\text{NR}_i^\uparrow := \bar{C}_i^\uparrow / N_i^\uparrow \qquad \text{load}_i^\uparrow := 1 - (1/\text{NR}_i^\downarrow) = 1 - \left(N_i^\uparrow / \bar{C}_i^\uparrow\right) \qquad (6.41)$$

With this definition, the load of an empty system is zero, while the load is 1 (or 100%) for infinite noise rise. In real systems, there is usually a *maximum noise rise* that is not to be exceeded at any base station in order to keep the system in a stable state.

Another performance figure derived directly from power that sheds light on network planning is *other-to-own cell received power ratio*, short 'Little i' (sometimes this is called intra-to-intra cell interference power factor). It is measured at the receiver. In the uplink, it is calculated per cell j, whereas in the downlink it is measured at each mobile m:

$$i^\uparrow(j) = \frac{\sum_{n \notin \mathcal{M}_j} v_n^\uparrow L_{nj}^\uparrow P_n^\uparrow}{\sum_{m \in \mathcal{M}_j} v_m^\uparrow L_{mj}^\uparrow P_m^\uparrow} \quad \forall j \in \mathcal{N} \qquad i^\downarrow(m) = \frac{\sum_{k \neq j} L_{km}^\downarrow \bar{P}_k^\downarrow}{L_{jm}^\downarrow \bar{P}_j^\downarrow} \quad \forall m \in \mathcal{M} \qquad (6.42)$$

It is assumed here that mobile m is being served only by cell j, such that the signals from this cell count as own cell interference.

As the allocated bearers and their spreading factors are determined for each user during the snapshot analysis, *code* and *channel element* consumption can also be estimated. The *throughput* of a cell is determined by adding the individual user's data rates. However, exact determination of these performance measures is beyond the scope of a static model.

Missed traffic
An essential part of the UMTS radio network evaluation is estimation of *missed traffic*. Missed traffic occurs if a user in the snapshot is not served at all or not with the desired degree of service. In evaluating a snapshot, it is crucial to record where and how much traffic has been missed for what reason. Some of the reasons for outage (e.g. failed E_c/I_0 coverage or UE power limits) are straightforward to evaluate in a static simulation. Outage and blocking due to depletion of radio resources touches the domain of load/call admission control and is therefore implementation dependent. Some reasons for unsatisfactory service for users can hardly be modelled in a static simulation, see Section 6.4.8.3.

Pilot pollution
In general, it is beneficial for a mobile to receive several cells' signals equally strongly because the diversity can be used in SHO. However, if the number of signals of similar strength is too large (notably if it exceeds the number of rake fingers in the receiver unit), one speaks of *pilot pollution*. For the exact definition, a number $n^{(\text{CPICH})}$ and a margin of $m^{(\text{CPICH})}$ (in dB) have to be fixed. A mobile suffers from pilot pollution, if the pilot signals of more than $n^{(\text{CPICH})}$ cells are received within a window of $m^{(\text{CPICH})}$ below the strongest signals, i.e. if

$$\left| \left\{ j : L_{jm}^\downarrow P_j^{\text{CPICH}} \geq \left(\max_{i \in \mathcal{N}} L_{im}^\downarrow P_i^{\text{CPICH}} \right) - m^{(\text{CPICH})} \right\} \right| \geq n^{(\text{CPICH})}$$

Coverage
For a snapshot, a mobile m has E_c-coverage if the received power of any cell i's pilot signal is above the threshold value:

$$L_{im}^\downarrow \cdot P_i^{\text{CPICH}} \geq P_{\text{limit}}^{\text{RSCP}}$$

This condition can be checked without any traffic considerations once the CPICH signal strength is known. For E_c/I_0-coverage, however, traffic is indispensable, as the power on the dedicated links adds to the interference I_0. A mobile m has E_c/I_0 coverage if for at least one installation (Equation 6.39) is satisfied.

6.4.8.2 The Monte-Carlo Approach

The performance of the network on a single snapshot is usually of little general interest. For network planning, the network's *expected* performance is important. A common technique to determine the expected performance is *Monte-Carlo simulation*. The principle is to evaluate the network on a large number of snapshots drawn at random from a probability distribution describing the *expected* traffic intensity. The different performance figures are then aggregated. A generic flow chart for Monte-Carlo simulation is shown in Figure 6.15.

Random influences in a snapshot
The main source of randomness for a snapshot is the *traffic intensity,* the number of users and their spatial distribution across the planning area. The distribution of users in a snapshot is usually the realisation of a (spatial) Poisson point process. Another random parameter is the users' *mobility* and *service usage*. The main influence of mobility on static analysis is the user's current speed, which influences the CIR target and the effect of fading. *Shadowing* may be included as a random parameter and drawn at random for each link. However, links to the same mobile from different antennas should have angular correlation. The assignment of a mobile to a cell can also be considered a random process using cell assignment probabilities.

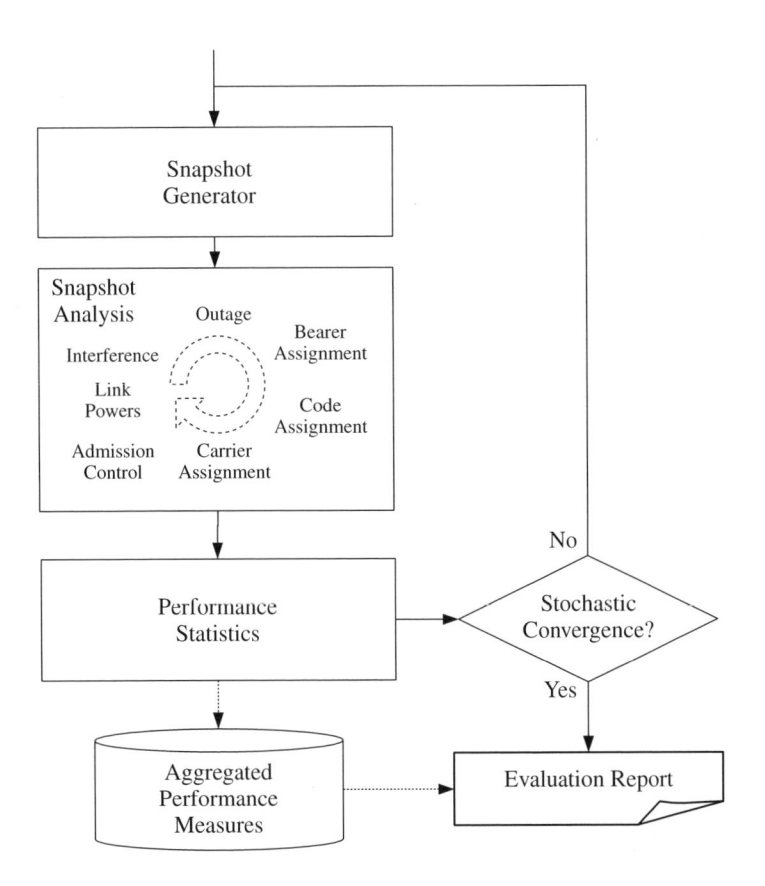

Figure 6.15 Performance evaluation using Monte-Carlo simulation.

Aggregation of performance figures
There are basically two ways of aggregating performance figures: (1) statistics on a per- cell- basis and (2) spatial distributions. The first class comprises all figures that are measured per cell, such as the uplink other-to-own cell interference ratio or average output power. The second class includes missed traffic maps, E_c/I_0 coverage and the downlink other-to-own cell interference ratio.

Care has to be taken in order to ensure *stochastic reliability* of results. For solid results, it is not sufficient to analyse a seemingly 'large' number of snapshots. Statistical tools such as confidence intervals should be used instead [38]. Spatial performance measures (such as outage probability plots with a high resolution) converge slowly with the Monte-Carlo methodology. It is thus necessary to analyse substantially more snapshots for spatial performance measures than for cell-based measures (e.g. average transmit power).

6.4.8.3 Limitations of Static Approaches

A static model is far easier to handle than a dynamic one, but it inevitably falls short of capturing dynamic aspects. In some cases, however, the shortcomings are not essential, because dynamic effects can be compensated by adjusting the static model's parameters.

Small scale dynamic effects
This category mainly comprises effects that are related to fast fading, which takes place on a 'microscopic' time scale, and power control. Throughout this section, it has been indicated how to take into account the most important effects and consequences of fading and power control into the static model, most prominently in Section 6.4.4.3.

In general, these effects can be reflected by parameter adjustment. The adjustment is adapted to the user's speed, where necessary. It is a task for link-level simulators to determine appropriate settings for these parameters. Even if the static approximation can never be exact, estimates on system performance figures are possible.

Large scale dynamic effects
Dynamic events on a larger timescale basically consist of the impact of users' behaviour on the system. In a static model, there is merely a set of users with no memory or history information. This introduces substantial shortcomings in the following cases:

- Load control. Realistic load and call admission control algorithms are based on historic knowledge. For example, users that want to set up a new call are usually prioritised lower than users that have a running connection. Also the evolution of code trees is beyond a static model.
- Handover. The effect of handover settings and algorithms can only be studied with users moving from one cell to another. Especially for soft handover this is a serious shortcoming. Some soft handover parameters do play a role in static evaluations (see Section 6.4.6), but their impact on the network is not fully reflected.
- Data services and shared channels. The impact and behaviour of data users – especially with 'background' and 'interactive' QoS requirements – can only be studied poorly in a static model. This begins with the bearer allocation. In a real system, the connection speed might vary over time according to the current situation. Scheduling policies, e.g. in HSDPA, and the development of buffer states over time, jitter and latency cannot be considered.

For studying the above parameters and situation, dynamic models described in Section 6.4 have to be used.

6.4.9 POWER CONTROL AT CELL LEVEL

In each direction (uplink and downlink), the static system level model described in Section 6.4.7 consists of $|\mathcal{M}|$ linear equations involving $|\mathcal{M}|$ variables. This number can be very big in a large scenario. Another problem is the large dynamic range of input parameters, notably of the attenuation values. State-of-the-art tools for snapshot evaluation use a compression 'trick' described in this section to circumvent these problems and significantly speed up simulations.

The main idea is to perform calculations on a per cell basis and to treat individual users only implicitly. Analytically, the equation system is formulated in terms of the dependent variables $\bar{C}_i^\uparrow$ and $\bar{P}_i^\downarrow$ defined in Equations (6.35) and (6.37). The variables $P_m^\uparrow$ and $P_{im}^\downarrow$ can be eliminated. The result is an equation system of dimension $|\mathcal{N}| \times |\mathcal{N}|$. This system is smaller and significantly easier to solve. These coupling systems are introduced and extended to WCDMA in [39–42]. We start from a network design with sectors $i \in \mathcal{N}$ and a traffic snapshot with mobiles $m \in \mathcal{M}$, using the notation from Table 6.1.

6.4.9.1 Uplink

Concerning the uplink at antenna i, recall that $\bar{C}_i^\uparrow$ is the total amount of received power including thermal and other noise. Under the above assumptions we use elementary transformation of the equality version of Equation (6.36) to derive two quantities for every mobile m served by cell i: First, the transmission power $P_m^\uparrow$ of mobile m given the total received power $\bar{C}_i^\uparrow$ at the serving installation. Second, the fraction of the total received power at the installation i originating in mobile m.

$$P_m^\uparrow = \frac{1}{L_{mi}^\uparrow} \frac{CIR_m^\uparrow}{1 + \nu_m^\uparrow CIR_m^\uparrow} \bar{C}_i^\uparrow \quad (6.43)$$

$$\frac{\nu_m^\uparrow L_{mi}^\uparrow P_m^\uparrow}{\bar{C}_i^\uparrow} = \frac{\nu_m^\uparrow CIR_m^\uparrow}{1 + \nu_m^\uparrow CIR_m^\uparrow} \quad (6.44)$$

We define the *uplink user load* $l_m^\uparrow$ of a mobile m as the right-hand side of (6.44).

$$l_m^\uparrow := \frac{\nu_m^\uparrow CIR_m^\uparrow}{1 + \nu_m^\uparrow CIR_m^\uparrow} \quad (6.45)$$

The contributions to the total received power $\bar{C}_i^\uparrow$ at cell i are broken down to all uplink connections (not just those served by i). Let $\mathcal{M}_i \subseteq \mathcal{M}$ denote the set of users served by cell i. Then Equation (6.35) reads as

$$\bar{C}_i^\uparrow = \sum_{m \in \mathcal{M}_i} L_{mi}^\uparrow \nu_m^\uparrow P_m^\uparrow + \sum_{j \neq i} \sum_{m \in \mathcal{M}_j} L_{mj}^\uparrow \nu_m^\uparrow P_m^\uparrow + N_i \quad (6.46)$$

Defining the installation *uplink coupling factors* $M_{ii}^\uparrow$ and $M_{ij}^\uparrow$ (where $i \neq j$) as

$$M_{ii}^\uparrow := \sum_{m \in \mathcal{M}_i} l_m^\uparrow \quad \text{and} \quad M_{ij}^\uparrow := \sum_{m \in \mathcal{M}_j} \frac{L_{mi}^\uparrow}{L_{mj}^\uparrow} l_m^\uparrow \quad (6.47)$$

and substituting Equation (6.43), the uplink transmission powers can be expressed as

$$\bar{C}_i^\uparrow = M_{ii}^\uparrow \bar{C}_i^\uparrow + \sum_{i \neq j} M_{ij}^\uparrow \bar{C}_j^\uparrow + N_i \quad (6.48)$$

The quantity $M_{ii}^\uparrow$ measures the contribution from the own users, and $M_{ij}^\uparrow$ scales the contribution from installation i. The matrix

$$M^\uparrow := \left(M_{ij}^\uparrow \right)_{1 \leq i,j \leq |\mathcal{N}|} \quad (6.49)$$

is called the *uplink coupling matrix*. An example for an uplink coupling matrix and its computation can be found in Section 6.4.9.4.

Collecting Equation (6.48) for all installations, we obtain a system of linear equations governing the uplink cell reception powers:

$$\bar{C}^\uparrow = M^\uparrow \bar{C}^\uparrow + N^\uparrow \tag{6.50}$$

Writing *I* for the identity matrix, this system becomes

$$\left[I - \begin{pmatrix} \sum_{m \in \mathcal{M}_1} l_m^\uparrow & \sum_{m \in \mathcal{M}_2} \frac{L_{m1}^\uparrow}{L_{m2}^\uparrow} l_m^\uparrow & \cdots & \sum_{m \in \mathcal{M}_{|\mathcal{N}|}} \frac{L_{m1}^\uparrow}{L_{m|\mathcal{N}|}^\uparrow} l_m^\uparrow \\ \sum_{m \in \mathcal{M}_1} \frac{L_{m2}^\uparrow}{L_{m1}^\uparrow} l_m^\uparrow & \sum_{m \in \mathcal{M}_2} l_m^\uparrow & \cdots & \sum_{m \in \mathcal{M}_{|\mathcal{N}|}} \frac{L_{m2}^\uparrow}{L_{m|\mathcal{N}|}^\uparrow} l_m^\uparrow \\ \vdots & \vdots & \ddots & \vdots \\ \sum_{m \in \mathcal{M}_1} \frac{L_{m|\mathcal{N}|}^\uparrow}{L_{m1}^\uparrow} l_m^\uparrow & \sum_{m \in \mathcal{M}_2} \frac{L_{m|\mathcal{N}|}^\uparrow}{L_{m2}^\uparrow} l_m^\uparrow & \cdots & \sum_{m \in \mathcal{M}_{|\mathcal{N}|}} l_m^\uparrow \end{pmatrix} \right] \cdot \begin{bmatrix} \bar{C}_1^\uparrow \\ \bar{C}_2^\uparrow \\ \vdots \\ \bar{C}_{|\mathcal{N}|}^\uparrow \end{bmatrix} = \begin{pmatrix} N_1^\uparrow \\ N_2^\uparrow \\ \vdots \\ N_{|\mathcal{N}|}^\uparrow \end{pmatrix}$$

Under the assumptions stated in the beginning of this section, the solution of Equation (6.50) is the received powers at each installation. Necessary and sufficient conditions on $M^\uparrow$ for the existence of positive and bounded solutions to Equation (6.50) are given in [41,42].

6.4.9.2 Downlink

In the downlink case, we basically repeat what has just been done for the uplink. The starting point is the CIR equality (Equation 6.38). It can be rewritten as

$$\frac{1 + \bar{\alpha}_m \nu_m^\downarrow CIR_m^\downarrow}{\nu_m^\downarrow CIR_m^\downarrow} \nu_m^\downarrow \bar{P}_{im}^\downarrow = \bar{\alpha}_m \bar{P}_i^\downarrow + \sum_{j \neq i} \frac{L_{jm}^\downarrow}{L_{im}^\downarrow} \bar{P}_j^\downarrow + \frac{N_m}{L_{im}^\downarrow} \tag{6.51}$$

We define the *downlink user load* of serving mobile *m* as:

$$l_m^\downarrow := \frac{\nu_m^\downarrow CIR_m^\downarrow}{1 + \bar{\alpha}_m \nu_m^\downarrow CIR_m^\downarrow} \tag{6.52}$$

Similar to the uplink case, further notation is helpful to express the dependency of the power $\bar{P}_i^\downarrow$ on the downlink transmission power at all cells. We introduce the downlink *coupling factors*

$$M_{ii}^\downarrow := \sum_{m \in \mathcal{M}_i} \bar{\alpha}_m l_m^\downarrow \quad \text{and} \quad M_{ij}^\downarrow := \sum_{m \in \mathcal{M}_i} \frac{L_{jm}^\downarrow}{L_{im}^\downarrow} l_m^\downarrow \quad (j \neq i) \tag{6.53}$$

for cells *i* and *j* as well as an cell's *traffic noise power*

$$P_i^{(N)} := \sum_{m \in \mathcal{M}_i} \frac{N_m}{L_{im}^\downarrow} l_m^\downarrow \tag{6.54}$$

The transmit powers at the cells satisfy the expression

$$\bar{P}_i^\downarrow = M_{ii}^\downarrow \bar{P}_i^\downarrow + \sum_{j \neq i} M_{ij}^\downarrow \bar{P}_j^\downarrow + P_i^{(N)} + P_i^{CCH} \tag{6.55}$$

We define the *downlink coupling matrix* as

$$M^\downarrow := \left(M_{ij}^\downarrow\right)_{1 \leq i,j \leq |\mathcal{N}|} \qquad (6.56)$$

The Equation (6.55) for all antennas in the network form a linear equation system that describes the downlink transmit power in each cell:

$$\bar{P}^\downarrow = M^\downarrow \bar{P}^\downarrow + P^{(N)} + P^{CCH} \qquad (6.57)$$

In the downlink, the explicit coupling matrix is

$$M^\downarrow = \begin{pmatrix} \sum_{m \in \mathcal{M}_1} l_m^\downarrow & \sum_{m \in \mathcal{M}_1} \frac{L_{m1}^\downarrow}{L_{m2}^\downarrow} l_m^\downarrow & \cdots & \sum_{m \in \mathcal{M}_1} \frac{L_{m1}^\downarrow}{L_{m|\mathcal{N}|}^\downarrow} l_m^\downarrow \\ \sum_{m \in \mathcal{M}_2} \frac{L_{m2}^\downarrow}{L_{m1}^\downarrow} l_m^\downarrow & \sum_{m \in \mathcal{M}_2} l_m^\downarrow & \cdots & \sum_{m \in \mathcal{M}_2} \frac{L_{m2}^\downarrow}{L_{m|\mathcal{N}|}^\downarrow} l_m^\downarrow \\ \vdots & \vdots & \ddots & \vdots \\ \sum_{m \in \mathcal{M}_{|\mathcal{N}|}} \frac{L_{m|\mathcal{N}|}^\downarrow}{L_{m1}^\downarrow} l_m^\downarrow & \sum_{m \in \mathcal{M}_{|\mathcal{N}|}} \frac{L_{m|\mathcal{N}|}^\downarrow}{L_{m2}^\downarrow} l_m^\downarrow & \cdots & \sum_{m \in \mathcal{M}_{|\mathcal{N}|}} l_m^\downarrow \end{pmatrix}$$

Note that a cell's mobiles appear only in the corresponding *row* of the downlink coupling matrix, whereas in the uplink, they appear in the referring *column*.

In practical implementations, the cell-based formulations Equations (6.50) and (6.57) converge faster than their explicit counterparts Equations (6.36) and (6.38). This is demonstrated empirically in [43]. The reason for this – besides the effect of dimension-reduction – is that the equation systems are better conditioned, as path loss coefficients do not appear as coefficients anymore, but only ratios of them, which have a smaller dynamic range.

6.4.9.3 Link Powers

The powers needed for the individual links can easily be deduced from the cell powers. Suppose mobile m is being served by antenna i, the antenna's total received power being $\bar{C}_i^\uparrow$, its total emitted power is $\bar{P}_i^\downarrow$. In the uplink, the mobile's emitted power is calculated via an easy transformation of Equation (6.44):

$$P_m^\uparrow = \frac{l_m^\uparrow}{\nu_m^\uparrow L_{mi}^\uparrow} \cdot \bar{C}_i^\uparrow \qquad (6.58)$$

In the downlink, we rearrange Equation (6.51) and obtain

$$P_{im}^\downarrow = \frac{l_m^\downarrow}{\nu_m^\downarrow} \left(\bar{\alpha}_m \bar{P}_i^\downarrow + \sum_{j \neq i} \frac{L_{jm}^\downarrow}{L_{im}^\downarrow} \bar{P}_j^\downarrow + \frac{N_m}{L_{im}^\downarrow} \right) \qquad (6.59)$$

6.4.9.4 Example

The following small example shows how the above computation scheme works. For better readability, we mix numbers in logarithmic and linear scale. Numbers in logarithmic scale carry one of the units dB or dBm. All computing operations are in linear scale. The figures given are rounded; a higher precision has been used in the underlying computations.

Table 6.3 Data for example calculations.

m	$L^\downarrow_{am} = L^\uparrow_{ma}[\text{dB}]$	$L^\downarrow_{bm} = L^\uparrow_{mb}[\text{dB}]$	$CIR^\downarrow_m[\text{dB}]$	$CIR^\uparrow_m[\text{dB}]$	$\nu^\downarrow_m = \nu^\uparrow_m$
1	−80.0	−85.0	−17.5	−19.5	0.5
2	−70.0	−80.0	−17.5	−19.5	0.5
3	−80.0	−60.0	−12.1	−14.6	1.0

Suppose there are three mobiles, $\mathcal{M} = \{1, 2, 3\}$, and two cells, $\mathcal{N} = \{a, b\}$. The first two users are speech users, the third one uses video telephony. The respective CIR targets (see Section 6.4.1), activity values and attenuation values are given in Table 6.3. In addition, we assume that the noise at all receivers is $N^\uparrow = N^\downarrow = -102.0\,\text{dBm}$, the global orthogonality factor is $\bar{\alpha} = 0.673$, and that the common channel powers in both cells are $P^{CCH}_a = P^{CCH}_b = 30.0\,\text{dBm}$.

The corresponding user loads, as calculated in Equations (6.45) and (6.52), are

$$l^\downarrow_1 = l^\downarrow_2 = 0.009, \quad l^\downarrow_3 = 0.059$$
$$l^\uparrow_1 = l^\uparrow_2 = 0.006, \quad l^\uparrow_3 = 0.034$$

Assuming a best-server assignment, mobiles 1 and 2 are served by cell a, while mobile 3 is served by cell b. The coupling matrices are thus calculated as

$$M^\downarrow = \begin{pmatrix} 0.673 \cdot 0.009 + 0.673 \cdot 0.009 & (-5.0\,\text{dB}) \cdot 0.009 + (-10.0\,\text{dB}) \cdot 0.009 \\ (-20.0\,\text{dB}) \cdot 0.059 & 0.673 \cdot 0.059 \end{pmatrix}$$
$$= \begin{pmatrix} 0.012 & 0.004 \\ 0.001 & 0.040 \end{pmatrix}$$

and

$$M^\uparrow = \begin{pmatrix} 0.006 + 0.006 & (-20.0\,\text{dB}) \cdot 0.034 \\ (-5.0\,\text{dB}) \cdot 0.006 + (-10.0\,\text{dB}) \cdot 0.006 & 0.034 \end{pmatrix}$$
$$= \begin{pmatrix} 0.011 & 0.000 \\ 0.002 & 0.034 \end{pmatrix}$$

Furthermore, the traffic noise power in the downlink is

$$P^{(N)} = \begin{pmatrix} (-12.0\,\text{dBm}) \cdot 0.009 + (-22.0\,\text{dBm}) \cdot 0.009 \\ (-32.0\,\text{dBm}) \cdot 0.059 \end{pmatrix} = \begin{pmatrix} -32.1\,\text{dBm} \\ -44.3\,\text{dBm} \end{pmatrix}$$

The downlink transmission powers of the two cells are calculated by transforming Equation (6.57). This yields

$$\bar{P}^\downarrow = (I - M^\downarrow)^{-1}(P^{(N)} + P^{CCH}) = \begin{pmatrix} 30.07\,\text{dBm} \\ 30.18\,\text{dBm} \end{pmatrix}$$

By using Equation (6.58) we determine the powers on the individual links:

$$P^\downarrow_{a1} = 12.53\,\text{dBm} \quad P^\downarrow_{a2} = 11.44\,\text{dBm}, \quad P^\downarrow_{b3} = 16.25\,\text{dBm}$$

These power values ensure that all referring CIR equalities (Equation 6.38) are exactly met.

In the uplink, we calculate the received powers at the cells by transforming Equation (6.50) and obtain

$$\bar{C}^\uparrow = (I - M^\downarrow)^{-1} N^\uparrow = \begin{pmatrix} -101.95\,\text{dBm} \\ -101.84\,\text{dBm} \end{pmatrix}$$

($N^\uparrow$ is a vector with two components). Using Equation (6.59), one deduces for the mobiles' transmit powers:

$$P_1^\uparrow = -31.47\,\text{dBm} \qquad P_2^\uparrow = -41.47\,\text{dBm} \qquad P_3^\uparrow = -46.59\,\text{dBm}$$

Again, with these powers the CIR inequalities (Equation 6.36) are met exactly.

Note that in this example there are few users, so the cells are almost empty. The equation system reduction in dimension from $|\mathcal{M}| \times |\mathcal{M}|$ to $|\mathcal{N}| \times |\mathcal{N}|$ removes just one dimension as $|\mathcal{M}| = 3$ and $|\mathcal{N}| = 2$. However, as in realistic settings tens of users are served by one cell, the reduction is considerable.

6.4.10 EQUATION SYSTEM SOLVING

Computational linear algebra has produced a variety of methods for solving linear equation systems. The reduced systems are comparatively small in practical applications, so a simple iterative scheme is sufficient in most cases. This scheme can also be extended for incorporating load control, as is described below. In some circumstances, the reduction cannot be applied (see below). In this case, the efficient computational solution of the resulting large equation systems is more involved. Some advanced methods are therefore presented and evaluated empirically in their fitness for solving the specific tasks related to static system-level models of UMTS systems.

6.4.10.1 Iterative Solving

The dimension of the equation systems (6.50) and (6.57) is $\mathcal{N}$. In a planning scenario, this value does not typically exceed a couple of hundreds or a few thousands. In this situation, the use of sophisticated linear equation system solving software does not pay off. If there is a feasible power allocation, the following simple iterative Jacobi scheme converges to the correct power values for $t \to \infty$. Convergence has been observed to be quite fast in practice.

$$\bar{C}_i^{(0)\uparrow} := N_i^\uparrow \qquad \bar{C}_i^{(t+1)\uparrow} := \sum_{j=1}^{|\mathcal{N}|} M_{ij}^\uparrow \bar{C}_i^{(t)\uparrow} + N_i^\uparrow \qquad i = 1\ldots|\mathcal{N}| \qquad (6.60\text{a})$$

$$\bar{P}_i^{(0)\downarrow} := P_i^{CCH} \qquad \bar{P}_i^{(t+1)\downarrow} := \sum_{j=1}^{|\mathcal{N}|} M_{ij}^\downarrow \bar{P}_i^{(t)\downarrow} + P_i^{CCH} \qquad i = 1\ldots|\mathcal{N}| \qquad (6.60\text{b})$$

The iteration converges if and only if the solution to the equation system (6.57) or (6.50), respectively, is positive in all components. (If the analytical solution has negative components, there are too many users in the system and no feasible power allocation exists.) This is the case if and only if the spectral radius (the absolute value of the largest eigenvalue) of the referring coupling matrix $M^\downarrow$ or $M^\uparrow$ is smaller than one. For more mathematical details and results on nonnegative matrices, see [44].

6.4.10.2 Load Control

If radio resources do not suffice for serving all users, it is not trivial to determine which users are denied service. This process is referred to as *admission control* and *load control*; it heavily depends on the equipment vendor's implementation and on the operator's priorities. Detailed modelling of call admission and load control as well as some radio resources in question (e.g. code trees) are beyond the scope of a static model as downright dynamic aspects of the system are touched. Power values, however, are calculated. These are used to determine the cell loading factors. It can thus be determined whether cell load stays within the allowed range, and load control can be mimicked in this aspect.

In a real system, call admission and load control mechanisms keep load and transmit powers within the allowed range. The system denies service to some users either partially (lower data rate than desired) or altogether. If there are too many users in the static system model, resulting power values are either too high (e.g. $\bar{P}_i^\downarrow > 20\,\text{W}$) or even negative (if there is no feasible power assignment at all).

In a simple implementation, users might be added in a random order to the system. At each step, the system is evaluated. If the system's state becomes infeasible when adding a certain user, the user is blocked. The decision whether or not to block a user can be made quickly and on the basis of distributed data, e.g. using methods in [45]. Other schemes [43] have a notion of quality of service and assign priorities to users. They then try to serve users such that the overall 'satisfaction' is maximised.

Quickly estimating blocking rates
When there is no feasible load allocation for all users, the iterations (6.60) do not converge, the power values tend to infinity. If there is a feasible allocation, the resulting values might become too large. In this case, they cannot be interpreted as power values in a real system. Both cases can be avoided by solving the (cell-level) equation system with some modifications. Scaling factors λ can be used per cell to reduce traffic when the cell's maximum load is reached.

In the downlink, this requires only a slight modification of the iteration (6.60b) involving an additional value $\lambda_i^{(0)}$ for each cell:

$$\lambda_i^{(0)} := 1 \qquad \lambda_i^{(t+1)} := \min\left\{\lambda_i^{(t)},\, \frac{P_{\max,i}^{\text{BS}} - P_i^{\text{CCH}}}{M_{ii}^\downarrow P_{\max,i}^{\text{BS}} + P_i^{(N)} + \sum_{j\neq i} M_{ij}^\downarrow \bar{P}_j^{(t)\downarrow}}\right\}$$

$$\bar{P}_i^{(0)\downarrow} := P_i^{\text{CCH}} \qquad \bar{P}_i^{(t+1)\downarrow} := \frac{P_i^{\text{CCH}} + \lambda_i^{(t+1)}\left(P_i^{(N)} + \sum_{j\neq i} M_{ij}^\downarrow \bar{P}_j^{(t)\downarrow}\right)}{\left(1 - \lambda_i^{(t+1)} M_{jj}^\downarrow\right)}$$

This iteration scheme always converges. The resulting power values $\bar{P}^\downarrow$ never exceed the power limit $P_{\max,i}^{\text{BS}}$. If the power limit is reached, however, the resulting scaling factor λ_i is below 1 and can be interpreted as an approximation of the cell's blocking rate. For details and a similar solution for the uplink, see [46].

6.4.10.3 Solution without Dimension Reduction

The Smart Antenna (SA) technique dynamically adjusts a radiation pattern of the antenna in a base station to served users, which leads to more efficient power control. The details on this are given, e.g., in [47]. The model for a UMTS network with SAs can also be expressed as a system of linear equations [48]. However, the dimension reduction given by Equations (6.43–6.58) cannot be applied for the SA as attenuation values of interferers depend on the SAs. We thus discuss here the selected methods for solving large systems of linear equations. We shortly refer to these methods as Linear Solvers. In general, given a power vector P, let us consider the following system of equations

$$AP = b \qquad A \in \mathbb{R}^{\mathcal{M} \times \mathcal{M}} \qquad P \in \mathbb{R}^{\mathcal{M}} \qquad b \in \mathbb{R}^{\mathcal{M}}. \qquad (6.61)$$

The system matrix A in a network with SAs is large, square, non-symmetric, non-singular and sparse for a very large network. After a suitable scaling, A is also quite well conditioned. By its construction A is irreducible, diagonally dominant, and hence, from the Gershgorin's theorem results that A is positive-definite, which assures the positive solution. Moreover, both in the uplink and the downlink, the equations are consistent.

Many methods [49,50] for solving such systems of equations could be applied here, but unfortunately the area of application is severely limited if we take into account the computational cost that should be kept as low as possible. This excludes usage of many direct methods [49], especially as A is different in each snapshot. For example, the cost of the Gaussian elimination is $O\left(\frac{2}{3}\mathcal{M}^3\right)$, which is prohibitive. Thus iterative methods seem to be the most preferable in our application. In this respect, we discuss the class of stationary iterative methods [49] that have the form

$$\mathbf{S}P^{(k+1)} = TP^{(k)} + b \qquad k = 0, 1, \ldots \qquad (6.62)$$

where $A = \mathbf{S} - T$, and $P^{(0)}$ is an initial guess.

The conditions under which iterations (Equation 6.62) are convergent can be found, e.g., in [49]. The matrix $\mathbf{S}$ can be regarded as a left preconditioner, and its selection determines the method as well as its convergence rate. Basically, we consider the Richardson, Jacobi, Gauss-Seidel, and Successive Over-Relaxation (SOR) methods. The computational costs per one iteration in these methods can be roughly estimated as $2\mathcal{M}^2 + 3\mathcal{M}, 2\mathcal{M}^2 + 3\mathcal{M}, 3\mathcal{M}^2$ and $3\mathcal{M}^2 + \mathcal{M}$ arithmetic operations, respectively. For the SOR, one must add to this the additional off-line cost of creating lower and upper triangular matrices L and U, which amounts to about $\mathcal{M}^2 + \mathcal{M}$ arithmetic operations.

The standard Richardson method is known in the literature to be very slow-convergent; however, with the appropriate preconditioning this method can be competitive. Since the eigenvalues of A for a typical WCDMA network configuration are very small, it is obvious that the right scaling must be applied to matrix A. This can be achieved by a right-side preconditioning, i.e. $(AD^{-1})(DP) = b$. Let $AD^{-1} = \tilde{A}$ and $DP = \tilde{P}$, thus we have $\tilde{A}\tilde{P} = b$. Matrix D should be defined in this way in order that eigenvalues of $\tilde{A}$ would be close to one, and the system $DP = \tilde{P}$ should be very easy to solve. Taking into account this, we may use column scaling with $D^{-1} = \mathrm{diag}\{\|a_1\|_2^{-1}, \ldots, \|a_\mathcal{M}\|_2^{-1}\}$, where $a_i (1 \leq i \leq \mathcal{M})$ is the i-th column of A. The right-side preconditioning requires about $2\mathcal{M}^2 + \mathcal{M}$ arithmetic operations.

A meaningful acceleration of the Richardson iterations can be also obtained with a polynomial approach that in most cases leads to the Krylov subspace projection methods. They belong to a very numerous and robust class of linear solvers (the polynomial Q can be defined in many ways, which is equivalent to various linear combinations of the vectors that span the Krylov subspace). Here, we only mention the exemplary methods which are known as very efficient for solving the systems with above-mentioned properties. There are the methods such as Conjugate Gradient Square (CGS) [51], Quasi-Minimal Residual (QMR) [52], Bi-Conjugate Gradient (BiCG) [53] and Bi-Conjugate Gradient STABilized (BiCGSTAB) [54], which belong to the Petrov-Galerkin class. Another robust method is the Generalized Minimum RESidual (GMRES) [55] and its modifications [50]. For a comparison, one iteration of the CGS needs about $4\mathcal{M}^2 + 22\mathcal{M}$ arithmetic operations.

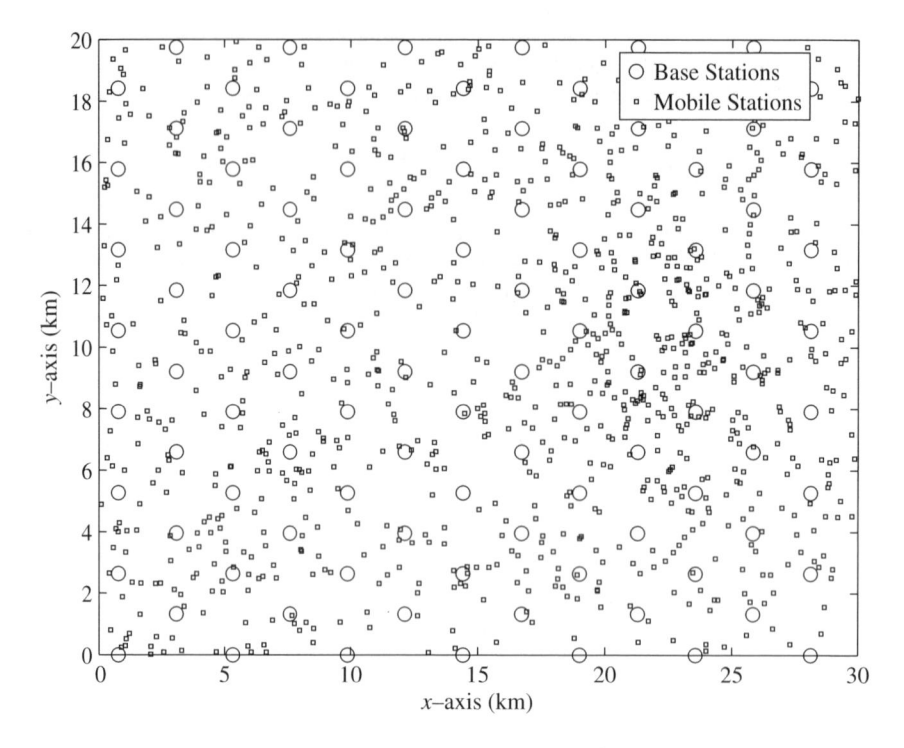

Figure 6.16 Layout of simulated network.

6.4.10.4 Computational Test for Solution Without Dimension Reduction

In the following, the performance of the selected linear solvers is tested for one randomly selected snapshot in the uplink in the WCDMA network with both omnidirectional antenna and SA. We assumed 1000 users randomly distributed in 104 cells* with a mixture of the uniform and screw-Gaussian distributions. Hence we have $A \in \mathbb{R}^{1000 \times 1000}$. The layout of base stations and mobile stations is presented in Figure 6.16. Half of the users work with a voice service ($CIR_m^\uparrow = -19.5\,dB$) and the other half with a data service ($CIR_m^\uparrow = -14.6\,dB$). Site-to-site distance equals 2.67 km. In the case of SA$_S$, the attenuation values are specific not only to a pair of sender/receiver, but to the triple sender/receiver/interferer. For this snapshot and the traditional antennas $\max_i\{|\lambda_i(A)|\} = 2.1 \times 10^{-7}$ and $\min_i\{|\lambda_i(A)|\} = 8.7 \times 10^{-13}$, and for the SA $\max_i\{|\lambda_i(A)|\} = 2.1 \times 10^{-6}$ and $\min_i\{|\lambda_i(A)|\} = 9.2 \times 10^{-12}$. Thus the convergence of the Richardson, Jacobi, Gauss-Seidel and SOR methods is definitely guaranteed. For the scaled matrix $\tilde{A}$, $\lambda \in [1.032 \quad 0.1813]$ for the case of traditional antennas and $\lambda \in [1.0536 \quad 0.9431]$ for SAs. Since the eigenvalues of the scaled matrix are slightly greater than one, the relaxation in the Richardson must be used.

All the iterative algorithms are run until the stopping criterion $e^k = \|P^k - P^{k-1}\|_\infty \geq \epsilon$ is met, where for arbitrary u: $\|u\|_\infty = \max_i\{|u_i|\}$, and ϵ is a small positive number. We assume that the solution

---

*In real applications the number of users per cell is usually much larger than 10. However, in the experiments our small problem has a similar spectral characteristics and much smaller complexity than a bigger one.

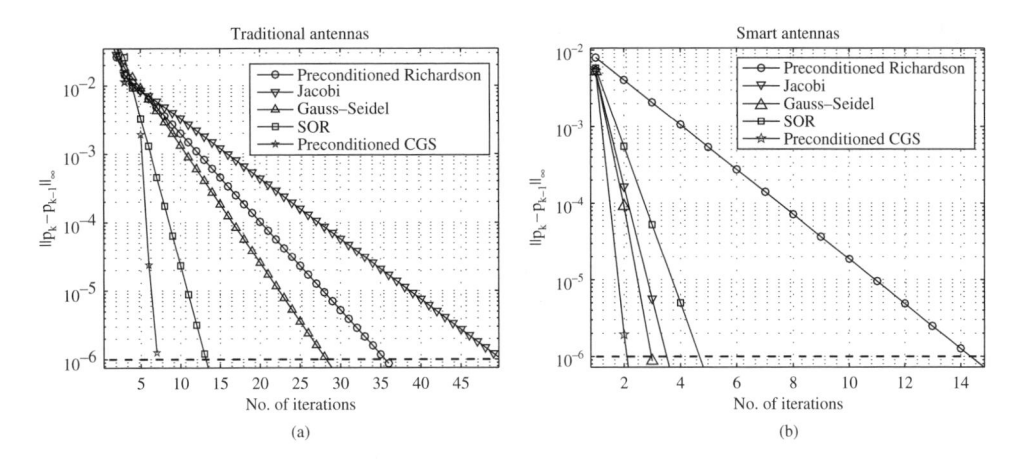

Figure 6.17 History of error e^k versus iterations for traditional antennas (left) and SAs (right).

should be computed with the accuracy up to the fifth significant digit, thus $\epsilon = 10^{-6}$. The plots of e^k versus iterations are illustrated in Figure 6.17 for the cases of traditional antennas (a) and SAs (b), respectively.

The dashed horizontal lines in Figure 6.17 mark the level 10^{-6} at which the iterations are stopped. It follows from Figure 6.17a that this level or lower is reached by the Richardson, Jacobi, Gauss-Seidel, SOR and Preconditioned CGS after performing 36, 50, 29, 14 and 7 iterations, respectively. For SAs (see Figure 6.17b), this level is reached within 15, 4, 3, 5 and 3 iterations for the Richardson, Jacobi, Gauss-Seidel, SOR and Preconditioned CGS methods, respectively.

For the traditional antennas, the computational cost of performing 36 iterations with the Richardson method (including the preconditioning) is about $72\mathcal{M}^2 + 108\mathcal{M}$. For the Jacobi's, Gauss-Seidel and SOR methods we have $100\mathcal{M}^2 + 150\mathcal{M}$, $87\mathcal{M}^2$ and $44\mathcal{M}^2 + 16\mathcal{M}$, respectively. The 7 iterations with the Preconditioned CGS cost about $28\mathcal{M}^2 + 154\mathcal{M}$.

The similar analysis for the SAs leads to the following estimations of the costs: $32\mathcal{M}^2 + 46\mathcal{M}$, $8\mathcal{M}^2 + 12\mathcal{M}$, $9\mathcal{M}^2$, $16\mathcal{M}^2 + 6\mathcal{M}$ and $14\mathcal{M}^2 + 67\mathcal{M}$ for the Richardson, Jacobi, Gauss-Seidel, SOR and Preconditioned CGS methods, respectively.

This comparison shows that the Gauss-Seidel method is the most promising for the SA. For the traditional antennas the CGS takes the first place, and then the SOR. The costs presented for the Richardson and SOR do not include the cost of determining the optimal value of relaxation parameters. This requires computation of at least the highest eigenvalue, which is generally very expensive. More details on the computational costs can be found in [48].

This shall finish the section on static system level modelling, and we will continue now on the issue of modelling the dynamics of the UMTS system.

6.5 DYNAMIC SYSTEM LEVEL MODEL

6.5.1 SIMILARITIES AND DIFFERENCES BETWEEN STATIC AND DYNAMIC MODELS

Static system level simulators are powerful tools to assess wireless networks, and are often used in planning cellular systems. As the name suggests, static simulators truncate the system changes over time. This, on the one hand, facilitates computations and enables quick evaluations. However, this limits the applicability of static simulators to the case when a snapshot solely integrates the essential features of the system. When the system is packet switched, such as in HSDPA, static simulators are insufficient due to the time division foundation, which causes significant changes in the transmission power and the consequent interference, as well as the instantaneous data rate of each user over time. Moreover, when the aim is to evaluate the packet delay, handoff frequency or flow control mechanisms (that require measurements over time), static simulators fall short of capability.

To mitigate such weaknesses of the static type, dynamic models can be configured. The fundamental structure of dynamic simulators is in many ways similar to static simulators. As in static models, multi-cell layout and user distribution, as well as techniques such as wrap around, can still be applied to dynamic models. However, dynamic simulators incorporate system mechanisms in greater detail, with the most important difference being the actual simulation of the time scale. This broadens the evaluation capability, at the cost of complexity in developing efforts and computational power. Additional features of a dynamic model may include but are not limited to:

- call arrivals and departures;
- mobility of users;
- transmission control mechanisms;
- radio resource management mechanisms;
- application layer mechanisms.

6.5.2 GENERIC SYSTEM MODEL

As in a static simulator, various levels of detail can be modelled in a dynamic simulator. The effective model and depth of detail depend on the system scenario and the required measurements. For example, when the call duration is relatively short in comparison with the mobility of users, as in a text page download, the user location is virtually the same during a call. Then mobility, as well as handoff mechanisms, can be omitted for simplicity. If the aim is to evaluate handoff frequency, then mobility and handoff procedures must be modelled in detail, whereas slot-by-slot transmission can be simplified into a stochastic model.

A typical flow of a dynamic simulator is shown in Figure 6.18. The model incorporates call arrivals and departures, with the next call arrival time denoted by t_{ca}. Hence, the number of concurrent users, namely N_{user}, changes through the simulation course. The time scale is simulated with a discrete step T, which can typically be the slot or frame interval. The scale of choice depends on the expected system variations (e.g. small scale fading) within the interval T, and the computational amount that can be afforded. From the calculated SIR, block errors are generated randomly by looking up a BLER (or BER) table. The table must be prepared to map the instantaneous SIR (or the SIR statistics, e.g. mean and variance, within the interval T), not the long-term average SIR as in static simulators. The observation time t_{obs} must be large enough to obtain the relevant statistics. Moreover, to avoid data collection during the initial transition from an empty system, a preliminary run is often required. This can be implemented simply by resetting the measurements when the simulation time t reaches the predefined time $t_{pre}(< t_{obs})$, or the number of arrived (or departed) calls reaches a predefined number.

To facilitate the laborious process of dynamic system level simulations, simplified models are appreciated where possible. For instance, the user locations, as well as the path loss and shadowing values, can be updated in a slower cycle than SIR calculations, since the location can be seen as virtually static within a frame interval. The small scale fading must still be updated at short intervals, if the mobility implies a large change.

Theoretical Models for UMTS Radio Networks

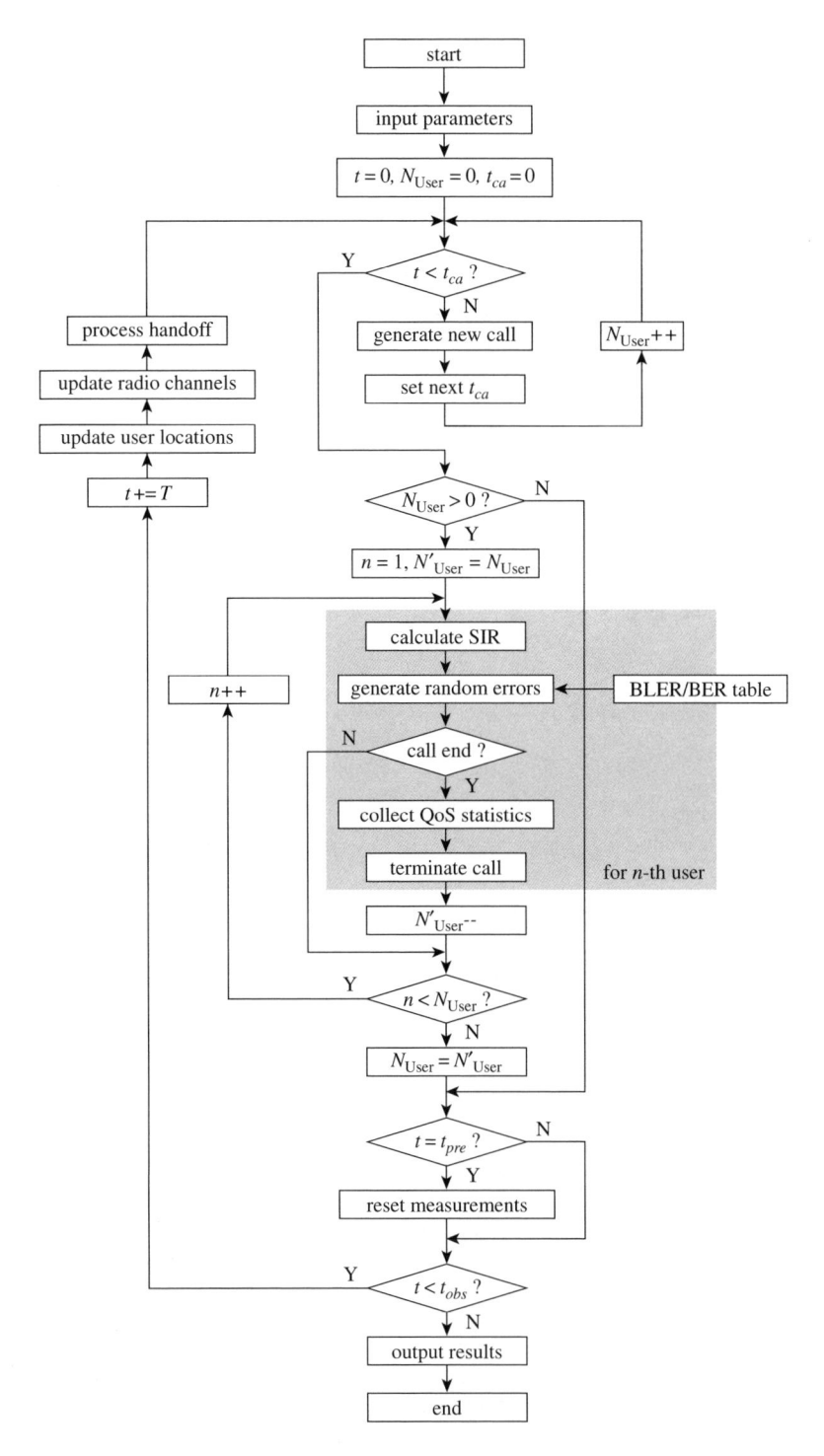

Figure 6.18 Generic flow of a dynamic system level simulator.

6.5.3 INPUT/OUTPUT PARAMETERS

The input parameter of a dynamic model ranges from the network layout, as in static models, to traffic and mobility models, as well as radio resource management strategies. Following is a list of some input parameters:

- time scale granularity;
- network layout (e.g. site locations, sectorisation, antenna pattern);
- user distribution and mobility;
- propagation models (e.g. path loss, shadowing, multipath fading);
- traffic models (e.g. call arrival pattern, packet arrival pattern, transport protocol);
- traffic QoS (e.g. required throughput, packet delay, packet loss rate);
- radio bearer (e.g. data rate, transmission time interval, maximum transmission power, dynamic range, BLER table);
- handoff mechanism (e.g. handoff thresholds, hysteresis, time to trigger)
- radio resource management strategies (e.g. call admission control, transmission power control, packet scheduling discipline).

The dynamic model is then capable of producing wide variety of measurements. Below are some typical outputs:

- user and system throughputs,
- packet delay,
- packet loss rate,
- call sojourn time,
- handoff frequency,
- transmission power,
- radio resource usage,
- call blocking probability,
- call dropping probability,
- system capacity.

These measurements can be obtained as mean values or as distributions. An inline processing and/or post-processing can further arrange data into various meaningful results. An example would be to collect user throughput as a function of location, to evaluate the location dependency.

6.5.4 MOBILITY MODELS

A plethora of mobility models have been proposed for various environments. The simplest model is the user travelling in a straight line at a constant speed. A more realistic model incorporates changes in the direction of travel. The mobility trace then becomes a series of straight segments articulated at random angles. For example in [56], an exponential distribution (mean value equals 25 s) is suggested for the duration of a straight travel, and a 1:2 blend of normal (mean value equal 82°, standard deviation value equals 10°) and exponential distribution (mean value equals 10°) are suggested for the grafting angle for macrocell environments, based on an empirical analysis of taxi traces in urban areas. Figure 6.19 shows a trace generated by this model. A more sophisticated model further incorporates changes in the velocity, given a certain transition function. Note that changes in velocity impose relevant changes in the fast fading (e.g. Doppler frequency). Stochastic models for Manhattan environment and indoor office environment are found in [57,58], for example.

Another approach would be to feed empirical data into the simulator. A noteworthy remark is that in such a case, the essential input, if available, is the radio propagation data rather than the location itself.

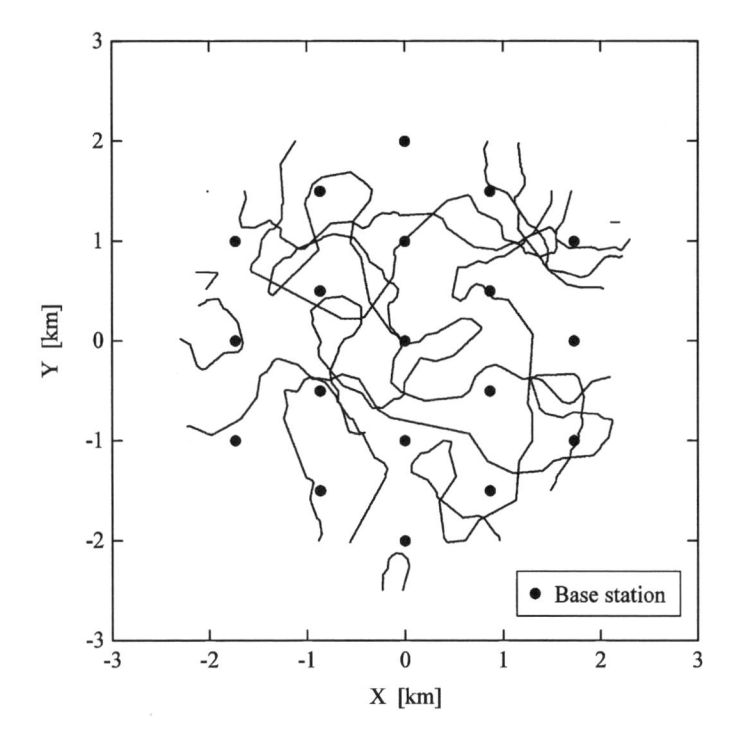

Figure 6.19 Example of mobility trace using a stochastic model.

A practical network is unaware of the geometric location of the user, but is able to obtain propagation estimates. Hence, radio resource management (such as handoff and transmission power control) is generally performed according to the radio propagation condition, not the location itself. As such, the location data is essentially unnecessary, unless the intension is to evaluate certain metrics by location (e.g. to evaluate throughput as a function of the distance from the serving base station, or to show SIR on a geographical map).

6.5.5 TRAFFIC MODELS

The traffic model largely depends on the service profile, and can be segregated into two parts: call arrival model and data (packet) arrival model within each flow. The simplest model for call arrivals is the fixed interval model; this applies when the traffic is generated by machines, as in sensor networks where measurement reports are sent periodically. The most widely appreciated model for call arrivals is the Poisson process [59], in which the interval is generated by an exponential distribution. When the sequential arrivals are independent and the arrival rate is fixed, the process is called a stationary (or homogeneous) Poisson process. This applies when the number of subscribers in a cell is very large, while the probability of a user making a call is small. If this does not apply, the process becomes non-stationary (or inhomogeneous), in which the arrival rate varies over time. This is due to the fact that as more users become concurrently active, the probability of a new call arrival reduces with a fixed number of subscribers. In a multi-service scenario, a Poisson process can be split into the multiple services. That is, a single Poisson process can be used to generate call arrival instances, and an independent random variable can be drawn at each arrival to determine the service (i.e. Bernoulli process). The resulting process for each service is also known to be Poisson [59].

The data arrival model within each flow depends largely on the application. For example, conversational speech traffic exhibits constant rate with (or without) activity, whereas interactive traffic has more 'bursty' characteristics. Furthermore, in flows that cross a packet switched network, the data (packet) arrival pattern depends on the transport protocol, network load and the packet forwarding behaviour at routers. The simplest model would be to consider fixed size bulk data transfer with constant rate. Although such model is unrealistic, the model is still useful to assess fundamental performance of the radio access network. To evaluate the traffic QoS in detail, relevant models that capture the traffic behaviour are necessary. Applicable stochastic models for conversational speech, video streaming and web browsing traffic are described in the sequel.

For conversational speech traffic, an exponentially distributed call duration model with voice activity has been widely used [57,58]. The mean call duration typically ranges around 60 to 90 s. The talk spurts and gaps also follow an exponential distribution with typical mean values of 1.0 and 1.5 s, respectively. Each talk spurt is to be divided into a number of codec/transport blocks, and arrive to the access point at constant intervals (or with slight jitter, more realistically, with a packet-switched backbone). Note that during silence, pilot symbols may still be transmitted on the dedicated channel to maintain radio synchronisation in UMTS. The transmission power allocated to such symbols should be taken into account in the interference calculation for increased precision.

For video streaming traffic, a different model is necessary to reflect the inherent variable bit rate (VBR) behaviour. The VBR video traffic has been shown to exhibit autocorrelation functions that decrease hyperbolically as the lag increases, hence exhibiting short-range dependence (SRD) and long-range dependence (LRD) [60]. The SRD and LRD characteristics are cumbersome to model, resulting in numerous models being reported, e.g. histogram-based models, Markov chain models, autoregressive processes and self-similar models [61,62]. Of the various schemes, the wavelet-based model [63,64] has been shown to provide adequate accuracy in capturing the marginal distributions and the autocorrelation structure, without needing significant amount of data storage and laborious computations. The details of the wavelet-based model can be found in [64], for example. If the traffic is constant bit rate (CBR), a model can be used, in which fixed size packets arrive at constant intervals.

For general web browsing traffic, the ETSI WWW model [57] with modified parameters has been widely used. Figure 6.20 illustrates the traffic pattern of a web browsing session in this model. The session comprises a number of packet calls, where each packet call corresponds to downloading a web page, an email message or a file. The reading time between consecutive packet calls represents the time required by the user to read the page and to trigger downloading of the next page. The packet call size (data size) is derived by a truncated Pareto distribution, the pdf of which is given by

$$pdf(x) = \begin{cases} \dfrac{\alpha_p \cdot k_p^{\alpha_p}}{x^{\alpha_p+1}} + \left(\dfrac{k_p}{m_p}\right)^{\alpha_p} \cdot \delta(x - m_p) & k_p \leq x < m_p \\ 0, & \text{otherwise} \end{cases} \qquad (6.63)$$

where the parameters α_p, k_p and m_p are set to $\alpha_p = 1.1$, $k_p = 4.5$ KB and $m_p = 2$ MB, yielding a mean packet call size of 25 KB. As shown in Figure 6.20 a packet call consists of a series of IP packets, which have a size of 1.5 KB (the maximum transmission unit (MTU) in the widely used Ethernet) except for the last IP packet to complement the packet call size. These IP packets arrive to the serving base station at geometrically distributed intervals.

In terms of QoS, the time taken to retrieve a packet call, or the throughput per packet call, is often of the main interest. Then, the throughputs of individual packet calls are measured by dividing the source data size (packet call size) by the sojourn time per packet call. If the QoS constraint only applies to

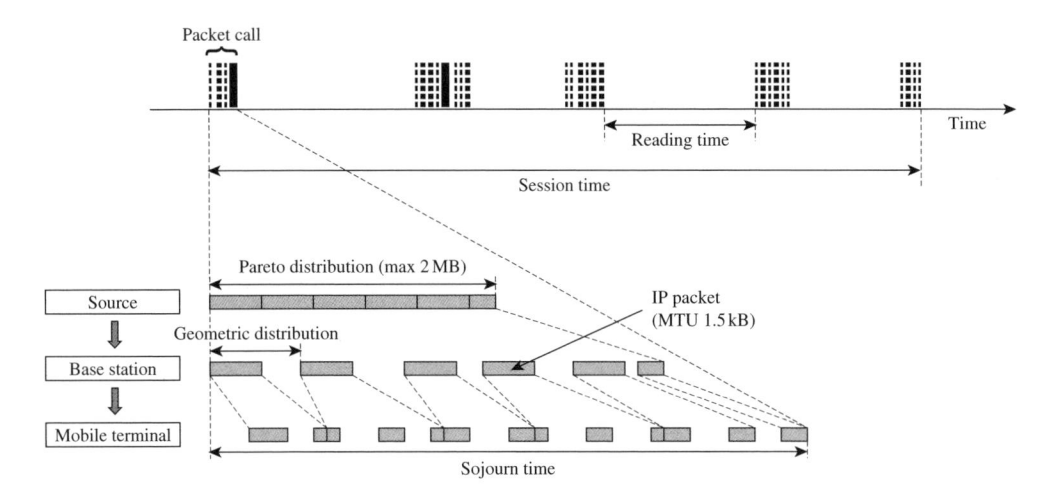

Figure 6.20 Modified ETSI WWW traffic model.

each packet call, the entire session does not necessarily have to be simulated. Instead, Poisson arrivals of packet calls can be assumed to facilitate simulations. If a QoS constraint applies to the entire session, an explicit model is necessary for the reading time, with an example being the exponential distribution.

A more direct and accurate approach would be to implement the actual transport and application layer protocols, or their simplified versions, in the simulator. This will enable evaluations of the impact of TCP slow start, for example.

6.5.6 PATH LOSS MODELS

The radio propagation channel can be characterised as a concatenation of the path loss, shadowing and multipath fading [65]. Figure 6.21 depicts this three 'layer' model. The path loss represents the mean (median) attenuation due to the distance between the base and mobile stations, whereas the shadowing represents the random fading of the received power, typically caused by the changes in the surrounding building heights. The multipath fading represents the rapid fluctuation caused by a number of propagation paths interfering one another, and causes large variations with deep fades in the received power.

The path loss represents the mean (median) signal attenuation at a certain distance from the transmitter, and can be predicted by the distance and other macroscopic parameters such as carrier frequency, transmitter and receiver antenna heights, terrain contour and building concentration [66,67]. The simplest form of path loss is the free space loss, which applies to the extreme case when nothing obstructs the propagation path. An observation that the power density integrated over a spherical surface equidistant from the transmitter is constant in free space yields that the transmitted signal decays by the squared distance [66]. However, in practice, a propagation path is almost always obstructed, by surrounding terrain, buildings or the atmosphere that causes a refraction loss [66].

To predict the path loss considering these practical effects, Okumura *et al.* developed an empirical method based on extensive measurements in an urban area [68]. The principle of the Okumura method is to compensate the free space loss by empirically obtained factors (that are read from graphical data). For more details on path loss propagation modelling please see Chapter 5.

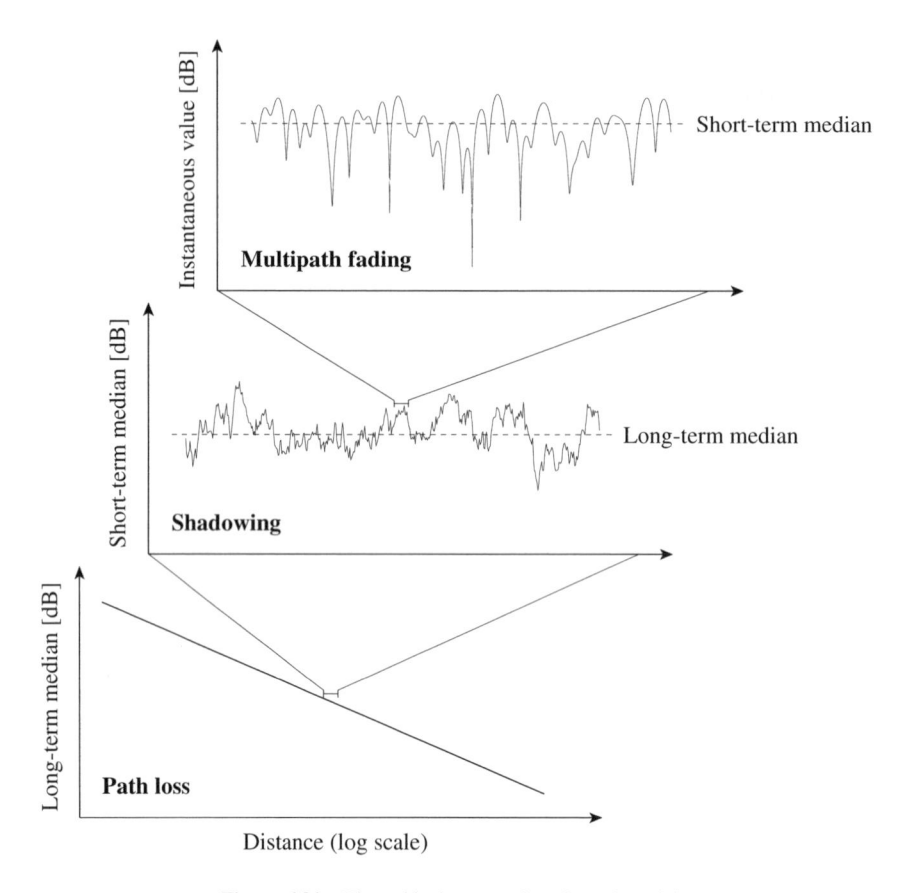

Figure 6.21 Hierarchical propagation channel model.

6.5.7 SHADOWING MODELS

Shadowing characterises variations of the local mean (median), caused by mobile terminals travelling through 'shadows' of surrounding obstacles. A mobile terminal may be severely shadowed by large obstacles or occasionally be in line-of-sight (LOS) to a base station. Empirical studies in [69] have shown that the shadowing can be modelled by lognormal distribution, having a standard deviation typically ranging between 4 and 12 dB [67,70], depending on the environment. Note that the mean of the lognormal distribution is usually absorbed into the path loss model, thereby making it a zero mean lognormal shadowing distribution.

In general, shadowing values are spatially correlated because of the geometrical nature. Gudmundson [71] has shown that the autocorrelation function $R_S(\Delta d)$ of shadowing can be described with sufficient accuracy by an exponential function, given by

$$R_S(\Delta d) = \exp\left(-\frac{|\Delta d|}{d_d}\ln 2\right) \qquad (6.64)$$

where d_d is the decorrelation length, i.e. the distance at which the autocorrelation is reduced to one half, that depends on the environment (e.g. $d_d = 20\,\text{m}$ is suggested for macrocells in [58]). Hence, denoting the shadowing L_S as a function of the location d, the shadowing value in decibels can be derived auto-regressively as

$$L_S(d + \Delta d) = R_S(\Delta d) \cdot L_S(d) + \sqrt{1 - R_S^2(\Delta d)} \cdot N(0, \sigma_S), \qquad (6.65)$$

where $N(0, \sigma_S)$ is a normally distributed random variable having zero mean and standard deviation of σ_S.

In cellular systems, the propagation from multiple cells needs to be considered. The cross-correlation coefficients between the shadowing of different base stations are generally non-zero, since shadowing is affected by obstacles in the mobile terminal's vicinity. In Monte-Carlo simulations, a set of correlated shadowing values $\{L_{S,1}, L_{S,2}, \dots\}$ can be generated by using correlated lognormal random variables [72]. Using a common seed L_0 given by

$$L_0 = N(0, \sigma_S) \qquad (6.66)$$

the shadowing value for the i-th site ($i = 1, 2, \dots$) can be derived by

$$L_{S,i} = \sqrt{\rho} \cdot L_0 + \sqrt{1 - \rho} \cdot N(0, \sigma_S) \qquad (6.67)$$

where ρ is the correlation coefficient, which depends on the environment. For macrocell environments, $\rho = 0.5$ is suggested in [58].

6.5.8 MODELLING OF SMALL SCALE FADING

The multipath fading is caused by a number of scattered waves arriving at the receiver, each with a different amplitude, phase and arrival angle. Consequently, the received signal strength varies rapidly, with successive deep minima occurring about every half wavelength of the carrier frequency. The multipath fading can be modelled as a linear superposition of plane waves having random phase and a Doppler shift [67]. Assuming N_w waves arrive at the receiver with amplitudes a_j and angles of arrival ζ_j ($j = 1, 2, \dots, N_w$), the received signal $r(t)$ can be given by

$$r(t) = \sum_{j=1}^{N_w} a_j \cdot \cos\left\{2\pi f_c t + \vartheta_j + \frac{2\pi v}{\lambda} \cos(\zeta_j) \cdot t\right\} \qquad (6.68)$$

where ϑ_j is the random phase of each wave, introduced as a result of the small difference in the path length of each wave, v is the mobile velocity and λ is the wavelength of the carrier. The coefficient of the third term in the bracket represents the Doppler shift, which depends on the arrival angle and speed.

Equation (6.68) can be rewritten as

$$r(t) = A_I(t) \cos 2\pi f_c t - A_Q(t) \sin 2\pi f_c t \qquad (6.69)$$

where

$$\begin{aligned} A_I(t) &= \sum_{j=1}^{N_w} a_j \cdot \cos\left\{\vartheta_j + \frac{2\pi v}{\lambda}(\cos \zeta_j) \cdot t\right\} \\ A_Q(t) &= \sum_{j=1}^{N_w} a_j \cdot \sin\left\{\vartheta_j + \frac{2\pi v}{\lambda}(\cos \zeta_j) \cdot t\right\} \end{aligned} \qquad (6.70)$$

If N_w is sufficiently large ($N_w \geq 6$, according to [73,74]) and the comprising waves have similar amplitudes (as in the case of NLOS), the central limit theorem invokes that $A_I(t)$ and $A_Q(t)$ are Gaussian processes having zero mean with the same variance, namely σ^2. Consequently, if we rewrite Equation (6.69) as

$$r(t) = A(t) \cdot \cos\{2\pi f_c t + \theta^{ph}(t)\} \qquad (6.71)$$

where

$$A(t) = \sqrt{A_I^2(t) + A_Q^2(t)} \qquad (6.72)$$

$$\theta^{ph}(t) = \tan^{-1}\{A_Q(t)/A_I(t)\} \qquad (6.73)$$

the envelope $A(t)$ is Rayleigh distributed and the phase $\theta^{ph}(t)$ is uniformly distributed over $(0, 2\pi]$. The pdf of $A(t)$ is thus given by

$$pdf(A) = \frac{A}{\sigma^2} \cdot \exp\left(-\frac{A^2}{2\sigma^2}\right) \qquad (6.74)$$

If a dominant path exists (as in the case of LOS) the envelope follows Ricean distribution [75], which is given by

$$pdf(A) = \frac{A}{\sigma^2} \cdot \exp\left(-\frac{A^2}{2\sigma^2} - K\right) \cdot I_0^{Bessel}\left(\frac{A\sqrt{2K}}{\sigma}\right) \qquad (6.75)$$

where K is the power ratio of the dominant wave and the sum of the other waves, and I_0^{Bessel} is the first kind Bessel function of the zero-th order. The Ricean pdf approaches Rayleigh as K is decreased (the dominant path is diminished).

In dynamic simulators, Equation (6.68) can be used to generate a Rayleigh distributed trace. The amplitudes a_j can be set uniform, and ϑ_j as a random variable $U(0, 2\pi)$, i.e. uniform distribution. The angles of arrival can be set as $\zeta_j = \zeta_1 + \frac{2\pi(j-1)}{N_w}$, where ζ_1 is $U(0, 2\pi)$. Note that N_w should be an odd integer to avoid a wave having another wave coming from exactly the opposite direction, which would diminish their effect in pairs. Alternatively, ζ_j can also be $U(0, 2\pi)$. For the Ricean model, the parameter K needs to be defined, which can be static or dynamic with some transition function.

In cellular systems, the multipath spread is typically about a few micro seconds, and can be up to about 20 μs in open areas where heavily delayed paths arrive to the receiver. If the multiple paths are distributed over a delay larger than the symbol duration (or chip duration in DS CDMA), the receiver is able to distinguish different paths. In such a (frequency selective) case, the multipath channel can be modelled as a tapped delay line [75], with each tap having an independent Rayleigh (or Ricean) distributed envelope. The sum of the autocorrelation coefficients of each tap is equal to one, such that the received power is distributed among the paths.

6.5.9 SIR CALCULATION

In static simulators the orthogonality factor is used to simplify SIR calculations under multipath interference. However, the orthogonality factor is based on the long-term average effect of multipath interference. Since fast multipath fading is simulated in dynamic simulators, the SIR calculation needs more detailed modelling. The SIR can be calculated per slot, and the SIR statistics (e.g. mean) over a frame (i.e. the transmission time interval or the coding block size) can be used to look up a BLER table.

Alternatively, the SIR can be calculated per frame, if slot-by-slot dynamics is not of main interest. The BLER is used to generate random errors. Details of the SIR calculation are described in the sequel.

The received signal of the k-th path for a data channel after despreading is given by

$$r_{d,k} = A_d \cdot c_k \cdot s_d + n_{d,k} \tag{6.76}$$

where A_d, c_k, s_d and $n_{d,k}$ denote the received signal amplitude, channel vector, data symbol ($|s_d|^2 = 1$) and noise/interference vector respectively on the k-th path after despreading. Using similar notations, the received signal for the CPICH is given by,

$$r_{c,k} = A_c \cdot c_k \cdot s_c + n_{c,k} \tag{6.77}$$

where A_c, s_c and $\mathbf{n}_{c,k}$ denote the received CPICH amplitude, deterministic common pilot symbol ($|s_c|^2 = 1$) and noise/interference vector respectively. Note that all variables are functions of time, although the time representation is omitted for simplicity. The channel vectors $\mathbf{c}_k$, ($k = 1, 2, \ldots$) are defined such that the sum of their autocorrelations for all k is equal to one. The signal amplitudes A_d and A_c are given by

$$A_d = \sqrt{Pt_d/L} \tag{6.78}$$

$$A_c = \sqrt{Pt_c/L} \tag{6.79}$$

where Pt_d and Pt_c represent the transmission power of the data channel and CPICH, respectively, and L is the short term average path loss for the serving cell including the shadowing and antenna gain. The channel estimate for the k-th path to demodulate $\mathbf{r}_{d,k}$ is derived by averaging $\mathbf{r}_{c,k}$ in the time vicinity. Assuming that the channel is constant over this averaging interval, the estimated channel vector $\hat{\mathbf{c}}_k$ is given by

$$\hat{\mathbf{c}}_k = E\left[\mathbf{r}_{c,k} \cdot \mathbf{s}_c^*\right] = A_c \cdot \mathbf{c}_k + \bar{\mathbf{n}}_{c,k} \tag{6.80}$$

where $E[\cdot]$ indicates ensemble average. The residual noise/interference component $\bar{\mathbf{n}}_{c,k}$ is the channel estimation error. By applying standard Gaussian approximations on the additive noise and interference, the variances of the vectors $\mathbf{n}_{d,k}$, $\mathbf{n}_{c,k}$ and $\bar{\mathbf{n}}_{c,k}$ are given by

$$\sigma_{d,k}^2 \equiv Var\left[\mathbf{n}_{d,k}\right] = \frac{I_k}{SF_d} \tag{6.81}$$

$$\sigma_{c,k}^2 \equiv Var\left[\mathbf{n}_{c,k}\right] = \frac{I_k}{SF_c} \tag{6.82}$$

$$\bar{\sigma}_{c,k}^2 \equiv Var\left[\bar{\mathbf{n}}_{c,k}\right] = \frac{I_k}{SF_c \cdot m_{\text{CPICH}}} \tag{6.83}$$

respectively, where SF_d and SF_c are the spreading factors of the data channel and CPICH, respectively, m_{CPICH} is the number of CPICH symbols in the ensemble of Equation (6.80) and I_k is the effective noise/interference power on the k-th path, which, considering the orthogonality of the codes from the serving cell on the same path, is given by

$$I_k = I - \frac{Pt_{or}}{L} \cdot |\mathbf{c}_k|^2 \tag{6.84}$$

In Equation (6.84), I is the aggregate power of the thermal noise, interference and own signals at the rake input, and Pt_{or} is the aggregate transmission power of the set of orthogonal codes from the serving cell. The aggregate power I is given by

$$I = N_0 W + \sum_{i}^{\text{all cells}} \sum_{k}^{\text{all paths}} \frac{Pt_i}{L_i} \cdot |c_{i,k}|^2 \qquad (6.85)$$

where N_0 is the thermal noise power spectral density (including the receiver noise figure), W is the chip rate and Pt_i is the total transmission power of the i-th cell.

Using the signal notations above, the output of the Rake receiver is given by

$$\begin{aligned} z &= \sum_k \mathbf{r}_{d,k} \cdot \hat{\mathbf{c}}_k^* \\ &= A_d A_c \sum_k |\mathbf{c}_k|^2 \mathbf{s}_d + A_c \sum_k \mathbf{c}_k^* \cdot \mathbf{n}_{d,k} + A_d \sum_k \mathbf{c}_k \cdot \overline{\mathbf{n}}_{c,k}^* + \sum_k \mathbf{n}_{d,k} \cdot \overline{\mathbf{n}}_{c,k}^* \end{aligned} \qquad (6.86)$$

The first term indicates the desired signal and the second term represents the noise and interference. The third and fourth terms represent the additional noise caused by the channel estimation error. By calculating the energies of the first term and the sum of the remaining terms, and taking their ratio, the received instantaneous SIR, namely γ_d, that includes the effect of the channel estimation error, is obtained as

$$\gamma_d = \frac{A_d^2 A_c^2 \left(\sum_k |\mathbf{c}_k|^2 \right)^2}{A_c^2 \sum_k |\mathbf{c}_k|^2 \sigma_{d,k}^2 + A_d^2 \sum_k |\mathbf{c}_k|^2 \overline{\sigma}_{c,k}^2 + \sum_k \sigma_{d,k}^2 \overline{\sigma}_{c,k}^2}. \qquad (6.87)$$

This γ_d can be used to look up a BLER table, and frame errors can be generated randomly according to the BLER.

This analysis completes the chapter on theoretical modelling aspects across the topics of antennas, point-to-point link level and both static and dynamic system level characterisation. We will now move on to business related planning exercises, before utilising above models for planning and optimisation of the UMTS network.

REFERENCES

[1] The IEEE Standard Definitions of Terms for Antennas, *IEEE Trans. Antennas Propag.*, vol. 31, no. 6, November 1983.
[2] R.F. Harrington, *Time-Harmonic Electromagnetic Fields*, McGraw-Hill, Inc., New York, 1961.
[3] C.A. Balanis, *Advanced Engineering Electromagnetics*, John Wiley & Sons, Ltd/Inc., New York, 1989.
[4] CCIR, *Broadcasting satellite service (sound and television)*, Geneva, 1983.
[5] C.A. Balanis, *Antenna Theory. Analysis and Design*, 2nd ed., John Wiley & Sons, Ltd/Inc., New York, 1997.
[6] G.E. Evans, *Antenna Measurement Techniques*, Artech House, Inc., Boston, 1990.
[7] K.-L. Wong, *Compact and Broadband Microstrip Antennas*, John Wiley & Sons, Ltd/Inc., New York, 2002.
[8] P. Słobodzian, R. Borowiec, Microstrip Antennas for Cellular and Wireless Communication Systems, *Microwave and Optical Technology Letters*, vol. 34, no. 5, pp. 380–384, September 2002.
[9] 3GPP working group specifications, www.3gpp.org.
[10] 3GPP 25.201 Physical layer – general description.
[11] 3GPP 25.211 Physical channels and mapping of transport channels onto physical channels (FDD).
[12] 3GPP 25.212 Multiplexing and channel coding (FDD).
[13] 3GPP 25.213 Spreading and modulation (FDD).
[14] 3GPP 25.214 Physical layer procedures (FDD).

[15] 3GPP 25.215 Physical layer; Measurements (FDD).
[16] Harri Holma, *WCDMA for UMTS: Radio Access for Third Generation Mobile Communications*, John Wiley and Sons Ltd, 2004.
[17] H. Vincent Poor, *Introduction to Signal Detection and Estimation,* Springer-Verlag New York, LLC, 1994.
[18] Rodney Vaughan, 'On Optimum Combining at the Mobile', *IEEE Trans on VT*, vol. 43, no. 4, November 1988.
[19] Francesco Ostuni, Iterative Processing For Space-Time Multiuser Wireless Communications, PhD Thesis, King's College London, University of London, 2004.
[20] IST METRA, deliverable 5.1, 'Architectural design and cost impact'.
[21] E. Hardouin, *Egalisation au niveau chip pour la liaison descendante des systèmes de communications mobiles DS-CDMA*, PhD Thesis, Université de Rennes 1, May 2004 (in French).
[22] K. HIGUCHI, A. FUJIWARA and M. SAWAHASHI, 'Multipath interference canceller for high-speed packet transmission with adaptive modulation and coding scheme in W-CDMA forward link', *IEEE Journal on Selected Areas in Communications*, vol. 20, no. 2, pp. 419–432, 2002.
[23] K.S. Gilhousen, I.M. Jacobs, R. Padovani, A.J. Viterbi, L.A. Weaver, C.E. Wheatley, 'On the Capacity of a Cellular CDMA System', *IEEE Trans. Veh. Technol.*, vol. 40, no. 2, pp. 303–311, May 1991.
[24] K. Hiltunen, R. Bernardi, 'WCDMA downlink capacity estimation', *in Proc. IEEE Veh. Technol. Conf.*, VTC-2000, part 2, vol. 2, pp. 992–6, Piscataway, NJ, USA, 2000.
[25] J.M. Aein, 'Power balancing in systems employing frequency reuse', *COMSAT Tech. Rev.*, vol. 3, no. 2, Fall 1973.
[26] R.W. Nettleton, H. Alavi, 'Power control for spread-spectrum cellular mobile radio system', in *Proc. IEEE Veh. Technol. Conf.*, VTC-83, pp. 242–246, 1983.
[27] J. Zanders, 'Performance of optimum transmitter power control in cellular radio systems,' *IEEE Trans. Veh. Technol.*, vol. 41, no. 1, pp. 57–62, February 1992.
[28] Momentum Project. Momentum public UMTS planning scenarios. Avaliable online at http://momentum.zib.de/data.php, 2003. IST-2000-28088.
[29] S. Burger, H. Buddendick, G. Wölfe and P. Wertz, 'Location dependent CDMA orthogonality in system level simulations'. *In Proc. VTC-Spring 2005. IEEE*, Stockholm, Sweden, 2005.
[30] K. Pedersen and P. Mogensen, 'The downlink orthogonality factors influence on WCDMA system performance'. *In Proc. VTC-Fall 2002. IEEE*, Vancouver, Canada, 2002.
[31] 3rd Generation Partnership Project: Technical Specification Group Radio Access Network, 'User equipment (UE) radio transmission and reception (FDD) (Release 5)', 3G TS 25.101, V5.16.0, September 2005.
[32] J.G. Proakis, *Digital Communications – Third Edition*, McGraw-Hill, Singapore, 1995.
[33] R. Steele, C-C Lee and P. Gould, *GSM, cdmaOne and 3G Systems*, John Wiley & Sons, Ltd/Inc., Chichester, 2001.
[34] 3rd Generation Partnership Project: Technical Specification Group Radio Access Network, 'Base station (BS) radio transmission and reception (FDD) (Release 5)', 3G TS 25.104, V5.11.0, September 2005.
[35] Multiple Access Communications Limited, 'Research into the Impact of Dead Zones on the Performance of 3G Cellular Networks', Final Report, RA0703DZ/R/18/008/1, January 2004, available at http://www.ofcom.org.uk/ research/technology/archive/.
[36] U. Türke, M. Koonert, R. Schelb, and C. Görg, 'Advanced site configuration techniques for automatic UMTS radio network design.' *In Proc. VTC-2004 Spring. IEEE*, Milan, Italy, 2004.
[37] J. Laiho, A. Wacker and T. Novosad (eds). *Radio Network Planning and Optimization for UMTS*. John Wiley & Sons, Ltd/Inc., 2001. ISBN 0-471-48653-1.
[38] R. Jain. *The Art of Computer Systems Performance Analysis: Techniques for Experimental Design, Measurement, Simulation and Modeling*. John Wiley & Sons, Ltd/Inc., New York, 1991.
[39] L. Mendo and J.M. Hernando. 'On dimension reduction for the power control problem'. *IEEE Trans. Comm.*, vol. 49, no. 2 pp. 243–248, February 2001.
[40] U. Türke, R. Perreira, E. Lamers, T. Winter and C. Görg. 'Snapshot based simulation techniques for UMTS network planning'. *In Proc. IST-Mobile Summit*. Aveiro, Portugal, 2003.
[41] D. Catrein and R. Mathar. 'On the existence and e-cient computation of feasible power control for cdma cellular radio'. *In Proc. of ATNAC 2003*. Melbourne, Australia, December 2003.
[42] D. Catrein, L. Imhof and R. Mathar. 'Power control, capacity, and duality of up and downlink in cellular CDMA systems'. *IEEE Trans. Comm.*, vol. 52, no. 10, pp. 1777–1785, 2004.
[43] U. Türke, R. Perera, E. Lamers, T. Winter and C. Görg. 'An advanced approach for QoS analysis in UMTS radio network planning'. *In Proc. 18th ITC*, pp. 91–100. VDE, 2003.

[44] A. Berman and R.J. Plemmons. 'Nonnegative matrices in the mathematical sciences. Classics in Applied Mathematics'. 9. Philadelphia, PA: *SIAM*,. xx, 340 p., 2nd edition, 1994.
[45] D. Catrein, A. Feiten and R. Mathar. 'Uplink interference based call admission control for W-CDMA mobile communication systems'. *In Proc. VTC-2005 Spring. IEEE*, Stockholm, Sweden, 2005.
[46] A. Eisenblätter, H.-F. Geerdes, and N. Rochau. 'Analytical approximate load control in WCDMA radio networks'. *In Proc. VTC-2005 Fall. IEEE*, Dallas, TX, September 2005.
[47] Nawrocki M.J., Slobodzian P.M. and Borowiec R.: 'Smart Antenna Techniques for WCDMA systems,' *In Proc. of ATAMS 2002*, Krakow, Poland, December 2002.
[48] Zdunek R., Nawrocki M.J., Dohler M. and Aghvami A.H.: 'Application of Linear Solvers to UMTS Network Optimisation without and with Smart Antennas,' *In Proc. of PIMRC 2005*, Berlin, Germany, September 2005.
[49] Björck Å., *Numerical Methods for Least Squares Problems*, SIAM, Philadelphia, 1996.
[50] Saad Y. and Van der Vorst A.H. 'Iterative Solution of Linear Systems in the 20-th Century', *J. Comput. Appl. Math.*, vol. 123, no. 1–2, pp. 1–33, 2000.
[51] Sonneveld P.: CGS: 'A Fast Lanczos-Type Solver for Nonsymmetric Linear Systems,' *SIAM J. Sci. Statist. Comput.*, vol. 10, pp. 36–52, 1989.
[52] Freund R.W. and Nachtigal N.M.: QMR: A Quasi-Minimal Residual Method for Non-Hermitian Linear Systems, *Numer. Math.*, vol. 60, pp. 315–339, 1991.
[53] Fletcher R., *Conjugate Gradient Methods for Indefinite Systems*, Lecture Notes, Math., vol. 506, pp. 73–89, Springer-Verlag, Berlin-Heidelberg-New York, 1976.
[54] Van der Vorst A.H.: Bi-CGSTAB: 'A Fast and Smoothly Converging Variant of Bi-CG for the Solution of Non-symmetric Linear Systems,' *SIAM J. Sci. Statist. Comput.*, vol. 13, pp. 631–644, 1992.
[55] Saad Y. and Schultz M.H.: GMRES: 'A Generalized Minimal Residual Algorithm for Solving Nonsymmetic Linear Systems,' *SIAM J. Sci. Statist. Comput.*, vol. 7, pp. 856–869, 1986.
[56] S. Nagatsuka, M. Sengoku, Y. Yamaguchi, and T. Abe, 'An evaluation of telephone traffic characteristics of various channel assignment in a mobile radio communication system,' *IEICE Technical Report*, vol. J71-B, no. 10, pp. 1167–1170, October 1988 (in Japanese).
[57] ETSI, TR 101 112, 'Selection procedures for the choice of radio transmission technologies of the UMTS,' UMTS 30.03, April 1998.
[58] ARIB IMT-2000 Study Committee, 'Evaluation Methodology for IMT-2000 Radio Transmission Technologies,' ARIB, September 1998.
[59] A. Papoulis, *Probability, Random Variables, and Stochastic Processes*, 2nd edition, McGraw-Hill, New York, 1984.
[60] J. Beran, R. Sherman, M. S. Taqqu and W. Willinger, 'Long range dependence in variable bit rate video traffic,' *IEEE Trans. Commun.*, vol. 43, no. 2/3/4. February/March/April 1995.
[61] V.S. Frost and B. Melamed, 'Traffic modelling for telecommunication networks,' *IEEE Commun. Mag.*, vol. 32, no. 3, pp. 70–81, March 1994.
[62] Adas, 'Traffic models in broadband networks,' *IEEE Commun. Mag.*, vol. 35, pp. 82–89, July 1997.
[63] S. Ma and C. Ji, 'Modelling video traffic using wavelets,' IEEE Proc. ICC 1998, 1998.
[64] O. Lazaro, D. Girma and J. Dunlop, 'Real-time generation of synthetic MPEG-4 video traffic using wavelets,' *IEEE Proc. VTC 2001 Fall*, vol. 1 pp. 418–422, Atlantic City, October 2001.
[65] ITU-R Recommendation, 'Guidelines for Evaluation of Radio Transmission Technologies for IMT-2000,' Rec. ITU-R M.1225, 1997.
[66] J.D. Parsons, *The Mobile Radio Propagation Channel*, 2nd edition, John Wiley & Sons, Ltd/Inc., Chichester, 2000.
[67] W.C. Jakes, *Microwave Mobile Communications*, John Wiley & Sons, Ltd/Inc., New York, 1974.
[68] Y. Okumura, E. Ohmori, T. Kawano and K. Fukuda, 'Field strength and its variability in the VHF and UHF land mobile radio service,' *Review of the Electrical Communications Laboratories,* vol. 16, pp. 825–873, September 1968.
[69] S. Kozono and K. Watanabe, 'Influence of environmental buildings on UHF land mobile radio propagation,' *IEEE Trans. Commun.*, vol. 25, no. 10, pp. 1133–1143, October 1977.
[70] S. Tabbane, *Handbook of Mobile Radio Networks*, Artech House, London, 2000.
[71] M. Gudmundson, 'Correlation model for shadow fading in mobile radio systems,' *IEE Electronics Letters*, vol. 27, no. 23, pp. 2145–2146, November 1991.
[72] F. Graziosi and F. Santucci, 'A general correlation model for shadow fading in mobile radio systems,' *IEEE Commun. Letters*, vol. 6, no. 3, pp. 102–104, March 2002.

[73] W.R. Bennett, 'Distribution of the sum of randomly phased components,' *Quart. Appl. Math.*, vol. 5, pp. 385–393, January 1948.
[74] M. Slack, 'The probability of sinusoidal oscillations combined in random phase,' *J. Inst. Elec. Eng.*, vol. 93, part III, pp. 76–86, 1946.
[75] J.G. Proakis, *Digital Communications*, 4th edition, McGraw-Hill, Singapore, 2001.

7

Business Modelling Goals and Methods

Marcin Ney

In this chapter, we will expose the prime business modelling goals and typical methods to achieve them. We will start with the discussion of issues related to business plans, infrastructure development and associated budgeting processes. We will then dwell on some business modelling methods, such as statistical or detailed quantitative methods. This chapter hence constitutes a basis for the issues raised in Chapter 8.

7.1 BUSINESS MODELLING GOALS

Since the late nineties, the approach to business planning has been significantly transformed. In the era of dot.com, Internet and new-tech hype, it was very easy to convince potential investors, even while not having a reasonable business plan. The *idea* of a new service or system itself was enough to acquire capital.

Nowadays, all the companies and enterprises, especially in telecommunications, concentrate on profitability. For many potentially interesting inventions and projects a large-scale deployment will never happen. Thus the key factor for 'go/no go' decision is a carefully prepared business plan. Such business plan should be based on strong assumptions and should have a relatively short payback period.

7.1.1 NEW BUSINESS PLANNING

The first area with a strong business modelling need is the planning of a new business. It can be connected with running new entities, looking for new positions in the value chain or just extensions of current products and service portfolios. During the past years, there were many examples of such situations; UMTS license auctions were one of them.

While applying for a new license (and the licensing process is auction based), all new entrants had to perform a very detailed and meticulous business analysis, which gave information about investment feasibility together with maximum license fees they could offer. The experience of capital groups like the biggest pan-European operators (Orange/France Telecom, T-Mobile and Vodafone) gave them significant advantage while performing basic assumptions for the business plan (network roll-out plans, coverage percentages, equipment prices, target customer segments, possible tariffs etc.). But the key factor for a business success may be the proper approach to local constraints. Parameters, like customs fee for imported equipment, local prices of civil work and legal permissions to build new sites, legal permissions acquisition time, local support from equipment vendors, maintenance fees, etc. may vary from one country to another. And what is also important is the wording of the license itself. Many issues defined in the license, like obligation or permissions concerning national roaming, infrastructure sharing, Mobile Virtual Network Operator (MVNO), influence the business plan significantly.

The key significance for the accuracy of the decision to be taken (bidding, proposed license fee etc.) requires a proper modelling of all the factors above and many others not mentioned here.

For the incumbent operators and all entities already present on a particular market (e.g. GSM operators), a business plan preparation is much easier. UMTS license acquisition for these operators means extension of the current network. Because they know well about local constraints and they most probably own a number of sites already that can be reused for Node B installations, they have a much better basis for a reliable business plan. The prime goal of business planning for them is not only profitability of the new investment, but they also have to assess potential threats and losses if not participating in a new bid (potential new competitors on the market, competitive advantage increase of current competitors, etc.).

It is worth mentioning that a proposed license fee is not the only thing evaluated by the regulator from a submitted offer, even though it usually has the highest rank. Regulators usually evaluate the proposed steps for the network roll-out, as well as the technical project and the credibility of the presented business plan.

7.1.2 INFRASTRUCTURE DEVELOPMENT

The other important issue with a strong business modelling demand is the development of the already existing infrastructure; a network upgrade from GSM to UMTS, as exemplified in the previous section, may be an example of such issue.

A business model is needed to assess the reasonability of every new platform or system introduction or extension. From the UMTS operator's perspective, it may be necessary to consider technical platform upgrades, e.g. the introduction of new UMTS releases (Rel. 4, Rel. 5), Intelligent Multimedia Systems (IMS), the Time Division Duplex (TDD) mode, High Speed Downlink Packet Access (HSDPA), value added services platform extensions (SMSC, MMSC, WAP Gateway, GMLC, IN etc.) and many others. While evaluating such possibilities and preparing appropriate business scenarios, it is necessary to take into account market demands, possible movements of competitors and new products and services that can be built on top of these platforms.

Prior to introducing new platforms, however, the operator should also monitor any capacity needs of its existing systems. It is usually done by monitoring certain interface saturations, calculating trends of traffic increases, forecasting the number of subscribers and their traffic profiles and using this information as an input to equipment dimensioning rules. The result of such dimensioning reveals the time when a new module ought to be added to a particular platform and its configuration. While not directly related to business planning, this process has a big impact onto the overall network development plan and is hence needed for budgeting purposes, which will be the subject of the subsequent section.

Also, the infrastructure development related to the introduction of a new system or technology, which is not positively justified by a business plan, can still be decided for implementation. Some reasons can be conscious decision about entering new development directions, which can possibly give competitive advantages in the future, or capital group global strategies etc.

7.1.3 BUDGETING

The last important goal of business modelling is the budgeting process. In modern telecommunication companies, it is nearly impossible to undertake any new initiative without having some budgetary resources planned for it beforehand. Because of this, budgeting is a process with iterative characteristics; a base budget version is usually followed by many revisions.

Three main time perspectives of a budgeting process can be distinguished:

1. yearly budget plan;
2. mid-term plan (MTP) – three to five years;
3. long-term plan (LTP) – usually up to ten years.

The differences between them are not focused on a time perspective only. The other differentiating factor is the level of detail: in a year's perspective, the assumptions are strong and so is the confidence of the plan itself; however, as the time perspective is becoming longer, the plan starts to be based more and more on assumptions and forecasts; the probability of false predictions hence rises.

In an yearly plan, the usual time quantum is thus one month, in MTP it is a quarter or half a year and in LTP it is one year. In fact, it is possible to prepare, e.g., LTP on a monthly basis, but it is unreasonable and does not give any added value compared to an yearly one. The yearly plan and MTP are mainly used on an operational level, while LTP is usually needed when working on a company strategy, i.e. to prepare scenarios for very important and expensive bids (like the UMTS spectrum license) and for the company to acquire additional capital.

A typical budgeting process consists of the following phases:

1. Forecasting and preparation of assumptions, such as market, clients' needs, tariffs, offered products, distribution channels – marketing and sales departments;
2. Technical equipment dimensioning and its CAPEX/OPEX calculation – technical department;
3. Financing sources, non-technical OPEX calculation and business plan financial index calculations – financial department;
4. Plan acceptance – management board/supervisory board.

Phases one to three have iterative characteristics and in most cases many different scenarios have to be prepared prior to a final acceptance of the plan.

7.2 BUSINESS MODELLING METHODS

There are many business modelling methods that can be distinguished. Which method or approach is best suited to a particular need strongly depends on the goal that should be achieved, time that can be allocated to the process and input data available (it can also be historical data). Different methods have different levels of accuracy, but the key factor for the accuracy of the resulting business plan is not the error induced by the method, but by the underlying assumptions. The weight of the error of input assumptions related to the overall error is the biggest and most significant; however, the choice of method should not be neglected either.

In the subsequent section, we present an overview of different business modelling methods and approaches.

7.2.1 TRENDS AND STATISTICAL APPROACH

One usage of the statistical business modelling method is when historical data from the network or from the new potentially attractive market is available. The goal is to model the market or network behaviour based on previous experiences related to similar systems, markets or even just assessing it, while not having any direct relation available.

When focusing on the market and revenue part, all the existing and new service demands can be modelled in a statistical fashion; example parameters are a country's population, clients' purchasing power connected with GDP per capita, country's mobile subscriber penetrations, number of mobile network operators, total network voice traffic, total network data traffic, Minutes of Usage (MoU), Average Revenue Per User (ARPU) etc. While starting with some empirical data taken from Technical/Management Information Systems (TIS/MIS), the first approximated projection can be a linear trend. It can be stated directly, as e.g. 100 % network traffic increase every two years; alternatively, it can be calculated precisely in any analytical tool like Microsoft Excel, as exemplified by means of Figure 7.1.

The second approximation takes into account a usual inflexion point of evaluated curves. That point is related to the market (and curve) saturation, while looking on penetrations, number of subscriber growth etc. On the other hand, when related to service revenues, the point divides the curve between the service infancy and maturity, with traditional take-up rates for cellular traffic. According to Tomy Ahonen [1], such curves can be named *hockey stick curves* and are exemplified in Figure 7.2.

The exact place where such inflexion point should be put and what will be the slope of the second curve segment strongly depends on the amount of empirical data processed earlier and on the experience of the modeller.

When performing such modelling process to important subsystems of the UMTS network, it can be easily related to the gradual CAPEX/OPEX increase.

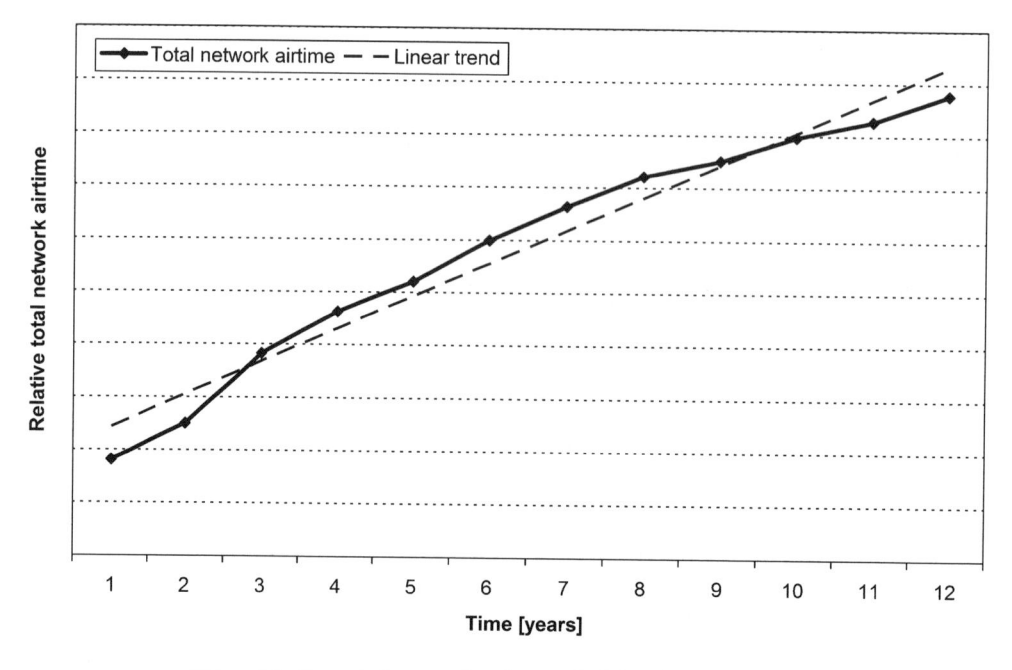

Figure 7.1 Example forecast of total network airtime (with linear trend).

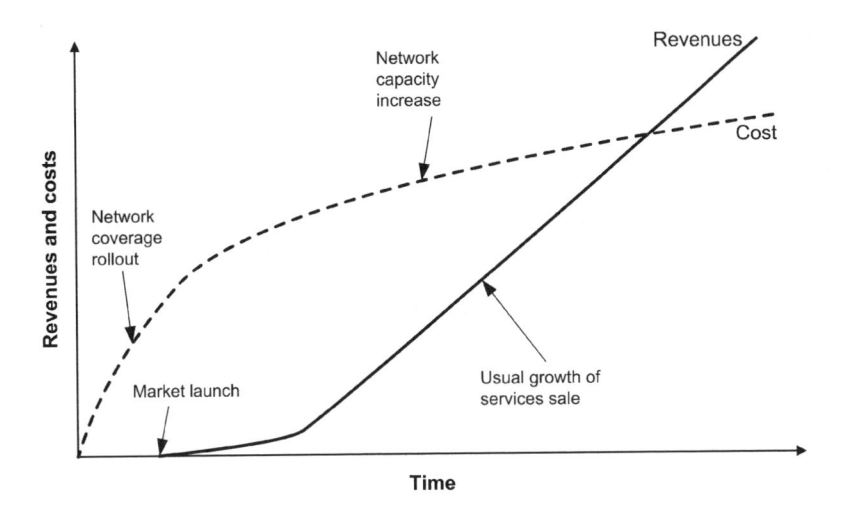

Figure 7.2 Example of hockey stick curves.

7.2.2 BENCHMARKING AND DRIVERS

As a first approximation to the particular business plan, the data from a similar case can be taken. Such an approach is called benchmarking. It can also be used for a business plan credibility evaluation.

The data from similar projects for benchmarking can be obtained from empirical data obtained from within the same company, or from other companies within the same Capital Group, or from competitors (practically unavailable) or from professional consultants. Benchmarking can significantly ease a business plan creation while having incomplete data (e.g. lack of equipment prices) and can be used as a first approximation. It is also useful when there is no time for detailed analysis.

Drivers related business modelling can be treated as a kind of benchmarking. The drivers are also inherited from empirical data or from other companies with similar projects. The difference between the drivers and typical benchmarking refers to the level of detail; whereas benchmarking acts as a general business plan approximation or sanity checking, the drivers refer to detailed business models. They can be used to bridge incomplete data, or to do the process the fastest possible way with acceptable accuracy (even if a detailed, albeit time-consuming, calculation is possible); they can also be useful in evaluating the credibility of the business plan. Examples of drivers are OPEX as CAPEX percentage, Civil Work CAPEX as UTRAN CAPEX percentage, Operations and Maintenance (O&M) or Project Management CAPEX as overall CAPEX percentage, CAPEX per subscriber, CAPEX per Node B etc. More practical applications of drivers are presented in Section 8.2.

7.2.3 DETAILED QUANTITATIVE MODELS

Detailed quantitative business modelling approaches yield most details and the best accuracy, but they are usually very time-consuming. The time allocated for that purpose can be significantly shortened by having a universal business model (fairly easily) adopted to new projects. However, it is even more time-consuming to develop such a model from scratch and to keep its developments ongoing in parallel with new technologies and system developments.

The detailed modelling is related to both: CAPEX/OPEX and revenue calculations. When considering CAPEX/OPEX, the prerequisite for commercial calculation is a very detailed network or system dimensioning process. At the same level of detail, the revenue forecasting should be done, where we present more information about this process and business modelling methods in Chapter 8.

7.2.4 OTHER NON-QUANTITATIVE METHODS

As one of the business planning goals is to help the decision-making process, not only financial methods apply. For the sake of a better understanding of the environment or presentation of a business plan itself, some descriptive methods can also be very useful.

One of them is the well established SWOT analysis. The acronym stands for Strengths, Weaknesses, Opportunities and Threats. To perform a SWOT analysis, one clearly needs to identify the most significant factors which can act as a particular project's strength, weakness, opportunity or threat. Then, all the factors identified should be jotted down in a table as presented in Figure 7.3. The table can then be added to the final presentation in order to highlight the context of the initiative and project.

Another method for non-quantitative business modelling is positioning on a value chain. It shows the relations and dependencies between the entities involved in business processes. As a new system, technology or project is implemented, a change of the value paradigm may occur, resulting in changes of the value chain itself, as presented in Figure 7.4. In the context of business modelling, it is important to predict such changes prior to the project implementation.

Figure 7.3 Example of SWOT analysis table.

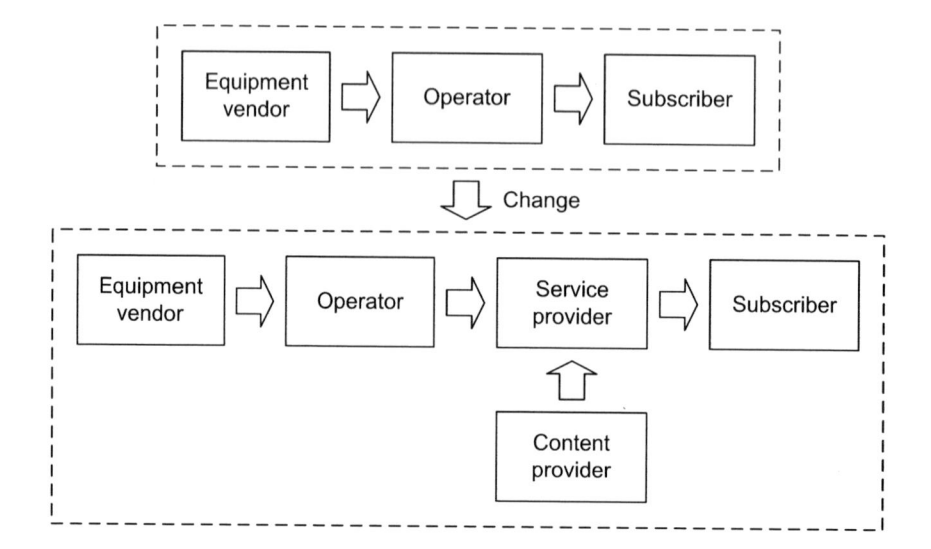

Figure 7.4 Example change in value chain.

These were two examples of non-quantitative business modelling methods. More details on them and others, as well as top-down/bottom-up classification can be found in [2,3].

REFERENCES

[1] Tomi T. Ahonen, Timo Kasper, Sara Melkko, *3G Marketing: Communities and Strategic Partnerships*, John Wiley & Sons, Ltd/Inc., 2004.
[2] John Tennent, Graham Friend, *Guide to Business Modelling*, Profile Books Ltd, 2001.
[3] Graham Friend, Stefan Zehle, *Guide to Business Planning*, Profile Books Ltd, 2004.

Part III

Planning

8

Fundamentals of Business Planning for Mobile Networks

Marcin Ney

After having exposed typical business modelling goals and methods in Chapter 7, we will now concentrate on some more detailed planning issues related to the business side of mobile 3G networks. We will commence with a description of the various planning processes, such as market analysis and forecasting etc. We will then dwell on the main constituents for calculating CAPEX and OPEX, before going into details related to revenue and non-technical related investment calculations. We will finish the exposure by discussing the results one ought to expect from a proper business planning approach.

8.1 PROCESS DESCRIPTION

The main phases of a business planning process in the context of mobile networks are related to market forecasting and system modelling. Depending on time constraints and desired level of accuracy, detailed quantitative models or drivers related methods (described in Chapter 7) should be used. The different phases of the process are presented in Figure 8.1. This section describes the business *planning* process and hence constitutes a basis for more detailed information provided in Sections 8.2, 8.3 and 8.4.

8.1.1 MARKET ANALYSIS AND FORECASTING

As shown in Figure 8.1, the part of a business planning process related to the market commences with the forecasts. Since it influences all of the subsequent phases, it is important to minimise the forecasting error. The forecast is giving general information about the mobile market, like penetration,

---

Understanding UMTS Radio Network Modelling, Planning and Automated Optimisation   Edited by Maciej J. Nawrocki, Mischa Dohler and A. Hamid Aghvami   © 2006 John Wiley & Sons, Ltd

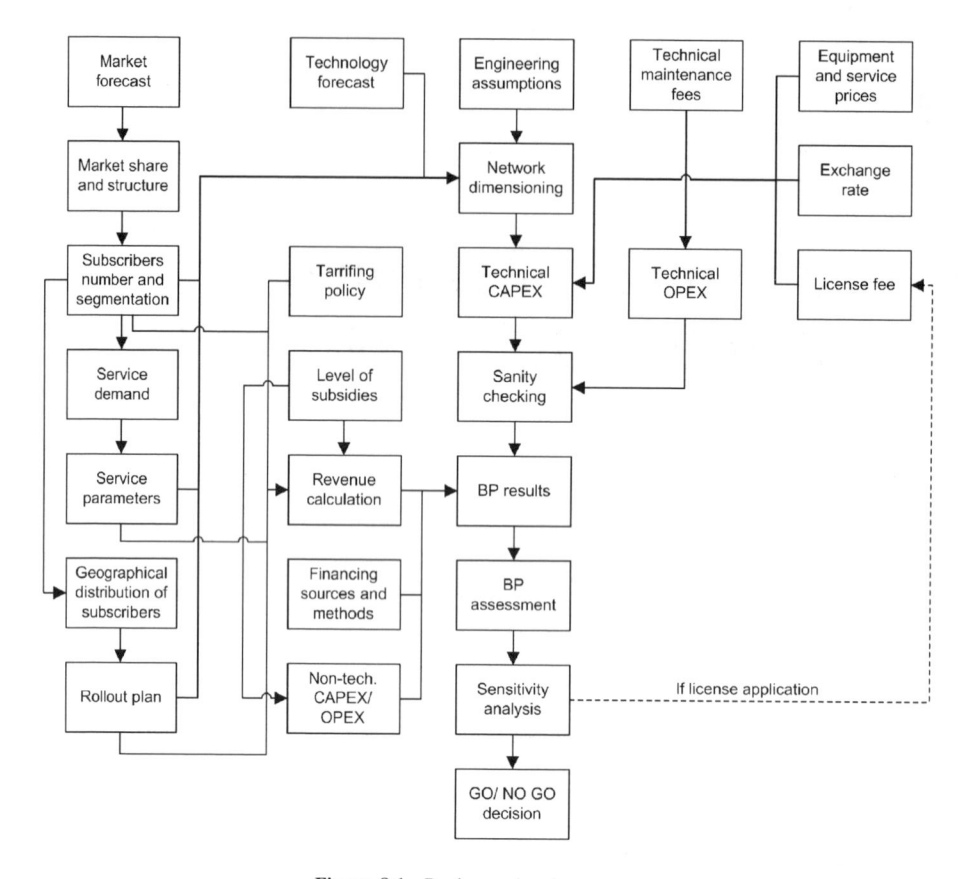

Figure 8.1 Business planning process.

revenue structure, growth, data usage etc. Three general types of forecasting methods can be distinguished [1]:

1. Extrapolative models

   - Moving averages, decomposition
   - Trend curves
   - Smoothing
   - Box-Jenkins ARIMA
   - Neural networks.

2. Causal models

   - Single equation
   - System model.

3. Judgemental forecasts

   - Individual
   - Survey, market research
   - Panel opinion, Delphi.

The details on the above methods can be found in [2] and [3].

While having a general view on the expected future market, the next phase is to develop an evolution of market shares. The operator market share can be valuable, quantitative and should be derived from both strategic goals (e.g. to be market leader in three years time) and realistic abilities to achieve them. The other dimension of the market share is technology, i.e. migration of subscribers from 2G to 3G network. The important factor for that is also 3G handset market penetration.

The next phase of the process is the split of subscriber types. According to customer behaviour, service profile, ARPU generated, etc., the potential subscriber base should be split into a number of different types. The usual types are prepaid, postpaid and business. As usual, it is easier for incumbent operators to build the split on historical data. Greenfield operators, not having such historical data, should assume the values based on market research.

The last major phase of the process is related to voice and data service parameters. The parameters cover service usage profiles (e.g. throughput, number of sessions per month, daily usage profile etc.) together with service related tariffing policies.

In parallel to market shares and segmentation forecasting, the values obtained should be related to particular geographical country areas, thus defining network rollout. Usually, marketing assumptions divide rollout into a number of phases, e.g. 1st year: X biggest cities with number of inhabitants above Y within administrative boundaries; 2nd year: extension of previous X cities to agglomeration boundaries and number of additional cities with number of inhabitants above Z; 3rd year: coverage extension to some rural areas etc. Based on GIS data and geomarketing statistics, these assumptions are converted into an area to be covered (in square kilometres) and the contents of different clutter types (e.g. dense urban, urban, suburban and rural) are connected with forecasted numbers of subscribers within the area. Based on this, the percentage of country area and population covered can also be calculated. The example of such data is presented in Table 8.1.

It is also important to know that the rollout planning process can have iterative characteristics and is closely connected with some system modelling parameters, e.g. if the number of sites to be build to reach some rollout targets is unrealistic, then some rollout modifications will be necessary.

8.1.2 MODELLING THE SYSTEM

Having all the market input and output parameters available, the next step in a business planning process is system modelling; such modelling would typically start with equipment numbers dimensioning, and finish with having the technical CAPEX and OPEX calculated.

As presented in Figure 8.1, the process starts with all available market input parameters, as well as underlying engineering assumptions. The first phase is related to dimensioning the number of sites

Table 8.1 Rollout plan example.

Year	Coverage Population (%)	Coverage Area (%)	Criteria
1	36	7	Agglomerations with more than 200 000 citizens, motorways
2	71	32	Cities with more than 50 000 citizens, major railroads, major tourist areas
3	79	44	Cities with more than 10 000 citizens, major roads, all tourist areas
4	90	70	Cities with more than 10 000 citizens

from both coverage and capacity perspective. Then, all the other network subsystems have to be dimensioned. Having all the equipment numbers and configurations calculated, the next step is to convert them to CAPEX and OPEX figures. The details on system modelling issues are presented in Section 8.2.

Afterwards, when financial values of all technical and marketing sides are available, the business plan itself is finally constructed. Its parameters and their evaluation are described in Section 8.4.

8.1.3 FINANCIAL ISSUES

Financial inputs and assumptions have a great influence on the accuracy of the business plan. The most important financial parameters in business planning are:

- equipment and services prices;
- investment financing sources and methods;
- exchange rate.

The influence of assumed equipment and service prices can be multiplied by the drivers used in the business plan. Thus, their accuracy should be as good as possible. The incumbent operators or the ones within capital groups can simply apply the prices from framework agreements or assume the ones from similar technologies. Greenfield operators, however, have to rely on some budgetary offers from the equipment vendors, with a risk of CAPEX and OPEX overdimensioning.

The investment financing sources can be own company capital, shareholders, bank loan etc. To have a full picture of a business plan, also the cost of capital has to be assumed – WACC (Weighted Average Cost of Capital). On the other hand, the increase of competition between vendors, new vendors' entrances to the market and their need to have operators' references enabled the usage of other financing methods. It is now possible to build a network based on pay-as-you-grow mechanisms. This means that payment time is shifted in months or years into the future, when revenue is becoming bigger. Whereas overall CAPEX (cumulative in long-term perspective) can be bigger, it has a very positive influence on the business plan.

Since different currencies apply in different countries, the accurate forecast of the exchange rate of the currency in which the contract is or will be signed is crucial for the success of a business plan. Note that even with a conservative exchange rate forecast, it is advisable to apply some additional safety margins. It should be remembered that any revenues are in home currency, whereas the CAPEX in the foreign one.

8.1.4 RECOMMENDATIONS

As was stated before, the business planning process has an iterative nature. A business planner usually has to evaluate many scenarios before a final recommendation. The differences can cover rollout modifications, different tariffing strategies, level of change of subsidies, license fees to be declared to pay any change etc. The details of such evaluation and sensitivity analysis are presented in Section 8.4.

But not only pure financial issues should be examined. Every business plan can have various traps in which the operator can fall. Thus, all the other factors with a potential influence should be extracted and the entire environment should be analysed. For example, the cheapest equipment can be unstable and simply not work as it should, prohibiting the network launch at a planned date; or, the 'low CAPEX' vendor can provide very costly equipment maintenance, thus making the OPEX very high etc. Important issues could also arise due changes in the regulatory environment.

8.2 TECHNICAL INVESTMENT CALCULATION

This section presents a guideline to the reader, how to calculate the total cost of investment needed for the network or system from a technical perspective. It mainly shows detailed quantitative and drivers-related methods. The reason is practice – these methods are most often used for that purpose.

8.2.1 CAPEX CALCULATION METHODS

To calculate the overall technical CAPEX, all the involved sub-systems and possible vendor services have to be taken into account. From a practical perspective, the following categories can be distinguished:

- Node Bs
- UTRAN transmission network
- Backbone
- UMTS core
- Supervision (NMS)
- VAS platforms
- Sites
- Radio planning
- Other services.

We will discuss each of them subsequently.

8.2.1.1 Node Bs

A full network dimensioning process has to be done according to technical rollout demands, forecasted subscriber numbers and service usage characteristics. The details to the most important process phases are presented in Section 10.3, whereas a process overview can be found below.

Generally, the process may consist of two main phases: coverage and capacity dimensioning. The coverage dimensioning phase is strictly connected with rollout targets (as described in Section 8.1.1). Having rollout targets defined, to obtain a total number of coverage sites, the cell range and area should be calculated. From the total pathloss, which can be taken from a previously calculated link budget using an appropriate propagation model and assuming a particular site configuration, the desired cell range and area can be calculated (different per clutter type). With this data, the total area can be divided into fractions of a particular clutter type, from which the number of sites per clutter type is obtained. To avoid rounding errors, such calculation should be done separately per connected area.

An easier way that can be used as a first approximation is to calculate the number of Node Bs being equal to the number of existing 2G sites within a desired area multiplied by the ratio between 2G and UMTS cell areas (where the choice on the ratio can sometimes be fairly arbitrary).

Having the number of sites available, the next phase is to compare the network capacity with a calculated traffic demand. If the capacity is too small, the site capacity configuration should be changed or, if it is not possible, additional sites should be added.

When the overall number of Node Bs is finally available, it is necessary to review their configuration. For this, it is very important to take all the needed elements into account. Usually, vendors present the configurations of Node Bs as a number of predefined kits. The kit that best corresponds to the defined needs should be chosen (or a mix of such kits, e.g. indoor vs. outdoor hardware cabinets) and then all the additional elements have to be added (e.g. additional Power Amplifiers, Channel Elements, Remote Electrical Tilt (RET) units, Mast Head Amplifiers (MHAs) etc.).

The Node B hardware configuration directly corresponds to its pricing. The other elements that should be taken into account are the Node B software license (which can be different for different types of configuration) and additional software features that should be implemented (e.g. UMTS to 2G handover etc.).

The overall Node Bs CAPEX consists of a number of Node Bs multiplied by a unit price and the price of all additional features. It is worth to remember that the Node Bs CAPEX is one of the most significant constituents in the overall network CAPEX, usually reaching between 20 and 35 % of it, and the number of Node Bs itself influence the UTRAN transmission network, the supervision (see Section 8.2.1.5), as well as the site and radio planning CAPEX.

8.2.1.2 UTRAN transmission network

What is generally called the UTRAN transmission network consists of two elements: the transmission links between Node Bs and RNC and the RNCs themselves. To include the cost of the transmission links into the overall CAPEX depends on the decision whether to build or reuse the operators' own infrastructure. The transmission links can use microwave point-to-point links, a LMDS network, fibre-optics network etc. The pros and cons of different solutions together with the guidelines to choose the best one are presented in Chapter 17.

To calculate the cost of the transmission links, the Node Bs' capacity configuration (throughput needs) and transmission network topology should be known. From a generic approach, the capacity configurations can be differentiated on a per clutter type basis. Thus, different transmission link capacity types (e.g. microwave links) can be directly assigned to them. Having the numbers of transmission links calculated and its unit price, the related CAPEX is obtained by simple multiplication.

The other important thing is the transmission network topology. In a real network, some concentration and aggregation nodes are often needed. As a result, the costs of these additional nodes together with additional links have to be included into CAPEX figure.

In the environment, where 2G to 3G network upgrades are considered, the optimal way is to reuse the current transmission infrastructure as much as possible. For that purpose, the existing network's spare transmission capacity has to be calculated and compared to 3G capacity needs. Instead of new elements cost, the upgrade cost can then be calculated.

The remaining part of the UTRAN transmission network CAPEX is RNC related. To calculate this CAPEX, the number and configuration of RNCs has to be known. The limiting factors in RNC dimensioning are:

- I_{ub} traffic capacity (in Mbps, in number of channels and in number of carriers per RNC);
- interface capacities (e.g. STM-1 and E1);
- maximum number of cells per RNC.

According to the criteria above, the vendors provide different RNC configurations with a different range of parameters; some example RNC configurations are presented in the Table 8.2.

The choice of a particular configuration is in the hands of the network planner and has to be considered as a trade-off between RNC and transmission CAPEX/OPEX. From a practical point of view, even when the initial network capacity need is relatively small, it is better to build a greater number of RNCs with a lower configuration (e.g. at least one RNC per big city) – it is much easier to upgrade an existing RNC than to build a new one in the future.

Finally, taking into account all the assumptions and factors above, the RNC related CAPEX can be easily calculated.

8.2.1.3 Backbone

The backbone CAPEX strongly depends on the chosen transport network topology and technology. It mainly consists of transmission links and switching/aggregating nodes CAPEX.

Table 8.2 Example RNC configurations.

Configuration Number	Maximum capacity in different configurations					
	I_{ub} traffic capacity			Interfaces	
	Mbps	Channels	Carriers	STM-1	E1
1	48	3000	384	16	96
2	85	5313	576	16	128
3	122	7625	768	16	160
4	159	9938	960	16	192
5	196	12250	1152	16	224

The number and capacity of transmission links depends on the number of network elements that have to be connected, as well as the overall core network traffic. The number of ATM switches may be equal to the number of MSCs (large switches), plus the number of RNCs (small/medium ones). The number of other transmission nodes (e.g. SDH add-drop multiplexers) depends on the particular network topology.

Having unit prices from the vendors, the overall backbone CAPEX can easily be calculated.

8.2.1.4 UMTS Core

The dimensioning of the UMTS core and its CAPEX strongly relates to the core revision that is going to be implemented. The general rule should be that when building a new UMTS network the latest available stable core release from a particular vendor should be considered. Implementing the old core version will make the new network out of date from the very beginning, whereas implementing the newest one without any market operation period may lead to network instability. For this example the core release 4 is assumed. This core release consists of the following elements:

- MSS Server
- Media Gateway
- SGSN
- GGSN
- HLRi/AuC
- Others (Charging Gateway, Border Gateway, Lawful Interception Gateway, Firewall, DNS/DHCP, EIR etc.).

Like with the RNCs, the different configurations of all the elements above are available. The limiting parameters for these elements are:

- MSS Servers: number of subscribers, Busy Hour Call Attempts (BHCA);
- Media Gateway: number of speech channels, BHCA, Iu interface throughput, number of RNCs, number of ATM ports, number of TDM ports;
- SGSN: maximum capacity in Mbps and in PDP contexts;
- GGSN: maximum capacity in Mbps and in PDP contexts;
- HLRi/AuC: number of subscribers.

Additionally, according to the planner's decision, a core network rollout strategy has to be defined. As for the UTRAN transmission case, the decision covers the choice between a smaller number of core elements with a large configuration and a greater number with a small configuration. For that purpose, transmission costs for the core traffic transport and aggregation, as well as future proofness of the solution should also be considered.

After implementing all these configuration options together with their pricing – and rules for numbers of elements calculation – the CAPEX figure can be obtained.

8.2.1.5 Supervision (NMS)

To assess the supervision system cost, both detailed quantitative and drivers related methods can be used. Usually, supervision platforms are divided into national and regional, and regional ones that are dedicated per radio access, circuit switched and packet switched domains. Since these systems are usually software applications installed on UNIX servers, the hardware can be acquired from UMTS system vendors or directly from server vendors. In order to properly dimension the server capacity and performance, all UMTS vendor dimensioning rules and guidelines have to be taken into account. Furthermore, supervision software can be licensed per number of carriers, number of Channel Elements, number of subscribers, number of speech channels etc. Having calculated the number of elements and software license options, the supervision CAPEX is available.

Another method for calculating the supervision CAPEX is a driver-based method. The driver in this case can be a percentage of all equipment CAPEX, and can reach the value between 1 and 7%. Compared with the detailed method, it is only an approximation, however, while calculating business scenarios in very time-critical constraints, it can be the only option available.

8.2.1.6 Value Added Services platforms

Value Added Service (VAS) platforms are generally independent from the mobile network generation. The same platforms used for 2G networks by incumbent operators can be reused for UMTS as well. The only differences are related to the application types with high throughput demand, which could not be used for 2G network because of system limitations; an example of such application is video streaming. On the other hand, Greenfield operators, who build their 3G network not having a 2G one, have to build their VAS systems from scratch; this can mount to a significant cost.

Due to its complexity and only loose relation to UMTS CAPEX, the VAS CAPEX calculation methods will not be covered here. For more information on that, a good reference is [4].

8.2.1.7 Sites

Site CAPEX is directly related to the number of physical sites in a UMTS network. The biggest constituent here is the cost of the Node B sites, whereas the remaining part covers all the other UTRAN and core elements sites. The Node B site CAPEX consists of the following components:

- site acquisition;
- site legalisation (usually some administrative fees for permission for radio emission etc.);
- site preparation (container, tower/mast, power supply system, battery backup system, air conditioning system etc.) including civil work;
- antenna system (antennas, Mast Head Amplifier, Remote Electrical Tilt and installation).

The cost of sites related to the other network elements (like RNC, SGSN etc.) covers site acquisition and preparation only.

Every component can be priced by the operator's subcontractor or equipment vendor directly. The difference is in strategy – whether the operator would like to perform network planning with his own resources and site deployment with local subcontractors or he would like to have a turnkey project from the equipment vendor. The second option is usually more expensive, but can be the only solution for a Greenfield operator without strong engineering expertise and strong local presence (i.e. the network of experienced subcontractors). It is also utmost important to know the exact Node Bs configurations, as well as the configurations of the other elements and also the type (e.g. indoor/outdoor hardware cabinet), because it can influence the site price significantly.

Again, the incumbent operators are in a privileged position here, because reuse of existing 2G sites seems to be a natural strategy for them. The factor that should be considered here is that not every 2G site can bear additional 3G equipment. There are many sites where 2G equipment is co-located with other systems or with 2G equipment of competitors (site sharing). In such a case, there can simply be too little room in the container to put new equipment in, or the mast can be overloaded with the existing antennas. But generally speaking, having already a big network of existing sites gives the operator a possibility of significant CAPEX savings.

But the cost is not the only important factor for sites: in many countries, the process of site legalisation can be a very time-consuming issue, which may take sometimes twelve months or more. It should hence be taken into account while preparing business scenarios for UMTS.

8.2.1.8 Radio Planning

What is called radio planning here is effectively the cost of the radio planning, the measurement tools and the UMTS testbed built on the operator's premises. As mentioned later in this chapter, the real cost in man-working-days dedicated to the project will not be covered here, as it would be very difficult to have detailed prices and calculations at business planning stage. Thus, the method most often used for radio planning cost calculation is a driver. The cost can be assessed at the level of 2% of the cumulative CAPEX.

8.2.1.9 Other Services

While calculating the total network CAPEX, some other services should also be counted. These are:

- assistance – emergency support
- project management
- training investments.

The above listed services can be provided by an equipment vendor or by external companies. The recommended ways to have their values obtained are drivers too. The drivers could take the range of 1 to 3% of the cumulative CAPEX each.

8.2.1.10 Additional Parameters and Overall CAPEX

To compose the overall CAPEX some additional factors and parameters also have to be considered. These are:

- shipping, installation and commissioning
- software licenses
- spare parts
- transport and delivery
- hardware replacement
- equipment price erosion
- UMTS license fee.

Having a contract with an equipment vendor signed, or at least available within a capital group framework, many of these parameters can be directly calculated from the price list. On the contrary, for a Greenfield operator, the only solution to prepare a business case for a UMTS license bid is to apply relevant drivers.

Shipping, installation and commissioning costs can be calculated as a certain percentage of the yearly CAPEX investments. The value depends mainly on local constraints and the type of equipment under evaluation. It can vary from 4 to 5% (core network equipment), through 9 to 13% (UTRAN), to around 20% (backbone).

Spare parts and software licenses cost can be calculated as a percentage of the cumulative CAPEX investments instead. The values usually have the range from 3 to 6% each.

Transport and delivery duties have to be added to equipment prices depending on the way these prices were set up. The equipment price can be defined as DDP (Delivery Duty Paid) or DDU (Delivery Duty Unpaid). While the prices are defined as DDU, the additional driver of 1 to 2% should be applied.

The remaining two factors for the overall CAPEX calculation in a long-term perspective (LTP) are equipment price erosion and hardware replacement. The reasons of price erosion are the increase of equipment production volumes, the maturity of products on the market, the introduction of newer equipment releases etc. The usual level of price erosion is from 8 to 12% per year. Furthermore, the equipment has also its lifetime; for LTP, the usual value assumed is 7 to 9 years. Therefore, the business case should be constructed in a way that after 5 to 6 years the equipment should be replaced with some new one.

The last thing that should be counted in CAPEX, whilst not directly related to any technical investment, is the UMTS license fee itself.

After calculating and summing up all the CAPEX constituents, the overall value is obtained. For further business plan processing and evaluation, it is very useful to present CAPEX values in both yearly and cumulative ways.

8.2.2 OPEX CALCULATION METHODS

To have the overall cost in the UMTS business plan it is very important to have a good estimation of OPEX. Whereas in the first network deployment years it is much lower than CAPEX, for LTP perspective the cumulative OPEX has a much greater significance. This subchapter presents two approaches for OPEX calculation.

8.2.2.1 Detailed OPEX Calculation

The detailed method of the OPEX calculation has similar principles to the CAPEX one. All the OPEX contents should be distinguished, quantified and calculated. The overall value consists of the following factors:

- network maintenance expenses
- frequency fees
- backhaul transmission (leased lines) fees
- interconnection fees
- energy and site rental fees
- project related expenditures (engineering staff).

The most significant factor in the overall OPEX is network maintenance. It is directly related to the equipment deployed and its value can directly be taken from the vendor's pricelist or the signed maintenance contract. It can cover repair services, help lines, new software versions and patch subscriptions, reactions to critical failures procedures etc.

The values of frequency fees can directly be taken from the country's regulator pricelist. They are not about the UMTS license itself, but rather the fees for using particular frequency bands on a certain area (e.g. the whole country or some areas/sites only). Such fees are related to both UMTS frequencies and backhaul microwave ones (point-to-point, point-to-multipoint). The other OPEX expenses directly connected with backhaul transmission are the fees of leased lines. These are directly connected to the provider's pricelist and, while used on a large scale, can be the most significant portion of the total OPEX. The details on backhaul transmission options are presented within Chapter 17.

As every network is connected with other networks (competitor's, fixed ones, international ones etc.) via interconnection points, the relevant fee has to be paid. The fee which should be paid is related to the volume of outgoing voice traffic. Furthermore, the fee for data (and Internet) connectivity has to be paid as well.

The OPEX portions related to physical sites are site rental and energy fees. The value can be calculated by summing up all the contractual values site-by-site.

If an existing 2G operator is willing to extend his portfolio of systems and services to UMTS, it can be assumed that the existing engineering resources will be reused for UMTS planning and optimisation tasks. However, due to the system complexity, it is good to estimate the need for additional staff and consider it from the beginning to avoid serious resource problems in the future. For the sake of conciseness, no direct method of calculating the man-working-days numbers is presented in this section. The cost of engineering has been covered in Sections 8.2.1.8 and 8.2.1.9.

8.2.2.2 Drivers-based OPEX Calculation

Most of the detailed OPEX calculation methods presented in the previous section are available for the incumbent operator only. The Greenfield operator, without any experience in a particular market, should take more assumptions and apply some drivers to have an OPEX estimation.

The first estimation of the total yearly OPEX value can be from 10 to 12% of the cumulative CAPEX; however, the driver strongly depends on the operator strategy (e.g. backhauling method) and his maturity on the market (own infrastructure availability). For a Greenfield operator, who is building his infrastructure from the very beginning, the OPEX driver can reach the value of 20% for the initial years of deployment, whereas for the operator using leased lines as a main backhauling solution, it can converge to the value of 30%.

8.2.3 THE ROLE OF DRIVERS: SANITY CHECKING

Whilst Section 8.4.2 presents the method of business plan results assessment, an initial check on the correctness of CAPEX/ OPEX calculations should be done at a much earlier stage. Having technical CAPEX and OPEX calculated, the results can be evaluated and validated using many available drivers and formulas. This process is called sanity checking.

The main reason for that is to check whether the CAPEX and OPEX calculation was error free and the values obtained are realistic. The method enables the discovery of the most significant errors only, but it is sufficient with respect to the accuracy of the CAPEX and OPEX calculations.

Some example drivers used for sanity checking are:

- average channels per site
- average cells per site

- network load
- average number of subscribers per site
- cumulative CAPEX per subscriber
- CAPEX per subscriber
- CAPEX per site
- OPEX per subscriber
- OPEX per site
- ratio of OPEX to cumulative CAPEX.

The range of values of above drivers can be determined empirically. It can be different for different phases of the network deployment, or for incumbent versus Greenfield operator etc. To have the possible range of drivers related to a particular environment, the number of sample business case calculations should be performed and the results should be collected.

8.3 REVENUE AND NON-TECHNICAL RELATED INVESTMENT CALCULATION

In order to properly close the business plan, information from the full investment together with the revenue calculation is needed. To have the full investment calculated it is necessary to estimate the technical CAPEX and OPEX together with other expenses (marketing, administrative etc.). Furthermore, the revenues should be forecasted as well. Although it is not the main focus of this book, for the sake of a better understanding of the business planning process, this section will briefly deal with all these non-technical investments and revenue calculation issues. For more detailed guidelines, the reader should refer to [5] and [6].

8.3.1 INPUT PARAMETERS AND ASSUMPTIONS

As is done with technical CAPEX and OPEX, very similar input data is needed when estimating revenue and marketing investments. To start the forecasting process, some market research campaigns are required in order to obtain market needs, the potential of new service (system) introduction, possible user segments etc. The second source can be the analysis of market reports and forecasts, showing market growth, segmentation, possible development of existing infrastructure, current market penetration and share of present operators etc. For incumbent operators, the important source may also be a current subscriber base and their statistical behaviour, i.e. peak traffic hours, traffic profile (voice and data), service usage, geographical distribution etc.

To start the calculation process, it is also important to make some critical assumptions: They should cover:

- market segmentations and decisions related to which segment should be addressed with a new offer and how (e.g. prepaid, postpaid mass, postpaid business etc.);
- the tariffing policy (it will strongly affect revenues, but also traffic figures);
- the level of subsidies;
- the service portfolio to be offered.

As a result of the above assumptions and the data analysed, the numbers of forecasted users and their traffic profiles can be calculated using some statistical models. Such data will act as an input for both revenue and marketing investment calculations.

8.3.2 REVENUE CALCULATION METHODS

Having the segmentation, the numbers of subscribers, service portfolio and usage profiles, together with tariffing policy, it is only 'spreadsheet work' to obtain the total revenue. The most widely used parameter in such context is called ARPU (Average Revenue per User). The ARPU value is usually different across different market segments and, together with valuable and quantitative market shares, can be used as a very good benchmark of the operator market position.

8.3.3 NON-TECHNICAL RELATED INVESTMENTS

Many marketing and administrative CAPEX and OPEX related expenses occur on the non-technical cost side. To get a business plan finalised, these should also be accounted for. The most important components of such investments cover:

- Marketing
  - mobile handset subsidies
  - advertising campaigns (ATL and BTL)
  - promotions
  - dealer commission.
- Customer management
  - Customer care
    a. SAC (Subscriber Acquisition Cost)
    b. SRC (Subscriber Retention Cost)
    c. loyalty programmes.
  - Billing.
- Personnel cost
  - The cost of running a new firm (for a Greenfield operator) or extension of existing one (incumbent operator – additional staff etc.).
- Content fees
- Others (insurance, management, finance and administration).

The way of estimating the costs for most of the factors above is pricelist based. However, the pricelist itself is very market dependent and should be acquired per market and country. The sum of these additional investments should be added to the total CAPEX and OPEX figures respectively, enabling a very accurate business planning.

8.4 BUSINESS PLANNING RESULTS

The finished business plan, or rather one of its scenarios, has to be evaluated in the time domain. From a practical point of view, the minimum time period of such evaluation should be several years. It is necessary to evaluate the profitability of a certain business, the amount of funds that have to be acquired to run the business and the sensitivity of such business plan to changes of input factors and assumptions. While this section presents the most important parameters and methods from a mobile network business planner perspective, more detailed information on a general evaluation of business plan results can be found in [7,8], and an interpretation of economic indicators can be found in [9].

8.4.1 BUSINESS PLAN OUTPUT PARAMETERS

The best way to assess business plan profitability and to compare it with other scenarios is to start with a cumulative cash flow calculation. An operating cash flow is the sum of net profit, depreciation, change in accruals and change in accounts payable, minus change in accounts receivable, minus change in inventories. The operating cash flow minus CAPEX, minus changes in working capital gives the value of undiscounted cash flow, which converted to cumulative values constitutes the basis for a business plan evaluation.

Other important parameters used for a business plan evaluation are:

- Payback Period
- Breakeven Point
- Return of Investment (ROI)
- Net Present Value (NPV)
- Internal Rate of Return (IRR)
- Peak Financing Need.

The payback period is the amount of time taken to break even on an investment. Since this method ignores the time value of money and cash flows after the payback period, it can provide only a partial picture of whether the investment is worthwhile. The factor closely connected with the payback period is the breakeven point. It is the price at which a transaction produces neither a gain nor a loss. Thus, on the graph presenting the business plan result, the breakeven point corresponds to a point in time of the payback period. The other important parameter is ROI. It is a measure of a corporation's (and particular business plan's) profitability, equal to a fiscal year's income divided by common stock and preferred stock equity plus long-term debt. ROI measures how effectively the firm uses its capital to generate profit; the higher the ROI, the better.

For the initial decision about investing and starting the business, the best parameter to start with is the Net Present Value. It is the present value of an investment's future net cash flows minus the initial investment. If positive, the investment should be made (unless an even better investment exists), otherwise it should not be made. Furthermore, the Internal Rate of Return should also be calculated. The value of IRR is the rate of return that would make the present value of future cash flows plus the final market value of an investment or business opportunity equal the current market price of the investment or opportunity. The last parameter important for investment financing is the peak financing need.

The presentation of business plan output parameters in the form of a graph seems to be the best way to see its results. Figure 8.2 presents example business plan results. The curve on the graph presents a cumulative cash flow value. On such a graph, some parameters like the breakeven point, payback period and peak financing need are easily visible without further analysis.

8.4.2 BUSINESS PLAN ASSESSMENT METHODS

As shown in the previous section, many financial parameters have to be calculated to properly assess business plan results. Having them available, the analysis and comparison between different scenarios should be done.

Generally, to accept the business case and start the investment (go/no go decision), there are some hard requirements that have to be fulfilled. Two example requirements are the NPV to be positive and the payback period to be no longer than three years. Some other requirements may be some direct constraints from potential investors or shareholders. They can affect the value of IRR, free cash flow or EBIDTA (Earnings Before Interest, Taxes, Depreciation and Amortisation). However, a business planner should not only look on hard requirements. Some other factors, potentially not connected

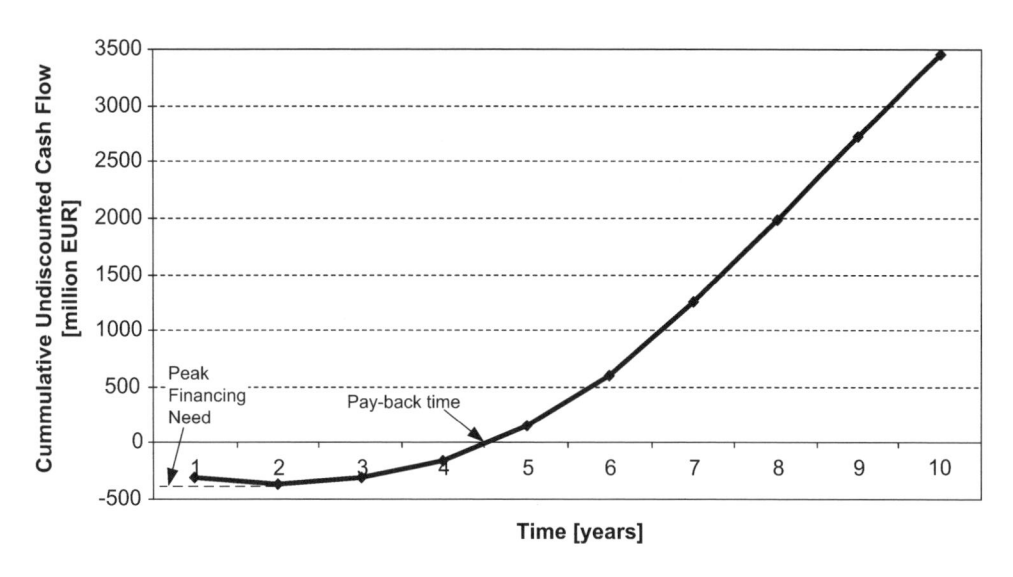

Figure 8.2 Example business plan results.

with the business plan, may enable running not profitable business. This can be the case when the abandonment cost is higher than losses from a new business implementation. A good example is a 3G operator evaluating the introduction of a new service, such as video conferencing. As the service seems to be unprofitable, the operator may refuse implementation; however, looking from a long term perspective, when all the competitors will implement such service, the outflow of best business customers may occur, thus decreasing revenue significantly.

It should be remembered that the changes of input parameters and assumptions can change business plan results significantly. Thus, the usual last step in business planning is a sensitivity analysis. The idea here is to change only one parameter at a time, when all the others remain unchanged. It should be done making several calculations with different parameter values within an assumed range and presenting the resulting extreme output values together. The reason for sensitivity analyses is a simulation of environmental changes (e.g. exchange rates, WACC etc.) and potential mistakes in assumptions (e.g. UMTS cell range, number of subscribers, equipment prices, ARPU etc.). The primary usage of sensitivity analyses is related to market forecasts, being the input data with the highest level of uncertainty. The main goal is to determine the factors critical for business plan closure (e.g. NPV positive) and further analysis of them (if needed). The finalised business plan presentation covers the base case scenario together with sensitivity analysis results.

This finishes our chapter on business oriented planning methods and underlying mechanisms. The aim has been to give the reader a grander picture on the 3G network planning procedure, and hence path the way for the discussion of more technical planning issues as exposed in subsequent chapters.

REFERENCES

[1] Robert Fildes, *Forecasting Techniques Workshop*, IBC Global Conferences, 20th March 2002, London.
[2] John E. Hanke, Dean W. Wichern, *Business Forecasting* (8th edition), Prentice-Hall, 2004.
[3] Bos, Newbold, *Introductory Business and Economic Forecasting*, South Western College Publishing, 1994.
[4] Tomi T. Ahonen, Joe Barret, *Services for UMTS: Creating Killer Applications in 3G*, John Wiley & Sons Ltd, Inc., 2002.

[5] Costas Courcoubetis, Richard Weber, *Pricing Communication Networks: Economics, Technology and Modelling*, John Wiley & Sons Ltd, Inc., 2003.
[6] Tomi T. Ahonen, *m-Profits: Making Money from 3G Services*, John Wiley & Sons Ltd, Inc., 2002.
[7] Graham Friend, Stefan Zehle, *Guide to Business Planning*, Profile Books Ltd, 2004.
[8] John Tennent, Graham Friend, *Guide to Business Modelling*, Profile Books Ltd, 2001.
[9] *Guide to Economic Indicators: Making Sense of Economics*, 5th edition, Profile Books Ltd, 2003.

9

Fundamentals of Network Characteristics

Maciej J. Nawrocki

The aim of this chapter is to shed some light onto the behaviour of power limited WCDMA networks. Knowledge about the network's power characteristics is hence the key for a proper understanding of its behaviour, especially under extreme load conditions. The following sections address various aspects related to power, including the power dependency on the distance between mobile and base station, the load, any irregularities of cell layout as well as the size of the actual UMTS network [1]. Most of the presented results concern simple cases with a uniform cell grid and a uniform traffic distribution, but they can easily be extended to more complex cases since tendencies in WCDMA network behaviour generally remain the same. The intention of this chapter is hence to expose general tendencies, and not the detailed real-life network behaviour, so as to smoothly prepare the reader for more complex scenarios in subsequent chapters.

9.1 POWER CHARACTERISTICS ESTIMATION

This section addresses the basics of power estimation and behaviour, these being the main resource in a WCDMA network. Relatively simple models were used in simulations to highlight the nature of the WCDMA network; advanced Radio Resource Management (RRM) algorithms, however, were not used in the simulations since they usually counteract with nonlinear power-dependent effects, which were intentionally emphasised in this chapter.

9.1.1 DISTANCE TO HOME BASE STATION DEPENDENCY

9.1.1.1 Power Transmitted by the Mobile Station (MS)

The first characteristic illustrating the fundamentals of WCDMA network behaviour is the MS's transmit power as a function of the distance to the serving base station. This power also depends

Understanding UMTS Radio Network Modelling, Planning and Automated Optimisation   Edited by Maciej J. Nawrocki, Mischa Dohler and A. Hamid Aghvami   © 2006 John Wiley & Sons, Ltd

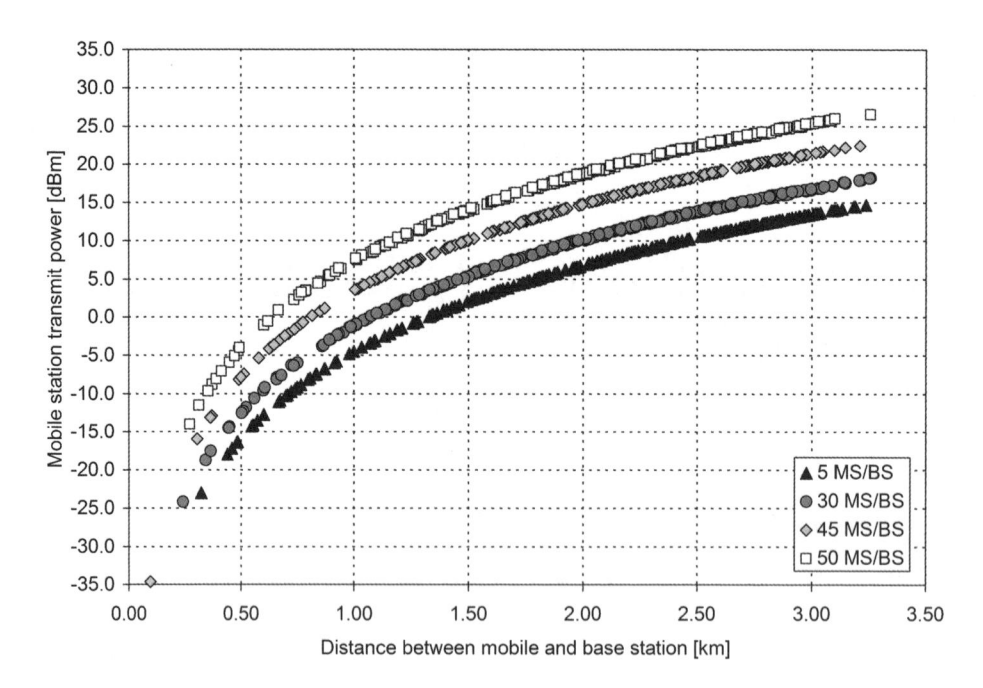

Figure 9.1 Mobile station transmit power as a function of distance to serving base station for voice service and various traffic loads.

on the network load as shown in Figure 9.1 and already described in Section 2.3.1, as well as in Chapter 10. Assuming a uniform terminal distribution within the cells, the MS transmit powers have been computed for voice service (12.2 kbps, $E_b/N_0 = 6$ dB). The cell radius was around 3 km excluding shadow fading to clearly show the tendencies in the curve shape.

Figure 9.1 presents the power transmitted by the mobile terminal as a function of the distance to the serving base station. The curves were computed for four different loads ranging from 5 to 50 MSs per cell. Taking into account the distribution of interference sources, the curve shapes mimic propagation curves when drawn in logarithmic scale. An increase of the number of active terminals in a cell does not change the depicted tendency; thus, more power is always required to overcome propagation losses. It can also be noted that the required power increase to compensate the increased traffic from 5 to 30 MS per BS is the same as for the case of 45 to 50 MS per BS, clearly showing a nonlinear power dependency caused by the load. Note that such an increase is the same for any location of the terminal within a cell; also, the trend does not change when shadow fading is assumed. Note further that these important tendencies in the uplink are very different from the downlink, as will be discussed in the subsequent section. Figure 9.1 also clearly shows the need for a large dynamic range of power control mechanisms in the uplink, since the terminal is mostly forced to compensate propagation losses which, by their nature, have a high dynamic range. The interference level rise is relatively small compared to the path loss increase. The interference level rise is also smaller than the interference rise in the downlink, but of course becomes a definite component of increased power control dynamic range requirement for the uplink.

9.1.1.2 Power Transmitted by the Base Station

Similar to the uplink, the same characteristics were computed for the downlink for a voice service (12.2 kbps, $E_b/N_0 = 8$ dB) as shown in Figure 9.2. The transmit power in the traffic channel

Fundamentals of Network Characteristics

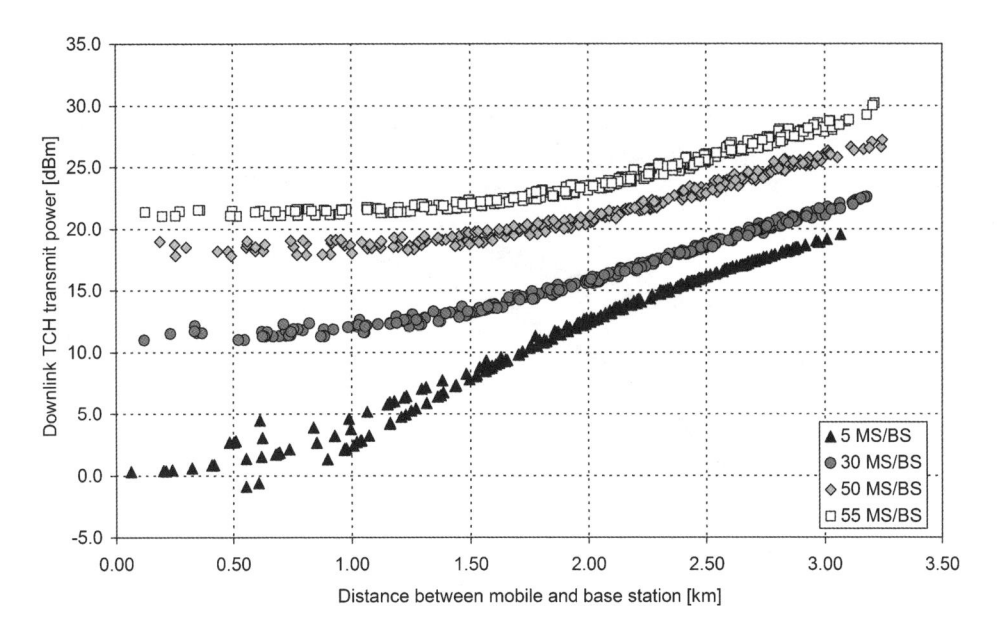

Figure 9.2 Base station transmit power in given TCH as a function of distance to active mobile for voice service and various traffic loads.

(TCH) is comparable to the uplink case, but only when the traffic load is low. An increased number of active terminals in the cells causes a significant increase of the required transmit power for terminals close to the base station, while the terminals near the cell edge are subject to a much smaller rise. For a fully loaded system, there is only a minor dependency between the terminal location within a cell and the base station transmit power in traffic channel since terminals close to the base station are very close to the source of intra-cell interference (Figure 9.4). As shown in Figure 9.2, there is only about 7 dB of difference in transmit power for terminals placed 100 m and 3 km from the serving BS. Adding shadow fading to the presented scenario scatters this effect, but the tendency remains the same. The curve for a fully loaded network would ideally be flat for the single cell case; however, for a multi cell network, terminals close to the cell edge are closer to other base stations, which increases inter-cell interference and the convexity of the presented curve. All these effects are typical to CDMA based networks and very different from FDMA/TDMA based.

Interference from the serving base station is the main reason of a reduced dynamic range of power control in the downlink compared to the uplink. Furthermore, propagation loss and intra-cell interference compensate each other's influence on the dynamic range of power control; thus the dynamic range is much smaller for downlink than for the uplink. A direct comparison between the uplink and downlink can be seen in Figure 9.3 for both lightly and heavily loaded networks.

Figure 9.4 presents an example of the interference level in the downlink as a function of the terminal distance to the serving base station. Intra- and inter-cell interference levels are shown, as well as the thermal noise component. Intra-cell interference was weighted by an orthogonality factor ($\alpha = 0.6$) to show the influence on the performance of a more realistic system. In the presented case, around 60% of the cell area is dominated by intra-cell interference, while the remaining 40% of the cell (outer region) by inter-cell interference. In cases other than this example, the considered values could vary depending on cell size and propagation environment, but tendencies remain the same.

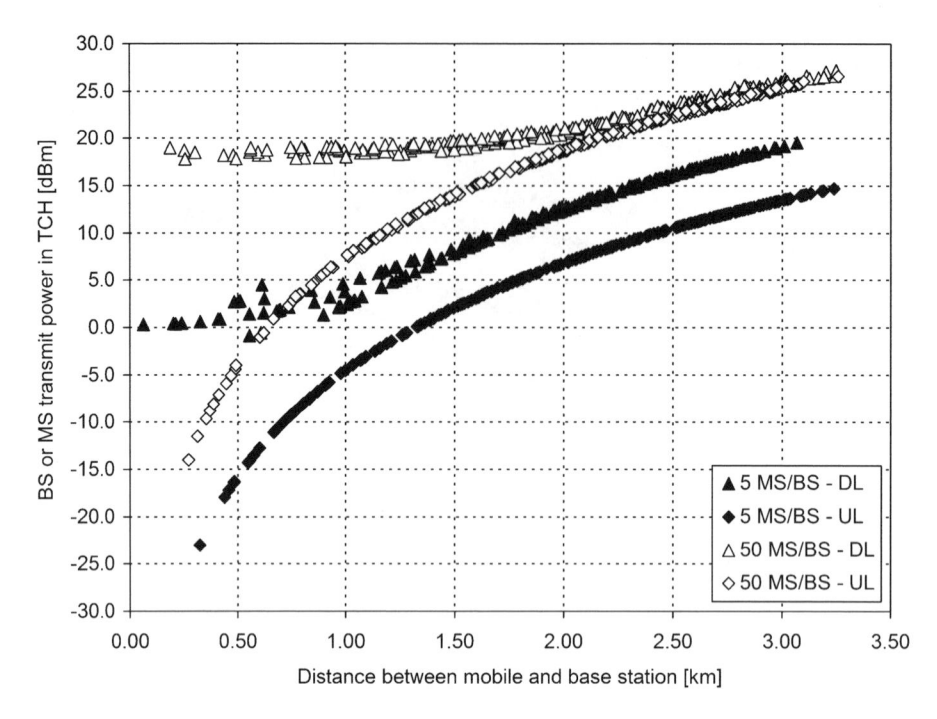

Figure 9.3 Comparison of TCH transmit power for both link directions for lightly and heavily loaded networks. Dynamic range of power control is higher for uplink compared to downlink direction.

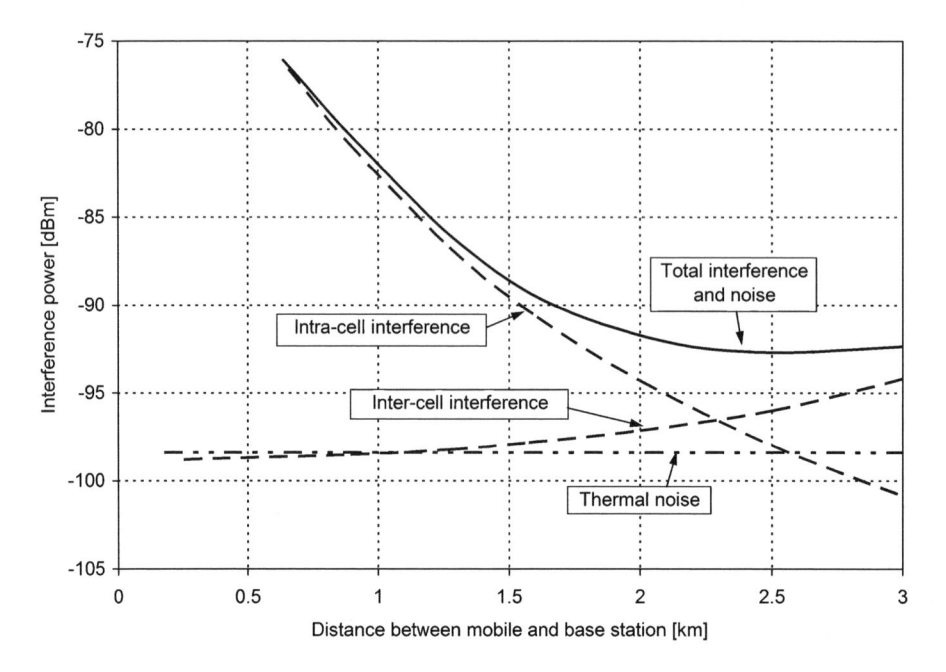

Figure 9.4 Interference power received by mobile terminal as a function of the distance to serving base station.

9.1.2 TRAFFIC LOAD DEPENDENCY

9.1.2.1 Base Station Received Interference Power

Figure 9.5 presents the received interference power at the base station as a function of the number of active terminals for a voice service (other services are not included but tendencies of the network behaviour remains the same). The network was computed as a static model with 1000 Monte-Carlo simulations assuming a bit rate of 12.2 kbps and a target $E_b/N_0 = 6$ dB. The cell layout was hexagonal with a uniform terminal distribution. The log-normal shadow fading was included with a standard deviation of $\sigma = 4$ dB in a macrocell environment with an approximate cell radius of 3 km. Figure 9.5 presents curves for maximum, average and minimum values obtained during simulations as well as values for 1 % probability of exceeding given interference power values.

The interference presented in Figure 9.5 can be split to intra- and inter-cell interference; thermal noise as such plays a minor role in the network performance analysis and is often neglected for loaded systems. It is interesting to note, however, that the thermal noise is the only reason for requiring power control procedures in the first place. In the absence of thermal noise, the system of linear equations presented in Section 6.4.9, Equation (6.50), would have zero power solutions, thus no transmit power would be required at all. Since thermal noise is present, however, a given transmit power is required to overcome this thermal noise and, in consequence, also the interference generated by other sources which try to overcome the thermal noise as well.

Figure 9.6 shows all three components of the major interference: intra-cell, inter-cell and thermal noise as well as their sum as a function of traffic load. (Here, we have neglected other sources of interference, such as adjacent channel interference – see Section 6.4.3.5.) It can be noted that the difference between intra- and inter-cell interference is almost constant and does not depend on the traffic load. A more detailed example is presented in Section 9.3.

Obviously, the curves from Figure 9.6 can have other relations to each other when the network layout and traffic distribution is different from the homogenous case.

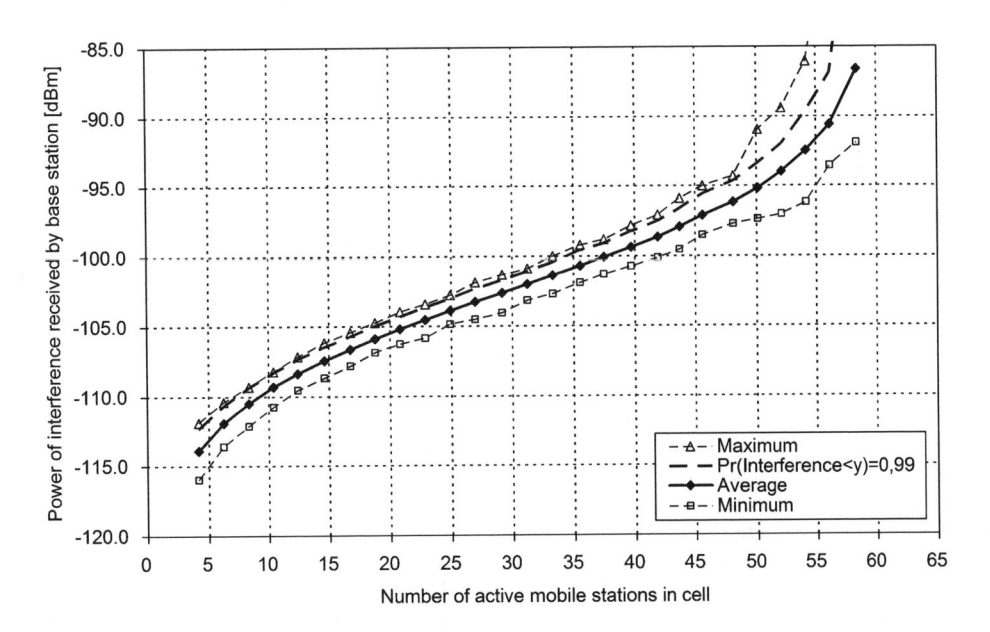

Figure 9.5 Interference power received by base station as a function of the number of active terminals in a cell.

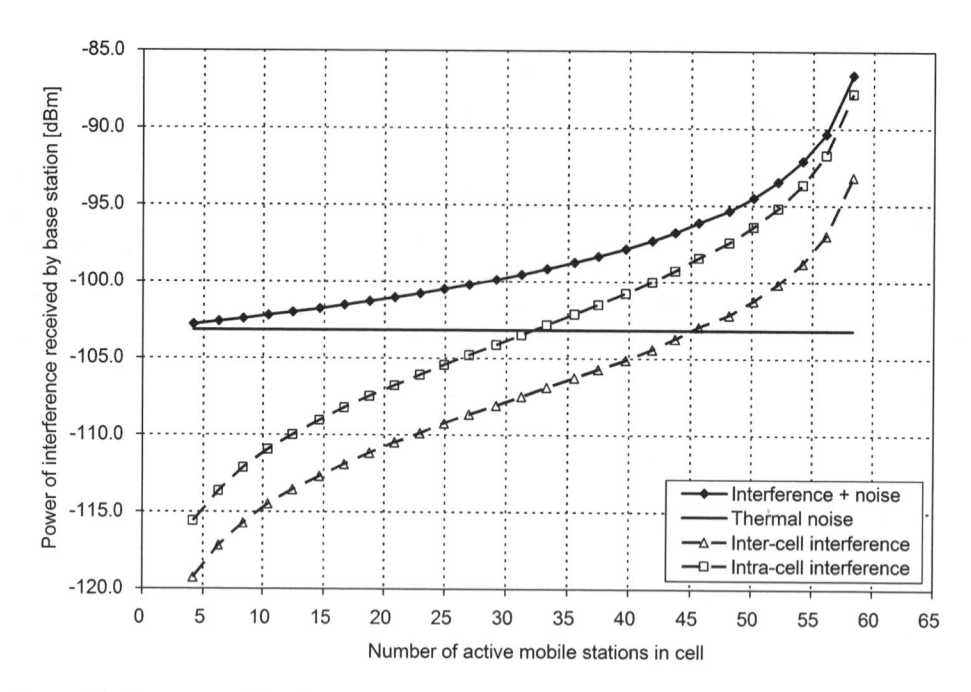

Figure 9.6 Components of interference power received by base station as a function of the number of active terminals in a cell.

9.1.2.2 Total Power Transmitted by the Base Station

The most important characteristic influencing the general network behaviour in the downlink is the relation of the total power transmitted by the base station as a function of the traffic load, i.e. the number of active terminals in every cell. Based on this characteristic, the maximum number of theoretically available channels can be estimated, i.e. the pole capacity, as well as more realistic values, which take into account finite power of the base station amplifiers.

Figure 9.7 presents above characteristics for a hexagonal network with a uniform traffic distribution and 12.2 kbps voice service (100 % activity), similar to the example in previous sections of this chapter (3 km of cell radius). Calculations were done for 1000 Monte-Carlo snapshots and are represented by maximum, average and minimum values as well as values for 1 % probability of exceeding given powers. It has been assumed that the macrocell base stations transmit with 20 W (43 dBm) of maximum transmit power. Assuming 20 % of this power is used by Control Channels (CCH), there is about 42 dBm left for TCHs; this value is marked in Figure 9.7. It corresponds to around 60 available TCHs for the average curve and around 55 channels for the 99 % probability curve.

The pole capacity equals about 69 channels (vertical asymptote) where a rapid growth of the required transmit powers occurs. Above this value, similar to the uplink, linear equations (Section 6.4) lose their physical meaning and the power values would need to be negative to satisfy the system of linear equations (which is clearly impossible in real situation). Adding one active terminal to the cell for loads over 90 % causes rapid increase of required transmit power. This nonlinearity is similar to the uplink growth of interference and MSs' transmit power in highly loaded networks, and is characteristic to all CDMA networks where the same frequency band is used in neighbouring cells.

To complement Figures 9.5 and 9.7, the standard deviation of the received interference power (uplink) and transmit power (downlink) over 1000 snapshots as a function of traffic load was computed and has been depicted in Figure 9.8. It exposes the increased variations of the results obtained from

Fundamentals of Network Characteristics

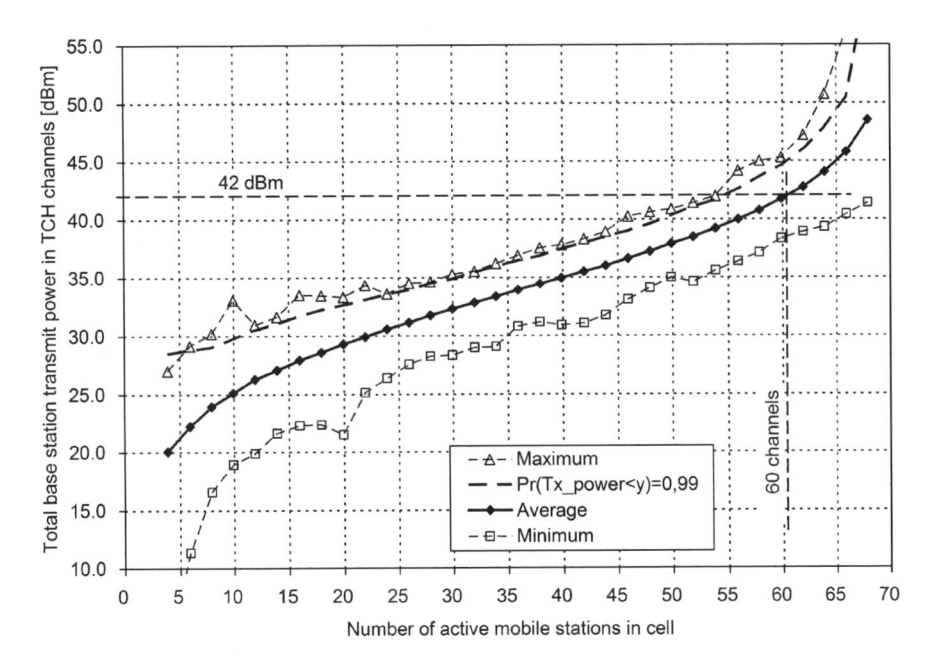

Figure 9.7 Total transmit power in TCH channels as a function of the number of active mobile stations in every cell.

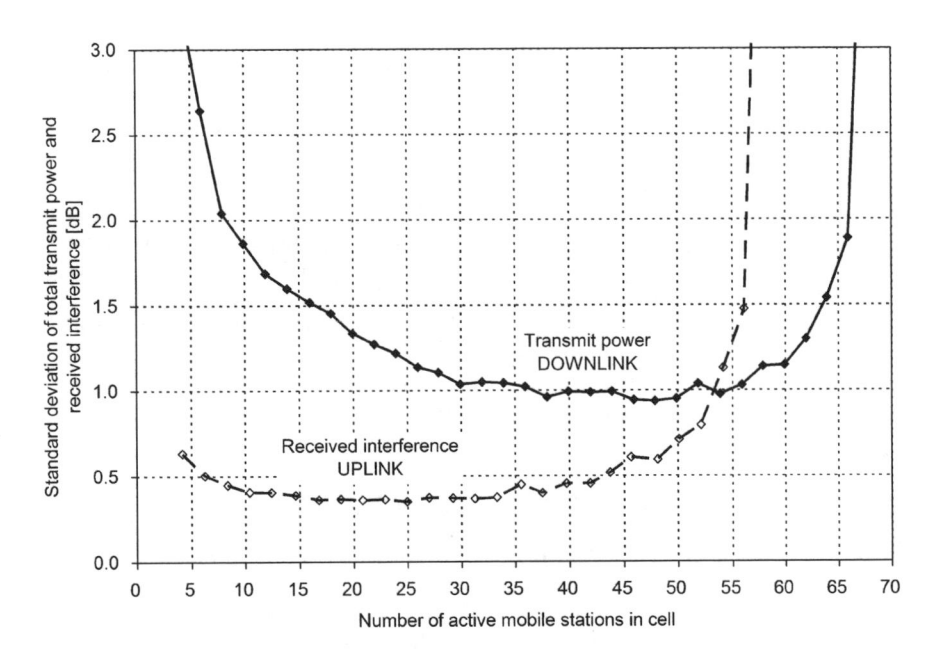

Figure 9.8 Standard deviation of the interference power received by base station (UL) as well as total base station transmit power (DL) as a function of the number of active terminals in a cell for 1000 Monte-Carlo snapshots.

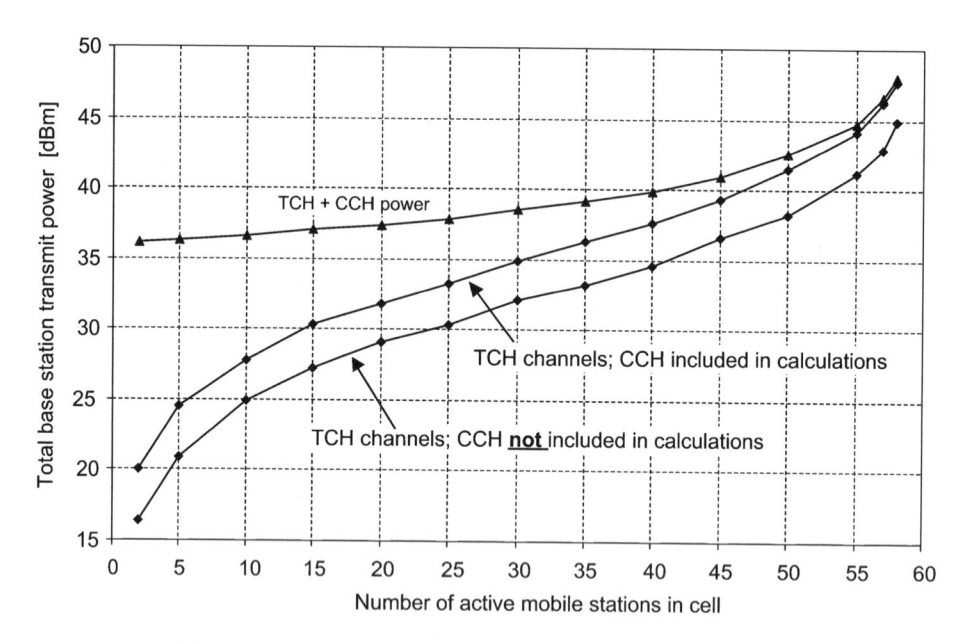

Figure 9.9 Influence of CCH channels existence into power characteristics in downlink direction.

Monte-Carlo simulations, not only for low loads but also for higher loads (around 85 %). This is of special importance when supporting simulations are made by network planning tools: an increased number of snapshots is hence recommended to be used for highly loaded and overloaded networks, as well as networks at low loads, but not necessarily at medium loads. Operators and equipment vendors usually assume that for stable real-world networks the load should not exceed about 75 %, but during the network planning process physically non-achievable cases (e.g. extremely high loads) are also considered and wrong conclusion could be drawn from their analysis. Furthermore, it can be seen that the downlink experiences a much higher dispersion of the obtained results and is consequently more demanding in terms of number of Monte-Carlo snapshots.

Results of the simulations presented in Figure 9.9 clearly show that common channels (most notably CPICH) do not influence the pole capacity value, which is in contrast to bit rate or target E_b/N_0. The presence of these channels causes increased transmit powers in TCH channels. This increase is constant in traffic load; thus, an increased power of the CCH channels has comparable effects as an increased level of thermal noise. Remember that the latter statement only holds for the transmission powers as such, without taking other effects into account that might be triggered by a change of the power allocated to the CCHs; for example, an increased CPICH power will change many other aspects of the network behaviour (e.g. best-server assignment).

9.2 NETWORK CAPACITY CONSIDERATIONS

9.2.1 IRREGULAR BASE STATION DISTRIBUTION GRID

An irregular base station distribution can cause a noticeable capacity loss when the traffic distribution is uniform. On the other hand, in real situations, traffic is usually very uneven and an irregular, albeit coordinated, base station placement can improve the network capacity compared to homogenous network roll-outs. Base station placements, matching traffic requirements optimally, are analysed in

Fundamentals of Network Characteristics 211

Chapter 14, where their locations are among the parameters required for an automated network layout optimisation. In contrast to these chapters and hence addressing more basic tendencies, this section assumes uniform traffic only.

A triangular grid for base station locations is the main theoretical layout used to create homogenous wireless networks when omnidirectional or three sector sites are used. This ensures optimal coverage of a given surface, generating hexagonal cell shapes. In real situations, cell sites often cannot be located where required due to electromagnetic compatibility issues, terrain ownership, lack of infrastructure etc., and must be moved to other locations, thereby violating the ideal roll-out.

Modelling irregularities of base station locations can be made in many ways. In the case considered in this section, a constant deviation to the standard triangular grid is assumed for all base stations. For instance, when the assumed deviation equals 20% of the cell radius, the base stations are randomly placed on circles with the centres in a triangular reference grid and a radius equal to 0.2 the cell radius for all snapshots; this has been illustrated in Figure 9.10. Subsequent results presented in this section were obtained for network irregularities of 0–30% assuming an average cell radius of 3 km and supported voice service. It must be noted that site location irregularities can be bigger in real situations than presented here.

9.2.1.1 Uplink Direction

The limited terminal transmit power is the main factor restraining the network capacity. Class 3 (24 dBm) terminals were assumed for calculations with 1% probability of exceeding this value as the capacity limit. It was found that the pole capacity experiences only a minor drop over the entire range from 0 to 30% irregularity, but it must be noted that the pole capacity is not limited by the terminal maximum transmission power. For irregularities equal to 30% of the cell radius, there are about 14% less voice channels available compared to a regular network layout (assuming 1% probability of exceeding 24 dBm).

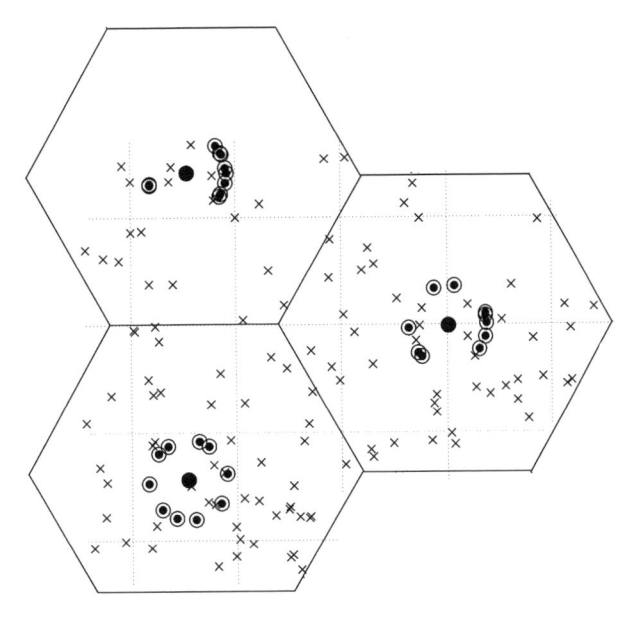

Figure 9.10 Irregular base station placement used in simulations.

9.2.1.2 Downlink Direction

The total transmit power available at the base station is the main capacity limiting factor in the downlink. It was assumed that this power equals 43 dBm, thereby clearly limiting the number of available TCH channels (CCHs were also included in the simulations). It was found that the capacity loss is slightly bigger in the downlink compared to the uplink. The experienced loss in capacity for both links is presented in Figure 9.11; it varies up to 20 % in the studied case. For a uniform traffic distribution (as here) the capacity loss is mainly caused by a non-uniform load within the considered cells when the base station location is not within the triangular grid, moving the problem from an irregular site location to a non-uniform traffic distribution (these two cases are replaceable in many situations). On the other hand, real life conditions can compensate this loss by the operator moving sites closer to intensive traffic areas (but, if not done properly, can also exaggerate it).

A similar case is presented in [2,3], but for more realistic network and propagation data and smaller cell sizes. The authors there analyse the network consisting of 17 three-sector base stations with 1.5 km or 3.0 km site-to-site distance, depending on the studied case. Only voice traffic is simulated, and indoor users are also considered. The base station location deviation of one-quarter of the site spacing was assumed. The conclusions from the Monte-Carlo simulation for this kind of irregular network have been that the network performance in terms of service probability, SHO probability and throughput remains almost untouched. It has been shown that hexagonal grids do not need to be optimal when terrain topography and morphology are taken into account.

9.2.2 IMPROPER ANTENNA AZIMUTH ARRANGEMENT

Locating sites on a triangular grid, two opposite cell layouts can be obtained, depending on the direction of the sector antennas, where both layouts differ by a 30° of all antennas in azimuth; both cases have

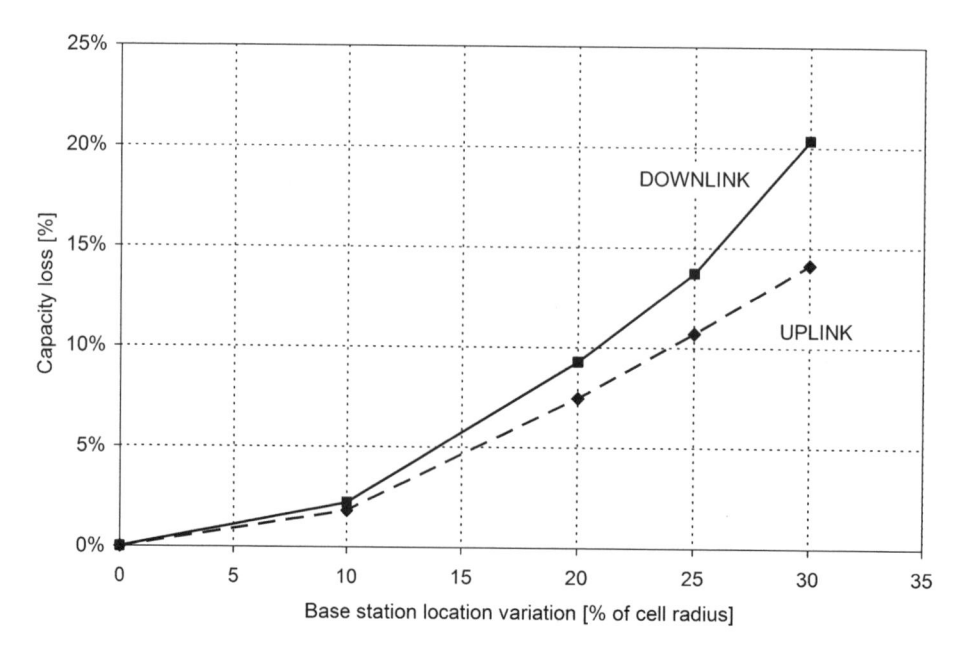

Figure 9.11 Capacity loss as a function of base station location variation.

Fundamentals of Network Characteristics

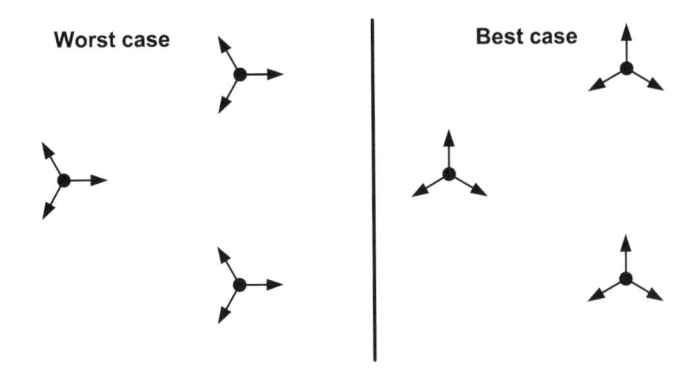

Figure 9.12 The worst and the best case of antenna orientation for triangular grid site placement.

been presented in Figure 9.12. Note that the term worst/best relates to sites with 120° of angular separation between neighbouring sectors.

Simulations were made for the worst and the best cases with uniform voice traffic and the same network specification as in previous examples. Sector antennas were used with a beamwidth equal to 65°(-3 dB) located on 19 three-sector sites [4].

Figure 9.13 presents the total power transmitted in the TCH channels as a function of the network load for both cases. An improper azimuth assignment of sector antennas can cause a capacity degradation exceeding 25 % and requires an increase of base station transmit powers of 3 to 6 dB. The decrease

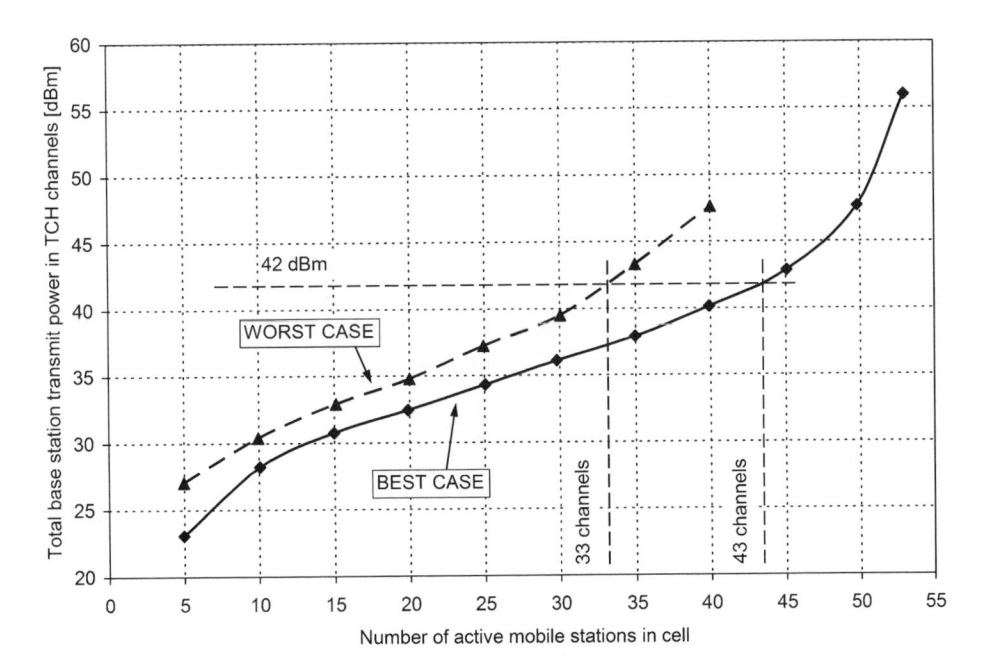

Figure 9.13 Total transmit power in TCH channels as a function of the number of active mobile stations in every cell for the worst and the best cases of site antenna orientation.

pertains not only to the capacity depending on the finite maximum power of the transmitter (in this case 42 dBm), but also to the pole capacity. The graph shown in Figure 9.13 represent average values over 100 Monte-Carlo simulations. The worst case scenario includes coverage holes (i.e. places with increased propagation attenuation) which make the network require more power resources to close them [5]. This effect is intentionally amplified by assuming a macrocellular environment with higher power demands. The author of [5] proposes an antenna azimuth optimisation algorithm based on attempts of avoiding these holes by properly adjusting the azimuth settings.

The authors of [2,3] also consider random azimuth variations in an ideal hexagonal grid but change each sector antenna direction independently, given the same network conditions as for the irregular base station distribution presented in Section 9.2.1. The resulting changes have a normal distribution with average deviation of 9.1° or 18.2° depending on the assumed scenario. The results show only small degradations in the network performance when azimuths are randomly varied. The softer handover probability, clearly, is the most influenced performance measure in that case. And adding indoor users does not change the situation significantly. It is worth mentioning that the results were obtained for sites located in a triangular grid. For highly deviated base station locations, as in real life situations, improper azimuth assignments might cause a significant network performance drop (see Chapter 15).

9.3 REQUIRED MINIMUM NETWORK SIZE FOR CALCULATIONS

In any CDMA network, the working frequency band is usually repeated in neighbouring cells, i.e. the frequency reuse factor is one. This causes cell coupling in terms of interference and hence transmission powers, even between quite distant areas of the same network. This property requires careful simulations, when it comes to the network planning of a real network with use of network planning tools. The question of the size of the network taken into account arises when the network planner wants to be sure of receiving proper simulation results.

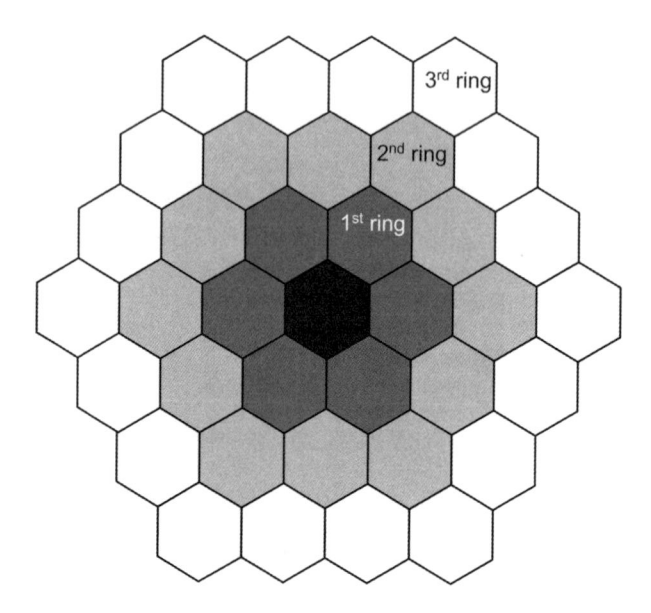

Figure 9.14 Structure of simulated network with three rings of cells around the central cell.

When having to analyse one part of the network for a given city, extended fragment must be taken into account to avoid border effects. From a simulator point of view, the rest of the network does not exist in this case, which obviously does not hold true in reality. For more academically inclined simulations, so-called *wrap-around* techniques are used to overcome this problem; however, for real world situations, this cannot be used.

For finding an answer to this problem, the following scenario was considered. There is one cell with an omnidirectional antenna, for which network performance characteristics need to be found. This cell is surrounded by consecutive rings of neighbouring cells, as shown in Figure 9.14. Three cases were considered, where the network consisted of 7 cells (one ring), 19 cells (two rings) and 37 cells (three rings).

Various cell sizes were assumed for calculations, ranging from 1 to 6 km for site-to-site distances. Two services were included in calculations: voice 12.2 kbps and circuit switched data transmission 64 kbps. Log-normal shadow fading was included with a standard deviation of $\sigma = 4$ dB. Terminals were uniformly distributed within the cells in 100 Monte-Carlo snapshots at every load point. Wrap-around techniques were not used. Several criteria were used to determine the cell capacity:

- Uplink

  - Pole capacity;
  - Capacity for a probability of exceeding the maximum terminal transmit power (21 dBm for voice, 24 dBm for data) equal to 1 %;
  - Capacity for a probability of exceeding the maximum terminal transmit power (21 dBm for voice, 24 dBm for data) equal to 5 %.

- Downlink

  - Pole capacity;
  - Capacity for 42 dBm average power available for TCH channels;
  - Capacity for a probability of exceeding the maximum transmit power per code channel (30 dBm) equal to 1 %;
  - Capacity for a probability of exceeding the maximum transmit power per code channel (30 dBm) equal to 5 %.

An example power characteristic for different number of cell rings for downlink direction is presented in Figure 9.15.

When the 37-cell scenario is considered as the reference, an error in the capacity estimation compared to the reference scenario can be calculated for various capacity criteria and both link direction. The error is presented in Figures 9.16 and 9.17; it virtually does not depend on the site-to-site distance, nor the service used when terrain topography and morphology is homogenous. The first noticeable observation is about the error in the downlink compared to the uplink being twice the value. The capacity error for downlink and two cell rings (19 cells) is around 5 %, while reaching 20 % in the one cell ring case (7 cells). The uplink direction reaches lower error values, on average 10 % for one cell ring (6 to 16 %) and on 5 % average (2.5 to 6 %) for the two cell rings.

It can be concluded that a single ring of cells around an analysed cell is not sufficient for downlink simulations and the second ring should be included to receive reasonable results. In the case of the uplink, the second ring of cells is also recommended; however, single ring simulations can be accepted when a reduced simulation time is required. This option can also be used for the downlink, but the planning engineer must be aware of overestimating the network performance measures since significant parts of the inter-cell interference are neglected. It must also be remembered that traffic non-uniformity can amplify this error when the important part of the traffic lies out of the considered area but close to its border or when the terrain becomes heterogeneous. The factor of inter-to-intra cell interference power, often used in network dimensioning (see Chapter 10.2.4.6), was also computed for the analysed multi ring scenarios and is presented in Figure 9.18 as a function of cell load for the

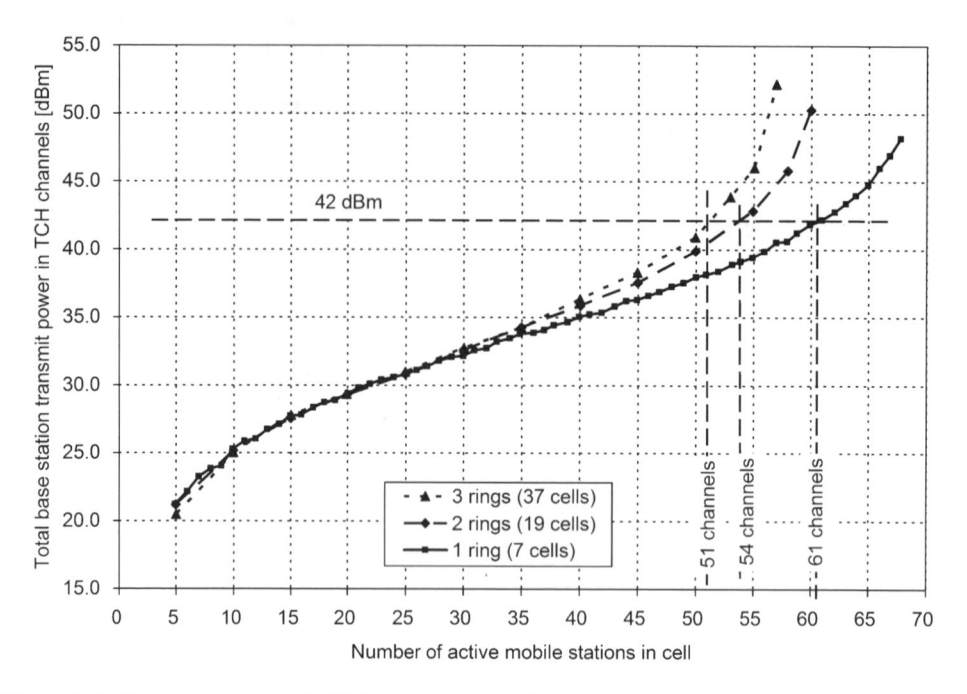

Figure 9.15 Total transmit power in TCH channels as a function of the number of active mobile stations in every cell for different number of cells assumed in simulations.

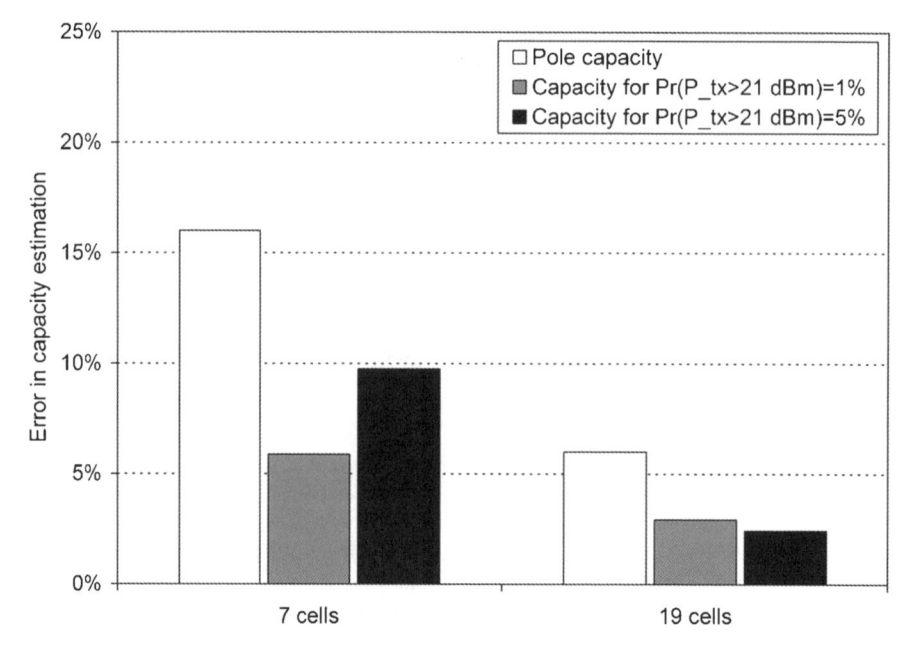

Figure 9.16 Error in capacity estimation for different number of cells assumed in simulations in the uplink.

Fundamentals of Network Characteristics 217

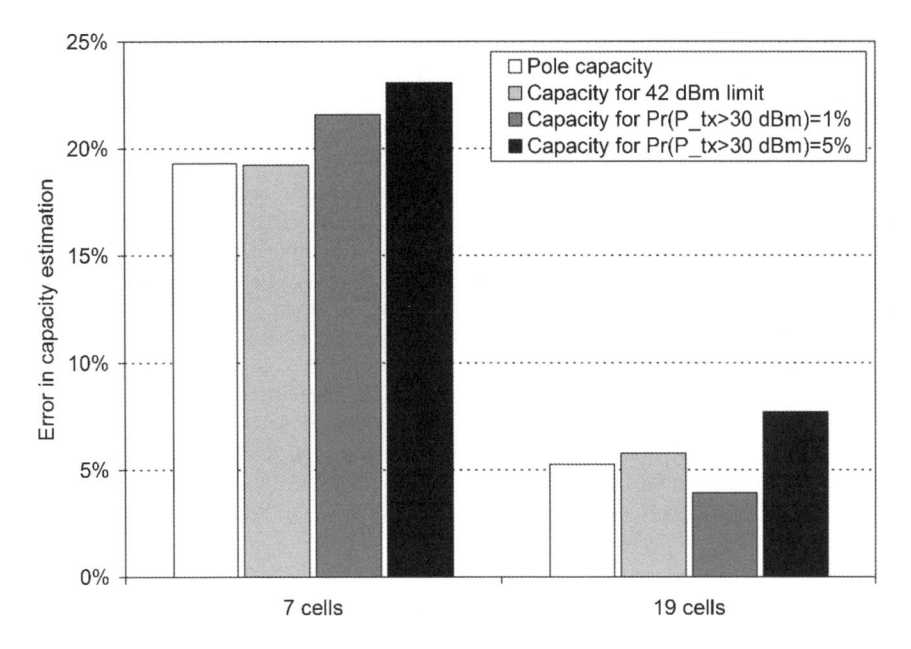

Figure 9.17 Error in capacity estimation for different number of cells assumed in simulations in the downlink.

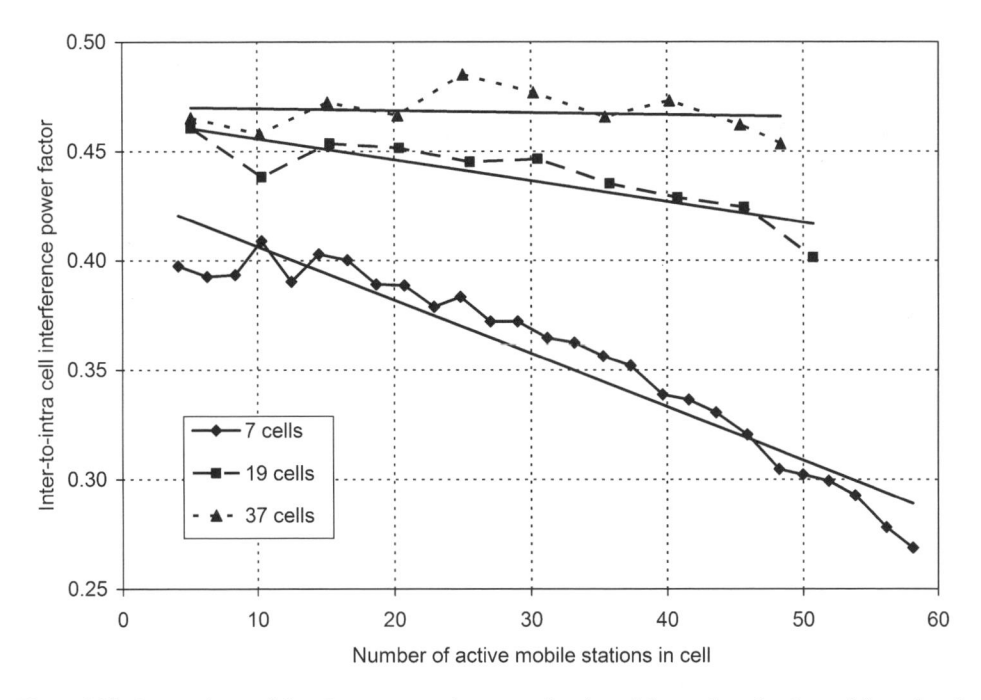

Figure 9.18 Inter-to-intra cell interference power factor as a function of the number of active mobile stations in a cell for uplink direction and three different cell configurations.

uplink (voice service and 100 Monte-Carlo simulations per point). The factor was computed for the central cell.

The influence of the number of cells assumed for simulations is clearly visible for both parameter values as well as its dependency on the cell load. Triple and dual ring models are quite similar, while the single ring model becomes significantly different; this emphasises the importance of the size of the simulated area for network planning purposes even stronger.

The problem of choosing the correct network size for system level simulations has also been studied in [6] for the uplink by analysing the CDF curves of the received interference. The authors considered a network consisting of one and two rings of three sector sites around the central site and used an advanced Monte-Carlo simulator with 1000 snapshots. The traffic was uniform with a mix of voice and 64/144/384 kbps data transmissions. Log-normal shadow fading was included with a standard deviation of $\sigma = 6$ dB. The authors discovered that the 7-site scenario leads to a gross underestimation of the total interference at the central site, thereby overestimating the cell capacity. The 19-site network becomes sufficient for system level simulations in the uplink direction.

To conclude, for reliable system level simulations used in network planning, the size of the simulated area should be extended by two additional tiers of surrounding sites. Such a structure was considered for simulations presented in this section, but only the inner cells were used for performance analysis, while the outer cells just represented sources of interference which exist in the real network.

REFERENCES

[1] Bem, D.J., Nawrocki, M.J., Wieckowski, T.W., Zielinski, R.J., *Modeling methods for WCDMA network planning*; Vehicular Technology Conference, 2001. VTC 2001 Spring. IEEE VTS 53rd vol. 2, pp. 962–966, 6–9 May 2001.

[2] Jukka Lempiäinen, Matti Manninen (eds), *UMTS radio network planning, optimization and QoS Management, For practical engineering tasks*, Kluwer Academic Publishers, 2003.

[3] Jarno Niemelä, Jukka Lempiäinen, *Impact of Base Station Locations and Antenna Orientations on UMTS Radio Network Capacity and Coverage Evolution*, IEEE 6th Wireless Personal Multimedia Communications Conference, WPMC, vol. 2, pp. 82–86, 2003.

[4] Maciej J. Nawrocki, Tadeusz W. Wieckowski, *Optimal Site and Antenna Location for UMTS – Output Results of 3G Network Simulation Software*, 14th International Conference on Microwaves, Radar and Wireless Communications, MIKON-2002, vol. 3, 2002.

[5] S. Jakl, *Evolutionary Algorithms for UMTS Network Optimization*, PhD thesis, Technische Universität Wien, 2004.

[6] Thomas Neubauer, Thomas Baumgartner, Ernst Bonek, *Required network size for system simulations in UMTS FDD uplink*, IEEE Sixth International Symposium on Spread Spectrum Techniques & Applications ISSSTA 2000, Newark, USA / September 6–8, 2000.

10

Fundamentals of Practical Radio Access Network Design

Ziemowit Neyman and Mischa Dohler

In this chapter, we aim to acquaint the reader with the fundamentals of practical Radio Access Network (RAN) design. In particular, we will start with a description of the input parameters required for RAN design. We will then discuss important network dimensioning metrics, such as coverage, capacity, their trade-off etc. The remaining part of the chapter is dedicated to some detailed approaches related to the actual network planning exercise.

10.1 INTRODUCTION

The life cycle of RAN can be roughly divided into six phases (Figure 10.1):

1. Definition
2. Dimensioning
3. Planning
4. Roll-out and Initial Optimisation
5. Operation and Ongoing Optimisation
6. Extension.

10.1.1 DEFINITION

The objective of the network definition phase is a clear description of the targets for the RAN, what kind of resources can be used to achieve them and what is the time frame for each of the activities involved. It also includes a specification of the resulting radio access network technology. The definition comprises the key actions to build a business plan and hence to secure the network's existence over a long period of time. The output of the definition yields the network requirements for

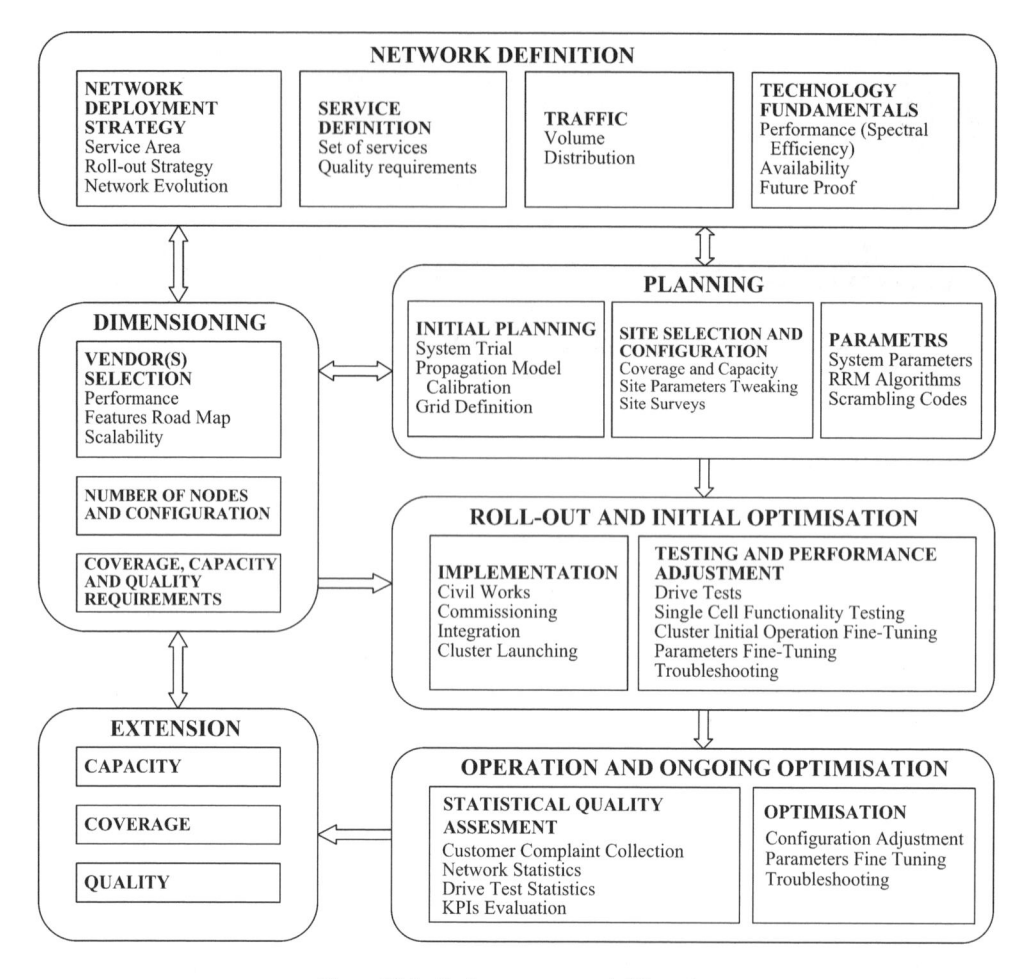

Figure 10.1 Radio access network life cycle.

coverage, capacity and quality for specified user services in a given area and over a given time frame. Typically, the network deployment would be defined in a number of phases or roll-out steps, where specific regions (markets) would be offered to potential customers. The service quality and variety may differ depending on phase and service area due to system availability and market development. Additionally, the future network evolution is defined to adapt the changes in technology and market progress.

10.1.2 DIMENSIONING

The process of dimensioning delivers a lower bound on the number of network elements and its configurations required to provide given services for a number of customers in a service area with certain quality and within a specific period of time. During dimensioning, a vendor selection is performed. This will significantly influence the dimensioning output, since the vendors' system performance and feature sets directly influence the parameters in the dimensioning. Hence, the number of required network nodes may differ between different vendors or the offered service can be obtained with different quality. Beside the performance criterion, the vendors' solution scalability and system development road map are important. Both are required to flexibly extend network capacity, coverage or quality and adjust the network to future needs and progress. Dimensioning delivers principal information for system planning, deployment and optimisation: topology of RAN, coverage, capacity and quality requirements, usually given within certain thresholds or key performance indicators (KPIs). The dimensioning stage may be supported by field trials and initial propagation model adjustments for a better fine-tuning of the dimensioning parameters. The dimensioning output can be fed back into the network definition to fine-tune the business case.

10.1.3 PLANNING

The system planning typically begins with a field trial of available technology solutions. It will help to verify the system behaviour in a real environment. In parallel, the propagation model calibration is performed and the grid is defined in the service area. Next, the site selection together with site configuration takes place. It is strongly supported by an advanced planning and optimisation tool, since, especially for UMTS, the trade-off between coverage and capacity cannot be described by a static coverage prediction only. Finally, the network data base's parameter fill is determined, which parameterises the system. The more realistic number of network elements and its cost of deployment are fed back to adjust the business case.

10.1.4 ROLL-OUT AND INITIAL OPTIMISATION

After the planning is finished, sometimes even in parallel, the system is implemented. On selected locations the necessary physical construction is done to support Node Bs, its configuration and radio network controllers. The particular system elements are then commissioned and integrated into the whole network and finally tested for basic functionality. A number of base stations in a selected area, typically between 10 and 15, are combined into clusters to assess the network performance regarding coverage, capacity and quality requirements, as well the user mobility. When the performance meets the requirements, a cluster is ready for launch.

10.1.5 OPERATION AND ONGOING OPTIMISATION

Following the system launching, there are mainly three sources of system performance verification: customer complaints, network and drive test statistics. This is used to assess prevailing network deficiencies and system trends. The first will be used to perform typically ad hoc optimisation actions, the second to identify any future bottlenecks or drawbacks.

10.1.6 EXTENSION

At a certain point, a given system does not offer any more head room for optimisation; the system resources are exhausted. Thus, additional traffic growth may cause unacceptable damage of QoS. The network needs to be extended. This may be performed by already given technology due to hardware or software extension or through service area extension or by introduction of a new technology.

10.2 INPUT PARAMETERS

In the following section, the parameters for network dimensioning and planning are described. After a short description, the average values for each parameter are presented to allow fast use for dimensioning purpose and proper parameter settings for simulation.

10.2.1 BASE STATION CLASSIFICATION

Taking into account the cell deployment, there are basically three types of base stations (BSs):

1. macro
2. micro
3. pico station.

The main application of macro base stations is to provide wide area coverage; a deployment assumes a hexagonal cell layout. The antennas of a macro BS are typically mounted above the roof-top level or on towers. The range of macrocells varies from a couple of hundred meters in dense urban areas to couple of kilometres in open areas; note that remote rural areas, desert or cells over water may range up to tens of kilometres. The macro base stations are dedicated to provide service to fast moving mobiles.

The micro base stations' main functionality is to provide capacity in densely populated areas or to improve both coverage and capacity, if not already achieved by the macro layer. The antennas of micro base stations are mounted usually below the roof-top level, on the wall of buildings. The range is significantly lower than for macro base stations (up to several hundreds meters), since the antenna height and output power are lower. The cell layout modelling follows in most of the cases the Manhattan-grid model. The micro base station may be used in a hierarchical cell structure, relieving the macro layer from stationary, low and medium moving users.

Pico base stations are realised usually as indoor cells and used to serve the traffic in hot spots, like offices, shopping malls and sport arenas. They are dedicated to provide coverage and capacity to slowly moving or stationary users, freeing the other layers from occasionally and permanently excessive loads. The other application scenario for pico cells is to provide indoor coverage, where the coverage quality cannot be controlled by macro or microcells, e.g. on higher storeys of tall buildings in urban areas or in deep ground metro stations and tunnels. The range of a pico base station is significantly limited by its low output power and the indoor environment, reaching in most cases only dozens of meters.

10.2.2 HARDWARE PARAMETERS

10.2.2.1 Receiver Sensitivity Level

The receiver sensitivity level is mainly limited by thermal noise. The thermal noise power spectral density N_0, normalised to 1 Hz bandwidth, is defined by the following equation:

$$N_0 = k \cdot T \quad (10.1)$$

where k is the Boltzman constant (1.38 E-23 K/J) and T is the temperature of the conductor (in Kelvin, typically 290 or 293 K, which refer to 17 or 20 °C respectively). The total thermal noise power N at the detector is limited by its filter bandwidth. For UMTS, the chip-matched filter bandwidth B equals to the frequency band occupied by the scrambling code, i.e. approximately 3.84 MHz; thus, the thermal noise power equals:

$$N = N_0 \cdot B = k \cdot T \cdot B \quad (10.2)$$

The real receiver noise floor is further limited by the quality of the internal components like Low Noise Amplifier (LNA), filter, synthesizer etc., which generate additive noise. This noise contribution is expressed as a noise figure, F, and describes how much of additional noise is added to the received signal within the receiver bandwidth. Thus, the total noise floor of a receiver is given by (10.3):

$$N \cdot F = N_0 \cdot B \cdot F = k \cdot T \cdot B \cdot F \tag{10.3}$$

This value can be used to calculate the receiver sensitivity level, which is defined as a minimum level of a RF signal at the receiver input point to provide at the output, after data demodulation, the useful signal at minimum acceptable quality. The quality of the receiver input signal is expressed by E_b/N_0 (bit energy divided by the noise spectral power density) value and the quality of the receiver output signal, in case of digital communications, by BER (Bit Error Rate) or BLER (Block Error Rate). The basic formula for E_b/N_0 is given by:

$$\frac{E_b}{N_0} = \frac{C}{N} \cdot \frac{R_c}{R_b} \tag{10.4}$$

where C is the minimum required signal power at the receiver input, R_c the chip rate and R_b the user bit rate. Therefore, the receiver sensitivity level can be defined as:

$$\text{receiver\_sensitivity\_level} = \frac{E_b}{N_0} \cdot \frac{R_b}{R_c} \cdot N \cdot F = \frac{E_b}{N_0} \cdot \frac{R_b}{R_c} \cdot k \cdot T \cdot B \cdot F \tag{10.5}$$

Assuming $R_c = B$, above formula can be rewritten as:

$$\text{receiver\_sensitivity\_level} = \frac{E_b}{N_0} \cdot R_b \cdot k \cdot T \cdot F \tag{10.6}$$

It should be noted that the receiver sensitivity level does not depend on the signal bandwidth, but on the quality requirements of the service (E_b/N_0), user data rate and quality of analogue components of the receiver (F). It can further be affected by interference coming from different sources, like own cell users, other cell users and any other interference sources, which generate interference within the receiver frequency bandwidth.

Usually, the receiver sensitivity level is expressed in dBm (decibels referred to 1 mW). Therefore, the above formula can be adjusted as:

$$\text{receiver\_sensitivity\_level (dBm)} = 10 \cdot \log_{10}\left(\frac{\frac{E_b}{N_0} \cdot R_b \cdot k \cdot T \cdot F}{0.001[\text{W}]}\right) \tag{10.7}$$

$$= \text{receiver\_sensitivity\_level (dBW)} + 30\,\text{dB}$$

10.2.2.2 Base Station Parameters

The typical configuration of a base station (or Node B) consists of (Figure 10.2):

- Base station hardware (built in a cabinet), namely BTS (base transceiver station);
- *Cable system* to connect the BS to the antenna(s);
- Antennas.

As for the BS, its performance is mainly characterised by the *receiver noise figure* and the *maximum output power of the power amplifier*.

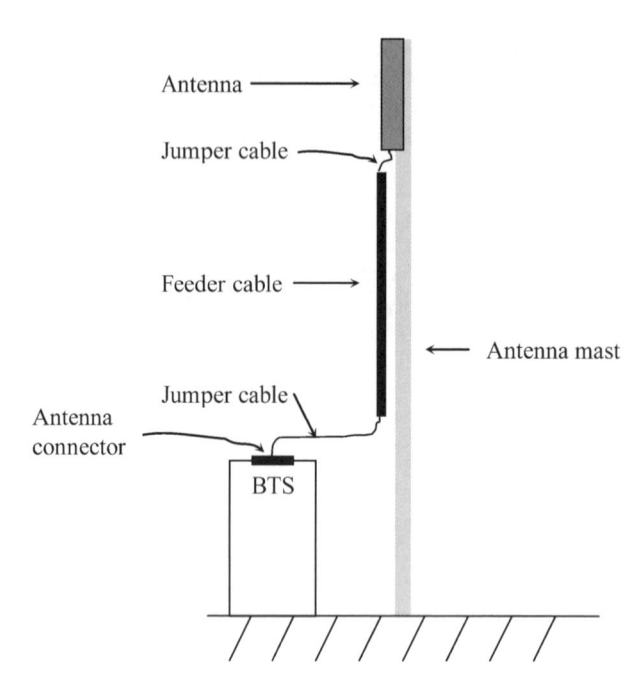

Figure 10.2 Example base station configuration.

Receiver noise figure
The typical range of base station receiver noise figure varies from 2.5 to 4 dB for a macro or wide range base station. For micro and pico base stations (medium range or local area BS), the noise figure could be higher (like for a UE), since the capacity would be the limiting factor rather than the coverage range of the cell.

The 3GPP TS 25.104 document [1] does not directly specify the requirements for base station receiver noise figures. It defines the minimum requirements for the receiver sensitivity level for the 12.2 kbps channel, measured at the base station antenna connector, if no other passive or active elements are applied to the antenna (Table 10.1). The minimum requirements for the sensitivity level should apply to receivers without receive diversity. In the case that diversity is implemented, the minimum requirements apply to each of the antenna branches separately, under the assumption that the other branches are disabled during the measurements.

Additionally [1], defines numerous performance requirements for a BS receiver. The core part of the requirements is focusing on the performance of DTCH (12.2, 64, 144 and 384 kbps) under different propagation conditions. The quality of a demodulated DTCH is indicated with a maximum value of BLER measured at a specific E_b/N_0 at the receiver input. The requirements are specified for BS with and without Rx diversity.

Table 10.1 Base station receiver sensitivity levels for the 12.2 kbps service.

Base station	Reference channel quality (BER)	Sensitivity level (dBm)
Macro	<1 in 10^3	−121
Micro	<1 in 10^3	−111
Pico	<1 in 10^3	−107

The maximum output power of amplifier
This is the mean power of one carrier measured at the antenna connector, given that no other passive or active elements are applied to the antenna. The minimum requirements apply to the rated output power, which is the mean power level per carrier declared by manufacturer and measured at the antenna connector.

In Table 10.2, the 3GPP specific values are depicted. For micro and pico base stations, there are clearly defined upper limits. For the macro BS, there is no definition of an upper limit of output power; typically, it varies between 10 and 45 W, with a step of 5 W, depending on manufacturer and product line. The tolerance of the output power could vary in a range between ±2.5 dB from the nominal level.

The maximum output power of any UMTS cell is limited in [2] to the value of 50 dBm by the range of Maximum Transmission Power parameter. It describes the maximal summarised power of all physical channels in a cell for downlink.

The total output power of the power amplifier (PA) per carrier is shared by all physical channels supported by the BS. There are channels with fixed power allocation, i.e. the Common Control Channels (CCCHs), and with flexible power allocation obeying power control algorithms. The power of CPICH* is defined as the percentage of the total power (typically 8–10%), and the other CCCH's output power is related to the CPICH [2]. In Table 10.3, some example power allocations are presented.

The CPICH power indicates the cell range; consequently, if a significant portion of the total power is assigned to the pilot signal, the coverage of the cell is considerably enlarged, but the remaining power allocated to the traffic channels decreases and hence the capacity is low. In addition, if more power is assigned to the other common control channels, the signalling reliability rises, but again less power would remain for traffic channels.

Table 10.2 BS rated output power levels [1].

Base station	Mean output power
Macro	—
Micro	$\leq$38 dBm
Pico	$\leq$24 dBm

Table 10.3 Example power allocation for CCCH.

CCCH	Vendor 1	Vendor 2	Vendor 3
CPICH	10% of total power	8% of total power	10% of total power
Primary SCH	−3 dB	−5 dB	−3 dB
Secondary SCH	−3 dB	−5 dB	−3 dB
Primary CCPCH (BCH)	−5 dB	−2 dB	−3 dB
Secondary CCPCH (FACH, PCH)	SCCPCH1 (only FACH) 0 dB SCCPCH2 (only PCH) −5 dB	−3 dB	−3 dB
AICH	−8 dB	−7 dB	−6 dB
PICH	−8 dB	−3 dB	−6 dB

---

* If no extra mentioned the CPICH term represents the P-CPICH

Cable system

A cable system of base station consists of:

- feeder cable, i.e. the central element of the signal transmission between BTS and antenna;
- jumpers, i.e. the flexible components connecting BTS with much less flexible feeder cables and the feeder cable with the antenna;
- connectors, i.e. the physical interface between all elements (BTS, jumpers, feeder cable and antenna).

The key parameter of the cable system is the RF signal attenuation or cable loss. It depends on the signal frequency, the length and diameter of jumpers and feeder cables. Since a jumper should be flexible, its diameter is relatively small when compared with the feeder cable. On the other hand, the small diameter causes higher signal attenuations; hence its length should be limited. Typically, the overall cable system attenuation should be less than 3 dB for a macro base station. For instance, assuming 6.1 dB attenuation per 100 m of feeder cable with 7/8″ diameter, 21 dB attenuation per 100 m of jumper cable and 2 jumpers each 1.5 m length, 0.02 dB attenuation per connector and 4 connectors in the cable system, the length of the feeder cable should not exceed 47 m [3].

Antenna

The antenna is used in the downlink (DL) direction to transform the EM waves propagating in a transmission line (e.g. a coaxial cable) into the EM waves propagating in free space. On the other hand, in the uplink (UL), the transformation occurs in the reverse direction, i.e. free space waves are transformed into waves propagating in the cable.

Antenna key parameters are:

- frequency range of operation;
- power gain (may be expressed in relation to a reference antenna, most commonly the isotropic radiator [dBi] or the half-wave dipole [dBd]);
- vertical and horizontal radiation patterns and their characteristics (half power beam width, side lobes level, front-to-back ratio, etc.);
- polarisation (primarily cross-polar, vertical and horizontal) – the direction of the electrical field vector oscillation;
- ability and range of electrical or mechanical tilt and RCT (remote controlled tilt);
- network (circuit) parameters (input impedance, frequency bandwidth and return loss);
- physical dimensions (height/width/depth and weight).

In Table 10.4 selected antennas and its main parameters are presented.

10.2.2.3 User Equipment Parameters

The performance of the User Equipment (UE) is mainly characterised by the receiver noise figure and the output power. The UE is usually a less complicated unit than the base station; therefore, the quality of its components may be lower. Typical values for the noise figure of a UE receiver can vary between 7 and 8 dB. The authors of [5] recommend that the noise figure of a UE cannot be higher than 9 dB, assuming 12.2 kbps DTCH as a reference channel, which requires $E_b/N_0 = 5.2$ dB for a BER less than 10^{-3} and a 1.8 dB implementation margin of a real receiver. The 3GPP TS 25.101 [6] specifies several tests for type approval purposes, similar to the base station performance tests.

3GPP defines four power classes with the following nominal power and tolerance values [6]:

1. Power Class 1: 33 dBm (2 W), +1/−3 dB;
2. Power Class 2: 27 dBm (0.5 W), +1/−3 dB;
3. Power Class 3: 24 dBm (0.25 W), +1/−3 dB;
4. Power Class 4: 21 dBm (0.125 W), +2/−2 dB.

Table 10.4 Selected antennas for 1920–2170 MHz and its main parameters [4].

#	Gain (dBi)	Horizontal beam width	Vertical beam width	Electrical tilt range	Mechanical tilt range	Polarisation	RET availability	Example application
1	19.8	33°	8.5°	0°–12°	0°–19°	+45°, −45°	yes	High capacity or long range sector macrocells
2	19.6	44°	6.7°	0°–8°	0°–15°	+45°, −45°	yes	High/middle capacity or long range sector macrocells
3	18	65°	6.2°	0°–10°	0°–10°	+45°, −45°	yes	Middle capacity, regular coverage urban/suburban sector cells
4	16.7	88°	6.5°	0°–8°	0°–10°	+45°, −45°	yes	Regular coverage suburban/rural sector cells
5	10	360° (omnidirectional)	9°	—	—	Vertical	—	Coverage of short range rural or isolated omnidirectional cells
6	2	360° (omnidirectional)	—	—	—	Vertical	—	Indoor/pico cells

Note that most handsets use Power Class 4.

The gain of a mobile antenna depends mostly on its size and the application of the UE; its directional gain can vary between −10 and 2 dB. In a typical handset application, such as voice, it is assumed that the antenna gain is equal to 0 dBi. UE used as data cards (PCMCIA) can contain an antenna with 2 dBi gain. The UEs may be connected to external antennas, e.g. on a car or house roof-top, to extend the coverage range in remote rural areas, leading to antenna gains of up to several dBi. In this case, a cable (with its loss) is required to connect both antenna and the UE. Therefore, the antenna gain and cable loss need to be accounted for when calculating the total gain.

Due to usage of the UE near the user's body, a body loss parameter is introduced. It defines the additional loss in the transmitting and receiving path. For instance, if a UE is used for a speech connection and there are no head speakers or external microphone, then one can assume that the user holds the UE in the hand and close to the head; the resulting loss typically mounts to an additional 2–3 dB on average. If a UE is used for data connection (e.g. as modem between a laptop and RAN for web browsing), then it is placed away from the user, hence no additional loss exists.

10.2.2.4 Masthead Amplifier (MHA)

The MHA is a low noise amplifier, which amplifies the signal in the uplink direction, minimising the negative effect of cable losses and improving the overall noise figure of the BS (BTS with cable system). It should be placed right after the antenna to increase the SNR (signal to noise ratio) before the cable loss reduces the signal coming into the receiver input, and hence the SNR itself.

The cable loss does not decrease the thermal noise level, which is on the same level at both ends of the cable.

The gain of using MHA can be calculated using Friis formula, which describes the calculation of the total noise figure of a chain of elements. The total nose figure $F_{\text{without\_MHA}}$ of a BTS with cable system and without MHA can be formulated as:

$$F_{\text{without\_MHA}} = F_{\text{cable}} + F_{\text{cable}} \cdot (F_{\text{BTS}} - 1) = F_{\text{cable}} \cdot F_{\text{BTS}} \qquad (10.8)$$

where F_{cable} is the cable system noise figure (equal to the cable loss) and F_{BTS} is the BTS noise figure. Similarly, the total noise figure ($F_{\text{with\_MHA}}$) of the BTS with MHA and cable system can be derived as:

$$F_{\text{with\_MHA}} = F_{\text{MHA}} + \frac{(F_{\text{cable}} - 1)}{G_{\text{MHA}}} + \frac{F_{\text{cable}} \cdot (F_{\text{BTS}} - 1)}{G_{\text{MHA}}} = F_{\text{MHA}} + \frac{F_{\text{cable}} \cdot F_{\text{BTS}} - 1}{G_{\text{MHA}}} \qquad (10.9)$$

where F_{MHA} and G_{MHA} are the noise figure and gain of the MHA, respectively.

In Figure 10.3, the MHA improvement on the BS noise figure is depicted. In the depicted example, two MHAs have been used, both with noise figure of 1.5 dB but with different gains, i.e. 12 (MHA1) and 24 dB (MHA2) respectively. The total noise figure of the BS system with cable rises linearly with respect to the cable loss (when expressed in dB). As expected, the greater improvement can be achieved with a higher gain MHA. The improvement of using MHA with a lower gain reaches its saturation much quicker than the MHA with a higher gain. The noise figure of a BS with MHA2 is almost constant and independent of cable loss. Hence, the improvement curve of MHA2 is almost parallel to the noise figure of the BS without MHA. One should notice that, if the MHA gain is high

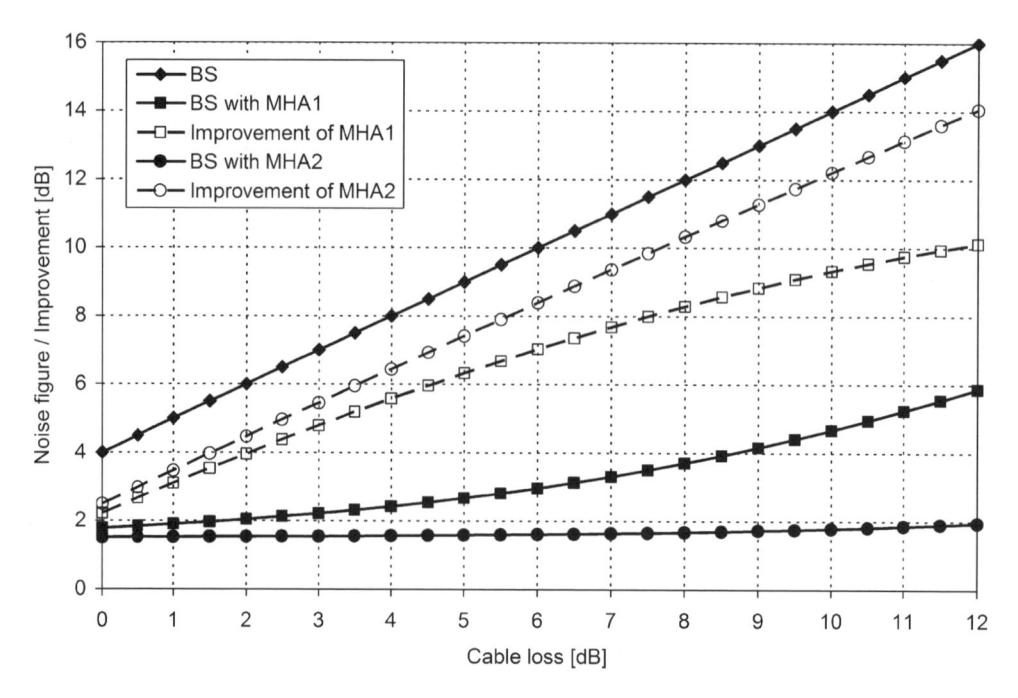

Figure 10.3 MHA noise figure improvement in the BS configuration ($F_{\text{BTS}} = 4$ dB, $F_{\text{MHA1}} = 1.5$ dB, $G_{\text{MHA1}} = 12$ dB, $F_{\text{MHA2}} = 1.5$ dB, $G_{\text{MHA2}} = 24$ dB).

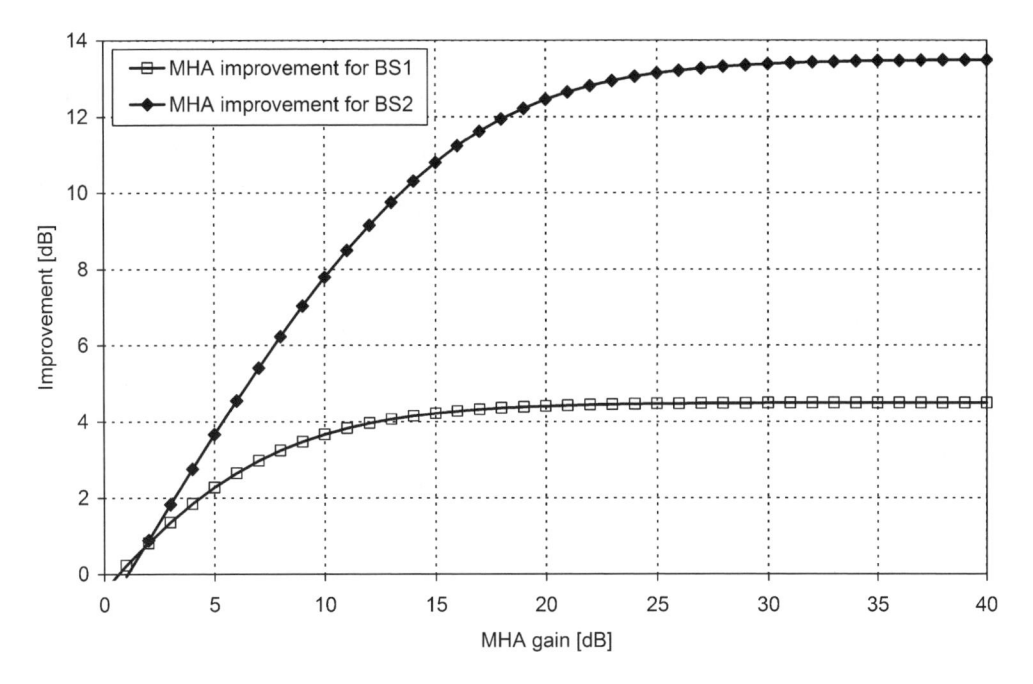

[Figure shows two curves of MHA improvement vs MHA gain]

Figure 10.4 MHA noise figure improvement in 2 BS configurations (BS1: $F_{BTS} = 3\,dB$, cable loss $= 3\,dB$, BS2: $F_{BTS} = 5\,dB$, cable loss $= 10\,dB$).

enough, the cable loss does not have a significant impact on the overall noise figure; therefore, to achieve the best SNR in the uplink, it would be practical to use a MHA for each BS.

Figure 10.4 presents the noise figure improvement as a function of the MHA gain for two BS configurations: first, where the total noise figure of the BS and cable is very low ($F_{BTS} = 3\,dB$, cable loss $= 3\,dB$), and second, where the total noise figure of BS and cable is very high ($F_{BTS} = 5\,dB$, cable loss $= 10\,dB$). For the same MHA gain, the overall noise figure improvement is higher for the configuration with a higher noise figure. On the other hand, each of the BS configurations reaches a saturation point where additional MHA gain does not lead to significant increase of the total noise figure. Consequently, in a real cell deployment, each of the BS configurations may require different MHA. Practically, it may be useful to use 2 or 3 types of MHA in a network, depending on the noise figure of a BS and cable systems.

It should be noted that the MHA brings an additional loss – the insertion loss in the downlink (roughly 0.3 dB). Consequently, it needs to be connected to an antenna with a supplementary jumper cable, which again results in extra loss in both directions, uplink and downlink. Therefore, both losses should be taken into account when calculating the total cable system loss.

10.2.3 ENVIRONMENTAL SPECIFICS

The propagation environment of a mobile can be roughly described by the following parameters: clutter class, penetration loss, channel profile, speed of mobile terminal and orthogonality factor. The clutter class defines the morphological type of a physical area. The other parameters quantify the nature of propagation in this specific area. As a consequence, the attributes of penetration loss, channel profile with given speed and orthogonality factor are used to illustrate a clutter's class.

10.2.3.1 Clutter Classes

The clutters represent the morphological structure of the communication environment. Multiple clutters may be defined, reaching very detailed description of the environment. Due to practical reasons, however, the number of clutters should be limited for dimensioning purpose to obtain fast results. In a real-world planning process, the detailed information of an environment is necessary, but it requires significantly more effort to adjust all related parameters (see Chapter 1). Typical clutter classes for dimensioning are described below:

- *Rural or Open*   It is characterised by remote areas without significant natural obstructions or buildings, with low or no vegetation, very sparsely populated.
- *Motorway or Quasi Open*   It contains road areas outside of urban, small and middle vegetations.
- *Suburban*   Suburban areas typically highlight sparse construction density and some vegetation. The class consists of residential, commercial and industrial zones in suburban environment. The mean building height is limited between 2 and 5 storeys, between 6 and 15 m.
- *Urban*   The area is limited by urban periphery. It encloses considerable building or building groups found in major cities or towns, like offices, big shopping moles, substantial residential houses etc. The density of construction is considerably higher than in suburban areas, but borders in-between are clear. The major streets are distinguishable, but the density of mean streets appears without outline. A small amount of vegetation, like trees or small gardens could be present. Average height is restricted between 10 and 14 storeys, thus the mean building height should be no bigger than 40 m.
- *Dense Urban*   It is a very dense populated and constructed region within urban perimeter. The streets are narrow, thus the differentiation of several buildings is difficult using, e.g. satellite photography. It can contain the down-town areas as well, thus the mean building height is constrained between 4 and 8 storeys, reaching 25 m.

For evaluation purposes of different 3G candidate technologies, more specific test environments, defined by [7], were used:

- Indoor office
- Outdoor to indoor and pedestrian
- Vehicular
- Mixed cell pedestrian/vehicular.

The test environments are mostly characterised by a path loss model and channel profile (impulse response model) of the RF signal, the speed of the mobile terminal and the corresponding link budget parameters. The channel profile, together with the speed, is used to estimate the lowest possible quality of signal reception E_b/N_0, at a given quality BER or BLER. The channel impulse response model in a specific environment is characterised by several signal echoes (taps), its delays are related to the first tap, its average powers are coupled with the strongest tap and the Doppler spectrum of each tap. In Table 10.5, a mapping between defined environments and the real environments is depicted.

10.2.3.2 Penetration Loss

In most of the cases, the mobile station is positioned indoors. The signal needs to penetrate through walls or windows to reach the mobile and hence arrives weaker if the terminal was positioned outdoors. The building attenuation needs to be considered to make sure that indoor users are served properly. In a typical planning application, the building penetration loss is added to the path loss of an outdoor user, with a mobile station height of 1.5–2 m.

The penetration loss varies to a large extent between different buildings and is mainly influenced by construction material, wall width, position of the mobile relative to signal entrance into the building,

Table 10.5 Mapping of ETSI environments to real network scenarios.

ETSI environment	Channel profile occurrence in a time		Real network scenario
	Channel A	Channel B	
Indoor office	50 %	45 %	Pico indoor cell
Outdoor to Indoor and Pedestrian, 3 km/h	40 %	55 %	Microcell/macrocell in dense urban or urban
Outdoor to Indoor, 50 km/h	40 %	55 %	Microcell/macrocell in urban or suburban
Vehicular, 120 km/h	40 %	55 %	Macrocell in rural

Table 10.6 Typical penetration loss related to clutter classes.

	In-car	Suburban	Urban	Dense urban
Penetration loss (dB)	5–8	10–15	16–20	20–30
Standard deviation (dB)	2	4	6	8

angle of signal incidence, number of glasses in the window, presence of infrared reflection material on the window glass, existence of LOS visibility and position of the mobile relative to the ground level of the building. Some example penetration losses are provided in the Table 10.6.

10.2.4 TECHNOLOGY ESSENTIALS

10.2.4.1 RAKE Receiver

On the way between transmitter and receiver, the RF signal experiences reflections, diffractions and free-space attenuation. The signal arrives at the destination by means of multiple and different paths. As a consequence, the mutually delayed signal echoes are time or phase shifted and have different energies. The instantaneous sum of these echoes leads to mutual additions and cancellations, called small-scale fading, and consequently to a degradation of the connection quality. Therefore, it is necessary to distinguish and ideally extract all possible multipath components and combine the energy of each path to maximise the effectiveness of demodulation. This is achieved by means of a RAKE receiver. Its detailed description can be found in Section 6.1.

In the case that four fingers are assigned to one antenna, about 98 % of the signal energy can be received on average for ITU vehicular channel A [8]. The gain of such a signal combining typically ranges from 0.7 to 4 dB and is included in the E_b/N_0 calculation. The highest gains will be obtained by relatively slow moving mobiles and without antenna diversity, the lowest for high speed mobiles and with antenna diversity. The influence of speed on the results is that faster moving mobiles have a higher chance to move out from an instantaneous fading deep and hence the channel coding and interleaving is able to recover the signal information at relative lower transmit power. On the other hand, antenna diversity dominates over multipath diversity, since for each antenna receiving path there is one receiver assigned and the signal energy collected with two receivers (antenna diversity) is more significant than in one (multipath diversity).

10.2.4.2 Orthogonality Factor

The connections in downlink within one cell are separated between users with synchronised orthogonal variable spreading factor (OVSF) codes, which are perfectly orthogonal when transmitted by the base station, assuming they are under the same scrambling code. Due to multipath propagation, however, the orthogonality is partly lost (see Chapter 2). In the uplink, all mobiles transmit in an asynchronous manner; it is thus assumed the orthogonality is almost completely lost.

The loss of orthogonality due to multipath propagation is expressed by the orthogonality factor a. It is used to scale the intra-cell interference in the SIR estimation. It therefore has a significant influence on the coverage and capacity calculation. Perfect orthogonality, where unwanted signals are separated entirely at the receiver from useful information, is described with value of one. The value of zero represents fully lost orthogonality. Similarly, the orthogonality loss factor can be defined as $(1-a)$. It should be noted that in some previous publications, e.g. [9], the opposite notification is used, i.e. zero corresponds to a completely orthogonal state and one for a complete loss in orthogonality. In other reference literature, it was also called the non-orthogonality factor.

Due to limitations or simplifications in the system parameters modelling, a typical application for UTRAN dimensioning and planning assumes an average and constant orthogonality factor, depending on:

- Clutter classes or
- Base station classes or
- Channel profile or
- Channel profile and RAKE receiver implementation.

The clutter classes (land usage) are commonly used in the planning and dimensioning tools to characterise the propagation environment. The conditions for open area (quasi line of sight) lead to existence of one dominant signal and relatively low power interferers between the base station and mobile. As a consequence, the orthogonality factor is the highest among all other clutter classes. On the other hand, the sealed area (open in urban) is surrounded by buildings, thus there is no line of sight connection. The signal arrives at the mobile through reflections or diffraction at near obstacles, leading to many similar power echoes. As a result the orthogonality factor is the lowest. In the urban area, there is typically no line of sight connection, but due to less possible propagation routes between base station and mobile the orthogonality factor is higher than in a sealed area. Fewer obstacles are present in the suburban land use than in an urban area. This leads to a higher chance of one dominant signal in the downlink and less signal reflections. Subsequently, the orthogonality factor is higher than in urban areas, but lower than in open area.

The dependency of the orthogonality factor on the base station class reflects the situation in a town location. For the macro base station, the orthogonality factor is used as an average among all clutters classes as depicted in Table 10.7. In the microcell, the antenna is located below roof top level, so the

Table 10.7 Mean orthogonality factor related to clutter classes for macrocells (based on results from [10]).

		Clutter class		
All	LOS (open)	Suburban	Urban	Sealed area
0.6	0.825	0.65	0.525	0.4

cell range is significantly lower than in the macro base station. There is hence a higher chance that the useful signal will not experience much multipath propagation, since fewer diffraction and reflections are possible. As a result, the orthogonality factor is higher in a microcell than in a macrocell. In the pico cell, the cell range is considerably smaller than in the microcell; hence, there is a lower probability of many significant paths with high delays. The mean orthogonality factor for pico cells therefore should be highest among all base station classes (Table 10.8).

The TU (typical urban) channel is characterised by many similar power echoes and a lack of one dominant signal. It is caused by many obstacles between transmitter and receiver, thus almost no line-of-sight connections exist. For that reason, the orthogonality is significantly degraded. In the hilly terrain (HT) environment, the channel profile is described by strong and significantly delayed, but few echoes, due to the possibility of the signal being reflected by distant hills or mountains. As a consequence, the orthogonality is corrupted, but not to the same extent as for TU. The best orthogonality factor among all channel profiles is provided by rural area (RA). It is possible due to a high probability of line-of-sight connections, as well as a low likelihood of signal reflections or diffraction. Typical values based on [9] are summarised in Table 10.9.

The number of RAKE fingers can have a considerable impact on the orthogonality factor. An increased number of RAKE fingers leads to a better resolution of different echoes of the useful signal. The SIR is hence better than at a receiver with less fingers; however, this clearly requires an enhanced and extended receiver structure (Table 10.10).

Table 10.8 Mean orthogonality factor related to base station classes (based on results from [10] and [11]).

Base station class		
Macro	Micro	Pico
0.5–0.6	0.8	0.9

Table 10.9 Mean orthogonality factor related to channel profile for macrocell (based on results from [9]).

Channel profile		
TU	HT	RA
0.347	0.577	0.669

Table 10.10 Mean orthogonality factor related to channel profile and number of Rake receiver fingers for macrocell (based on results from [9], the number x of Rake fingers corresponds to the full paths being resolved by fingers spaced by half length of the chip duration).

Number of RAKE fingers	Channel profile		
	TU	RA	HT
1	0.284	0.652	0.568
2	0.368	0.67	0.574
3	0.377	0.655	0.563
x	0.332	0.654	0.543
	($x=10$)	($x=4$)	($x=5$)

The above consideration of orthogonality is suitable for dimensioning purposes. For the real UTRAN planning, however, different approaches can be more adequate since the orthogonality factor is not constant in an entire cell or network. It depends on the particular location, identified by the channel profile, as well as on the receiver implementation. Theoretically and practically, the orthogonality factor is hence clearly separate for each user in the network. It seems to be reasonable to compute the channel's Power Delay Profile (PDP) with use of e.g. ray tracing propagation methods and then compute the orthogonality factor for each terminal separately [12,13]. The degree of orthogonality decreases on average with an increasing distance between base station and mobile and between the centre of the cell and its peripheral areas. The orthogonality is also affected by the dominance of own cell interference to the other cell interference. At the base station, a is approaching the highest values, since the delay spread of the signal increases with an increasing distance between base stations and mobiles and the own cell interference overshadows the other cell interference.

10.2.4.3 Soft Handover Gain

Soft handover occurs if a mobile is connected to at least two cells at a time. In the case that the cells belong to different Node Bs, the uplink signal combining is performed at RNC level. If the cells belong to the same Node B, then the handover is called *softer* and the uplink signal combining is performed at the same Node B. In the downlink, the combing is done by the mobile's RAKE receiver independently of the source of incoming signals. A soft handover gain is achieved due to macro diversity and hence reduces the negative effects of shadow and small scale (fast) fading.

In a real network, most of the cells' coverage areas overlap. A macro diversity gain combating shadow fading occurs, because the mobile at the cell edge may choose a better serving cell from a set of cells and hence is not limited to only one connection. This leads to a reduction of the shadowing fading margin, and the cell edge dimensioning limits may be more moderate. Also, for small scale fading, the instantaneous required E_b/N_0 or transmit power can be lower in a connection with at least two cells. In both cases, the gain increases when the decorrelation between the signals increases but their amplitudes are in the same or similar order.

The gain in the case of small-scale fading has been simulated and measured in [8], [14] and [15]. The simulation results give almost always positive gains. For uplink and downlink, the gain is around 2 dB, when the path loss from mobile to two different cells is equal. If the relative path loss is different from zero, then the gain decreases with the magnitude of the path loss difference. From the simulation results one can observe that the gain reaches 0 dB if the relative path loss reaches a little bit more than 4.5 dB. This indicates the maximum window size for the soft handover parameter.

From measurements with real UTRAN equipment, different observations have been made. Surprisingly, the net gain in the downlink is negative for both the softer and soft handover. The average gain is −1.9 and −1.2 dB respectively. The soft handover impairments come from the power control algorithm applied. After estimation of the link quality, the mobile sends the power control command to all cells in the active set. Due to errors in the radio transmission, the power control command may be interpreted differently at each base station. For instance, one base station may interpret a command as power up, the other as power down. Additionally, the power command is not combined by the RNC for common power management due to excessive signalling delays. The possible misinterpretation of power commands, together with a lack of its central management, may lead to power drifting and hence to signal degradation. In the uplink, where the power commands are combined, the average gain of soft handover is around 0.5 dB. In the other cases, when the power commands are not combined, similar losses as in the downlink of around 1.9 dB have been measured.

Theoretically, the macro diversity gain from shadow fading can be calculated using Jake's formula [16] and the theory of multiserver coverage [17]. The details will be presented in Section 10.3.2.1.

10.2.4.4 Service Quality Requirements

For the purpose of the link range and capacity calculations, the quality of different services is often described with only one parameter: target or minimum allowable E_b/N_0 value. It expresses the quality of a traffic channel before data demodulation, determined by maximum or target BER or BLER. Unfortunately, according to the standards, the BER and BLER are not the only parameters to guarantee the quality for a specific service. Rather the QoS attributes are associated to a service class (Section 3.2). On the other hand, there are many aspects influencing the instantaneous target of E_b/N_0 for a specific service, like propagation conditions in multipath environment, soft handover, RAKE receiver usage, antenna Rx and Tx diversity, velocity of the receiver, type of power control algorithm, voice activity factor, coding etc. It is thus ambiguous to provide one fixed value of E_b/N_0 per service for computing coverage or capacity figures; rather a set of E_b/N_0 values should be available to cover most reasonable dimensioning or planning exercises. If target E_b/N_0 are provided, however, then it is paramount to include the conditions under which they are required, i.e. PDP, terminal speed etc. Table 10.11 shows example E_b/N_0 values for 64 kbps DTCH. The data have been calculated using link level simulations and assuming no Rx diversity.

The target BER or BLER depends on user service, thus following minimum requirements may apply:

- Speech conversation: BER not greater than 10^{-3};
- Video conversation: BER not greater than 10^{-4};
- Real Time (RT) data transfer: BER not greater than 10^{-6};
- Non Real Time (NRT) data transfer: BLER not greater than 10^{-1}.

Obviously, the proposed minimum requirements can be made stricter, thereby improving the service quality; however, as a consequence, the E_b/N_0 requirements will rise and the dimensioning and planning will be more constrained. Prior to making any requirements more stringent, however, it is important to determine if an improvement of BER or BLER will really be noticed by customers in the expected manner.

10.2.4.5 Power Control Effects

Transmit power rise
The fast power control algorithm applied in the uplink tries to overcome fast fading notches. This leads to an increase in the average required transmit power to maintain a target E_b/N_0 as compared with no fading. As a result, the interference in other cell rises [18]. The negative influence of the interference

Table 10.11 Example of E_b/N_0 values for 64 kbps DTCH.

Propagation model		Required E_b/N_0 (dB)			
Speed (km/h)		ITU pedestrian A	ITU vehicular A		
		3	3	50	120
BER	1%	6.2	5.6	6.3	6.7
	0.1%	7.1	6.3	7.0	7.4
	0.01%	7.4	6.5	7.3	7.8
	0.001%	8.5	7.3	8.1	8.5
BLER	10%	5.3	5.1	5.4	5.9
	1%	6.3	5.8	6.5	6.8
	0.1%	6.8	6.1	6.8	7.2

rise is particularly notable at the cell edges, where mobiles transmit with maximum power. Also, there are less power dynamics in the downlink than in the uplink (see Chapter 9); thus, the effect of power rise is smaller in the downlink and typically included into E_b/N_0 calculations. The average power rise, and hence the received interference, depends on the number of multipath components and the application of receive diversity.

Power control headroom or fast fading margin
The fast power control algorithm has been introduced in UMTS to maintain a target E_b/N_0 during small scale fades of the communication channel, which may be in the order of up to 30 dB. This is especially important for slowly moving mobiles, since they cannot change their positions rapidly enough to compensate for any deep fades. At the cell edge, the mobiles can transmit almost at the maximum available power; thus, there is no more headroom to follow and reduce the negative influence of the small-scale or fast fading. In order to include this phenomenon in the dimensioning process, a power control headroom (or fast fading margin) needs to be applied. The fast fading margin depends on the mobile speed and its typical value:

- varies between 4 and 5 dB for slowly moving mobiles (3 km/h);
- ranges between 1 and 2 dB for mobiles moving with a velocity of 50 km/h;
- is marginal and may be assumed at 0.1 dB for high speed mobiles (120 km/h).

10.2.4.6 Total Interference, Interference Margin and Noise Rise

The limiting factor of the Node B link performance and capacity is the received total interference I_{total}. It consists of own cell interference I_{own}, other cell interference I_{other}, as well as other system interference I_{other_system} and thermal noise N:

$$I_{total} = I_{own} + I_{other} + I_{other\_system} + N \tag{10.10}$$

The own cell interference is caused by own cell users and can be minimised by advanced features, such as multi-user detection or intelligent beamforming antennas. The other cell interference is caused by other cell users and their serving base stations. The key role of a proper radio network planning and optimisation is to reduce any negative influence of interference and hence to achieve a maximum cell decoupling or isolation. The planning and optimisation ought to leaving some headroom for soft handovers; typically, the soft handover areas should be limited to 25–35 % of the total coverage area. The other system interference comes from different systems. It arrives from adjacent channels, as the adjacency power leakage into the operating frequency band (see Section 6.4.3.5 and 12.2), or neighbour UMTS networks or systems transmitting at significant power out of its operating frequency, the spurious emission in the UTRAN band. The latter issue may apply to older GSM technology, which has been manufactured according to specifications not taking into account the existence of 3G networks. In the real UTRAN dimensioning and planning approach, mainly the own and other cell interferences are taken into account, since other sources of interference do not occur generally. They may have a temporal effect; their minimisation is hence rather a local than a global optimisation issue.

The ratio of other-to-own cell interference is called the *i*-factor:

$$i = \frac{I_{other}}{I_{own}} \tag{10.11}$$

The value of the *i*-factor depends on the propagation environment, overlapping cell areas, the number of sectors in the Node B, the traffic intensity and its distribution, as well as on the distance to the serving and interfering cells.

Table 10.12 Typical values for *i*-factor.

Pico indoor	Micro omni	Macro omni	Macro two sectors	Macro three sectors	Macro six sectors
0.1–0.2	0.25–0.55	0.45	0.55	0.65	0.85

In the uplink, the value is the same for all mobiles, since it is calculated at the base station receiver. For well isolated cells, e.g. indoor pico cells, the value of *i* is small, ranging around 0.1. For cells with bad cell decoupling, e.g. macrocells, it may be 10 times or more the value of indoor cells. This implies a very high influence of other cells and hence loss in capacity and coverage of the own cell. Such situation should be avoided by means of proper radio planning or optimisation. The typical values of the *i*-factor in the uplink are shown in the Table 10.12, taking into account a uniform traffic distribution. In the locations closer to the serving cell, the *i*-factor decreases and the cell capacity and coverage rise. In the locations away from the serving cell, the *i*-factor rises and the cell capacity and coverage accordingly decrease. For some other details about the *i*-factor, please consult Chapter 9.

In the downlink, the *i*-factor depends on the location of the mobile, thus its value is different for each user. On the other hand, for dimensioning purposes, it is usually assumed that the traffic is uniformly distributed over the entire network; thus, each cell serves the same amount of traffic.

Taking into account Equation (10.11), the total inference can expressed as:

$$I_{total} = I_{own}(1+i) + I_{other\_system} + N \qquad (10.12)$$

The interference margin, IM, in a cell is defined as the ratio of the total interference over the thermal noise, i.e.:

$$IM = \frac{I_{total}}{N} \qquad (10.13)$$

This parameter limits the receiver sensitivity, thereby being significant for the coverage estimation. For the purpose of power budget calculation it is expressed as a noise rise *NR*:

$$NR = 10 \cdot \log(IM) \qquad (10.14)$$

and represents the portion of received total power in dB over the thermal noise level. Typically the interference margin should be between 2 and 4 dB for coverage limited cells and between 4 and 7 dB for capacity limited cells.

The noise rise is also often described by means of the cell load, η:

$$NR = 10 \cdot \log(\frac{1}{1-\eta}) \qquad (10.15)$$

where the cell load, η is defined as:

$$\eta = \frac{I_{own} + I_{other}}{I_{own} + I_{other} + N} = \frac{I_{own} + I_{other}}{I_{total}} \qquad (10.16)$$

Comparing Equation (10.14) to (10.15), the dependency between interference margin and cell load can easily be shown to be:

$$IM = \frac{1}{1-\eta} \qquad (10.17)$$

10.3 NETWORK DIMENSIONING

10.3.1 COVERAGE VERSUS CAPACITY

An inherent phenomenon of CDMA technology is that, by increasing the number of active users, the interference and consequently the total noise level also increases. This implies a decrease of the sensitivity level of a receiver, as well as the need to consume more power to serve existing users. If the transmission power limit is reached, then, whilst trying to maintain the link quality of existing users, this leads to service degradation for distant users where the power requirements are the largest. As a result, the coverage shrinks and the capacity is degraded, i.e. even the existing users cannot be maintained and less requesting users may be served. On the other hand, if the service requirements decrease or some of the users leave the cell, then there is clearly less interference and more power available. Consequently, the cell coverage rises and the capacity potential increases. It is often referred to as cell breathing.

The increase of interference can be caused by customers, when they move away from a serving Node B. The increased distance will lead to increased path loss, and therefore the transmit power needs to be adjusted. Even users moving not far away from the base station, but moving into more difficult radio environments with a shrinking orthogonality factor, can cause more transmitted power and hence more interference. The increase of transmitted power cannot be infinite; especially in the downlink it saturates much faster than in the uplink, since the total power of the Node B is shared among all users in a cell. The power consumption of a base station is dependent on the user distribution as a function of the distance to base station, the orthogonality factor (which in turns depends on the distance to the BS) and the required service quality. On the other hand, the number of covered users depends on the cell range.

In Figure 10.5, the example of the coverage and capacity trade-off for a macrocell is depicted. Going towards higher loads, the allowable path loss decreases for uplink and downlink, albeit in a different way. The deterioration of the downlink is faster. For low values of the load, the downlink is superior to the uplink regarding the cell range; system coverage is hence uplink limited. At a certain load point, both of the links become equal. Thereafter, the downlink direction deteriorates very quickly towards its limit; the system capacity is hence downlink limited.

The noise rise due to the prevailing traffic is strongly dependent on the link direction. The load asymmetry indicates a link range asymmetry as well. For asymmetrical services there is typically more traffic in the downlink than in the uplink, e.g. web browsing. Thus, the cell capacity will be downlink limited. On the other hand, in a cell with significantly higher uploads than downloads, e.g. FTP file uploading, its capacity and coverage will be uplink limited.

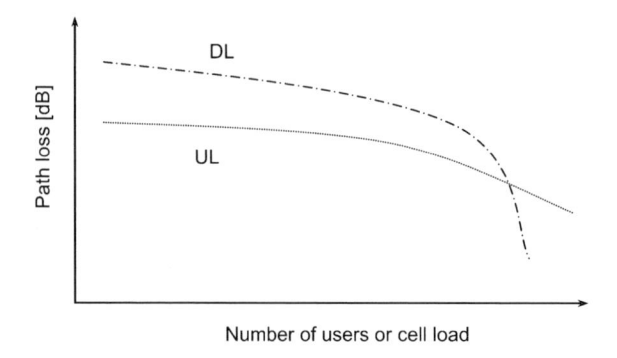

Figure 10.5 Example trade-off between coverage and capacity for macrocell.

Coverage and capacity depend on the ratio of the received signal to total noise, expressed by the E_b/N_0 (or better $E_b/(N_0+I_0)$, where I_0 is the spectral density of the total received interference power) requirements per service. Therefore, any measure which increases the received signal level, or decreases interference or minimises the service quality requirements will directly influence the system capacity and coverage.

The phenomenon of cell breathing is controlled by RRM procedures and requires the soft capacity limits to be set-up in advance. The trade-off between capacity and coverage enforces a dimensioning of both parameters together and in parallel. Final and hopefully satisfactory results are then achieved in a recursive approach.

The dimensioning of a UMTS radio access network in terms of coverage and capacity will involve the following steps:

1. *Uplink cell load estimation and power budget computation.* Since the coverage limiting factor for macrocells is the uplink direction, the corresponding uplink loading needs to be calculated. It will be used to estimate the total interference for power budget calculations. The power budget is necessary to calculate the maximum allowable path loss, which a signal may experience between the mobile and base stations for a certain service. It includes the total interference, a sum of all possible unwanted signals, margins to take into account signal coverage reliability, environment or system losses and gains and the hardware parameters of Node B and UE.
2. *Downlink power budget assessment.* Taking into account the uplink cell load and the asymmetry factor for each service, the downlink power allocation for the dedicated channels is calculated, being a portion of the total power. Consequently, it is used to calculate the downlink noise rise to verify that the downlink maximum allowable path loss is not less than that in the uplink (also accounting for the power allocated to common channels). If in discrepancy, the uplink loading from the first step needs to be adjusted accordingly or additional features for capacity or coverage enhancements need to be applied.
3. *Cell coverage assessment.* Using standard empirical propagation models (e.g. COST231-Hata) and the maximum allowable path loss, the cell range and coverage can be estimated for a certain service. The cell coverage area is usually calculated using the approximation of hexagonal cells.
4. *Cell capacity calculation.* For a given Grade of Service (GoS), the cell capacity should be calculated for the uplink taking into account the uplink loading, and for the downlink taking into account the downlink asymmetry factor and subsequently the loading and soft handover overhead. The analysis needs to consider the stochastic process of call arrivals and the given distributions of the call duration.

At the end of the dimensioning phase, the single cell coverage and capacity will be known for underlying assumptions. This can be used to estimate the total number of cells in the given environment.

10.3.2 CELL COVERAGE

10.3.2.1 Coverage Reliability

The design of the link budget assumes a certain signal level to be received at the cell edge, which is usually given in terms of a maximum allowable path loss together with a margin to provide a certain Quality of Service (QoS). The QoS is expressed with a cell edge reliability or cell area reliability.

A statistical analysis of the radio signal measurements exposed that the path loss L at any particular location between base and mobile stations at any distance d can be expressed as a random variable which follows a lognormal distribution (normal distribution in dB) about the mean path loss value [19]:

$$L(d)[\text{dB}] = L_{mean}(d)[\text{dB}] + X_\delta[\text{dB}] \qquad (10.18)$$

where X_δ is zero-mean Gaussian distributed random variable with a standard deviation of σ, both in dB. Since the received power $P_{Rx}(d)$ at a distance d is equal to the output power at the transmitter P_{Tx} decreased by the path loss, i.e.

$$P_{Rx}(d)[\text{dBm}] = P_{Tx}[\text{dBm}] - L(d)[\text{dB}], \qquad (10.19)$$

the distribution of the received power follows as well a Gaussian distribution with a local mean measured in dBm and standard deviation σ in dB. The lognormal distribution describes the random shadowing phenomena, which is often referred to as lognormal shadowing and sometimes also as slow fading.

The shadowing standard deviation [20] varies from 5 to 12 dB for macrocells, with a typical value of 7 or 8 dB. It rises a little with the increase of the carrier frequency, e.g. there is a 0.8 dB difference between 900 and 1800 MHz. It is virtually independent of the distance between terminal and base station. The standard deviation, however, differs among clutters and increases with a decreasing level of urbanisation or amount of scatters, e.g. in urban environment it is lower by 1.3–1.8 dB than in suburban areas. For microcells, the standard deviation of shadowing ranges from 4 to 13 dB.

Due to random effects of shadowing, the received signal level in a particular cell location can be below a desired target with a non-zero probability. It is therefore interesting to find this probability at the cell edge, often referred to as the outage probability $p_{out,edge}$. It quantifies the likelihood that the received signal level P_{Rx} falls below a required signal threshold P_{thr} and is given as:

$$p_{out,edge} = \frac{1}{2}\left[1 - erf\left(\frac{P_{Rx} - P_{thr}}{\sigma \cdot \sqrt{2}}\right)\right] \qquad (10.20)$$

where erf is the error function. An opposite quantity, i.e. the probability $p_{cov,edge}$ that the received signal at the cell edge is greater than a required signal threshold, is given by:

$$p_{cov,edge} = \frac{1}{2}\left[1 - erf\left(\frac{P_{thr} - P_{Rx}}{\sigma \cdot \sqrt{2}}\right)\right] \qquad (10.21)$$

The corresponding formula for the cell area or location probability A_{cell} is given as:

$$A_{cell} = \frac{1}{2}\left[1 - erf(a) + \exp\left(\frac{1 - 2 \cdot a \cdot b}{b^2}\right) \cdot \left(1 - erf\left(\frac{1 - a \cdot b}{b}\right)\right)\right] \qquad (10.22)$$

where

$$a = \frac{P_{thr} - P_{Rx}}{\sigma \cdot \sqrt{2}} \qquad (10.23)$$

$$b = \frac{10 \cdot n \cdot \log_{10} e}{\sigma \cdot \sqrt{2}} \qquad (10.24)$$

and n is the propagation constant (path loss coefficient). The difference between the received signal at the cell edge and the required threshold is called the shadow fading margin and will be used in the link budget calculation. Equation (10.22) is well known as 'Jake's formula', and has been introduced in [16]. For the same margins, the cell area probability is greater than the cell edge probability. For the same probabilities, the corresponding margin is greater for the cell edge probability than for cell area probability. As a result, the cell edge coverage reliability is more constrained than the cell area reliability.

The formulas above describe the probability of coverage for a single cell. In reality, cells will also overlap. This implies an increase of the coverage probability since in CDMA networks soft handover is used to connect the mobile simultaneously to different cells. Due to the resulting macro diversity combining, the transmitted power from the base station and the mobile can be decreased. Consequently, the coverage reliability is defined as the probability that the received signal level from any cell involved in the soft handover is greater than the coverage threshold, i.e.:

$$p_{\text{out,edge,softhandover}} = \frac{1}{\sqrt{2\pi}} \int_{-\infty}^{+\infty} e^{-\frac{x^2}{2}} \cdot \left[Q\left(\frac{P_{\text{Rx}} - P_{\text{thr}} - a \cdot \sigma \cdot x}{b \cdot \sigma}\right) \right] dx \qquad (10.25)$$

where Q is the cumulative distribution function of the normal distribution.

The soft handover gain with respect to shadow fading, obtained by combining the signals from different sources, is defined as the difference between the shadow fading margins and the coverage signal thresholds required for achieving a certain QoS in single cell and multiple cells scenarios. The soft handover gain depends on the propagation environment, i.e. the standard deviation, the propagation constant and the correlation between signals. In urban areas, where the fading is significant, the correlation between signals coming from different sources is small. As a result, the soft handover gain is greater. On the other hand, in rural areas, where fading does not significantly affect the signal, the correlation is accordingly higher and the achieved soft handover gain is smaller. The value of the soft handover gain may vary between 2 and 5 dB, with typical conservative values for the link budget of 2–3 dB. The highest gain will be obtained at cell edges, where the difference between signals from different sources is equal to zero and the macrodiversity gain is the greatest. The soft handover gain rises with an increased coverage probability, since higher received signals at cell edges will further outperform the signal threshold. For instance, for $n = 3.5$, $\sigma = 8$ dB, a signal correlation of 0.5 and a coverage probability for the cell area of 95 %, the gain is equal to 4 dB; for $n = 3.2$, $\sigma = 7$ dB, a signal correlation of 0.7 and a coverage probability for the cell area of 90 %, the gain is around 3 dB (see Table 10.13).

In the case that an outdoor cell needs also to provide indoor coverage, the combined standard deviation of the fading phenomena should be considered and, as already mentioned Chapter 5, the combined total standard deviation σ_{total} needs to be used, i.e.:

$$\sigma_{\text{total}} = \sqrt{\sigma_{\text{outdoor}}^2 + \sigma_{\text{indoor}}^2} \qquad (10.26)$$

where σ_{outdoor} and σ_{indoor} are the outdoor and indoor standard deviations, respectively. If other statistical variations of the link budget design should be considered, e.g. transmit power or indoor penetration loss deviations, Equation (10.26) may be extended with appropriate standard deviations.

Table 10.13 Typical values for coverage reliability calculation for cells in different clutter classes.

	n	σ_{outdoor} (dB)	σ_{indoor} (dB)	Signal correlation
Rural	3.2	8	—	0.7
Motorway	3.2	8	—	0.7
Suburban	3.5	8	4	0.5
Urban	3.5	7	5	0.5
Dense urban	4	7	6	0.5

10.3.2.2 Loading Formulas for Link Budget Analysis

Uplink
The WCDMA uplink load in a cell is the sum of loadings of a particular user. To obtain the loadings, one needs to consider both the thermal noise and the additional noise due to traffic. The noise rise can be used in the link budget calculations as one of the parameters to estimate the path loss. The uplink load η_{UL} in a cell can be described with the following equation [3], [21]:

$$\eta_{UL} = \sum_{j=1}^{N_{UL}} \frac{1}{1 + \frac{B}{(E_b/N_0)_j \cdot R_j \cdot v_j}} \cdot (1+i) \qquad (10.27)$$

where i is the i-factor, N_{UL} is the number of active users in uplink in the cell, B is the chip rate equal to 3.84 Mcps for UTRAN, $(E_b/N_0)_j$ is the required service quality, R_j the service data rate and v_j is service activity factor of the j-th user, respectively. The service data rate may vary according to Radio Access Bearer (RAB) settings and has quantified values, e.g. 12.2, 32, 64, 128, 144, 256 and 384 kbps. The activity factor describes the portion of the time used by a speech connection for speech frames with respect to the total time of the connection, in case that a discontinuous transmission (DTX) is applied. Typical speech frames may occupy 50 % of the traffic channel's connection time. For dimensioning purposes, the voice activity factor in the uplink can be assumed to be 66 or 67 %, which takes into account the overhead due to the Dedicate Physical Control Channel (DPCCH) equal to about 16–17 %. For data connections, the activity factor is assumed to be 100 %, indicating that resources are reserved and used only when there is information to be sent.

The corresponding noise rise due to the various traffic sources can be expressed as:

$$NR = -10 \cdot \log_{10}(1 - \eta_{UL}) \qquad (10.28)$$

The noise rise will limit the maximum path loss between the mobile and base stations. If the load approaches 100 %, the noise rise reaches infinity and hence the cell range shrinks to zero. This would require infinite power at the transmitters to maintain all the connections. It is therefore necessary to constrain the load, since the available power is limited. Depending on the assumed scenario, the loading can be limited to 50 or 60 % for coverage constrained deployment and to about 75 or 80 % for capacity constrained deployment. It should be noted that cell range variations due to low and high loading need to be compensated by an appropriate cell planning, so as to minimise extensive cell breathing when higher loads are experienced in the network. On the other hand, under conditions of lower loading, the extensive noise rise can be controlled by Radio Resource Management (RRM) algorithms.

Assuming all users in a cell follow the same average traffic and propagation condition profile (E_b/N_0, R and v are averaged for the whole cell), then Equation (10.28) can be rearranged to calculate the average number of subscribers per cell, $\overline{N}_{UL}$, to be used for dimensioning purposes:

$$\overline{N}_{UL} = \frac{\eta_{UL}}{(1+i)} \cdot \left(1 + \frac{B}{(E_b/N_0) \cdot R \cdot v}\right) \qquad (10.29)$$

Downlink
The loading equation for downlink differs from the one for uplink. The capacity limiting factor in the downlink is the total available power at the base station, which is known as the soft limit. (There are hard limits as well, but not considered here: like number of OVSF codes or channel elements.) The total power P_{TX_total} needs to be shared among the Common Control Channels (CCCHs), denoted as P_{TX_CCCH}, and the dedicated traffic channels, denoted as $P_{TX\_traffic}$, i.e.

$$P_{TX\_total} = P_{TX\_traffic} + P_{TX\_CCCH} \qquad (10.30)$$

The CCCHs are transmitted on fixed powers and with predefined activities*, without power control. Thus, its contribution to the noise rise is predetermined. The dedicated channels obey power control routines and their contribution to the noise rise is dependent on the user distribution and service profile, meaning it is not deterministic. In a given cell, all traffic connections in the downlink are originating from one point using fully orthogonal codes. As already mentioned before, due to multi path propagation, this orthogonality is diminished, resulting in the increase of interference, i.e. more power is needed to maintain the same connection quality as compared to the fully orthogonal case. The total noise rise should then be calculated as the sum of the noise sources from dedicated and common channels.

Taking into account Equation (10.13) and assuming that traffic is uniformly distributed in a homogenous environment, thus all base stations transmit at the same output power P_{TX_total}, the interference margin received by a j-th user connected to cell l in the downlink can be expressed as:

$$IM = \frac{I_{total}}{N} = \frac{\frac{(1-\alpha) \cdot P_{TX\_total}}{L_{j,l}} + P_{TX\_total} \cdot \sum_{k=1, k \neq j}^{K} \frac{1}{L_{k,l}} + N}{N} \qquad (10.31)$$

where the first part of the denominator is the own cell interference, the second one represents the other cell interference, N stands for noise power, L is the path loss between base station antenna output and mobile station receiver input and a is the orthogonality factor. Assuming that the i-factor is equal to (based on [22]):

$$i_j = \sum_{k=1, k \neq j}^{K} \frac{L_{j,l}}{L_{k,l}} \qquad (10.32)$$

Equation (10.31) can be rewritten as:

$$IM = 1 + \frac{((1-\alpha) + i) \cdot P_{TX\_total}}{N \cdot L_{j,l}} \qquad (10.33)$$

For the link budget calculation, it is interesting to calculate the noise rise at the cell edge, hence $L_{j,l}$ needs to reflect the total path loss between base station antenna output and the mobile station receiver input. Equation (10.33) can be used to estimate the total required power of a cell for a given noise rise, i.e.:

$$P_{TX\_total} = \frac{N \cdot \overline{L} \cdot (IM - 1)}{((1-\alpha) + i)} \qquad (10.34)$$

where $\overline{L}$ is the average path loss between the base station transmitter and the mobile station receiver. The requirement on the average path loss (not the cell edge path loss) comes from the power requirements of users distributed in a cell [21], i.e. users located closer to the base station will require less power than users located at the cell edge. The typical difference between the average and maximum path loss can be assumed between 6 and 8 dB.

---

* The assumption of constant activities of CCCHs holds only when the traffic in a cell is constant. It does not hold, for instance, if more call attempts occur, because more signalling will be used for call establishment, or if less call attempts occur, because less signalling between mobiles and the base site will be exchanged.

Assuming the basic formula for the link quality given in Equation (10.35), the traffic channel powers can be estimated at a given location of mobile j with dedicated power from a cell $P_{TX,j}$ [22]:

$$\left(\frac{E_b}{N_0}\right)_j = \frac{B \cdot C_j}{R_j \cdot I_{total}} = \frac{P_{TX,j} \cdot B/L_{j,l}}{R_j \cdot \left((1-\alpha_j) \cdot (P_{TX\_traffic} + P_{TX\_CCCH})/L_{j,l} + (P_{TX\_traffic} + P_{TX\_CCCH}) \cdot \sum_{k=1, k\neq j}^{K} \frac{1}{L_{k,l}} + N\right)} \qquad (10.35)$$

where C_j is the received signal power at mobile j. Solving the equation for $P_{TX,j}$, multiplying by activity factor v_j and summing up for all connection in the cell in the downlink N_{DL}, the total traffic power $P_{TX\_traffic}$ can be estimated:

$$P_{TX\_traffic} = (P_{TX\_traffic} + P_{TX\_CCCH}) \cdot \sum_{j=1}^{N_{DL}} \left[\frac{E_b/N_0 \cdot R_j \cdot v_j}{B}\left((1-\alpha_j) + \sum_{k=1, k\neq j}^{K}\frac{L_{j,l}}{L_{k,l}}\right)\right]$$

$$+ N \cdot \sum_{j=1}^{N_{DL}} \frac{E_b/N_0 \cdot R_j \cdot v_j}{B} \cdot L_{j,l} \qquad (10.36)$$

Taking into account Equation (10.32) and rearranging Equation (10.36) the total traffic power $P_{TX\_traffic}$ can be given by [22]:

$$P_{TX\_traffic} = \frac{\eta_{DL} \cdot P_{TX\_CCCH} + N \cdot \sum_{j=1}^{N_{DL}} \frac{E_b/N_0 \cdot R_j \cdot v_j}{B} \cdot L_{j,l}}{1 - \eta_{DL}} \qquad (10.37)$$

where η_{DL} is the downlink load caused by the dedicated channels [22]:

$$\eta_{DL} = \sum_{j=1}^{N_{DL}} \frac{(E_b/N_0)_j \cdot R_j \cdot v_j}{B} \cdot ((1-\alpha_j) + i) \qquad (10.38)$$

The other difference to the uplink is the contribution of DPCCH to the activity factor, which is about 8 %; the speech activity factor in the downlink is hence equal to 58 %.

Assuming only one service is supported and a flat traffic distribution is present, Equation (10.37) can be simplified to:

$$P_{TX\_traffic} = \frac{\eta_{DL} \cdot \left(P_{TX\_CCCH} + \frac{N \cdot \overline{L}}{(1-\alpha)+i}\right)}{1 - \eta_{DL}} \qquad (10.39)$$

and Equation (10.38) can be used to calculate the average number of users in the downlink:

$$\overline{N_{DL}} = \frac{\eta_{DL}}{(1-\alpha)+i} \cdot \frac{B}{(E_b/N_0) \cdot R \cdot v} \qquad (10.40)$$

It should be noted that a part of the CCCHs, i.e. the primary synchronisation channel (P-SCH) and secondary synchronisation channel (S-SCH), are not scrambled with the primary scrambling code; thus, the synchronisation channels are not orthogonal to other CCCHs and the dedicated channels. Consequently, the SCH power is seen by the user as interference, independent of channel-induced orthogonality factor. This has not been considered in the formulas above. In average, the contribution of the SCHs' power to total output power of a cell is around 1 % or less; the effect is thus usually neglected.

10.3.2.3 Link Budget Analysis

Albeit available in many books, for the sake of completeness, we shall review the UTRAN link budget analysis here in a compact form. Generally, the link budget analysis is used:

- to balance the uplink and downlink, since, due to different load dependencies, load asymmetries and available power, there is a significant difference in both directions; the uplink is coverage limited for macrocells, because the output power of the mobile station is considerably lower than that of a base station; and the downlink is capacity limited, because the total power available at the base station needs to be shared among all users in a cell and therefore, as the load rises, the power available for coverage per user is smaller, potentially making the uplink coverage larger than the downlink one;
- to calculate the cell range for a given cell load for the most demanding service;
- to calculate the corresponding coverage thresholds for network planning, roll-out and optimisation.

Table 10.14 displays an example link budget for a 64 kbps RT service in a typical urban environment macrocell with a voice activity factor of $v = 1$, an orthogonality factor of $\alpha = 0.6$, an other-to-own interference ratio of $i = 0.6$ and a soft handover overhead of 30 %. The transmitter end is considered first. The output power at the antenna connector for a dedicated connection is decreased by losses of the cable system and increased by antenna gain, if applicable. As the results, the *Equivalent Isotropic Radiated Power, EIRP*, can be obtained. The uplink transmission power has been set for UE class 3 and for the downlink adjusted to balance the uplink. The total power in the downlink has been assumed 20 W; thus, there is sufficient headroom for more traffic power increase in the downlink. The average power assigned to the CCCH has been limited to 3.6 W, including a CPICH power of 2 W. In the next step, before the receiver input, the signal must be corrected with the antenna gain, cable loss and MHA improvement, if applicable. This signal should fulfil the minimum requirement of *C/I – Carrier-to-Interference* ratio – which is the product of the quality required for a RT service, i.e. $E_b/N_0 = 8.1$ dB (in linear scale), the inverse of the processing gain G_p, as well as the ratio between the system chip rate $B = 3.84$ Mchps and the service bit rate $R_b = 64$ kbps. Consequently, the signal is received by the receiver, the sensitivity of which is determined by the noise power density of -174 dBm/Hz, the quality of the receiver hardware, i.e. the noise figure F, the E_b/N_0 and rate R_b requirements of the supported service. The margin section considers other possible signal gains and losses on the way between transmitter and receiver. Finally, the maximum allowable path loss is calculated that a signal may experience between transmitter and receiver, being the difference between *EIRP* and the receiver sensitivity level, including the margins. Due to the statistical approach used for the cell range estimation, the given coverage reliability for outdoor users needs to be considered next. This is done by taking the slow fading margin into account and consequently decreasing the allowable path loss. Indoor users experience less allowable path loss, which is due to indoor penetration losses and indoor shadow fading margins. Therefore, for the same coverage reliability, an indoor user requires a larger shadow fading margin than an outdoor user. Finally, the planning thresholds need to be calculated. Typically, the coverage reliability and its quality for a given *Data Traffic Channel* (DTCH) are expressed with CPICH RSCP, *Received Signal Code Power* and CPICH E_c/I_0. The required CPICH RSCP is calculated using the DTCH EIRP decreased by the appropriate maximum allowable indoor or outdoor path loss, and corrected with the power ratio between DTCH and CPICH. The estimation of the DTCH quality with CPICH E_c/I_0 is more complex, since the CPICH E_c/I_0 value itself does not provide enough information to conclude on the quality of the DTCH. The quality of the DTCH is determined by the orthogonality, which in turn depends on

Table 10.14 Example link budget for macrocell and 64 kbps RT service.

	Uplink	Downlink	Unit	Formula
Transmitter	Mobile station	Base station		
Max Tx power per connection at antenna connector point	0.25	1.50	W	a
	24	32	dBm	$b = 10\log(a/0.001W)$
Cable loss including MHA insertion loss	0	3.3	dB	c
Antenna Tx gain	0	18	dBi	d
EIRP	24	46	dBm	$e = b - c + d$
Receiver	Base station	Mobile station		
Antenna Rx diversity gain	3	0	dB	f
Antenna Rx gain	18	0	dBi	g
Cable loss	3	0	dB	h
MHA improvement	5.5	0	dB	i
Thermal noise density	−174	−174	dBm/Hz	j
Receiver noise figure	4	7	dB	k
Receiver noise power	−104.16	−101.16	dBm	$l = j + k + 10.\log(3840000)$
Service required E_b/N_0	8.1	8.1	dB	m
Data rate	64000	64000	bps	n
Receiver sensitivity level	−113.84	−110.84	dBm	$o = l + m - 10.\log(3840000/n)$
Margins (additional gains or losses)				
Noise rise (interference margin)	4.0	2.9	dB	p
Soft handover gain	3	3	dB	q
Fast fading margin (power control headroom)	2	0	dB	r
Body loss	0	0	dB	s
Transmitt power rise (due to power spikes)	1	0	dB	t
Maximum allowable path loss	157.34	157.40	dB	$u = f + g - h + l + e - o - p + q - r - s - t$

Coverage reliability

Outdoor probability of coverage at cell edge	90%		v
Outdoor slow fading standard deviation	7	dB	w
Outdoor slow fading margin	9.0	dB	$x = NORMSINV(v).w$
Max allowable path loss for outdoor coverage	148.37	dB	$y = u - x$
Indoor probability of coverage	90%		z
Indoor slow fading standard deviation	6	dB	aa
Indoor penetration loss	20	dB	ab
Indoor penetration loss standard deviation	4	dB	ac
Combined standard deviation	10.05	dB	$ad = SQRT(w.w + aa.aa + ac.ac)$
Intdoor slow fading margin	12.9	dB	$ae = NORMSINV(z).ad$
Max allowable path loss for indoor coverage	124.46	dB	$af = u - ab - ae$

Planning thresholds

CPICH power	2	W	ag
CPICH power	33	dBm	$ah = 10\log(ag/0.001W)$
Power ratio between CPICH and DTCH	1	dB	$ai = ah - b$
Outdoor CPICH RSCP threshold	−100.7	dBm	$aj = e - y(uplink) + ai$
Indoor CPICH RSCP threshold measured outdoor	−76.7	dBm	$ak = e - af(uplink) + ai$

Values in the right-side computed column: 90%, 7, 9.0, 148.43, 90%, 6, 20, 4, 10.05, 12.9, 124.52

the ratio between own and other interference (Section 10.2.4.2). The minimum required CPICH E_c/I_0 to guarantee a given DTCH quality can be obtained by means of Equation (10.41) [23]:

$$\left(\frac{E_c}{I_0}\right)_{min} = \frac{P_{TX\_CPICH}}{\max(P_{TX\_total}) \cdot \left(1 + \frac{1}{GF}\right)} \cdot \frac{E_c}{I_{0r}} \quad (10.41)$$

where P_{TX_CPICH} is the CPICH transmitted power, E_c/I_{0r} is the ratio of the chip energy of the DTCH to the power spectral density of P_{TX_total} and GF is the geometry factor defined as the ratio between own interference and the sum of other interference and noise power, i.e.:

$$GF = \frac{I_{own}}{I_{other} + N} \quad (10.42)$$

The geometry factor indirectly quantifies the location of mobile station in a cell. For bigger values of GF, the own interference is dominating; the mobile is thus located closer to the base station. For smaller values GF, the other interference is outperforming the own one; thus, the mobile is closer to the cell edge. The values for E_c/I_{or} in dependency of GF should be obtained from physical layer simulations and a calculation of the required minimum CPICH E_c/I_0 should consider the worst case scenario, i.e. the cell edge. Using Equation (10.41), the values for the minimum CPICH E_c/I_0 can be obtained for network roll-out or initial optimisation of lightly loaded network and ongoing optimisation for heavily loaded network. In lightly loaded cases, the total transmitted power would contain only the contribution of the CCCHs and the contribution of N in GF would be significant. For the loaded case, the traffic channel power will considerably influence the total output power and other interference may be much bigger than the noise power; the geometry factor will thus be almost equal to the inverse of the i-factor.

10.3.2.4 Coverage and Site-to-site Distance Calculation

The cell range can be obtained using the maximum allowable path loss taken from the link budget assessment and some standard propagation path loss models. Table 10.15 depicts an outline of the usability of certain propagation models according to the considered cell type; Sections 5.3 and 5.4 provide more details on the models.

After the cell range estimation, the site coverage area and site-to-site distance can be calculated. The cell area is used to calculate the number of sites required in a specific environment to fulfil the

Table 10.15 Standard propagation models and its application for cell range estimation.

	COST 231 Hata	COST 231 Walfish-Ikegami	Manhattan model
Environment	Macrocell	Regular microcell or small macrocell	Microcell in like Manhattan environment (e.g. street grid or canyons)
Application	Average cell with antenna above the roof top level and very simplified environment characterisation – purely statistical approach	Cell with antennas above the roof top level and some typical environment parameters – semi deterministic approach	Cell with antenna below roof top level and some typical environment parameters – semi deterministic approach

Table 10.16 Coverage area and site-to-site distance for sites with different number of sectors as a function of cell radius R.

	Omni	Two sectors	Three sectors	Six sectors
Site-to-site distance	$\dfrac{\sqrt{3}}{2} \cdot R$	$2 \cdot R$	$1.5 \cdot R$	$\dfrac{\sqrt{3}}{2} \cdot R$
Site area	$\dfrac{3 \cdot \sqrt{3}}{2} \cdot R^2$	$\dfrac{3 \cdot \sqrt{3}}{4} \cdot R^2$	$\dfrac{9 \cdot \sqrt{3}}{8} \cdot R^2$	$\dfrac{3 \cdot \sqrt{3}}{2} \cdot R^2$

coverage requirements. The site-to-site distance is necessary to define the planning grid and hence the areas where a potential site can be placed.

The coverage area and site-to-site distance can be calculated assuming that sites are located on crossings of a geometric grid and the cells assume shapes of some geometric figures. It has typically been assumed that cells take a shape of a hexagon, which not necessarily reflects reality well. For instance, a three sector site with antennas with a horizontal beam width of 65° is well reflected by a hexagonal structure; however, if the beam width is 90° or more, a better shape for a cell is a lozenge and all cells together build a clover leaf. Note that three sector antennas with a beam width of 90° or more would cause a high percentage of soft handovers; thus, reducing the downlink capacity and throughput on one hand, but improving the coverage on the other.

Table 10.16 provides example figures for coverage area and site-to-site distance calculations. For the omni site case, the cell shape would typically be a circle, but to fit into the hexagonal grid the cells have been assumed to be hexagons with some overlap. The two sector configuration is used for line coverage, e.g. road coverage in remote or rural areas; the cells are hexagons without overlap between neighbours. The three sectors pattern fits exactly into the hexagonal structure without overlapping between adjacent sites. The six sectors structure would hardly fit into a hexagon geometry, but due to the number of sectors it has been assumed that it is similar to the omni case.

10.3.3 CELL ERLANG CAPACITY

A telecommunications system is usually dimensioned for the busy hour, where the peak load (system traffic) is served with a required service quality and given resource availability, and it is quantified by its blocking probability [24]. The served traffic is given in Erlangs. The UTRAN capacity can be either hardware resource limited (hard limitation) or interference limited (soft limitation), or both.

10.3.3.1 Hard Capacity

For hardware limited systems, the supported traffic can be calculated using the well known Erlang B formula to obtain the blocking probability, where the traffic arrival process follows a Poisson distribution and the holding time is exponentially distributed. The Erlang B formula describes the blocking probability of a new call for given traffic and available resources (hardware channels), assuming that one channel can serve one user; such a system is called hard limited.

The standard Erlang B formula is valid only for single service dimensioning, but the nature of UMTS is to support multi-services. Using the formula for multi-service capacity, calculations leads to an underestimation of the capacity; hence the multidimensional Erlang B formula needs to be used. It defines the probability of j bandwidth units being occupied (busy) in a system with a given number of total bandwidth units [24].

Hardware limitations in UTRAN occur when, e.g., baseband resources, like channel elements or codes, are exhausted. Typically, in well isolated or low dispersive cells, a shortage of baseband resources occurs before the limitation due to interference. The first implies that the i-factor reaches zero, and the second that the orthogonality factor reaches one. The number of allowed users in a cell may therefore increase significantly without damaging the load equation.

10.3.3.2 Soft Capacity

In a CDMA system, the Erlang capacity calculation for a given blocking probability is more complex, since the number of available channels or radio resources is not fixed and can vary according to the load conditions and hence interference. As a result, the CDMA radio capacity is soft limited, since coverage can be traded against capacity and vice versa. As a consequence, the direct application of the Erlang B formula with an average number of available radio channels provides too pessimistic results; the Erlang capacity hence yields underestimated results. This underestimation rises with the increase of the service data rate [21].

In Section 10.3.2.2, the cell load was used to calculate the average number of supported users in a cell at a given time, often referred to as the channel capacity of a cell. The problem of this approach is that the traffic call arrival and departure are described by random processes, requiring the probability to be found that, when a new call arrives,

- in the uplink, the predefined noise rise will be kept (thus all users will be able to keep their connection at a desired E_b/N_0);
- in the downlink, there will still be enough radio resources (power) to maintain the existing connections and serve the new call.

As a consequence, if the channel capacity will be taken as an alternative to the Erlang soft capacity to estimate the cell capacity, then the network capacity will be under dimensioned.

There are number of other random parameters which need to be considered, when calculating the Erlang soft capacity, e.g.:

- given the nature of CDMA technology, the user activity factor depends on the user behaviour; intracell and intercell interferences depend on the user distribution and orthogonality factor; signal fading is also a random variable that determines the transmit power and hence the interference;
- caused by CDMA system impairments, imperfect power control algorithms cause the required E_b/N_0 to be increased; the receiver hardware quality determines the hardware noise figure, which might be different for each user and hence leading to a random distribution of noise figures in the system; randomness also occurs with the maximum output power of a mobile or the power amplifier of a base station, where power deviations are even allowed by the 3GPP specifications.

For the uplink there are a few analytical approaches to calculate the Erlang soft capacity, which:

1. Take into account a static and fixed i-factor* and the random process of traffic arrival and departure by using the classical Erlang B formula; scale up the number of available radio channels N_{UL} by $(1+i)$; divide the estimated isolated cell Erlang by $(1+i)$ to finally obtain the multicell Erlang capacity [21].
2. Take into account a static and fixed i-factor and the random process of traffic arrival and departure, as well as significant load parameters (voice activity factor and E_b/N_0 requirements); assume that the sum of all random contributions in the total interference equation is Gaussian distributed [25].
3. Take into account a static and fixed i-factor and the random process of traffic arrival and departure, as well as significant load parameters (voice activity factor and user power requirements, which are directly linked to E_b/N_0); assume that the sum of all random contributions in the total interference equation is either Gaussian or lognormally distributed [26].

---

* Thus other-cell interference is constant for the entire cell and is independent of the mobile distribution.

The first approach is relatively simple to implement and can be used for prompt dimensioning:

$$\text{Erlang\_soft\_capacity} = \frac{A\left(p_{\text{blocking}}, N_{\text{UL}} \cdot (1+i)\right)}{(1+i)} \qquad (10.43)$$

where $A\left(P_{\text{blocking}}, N_{\text{UL}} \cdot (1+i)\right)$ is the Erlang B capacity of an isolated cell calculated for a given blocking probability p_{blocking} and number of radio channels $N_{\text{UL}} \cdot (1+i)$.

The second and third approach for calculating the Erlang soft capacity are similar, and it has been shown that the Gaussian approximation provides almost the same results as the lognormal approximation for a blocking probability around and below 1 % [26]. For dimensioning purposes, the Gaussian approximation can be used, since in this case the expression for the soft blocking probability $p_{\text{blocking,soft}}$ is simpler [25].

Due to many random variables involved in the (W)CDMA soft capacity estimation, the probably exactest way to calculate the Erlang capacity for real systems would be to use the results from simulations, see e.g. [27]. Other possibilities comprise analytical approaches, as, e.g., proposed in [28] and [29]; they provide a fairly good matching to simulation results and yield probability figures versus offered capacity faster than simulations.

10.4 DETAILED NETWORK PLANNING

Most of the radio network planning activities are related to the cell layer that provides the wide area coverage, i.e. the network footprint into a market. It is thus typically related to the macrocell layer deployment. Before any planning activity can commence, the key parameters of the radio network layout need to be defined to find an optimum between coverage and capacity with a minimum investment; these include:

- *Site-to-site distance*, together with *antenna height* to define the network grid for planning;
- *Sectorisation*, including the number of sectors combined with proper antenna characteristics;
- *Antenna direction* tweaking;
- *Antenna tilt* adjustment.

These parameters are set a priori with certain assumptions; for instance, the site-to-site distance is calculated using a link budget, where parameters are averaged and the antenna height is fixed in advance. Consequently, the process of searching an adequate Node B candidate site can be initialised, resulting in a site location. In the classical approach, the assignment of Node Bs should follow an ideal geometry grid, where the neighbour grid nodes are elements of equilateral triangles. The corresponding cell shapes are hexagons or diamonds, depending on the antenna direction and antenna horizontal beamwidth. This approach is to some extend impractical, since the environment and traffic distribution are not homogenous (see Section 9.2). The actual Node B placement, configuration and its antennas tilt arrangements will hence differ from the ideal grid, and the initial settings will be adjusted according to the local environment and traffic requirements. Reality then may impair the prior assumed approximations very strongly; for instance, a suitable Node B candidate site is practically not available, or the assumed antenna direction cannot be set due to significant obstacles in the vicinity or due to other RF systems in the main antenna direction. Also, uptilting an antenna to achieve cell dominance causes strong interference to distant other cell users. All these real network planning issues will enforce additional measures to be taken to provide a certain coverage and capacity by means of further solutions, e.g. extra repeaters to amend the coverage gap, or additional Node Bs in a new location to secure coverage and capacity to a certain quality level, when neighbouring Node Bs have been shifted too far away from each other. The thus achieved network may vary considerably from the ideal initial plan.

Finally, note that any results presented here and in the open literature strongly depend on the assumed network and simulation parameters; results obtained in a different environment may differ in absolute numbers. However, the general tendencies of the presented solutions in this section are valid for the most typical configurations.

10.4.1 SITE-TO-SITE DISTANCE AND ANTENNA HEIGHT

The site-to-site distance determines the network density and, together with the antenna height, the achieved coverage and capacity; both parameters are contradictory, as explained below. Imagine that the antenna height is increased but the site-to-site distance remains constant, then the coverage probability increases and so does the area coverage per cell; hence, more potential traffic can be served. On the other hand, the interference range and strength to the surrounding cells rises as well. If the site-to-site distance increases but the antenna height remains constant, then the coverage probability per cell decreases; thus, less potential traffic can be served. However, the interference between cells also decreases, thereby leaving some headroom for a potential capacity increase. The proper approach would be to find the best trade-off between both parameters for coverage and capacity through iterative planning activities with a reference network and assumed site configuration*. This, unfortunately, implies an optimisation process before even any planning activities.

Another concern regarding the proper settings of antenna height and site-to-site distances is the quality of predicted data used to estimate these parameters, which include the received signal power, its quality and the traffic. As such, the prediction of the RF signal propagation depends on many factors. The most important ones are the resolution and number of dimensions of the environment data, the RF propagation model and the antenna model. This, unfortunately, is far from perfect in most cases, since applied approximations give relatively high errors; for instance, 2D versus 3D antenna pattern modelling (see Section 6.1), statistical/semi-empirical signal prediction models versus advanced and detailed deterministic techniques (see Chapter 5), low resolution 2D versus high resolution 3D geographical data etc.

In the following, one example is described in more detail. Given is an urban environment in an European city, 2D antenna models, 2D morphology data and an elevation model with a 50 m × 50 m pixel size, the standard propagation model from COST 231 includes correction factors for losses due to clutter, diffraction, effective antenna height and dual slope approach (see Chapter 5). The prediction model has been calibrated to the standard deviation of around 7 dB for the average predicted signal strength against the measured sample before any planning started on the selected locations. Unfortunately, the measurements of real implemented cells provided a local difference between predicted and measured signal strength of more than 10 dB for around 40 % of all measured points. For the prediction of the signal quality (E_c/I_0), these errors may add up and cause the term of the total interference in a dense network to be underestimated. Additionally, the real Node B antenna characteristic is far from ideal. After the half power point at −3 dB, the radiation power does not decrease smoothly to zero, but unfortunately contains significant side lobes. These are especially difficult to control by means of tilting in the vertical plane. Unwanted radiation power from the side lobes has a significant influence on the interference to other cells, which is crucial for the overall system performance. The traffic forecasting is not an easy task either, especially for a new network: not only the number of potential customers in certain time frame, the user profile and thus generated traffic with specific data rates are difficult to estimate, but also the geographical distribution. The number of customers and the service usage may be dependent on a positive or negative resonance to a marketing campaign, as well as the current penetration rate in the market, or the availability of handsets, or the

---

* Actually, the configuration should be optimised at the same time as well, since there are several different optimal configurations depending on antenna height, site-to-site distance and traffic. Please, refer to Sections 10.4.3 and 10.4.7.

availability of a killer application etc. A precise estimation of the traffic amount per cell is hence barely possible[†]. In summary, the above clearly shows the difficulty and imprecision of any prediction and planning activity (for an overview of other aspects, please, consult Chapter 1).

The uncertainties in the early planning phase enforce that the cells are planned for an a priori fixed and rather conservative antenna height[*]. This helps minimising and strictly controlling the interference to other cells. Additionally, a fixed maximum load is defined to guarantee a certain trade-off between capacity and coverage.

Subsequently, some average site-to-site distances are calculated from link budgets. To this end, Table 10.17 provides some examples on antenna height assessment and corresponding site-to-site distance for several clutter classes. It is obvious that a denser environment comes along with a denser user population, hence more potential customers, and more difficult propagation conditions for indoor users; this, in turn, requires smaller cell ranges to be used and hence lower antenna heights to be deployed. This demands more base stations to cover the same coverage area, e.g. in dense urban clutter compared to suburban clutter. Therefore, the potential capacity broken down per surface unit in sparse environment is smaller.

At the actual physical location, the antenna height is determined by the location height, close obstacles and break point distance. If the Node B was implemented on a building, its height would almost always determine the lowest possible antenna mounting position due to limitations of antenna placement on building walls. The close obstacles may shadow especially the antenna vertical beam; thus, the 1st Fresnel zone needs to be considered. The breakpoint distance affects the slope of the average path loss from the serving cell to served and interfered mobiles.

From the diffraction theory it is known that the 1st Fresnel zone is bounded by an ellipsoid, where the difference between the sum of path lengths from transmitter and receiver to a point on that ellipsoid and the distance from transmitter to receiver is equal to half of the wave length. For any reflected signal echo in the 1st Fresnel zone, the path difference with respect to the direct signal is equal or less

Table 10.17 Example site-to-site distances and antenna heights for macro layer for 64 kbps DTCH dimensioning and modified COST231 Hata propagation formula (based on assumption from Sections 10.2 and 10.3).

	Dense Urban		Urban		Suburban		Motorway		Rural (open)	
BS antenna height (m)	20	25	25	30	30	40	40	60	40	60
MS antenna height (m)	1.5	1.5	1.5	1.5	1.5	1.5	1.5	1.5	1.5	1.5
Building penetration loss (dB)	20	20	15	15	10	10	5	5	0	0
SD of building penetration loss (dB)	6	6	5	5	4	4	0	0	0	0
Outdoor slow fading SD (dB)	7	7	7	7	8	8	8	8	8	8
Indoor slow fading SD (dB)	6	6	6	6	6	6	0	0	0	0
Probability of coverage	95 %	95 %	95 %	95 %	95 %	95 %	95 %	95 %	95 %	95 %
Cell range (km)	0.248	0.281	0.502	0.556	1.665	1.892	5.793	7.287	8.168	10.397
Site-to-site distance (km)	0.372	0.422	0.752	0.834	2.497	2.837	11.586	14.575	12.251	15.596

[†] For the mature network, where new customers do not significantly change the average user profile, the traffic forecasting can be done in a less complicated fashion by simply scaling up the measured traffic.
[*] A different approach related to Ultra High Sites is presented in Section 12.3.

than the half of the wave length. Thus, at the receiver, the sum of both signals may lead in a worst case to signal cancellation. The radius of the 1st Fresnel zone $r_{\text{1st zone}}$ can be calculated by Equation (10.43):

$$r_{\text{1st zone}} = \sqrt{\frac{\lambda \cdot d_1 \cdot d_2}{d_1 + d_2}} \qquad (10.44)$$

where λ is the wave length, d_1 and d_2 are the distances from transmitter and receiver to the close obstacle, respectively. It is important that the 1st Fresnel zone is free from any obstacles close to the transmitter to minimise the losses due to reflection or diffraction. For cellular systems it is generally not possible to keep that zone totally free of obstacles, unlike the microwave link between base stations. It is, however, utmost important to have some clutter clearance close to the base station antenna. From the knife-edge diffraction theory, at least 50% of the 1st Fresnel zone radius needs to be cleared, thus the resulting diffraction loss is 0 dB. For instance, a site has been placed on a 20 m high building, the distance between antenna and the roof edge is 5 m and the vertical beamwidth of antenna is 7°; if the total tilt was 0°, 5°, 10° or 15°, then the corresponding antenna bottom height above ground level should be 21, 21.7, 22.4 or 22.9 m respectively.

Assuming a two ray model with a reflection off ground, the breakpoint distance for antennas placed above roof top level can be derived as:

$$\text{breakpoint} = 4 \cdot \frac{h_{MS} \cdot h_{BS}}{\lambda} \qquad (10.45)$$

where h_{MS} and h_{BS} are the heights of the antennas of base and mobile stations, respectively. Before the breakpoint distance, the LOS condition dominates and hence the path loss between base and mobile stations decreases with the slope of 20 dB per decade distance, similar to the free space but with fades. After the breakpoint distance, the corresponding path loss falls to 40 dB loss per decade distance (more complex environments than the two-ray model which typically yield values between 30 and 60 dB loss per decade). The dependency between Node B antenna height and the breakpoint distance is straightforward. The greater the antenna height the greater the distance where the breakpoint of the path loss slope occurs. With a large breakpoint distance, the signal attenuation from the serving cell to a potential user follows 20 dB per decade and also more interference to neighbouring cells is provided. The cell coverage should hence be inside the range of the breakpoint distance and the cell interference to other cells outside of it. Some cell range examples from Table 10.16 in dense urban, urban and suburban areas fit into the breakpoint range, e.g. for a mobile antenna height of 1.5 m and Node B antenna heights of 20, 25, 30 and 40 m, the corresponding breakpoint distances are 624, 780, 936 and 1247 m respectively.

10.4.2 SITE LOCATION

WCDMA systems are limited by interference; therefore, a new site should not only provide significant coverage to a predefined area, but also needs to guarantee a rigorous control of the created interference. Since perfect locations are often not possible, the only option to keep the negative influence of the transmitted power within limits is a proper Node B configuration by means of selecting a proper number of sectors, antennas, their tilts and orientations.

The Node B locations should be selected according to the predefined ideal grid. The grid should be referred to places, where the traffic would most probably be generated. The planned Node Bs should provide dominant coverage in the predicted traffic areas and minimise the areas where the dominance is weak. This will ensure a low pilot pollution. Spots with a high pilot pollution will experience a deficient network performance. Pilot pollution can be identified by a poor E_c/I_0 and when there are more than 3 pilots in a coverage point and the difference of the fourth pilot to the first pilot is less than 6 dB.

The average antenna height resulting from *a priori* settings needs to be maintained, especially in metropolitan areas, so as to allow a harmonised dominance of neighbouring cells and interference control. On the other hand, locations with significant RF dominance over potential neighbours should be avoided. The produced overlapping between cells should be kept to an average of 25–35 %, so that maximum 2 or 3 cells overlap. This facilitates a sufficient amount of connections in soft or softer handover. A higher percentage of either handover, however, leads to deficiency of the capacity. The soft handover demands an increase of both downlink and transmission capacities, where the softer handover requires only the downlink capacity to be increased. Cell shapes should be as regular as possible and cell areas should be fairly equal to ensure smooth transitions between the same layer cells and a uniform distribution of potential traffic. The Node Bs built on selected locations need to fulfil the coverage, capacity and quality criteria in a given service area. These criteria are expressed with appropriate probabilities. Since the capacity can be checked only under real load conditions, service points with coverage and quality probabilities not fulfilled should be spread out evenly among the service area or moved to lowly populated spots.

As mentioned before, the site location is determined by the search process of a suitable candidate. Based on a grid, an area is defined around the ideal location where the candidate site can be found. The span of the search area is determined by the network density and may vary between 50 and 250 m in metropolitan areas. The search area is then adapted to local conditions, both environment and traffic. Example situations for adapting a search area to the local circumstances are described below:

- An ideal grid node is placed on a small residential buildings spot, but the average clutter building height is greater that the chosen spot. From a coverage point of view, it is favourable to move the search area to the surrounding higher buildings.
- According to the grid, a new site would be better located on a building which is significantly higher than the surrounding constructions and the average antenna height in that area; however, mounting of the antenna on the wall is forbidden. The search area should therefore be moved to surrounding smaller buildings, since a Node B in a high location (but not on the wall) would dominate the area and hence there would be too much overlapping and interference[*].
- The location of the base station would be near the hot spot. From a traffic point of view, it could be advantageous to move the site location to the hot spot to allow uniform load sharing among all Node B cells.

In Section 9.2.1, two cases have been described showing the dependency of the site location variation with respect to a predefined grid, fixed antenna height and the system performance based on results from [30] and [31]. The first case uses results obtained from simulations of an ideal hexagonal network with a homogenous environment and uniform traffic distribution; the second case uses simulation results from a network with real environmental data and uniform traffic distribution. In [32], simulation results with similar real network data is provided, but three different traffic scenarios are presented: a homogenous distribution without indoor users, a homogenous distribution with indoor users and a non-homogenous distribution with an indoor/outdoor user mix of 70/30 %. In general, when the coverage probability is high enough, thus sufficient cell overlapping exists, small variations of the ideal site location are allowed and they only marginally influence the network performance. This means that there is very little impact on the site displacement in metropolitan areas, where the network is dense and the coverage probability is comparably high; this is also reflected in the location search process. On the other hand, if the coverage probability is relatively low, thus less cell overlapping exists, then the displacements may impact the network performance, especially for rural or motorway sites.

---

[*] If the new location was very close to the high building, then the high building would provide significant shadowing to one of the sectors. In any case, such a situation would be preferable in a dense network since, with building shadowing, there is less local network performance loss as with one dominant cell, which may destroy the overall capacity.

In reality, problems arise if the deviation is too big [30,32] so that, even with some site configuration tuning, a substantial network degradation or coverage gap can be locally observed. More serious, however, are situations where a suitable site candidate cannot be found at all. Consequently, the network requires to be redesigned and additional Node Bs may be required. The coverage gap could be closed with a repeater. Unfortunately, it does not bring further capacity to the network and may shift problems of pilot pollution to different areas; thus, for a macrocell layer in metropolitan areas this is not the preferred solution. Such situations appear quite often, causing the final site count and network cost to differ considerably from the initial plan.

For instance, the UTRAN network built in one of the markets of a big European country provides indoor coverage for 64 kbps RT services for all cities with more than 10 000 inhabitants, and also in-car coverage is provided along the main roads and motorways connecting the cities. The link budget analysis led to a number of sites, which have been required to fulfil the coverage and capacity constraints. The same constraints have been applied in the planning tool and, where possible, all existing GSM1800 sites have been re-used for initial RF planning. After this, the number of required sites rose by 4 %. Following that, the real RF planning started requiring new locations to be searched. The process had been finished with a site count increased by 19 % compared to the initial planning!

10.4.3 SECTORISATION

Sectorisation is commonly used technique to enhance capacity or coverage by equipping the site with (more) sectors; it is traditionally provided by means of directional high gain antennas. Each of the sectors creates a new cell. This is different from cell splitting, where the sectors may still belong to one logical cell. In the case of cell splitting, the enhancement leads only to a coverage extension in the downlink and uplink; it could also provide additional capacity improvements in the uplink[†], but due to power splitting leads to a decrease of the downlink capacity [11]. With sectorisation, due to the newly created cells, coverage and capacity can be improved in both directions.

Depending on the application, the site may be comprised of one sector (omnidirectional antenna) to provide services in an isolated area [11]. Such single sector approach is applicable to micro or pico cells, as well as macrocells; the cell requires being isolated from the remaining network by, e.g., terrain obstacles. The one sector solution may also be realised by a sectorised antenna to cover, e.g. a tunnel. Two sectors may be employed for line coverage at macro layer (e.g. sites along the motorway) or for coverage and capacity of dense areas at micro layer. Three sectors can be used to provide regular footprints of the network service at macro layer with low or medium load, and six sectors to provide additional capacity and improved coverage to regular macro layers. Other configurations are rarely used[*]; an example is the case of a four sector site to cover a road crossing, where each of the sectors is dedicated to one of the roads.

The performance of a sectorised site is mainly given by the horizontal (azimuth) characteristics of the used antennas and, obviously, the number of sectors. The more the sectors are used, the narrower the beamwidth of the antennas has to be; an optimum antenna beamwidth can be found for each configuration [11,32–35]. A narrower antenna provides usually a higher directional gain, and hence improves the link budget in both directions. In addition, less overlapping between neighbouring sites is provided; thus, less interference is received from other cells and less soft handover overhead is present. The improved directional link gain offers an improved coverage in the main direction, however requires the edges of the service area to be covered by the adjacent sectors. The directional gain can be used to

---

[†] This assumption holds true, when the loss due to power splitting of the original cell among all sectors is less than the difference between the gain of directional sector antenna and the antenna of the original cell. On the other hand, due to an increased overlapping, the soft handover area increases and the system capacity may hence be decreased.
[*] Different sectorisation approaches of Ultra High Sites are presented in Section 12.3.

increase the site-to-site distance and hence reduce the network density. Thus, the greater allowable path loss, together with less interference and less soft handover overhead, increases the network capacity. The number of connections for soft handover can be controlled by an appropriate soft handover parameter: number of pilots allowed in the active set, as well as add and drop thresholds. Unfortunately, soft handover parameters do not change the amount of received interference. Also, because the overlapping areas are decreased, the risk of less reliable coverage increases. The latter may be influenced to some extent by a proper adjustment of soft handover parameters, which need to reflect the number of pilots in the active set and soft handover gain achieved (see Section 10.2.4.3).

In general, the most typical configurations of three or six cell sector antennas with a horizontal beamwidth of 65° or 33° respectively provide the best performance. As an example, the results from [35] are presented. Figure 10.6 provides a comparison for the three sector cases with 65° and 90° antennas and some reference load. Figure 10.7 depicts the difference between 33° and 65°* for the six sector case with double load compared to the three sector case. For both sets of results, the site separation in simulation was 1.5 km and the antenna height 25 m.

The 65° antenna outperforms the 90° antenna in each of the considered performance parameters for the three sector site. The coverage probability with a 65° antenna is improved by 1.8%. Due to less overlapping among the cells, there is less percentage of soft and particularly softer handovers, as well as other-to-own cell interference. Consequently, less power and other network resources (throughput) are utilised to serve the users. Additionally, the resulting loading factor in the uplink is decreased by 4%.

A similar behaviour can be observed when comparing the performance between 33° and 65° wide antennas for the six sector case. The difference in the coverage probability is greater between 33° and

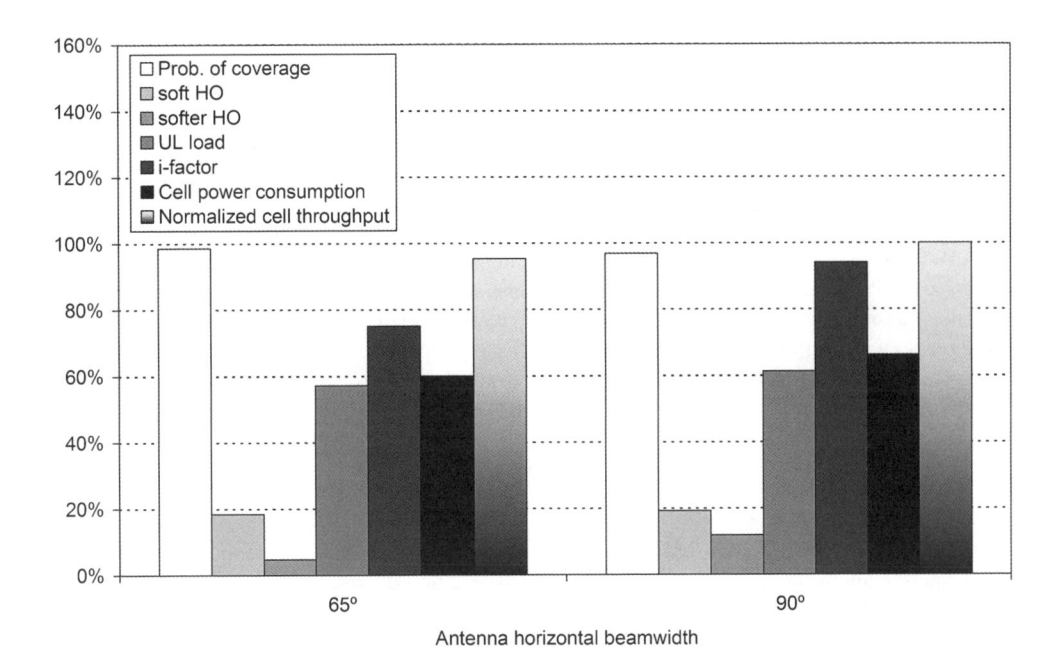

Figure 10.6 Example performance of sites with three sectors (based on results from [35]).

---

* The omnidirectional antenna, due to lack of directionality, and the 120° wide antenna, due to enormous overlapping, yield worst performance ([33]) and have been omitted for demonstration.

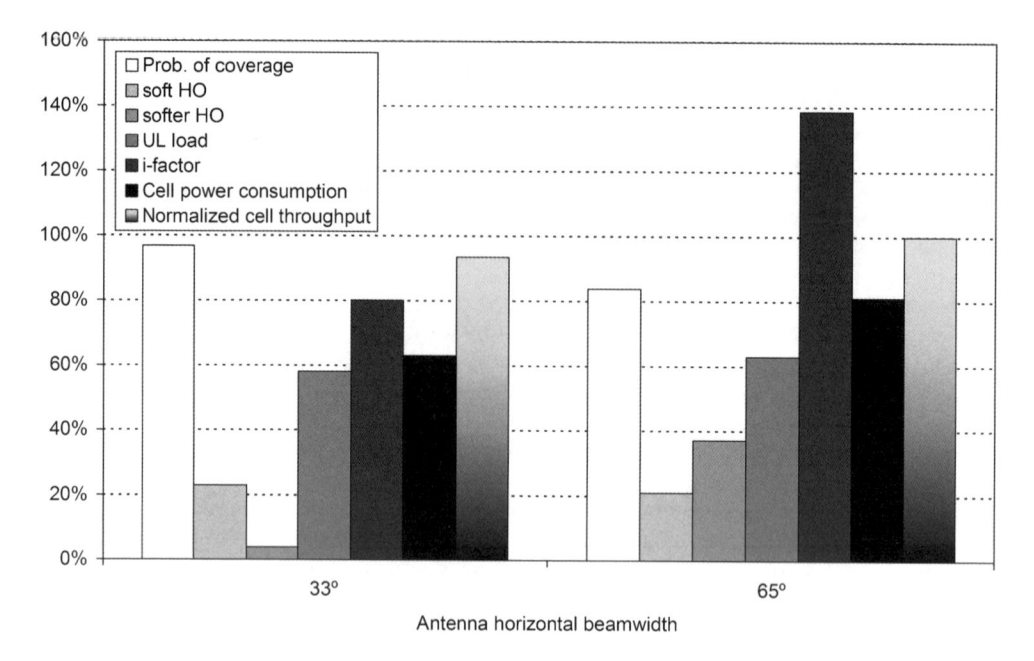

Figure 10.7 Example performance of sites with six sectors (based on results from [35]).

65° antennas for the six sector case, than between 65° and 90° for the three sector case. This is due to the significant overlapping among six cells with 65° antennas. Hence, the interference and the softer handover percentage are significantly higher than for the 33° antenna case. On the other hand, the high directivity of the 33° antenna causes a little higher soft handover overlap. As a result, the power consumption and throughput of a 33° cell sector is respectively almost 20 and 6.5 % lower than for a 65° cell sector. This implies considerable resource savings.

The capacity difference between the three and six sector configuration has been simulated in [34] in terms of the number of served users as a function of used transmit power. The transition from three sectors with 65° antennas to six sectors with 33° antennas provides around 113 % more capacity at the same transmit power (which was 2 dB lower than the maximum allowable 20 W). Moreover, when the three sector case runs out of power at a specific load, the six sectored site still has significant power headroom left for increasing the number of connections. The capacity increase between three and six sectors rises with an increasing load. The difference between 65° and 90° antennas in three and six sector configurations is rather minor and almost constant for a rising system load. The capacity of 65° antennas is greater than for the 90° antennas by around 1.5 and 1 % for three and six sector arrangements respectively.

During the life time of the network, the expected traffic will grow, requiring the side sectorisation to be adjusted; one has, of course also other possibilities [11], such as introducing multiple carriers, additional scrambling codes, receive and transmit diversity or opt for a microcell deployment. If the network will be launched with three sector sites, the next step could be six sector sites. In this case, the capacity is increased by nearly 80 % [11]. The transition requires the 65° antennas to be exchanged against a doubled number of 33° antennas. To save 50 % of antenna costs, a hybrid solution has been proposed [3]: three antennas of 33° are added to the original three sectors sites with 65° antennas and the narrower antennas are placed between the broader ones. The corresponding performance of the hybrid solution does not differ significantly from the pure 33° antennas configuration. In terms of coverage, soft handover probabilities and consumed power, the solutions are almost equal; the only significant difference obviously comes from the overlap between sectors, causing the softer handovers in the hybrid solution to increase by 7 % and the throughput by 5 %. Thus, the hybrid solution would require more transmission and baseband resources as well as transmit power.

Higher order sectorisations may also be employed. In [36], the 12 sector configuration with 20° antennas is proposed. The achieved average capacity gain with reference to the three sector configuration with 65° antennas and six sector configuration with 33° antennas is around 310 and 180 %, respectively. Naturally, the overlapping between neighbouring cells in the 12 sector case is noticeably greater than in the other arrangements. Thus, the softer handover probability for the 12 sector case is increased by about 7 and 9 % compared with the 6 and 3 sectors, respectively. The soft handover probability, however, remains almost the same among all the configurations. Practically, it could be very complex to implement such configuration, since the high capacity requirements occur mainly in metropolitan areas. First, the physical space for 12 different antennas and corresponding hardware (cabinets, masts, feeders, MHAs) need to be given. This may be a problem on already available sites in urban areas, where other technologies are present as well. Second, the load of the required construction and hardware need to be supported by the given location. Third, and likely the most important fact, is that a 'forest' of antennas may considerably scare the residents of surrounding building, thus the site could be blocked by protests of residents; for instance, due to serious aesthetic, albeit not technical objections, an operator had to dismount such an array in Barcelona in the summer of 2005.

Yet another higher order sectorisation solution has been proposed in [37]. Using beamforming techniques, 18 different sectors per site are employed. The approach consists of three uniform linear antenna arrays with four elements each. Each of the antenna arrays produces six different beams, which cover the hexagonal shape from a three sector arrangement. The six beams are symmetrical pair wise to each other with reference to broadside direction. For the performance evaluation, results from dynamic simulations are provided in [37], with the assumption that the power per site is fixed to 60 W. This indicates that for a comparable three sector configuration, each sector may use 20 W; for a six sector configuration, each sector may spend 10 W; and for each of the 18 sectors, only 3.33 W per cell is available. The capacity increase obtained by the 18 sectors and six sectors with reference to the three sectors is 190 and 150 % respectively. Thus, the difference between the 18 and the six and three sector configurations in terms of supported users is 27 and 90 %, respectively. The relative small capacity increase with reference to the six sector solution is mainly caused by the limited power per sector and the high soft handover overhead (67 and 120 % more than in the six and three sector configuration, respectively), which contains significant contributions from the softer handover. Additionally, the simulation results from [37] provide figures showing that fixed beamforming (three sectors, each with six fixed beams) outperforms the 18 sector arrangement by 26 % in terms of capacity.

10.4.4 ANTENNA AND SECTOR DIRECTION

Another key parameter influencing the performance of UTRAN is the direction of antennas (sectors). Already in Section 9.2.2, it has been shown that for a hexagonal structure there are significant consequences in the case of improper antenna direction adjustment. Figure 8.12 indicates that the best configuration for a three sectorised site would be 0°, 120° and 240°, and the worst 90°, 210° and 330°. But the situation may be diametrically opposite, when the reference nodes layout will be, e.g., turned clockwise by 60° as a whole; then the best performance will be given by 90°, 210° and 330°, and the worst by 0°, 120° and 240°. This example clarifies the dependency of the sector orientation on the Node B's relative arrangement, and it points out that a proper adjustment controls the soft handover overhead and interference leakage. Another important aspect is that the spacing between sectors ought to be as identical as possible, as this would allow an equal distribution of softer handover areas among all sectors.

In [32], the authors provide simulation results of the network performance when a deviation in antenna direction is allowed. The reference network is hexagonal, based on an ideal grid with 1.5 km site spacing, but with real morphological and topographical data. Each site consists of three sectors with 65° antennas. Basically, two cases are presented, where the direction deviation is relatively small (the average deviation is 9.1°) and where the direction deviation is fairly substantial (the average deviation is 18.1°). The network has been tested with uniform and non-uniform traffic distributions, as well as with and without indoor users, and three different load conditions. The interesting case is clearly the network with non-uniform traffic distribution, indoor users and the load for which highest coverage probability is provided. For the small deviation in sector direction, the network performance is almost the same as for the reference network; small deficiencies can be noticed in all performance parameters, but they are negligible. Even for higher traffic loads, as the probability of coverage shrinks, the changed network does not operate different than the original network. For the case of doubled deviation in antenna direction, there are no simulation results with non-uniform traffic and indoor users available, but the system performance for a uniform traffic distribution and indoor users is again almost the same as for the reference network. The softer handover overhead rose in this case by 1%, which is negligible too. Similar results are provided in [38].

As previously mentioned, in the real phase of the planning process, choosing the ideal sector direction is not always possible. First of all, the sites are not placed in the ideal locations, thus the grid and corresponding antenna orientation cannot be maintained. Second, the physical environment close to the site location may force changes in the sector direction due to significant obstacles. Third, the installation errors may lead to imprecise antenna placements. In the past, the angular difference between planned and practically installed configuration could be up to 20°. The deviation of the antenna direction may lead to increased softer handover areas, where the separation between two sectors from the same site is less than 120°. On the other hand, a separation bigger than 120° may lead to coverage holes. When the antennas of neighbour sites are directed to each other, then the soft handover overhead increases as well, leading to capacity deficiency. Generally, the antenna azimuth of a physically placed Node B should provide a clear best server coverage scenario, leading to dominant sectors and pilots in serving area. At the same time, the areas with insufficient pilot dominancy and extensive cell overlapping should be avoided. Two basic measures can be used to minimise the lack of dominant pilots: pointing sectors into an area in-between the sectors of surrounding sites and preventing adjacent site sectors to point at each other. In parallel, the sector orientation should be adjusted to traffic requirements, so as to evenly spread the expected load among all sectors. The situation of a street propagation canyon should be excluded by eliminating antennas pointing along straight long streets. The sector separation should be kept regularly. The physical antenna mounting should leave sufficient freedom for further optimisation processes, so that some future orientation (and tilt) changes can be performed.

10.4.5 ELECTRICAL AND MECHANICAL TILT

The antenna tilting technique is a very powerful measure to increase the network performance. The main aim of tilting is to increase local area cell dominance and reduce the cell overlapping, hence the inter cell interference. It is achieved by concentrating the radiation pattern of an antenna towards the anticipated cell serving area, thereby increasing the cell isolation to neighbouring cells. Basically there are two possibilities to tilt an antenna: mechanically or electrically [38].

10.4.5.1 Mechanical Tilt

The mechanical way tilts the antenna radome in reference to the ground plane. Unfortunately, the mechanical tilt affects predominantly the main direction and the tilt effect decreases from the main direction towards the side lobes in the azimuth. At azimuths of ±90° from the main direction, the influence of the mechanical tilt is zeroed. As a result, the gain reduction is a function of the azimuth and the downtilt angle: the horizontal half power beamwidth increases with an increasing mechanical downtilt angle (Figure 10.8). The mechanical tilt can be carried out by adjustable brackets.

10.4.5.2 Electrical Tilt

The electrical method is realised by adaptation of the relative phases between all elements of antenna array (dipoles), allowing the antenna pattern to be tilted evenly in all horizontal directions. Contrary to the mechanical approach, there is thus no dependency between azimuth and the tilt angle. Consequently, the gain reduction is independent of azimuth and the horizontal half power beamwidth remains unchanged as a function of the downtilt angle (Figure 10.8). The shape of vertical pattern, however, is modified in dependency of tilt angle. The phase adjustment is performed by means of different feeder lengths for each dipole, which results in a fixed electrical tilt, or by a phase shifter, which results in an adjustable electrical tilt.

Independent of the tilt scheme, there is always an optimum tilt angle* for a given configuration [3,39–42]. From the results and conclusions made in the literature, the following statements can be made. The optimum downtilt angle is mainly a function of the site-to-site distance, base station antenna height and the vertical beamwidth of that antenna, as well as the number of sectors and the tilt scheme utilised. Practically, the influence of the two last parameters is smaller than the first three. The optimum downtilt range varies between 3.5° and 10.5° for site-to-site distances between 1.5 and 2.5 km, base station antenna heights between 25 and 45 m, antenna vertical beamwidths of 6° and 12° and for a homogenous traffic distribution. The optimum downtilt angle decreases as the site separation increases and increases as the antenna height or antenna horizontal beamwidth increases. The change of optimum tilt angle is more sensitive to antenna height changes than to site-to-site distance modifications. This is because for greater antenna heights the cell overlapping increases more rapidly than with site

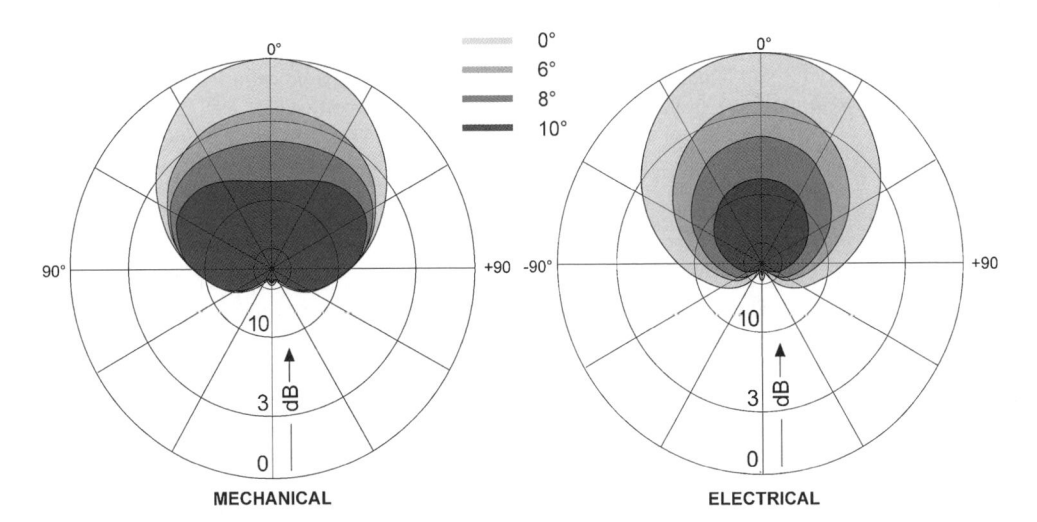

Figure 10.8 Horizontal radiation patterns in dependency of mechanical and electrical tilts [38]. (Reproduced by permission of KATHREIN-Werke KG Rosenheim).

---

* The optimum downtilt angle was defined in [39] taking into account downlink and uplink directions. For the range of the downtilt angle, the probability of coverage is not allowed to differ by more than 2% from the best possible case for a given configuration. From the tilt range, the minimum uplink and downlink power loads are considered and its corresponding tilt angles are averaged.

separation reduction. For three sector configurations, the optimum downtilt angle for broader vertical beamwidth antennas is greater for most of the cases than for narrower beamwidth, but the change of tilt for broader antennas is not so sensitive to the probability of coverage as for narrower antennas. Narrower vertical beamwidth antennas can be more precisely directed to the expected service area, thus the cell dominance and the isolation could be more evident. On the other hand, for non-tilted configurations, the broader antenna delivers a better probability of coverage, which results in a lower downlink load. The broader antennas also realise less directional gains, thus interference to other cell is reduced and coverage is spread more widely, especially for areas closer to the site. The electrically tilted antennas reduce the coverage cell area evenly, thus there is a monotone dependency of downtilt and the downlink load. This implies that with increased electrical downtilt angle, the service probability decreases. The mechanical tilt influences the downlink load slightly different; until a certain optimum angle, the load is reduced, but after that the load increases due to increased softer handover overhead. For narrower antennas and three sector configurations, the optimum downtilt angle of electrically controlled tilts is smaller than for mechanical tilt arrangements. For the broader antennas, the tendency is opposite. This can be justified by a less efficient coverage reduction for electrical tilt with wider antennas and limitation of optimum downtilt angle with mechanical arrangement by softer handover overhead. For six sector configurations, the optimum downtilt angles are similar to the three sector sites, but cell overlapping is increased. Consequently, the electrical tilt is a little more sensitive to antenna height and site separation than in the three sector case. Additionally, the performance of the mechanical tilt is further limited by the percentage of the softer handover probability. Thus the optimum downtilt angle achieved with the mechanical tilt is a little bit less than with the electrical one and the mechanical tilt adjustment is much more sensitive to the network performance than in three sector case, hence less capacity gain can be achieved. Generally, the system capacity gain with optimum downtilt angle, referred to the non-tilted scenario, rises as the antenna height and number of sectors rise and the site separation and antenna horizontal beamwidth decrease. A maximum gain of 58 % has been obtained for six sector configuration, with an antenna height of 45 m, site spacing of 1.5 km and a vertical beamwidth of 6°. This may be overestimated as the soft handover gain, reducing the negative effects of fast fading, was not considered and in the non-tilted scenario, there is a higher probability for soft handover. As suggested in [43], the antenna downtilt leads to a reduction of the delay spread, thus the orthogonality in downlink can be improved as well as the value of required E_b/N_0. As a result, the capacity can be further improved. The geometrical load distribution, service mix and indoor/outdoor spread do not noticeably influence the optimum downtilt angle [41,42]. However, the geographical load distribution and its volume may significantly impact the optimum downtilt angle. It has been shown in [44] that the tilt angle significantly influences the cell breathing phenomenon.

Based on simulation results, the following empirical formula has been developed in [42] to calculate the optimum downtilt angle v_{opt} in dependency of the base station antenna height h_{BS} in meters, cell range (sector dominance range) d in kilometres and the vertical beamwidth v_{-3dB}:

$$v_{opt} = 3 \cdot \left[\ln(h_{BS}) - d^{0.8}\right] \cdot \log_{10}(\theta_{-3dB}) \tag{10.46}$$

The formula is valid only for the parameters described above; for instance, for a base station antenna height of 40 m and a cell range of 6.45 km, the optimum tilt angle is less than zero, which suggests an uptilt. The results obtained with Equation (10.46) differ from the geometrical approach by a factor k:

$$v_{opt,geom} = \arctan\left(\frac{h_{BS} - h_{MS}}{d}\right) + k \tag{10.47}$$

where h_{MS} is the height of the mobile antenna; the unit for height and range is meter. In [42], the correction factor is equal to half of the vertical beamwidth, i.e. $v_{-3dB}/2$. In practical situations, the correction factor ranges from 1° to 3°, depending on the network density and antenna height.

From the above, the following practical considerations are highlighted, which shall conclude this section:

- Selection of downtilt angle is more sensitive in a denser network and significant antenna heights, where the dominant area is rather small; thus, for metropolitan areas, an optimum downtilt angle is of more importance than for a motorway or rural areas.
- Selection of downtilt angle for higher gain and narrower horizontal beamwidth antennas is more sensitive to network capacity than for lower gain and broader ones.
- Electrical tilt is more favourable against the mechanical tilt, since it allows more precise control of interference, cell dominance and isolation and pilot pollution.
- Mechanical tilt may be used as an extension to the electrical tilt to minimise the negative effect of back lobes and excessive overlapping between sectors pointing to each other.
- Application of adjustable electrical tilt, in contrast to a fixed electrical tilt, is useful due to a broader range of possible downtilt angles; this is because in real networks, cell ranges and antenna heights actually differ from cell to cell.
- The Remote Electrical Tilt (RET) solution would be desirable for the purpose of load balancing between the cells.

We will now move on to the temporal characterisation of Hierarchical Cell Structures (HCSs), which are known to significantly boost system capacity and hence to be very useful for the UMTS planning process.

10.4.6 TEMPORAL ASPECTS IN HCS

Hierarchical cell structures are a powerful network configuration technique to balance network traffic and hence increase the overall network performance. The main aim of HCSs is to provide a set of overlaying cells in the same geographic area, which facilitates traffic to be assigned to various layers according to some criteria. Interference is a paramount issue in such arrangements, thereby requiring stringent interference avoidance or cancellation mechanisms. Interference can be avoided (or at least minimised) if the cell layers are separated in frequency, time or space.

As already mentioned in Section 4.4.1, a separation in frequency is the most common approach, where the separation criteria is traditionally based on the terminal's speed, i.e. fast moving terminals are assigned to the macrocell and slow moving users to the microcell. In 2G systems, such a separation has been justified by the savings in signalling overhead related to handover procedures, i.e. a fast moving terminal would require very frequent handovers if assigned to the microcells. In 3G systems, a similar justification is appropriate; however, the additional factor here is that, according to gathered statistics, the majority of slow moving users require higher data rates which are easily delivered in short-range microcells. In any case, a separation in frequency between the layers is clearly a spectrally inefficient arrangements, and other separation mechanisms have been sought.

Notably, a separation in location, where natural or man-made constructions severely attenuate the signals between the cell layers, allows the hierarchical structure to communicate at the same time in the same bands. A typical example of such an arrangement is the deployment of an outdoor macrocell overlaying one or several well separated indoor hot spot cells. The network capacity in such an arrangement is clearly drastically higher than for a frequency separated HCS, with the caveat of increased design limits (i.e. one cannot place the layers where necessarily needed, but only where the attenuations between both layers are high enough).

The system introduced, analysed and enhanced in [45–49] and depicted in Figure 10.9 clearly mitigates above design limit, where one can place hot spot microcells under the overlay of a macrocell using the same frequency band. Such a deployment is clearly very advantageous from an operator's point of view, because cell planning can be performed with this increased capacity in mind. The deployment of such a system relies on the observation that interference varies over time, where

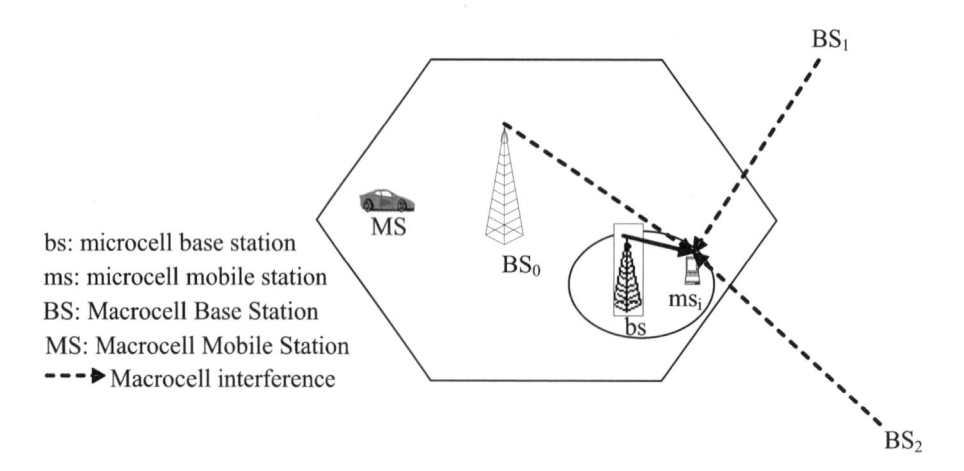

Figure 10.9 Two-layer HCS with a microcell being overlaid by a macrocell and interfered by several surrounding macrocell tiers.

the temporal drop of interference from the macrocell below a given threshold allows the users in the microcell to communicate; one speaks from an opportunistic transmission or interference window. Therefore, the microcell user reports the experienced macrocell interference level to the microcell base station, which transmits *packets* only during opportunistic transmission windows. The temporal duration of such opportunistic windows depends on many factors, however, most notably on the:

- temporal characteristics of the macrocell interference;
- PHY layer used in the mircocell.

As for the macrocell interference, it is caused by the macrocell base stations adjacent to the microcell hot spot. Three of them, BS0-BS2, have been depicted in Figure 10.9; however, there are clearly more interfering tiers. Their transmission power is known to vary in time due to the closed-loop power control to their associated mobile terminals. The frequency of change of the power control commands clearly depends on the macrocell propagation environment and the speed of the macrocell mobile terminals; i.e. if shadowing and fading are severe then the dynamic range of power control can be very large (10–20 dB in the downlink) and if the majority of the terminals moves very fast then the temporal variations are high (within ms).

A typical example temporal evolution of such an interference power due to 17 macrocell base stations, as observed in a given location within the microcell, is shown in Figure 10.10. The depicted interference power variations follow a Gaussian distribution in dB. We will use the Gaussian property to facilitate some closed form mathematical analysis for the purpose of cell planning; however, it ought to be noted that the interference power is by no means guaranteed to be Gaussian distributed. Indeed, it has been observed that the distribution differs from Gaussianity, particularly under heavy macrocell load. We will utilise the Gaussian assumption, nonetheless, because it yields some good performance indicators. We clearly observe from the figure that, at a given interference threshold, opportunistic transmission windows occur. They are random and can be statistically described by a frequency and duration of occurrence. We will deal with their description below, but will confine for the moment to the observation that the higher the threshold, the longer the opportunistic window.

In Figure 10.10, thresholds $I_{\text{Thr},1}$ and $I_{\text{Thr},2}$ have been depicted, where $I_{\text{Thr},1} > I_{\text{Thr},2}$. These thresholds are associated with two given PHY mechanisms (modulation + coding), i.e. PHY1 and PHY2. Clearly,

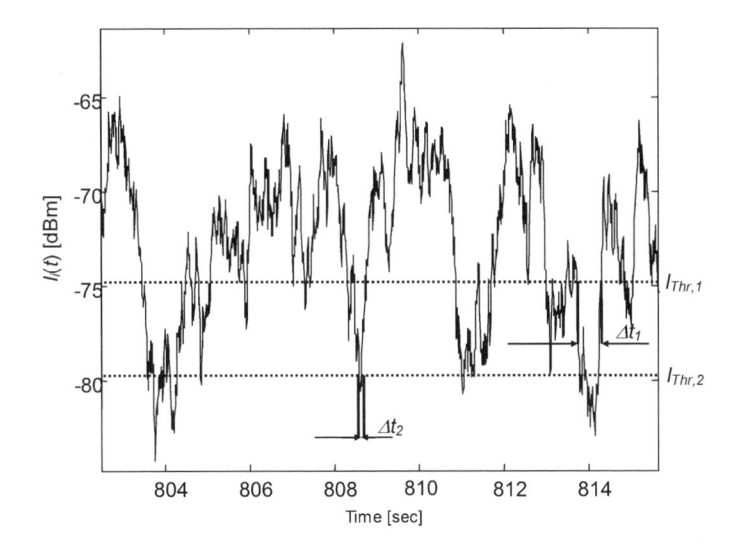

Figure 10.10 Macrocell interference profile, as measured at a microcell location.

PHY1 is a more robust mechanism which can tolerate a higher level of interference, but is likely of lower throughput; and PHY2 is the weaker mechanism but with a higher throughput. This demonstrates the PHY layer dependency on the performance in a HCS.

With this picture in mind, we can now proceed to the aims of a network planning engineer, which can be summarised as:

1. decide which traffic/service to associate to which HCS layer;
2. quantify the throughput in the macrocell, given the interference from the microcell;
3. quantify the throughput in the macrocell, given the interference from the macrocell; and then
4. perform respective macro and microcell planning exercises, as outlined in this book.

As for the first point, the introduced opportunistic scheduling approach only allows non-real-time (NRT) services to be serviced in the microcell, corresponding to the UMTS interactive and background traffic classes (see Section 3.2), and realtime (RT) and NRT services to be served by the macrocell. This is not a shortcoming, because, given that 3G systems need to support a mixture of RT and NRT circuit- and packet switched services, the importance of efficiently accommodating NRT data traffic in cellular systems has increased. NRT services (such as web browsing, e-mail, paging, fax, file transfer etc.), unlike the RT services, are delay insensitive and are able to change their transmission rates from a maximum to even zero in order to release resources for other users or wait for better channel conditions.

As for the second point, the interference from the microcell may cause harming interference to the macrocell terminals nearby. Comparing the miniscule coverage of a typical microcell with a radius of about hundred metres to the coverage of a macrocell of several kilometres, it is obvious that the likelihood and hence impact of such interference is fairly low. If it becomes a serious issue, e.g. due to a large number of microcells embedded into the macrocell, or because of a large macrocell terminal density, other interference mitigating mechanisms need to be catered for by the network planner. An example solution could be that the macrocell and microcell Node Bs are all connected to the same RNC, which in turn may disable the microcell Node Bs if the macrocell Node B is suffering from severe interference. Other mechanisms are also feasible, such as beamforming at the microcell site, but are not considered further.

As for the third point, given the characteristics of the PHY modes, i.e.:

- the respective over the air packet duration, TTI = 2, 10, 20, ... ms and
- the associated interference power thresholds,

the main issue here is to determine:

- the likelihood of the respective PHY mode to succeed;
- the average duration of success (transmission window);
- the hence achieved average throughput,

an analytical approach to which shall be demonstrated below. To this end, we will examine the average microcell signal and macrocell interference power. We shall also assume that the HCS system operates in the interference limited regime, i.e. we neglect the thermal noise in the following equations.

The microcell signal power is gathered by an L-fingered Rake receiver. It is a fairly complex stochastic process, depending on the statistics and temporal characteristics of possibly power controlled microcell transmission power, as well as shadowing and fading effects; a closed form description has hence evaded the research community's luck to date. Assuming, however, the shadowing process to obey a lognormal distribution and the combined Rake power to obey a central-χ^2 distribution with $2L$ degrees of freedom [50], it can be shown that the resultant distribution is well approximated with another lognormal distribution. In addition, it has been demonstrated that the power control error is also likely to obey a lognormal distribution [51]. This somehow lends itself to the assumption that the received microcell signal power obeys a lognormal distribution. The macrocell(s) interference power, as seen at a microcell location, has been shown also to obey a lognormal distribution [52].

In dB, the powers are hence Gaussian distributed, where the signal power at the receiver input $C_i(t) \sim N(\mu_s, \sigma_s^2)$ and the interference power $I_i(t) \sim N(\mu_i, \sigma_i^2)$. The resultant CIR, i.e.

$$CIR_i(t) = C_i(t) - I_i(t) \tag{10.48}$$

is hence also Gaussian distributed with mean $\mu_s - \mu_i$ and variance $\sigma_s^2 + \sigma_i^2$, i.e. $CIR_i(t) \sim N\left(\mu_s - \mu_i, \sigma_s^2 + \sigma_i^2\right)$. Reflecting above scenario of service and user separation, we will now assume that the umbrella macrocell is subjected to high temporal dynamics, whereas the microcell remains fairly static. This allows Equation (10.48) to be written as

$$CIR_i(t) = C_i - I_i(t) \tag{10.49}$$

where $I_i(t)$ is described by its temporal covariance function $R_i(t)$. This function can either be determined analytically for analysis/synthesis purposes or measured by the microcell terminal in the case of real-world implementation.

For a microcell packet transmission to be successful, it is not only imperative for the $CIR_i(t)$ to remain above a certain threshold but also to remain above it for at least τ_M seconds, where τ_M is the opportunistic transmission window duration. With respect to the interference level, we are hence interested in the likelihood that $I_i(t) < I_{\text{Thr}}$ for at least τ_M seconds. This likelihood is henceforth referred to as success probability of the opportunistic transmission window, P_{success}, and is well approximated by [53]:

$$\Pr(I_i(t) < I_{\text{Thr}} | \tau_M) = P_{\text{success}}(I_{\text{Thr}}, \tau_M) = \frac{\sqrt{\lambda_i}}{2\pi\sigma_i} e^{-\lambda_i \frac{I_M^2}{8}\left(\frac{\tau_M^2}{\sigma_i^2} + \frac{4}{\lambda}\right) + A\tau_M^2} \left(\tau_M e^{-A\tau_M^2} + \sqrt{\frac{\pi}{A}} Q\left(\sqrt{2A}\tau_M\right)\right) \tag{10.50}$$

where

$A = \lambda_i I_n^2/8$ per definition;
$I_n = (\mu_i - I_{Thr})/\sigma_i$ is the normalised difference between the interference mean and the threshold;
μ_i is the interference mean at the given location;
$\sigma_i^2 = R_i(0)$ is the interference variance;
$\lambda_i = -\dfrac{d^2}{dt^2} R_i(t) \bigg|_{t=0}$, $\lambda_i < \infty$ is the rate of change of the interference autocorrelation function;
τ_M is the opportunistic transmission window duration, which might be a multiple of the TTI; and
$Q(x) = \dfrac{1}{\sqrt{2\pi}} \int_x^\infty e^{-t^2/2} dt$, $x \geq 0$ is the Marcum Q-function.

In order to obtain the throughput of the NRT data traffic in the microcell, we will define the set of rates $\vec{R} = \{0, R_1, R_2, \ldots, R_K\}$ corresponding to set of interference threshold levels $I_{Thr} = \{I_{Thr,0}, I_{Thr,1}, \ldots, I_{Thr,K}\}$, where K is the maximum number of these feasible rates and at most a rate R_i is possible for $I_i(t) \leq I_{Thr,i}$. Relating energies and power, it is easy to show that $I_{Thr,i} \propto 1/R_i$, i.e. the achievable rate is inversely proportional to the interference threshold. This allows us finally to write for the average microcell throughput:

$$Throughput = \sum_{k=1}^{K} \Pr(I_{Thr,k+1} < I_i(t) \leq I_{Thr,k} | \tau_M) \cdot R_k$$

$$= \sum_{k=1}^{K} \left(P_{success}(I_{Thr,k}, \tau_M) - P_{success}(I_{Thr,k+1}, \tau_M) \right) \cdot R_k \quad (10.51)$$

The throughput, being a function of the average macrocell interference power μ_i, has been simulated by means of a sophisticated UMTS system level simulator [54] and is depicted in Figure 10.11. For the simulations, it has been assumed that standard deviation was $\sigma_i = 5.43$ dBm and $\tau_M = 10$ ms, and the set of feasible data rates was $\vec{R} = \{0, 8, 16, 32, 64, 128, 256, 512, 1024, 2048, 4096\}$ [kbps] and the associated set of macrocell interference levels was $I_{Thr} = \{\infty, -56.2, -59.2, -62.2, -65.2,$

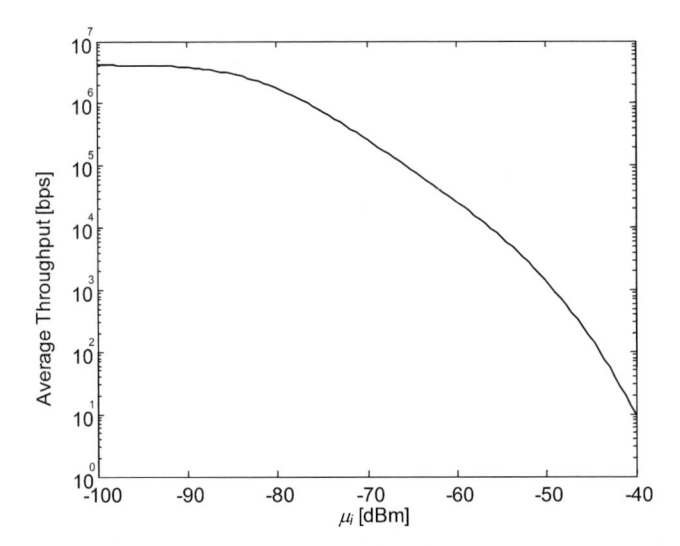

Figure 10.11 Average throughput at the microcell as a function of the average interference, μ_i.

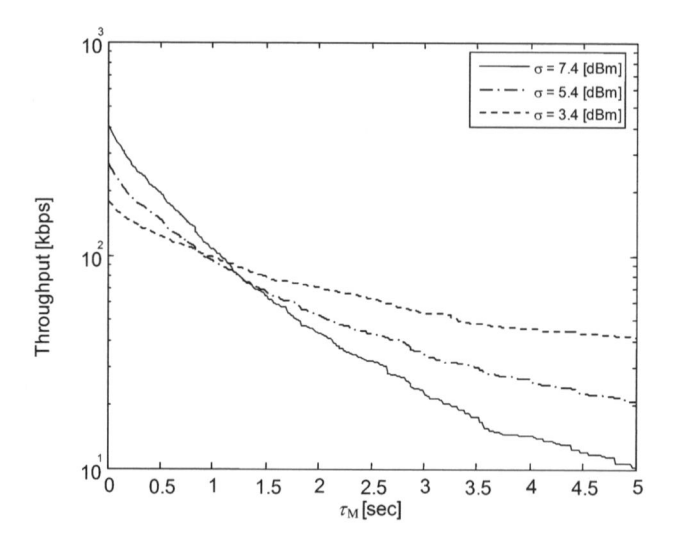

Figure 10.12 Average throughput at the microcell as a function of the opportunistic transmission window duration τ_M, parameterised on σ_i, and assuming $\mu_i = -70$ dBm.

$-68.2, -71.2, -74.2, -77.2, -80.2, -83.2, -\infty\}$ [dBm]. From the figure, one clearly observes that, as the average macrocell interference increases, the throughput in the microcell decreases. Furthermore, a macrocell interference average of $\mu_i < -90$ dBm allows data transmission with the maximum possible data rate (4096 kbps). It can also be seen that even when μ_i is around -50 dBm, a throughput of 1 kbps is still possible, which can prove vital in cell planning for NRT packet based services.

Finally, the impact of the assumed transmission window length τ_M onto the microcell throughput, parameterised on the macrocell interference variation, is depicted in Figure 10.12. It can be seen that, by increasing the interference variations, the slope of the average throughput increases; this means that the throughput under interference profiles with higher variations is more sensitive to the actual opportunistic transmission window.

As for the fourth point, given the above quantifications of the microcell opportunistic throughput, a planning of both macro and microcells can be performed according to the techniques outlined in this book. If, for above HCS arrangement, microcell to macrocell interference becomes an issue, then more sophisticated techniques have to be deployed at the microcell site. As mentioned before, this could include beamforming, multiuser detection, interference cancellation mechanisms etc.

To summarise this chapter, it was dedicated to planning and dimensioning issues of the UMTS radio access network. We have mainly discussed technical issues influencing capacity and coverage by tweaking some system related parameters. In the next chapter, we will discuss the impact of other systems – adjacent in frequency – on the overall planning process.

REFERENCES

[1] 3GPP TS 25.104, *Base Station (BS) radio transmission and reception (FDD)*.
[2] 3GPP TS 25.433, *UTRAN Iub interface NBAP signalling*.
[3] Jukka Lempiäinen, Matti Manninen, *UMTS Radio Network Planning, Optimization and QoS Management*, Kluwer Academic Publisher, 2003.
[4] Kathrein, *Technical Information and New Products* http://www.kathrein.de/de/mca/techn-infos/download/9987102.pdf.

[5] Il-Kyoo LEE, Dong-Han LEE, Jae-Young KIM, Hyun-Jin HONG, Seung-Hyeub OH, *Analysis of UE RF Parameters for 3GPP Specifications*, IEEE Proceedings of Vehicular Technology Conference, September 2002.
[6] 3GPP TS 25.101, *User Equipment (UE) radio transmission and reception (FDD)*.
[7] UMTS 30.03, *Selection procedures for the choice of radio transmission technologies of the UMTS*, version 3.2.0.
[8] Harri Holma, *A Study of UMTS Terrestrial Radio Access Performance*, Thesis for the degree of Doctor of Technology, Helsinki University of Technology Communications Laboratory, Technical Report T49, 2003.
[9] Neelesh B. Mehta, Larry J. Grenstein, Thomas M. Willis, Zoran Kostic, *Analysis and Results for the Orthogonality Factor in WCDMA Downlinks*, IEEE Transactions on Wireless Communications, vol. 2, no. 6, November 2003, pp. 1138–1149.
[10] Heinz Droste, Jürgen Beyer, *Distributions of orthogonality factor and multipath gain of the UMTS downlink obtained by measurement based simulations*, IEEE Proceedings of Vehicular Technology Conference, May–June 2005.
[11] Jaana Laiho, Achim Wacker, Tomas Novosad, *Radio Network Planning and Optimisation for UMTS,* John Wiley & Sons, Ltd/Inc., 2002.
[12] S. Burger, H. Buddendick, G. Wölfle, P. Wertz, *Location Dependent CDMA Orthogonality in System Level Simulations*, In Proc. of IEEE VTC, 2005, Stockholm, Sweden, September 2005.
[13] Klaus Ingemann Pedersen, Preben Elgaard Mogensen, *The Downlink Orthogonality Factors Influence on WCDMA System Performance*, In Proc. of VTC 2002 Fall, Vancouver, Canada, 24–28 September 2002.
[14] R.M. Joyce, T. Griparis, I.J. Osborne, B. Graves, T.M. Lee, *Soft Handover Gain Measurements and Optimisation of a WCDMA Network*, Fifth IEE International Conference on Mobile Communication Technologies (3G 2004), London, UK, 18–20 October 2004, pp. 658–662.
[15] Mario Da Silva, Yann Farmine, *W-CDMA Uplink Soft Handover Gain Measurements*, Proceedings of IEEE Vehicular Technology Conference VTC'05 Spring, Stockholm, Sweden, 2005.
[16] W. Jakes, Microwave Mobile Communications, John Wiley & Sons, Ltd/Inc., 1974.
[17] Andrew J. Viterbi, *CDMA Principles of Spread Spectrum Communication*, Addison-Wesley Wireless Communications, 1995.
[18] Kari Sipilä, Jaana Laiho-Steffens, Achim Wacker, Mika Jäsberg, *Modelling The Impact Of The Fast Power Control On The WCDMA Uplink*, Proceedings of IEEE Vehicular Technology Conference, VTC'99 Spring, Houston, USA, May 1999.
[19] Theodore S. Rappaport, *Wireless Communications, Principle and Practice*, Prentice Hall PTR, 1996.
[20] Gordon L. Stüber, *Principles of Mobile Communication*, Second Edition, Kluwer Academic Publisher, 2001.
[21] Harri Holma, Antti Toskala, *WCDMA for UMTS*, Third Edition, John Wiley & Sons, Ltd/Inc., 2004.
[22] Kari Sipilä, Zhi-Chun Honkasalo, Jaana Laiho-Steffens, Achim Wacker, *Estimation of Capacity and Required Transmission Power of WCDMA Downlink Based on a Downlink Pole Equation*, Proceedings of IEEE Vehicular Technology Conference 2000 Spring, Tokyo, Japan, May 2000, pp. 1002–1005.
[23] Stefan Brueck, *Performance Characterisation of WCDMA dedicated Traffic Channels based on Node B Transmit Power and Pilot Measurements*, Fifth IEE International Conference on 3G Mobile Communication Technologies (3G 2004), pp. 480–484.
[24] Toni Janevski, *Traffic Analysis and Design of Wireless IP Networks*, Artech House, 2003.
[25] A.M. Viterbi and A.J Viterbi, *Erlang Capacity of a Power Controlled CDMA System*, IEEE Journal on Selected Areas in Communications, vol. 11, no. 6, August 1993, pp. 892–900.
[26] J. Lee, L. Miller, *CDMA System Engineering Handbook*, Artech House, 1998.
[27] D. Molkdar, S. Burley, *On the Nature of UMTS Circuit-Switched Capacity Figures Obtained Through Monte Carlo System-Level Simulation*, fifth IEE International Conference on Mobile Communication Technologies (3G 2004), London, UK, 18–20 October 2004.
[28] Andreas Mäder, Dirk Staehle, *Analytical Modelling of the WCDMA Downlink Capacity in Multi-Service Environment*, Research Report Series, Report Number 330, April 2004, http://www3.informatik.uni-wuerzburg.de/TR/tr330.pdf.
[29] Andreas Mäder, Dirk Staehle, *Uplink Blocking Probabilities in Heterogenous WCDMA Networks considering Other-Cell Interference*, Research Report Series, Report Number 333, May 2004, http://www3.informatik.uni-wuerzburg.de/TR/tr333.pdf.
[30] Maciej J. Nawrocki, Tadeusz W. Wieckowski, *Optimal Site and Antenna Location for UMTS – Output Results of 3G Network Simulation Software*, 14th International Conference on Microwaves, Radar and Wireless Communications, MIKON-2002, 2002.

[31] Jarno Niemelä, Jukka Lempiäinen, *Impact of Base Station Location and Antenna Orientations on UMTS Radio Network Capacity and Coverage Evolution*, in Proc. IEEE 6th Int. Symp. on Wireless Personal Multimedia Communications Conf., Yokosuka, 2003, vol. 2, pp. 82–86.

[32] Jarno Niemelä, *Impact of Base Station Site and Antenna Configuration on Capacity in WCDMA Cellular Networks*, Master of Science Thesis, Tampere University of Technology, 2003.

[33] Jaana Laiho, Achim Wacker, Pauli Aikio, *The Impact of the Radio Network Planning and Site Configuration on the WCDMA Network Capacity and Quality of Service*, IEEE Proc. of Vehicular Technology Conference 2000 spring, Tokyo, Japan, May 2000, pp. 1006–1010.

[34] Jarno Niemelä, Jukka Lempiäinen, *Impact of the Base Station Antenna Beamwidth on Capacity in WCDMA Cellular Networks*, in Proc. IEEE 57th Vehicular Technology Conference, Jeju, April 2003, vol.1, pp. 80–84.

[35] Francesc Borràs Torà, *Impact of Antenna Beamwidth, Propagation Slope and Coverage Overlapping on Capacity in WCDMA Networks*, Master of Science Thesis, Tampere University of Technology, 2003.

[36] Afif Osseiran, Andrew Logothetis, *Impact of Angular Spread on Higher Order Sectorisation in WCDMA Systems*, in Proceedings of the 16th Annual IEEE International Symposium on Personal Indoor and Mobile Radio Communications September 11–14, 2005.

[37] Klaus Ingemann Pedersen, Preben Elgaard Mogensen, *Application and Performance of Downlink Beamforming Techniques in UMTS*, IEEE Communication Magazine, pp. 134–143, October 2003.

[38] Kathrein, Technical Information and New Products, Issue no. 3, September 2000, http://www.kathrein.de/de/mca/techn-infos/download/9986223.pdf.

[39] Tero Isotalo, Jarno Niemelä, Jukka Lempiäinen, *Electrical Antenna Downtilt in UMTS Network*, in Proc. 5th European Wireless Conference, Barcelona, Spain, February 2004, pp. 265–271.

[40] Jarno Niemelä and Jukka Lempiäinen, *Impact of Mechanical Antenna Downtilt on Performance of WCDMA Cellular Network*, in Proc. 59th IEEE Vehicular Technology Conference, Milano, Italy, May 2004, pp. 2091–2095.

[41] Jarno Niemelä, Tero Isotalo, Jakub Borkowski, and Jukka Lempiäinen, *Sensitivity of Optimum Downtilt Angle for Geographical Traffic Load Distribution in WCDMA*, in Proc. 62nd IEEE Vehicular Technology Conference, Dallas, USA, September 2005.

[42] Jarno Niemelä, Tero Isotalo, Jukka Lempiäinen, *Optimum Antenna Downtilt Angles for Macrocellular WCDMA network*, EURASIP Journal on Wireless Communications and Networking, 2005.

[43] E. Benner, A.B. Sesay, *Effects of Antenna Height, Antenna Gain, and Pattern Downtilting for Cellular Mobile Radio*, IEEE Transactions on Vehicular Technology, May 1996, vol. 45, No. 2, pp. 217–224.

[44] I. Forkel, A. Kemper, R. Pabst, R. Hermans, *The Effect of Electrical and Mechanical Antenna Down-Tilting in UMTS Networks*, in Proc. of IEE 3G Mobile Communication Technologies, London, UK, May 2002.

[45] S.A. Ghorashi, F. Said, A.H. Aghvami, 'Beamforming and Intelligent Scheduling for Layer Separation in HCS', U.K. Patent Application 0216-291.5, filed on 15th July 2002.

[46] Dong Hee Kim, Dong Do Lee, Ho Joon Kim, Keum Chan Whang, 'Capacity analysis of macro/microcellular CDMA with power ratio control and tilted antenna,' *IEEE Trans. Veh. Technol.*, vol. 49, no. 1, pp. 34–42, January 2000.

[47] Jung-Shyr Wu, Jen-Kung Chung, Yu-Chuan Yang, 'Performance study for a microcell hot-spot embedded in CDMA macrocell systems', *IEEE Trans. Veh. Technol.*, vol. 48, no. 1, pp. 47–59, January 1999.

[48] Cheolin Joh, Keunyoung Kim, Youngnam Han, 'Performance of a microcell with optimal power allocation for multiple class traffic in hierarchically structured cellular CDMA systems', *IEEE Veh. Tech. Conf. (VTC 2001-Spring)*, pp. 2818–2822, 2001.

[49] A. Catovic, S. Tekinay, 'Projection multiuser detectors for hierarchical cell structures in CDMA cellular systems', in *IEEE Veh. Tech. Conf. (VTC 2001-Spring)*, vol. 3, pp. 1853–1857, 2001.

[50] M.K. Simon, M-S. Alouini, *Digital Communication over fading channels*, John Wiely & Sons, Ltd/Inc., New York, 2000.

[51] A.J. Viterbi, 'CDMA: Principles of spread spectrum communications', Addison-Wesley, 1995, 2nd Edition.

[52] L. Wang, A.H. Aghvami, W.G. Chambers, 'Capacity Estimation of SIR-based Power Controlled CDMA Cellular Systems in Presence of Power Control Error', *IEICE Trans. Commun.* vol. E86-B, no. 9, September 2003.

[53] N.B. Mandayam, P. Chen, and J.M. Holtzman, 'Minimum duration outage for cellular systems: A level crossing analysis', in *Proc. IEEE VTC*, pp. 879–883, April 1996.

[54] S.A. Ghorashi, E. Homayounvala, F. Said, A.H. Aghvami, 'Dynamic simulator for studying WCDMA based hierarchical cell structures', *PIMRC 2001*, vol. 1, pp. 32–37, September 2001.

11

Compatibility of UMTS Systems

Maciej J. Grzybkowski

Spectrum resources are limited and generally lack sparse and unused frequency bands; therefore, new radio systems are introduced over previously reserved bands. The same situation prevails with the UMTS/IMT-2000 core (sometimes called initial) band, extended bands and other bands (see Section 3.1), which before and sometimes even after utilisation by 3G will be used by other systems and radio services.

In order to assure coexistence of the new system with other operating systems (sharing), whilst limiting mutual interference and assuring effective band usage, the conditions of such coexistence must be clearly defined. System coexistence conditions are determined by analysing spectrum usage and identifying the services and systems operating within the given frequency band. Next, appropriate technical evaluations take place, in order to determine the compatibility conditions for the individual services and systems. Based on this procedure, the introduction of UMTS/IMT-2000 in each of the earlier mentioned frequency bands required earlier compatibility research, which would define the coexistence conditions with respect to already existing systems. According to Recommendation ITU-R SM.1132 [1], such coexistence has been deemed possible, if at least one of the following system separation methods is possible: frequency and/or space and/or time and/or signal.

The introduction of each new public system, especially a digital one, generally is synonymous with creating a high level of interference in the EM environment. This is particularly true, when these are introduced into an environment already occupied by other systems (fixed radio-links, radio-access and mobile satellite services) where 3G systems must take into account the existing situation. A conflict-free coexistence of different telecommunication systems consists of limiting the mutual interference between them to acceptable levels, the later usually being defined by international bodies such as by the ITU internationally or CEPT and ETSI in Europe. Locating 3G alongside other existing systems was done with full awareness of the required compatibility assessments, because under certain conditions in one environment, the introduction of the new system can be 'pain-free', whilst under other conditions in another environment, this might require certain pre-existing systems to be removed from the bands of concern.

Another important aspect is assurance of compatibility within the UMTS system itself. Here, compatibility must be assured between stations operating in the terrestrial and satellite components,

---

Understanding UMTS Radio Network Modelling, Planning and Automated Optimisation   Edited by Maciej J. Nawrocki, Mischa Dohler and A. Hamid Aghvami   © 2006 John Wiley & Sons, Ltd

as well as between stations belonging to different operators – especially ones located on different sides of the country border. In each individual network, compatibility must also be assured for stations belonging to different HCS (Hierarchical Cell Structure) UTRA layers, as well as ones using different duplex modes (FDD/TDD).

11.1 SCENARIOS OF INTERFERENCE

11.1.1 INTERFERENCE BETWEEN UMTS AND OTHER SYSTEMS

Assurance of conflict-free operation of different radio services or systems using the same or adjacent frequency bands is possible by applying appropriate know-how backed by suitable national and international legal regulations. Spectrum usage is governed:

- on worldwide level – by ITU Radio Regulations [2];
- on continental level – in case of Europe, by Report 25 of European Radiocommunication Committee (The European Table of Frequency Allocations and Utilisations Covering the Frequency Range 9 kHz–275 GHz) [3];
- on national level – based on national frequency allocation tables, some of which are published for example in the EFIS (ERO Frequency Information System) system [4].

The introduction of UMTS/IMT-2000 required an analysis of its effect onto other radio system services, as well as the influence of those systems onto UMTS/IMT-2000. When evaluating the possibility of implementing UMTS/IMT-2000 at some chosen frequency bands, it is necessary to analyse the current spectrum allocation of these bands to various radio services (or systems). This analysis is performed taking into account current international and national regulations in force. This pertains to both the frequency band in question as well as the adjacent frequency bands. International regulations pertaining to spectrum allocation are drafted by the Radiocommunications Section of ITU, which are then approved at the World Radiocommunication Conferences (WRC) held every three (or four) years or the Regional Radiocommunication Conferences (RRC) (e.g. at continental level). Based on these regulations, appropriate national rules are defined, which however sometimes differ from the above.

Analysing the frequency allocation for the introduced system, as well as the adjacent frequency bands, it is possible to indicate other services and systems using these bands. Based on this analysis, an appropriate sharing matrix is created, clearly showing which services and systems require definition of inter-system compatibility conditions with respect to UMTS/IMT-2000.

Task group ERC TG1 (European Radiocommunications Committee, Task Group 1) performed such an analysis for the Euro-region before introduction of UMTS operation in the initial (core) band. Based on the previous version of European Table of Frequency Allocations and Utilisations [5], TG1 prepared the sharing matrix [6] shown in Table 11.1.

Similar actions were undertaken by project team ECC PT1 before the planned introduction of UMTS operation in the extended band. Another sharing matrix [7], presented in Table 11.2, was created after analysing the relevant part of the European Table of Frequency Allocations and Utilisations [3].

The resulting analysis of the possible influence of UMTS/IMT-2000 onto other systems operating within the same or adjacent bands in Europe has to take into consideration the following compatibility criteria:

- sharing channel compatibility in the initial band between UMTS-T and Fixed Service (FS) and in the extended band between UMTS-T and possible Mobile Satellite Service (MSS) as well as between UMTS and the Multipoint Multimedia Distribution Systems (MMDSs) used in some countries;

- adjacent channel compatibility in the initial band between UMTS-T and Digital Enhanced Cordless Telecommunication System (DECT) and FS and MSS/Space Service (both UL and DL) and in the extended band between UMTS-T and Radio Astronomy Service (RAS) as well as MMDSs.

In other regions, it is necessary to take into account the mutual interaction (both sharing and adjacent band compatibility) between IMT-2000 and other radio services/systems. For example, in the USA it is necessary to take into account adjacent band compatibility for IMT-2000/GSM in the initial band (GSM operates there at a different band as Europe), whilst in some Far-East countries the 2630–2655 MHz band is shared by terrestrial IMT-2000 and Broadcasting Satellite Service (BSS). The use of BSS in Europe is possible in accordance with ECC Recommendation [8], according to which the future use of the 2520–2670 MHz band in Europe will give priority to terrestrial UMTS/IMT-2000 systems over broadcasting satellite systems; however, 'broadcasting satellite systems may only operate over Europe provided that they do not cause harmful interference to UMTS/IMT-2000 and do not claim protection from them'.

Table 11.1 UMTS sharing/compatibility matrix. European initial band. Source [4].

Frequency bands	Allocated/assigned to...	Co-frequency band sharing with...	Adjacent band compatibility with... Lower band edge	Adjacent band compatibility with... Upper band edge
Below 1900 MHz	DECT	—	—	—
1900–1920 MHz	UMTS-T TDD	Fixed service[1]	DECT	Terrestrial UMTS FDD, Fixed service
1920–1980 MHz	UMTS-T FDD	Fixed service[1]	Terrestrial UMTS TDD, Fixed service	Mobile satellite service/UMTS-S, Fixed service
1980–2010 MHz	Mobile satellite service/UMTS-S	Fixed service[2]	Terrestrial UMTS FDD, Fixed service	Terrestrial UMTS TDD, Fixed service
2010–2025 MHz	UMTS-T TDD	Fixed service[1]	Fixed service, MSS/UMTS-S	Fixed service, Space services (E–s)
2025–2110 MHz	Fixed service, Space services (Earth–space, space–space)	—	—	—
2110–2170 MHz	UMTS-T FDD	Fixed service	Fixed service, Space services UL	Fixed service, MSS/UMTS-S
2170–2200 MHz	MSS/UMTS-S	Fixed service[2]	Terrestrial UMTS FDD, Fixed service	Fixed service, Space services (s–E)
Above 2200 MHz	Fixed service, Space services (space–Earth, space–space)	—	—	—

Notes
1. See ERC Report 64, *Frequency sharing between UMTS and existing fixed services*, Menton, May 1999.
2. See ITU-R Recommendations M.1141, M.1142, and M.1143 pertaining to sharing/compatibility between GSO/NGSO space stations operating in the mobile satellite service and stations in the fixed service.

Table 11.2 UMTS sharing/compatibility matrix. Extended band in Europe. Source [6].

Frequency bands	Allocated/assigned to...	Co-frequency band sharing with...	Adjacent band compatibility with...	
			Lower band edge	Upper band edge
Below 2500 MHz	Mobile satellite service (DL) Fixed service Mobile service Radiolocation	—	—	—
2500–2520 MHz	UMTS-T[1] Mobile satellite service (DL) (Fixed service)[2]	Mobile satellite service *Fixed service*[3]	Mobile satellite service Radiolocation[3] Fixed service[3]	*Fixed service*[3]
2520–2670 MHz	UMTS-T[1] (Fixed service)[2]	MMDS[4]	Mobile satellite service	Mobile satellite service
2670–2690 MHz	UMTS-T[1] Mobile satellite service (UL) (Fixed service)[2]	Mobile satellite service	*Fixed service*[3]	Radio astronomy service Space research service (passive)[3] Fixed service[3] Earth exploration satellite service (passive)[3]
Above 2690 MHz	Radio astronomy service[5] Space research service (passive)[5] Earth exploration satellite service (passive)[5]	—	—	—

Notes
1. In Europe the terrestrial component of UMTS/IMT-2000 will use the entire 2500–2690 MHz band, see Section 3.1.3.
2. With the implementation of UMTS/IMT-2000 in extended band in Europe, the fixed service will become secondary in appropriate parts of this band and transitional arrangements for the FS may be needed.
3. Scenarios in italics have not been examined in Report [6].
4. Within the band 2520–2670 MHz, MMDS is used in several European countries: Iceland, Ireland, Latvia and Lithuania. In some of these countries operation within 2500–2520 MHz and 2670–2690 MHz bands will be phased out.
5. Radio Regulation footnote 5.340 states: 'all emissions are prohibited' within the 2690–2700 MHz band.

The guidelines for a compatible operation of terrestrial UMTS (UTRA) systems and MSS can assume that these are separate radio services, because the UMTS satellite component is considered to be a Satellite Service (thus part of MSS), whilst UTRA can be considered as a Mobile Service [5]. However, because both the terrestrial and satellite segments belong to the IMT-2000 system family, the mutual interference between them will be treated as intra-system interference.

11.1.2 INTRA-SYSTEM INTERFERENCE

When designing the UMTS system, it is necessary to analyse possible interference between the terrestrial and satellite components, as well as intra-system interference in the terrestrial component. Analysis of the potential UTRA station locations shows that, at a given area or within a close neighbourhood, it is possible to have both base stations and mobile stations belonging to the same operator, but operating at different HCS layers or different duplex modes (FDD or TDD with its two modes – high chip rate, HCR, and low chip rate, LCR). This can lead to sharing and adjacent band and out of band interference, as well as intermodulations. In order to avoid such interference, it is important to carefully choose the location of base stations. Considering, however, that UMTS users are not necessarily stationary, it is not possible to totally exclude above interferences due to out of band and spurious emissions, intermodulation and blocking.

The problem of system compatibility is quite significant in the case of collocation, i.e. having different base station types installed at a single location (e.g. ones serving macro and micro cells or operating in FDD and TDD modes). Collocation is required in order to provide increased coverage and system capacity, which is synonymous with higher cell density. This can in turn cause receiver blocking.

When at the same or neighbouring areas there are networks belonging to different operators, it is likely that the base station locations will be totally uncoordinated. This easily leads to situations, where the base stations are collocated or located in close vicinity to each other, as it might be difficult to obtain more favourable locations for them. Considering this, base station locations and service areas should be mutually coordinated by the operators. Coordination is especially important at borderline regions, where appropriate coverage must be assured. This problem is considered in greater detail in Section 11.5.

A UTRA intra-system compatibility analysis should take into account all types of possible cell sizes: macro, micro and pico. It is, however, assumed that base station antennas for macro cells are located outdoors above rooftop, in micro cells outdoors below rooftop and in pico cells only indoors. It should also be noted that interference between stations operating in different modes usually occur in adjacent channels at a close adhering frequency f_a (Figure 11.1), which results from the frequency arrangement foreseen for UMTS. When the appropriate guard band between the adjacent TDD and FDD channels exists, this interference is reduced.

When considering compatibility between the UMTS terrestrial and satellite components, the analysis should concentrate on adjacent channel interference at the adhering frequencies $f_a = 1980\,\text{MHz}$ (FDD/MSS), $f_a = 2010\,\text{MHz}$ (TDD/MSS), $f_a = 2170\,\text{MHz}$ (FDD/MSS) in the initial band and within the range 2500–2520 MHz and 2670–2690 MHz (FDD/MSS) for the extended band. Note that this does not pertain to Europe; please, see Section 3.1.

11.2 APPROACHES TO COMPATIBILITY CALCULATIONS

The main objective of compatibility calculations is to determine the amount of interference caused by transmitters of one system into the receivers of the other system or by transmitters of a system belonging to one operator into the receivers of the same type radio system belonging to another operator. In UMTS, the intra-system compatibility problems pertain also to interaction between transmitters and receivers operating on different system levels, using adjacent channels or using different modes (FDD/TDD). In these situations, the approach consists usually of determining the probability of potential interference between the transmitters and receivers. However, in some situations it is necessary to take into account the worst-case scenario of actual interference and thus review all possible interference scenarios.

11.2.1 PRINCIPLES OF COMPATIBILITY CALCULATIONS

In order to perform calculations pertaining to compatibility between UMTS and other systems operating in the same or adjacent channels, it is necessary to create appropriate interference simulation models.

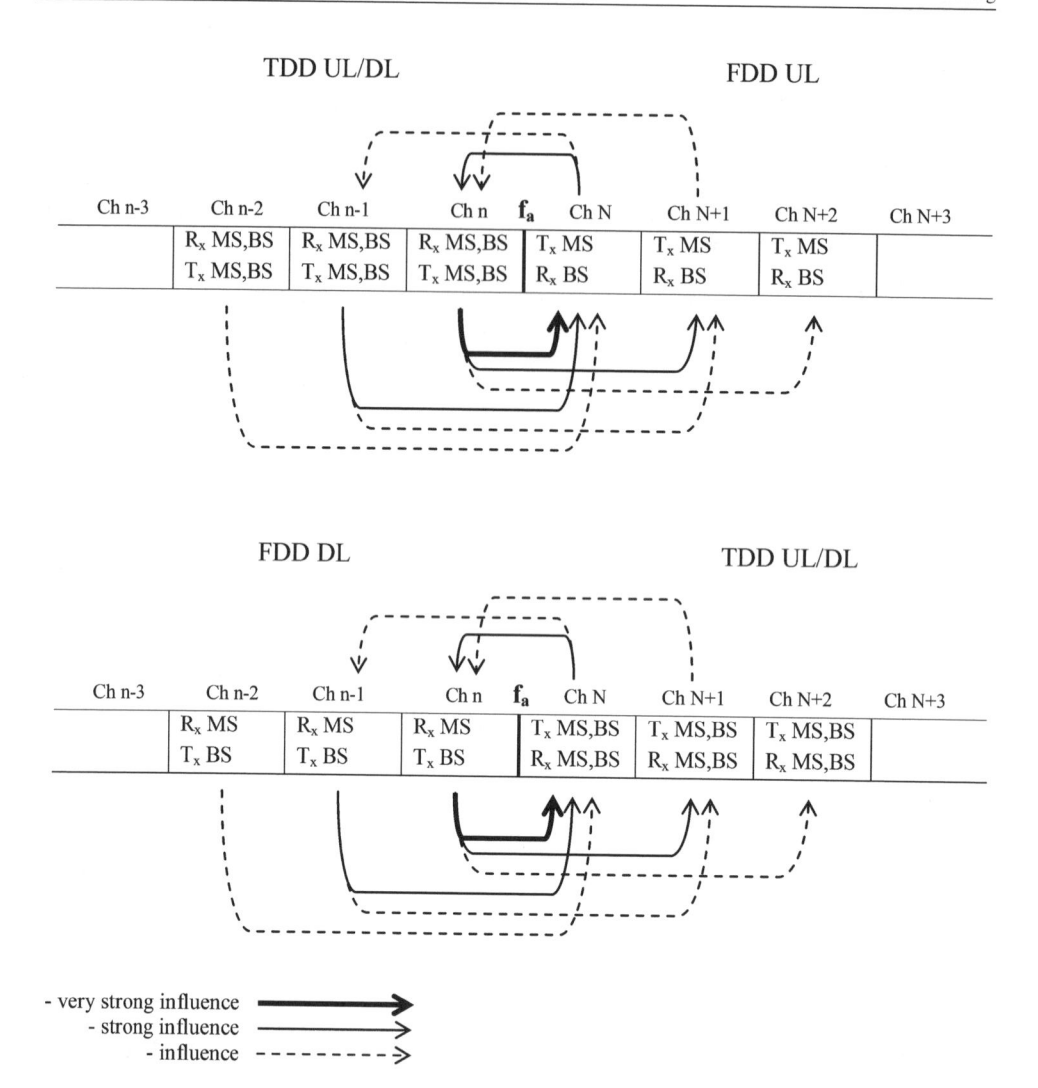

Figure 11.1 Possible interference between FDD and TDD channels at the adhering frequency f_a.

These models should take into account different path elements affecting the interference propagation. The analysis ought to include:

- transmitter causing the interference;
- receiver affected by the interference;
- transmitter and receiver antennas; and
- propagation path.

The concept of calculating the power spectral density at the receiver input takes into account different interference scenarios [9], such as combinations of a variety of signal types (having different allowable

Compatibility of UMTS Systems

interference levels) and different bandwidths. The power spectral density of a signal at the affected receiver can be determined based on the following relation:

$$I_R(f, p) = \frac{P_T(f) \cdot G_T(\varphi) \cdot G_R(\theta) \cdot PM_R(\theta) \cdot S(f)}{L_{\text{feeder,Tx}} \cdot L_{\text{feeder,Rx}} \cdot L_b(f, p)} \qquad (11.1)$$

where
$I_R(f, p)$ – power spectral density of interference signal at receiver input,
$P_T(f)$ – power spectral density at the transmitter output,
$G_T(\varphi)$ – transmitter antenna gain towards receiver,
$G_R(\theta)$ – receiver antenna gain towards transmitter,
$PM_R(\theta)$ – receiver antenna polarisation mismatch,
$S(f)$ – receiver selectivity,
$L_{\text{feeder,Tx}}$ – transmitter antenna feeder loss,
$L_{\text{feeder,Rx}}$ – receiver antenna feeder loss,
$L_b(f, p)$ – path loss based on propagation characteristics,
f – frequency,
p – time percentage,
φ – angle between transmitter antenna maximum radiation azimuth and the receiver antenna,
θ – angle between receiver antenna maximum radiation azimuth and the transmitter antenna.

Taking into account that

$$P_T(f) = P_{\text{out}} \cdot M_E(f) \qquad (11.2)$$

where
P_{out} – power level at transmitter output,
$M_E(f)$ – envelope of modulated signal at transmitter output.

Introducing the term *isolation*, $L_I(p)$, between the transmitter and receiver, formulated as:

$$L_I(p) = \frac{G_T(\varphi) \cdot G_R(\theta) \cdot PM_R(\theta)}{L_{\text{feeder,Tx}} \cdot L_{\text{feeder,Rx}} \cdot L_b(f, p)}, \qquad (11.3)$$

the power spectral density of an interference signal at the receiver input, which represents the overall level of interference and allows for assessment of all types of effects related to such interference, can be defined as:

$$P_T(f, p) = P_{\text{out}} \cdot M_E(f) \cdot L_I(p) \cdot S(f). \qquad (11.4)$$

If the required isolation value is known, it is possible to determine the required geographical separation (spacing) between the interfering transmitter and affected receiver. However, the calculated separation distance greatly depends on the chosen propagation model. The worst-case scenario with respect to interference (basis) loss occurs in free-space (basic) loss, where the attenuation $L_b(f, p)$ is the same as the free-space attenuation L_{bf}:

$$L_b(f, p) = L_{\text{bf}} = \left(\frac{4\pi d}{\lambda}\right)^2 \qquad [\text{dB}] \qquad (11.5)$$

where
d – distance between the antennas [m],
λ – wavelength (calculated for the mid-frequency of the interference signal band) [m].

In the above situation, the limit superior of the separation distance will be obtained. Interference scenarios, where many transmitters operate within the same frequency band and at the same geographical area, require an appropriate strategy for combining the interference signals reaching the affected receiver. The ITU Recommendation [9] states that in order to assess the resultant effects, it is necessary to sum the power of N interference signals and calculate the aggregated power spectral density of the interference at receiver input, given as:

$$I_1(f, p) = \sum_{I=1}^{N} I_R(f, p) \qquad (11.6)$$

Nevertheless, certain interference situations require the calculation of the value of the effective interference power level at predefined parts of the radio spectrum. In these cases, the effective interference power level at the receiver input is calculated by integrating the aggregated power spectral density over the required frequency range (f_1-f_2):

$$I_1(p) = \int_{f_1}^{f_2} I_1(f, p) \, df \qquad (11.7)$$

Furthermore, the average power spectral density of interference $P_{ds}(p)$ in the given frequency range can be calculated using the formula:

$$P_{ds}(p) = \frac{I_1(p)}{f_2 - f_1} \qquad (11.8)$$

A system is considered well designed and properly situated in a radio environment, if the effective interference power at the input of each of the receivers $I_1(p)$ is below (or equal to) the tolerable aggregated interference power in the receiver. This condition can be fulfilled by either using the above-mentioned spatial separation or by frequency separation.

Appropriate frequency separation between the transmitter and receiver bands is the key to establishing a guard band between different systems or between terminals belonging to different operators. In order to achieve maximum attenuation of interference signals within the receiver's operating bandwidth, the following relation should be fulfilled:

$$\Delta f = \frac{(2.5 \cdot B)}{2} + \frac{B_{R60}}{2} \qquad (11.9)$$

where
Δf – frequency separation,
B – bandwidth of emitted signal,
B_{R60} – width of 60 dB receiver I-F band.

In many cases, in order to achieve the required interference noise reduction, full frequency separation is not necessarily required. Depending on the spectral density of the interference signal and filter capabilities of the affected receiver, it is possible to use a smaller frequency separation Δf. This separation must fulfil the following requirement:

$$\int_{f_1}^{f_2} I_1(f + \Delta f, p) \, df \leq I_{tot} \qquad (11.10)$$

where

$I_1(f + \Delta f, p)$ – aggregated power spectral density of interference at receiver input,
f_1 – lower edge frequency 60 dB I-F bandwidth of receiver,
f_2 – upper edge frequency 60 dB I-F bandwidth of receiver,
Δf – frequency separation,
I_{tol} – allowable aggregated interference power at the receiver.

The peak interference level is determined in the cases where transmitters and receivers are equipped with either high-gain and/or rotating antennas. In this case, the calculations are simplified, as they assume transmitter and receiver antenna gains at an azimuth of the maximal radiation. The maximum power spectral density of interference at the receiver input can be determined in this case using Equation (11.1), but ignoring parameters $PM_R(\theta)$ and $S(f)$ and assuming the maximal transmitter and receiver antenna gains.

The transmitter models used for the calculations must take into account the fundamental emissions of out-of-band harmonic and non-harmonic as well as wideband parasitic emissions. The spectral density of power emitted by the transmitter is given in the form of a transmitter spectrum mask. However, considering the complexity of the signal emissions spectrum, the interference assessment process can use a simplified model of such a mask. According to the ITU Radio Regulations [2], fundamental emissions should be identified as the transmitter's emission spectrum within the frequency bounds covering 250 % of the so-called necessary bandwidth. Outside of this fundamental emissions-band, additional spurious emissions occur. The spectral power density attenuation distribution is defined as a function of frequency offset.

The receiver models should also take into account the susceptibility to co-channel, adjacent channel and out-of-band interference, as well as the spurious response rejection, resistance to blocking or desensitisation (reduction of the signal to noise ratio after exceeding power saturation) as well as intermodulation aspects. Receiver selectivity is also a crucial parameter (both the HF and IF rejection). If technical (or measurement) data is not available, a good way for defining the receiver selectivity characteristic is using the ratio of the so-called 60 dB band (outside of which input signals are attenuated by at least 60 dB) to the 3 dB band (within which the input signals are attenuated by at most 3 dB), referred to as the shape factor.

Antenna patterns used for interference calculations (e.g. for cross-border coordination) should be determined based on the following sources:

- Manufacturer information;
- Recommendations of ITU-R (and/or CEPT Recommendations in Europe);
- Technical standards (e.g. ETSI);
- Vilnius Agreement [10].

The Vilnius Agreement led to standardisation of antenna characteristics. Depending on their patterns, they have been classified into groups. A more detailed characteristic pattern is defined through a special set of codes developed for this purpose. This code allows, for example, estimating the original antenna pattern with good accuracy and for calculating the interference range at different azimuths. If necessary, polarisation effects should also be taken into account.

Compatibility calculations must also take into account deployment and interference scenarios. Radio-communication network deployment is especially important in view of possible concentration of local stations, which, if operating in the same radio channel, can lead to various unexpected interference situations. The target network density should also be defined in order to assess future interference levels. The interference scenarios should clearly distinguish between base stations (BS) and mobile stations (MS), because of their different operation specifics. The interference simulation procedures must take into account aspects such as the station being mobile or fixed, access times as well as different penetration (user saturation) levels at different regions, as this may require the application of different interference calculation procedures.

An operation of all radiocommunication stations using the same or adjacent frequency bands as the potentially affected (receiving) stations should be evaluated in terms of:

- calculation of isolation (separation) between the examined stations in all possible UMTS user activity combinations and the potentially affected system;
- calculation of aggregated interference power spectral density at the receiver input of the examined stations.

The aggregated interference power spectral density distribution at the receiver input allows for assessment of the probability of disturbances with respect to the maximum allowable interference levels. Depending on the requirements defined for the system, to which UMTS should be compatible, it is possible to use one of the concepts below for evaluation of interference at the receiver input [9]:

- I/N concept, where the allowable level of interference is considered with respect to the internal receiver noise, defining a certain C/N coefficient degradation;
- C/I concept, where the allowable level of interference is considered with respect to the required level of usable signal.

Among the multitude of methods used for evaluating interference between elements of different radio communication systems as well as between elements of the same radio communication system, two methods seem to be most important. They are respectively called Minimum Coupling Loss (MCL) and Monte Carlo (MC). The MCL method is usually applied for a preliminary assessment of frequency sharing possibilities and it is suitable for static interference situations, e.g. between radio link stations and base stations of a mobile system or between base stations of mobile systems. The results obtained with this method concern the worst cases of the interference scenarios. The MC method is applicable where there are moving radio terminals. This statistical method enables one to get more realistic results; however, in some cases they must be treated with precaution. Accurate results are obtained principally only when the probability distributions of all the input parameters are well known. The MC method is more complicated in terms of application than MCL.

11.2.2 MINIMUM COUPLING LOSS (MCL) METHOD

The Minimum Coupling Loss (MCL) determines the minimum attenuation (isolation) between a pair of radio terminals securing that the interference emitted by the interfering transmitter does not deteriorate reception of the victim receiver. That attenuation is evaluated on the radio path between the antennas of those terminals, taking into account the antenna gain measured between the antenna connectors (gain includes the feeder loss). If the value of the minimum attenuation is known, it is possible to evaluate, using a certain propagation method, the minimum separation distance. It is also possible, using appropriate protection coefficients, to calculate the minimum guard-band. Since this is the worst-case analysis, the results are inefficient with respect to both distance and spectrum. While evaluating the minimum attenuation, no fading in the propagation path and no statistics of interfering transmitter locations in the terrain are taken into account.

Computations with the MCL method are performed assuming that the receiver operates at 3 dB above its reference sensitivity. Here, interference cannot exceed the interference threshold. Formulas employed in the MCL analysis refer to two interference scenarios [11]:

- unwanted emission (Figure 11.2) and
- receiver blocking (Figure 11.3)

Compatibility of UMTS Systems 281

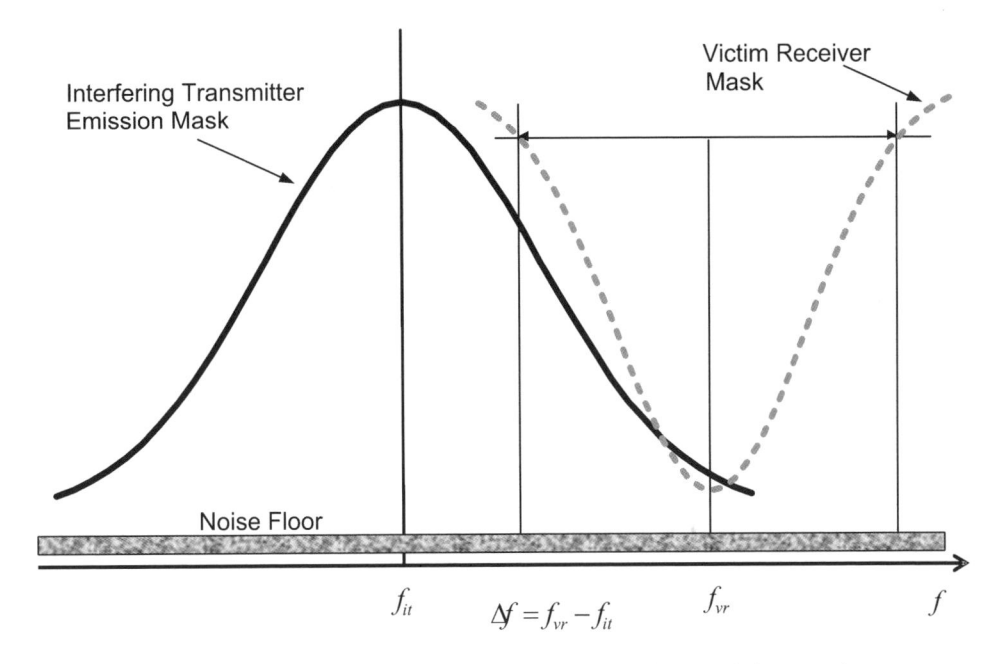

Figure 11.2 Victim receiver absorbs of unwanted emission from interfering transmitter.

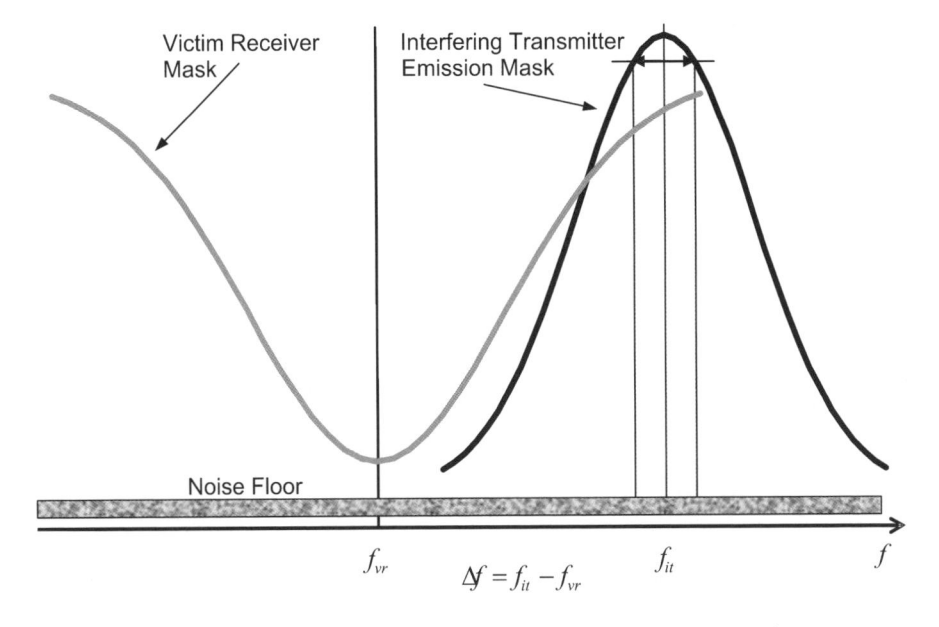

(f_{it}, $\Delta f = f_{vr} - f_{it}$, f_{vr})

Figure 11.3 Receiver blocking by a strong signal of the interfering transmitter.

(f_{vr}, $\Delta f = f_{it} - f_{vr}$, f_{it})

In the case of analysing the unwanted emission being absorbed by a receiver, the isolation loss L_i [dB] between an interfering transmitter (Ti) and a victim receiver (Rv) is expressed as:

$$L_i = P_{Ti} + \text{BWc} + M_{mc} + G_{Ti} + G_{Rv} - I_{TRv} + \text{PWo} \qquad (11.11)$$

where

P_{Ti} – maximum transmitted power of the interfering transmitter [dBm],
BWc – bandwidth conversion factor between interfering transmitter and victim receiver [dB],
M_{mc} – multiple carrier margin in case when the interferer is a base transmitter and transmits more than a single carrier [dB],
G_{Ti} – gain of the interfering transmitter antenna (including feeder loss) [dBi],
G_{Rv} – gain of the victim receiver antenna (including feeder loss) [dBi],
I_{TRv} – interference threshold of the victim receiver [dBm],
PWo – function defining the power of the wideband noise at the frequency offset being considered relative to the carrier power of the interfering transmitter [dB].

The bandwidth conversion factor and the function defining the power of the wideband noise are determined by spectrum masks of the interfering transmitter and victim receiver. Interference thresholds may be determined as the difference between sensitivity and protection ratio C/I of the victim receiver. In the case that the interfering transmitter (system) and the victim receiver (system) operate within the same bandwidth, BWc $= 0$ and the adjacent band isolation is expressed by the adjacent channel interference ratio (ACIR). ACIR is defined as the ratio of the total power transmitted from the interferer to the total interference power affecting the victim receiver, resulting from filtering imperfections both of transmitter and receiver. In linear terms [12–14], we obtain:

$$\text{ACIR} = \frac{1}{\frac{1}{\text{ACLR}} + \frac{1}{\text{ACS}}} \qquad (11.12)$$

where

ACLR – adjacent channel radio leakage ratio of interfering transmitter,
ACS – adjacent channel selectivity.

The ACIR is determined only in cases of standard frequency carrier separations (e.g. 5, 10, 15 MHz). Then,

$$L_i = P_{Ti} + M_{mc} + G_{Ti} + G_{Rv} - I_{TRv} - \text{ACIR} \qquad (11.13)$$

After computing the isolation loss, one can evaluate the minimum separation distance between the interfering transmitter and the victim receiver:

$$L_i \rightarrow \text{Propagation model} \rightarrow d_{\min},$$

or minimum guardband separation

$$L_i \rightarrow \text{Propagation model} \rightarrow d_{\min} \rightarrow f_{\min}.$$

Blocking is a measure of the victim receiver's capability of correctly receiving the desired signal in the presence of (strong) signals emitted by the interfering transmitter at frequencies other than those of spurious response or adjacent channels. In the case of blocking of the victim receiver, the isolation loss L_i may be obtained from:

$$L_i = P_{Ti} + M_{mc} + G_{Ti} + G_{Rv} - b_{Ro} \qquad (11.14)$$

where

$P_{Ti}, M_{mc}, G_{Ti}, G_{Rv}$ – parameters same as in (11.11),
b_{Ro} – blocking performance of the victim receiver at specified frequency offset.

Also, in the following case, after computing the isolation loss, the minimum separation distance or the necessary frequency separation between the interfering transmitter and the victim receiver can be evaluated. Nevertheless, the separation values calculated with the MCL method are often overestimated (corresponding to the worst case) and in many cases unrealistic.

11.2.3 MONTE CARLO (MC) METHOD

The Monte Carlo approach is a statistical technique, the concept of which is based on the interference scenarios modelling while calculating the probability of the reception interference. The term 'Monte Carlo' was used for the first time by the American scientists – Janos von Neumann coming from Hungary and Stanislaw Ulam coming from Poland, during the World War II, as they made research towards the development of the atomic bomb. The Monte Carlo method employs the simulation of random processes by taking samples of random variables with previously defined probability density functions. In analysing the interference to radio reception, the statistics of both the desired and the interfering signals are determined. Here, the straightforward analysis of the minimum-carrier-to-interference ratio at the receiver input is insufficient.

A population of interfering transmitters randomly located around the victim receiver is under consideration. However, only some of these transmitters can appear intrusive to the reception of the receiver in question. These are the ones, the unwanted emissions of, which invade into the receiving band of the victim receiver. The transmitters must have big enough radiated power at their disposal and the attenuation of radio paths (transmitter–receiver) selected from the whole population must be sufficiently small. An example of the Monte Carlo simulation process, where isolation loss is being calculated when a number of transmitters interfere a receiver, is shown in Figure 11.4 [15,16].

In this case, the isolation loss evaluated for each event (random sample) depends on the transmitter and the receiver location with respect to each other, on the radio wave propagation and on

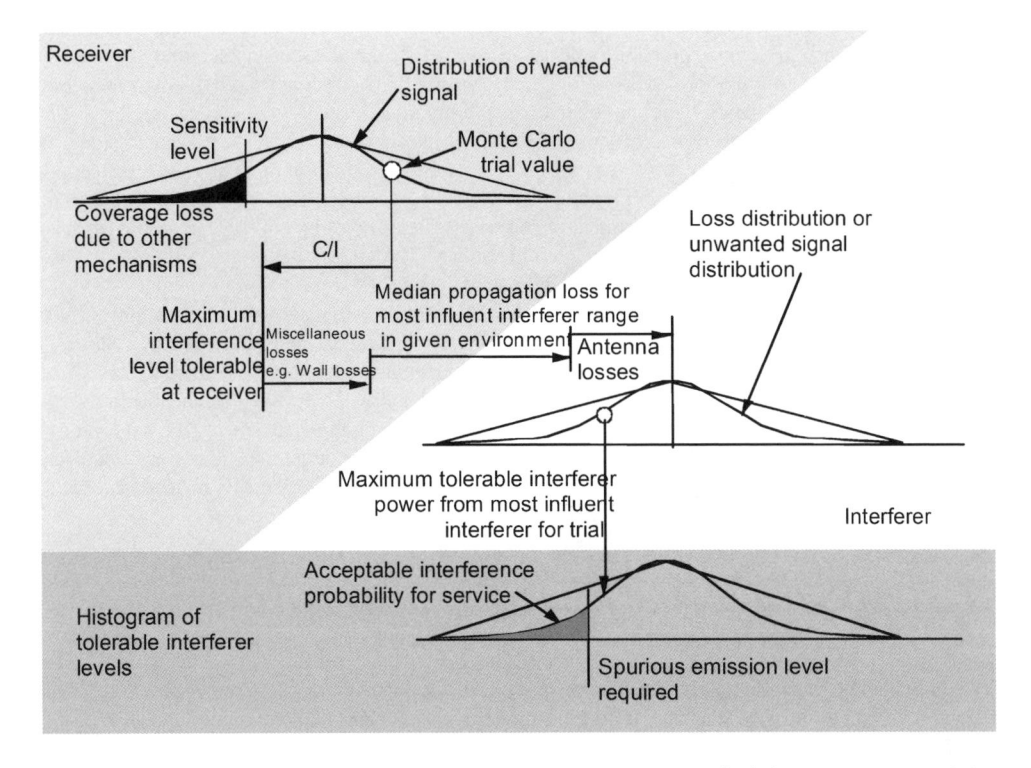

Figure 11.4 An example formulation of the Monte-Carlo simulation process [15] (Reproduced by permission of ITU).

positioning with respect to each other of the transmitting and receiving antennas (antennas directional patterns). The minimum attenuation obtained determines the maximum power of an interfering signal generated by one of the transmitters. Next, for many events, a histogram of the acquired levels of the interfering signals is determined. With the given interference probability and with the known receiver's endurable interference level, it is possible to evaluate, for the receiver, the maximum tolerable interfering signal power. Power control of the interfering transmitters and soft handover gains at the cell borders (where interference are the strongest) are taken into consideration here.

The acceptable interference probability depends on the chosen model of interference scenario. For example, 2 % interference probability between the DECT and UMTS systems is considered as the maximum value. The Monte Carlo method of interference simulation yields the interference probability which is a measure in the performance evaluation of radio systems.

The state administration group of CEPT, together with the members of ETSI and international scientific bodies, worked out and inured the implementation of the simulation model of the Monte Carlo method in the form of a computer tool called SEAMCAT® (Spectrum Engineering Advanced Monte Carlo Analysis Tool) [15,16]. SEAMCAT is being distributed by ERO (European Radiocommunications Office) [17]. The recently used version is SEAMCAT-2 and the beta-version SEAMCAT-3 is available for tests as of 2005.

The SEAMCAT tool consists of four processing engines: the event generation engine, the distribution evaluation engine, the interference calculation engine and the limits evaluation engine. Particular engines generate trials for the desired and interfering propagation paths. It takes into account such effects as unwanted emission generation (in transmitters), their elimination (in receivers), wideband noise, background noise (antenna and man-made), intermodulation, receiver blocking, generation and elimination of co-channel and adjacent channel interference. These effects are ascribed to three main categories of interference mechanisms: unwanted emission, intermodulation and receiver sensitivity. The quantity and statistics of samples is determined by means of establishing the distribution function as well as correlation between the desired signal and various kinds of interfering signals and phenomena. All the interference probability distributions are calculated with consideration of site interfering signal, related to emission masks of interfering transmitters, receiver mask (blocking process) or intermodulation attenuation. The cumulative probability functions of interference can be evaluated for the following parameters: C/I (wanted signal to interference), $C/(N+I)$ (wanted signal to noise + interference) and I/N or $N/(N+I)$. Finally, there is an evaluation of limit values for parameters, such as unwanted blocking or intermodulation levels, by applying an optimisation algorithm. These limit values are analysed in the context of spectrum efficiency, radio coverage or traffic capacity for the studies of internal and external system compatibility. The main criterion of optimisation is the cost function, which depends upon the used radio parameters.

The outcome of simulations executed using the SEAMCAT tool is the interference probability which is a measure of the degradation of system range and capacity. However, the results must be cautiously interpreted in view of the accuracy of modelling the interference situation. Not all receivers under consideration are and can be interfered with the calculated probability. Neither is it reasonable to believe that, determined by the probability function, a percentage of receivers shall always be subjected to interference. Nevertheless, calculation results facilitate to assess, for example, the necessary frequency separation between two systems operating in the same area. Unlike MCL, the MC method enables to consider the shadow fading in radio propagation environments.

11.2.4 PROPAGATION MODELS FOR COMPATIBILITY CALCULATIONS

The radio path propagation loss between a transmitter causing interference and the potentially affected receiver (as well as interference field strength) is the basic component of all compatibility calculations, especially when it comes to coordination aspects. When creating the scenario of potential interference it is important to use an appropriate, standardised propagation model. Such a model should be based on topographic data (terrain) and coverage, which allows for accurate assessment of the level of interference at the receiver.

In the case of compatibility calculations between the UMTS terrestrial and satellite components (not covered in this book) several other factors have to be taken into account considering the earth–space propagation specifics.

Depending on the scenario of potential interference (location of transmitters and receivers, which are subject to coordination), there are different application-dependent propagation models possible:

- point to point;
- point-to-multipoint (area);
- signal penetration inside buildings.

In general, the propagation models used in the calculations related to compatibility should consider the instability of propagation phenomena. Then, the statistical methods must be used (see Section 5.5). However, statistical models not always have to be used. In static interference situations, e.g. in a radio path between base stations, one can use deterministic models. In particular, the deterministic models can be used in calculations made with the MCL method. As is it well known, interference level calculations employing the free-space model lead to the 'worst case', resulting in the biggest location and frequency separations.

The selection of propagation models for calculations is very important and it should be made individually in each case of analysis. However, in the most general situations, application of the following models is recommended.

(a) **Point-to-point models**

If minute resolution data concerning terrain are available and if the distance between the interfering transmitter and the victim receiver exceeds 20 km, then the propagation loss is to be calculated using the method ITU-R Rec. P. 452 [18], please see also Section 5.5.2 of this book, or by applying a combination of the methods ITU-R Rec. P. 452, P. 526 [19] and P. 676 [20] Spherical Diffraction Model used in SEAMCAT. These models are applicable in the open terrain and they take into account the curvature of the earth surface. If the distances are smaller than 20 km, one may use the models COST Okumura-Hata or COST Walfisch Ikegami (restricted to 2000 MHz [21]) or implemented in SEAMCAT model Modified Hata [13,15,16].

(b) **Point-to-area models**

A typical propagation model, which may be recommended in this case, is the model ITU-R Rec. P. 1546 [22]. Extensive information about its application is given in Section 5.5.1.

(c) **Lodging-penetration models**

If a base station or a UMTS mobile station or a station of any other radio service is placed inside a building, additional path attenuations are to be expected because the radio waves have to cross the main walls, the non-bearing walls and the floors. Here, the attenuation value is strictly dependant on the building's structure. Because of the diversity of the building structures, it is not possible to determine exact values of all the possible parameters concerning the radio waves building penetration.

In the literature, e.g. in the documents of COST [21], CEPT [6,13,16], ETSI [23] or ITU [15], standard propagation scenarios are recommended to be used. This enables a fairly realistic representation of the radio path attenuations in a building development environment. In each particular situation, one of the following models is to be used: combined indoor–outdoor models or indoor–indoor models for the same or different buildings.

(d) **Space-to-Earth and Earth-to-space models**

For the assessment of interference between terrestrial and satellite stations, the propagation model may be used as presented in the Recommendation ITU-R P. 619 [24]. This model, which in fact is a supplement of the free-space propagation model, comprises three propagation mechanisms: propagation in the clear air (taking into account, e.g. gaseous absorption, cloud attenuation, tropospheric scintillations, ducting and ground/building reflections), precipitation scatter and differential attenuations on the adjacent radio paths Earth–space. In calculations with minor accuracy, one may use the attenuation model of free space neglecting the other factors.

In compatibility related calculations one may use other (and similar) than above-mentioned propagation models, e.g. given in ITU-R Rec. M.1225 [25]. However, in the cases of intersystem compatibility analyses or the compatibility analyses of systems belonging to different operators, the methods commonly recognised and accepted in the international forum should be used.

11.2.5 CHARACTERISTICS OF UTRA STATIONS FOR THE COMPATIBILITY CALCULATIONS

The characteristics of UTRA stations which are shown in this section (Table 11.3) should be used in frequency sharing and interference analysis studies for UMTS and between UMTS and other systems. These characteristics are taken from the ITU Report ITU-R M.2039 [26].

11.3 INTERNAL ELECTROMAGNETIC COMPATIBILITY

Internal compatibility in UMTS is important if the operator, or several operators, uses either the FDD or TDD interface in a certain location or in nearby locations, and yet operate in the same or adjacent frequency bands. The system planning must then be done very carefully. One has to consider the necessity of applying the geographical separation (in distance) or frequency separation between the stations. In the case of systems installed near the country border it is necessary to take into account the possibility of mutual impact with the foreign stations. The cross-border coordination of locations and station parameters is then indispensable (see Section 11.5).

The requirements of internal compatibility concern equally the terrestrial as the satellite components of UMTS/IMT-2000. Here, only the terrestrial component is considered. However, if in some area both components, the terrestrial and the satellite, are developed concurrently then the requirements of their mutual compatibility must be considered. Shared as well as adjacent band compatibilities are to be considered. These problems are dealt with in e.g. ERC Report 65 [6] for the core band 2 GHz, and in ITU-R Report M. 2041 [13] for the extended band 2.5 GHz. In cases when the UMTS/IMT-2000 systems will use high-altitude platform stations (HAPSs), the methodology of evaluation of co-channel interference, as well as a separation distance between base stations is presented in the ITU-R Recommendation M.1641 [34]. Calculations performed according to this methodology have revealed the required sizes of separation distances. They can be just a few kilometres in the case if the number of users per cell is small (less than 100) or if the transmission power of the HAPS is small (less than 50 mW), and they can reach some tens of kilometres in the case if there are some hundreds of users or the transmission power higher than some tens of mW.

Concerning problems of internal compatibility of UMTS terrestrial component, for both TDD or FDD modes, the following impact scenarios exist MS ↔ MS, MS → BS, BS → MS and BS ↔ BS. It is also essential in which HCS (macro/micro/pico) layer the analysed base or mobile stations are placed, and whether the stations are located indoors or outdoors. However, for co-existence studies not all scenarios are essential. Out of the contents of the document 3GPP TR 25.942 [35], it follows

Table 11.3 Characteristics of terrestrial UMTS/IMT-2000 base and mobile stations for frequency sharing/interference analysis. Source [26].

Parameter	UMTS/IMT-2000 FDD Base station See [27,28]			UMTS/IMT-2000 FDD Mobile station See [29]	UMTS/IMT-2000 CDMA TDD Mobile station						UMTS/IMT-2000 CDMA TDD Mobile station	
					1.28 Mchip/s low chip rate See [30,31]			3.84 Mchip/s high chip rate See [30,31]			1.28 Mchip/s low chip rate See [32]	3.84 Mchip/s high chip rate See [32]
Carrier spacing	$5\,\text{MHz} \pm n \times 0.2\,\text{MHz}$			$5\,\text{MHz} \pm n \times 0.2\,\text{MHz}$	$1.6\,\text{MHz} \pm n \times 0.2\,\text{MHz}$			$5\,\text{MHz} \pm n \times 0.2\,\text{MHz}$			$1.6\,\text{MHz} \pm n \times 0.2\,\text{MHz}$	$5\,\text{MHz} \pm n \times 0.2\,\text{MHz}$
Duplex method	FDD			FDD	TDD			TDD			TDD	TDD
Cell type	Macro	Micro	Pico		Macro	Micro	Pico	Macro	Micro	Pico		
Transmitter power, dBm (typical)[1]	43	38	24	20	43	tbd	tbd	43	tbd	tbd	20	20
Transmitter power, dBm (maximum)				24 or 21							24 or 21	24 or 21
Antenna gain[2] (dBi/120° sect.)	17	5	0	0	17	5	0	17	5	0	0	0
Antenna height (m)	30	5	1.5	1.5	30	5	1.5	30	5	1.5	1.5	1.5
Tilt of antenna (deg. down)	2.5	0	0	0	2.5	0	0	2.5	0	0		
Access techniques	CDMA			CDMA	TDMA/CDMA			TDMA/CDMA			TDMA/CDMA	TDMA/CDMA

Table 11.3 (continued)

Parameter	UMTS/IMT-2000 FDD Base station See [27,28]	UMTS/IMT-2000 FDD Mobile station See [29]	UMTS/IMT-2000 CDMA TDD Mobile station		UMTS/IMT-2000 CDMA TDD Mobile station	
			1.28 Mchip/s low chip rate See [30,31]	3.84 Mchip/s high chip rate See [30,31]	1.28 Mchip/s low chip rate See [32]	3.84 Mchip/s high chip rate See [32]
Data rates supported	Pedestrian: 384 kbit/s, Vehicular: 144 kbit/s, Indoors: 2 Mbit/s Higher data rates up to 10 Mbit/s are supported by technology enhancements (HSDPA), See [33]	Pedestrian: 384 kbit/s, Vehicular: 144 kbit/s, Indoors: 2 Mbit/s Higher data rates up to 10 Mbit/s are supported by technology enhancements (HSDPA), See [33]	Pedestrian: 384 kbit/s, Vehicular: 144 kbit/s, Indoors: 2 Mbit/s Higher data rates up to 2.8 Mbit/s are supported by technology enhancements (HSDPA), See [33]	Pedestrian: 384 kbit/s, Vehicular: 144 kbit/s, Indoors: 2 Mbit/s Higher data rates up to 10.2 Mbit/s are supported by technology enhancements (HSDPA), See [33]	Pedestrian: 384 kbit/s, Vehicular: 144 kbit/s, Indoors: 2 Mbit/s Higher data rates up to 2.8 Mbit/s are supported by technology enhancements (HSDPA), See [33]	Pedestrian: 384 kbit/s, Vehicular: 144 kbit/s, Indoors: 2 Mbit/s Higher data rates up to 10.2 Mbit/s are supported by technology enhancements (HSDPA), See [33]
Modulation type	QPSK	HPSK[3]	QPSK/8PSK	QPSK	QPSK/8PSK	QPSK
Emission bandwidth	See [27]	See [29]	See [30]	See [30]	See [32]	See [32]
Receiver noise figure (worst case) (dB)	5 for macro BS	9	7 for macro BS	5 for macro BS	9	9
Thermal noise in specified bandwidth[4] (dBm)	−103 in 3.84 MHz for macro BS	−108 in 3.84 MHz	−106 in 1.28 MHz for macro BS	−103 in 3.84 MHz for macro BS	−113 in 1.28 MHz	−108 in 3.84 MHz
Receiver thermal noise level (dBm)	−98 in 3.84 MHz for macro BS	−99 in 3.84 MHz	−99 in 1.28 MHz for macro BS	−98 in 3.84 MHz for macro BS	−104 in 1.28 MHz	−99 in 3.84 MHz
Receiver bandwidth	<5 MHz (See [27])	See [29]	<1.6 MHz (See [30])	<5 MHz (See [30])	See [32]	See [32]
E_b/N_o for $P_e = 0.001$	See [27]					
Receiver reference sensitivity (dBm)	−121 (See notes 5,6)	−111 (See note 5)	−111 (See note 5)	−117 in 3.84 MHz[7]	−110 (See note 5)	−96 (See note 5)	−96 (See note 5)	−109 (See note 5)	−95 (See note 5)	−95 (See note 5)	−108 in 1.28 MHz[7]	−105 in 3.84 MHz[7]

	−109 in 3.84 MHz for macro BS[9]	−105 in 3.84 MHz	−112 in 1.28 MHz for macro BS	−109 in 3.84 MHz for macro BS	−110 in 1.28 MHz	−105 in 3.84 MHz
Interference threshold[8] (dBm)						
Transmitter ACLR	See [27,28]	See [29]	See [30]	See [30]	See [32]	See [32]
1st adjacent channel	45 dB @ ±5 MHz	33 dB @ ±5 MHz	40 dB @ ±1.6 MHz	45 dB @ ±5 MHz	33 dB @ ±1.6 MHz	33 dB @ ±5 MHz
2nd adjacent channel	50 dB @ ±10 MHz	43 dB @ ±10 MHz	45 dB @ ±3.2 MHz	55 dB @ ±10 MHz	43 dB @ ±3.2 MHz	43 dB @ ±10 MHz
Transmitter spurious emissions	See [27]	See [29]	See [30]	See [30]	See [32]	See [32]
Receiver ACS (dBm) (relative ACS) (dB)	−52 (46)[10] −42 (46)[10]	(33)	−52 (46)[10] −41 (46)[10]	−52 (46)[10] −38 (46)[10]	(33)	(33)
Receiver blocking levels	See [27,28]	See [29]	See [30,31]	See [30,31]	See [32]	See [32]

Notes (See [26])
1. May not be appropriate for all scenarios, for example when calculating aggregate interference from all users in a cell.
2. Feeder losses are not included in the values and should be considered in the sharing/compatibility issues.
3. Hybrid Phase Shift Keying: a method peculiar to UMTS/IMT-2000 FDD in which the peak to average ratio is reduced in comparison to a QPSK signal by mixing the orthogonal variable spreading factor (OSVF) with both information sources as real signals, i.e. those destined for I and Q modulation components, and then shifting one component by 90° to produce an equivalent imaginary signal and then utilising gain control on the Q channel to preserve orthogonality.
4. $10\log(kTb) + 30$ (dBm), where k = Boltzman's constant = $1.3807 \cdot 10^{-23}$ J/K, T = reference temperature = 277 K, b = noise equivalent bandwidth (Hz).
5. For a 10^{-3} raw bit error rate, theoretical E_b/N_0.
6. The thermal noise figure for a WCDMA receiver is −108 dBm based on kTf, where k is Boltzman's constant ($1.3807 \cdot 10^{-23}$ J/K), T(K) is the temperature in Kelvin, and f is the bandwidth in Hertz. For a noise figure of 4 dB (typical value for a base station receiver), the thermal noise becomes −104 dBm. However, receiver sensitivity depends on the service (i.e. voice, packet, etc.). For example, the voice (DTCH 32) sensitivity for the base station receiver is −121 dBm for BER <0.001.
7. For a 10^{-3} raw bit error rate, $\hat{I}_{or}$, the received power spectral density (integrated in a bandwidth of $(1+a)$ times the chip rate and normalized to the chip rate) of the downlink signal as measured at the MS antenna connector.
8. $I/N = -6$ dB for a 10 % loss in range applicable to cases where interference effects a limited number of cells, e.g. sharing (international coordination) with BSS (sound) in the 2630–2655 MHz band, a value of $I/N = -10$ dB is appropriate.
9. The tolerable I/N thresholds are as follows: coordinated use (−6 dB), agreement trigger (−10 dB), licence exempt (−20 dB).
10. The absolute ACS values are the test values as specified in [27] and [30]. The following conversion formula: ACS_relative = ACS_test − Noise_floor − $10 \ast \log_{10}(10^{M/10} - 1)$, can be used to derive relative ACS values, where M is the margin expressed in dB used in the ACS test, which is the useful signal level above the reference sensitivity level. For both UMTS/IMT-2000 FDD and UMTS/IMT-2000 TDD, $M = 6$ dB. ACS relative values are often used in sharing studies.

that in the core band (where one adhering frequency, $f_{a1} = 1920\,\text{MHz}$, is being only) one has to consider:

1. For scenario MS ↔ MS

   a. stations located 'near-far'

   - FDD → TDD at adhering frequency (macro/micro, macro/pico)
   - TDD → FDD at adhering frequency (micro/micro, pico/pico)
   - TDD → TDD (micro/micro, pico/pico) for nonsynchronised networks

   b. all stations co-located

2. For scenario MS → BS

   a. Inter-operator guard band (uncoordinated deployment)

   - FDD macro → FDD macro/micro/pico (indoors)
   - FDD micro → FDD pico (indoors)
   - TDD macro → TDD macro/micro/pico (indoors)
   - TDD micro → TDD pico (indoors)
   - FDD macro → TDD macro/micro/pico at adhering frequency
   - FDD micro → TDD micro/pico at adhering frequency

   b. Intra-operator guard bands

   - FDD macro → FDD macro (co-located)/micro/pico (indoors)
   - FDD micro → FDD pico (indoors)
   - TDD macro → TDD macro/micro/pico (indoors)
   - TDD micro → TDD pico (indoors)
   - FDD macro → TDD macro/micro/pico at adhering frequency
   - FDD micro → TDD micro/pico at adhering frequency

3. For scenario BS → MS

   a. stations located 'near-far'

      i. Inter-operator guard band (uncoordinated deployment)

      - FDD macro → FDD macro
      - TDD macro → TDD macro
      - TDD macro → FDD macro at adhering frequency

      ii. Intra-operator guard bands

      - FDD macro → FDD micro
      - TDD macro → TDD micro
      - TDD macro → FDD macro at adhering frequency

   b. all stations co-located

4. For scenario BS → BS

   - TDD → FDD at adhering frequency (macro/micro, macro/pico)
   - TDD → TDD for nonsynchronised networks (micro/micro, pico/pico).

In all of the above-mentioned impact scenarios, a maximum of 2% acceptable probability of interference is assumed. Only for scenario BS → MS, a 10% maximum acceptable loss of system capacity is determined. The analysis of mutual impact between the particular stations must contain the study of all the most essential parameters of compatibility. They are: out-of-band and spurious emissions, intermodulations (between MS or between BS), reference interference level and blocking. The propagation methods used in compatibility calculations are selected according to environments, and level in the hierarchical structure of cells (single or multi-operator case) in which both, the victim and the interfering stations, are placed. If necessary, the handover and power control should be used in this analysis.

The most important conclusions of simulations presented in Report [35] concern the adhering frequency 1920 MHz. The research revealed that co-location of wide area FDD and TDD base stations (operate in adjacent bands) with 30 dB coupling loss is not possible. But co-location of FDD and TDD base stations, operating not in adjacent bands, is possible if an external filter (with minimum attenuation of 56 dB) will be added in the FDD uplink chains. However, if the TDD base station works in a local area, then the adjacent channel operation of both TDD and FDD systems is possible under the conditions stated in this Report – only minor capacity losses may be observed when TDD stations are too close to FDD BSs.

Further, in extended bands, where the adhering frequencies f_{a1} and f_{a2} may appear, Report [12] suggests to consider all the possible scenarios BS FDD ↔ BS TDD (macro/micro/pico, outdoors/indoors). In these cases, the interference may be essential, and the compatibility calculations should be conducted using deterministic methods (MCL). In other scenarios (BS → MS, MS → BS, MS ↔ MS), the Monte Carlo calculations have shown that those interference have very little or negligible impact on the system's capacity. However, the calculations conducted with deterministic methods for scenario MS ↔ MS indicated that critical situations may exist when the MSs are closely located and operating on proximate frequencies. Then, mobile terminals will cause strong mutual interference.

In Report [12], for the assumed interference scenarios and base stations parameters, the separation distances between BS stations have been calculated. The simulation procedures have shown that there are no significant differences in interference magnitude considering FDD DL to TDD UL interference or TDD DL to FDD UL interference. The required separation distance, depending on frequency separation (15–5 MHz) reaches from single to teens of kilometres for BS FDD ↔ BS TDD macro/macro scenario, some tens to hundreds meters for BS FDD ↔ BS TDD macro/micro or micro/micro scenarios, and from single to some tens of meters for BS FDD ↔ BS TDD micro/pico or pico/pico scenarios. These values have been calculated for 3.84 Mchip/s TDD. In the case of 1.28 Mchip/s TDD (TD – SCDMA), with 3.5 MHz carrier separation, the required separation distance decreases significantly. However, the simulations have shown that co-locating FDD and TDD base stations without providing additional isolation is impossible, and even the application of a guard band of 5 and 10 MHz give no satisfactory results. Therefore, this is a situation similar to the one presented in Report [35] for the frequency at 1920 MHz.

Reduction of mutual interference between the base stations FDD and TDD is possible only after introduction of additional highly effective filtration, careful station location preserving the necessary distances between them, and application of appropriate guard bands. The transmitter power reduction is a partial solution; however, the coverage of base stations will also be reduced.

The problems of inter-operator interference are analysed in the book by Holma and Toskala, 'WCDMA for UMTS' [36], Section 8.5. The uplink and downlink effects are described, in particular the local and average downlink interference. The authors demonstrate the calculation of sizes of 'death zones', caused by adjacent channel interference, and they give an analysis of solutions to avoid the adjacent channel interference. These problems and potential solutions are also presented in Chapter 12 of this book.

11.4 EXTERNAL ELECTROMAGNETIC COMPATIBILITY

Considerations of external compatibility of UMTS presented in this section refer only to the terrestrial systems. However, it is commonly known that there are countries where the broadcasting satellite services (BSS sound) are used in the 2630–2655 MHz band; therefore, the compatibility in this band must be secured by appropriate planning of an UMTS system.

Information concerning compatibility studies and a methodology to assess interference between terrestrial UMTS and BSS (sound) are contained in ITU Recommendations M.1646 [37] and M.1654 [38]. The results of compatibility studies between UMTS and Space Services in the form of evaluated guard bands, in proximity of the frequency 2025 and 2110 MHz, are given in ERC Report 65 [6].

11.4.1 UMTS TDD VERSUS DECT WLL

Near the lower limits of the core band of UMTS in Europe, there exists the unpaired band in which only the operation in TDD mode is possible. For this reason, the mutual interference of DECT and UMTS are assessed for this mode of operation. Calculations of the scale of the impact of the mutual interference of those two systems are to be made with one of the two methods, MCL or MC, according to the interference scenario.

Analysis of possible interference scenarios conducted by specialised CEPT working groups (and UMTS sharing matrix in European initial band, see Table 11.1) has shown that an introduction of the UMTS system in the 1900–1920 MHz band can cause its strong mutual interference with DECT Wireless Local Loop (WLL) systems. This interference mainly arises due to spurious emissions of transmitters and due to receiver blocking.

While looking into the matter of mutual compatibility of these two systems, three profiles of the DECT are considered: the access radio fixed profile WLL RFP (Radio Fixed Profile), the profile of fixed subscriber unit (CTA – Cordless Terminal Adapter) and the portable profile (PP); all with respect to base stations (BS) and mobile stations (MS) used in the UMTS system. The basic parameters of UMTS base stations and mobile stations used for calculations, as well as the parameters of the System DECT (accepted according to the standard ETS 300 175-2 [39]) are contained in the ERC Report [6]. That report also presents the detailed basis of compatibility calculations between the systems DECT and UMTS.

The only situation in which the MCL method should be used is the case of interference between the base stations of both systems. This is due to the fact that the distance between them is fixed and constant. The free space propagation model used in this case refers to stations in outdoor locations. If the stations were indoors, then that model was valid only up to 10 m distance. Beyond the 10 m distance, the fourth power model was used ($L[\text{dB}] = 18 + 40 \log d$). The model did not fully comprise the specificity of radio waves propagation in the interior of buildings.

In scenarios with at least one mobile station, the Monte Carlo method approach is preferred. In the cases under consideration, the Monte Carlo simulation method was performed with the assumption of a density of 500 mobile objects (DECT and UMTS) per km^2. It is important to note that the Monte Carlo method simulations were performed using the propagation model adequate only in outdoor propagation scenarios and incorrect for indoor wave propagation. The indoor simulations require the proper model of indoor propagation through more than one storey, information about antenna heights, attenuations pertinent to the building construction and consideration of very small distance between the transmitter and victim receiver (a high user density). In order to avoid interference between UMTS MS and DECT MS, it was primarily assumed that the distance inside the buildings between the two system's mobile stations must exceed 5 m.

During the examination, different scenarios were considered of the mutual impact of the two systems terminals (UMTS base stations BS and mobile stations MS and the terminals with profiles DECT RFP,

CTA and PP). As a result of the conducted qualitative analysis, the following cases were said to be most dangerous: above roof-top WLL RFP ↔ BS (macro) and above roof-top CTA ↔ BS (micro) and indoor RFP ↔ indoor MS and indoor PP ↔ indoor MS. In the remaining cases (considered was the combination below roof-top WLL RFP, outdoor PP, below roof-top micro BS, indoor micro BS and outdoor MS), it was admitted that no dangerous interference exist.

For the cases of RFP ↔ BS and CTA ↔ BS, the minimum (necessary) levels of interference attenuation with respect to the central frequency of the UMTS channel have been calculated. For the remaining cases, the probability of interference between both system terminals has been calculated – again as a function of the central frequency of the UMTS channel. The results of these calculations facilitated the determination of minimum permissible distances in deployment of the UMTS and DECT terminals in the case of neighbourhood operation.

The calculation results given in ERC Report [6] indicate that if UMTS TDD system operates in the lowest accessible frequency channel (central frequency of 1902.5 MHz) and DECT system operates in the whole frequency range, then the minimum distances between the terminals should be:

- ca. 1500–1700 m in the case UMTS BS macro ↔ DECT FWA (Fixed Wireless Access);
- ca. 20–50 m in the case indoor UMTS MS ↔ indoor DECT.

If a UMTS System operates in the channel with a central frequency of 1912.5 MHz, these minimum distances diminish respectively to 150 m in the first and to few meters in the second case.

It follows from the compatibility analysis that the station deployment and planning of both systems ought to be done with appropriate care. The separation of base stations located above the roofs of buildings should be maximised. The use of directional antennas becomes necessary.

Decrease of coordination distances between terminals, having their antennas mounted high over the roofs of buildings (up to ca. 300 m for the UMTS channel central frequency 1902.5 MHz, and up to ca. 200 m for 1907.5 MHz) is permissible only when the power of the interfering signal has the level of background noise or if [6]:

- the highest carrier of DECT (1897.344 MHz) will not be used;
- the adjacent channel selectivity ACS of the UMTS base station will be increased to 55 dB (e.g. by applying an external filter);
- in situations of earlier installation of the UMTS TDD system than the DECT system, the limitation will be introduced for DECT FWA of its nominal wanted signal level down to 70 dBm (instead of 80 dBm);
- in situations of earlier installation of the DECT system than the UMTS TDD system, the level of emitted power of the UMTS TDD base station will be limited to 45 dBm.

In the case that UMTS and DECT systems are deployed indoors no additional organisational action is needed.

The results presented in [6] indicate that the minimum channel separation between the systems DECT and UMTS TDD should be equal to 5.2 MHz. If the UMTS TDD system is developed indoors no additional protection bands between DECT and UMTS are needed. Such bands may be necessary if both systems are developed outdoors and their antennas are placed above the roofs, then principally the frequency band 1900–1910 MHz should not be used. In order to reduce the level of interference to DECT coming from micro cell (external) UMTS TDD base stations, interference reduction techniques are to be used (as for example intra-cell handover and instant Dynamic Channel Selection, iDCS, in the time domain).

11.4.2 COMPATIBILITY BETWEEN UMTS AND RADIO ASTRONOMY SERVICE

The Radio Astronomy Service (RAS) uses the frequency range 2690–2700 MHz in which, in obedience to footnote 5.340 of Radio Regulations, any usage of radio transmitters is totally prohibited. But IMT-2000/UMTS operates in the frequency band adjacent to the lower edge of the band reserved for radio astronomy (see Section 11.1.1).

The ITU-R Recommendation RA.769 [40] presents the protection criteria for radio astronomy measurements. The determination of maximum power flux density S_H per 1 Hz or per 10 MHz is among these. For the frequency range 2690–2700 MHz, the respective values are: -247 dBm/Hz and -177 dBm/10 MHz. They apply to all the systems operating in the bands adjacent to the above-specified frequency range and in the geographical areas close to radio telescopes.

In accordance with Recommendation [40] 'administrations should afford all practicable protection to the frequencies and sites used by radio astronomers in their own and neighboring countries and when planning global systems, taking due account of the levels of interference ...'; hence, tough requirements should be secured to protect radio astronomy stations.

In order to determine the adjacent channel compatibility for IMT-2000/UMTS and radio astronomy, it is necessary to set the requirements concerning their geographical separation. This is done by defining coordination zones where it is allowed to transmit and exclusion zones where this is totally forbidden. These zones are assigned after determination of the minimum separating distances necessary for radio astronomy stations protection against interference.

A minimum separating distance is calculated based on the link balance, taking into account the requirements of radio astronomical stations protection, the maximum out-of-band emission in the frequency range reserved for RAS and applying an appropriate propagation model. In some cases, in determination of these zones, an additional isolation distance is required.

Coordination zones are the areas in which coordination is required of radio astronomy receiver locations and terrestrial base stations of IMT-2000/UMTS locations; in contrast, the exclusion zones are the areas where no operation of any stations is allowed.

Table 11.4 presents the required isolations between the radio astronomy stations and the base and mobile stations (FDD BS/MS) of the terrestrial component of IMT-2000/UMTS, calculated in ECC Report 45 [7]. These isolations have been calculated using the MCL method. The minimum isolation between RAS and mobile-earth stations of the satellite component of IMT-2000/UMTS (e.g. Satellite Digital Multimedia Broadcasting, S-DMB, Satellite Radio Interface – E, SRI-E) is evaluated in [7] as well.

The determination of coordination and exclusion zones in the case of terrestrial systems is based on the following rules [7]:

- The determination of size of coordination zones for base stations of the terrestrial component IMT-2000/UMTS should be based on the required isolation values. The zone's magnitude determines the telecommunication administration of a given country. Every base station planned to be set in operation within such a zone should be submitted to the process of coordination of its location and frequency with radio astronomy stations.
- The size of an exclusion zone will be determined also by national administrations. For the terrestrial mobile stations, the size of an exclusion zone will be determined in result of a coordination process of the base stations. Exclusion zones have to be determined depending upon the local terrain situation.
- The sizes of the exclusion and coordination zones will be determined considering the location of RAS and UMTS stations. Research conducted before the year 2004 has shown that typical coordination distances for base stations are in the range from 60 to 100 km. For a singular mobile station transmitting its maximum radiation power signal, the exclusion zone ranges from 30 to 50 km. These sizes of zones were determined neglecting the protection frequency bands. The addition of filters limiting the out-of-band radiation of the base stations can effectively reduce the zone's magnitude.

Table 11.4 Isolations required for the relation Radio Astronomy – terrestrial component of the IMT-2000/UMTS.

Station	Required isolation [dB]
WCDMA FDD, BS ($P = 43$ dBm)*	190
WCDMA FDD, MS ($P = 24$ dBm)*	174

* The maximum value of out-of-band (OOB) emissions acquired with consideration of the maximum power of BS/MS transmitters. Usually the power of base station or mobile terminal IMT-2000/UMTS transmitters is lower than declared.

The RAS stations are mostly located in rural areas in a manner facilitating to obtain the minimum of interference caused by terrestrial transmitters. The locations of these stations are chosen intentionally, e.g. in the valleys, to diminish the impact of significant permanent terrestrial interference sources. They may also be protected by natural terrain screens like hills or forests. The coordination or exclusion zones, determined for RAS stations, should be totally contained within the borders of their countries.

11.4.3 COMPATIBILITY BETWEEN UMTS AND MMDS

In some countries, within the Fixed Service with radio access, the Multipoint Multimedia Distribution System (MMDS) is developed. This system exploits the same extended frequency band as UMTS. Therefore, before any UMTS deployment, the conditions of co-channel and adjacent channel compatibility have to be determined.

Co-channel compatibility. The research results presented in [7] indicate that it is impossible to share the frequency ranges by IMT 2000/UMTS and MMDS, if they operate in the same geographical region.

These systems may co-exist operating in the same frequency range only preserving relatively large distance separations – above 70 km for macro-cells. Only then can their mutual interference be brought to a minimum. However, simulations conducted by the Working Group CEPT ECC WG SE revealed that sharing of frequency ranges by the devices of the two systems under consideration may prove difficult to be implemented because of the required large magnitudes of geographical separation (Table 11.5).

One ought to bear in mind that if co-channel interference appears in channel sharing MMDS receivers, there is a chance to reduce it due to large values of the front-to-back attenuation coefficient of the antennas of MMDS receivers. However, these receivers must be located outside the service area of a UMTS system.

Table 11.5 Magnitude of the geographical separation between MMDS and IMT-2000/UMTS systems for various interference scenarios.

Scenario of interference	Required distance separation (km)
UMTS MS → MMDS Rx	5
UMTS BS → MMDS Rx	5/25/70 – pico/micro/macro cell
MMDS Tx → UMTS BS	5/25/70 – pico/micro/macro cell
MMDS Tx → UMTS MS	5

Notes
MS – mobile station (user equipment)
BS – base station
Tx – transmitter
Rx – receiver

Table 11.6 Guard bands between IMT-2000/UMTS and MMDS systems.

Scenario of interference	Required frequency separation (MHz)
UMTS MS → MMDS Rx	0
UMTS BS → MMDS Rx	20/15 – macro/micro cell
MMDS Tx → UMTS BS	15 – macro cell
MMDS Tx → UMTS MS	10

Adjacent channel compatibility. Table 11.6 presents the required protection margins values between IMT-2000/UMTS and MMDS operating in adjacent channels with the geographical separation preserved. The research results presented in Report [7] indicate that with the assumed geographical separation, i.e. clearly separated service areas of MMDS and IMT-2000/UMTS, the minimum frequency separation in the case of micro-cells and macro-cells should be 10–20 MHz and for pico-cells no guard band consideration is required.

As in the case of co-channel interference, also here, when operating in adjacent channels, a significant reduction of interference to MMDS receivers is possible due to a high value of the antenna's front-to-back attenuation coefficient of MMDS receivers.

11.5 INTERNATIONAL CROSS-BORDER COORDINATION

11.5.1 PRINCIPLES OF COORDINATION

During UMTS implementation in individual countries, several problems have to be resolved with respect to compatibility of stations located by operators in borderline regions. The main goal of international cross-border coordination is to allow each of the countries a mutual and optimal use of the electromagnetic spectrum. Coordination of same frequency band of radiocommunication systems between neighbouring countries is conducted according to guidelines defined by bilateral or multilateral agreements between countries, following Radio Regulations [2]. The cross-border coordination considerations presented below pertain exclusively to systems using UMTS Terrestrial Radio Access (UTRA).

In general, coordination of UTRA systems located at borderline regions consists of an evaluation and analysis of signals reaching the other side of the border. The coordination process is then initiated, only if the signals coming from across the border can affect an operation of radio systems of the other country. Detailed coordination procedures and technical aspects of the coordination process (allowable interference's mean field strength, level of maximum interference, models of antenna characteristic, propagation calculation methods, etc.) are usually defined through international forums. In Europe, the general guidelines for UMTS cross-border coordination were defined, among others, by CEPT [41] and the member countries of the Berlin Agreement [10]. These guidelines were defined so far for the core bands (1900–1980, 2010–2025 and 2100–2170 MHz). Soon, these should also be defined for the extended band (2500–2690 MHz).

2G system cross-border coordination (e.g. for GSM) generally consists of an appointment (based on mutual agreements) to the neighbouring countries of so-called *preferential frequencies* (or to be exact, narrowband frequency channels), allocated within the available set of frequencies. This simplifies coordination activities, greatly reducing the number of coordination requests for the use of certain frequencies at borderline regions.

3G systems based on CDMA use wideband frequency channels. Furthermore, the number of these channels is very limited. It is generally known (please see Section 3.1), that during the first stage, a terrestrial UMTS component development foresees the use of only 12 paired FDD channels (1920–1980/2110–2170 MHz) and 7 unpaired TDD channels (1900–1920 and 2010–2025 MHz). The number of operators at each given country will basically be limited to 6 (the number of core band licences granted at particular countries at this time varied so far between 2 and 6 operators). Considering the above, the cross-border coordination of UTRA systems has to be resolved differently as compared to narrowband systems.

There are several methods for resolving cross-border coordinations, consisting mainly of:

- frequency coordination;
- special techniques for reducing interference between channels, for example:
  - coordination of scrambling codes, or
  - coordination of time slots (in case of TDD).

The performance or capacity of borderline systems will be greatly affected by the coordination method chosen by the operator or administrative bodies of the given country. Most important is the coordination of codes and frequencies at certain internationally agreed mean field strength levels, which should not be exceeded at the borderline areas of the given countries.

Considering the small number (2–3) of frequency channels usually assigned to each of the UTRA operators, a cross-border coordination based on the preferential frequency method is basically impossible. In Europe, there are many medium and large size cities located near national borders; therefore, operators will likely not accept to limit in any way the coverage of their UTRA systems. Considering this, each of the cross-border coordination methods must assure full coverage at the borderline regions, whilst also guaranteeing an appropriate and minimum level of interference to the foreign operator's as well as their own networks. This problem can be resolved by:

- applying the above-mentioned techniques for reducing interference between frequency channels;
- using a possibly more accurate method for predicting mean interference field strength levels.

However, it can be proven that it is impossible to develop a solution, which would simultaneously protect the bordering networks from interference and at the same time would assure good coverage. Cross-border coordination allows for significant reduction of the effects of interference onto operation of systems at both sides of the border. A cross-border coordination agreement between countries should define, among others:

- methods for harmonisation of frequency band usage and conditions for shared use;
- division of frequency bands into channels and code sets into blocks (uniform numbering);
- selection of propagation methods for calculation of interference range;
- techniques used for reduction of interference;
- allowable mean field strength levels with respect to code and frequency coordination;
- data exchange between administrative bodies of neighbouring countries (e.g. form of coordination requests).

11.5.2 PROPAGATION MODELS FOR COORDINATION CALCULATIONS

For coordination calculations, various propagation models may be used, depending upon the predicted location, as well as existing environment, of the coordinated radio stations. The selected propagation models ought to be commonly agreed with by the concerned countries.

In general, coordination calculations are performed in two cases. The first case pertains to a general evaluation of interference at the borderline as well as the territory of the neighbouring country. In the second case, detailed interference calculations are performed for a specific path between a transmitter causing the interference and the affected receiver.

The general evaluation is used to decide if cross-border coordination is necessary or not. The calculations are performed for areas along the borderline as well as for neighbouring countries. In this case, detailed terrain characteristics along the path are not always needed. The general evaluation shall use propagation calculations based on the point-to-multipoint (area) model. A model, given in ITU-R P.1546 [22] (please see Chapter 4.5 of this recommendation) or similar given in Annex 5 of the Vilnius Agreement [10], can be used for this purpose. In this case, calculations should typically be performed for 50 % of the locations and 10 % of time, assuming a receiver antenna height of 3 m. Concerned administrative bodies/operators from neighbouring countries can request an inclusion of terrain and coverage details (e.g. woodlands, buildings or use of a model taking into account propagation inside buildings) into the calculations, which will assure a better prediction accuracy. The quality of these calculations is significantly improved when an appropriate digital terrain model is used.

The detailed calculations shall use propagation calculations based on the point to point model. The model given in ITU-R P.452 [18] is appropriate for interference path loss calculations (see also Section 5.5.2). According to the Recommendation [41], path loss calculations based on this method can be made for predefined steps of distances from the given transmitter, along evenly distributed straight line beams coming from the transmitter. The detailed parameters pertaining to distance spacing/number of such beams are to be agreed between the concerned countries. The loss values calculated for predicted receiver locations shall be used for graphing a histogram showing excessive interference field strength levels. If more than 10 % of the field strength levels exceed the agreed maximum value, then the interfering station shall be subjected to coordination procedures.

11.5.3 APPLICATION OF PREFERENTIAL FREQUENCIES

Many national administrations granted licenses for four or more operators. UTRA networks will typically use three cell layers (macro, micro and pico), where it might happen that each requires channel separation at the given area.

Cross-border coordination of 3G systems can be resolved by using preferential channel assignments. This means that for borderline stations, UTRA operators would be assigned to certain preferential channels, i.e. frequency channels (from a limited set), which would be treated as privileged. This privilege means that signals using the preferential frequency can have a higher level at the borderline (as well as the area of the neighbouring country) as compared to the regular (non-preferential) frequencies. This also means that preferential frequencies provided to an operator for use at one side of the border, shall be treated as non-preferential frequencies on the other side of the border.

A use of preferential frequencies by an operator does not lead to significant reduction of cell capacity for the borderline cells with respect to the interference coming from cells operating on the same frequencies in the neighbouring country. However, border area coverage will be greatly reduced for each operator, as the number of frequency channels available to each operator is reduced to one-half in the case of two countries or even one-third in the case of coordination for areas where three borders meet. Considering this, any of the UMTS operators having only two channels, would not be able to provide continuous coverage along the border.

Assignment of a small number of frequency channels means that not all operators will be granted preferential frequencies for use at the borderline regions. Furthermore, operators using non-preferential frequencies must accept interference coming from across the border from systems using preferential frequencies at the neighbouring country. Considering this, preferential frequency channels should rather be used as a supplemental coordination method as compared to assignment of preferential codes and thus a term of 'neutral channels' should be introduced for channels treated equally at both sides of the border.

Nevertheless, in certain cases, frequency channel coordination might still be needed between the neighbouring countries and considering this, all frequency channel types have predefined maximum

Compatibility of UMTS Systems

allowable field strength levels for the borderline and beyond the border. This is required to prevent UTRA operators from increasing their base station power to increase their borderline coverage, but which would reduce coverage of the operator across the border.

The general assumption was that the maximum allowable field strength level at the border should be high enough to allow for voice communications (8 kbps) inside buildings, regardless of the compatibility between neighbouring networks operating on the same frequency. The maximum allowable borderline field strength level was determined, assuming that an increase of noise by 50 % at the mobile user terminal (MS) causes a coverage reduction of 10 %. In order to assure full coverage, an additional margin was also assumed for signal fading inside buildings as well as cell overlap at the border. In the cases where the foreseen (or measured) mean field strength levels at the border for each of the carriers generated by a given station do not exceed the allowable value, the given frequency does not require coordination.

As the result of works performed by CEPT TG1 and PT1, for all types of frequencies used in FDD systems, appropriate maximum field strength levels were established, which should not be exceeded at a height of 3 m above terrain at the border between the countries (and inside the neighbour country). These values are defined for downlink (with respect to field strength levels generated by the base stations), as these are more critical as compared to the uplink in the link budget. Using preferential frequencies, the predicted mean field strength level should not exceed $E_{max} = 65\,dB\mu V/m/5\,MHz$, whilst in case of non-preferential and neutral frequencies this shall be $E_{max} = 45\,dB\mu V/m/5\,MHz$. If any of these values are exceeded, cross-border frequency coordination is needed. Figures 11.5 and 11.6 present the possible combination of use of non-coordinated preferential vs. non-preferential frequencies as well as neutral frequencies.

The number of preferential frequencies for both sides of the borders should be equal. This means that at the low number of available UTRA channels this is only possible if at both sides of the border the UMTS terrestrial component channels have been divided among the same number of operators. In all

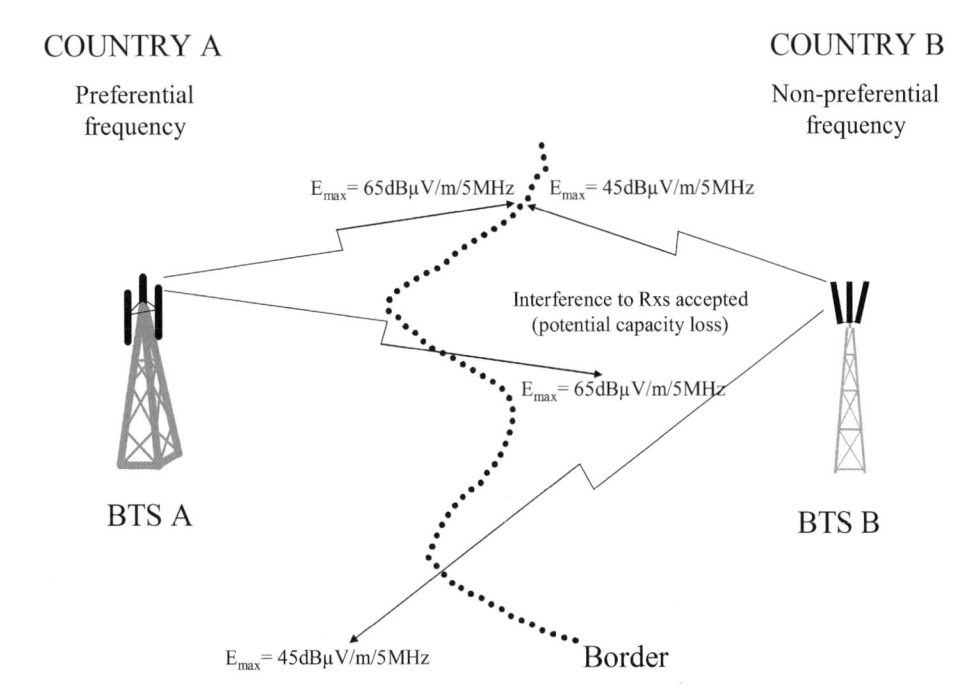

Figure 11.5 Sample border area interference situation, preferential vs. non-preferential frequency scenario.

COUNTRY A
Neutral frequency

COUNTRY B
Neutral frequency

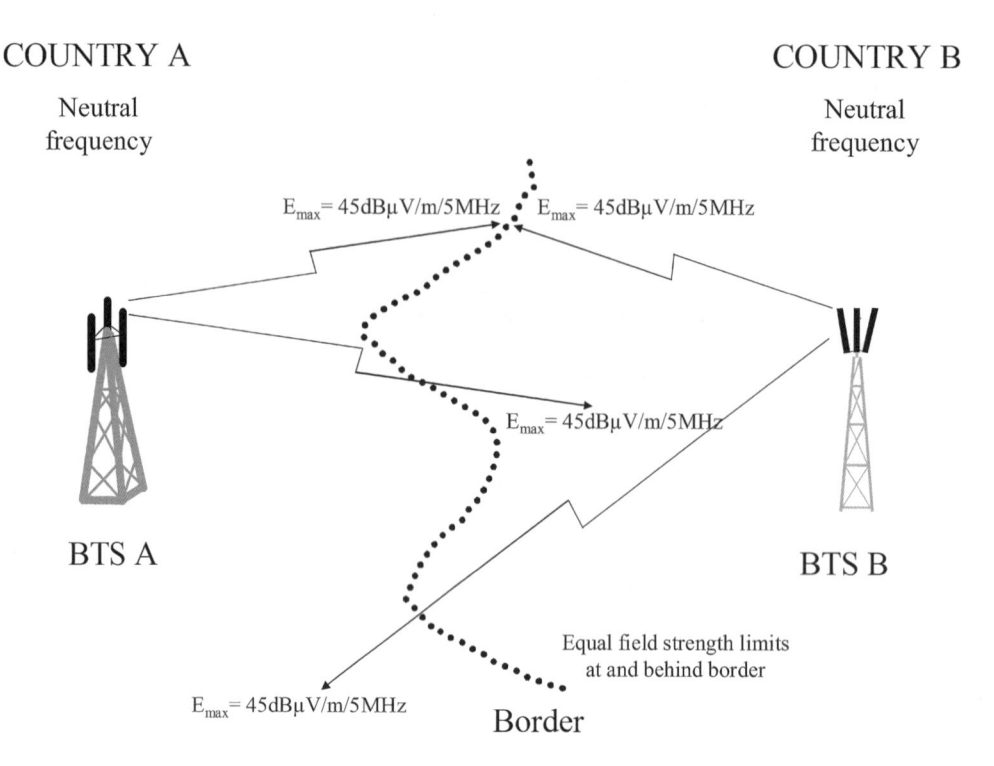

E_{max} = 45dBμV/m/5MHz
E_{max} = 45dBμV/m/5MHz
E_{max} = 45dBμV/m/5MHz
E_{max} = 45dBμV/m/5MHz

BTS A

BTS B

Equal field strength limits at and behind border

Border

Figure 11.6 Sample border area interference situation, neutral frequency scenario.

the other cases, one of the operators would have a higher number of preferential channels, which would breach the rule of equal treatment for both sides. Furthermore, the centre channel frequencies at both sides of the border should be the same. Considering this, code coordination becomes vital and it also allows for smooth handover when a subscriber crosses the border. Frequency coordination is possible only if, regardless of the number of issued licenses, there will be the same number of active operators on both sides of the border. It should be noted, however, that if only the initial band will be used, the maximum number of such operators cannot exceed 4, as there are only 12 paired FDD channels to be split among them.

11.5.4 USE OF PREFERENTIAL CODES

The use of unique codes for coordinating undisturbed operations of UMTS systems at borderline regions is based on the idea that information data within the UTRA is channelised by spreading, scrambling and modulation operations. Each bit pair of data (and control) is assigned to either I or Q branch. These orthogonal signals are then spread into the branches according to the predefined chip rate (3.84 Mchip/s) using the one channelisation code (for DL or two different codes for UL), selected from the Walsh-Hadamard code set. Then, the spread results are scrambled using a pseudo-random sequence (see Chapter 2 for more details). This method of coding can thus be profited from for cross-border coordination. For example, in FDD downlink, the scrambling code is specific for each cell and is selected from the set of 512 'initial' scrambling codes, divided into 64 subsets (each consisting of 8 codes) in order to facilitate easier selection. The scrambling code is unique for each connection. Considering this, the total number of available scrambling codes for both sides of the border should be divided into preferential and non-preferential codes to facilitate coordination.

The use of code coordination in order to minimise mutual interference of UTRA at borderline regions is considered to be a very effective solution. However, this technique alone is not sufficient for assuring effective spectrum usage and thus should be combined with limits imposed onto interference mean field strength levels at borderline regions.

The preferential code coordination method is not as easy to apply as e.g. the preferential frequency coordination method used in narrowband systems. In preferential frequency coordination, if the operators on both sides of the borders use exclusively their preferential channels, there should be basically no interference between their systems. On the other hand, a use of only preferential code coordination cannot assure such a level of protection against interference to the other operator. This arises from the fact, that even if different codes are used, there still remains some interference between two CDMA systems. This means that, in order to assure proper system operations based of preferential codes, limitations to signal levels for preferential codes of the neighbouring operator are still required. In other words, code coordination is not able to provide the same solution as preferential frequency channel coordination and an operator using his preferential codes must be protected against interference coming from systems using both the non-preferential as well as preferential codes. This is the reason why separate limits for field strength levels are also defined for preferential and non-preferential codes, along with areas where such limitations apply. It is assumed that if the concerned countries have aligned centre channel frequencies, they must agree on preferential code groups and/or preferential code group blocks.

CEPT recommends [41] that coordination with the neighbouring country is not required when using:

- FDD systems in downlink using preferential codes (or where centre channel frequencies are not aligned or not using CDMA UMTS interface) if the predicted mean field strength of each carrier produced by the base station does not exceed a value $E_{max} = 45\,\mathrm{dB\mu V/m/5\,MHz}$;
- TDD systems using preferential codes (or where centre channel frequencies are not aligned) if the predicted mean field strength of each carrier produced by the base station does not exceed a value $E_{max} = 35\,\mathrm{dB\mu V/m/5\,MHz}$;
- systems using non-preferential codes (and where the centre channel frequencies are aligned) if the predicted mean field strength of each carrier produced by the base station does not exceed a value $E_{max} = 21\,\mathrm{dB\mu V/m/5\,MHz}$;

at a height of 3 m above terrain at and beyond the border line between the concerned countries. This has been exemplified by means of Figure 11.7.

Limits imposed onto mean field strength levels at the borderline areas guarantee that:

- proper operation is assured for each of the operators at his preferential channels, by limiting non-preferential channel field strength levels of other operators in the borderline areas;
- the operator is able to provide full coverage up to the border, using his preferential channels, which are allowed to have higher maximum field strength beyond the border. At the same time, the appropriate limit level also guarantees that the foreign operator will be able to use his non-preferential channels further in the depth of his own country (several dozen of kilometres);
- the limits are to be treated as trigger levels, i.e. in the case when the trigger is exceeded, coordination will be required. This coordination is possible, thanks to the large number of available codes.

11.5.5 EXAMPLES OF COORDINATION AGREEMENTS

The split of codes/frequencies into preferential/non-preferential or other groups, which are later assigned to operators, is performed according to bilateral or multilateral agreements between countries. It should be noted that national administration bodies should seek an opinion of operators when undertaking any activities in this area. An example of such agreement is the Vilnius Agreement [10], which sets

COUNTRY A

Preferential code
group A
Non-preferential
code group B

COUNTRY B

Non-preferential
code group A
Preferential code
group B

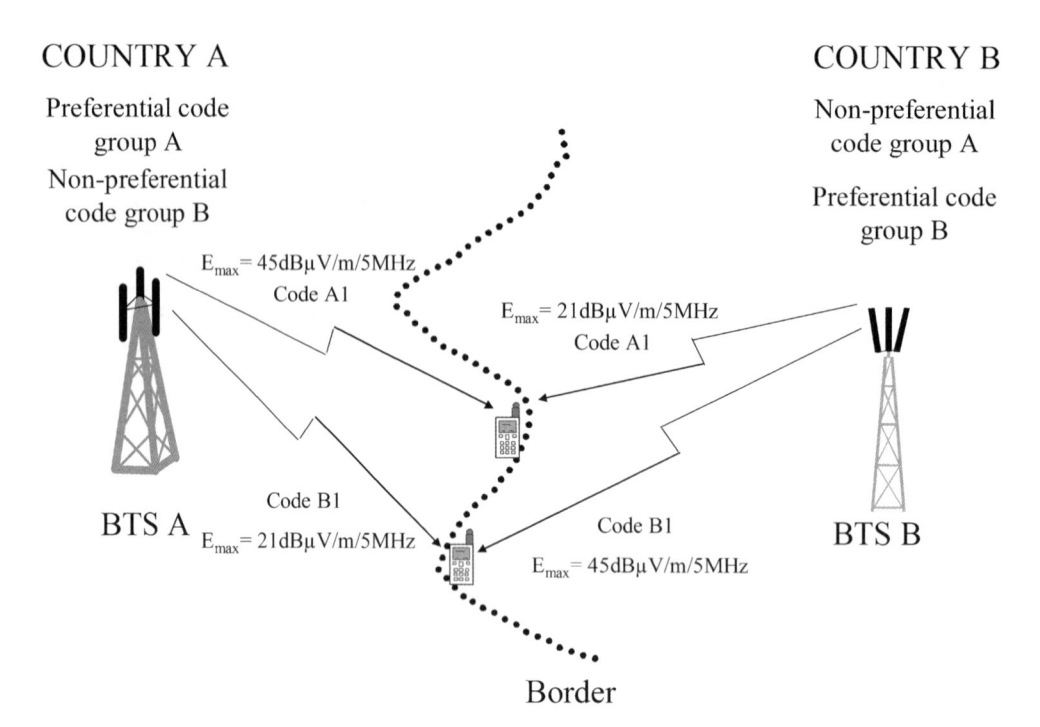

E_{max} = 45dBµV/m/5MHz
Code A1

E_{max} = 21dBµV/m/5MHz
Code A1

Code B1

BTS A E_{max} = 21dBµV/m/5MHz

Code B1

E_{max} = 45dBµV/m/5MHz

BTS B

Border

Figure 11.7 Sample border area interference situation, preferential vs. non-preferential code scenario.

out the guidelines for coordination, information exchange, form of coordination documents as well as the technical coordination conditions (coordination parameter values, allowable interference levels, propagation calculation methods).

The CEPT Recommendation [41], pertaining to cross-border coordination of UMTS systems, is also an excellent example of such international agreement. Considering that for the FDD mode, the 3GPP TS 25.213 document defines 64 groups (numbered from 0 to 63) of scrambling codes and for the TDD mode the 3GPP TS 25.223 document defines 32 (numbered from 0 to 31) groups of scrambling codes, these were divided into 6 subsets (each having 10/11 groups in FDD mode or 5/6 groups in TDD mode). Then, these subsets were assigned to the individual countries. The general assumption here was that each country should receive three groups of preferential codes to be used along any of the borders with another country, as well as two such groups for use at the borderline of three UMTS systems (i.e. when neighbouring two other countries). Table 11.7 shows the possible division of codes between countries.

In order to divide the codes between European countries (according to CEPT), they were grouped into four sets:

1. Set for countries 1 (SC1): BEL, CVA, CYP, CZE, DNK, E, FIN, GRC, IRL, ISL, LTU, MCO, SMR, SUI, SVN, UKR, YUG;
2. Set for countries 2 (SC2): AND, BIH, BLR, BUL, D, EST, G, HNG, I, MDA, RUS (Exclave);
3. Set for countries 3 (SC3): AUT, F, HOL, HRV, MKD, POL, POR, ROU, RUS, S, TUR;
4. Set for countries 4 (SC4): ALB, LIE, LUX, LVA, NOR, SVK.

Marking the available preferential code groups with letters A to F (subsequent code groups 0–10, 11–20, 21–31, 32–42, 43–52 and 53–63 for FDD or 0–4, 5–10, 11–15, 16–20, 21–26

Table 11.7 General code division for different modes of UMTS system operation among neighbouring countries.

	FDD mode	TDD mode
Borderline of two countries (countries A and B)	32 code groups are preferential for country A	16 code groups are preferential for country A
	32 code groups are preferential for country B	16 code groups are preferential for country B
Borderline of three countries (countries A, B and C)	22 code groups are preferential for country A	10 code groups are preferential for country A
	21 code groups are preferential for country B	11 code groups are preferential for country B
	21 code groups are preferential for country C	11 code groups are preferential for country C

Table 11.8 Detailed code group assignment for different UMTS system operation modes at borderlines between two and between three countries from different sets.

	Set for countries					
	Code Set A	Code Set B	Code Set C	Code Set D	Code Set E	Code Set F
	0..10 FDD	11..20 FDD	21..31 FDD	32..42 FDD	43..52 FDD	53..63 FDD
	0..4 TDD	5..10 TDD	11..15 TDD	16..20 TDD	21..26 TDD	27..31 TDD
Coordination area						
Border 1-2	SC1	SC1	SC2	SC2	SC2	SC1
Zone 1-2-3	SC1	SC1	SC2	SC2	SC3	SC3
Border 1-3	SC1	SC1	SC1	SC3	SC3	SC3
Zone 1-2-4	SC1	SC4	SC2	SC2	SC4	SC1
Border 1-4	SC1	SC4	SC1	SC4	SC4	SC1
Zone 1-3-4	SC1	SC4	SC1	SC4	SC3	SC3
Border 2-3	SC3	SC2	SC2	SC2	SC3	SC3
Border 2-4	SC4	SC4	SC2	SC2	SC4	SC2
Zone 2-3-4	SC4	SC4	SC2	SC2	SC3	SC3
Border 3-4	SC4	SC4	SC3	SC4	SC3	SC3

and 27–31 for TDD), these can be split among the countries for example as shown in Table 11.8.

It will be possible to define detailed division of codes among the concerned countries by using the above specified code grouping, as recommended by CEPT. Tables 11.9 and 11.10 show the recommended code usage along the borders of a selected European country (in this case for Poland) for both FDD and TDD operation.

The above proposals should serve as the basis for cross-border coordination agreements between the different European countries.

Table 11.9 Code coordination: recommended code group assignment around the borders of Poland (FDD mode).

Code group subset	A	B	C	D	E	F
Countries/preferential code groups	0–10	11–20	21–31	32–42	43–52	53–63
Number of groups	11	10	11	11	10	11
CZE/D/POL 21/22/21	CZE	CZE	D	D	POL	POL
CZE/POL 32/32	CZE	CZE	CZE	POL	POL	POL
D/POL 32/32	POL	D	D	D	POL	POL
DNK/D/POL 21/22/21	DNK	DNK	D	D	POL	POL
DNK/POL 32/32	DNK	DNK	DNK	POL	POL	POL
SVK/UKR/POL 21/22/21	UKR	SVK	UKR	SVK	POL	POL
SVK/POL 32/32	SVK	SVK	POL	POL	SVK	POL
SVK/CZE/POL 21/22/21	CZE	SVK	CZE	SVK	POL	POL
UKR/POL 32/32	UKR	UKR	UKR	POL	POL	POL
UKR/BLR/POL 21/22/21	UKR	UKR	BLR	BLR	POL	POL
BLR/POL 32/32	POL	BLR	BLR	BLR	POL	POL
LTU/BLR/POL 21/22/21	LTU	LTU	BLR	BLR	POL	POL
LTU/POL 32/32	LTU	LTU	LTU	POL	POL	POL
LTU/RUS/POL 21/22/21	LTU	LTU	RUS	RUS	POL	POL
RUS/POL 32/32	POL	RUS	RUS	RUS	POL	POL

Table 11.10 Code coordination: recommended code group assignment around the borders of Poland (TDD mode).

Code group	1	2	3	4	5	6
Countries/preferential code groups	0–4	5–10	11–15	16–20	21–26	27–31
Number of groups	5	6	5	5	6	5
CZE/D/POL 11/10/11	CZE	CZE	D	D	POL	POL
CZE/POL 16/16	CZE	CZE	CZE	POL	POL	POL

D/POL 16/16	POL	D	D	D	POL	POL
DNK/D/POL 11/10/11	DNK	DNK	D	D	POL	POL
DNK/POL 16/16	DNK	DNK	DNK	POL	POL	POL
SVK/UKR/POL 11/10/11	UKR	SVK	UKR	SVK	POL	POL
SVK/POL 16/16	SVK	SVK	POL	POL	SVK	POL
SVK/CZE/POL 11/10/11	CZE	SVK	CZE	SVK	POL	POL
UKR/POL 16/16	UKR	UKR	UKR	POL	POL	POL
UKR/BLR/POL 11/10/11	UKR	UKR	BLR	BLR	POL	POL
BLR/POL 16/16	POL	BLR	BLR	BLR	POL	POL
LTU/BLR/POL 11/10/11	LTU	LTU	BLR	BLR	POL	POL
LTU/POL 16/16	LTU	LTU	LTU	POL	POL	POL
LTU/RUS/POL 11/10/11	LTU	LTU	RUS	RUS	POL	POL
RUS/POL 16/16	POL	RUS	RUS	RUS	POL	POL

REFERENCES

[1] ITU, ITU-R Recommendation SM.1132, *General principles and methods for sharing between radiocommunication services or between radio stations.*
[2] ITU, ITU-R Radio Regulations, Geneva 2004.
[3] ECC, ERC Report 25, *The European table of frequency allocations and utilisations covering the frequency range 9 kHz to 275 GHz*, Lisboa January 2002 – Dublin 2003 – Turkey 2004 – Copenhagen 2004.
[4] http://www.efis.dk/search/general.
[5] ERC, ERC Report 25, *Frequency range 29.7 kHz to 105 GHz and associated European table of frequency allocations and utilisations,* ERC, Brussels, June 1994, revised in Bonn, March 1995 and in Brugge, February 1998.
[6] ERC, ERC Report 65, *Adjacent band compatibility between UMTS and other services in the 2 GHz Band*, Menton, May 1999, revised in Helsinki, November 1999.
[7] ECC, ECC Report 45, *Sharing and adjacent band compatibility between UMTS/IMT-2000 in the band 2500–2690 MHz and other services*, Granada, February 2004.
[8] ECC, ECC Recommendation (03)03, *Measures to safeguard the future use of terrestrial UMTS/IMT-2000 in the 2.5 GHz range with respect to broadcasting satellite systems.*
[9] ITU, ITU-R Recommendation M.1635, *General methodology for assessing the potential for interference between IMT-2000 or systems beyond IMT-2000 and other systems.*
[10] Agreement between the Administrations of Austria, Belgium, . . . , and Switzerland on the co-ordination of frequencies between 29.7 MHz and 39.5 GHz for the fixed service and the land mobile service. (HCM Agreement), Vilnius, 12 October 2005.
[11] ERC, ERC Report 101, *A comparison of the Minimum Coupling Loss Method, Enhanced Minimum Coupling Loss Method, and the Monte Carlo Simulation*, Menton, May 1999.

[12] ITU, ITU-R Report M.2030, *Coexistence between IMT-2000 time division duplex and frequency division duplex terrestrial radio interface technologies around 2600 MHz operating in adjacent bands and in the same geographical area*.
[13] ITU, ITU-R Report M.2041, *Sharing and adjacent band compatibility in the 2.5 GHz band between the terrestrial and satellite components of IMT-2000*.
[14] 3GPP, TR 25.942 v6.4.0, Technical Specification Group Radio Access Networks, *Radio Frequency (RF) system scenarios (Release 6)*, 2005–03.
[15] ITU, ITU-R Report SM.2028-1, *Monte Carlo simulation methodology for the use in sharing and compatibility studies between different radio services or systems*, 2002.
[16] ERC, ERC Report 68, *Monte Carlo simulation methodology for the use in sharing and compatibility studies between different radio services or systems*, Naples, February 2000, revised in Regensburg, May 2001 and Baden, June 2002.
[17] ERO, http//www.ero.dk.
[18] ITU, ITU-R Recommendation P.452, *Prediction procedure for the evaluation of microwave interference between stations on surface of the Earth at frequencies above about 0.7 GHz*.
[19] ITU, ITU-R Recommendation P.526, *Propagation by diffraction*.
[20] ITU, ITU-R Recommendation P.676, *Attenuation by atmospheric gases*.
[21] COST, COST 231 Final Report, Digital Mobile Radio Towards Future Generation Systems, 1998.
[22] ITU, ITU-R Recommendation P.1546, *Method for point-to-area predictions for terrestrial services in the frequency range 30 MHz to 3000 MHz*.
[23] ETSI TR 101.112 v3.2.0 (1998–04): Technical Report, Universal Mobile Telecommunications System (UMTS), *Selection procedures for the choice of radio transmission technologies of the UMTS*.
[24] ITU, ITU-R Recommendation P.619, *Propagation data required for the evaluation of interference between stations in space and those on the surface of the earth*.
[25] ITU, ITU-R Recommendation M.1225, *Guidelines for evaluation of radio transmission technologies for IMT-2000*.
[26] ITU, ITU-R Report M.2039, *Characteristics of terrestrial IMT-2000 systems for frequency sharing/interference analyses*.
[27] ETSI TS 125.104 v6.9.0 (2005–06): Technical Report, Universal Mobile Telecommunications System (UMTS), *Base Station (BS) Radio Transmission and Reception (FDD)*.
[28] ETSI TR 125.951 v6.2.0 (2003–09): Technical Report, Universal Mobile Telecommunications System (UMTS), *Base Station (BS) classification (FDD)*.
[29] ETSI TS 125.101 v6.8.0 (2005–06): Technical Report, Universal Mobile Telecommunications System (UMTS), *User Equipment (UE) Radio Transmission and Reception (FDD)*.
[30] ETSI TS 125.105 v6.2.0 (2004–12): Technical Report, Universal Mobile Telecommunications System (UMTS), *Base Station (BS) Radio Transmission and Reception (TDD)*.
[31] ETSI TR 125.952 v5.2.0 (2003–03): Technical Report, Universal Mobile Telecommunications System (UMTS), *Base Station (BS) classification (TDD)*.
[32] ETSI TS 125.102 v6.1.0 (2005–06): Technical Report, Universal Mobile Telecommunications System (UMTS), *User Equipment (UE) Radio Transmission and Reception (TDD)*.
[33] ETSI TS 125.308 v6.3.0 (2005–01): Technical Report, Universal Mobile Telecommunications System (UMTS), *UTRA High Speed Downlink Packet Access (HSDPA); Overall description; Stage 2 (Release 5)*.
[34] ITU, ITU-R Recommendation M.1641, *A methodology for co-channel interference evaluation to determine separation distance from a system using high-altitude platform stations to a cellular system to provide IMT-2000 service within the boundary of an administration*.
[35] 3GPP TR 25.942 v6.4.0 (2005–03): Technical Report, 3rd Generation Partnership Project, Technical Specification Group Radio Access Networks, *Radio Frequency (RF) system scenarios (Release 6)*.
[36] Holma H., Toskala A. (ed.), *WCDMA for UMTS, Radio Access for Third Generation Mobile Communications*, John Wiley & Sons, Ltd/Inc., 2004.
[37] ITU, ITU-R Recommendation M.1646, *Parameters to be used in co-frequency sharing and pfd threshold studies between terrestrial IMT-2000 and broadcasting-satellite service (sound) in the 2630–2655 MHz band*.

[38] ITU, ITU-R Recommendation M.1654, *A methodology to assess interference from broadcasting-satellite service (sound) into terrestrial IMT-2000 systems intending to use the band 2630–2655 MHz.*
[39] ETSI, ETS 300 175-2, Radio Equipment Systems (RES); Digital European Cordless Telecommunications (DECT) Common Interface Part 2: *Physical layer*, September 1996.
[40] ITU, ITU-R Recommendation RA.769, *Protection criteria used for radio astronomical measurements.*
[41] ECC, Revised ERC Recommendation 01–01, *Border Coordination of UMTS/IMT-2000 Systems.*

12

Network Design – Specialised Aspects

Marcin Ney, Peter Gould and Karsten Erlebach

This chapter deals with some specialised aspects in radio network design, which are often neglected in the 3G planning process. We will commence with issues related to network infrastructure sharing, including legal aspects of it, and then deal with adjacent channel interference control. The chapter is finalised by a section on ultra high sites, which have lately emerged as a network capacity booster and hence constitute an important element in future planning exercises.

12.1 NETWORK INFRASTRUCTURE SHARING

The UMTS network rollout is a very costly and time-consuming process. Especially for Greenfield operators, its complexity may significantly influence their ability to fulfil license requirements and to compete successfully with incumbent ones. Every operator should thus analyse any potential possibility to ease this process. This originated the concept of network infrastructure sharing. This section presents some methods of network sharing together with practical and legal guidelines concerning implementing such a sharing.

12.1.1 NETWORK SHARING METHODS

Four major network sharing methods can be distinguished:

1. site sharing;
2. RAN sharing;
3. national roaming – geographical split;
4. national roaming – common network.

Understanding UMTS Radio Network Modelling, Planning and Automated Optimisation   Edited by Maciej J. Nawrocki, Mischa Dohler and A. Hamid Aghvami   © 2006 John Wiley & Sons, Ltd

The choice of a particular method determines the speed of rollout, CAPEX savings, but also the level the network is bound with the other ones, which is very important if a divorce process needs to be done.

12.1.1.1 Site Sharing

The site sharing concept is very well known and commonly used among 2G network operators. It is a simple sharing of the physical site and basic installations. Such a solution leads to CAPEX savings (acquisition cost, civil work, air conditioning, mast, tower, etc.) and OPEX savings (site rental fee, site maintenance fee, etc.). The typical operator strategy of site sharing is a one-by-one site exchange, so the saving level depends on the number of sites shared and the number of operators in a sharing process. The other potential benefits include a greater choice of possible sites and improved public acceptance due to a reduced number of sites. The drawbacks are the need to coordinate site related operational aspects, to adjust radio planning according to competitors and the possible limited space for expansion on certain sites. In the case of a possible divorce, the 'guest' operator should sign a rental agreement with the site owner.

The second option of site sharing is the so-called ancillary sharing. In addition to sharing the physical site, power and backup systems hosted in a common cabinet (site support cabinet) can be shared (but powering two independent Node Bs). It gives the operator further CAPEX savings but leads to a further increase in site maintenance complexity.

Rack sharing is the last site sharing option; it can also be classified as a RAN sharing. It means sharing of the same physical Node B cabinet, but without sharing other Node B modules (e.g. Power Amplifier (PA), Channel Element (CE)), filters, transmission interface). Only cabinet, power supply, battery backup and cooling equipment are common. The benefits and drawbacks are the same as with an ancillary sharing option.

In addition to the above site sharing options, also antenna sharing is possible. The versions that can be implemented are: shared antenna radomes, shared antenna system by combining and shared antenna systems by receiver chaining. The major drawbacks are coordination of radio planning (the same antenna height, azimuth and tilt), combining loss in a combining option (has to be included into power budget calculation) and no RX diversity in the receiver chaining option (also has to be included in the power budget calculation). However despite these drawbacks, it can be the only option when e.g. the tower is overloaded and there is no possibility to place another antenna system on it.

12.1.1.2 RAN Sharing

The logical sharing of RAN infrastructure (Node B, backhaul transmission, RNC and O&M) is called RAN sharing. In that solution, each operator has a dedicated Public Land Mobile Network (PLMN), and its subscribers use the same Mobile Network Code (MNC). The shared RAN can cover the whole country or can be split geographically, e.g. on urban areas every operator builds its own network, but the rural part coverage is divided among them (one covers northern part and the second southern). From a technical perspective, each operator has its own core network and only the radio access network is shared (Figure 12.1). Generally, the options are that Node B cabinets can be shared (two logical Node Bs hosted in one physical Node B) and both Node Bs and RNCs are shared. The important prerequisites for RAN sharing are a choice of common vendor and software with a RAN sharing features availability.

Unfortunately RAN sharing solutions have many drawbacks. Practically, not more than two operators can share the same RAN and it requires a common technical policy and strong cooperation between partners. All the optional RAN features have to be the same for both operators. The next disadvantage is pooled capacity – one operator can exhaust the capacity of others. And finally, the O&M architecture

Network Design – Specialised Aspects

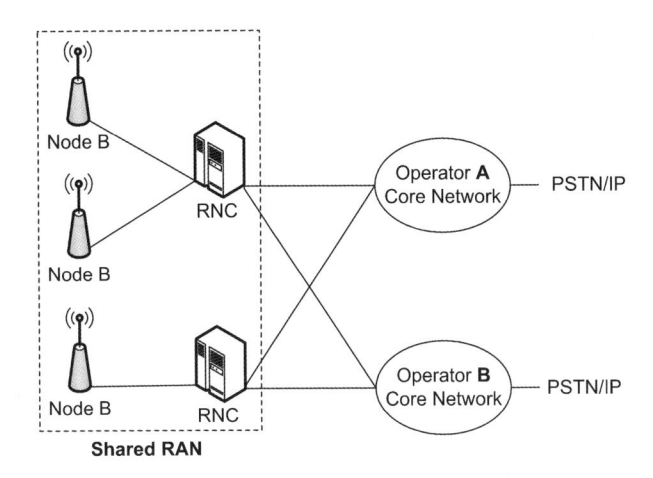

Figure 12.1 RAN sharing idea.

to manage shared and not shared items will be very complex. Also, while considering a divorce scenario, it practically means the rollout of a new network.

While considering Node B sharing, only channel element cards sharing should be considered. The reason comes from regulations – both operators have their dedicated band, so they have to use independent carriers. The other factors are 20 W output power per operator requirement, and the fact that the PA bandwidth is usually limited to 20 MHz. So only in the case of having adjacent bands by both operators, a PA can be shared.

Therefore having shared Node Bs, both operators pool the total number of channel element cards between them. It has a very positive trunking effect, but may also lead to a possible exhaustion of available capacity by one operator only. To prevent such a situation, it is necessary to have some capacity reservation mechanisms. It can be done by dedicating a pool of hardware (CE cards) to a particular operator only, or by limiting the use of CE cards at RRM admission control level. It is up to the equipment vendor if such functionalities are available. However, while using such features, the trunking effect is lost.

While moving towards full RAN sharing, the other element shared is RNC. The RNC capacity is usually defined with maximum throughput and maximum connectivity capacity (maximum number of Node Bs, sectors, etc.). Generally, all of these are pooled (hence can be shared). The exceptions in some vendors' implementations are I_u, I_{ur} and I_{ub} interface boards, which are not shared. To secure a fair access to RNC capacity resources by both operators, it is possible to let CAC algorithms separately control each I_u interface and guarantee that the maximum traffic per operator is not exceeded. The other way is to dimension the RNC to support maximum assumed traffic from both operators simultaneously.

Backhaul sharing is one of the most positive effects in RAN sharing, but the threat of blocking one operator by the other also applies here. It is related to both I_{ub} and I_{ur} interfaces. Concerning the I_{ur} interface, a mixed architecture implementation is possible (shared RNC connected to not shared RNC).

To have a full picture of RAN sharing, also RRM algorithms and parameters sharing should be evaluated (some of them should be the same for both vendors).

12.1.1.3 National Roaming – Geographical Split

In national roaming, the idea is that an operator builds its own network and covers some part of a country while making use of the competitor's network elsewhere (Figure 12.2). So each operator is

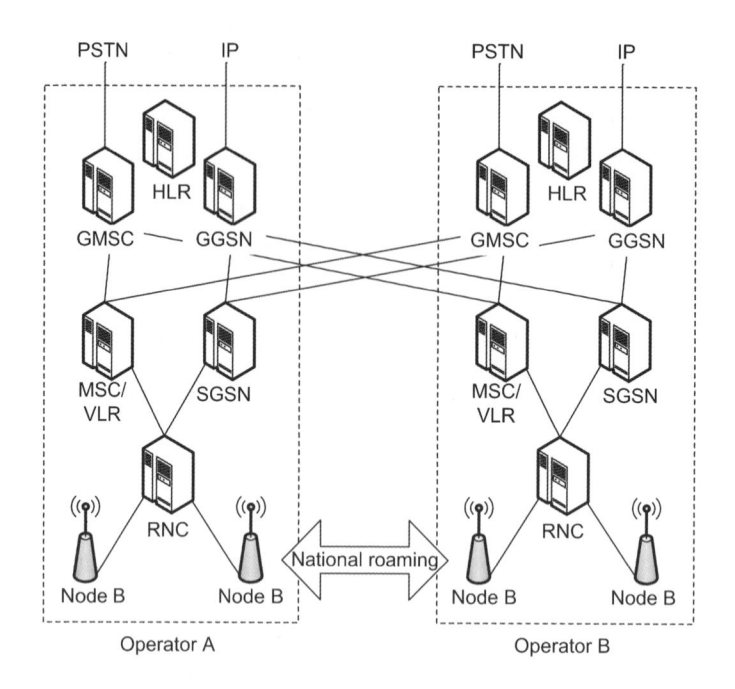

Figure 12.2 National roaming – geographical split.

using its own PLMN and frequency band only within its own coverage area. There is no coverage overlapping between two operators and subscribers are considered as roamers while leaving their home network and entering the partner's one.

From a technical point of view, the key point for such a national roaming scenario is the availability of inter PLMN mobility procedures. Furthermore, a coordination between operators concerning adjacent cells declaration should also be in place. The list of required RAN features relevant to national roaming includes:

- International Mobile Subscriber Identity (IMSI) based handover (HO) – cell reselection (selective inter PLMN Handover) 3G to 3G;
- Inter PLMN HO – cell reselection based on IMSI (3G to 2G and 2G to 3G) (support on the 3G side);
- Hard HO (Serving Radio Network System relocation);
- Inter frequency hard HO on dedicated channels;
- Neighbour list based on IMSI;
- Priority Handling (at the border);
- 2G to 3G Handover (support on 2G side).

From the core network side, the Equivalent PLMN feature is also required. From a legal perspective, both operators need a national roaming agreement, a Service Level Agreement (SLA) and an agreement from the regulator to implement a national roaming solution.

Such an approach to network sharing seems to be a good strategy to achieve a wide coverage in the early years of a network rollout – especially, when the license obligations are tough. Also later, migration to two independent networks is very easy, because there is no software and hardware impact of this solution. Thus, when the coverage of the remaining part of the country is complemented, only some changes in parameters will be required (adjacent cell and Location Areas (LAs) definition, etc.).

Network Design – Specialised Aspects 313

It is worth mentioning that there is no virtual limit for the number of operators sharing the network. The only limit is the existence of mutual roaming agreements and lawful regulations.

Nowadays, not only the speed of UMTS rollout is a coverage limiting factor, but also the strategy, that can be observed among many operators, not to build a UMTS country-wide coverage but to limit it to dense traffic areas only (e.g. cities and agglomerations). Thus, a national roaming scenario is also very interesting for Greenfield operators in the context of 3G to 2G roaming. Customers are used to having connectivity everywhere. So for new entrants into the mobile business, it is important to use a 2G network of a competitor. The way of doing so, is to become a Mobile Virtual Network Operator (MVNO) in a 2G network. In such a scenario, the subscriber perception is that he is using a 2G/3G network from one operator only. On the contrary, an incumbent 2G/3G operator should evaluate if it has a strategic and economical interest to let MVNO enter its 2G network. Details on MVNO possible scenarios and strategy are covered in [1].

12.1.1.4 National Roaming – Common Network

A common shared network solution (so-called NETCO) equals to a geographical split solution (inter-PLMN mobility is used) from a RAN point of view. The operators have common RANs and a partially shared core network (MSC and SGSN); only HLR and GGSN are separate (Figure 12.3). Thus, similar benefits and drawbacks (as in a geographical split) apply.

12.1.2 LEGAL ASPECTS

Even though some network sharing scenarios seem to be very interesting from a technical and economical perspective, it is also necessary to evaluate all the legal aspects. Some country regulators may

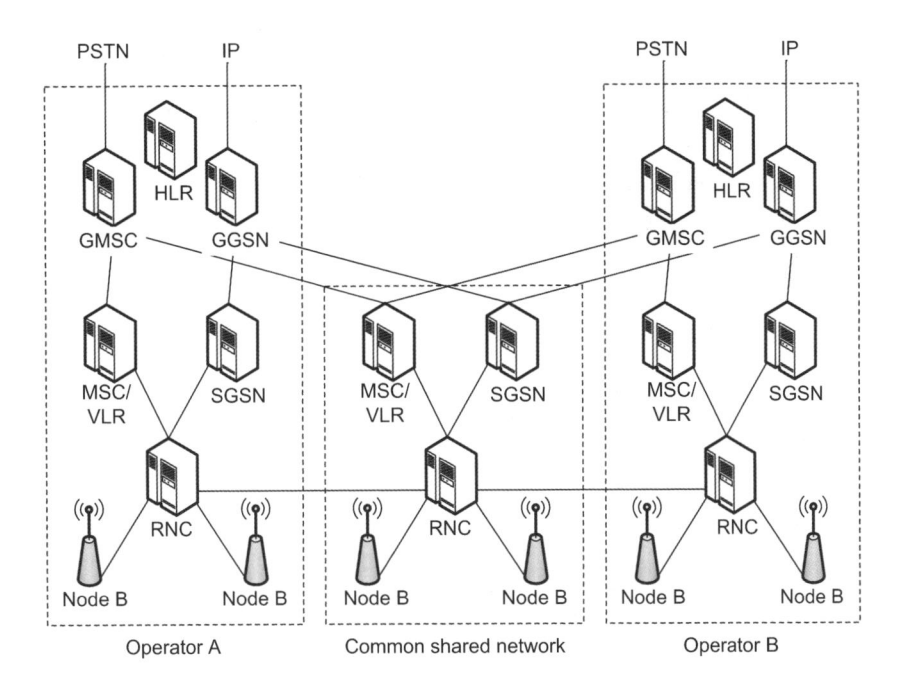

Figure 12.3 National roaming – common network.

simply not allow a particular sharing scenario and in such a case there is no use to take it into further considerations.

A site sharing scenario, being the easiest one, has no regulatory limitations. Therefore, it is up to the operators to use it or not. Even regulators have favourable opinions on that because of ecological aspects (less overall number of sites). Operators should have only a mutual site sharing strategy and agreement, as well as some operational arrangements on case-by-case basis.

A much more complicated situation is related to RAN sharing. First of all, every operator has its own UMTS frequency band assigned. So the usage of a competitor's band is generally not allowed. Therefore, the radio part of Node B should be separated for both operators. The other important issue is planning and maintenance of a shared RAN. To avoid elaboration on which part should take the lead in planning and responsibility in maintenance, the recommended solution is to create a joint company dedicated to these tasks. Both operators can have 50 % of shares in a joint company, and from a legal perspective it can be treated as a simple outsourcing process. To guarantee the quality and reliability of the network it is advisable to sign an SLA between two UMTS license holders and their joint company.

In national roaming with a geographical split scenario, there is no need of a joint company creation. However, it is also advisable to sign an SLA between the two roaming partners. From a regulations perspective, it is very important to check if a national roaming possibility exists and what can be counted to meet license obligations (own plus partner's network or own network only). The same aspect applies for national roaming with a common network and for RAN sharing combined with geographical split.

12.1.3 DRIVERS FOR SHARING

It is important to mention that, from a purely technical point of view, there is no advantage of network sharing at all. Every type of network sharing leads to more technical constraints and difficulties, both in the network design and maintenance phases. On the contrary, financial drivers make network sharing the factor that has to be considered while planning the business case for UMTS network rollout. The drivers that should be evaluated while considering network sharing include:

- License fee payback;
- License obligations fulfilment;
- Operator position on the market (incumbent or Greenfield);
- CAPEX savings;
- Early revenue;
- Optimal level of sharing;
- Hidden cost of sharing.

To evaluate a license fee payback, every operator should be aware of the country's conditions. The most important factors here are: the license fee itself, the number of operators and the country's population. Operators should consider network sharing if a great number of UMTS licenses awarded is combined with a high license fee and a small country population. In such a situation, building a shared network may significantly decrease license fee payback time.

A very similar situation occurs when license obligations are very high (e.g. percentage of the country's population or area to cover). Deploying the network in some rural areas with a small population density and low predicted traffic can be costly. And comparing it to possible revenues from that area may lead to the conclusion, that sharing network deployment costs with a partner is the best solution.

It is obvious that sharing is more profitable for a Greenfield operator than for an incumbent one, having already its own site base and infrastructure (e.g. from 2G network). The figures of possible

CAPEX savings from 15 to 40 % can be seen in market and strategic reports, but to have an idea of a particular case, the detailed calculation is necessary (long-term business plan including CAPEX, OPEX and revenue sides). As a result of such a calculation, the usual conclusion is that the optimal level of sharing is RAN sharing and the biggest savings are related to a coverage phase of rollout.

Furthermore, all the hidden costs and factors of a shared network should be considered:

- The cost of a joint company creation and operating, i.e. the company created to build and operate a shared network;
- Technical limitations, i.e. future capacity demands may lead to inefficient configuration of shared nodes;
- Corporate strategy, i.e. the need to have uniform vendors for a sharing solution and vendors shortlists chosen on a corporate level may not have intersection;
- Unbalanced union, i.e. it is not optimal to have an agreement with a much weaker partner;
- Divorce scenario, i.e. while reaching equipment capacity limits of shared networks it is necessary to divorce (split the equipment or stop a joint company activity).

It is extremely important to evaluate all the factors that can influence a network sharing decision and to prepare a detailed business case on a long-term perspective. Whereas site sharing is the most obvious and safe scenario, RAN sharing and national roaming may lead to a shift in a business case (lower CAPEX requirements at early years, early revenues), but cause many technical difficulties and big additional costs for a divorce scenario.

12.2 ADJACENT CHANNEL INTERFERENCE CONTROL

Given the amount of spectral overlap between the radio carriers in the UTRA FDD technology, interference between adjacent channels can become an issue without careful network design. In this section we examine the network design techniques that can be used to mitigate the effects of adjacent channel interference. For a more detailed analysis of adjacent channel interference the reader is referred to [2], which contains an assessment of the impact of adjacent channel interference on the performance of a UTRA FDD network and also an in-depth review of the possible interference mitigation techniques.

Let us first start by considering the conditions for adjacent channel interference problems to occur. The CDMA receiver has the ability to suppress the signal received on an adjacent channel to a certain degree using a filter centred on the wanted channel. The overall suppression of the adjacent channel interference is characterised by the adjacent channel protection (ACP), which was introduced in Section 6.4.3.5, and this includes contributions from the receiver performance and the transmitter performance. The processing gain achieved through the despreading of the received signal will also serve to decrease the level of the adjacent channel interference that enters the demodulator. Therefore, adjacent channel interference only becomes a significant issue in a CDMA system if the adjacent channel is much stronger than the wanted signal that the receiver is attempting to decode.

This means that adjacent channel interference is rarely a problem between radio channels belonging to the same network. The handover mechanism tends to mean that the UE always communicates with the Node B offering the highest signal strength. Therefore, as the received signal strength of the adjacent channel starts to exceed that of the current serving channel, the UE will naturally switch to the adjacent channel and the new adjacent channel will now have a lower received power than the new serving channel. Adjacent channel interference could become a problem in UTRA FDD networks that use different radio channels at different Node Bs. For example, an operator might choose to use one radio carrier frequency on its macrocells and another, adjacent carrier frequency on its microcells. A situation could arise where a UE is forced onto a microcellular Node B for capacity reasons, but it experiences interference from a nearby macrocellular Node B operating on an adjacent frequency. A scenario of this nature is likely to arise infrequently in practical networks and, even if it does, the network operator has the ability to address the problem, e.g. by enabling a handover to the macrocellular Node B.

The more serious form of adjacent channel interference occurs between radio carriers belonging to different network operators. The most likely scenario in which adjacent channel problems will occur in practice is where a UE from one network moves close to the Node B from another network, as shown in Figure 12.4. If the two Node Bs are operating on adjacent carrier frequencies and the serving Node B is some distance from the UE, the situation could arise where the received signal strength of the nearby adjacent channel Node B is significantly higher than the received signal strength of the serving Node B. In this situation, the UE may not be able to suppress the adjacent channel interference sufficiently to allow it to decode the information transmitted by its serving Node B. This means that a 'deadzone' will effectively occur around the interfering Node B within which it is not possible to make a call on the 'victim' network. If the UE is able to make a call, then it is also important to consider what happens on the uplink path. The UE is transmitting to a distant Node B and, as such, it is likely to be transmitting at a relatively high power. This power will be received at the nearby Node B as adjacent channel interference. This interference is unlikely to completely block the wanted signals at the victim Node B receiver, but it will have the effect of increasing the level of interference at this Node B and, in turn, the Node B will request more power from the UEs that it is serving. If a Node B is on the edge of its uplink coverage region, then this additional adjacent channel interference could have the effect of rendering the UE unable to make or continue a call. Therefore, on the uplink, adjacent channel interference from a nearby UE has the effect of decreasing the uplink coverage area of the Node B.

Having examined the scenarios in which adjacent channel interference can become a problem, let us now consider the ways in which the interference can be mitigated in a practical network. Returning to Figure 12.4, the main problem here is that the received power of the wanted signal is much lower than the received power of the adjacent signal. This situation could be eliminated by collocating the two Node Bs, as shown in Figure 12.5. In this case, as the UE moves closer to the adjacent channel Node B, it is also moving closer to its serving Node B and, therefore, the received power differential between the wanted and the interfering signals will remain roughly the same. On the uplink, as the UE moves closer to the adjacent channel Node B it will naturally decrease its transmitted power because it is also moving towards its serving Node B. Given the emphasis on site sharing as a means

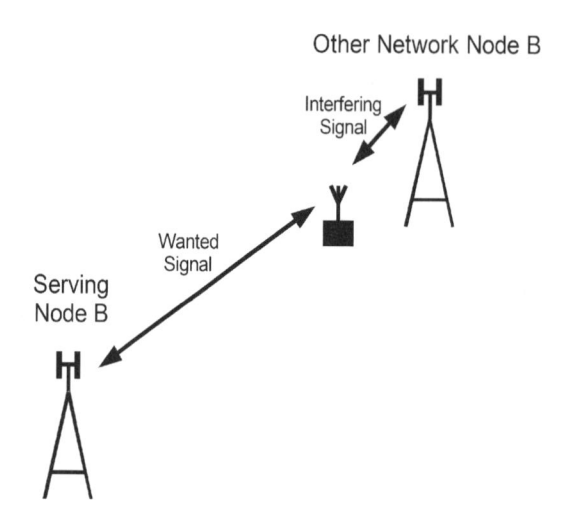

Figure 12.4 Adjacent channel interference scenario.

Network Design – Specialised Aspects 317

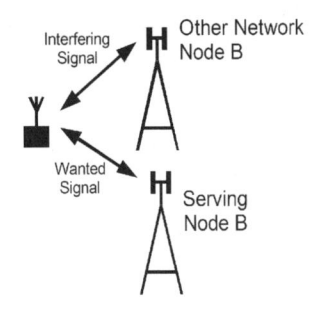

Figure 12.5 Node B collocation.

of decreasing the proliferation of cellular masts in many countries around the world, this collocation situation could occur quite frequently in practical networks and this will serve to decrease the impact of adjacent channel interference in areas where site sharing is used.

However, site collocation is by no means a universal feature of cellular networks and we must also consider other ways of mitigating adjacent channel interference in practical networks. One approach that could be used in specific areas where adjacent channel interference is a significant problem is to increase the power of the wanted signals in the vicinity of an adjacent channel Node B using an on-frequency, off-air repeater, as shown in Figure 12.6. This approach has a similar effect to Node B collocation in that the wanted and adjacent channel signals increase and decrease together and the difference between the two signals remains relatively constant. This approach is complicated by the fact that, in order to solve interference problems in one network, it requires the cooperation of another, competitive network operator, i.e. the competitor must agree to having a repeater installed at its Node B site.

The two mitigation techniques considered above rely on equalising the received power of the wanted and interfering signals. Another approach is to improve the suppression of adjacent channel interference by adding additional filtering at the Node B transmitter and receiver. Adding additional filtering at the Node B transmitter will improve the adjacent channel leakage ratio (ACLR) and this will decrease

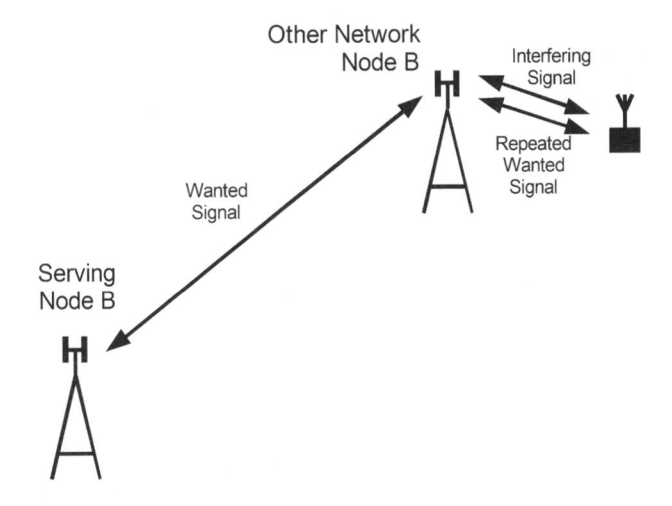

Figure 12.6 Using repeaters to mitigate adjacent channel interference.

the amount of adjacent channel interference experienced by nearby UEs operating on an adjacent channel. In common with the repeater approach, this technique is complicated by the fact that it relies on an action by one operator to decrease the amount of interference experienced by UEs belonging to a second, competitive operator.

An operator can improve the performance of a Node B receiver in the face of interference from nearby UEs operating on adjacent channels by installing additional filtering at its own Node B receiver. This technique has the advantage that an operator can improve the performance of its own network by taking action itself, but it has the disadvantage that it will only improve the performance of the Node B receiver, i.e. it is an uplink-only solution.

Other techniques that can be used to mitigate the effects of adjacent channel interference involve improving the ability of the receiver to suppress interference sources of any kind, i.e. both co-channel and adjacent channel. These techniques include the use of interference cancellation techniques at both the Node B and the UE, which allow the receiver to reject interference more effectively by attempting to characterise the interference and then cancel (or subtract) this from the received signal. Adaptive antenna techniques can also be used to suppress adjacent channel interference by steering antenna beams towards the wanted signal sources and steering nulls in the beam pattern towards the interference sources.

12.3 FUNDAMENTALS OF ULTRA HIGH SITE DEPLOYMENT

The design of macro cell layers using third generation CDMA technology offers a variety of options in regard of antenna pattern use. Most second generation macro-cellular networks were operated using either FDMA/TDMA or CDMA (IS-95) air interface technologies. These technologies limited the planner with critical constraints in either adjacent channel interference or PN-offset window sizes which, in most markets, led to a design of hexagonal grids using 3-sector antenna configuration per site (see Section 10.4.3). In CDMA based networks, the RF-planner is released from these constraints, and it enables him to design sites with a much higher amount of sectors. Economical circumstances, which forced many operators in the early 2000s to reduce the infrastructure cost led to the concept of using extraordinary elevated sites with a high numbers of sectors, called Ultra High Sites (UHS), to cover an entire medium sized European city.

Possible sites could be located on TV-towers as well as high office or industrial buildings. Figure 12.7 shows the scheme of the proposed site configuration. The German operator E-Plus has filed for a patent of this technology using a set of 24 sectors for an inner ring and 72 sectors for an outer ring deployed on two height levels, both of them higher than 120 m.

A test network in Erlangen, Germany, operating with 12 sectors has been successfully tested with throughputs of up to 384 kbps for a single user. More sites, like an 18 sector pattern on the Munich TV tower, have been deployed even in urban areas. However, up to now there is no information of successful test results using UHS-sites with 24 or 72 sectors, particularly in a loaded system.

The capacity N_{max} of UHS can be approximately derived from the common pole capacity formulas and for uplink can be expressed in form of:

$$N_{max}^{UHS} = \frac{W}{R_b} \cdot \frac{E_b}{N_0} \cdot \frac{FEU}{v_a} \cdot G_s \quad (12.1)$$

$$FEU = 1 - \frac{I_{total}}{C_{own}}$$

where W is channel bandwidth, R_b is data rate, E_b/N_0 is target bit energy to noise spectral density factor for given service, FEU stands for frequency reuse efficiency, I_{total} is the total interference power, C_{own} is the own cell signal power, v_a is voice/data activity factor and G_s stands for sectorisation gain.

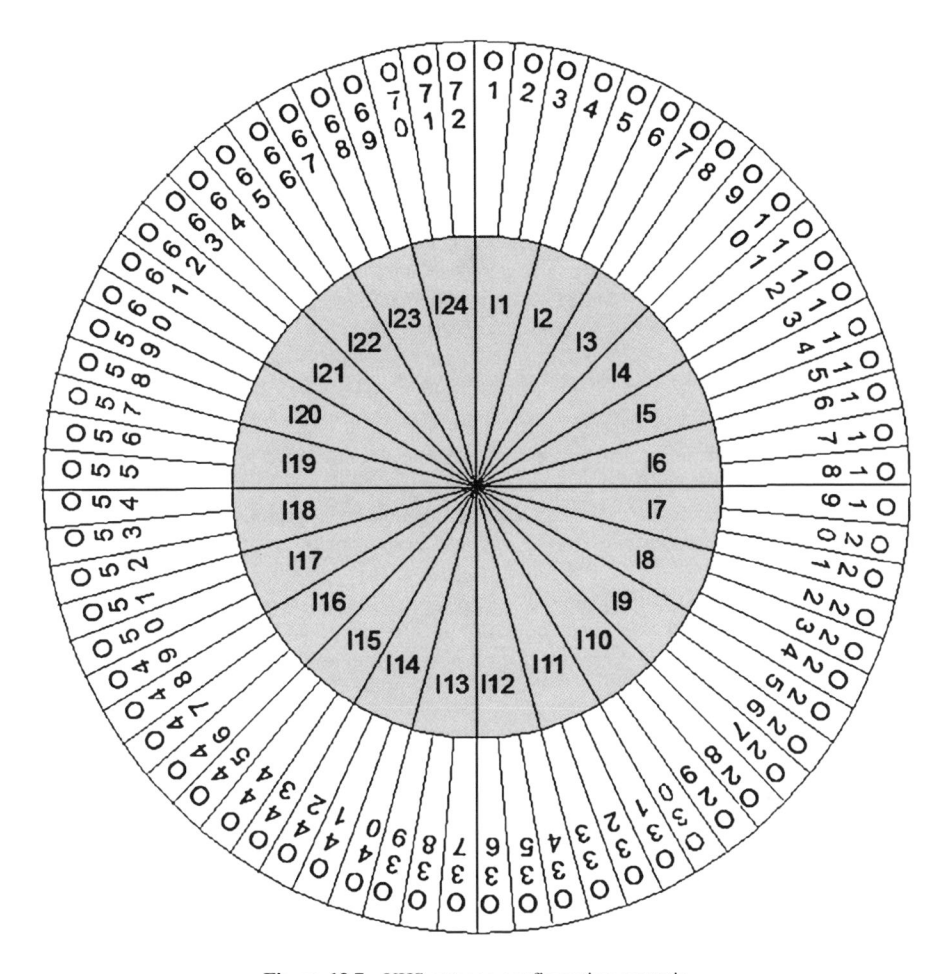

Figure 12.7 UHS antenna configuration scenario.

While W, R and v_a will not differ between a 3-sector site and an UHS, the FEU and s could deviate, where

- a standard value for G_s of a 3-sector site is commonly 2.55; this value can be different with more sectors due to higher side lobe interference;
- the FEU could experience a significant decrease in the uplink due to the higher chance that the antenna receives signals from mobiles in other cells, since it is placed on a high altitude.

Both factors will possibly lead to a significantly lower capacity compared to a network of 3-sector sites containing a comparable number of cells. In the downlink, however the conditions change, i.e.:

$$N_{\max}^{UHS} = \frac{W}{R_b} \cdot \frac{E_b}{N_0} \cdot \frac{FEU}{v_a} \cdot G_s \cdot \frac{1}{1-\alpha} \tag{12.2}$$

The factor α in Equation (12.2) describes the code orthogonality, which depends on the surrounding clutter. As a difference to the uplink, a UHS offers the chance that all downlink codes of an entire cluster can be transmitted from almost exactly the same point and time, which significantly increases the FEU and the code orthogonality. Considering this advantage, a higher downlink capacity compared to a network of several 3-sector sites can be assumed. Considering the fact that most data applications have a higher downlink volume, a UHS system could give an advantage here.

Economically, the significantly lower number of required sites for UHS deliver some strong economical arguments, e.g.:

- a significantly lower number of site leases;
- less problems and cost on EMC-issues;
- less backhaul transmission links;
- enables to get very good indoor CPICH coverage for a large urban area fast;
- good coverage can be provided faster with less investment.

These arguments have to be counted against strong technical concerns:

- Much higher cell boundary scattering, particularly in built-in areas. This leads to a much higher potential of soft handoff areas, which worsens with decreased distance to the site.
- More handovers, which bears the higher risk of call drops, particularly close to the site.
- More intra-cell interference due to the fact that the mobiles deviate significantly in their distance to the site, and thus in their output power. This can lead to sudden power changes causing a larger TX power headroom.
- Some vulnerability to wind load. Commonly antenna infrastructure is designed to a horizontal sway of 5° assuming wind speeds that occur on height levels <60 m. Given the required tolerances to design 72 sectors, which mean 5° per sector on 120 m altitude, the antenna mountings require a really solid built.

UHS is a straight contradiction to the very most of all CDMA optimisation approaches worldwide. On the other hand, it shows an innovative way to increase site capacity, and could pioneer the path away from hexagonal 3-sector designs, which are actually remains driven by previous 2G limitations (frequency reuse patterns), and not by actual 3G limitations. It should be noted that the use of UHS requires development of new propagation models since the antennas are located much above the height for typical base station [3]. It can seem reasonable that in the mid-term future a network capacity increase is rather feasible by a site capacity increase, and less by network densification. However, it remains to be seen, whether site capacity increase can be achieved by an increase of height and sectors or rather by advanced air interface techniques like HSDPA.

REFERENCES

[1] Eirwen Nichols, Ines Respini, Martin Garner, Virtyt Koshi, Carrie Pawsey, Ajay Ghambir, '3G survival strategies: build, buy or share', 2001, OVUM.
[2] Multiple Access Communications Limited, 'Research into the impact of dead zones on the performance of 3G cellular networks', RA0703DZ/R/18/008/1, January 2004, available at http://www.ofcom.org.uk/research/technology/archive/dzone.pdf.
[3] Andreas Hecker, Thomas Kürner, 'Analysis of propagation models for UMTS Ultra High Sites in urban areas', *Proceedings of 16th IEEE International Symposium on Personal, Indoor and Mobile Radio Communications PIMRC 2005*, 12–14 September 2005, Berlin.

Part IV
Optimisation

13

Introduction to Optimisation of the UMTS Radio Network

Roni Abiri and Maciej J. Nawrocki

What does it mean to optimise a UMTS network? Going through the books which have the word *optimisation* in the title, the reader can feel a bit lost since this term is usually not defined. Often the word *optimisation* is used for network planning activities and even more often authors imply *manual* optimisation. What is, therefore, the right approach? Optimisation theory is devoted to searching the coordinates of the highest (maximum) or the lowest value (minimum) of a function. This process can be very simple and conducted in analytical form when, e.g., the function derivatives can be computed in the entire domain. It can also turn out to be really complicated, usually when the function is described as a 'black box'; in this case, many iterations need to be performed to estimate the optimum. Therefore, optimisation theory as a whole additionally concentrates on methods which are able to find the optimum and do this as quickly as possible. Unfortunately, the majority of real life optimisation problems, including UMTS radio network optimisation, cannot be treated in a direct and simple way by means of analytical optimisation, since the mathematical model of the network is far too complicated (if possible to create at all).

The ideal situation would be a direct network *synthesis* to meet performance requirements, where the operator, using mathematical models, would be able to compute, e.g., exact coordinates of site locations or angles of antenna tilts. This seems to be impossible at the present state-of-the-art of network description; thus, other procedures need to be used. These comprise network performance *analysis* and, based on analytical results, making decisions about some parameter settings. This process is repeated until achieved results are acceptable, i.e. the network performance is good enough.

In the above described process, we have two important modules. The first one calculates the network performance (*analysis module*) and the second one, based on the results from the first module, makes decisions about parameter value changes (*decision module*). In traditional (i.e. manual) network planning and optimisation processes, a network planning software tool executes the *analysis module*, while the human becomes the *decision module*. Consequently, a radio planning engineer configures network parameters manually and the network planning tool analyses the given configuration. If the

---

Understanding UMTS Radio Network Modelling, Planning and Automated Optimisation   Edited by Maciej J. Nawrocki,
Mischa Dohler and A. Hamid Aghvami   © 2006 John Wiley & Sons, Ltd

obtained results are not acceptable, the analysis process has to be repeated several times, until the goal is achieved.

Modern cellular networks are large and many of their key parameters are interdependent. This is the main reason of an increased demand for automated optimisation solutions, since an engineer usually cannot cope with this level of complexity (traditionally involving an examination of coverage plots as colour images together with some statistical analysis). The computer, together with specialised software, becomes the *decision module* in this case. Consequently, the whole optimisation process can be performed *automatically* through iterative trimming of network parameters with only small assistance of a human. This process should correctly be referred to as *optimisation*. In the book, however, it is purposely called *automated optimisation* to emphasise this idea.

Several works and projects were carried on optimising wireless networks, specifically UMTS networks. The IST Momentum project, finalised in 2003, was the most extensive one [1]. Earlier example projects are ARNO [2] and STORMS [3], completed in 2000 and 1999 respectively. These projects focused in solving the optimisation problem solely based on planning data, because UMTS networks were not yet commercial. Currently, there are several commercial products available in the market that are offering optimisation solutions to UMTS and other cellular technologies.

13.1 AUTOMATION OF RADIO NETWORK OPTIMISATION

Wireless networks have become an essential part of the infrastructure that supports people's daily life all over the world. The usage of cellular phones has boomed in the last decade and the penetration rates are close to saturation in many countries. As competition is enforced by the different regulators, the wireless service costs went down considerably, thus becoming a commodity service. Today, there is a steady growth in the cheap air minutes due to the desire of the carriers to increase or at least sustain their Average Revenue Per Unit (ARPU). Also, the on-going introduction of new services, many of them requiring high throughputs, adds to the network loading. Although wireless networks become more and more mature, the optimisation work is not necessarily decreasing, because of the following reasons:

- Constant drive for performance improvement: As the wireless service replaces in many cases the landline service, users are looking for landline quality in terms of coverage (specifically in-building) and connection quality.
- Changes in users' profile: The introduction of new services puts additional stress on the infrastructure, causing additional optimisation efforts.
- Cost savings: The erosion of prices creates a pressure to lower both CAPEX and OPEX costs, both from the infrastructure side and the manpower expenses. For example, it can be translated into having less optimisation engineers or using less qualified people.
- Affect of permissions (often referred as zoning) and regulations: In most countries, there is a growing demand for limiting the location and transmitted power levels of the cellular sites; often, sites can be deployed at city borders only, thus complicating the optimisation, especially for CDMA based technologies.
- Multi-Radio Access Technologies (RAT) networks are reality: Multi-RAT networks such as UMTS/GSM/GPRS are common in many places and it adds a high complexity to the network planning and optimisation. In the near future, additional wireless access technologies, such as WiFi, WiMAX and others, may be available from the same service provider, adding a degree of complexity to the optimisation.
- Changing propagation conditions: The allocation of a higher frequency band for UMTS compared to GSM requires deployment of more sites compared to 2G networks, mostly in urban areas, thus increasing inter-cell interference.

Luckily, high performance technological tools, including computers, software, etc., have become more easily available at reasonable costs. For example, the costs of 3D building databases have fallen dramatically in the past years, in parallel to an improvement in resolution.

Furthermore, the relative complexity of 3G networks is considerably higher compared to 2G networks; many more parameters need to be optimised. These parameters are not anymore limited to basic RF requirements only, such as coverage, but also interact with and influence the data delivery performance and other new network Key Performance Indicators (KPIs). The culmination of these facts leads to the growing demand for automation of the optimisation process; this can also be induced from the strong market requirements for such optimisation products tried to be matched by both commercial companies and academic institutions in the field of optimisation.

13.2 WHAT SHOULD BE OPTIMISED AND WHY?

Generally speaking, one can say that every parameter that exists in the network may undergo optimisation. There are numerous ways to group these parameters under different categories, for example:

- remotely tuneable parameters versus those which require a physical site visit;
- parameters that influence only locally within the cell's service area versus parameters that may impact other cells as well;
- RF related parameters, mostly affecting coverage and interference;
- traffic related parameters, mostly affecting delivery of both voice and data;
- 'hard' parameters such as CPICH transmitted power level versus 'soft' parameters such as power control parameters;
- 'static' parameters which are planned according to the static behaviour of the users versus 'dynamic' parameters which govern the dynamics of calls such as handover from cell to cell.

We will, however, group them into parameter sets useful for automated optimisation. To this end, the first set of parameters is the one that directly governs inter-cell relations. Parameters within this set may be grouped into the following two categories and dealt with in sequential manner:

1. parameters that affect RF footprint, coverage and interference;
2. parameters that affect mobility between cells.

Chapter 15 elaborates on these parameters.

The next set of parameters that may be optimised automatically is described as follows. As wireless networks become more and more sophisticated, there are parameters that are adjusted automatically by various infrastructure controllers. For example, the order of and the allocation of scrambling codes for a specific call is done by the cell's controller, because it makes no sense to leave it to the optimisation engineer. However, this automatic setting of parameters is done in the cell only and does not take into account the inter-cell effect it may cause. The engineers thus regard these parameters per cell, but lack tools (and attention) to see their inter-cell influence. To illustrate this, let us consider the downlink power control: The controller sets the specific transmit power levels of a DCH according to:

- BER/BLER report from the User Equipment (UE);
- total transmitted power from this specific cell.

However, the following aspects are not being considered:

- Effect on the other UEs that are served by the same cell (e.g. cell breathing);
- Effect on UEs served by other cells, due to the increase of interference.

Now, the optimisation engineer may set a different limit to the DCH powers per Radio Access Bearer (RAB) and cell, but in reality these values are usually set to default as it seems to be impossible to monitor and optimise these parameters manually.

Regardless of the discussed parameters, there are several requirements to make them qualify for optimisation:

- The optimisation engineer can control this parameter, i.e. excluded are those parameters which are either set on a default value for the whole network/cluster of cells or which the engineer cannot access using network configuration management tools.
- There is a way of either predicting or measuring the results of the parameter change. If a specific parameter does change the network performance, but in a way that cannot be predicted or measured, there is no way of optimising this parameter – neither manually nor automatically. Such situation may arise due to a lack of predictive tools, algorithms or reports generated by the network.
- The parameter change has a measurable influence on the network performance, i.e. excluded are those parameters that hardly change the network performance; although it may technically be possible to tweak these parameters, an improvement in outcome is negligible, and hence investing into their optimisation is not warranted. An example of such a parameter is the minimal channel power level: studies have revealed that setting this parameter below a minimal value, so that it has a measurable effect on the total transmitted power of the cell, does neither improve the network performance nor the specific call performance. In US CDMA networks, based on the IS-95 standard, this value is about $-12\,dB$ compared to the pilot transmitted power; therefore, even if an engineer sets it to $-20\,dB$ below the pilot level, it will not have any affect on the network's performance.

13.3 HOW DO WE BENCHMARK THE OPTIMISATION RESULTS?

In past, when 2G networks were analysed, it was quite easy to see the difference between pre- and post-optimised networks just by looking at two major plots: coverage plot and C/I plot. This is unfortunately not the case for WCDMA networks, which require much more information to be used to assess the improvement achieved by means of optimisation. The situation might even be more complicated, where an optimisation procedure achieved some performance improvements locally, but globally led to deterioration in the network performance. This chapter discusses the minimal set of data that should be prepared and analysed in order to understand the effect optimisation has on the network. For this, we obviously do not want to use *a posteriori* network statistics as we may impair the service quality, but we prefer to know the merit of the proposed changes in advance. Additionally, as cellular operators need to minimise their operating expenses, there must be an analysis forecasting both the improvement and the investment in equipment and labour required for achieving this task.

When working with simulation tools to benchmark the optimisation gain (regardless if done manually or automatically), the engineer has to look at the following groups of data:

1. Location based, area distribution and traffic distribution plots as well as histograms;
2. Sectors (or network) statistical data and histograms;
3. Costs and optimisation efforts.

These are explained in more detail in subsequent subsections.

13.3.1 LOCATION BASED INFORMATION

The following plots are the minimal set for the analysis:

a. Pilot channel (CPICH) Received Signal Code Power (RSCP) level at each geographical bin;
b. CPICH E_c/I_0;
c. UE transmitted power level;
d. Handover state, i.e. 1-way, 2-way or 3-way;
e. Best server plot.

From these charts, most commercial simulation tools enable histograms of above data to be obtained. These histograms, when analysed correctly, supply valuable information on the network performance and can be used for benchmarking the benefit in performing the suggested changes.

To illustrate the suggested process, let us consider the analysis of CPICH E_c/I_0 as it is perhaps the single most significant parameter in benchmarking the network performance. Figure 13.1 shows the CPICH E_c/I_0 before and after optimisation, where:

a. it is very difficult to see whether the total change is positive or negative; and
b. there is no traffic weighting, so maybe the CPICH E_c/I_0 improvement was mainly received in low traffic areas, whereas the high traffic areas suffered from worse E_c/I_0 values.

Another and simpler way of looking at the network improvement is to look at the histogram of the traffic (or area) versus E_c/I_0 values, as shown in Figure 13.2.

It can clearly be seen that the optimisation activity met the goal of improving the E_c/I_0, mostly by taking traffic off both ends – with either too low or too high E_c/I_0 values, so that the accumulated percentage of traffic below -11 dB went down from 16 to 12%, leading to a quite significant improvement.

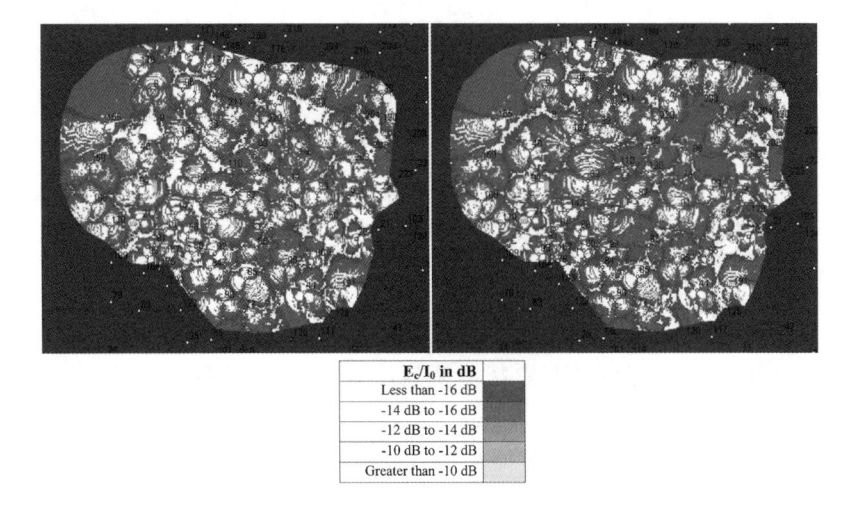

E_c/I_0 in dB
Less than -16 dB
-14 dB to -16 dB
-12 dB to -14 dB
-10 dB to -12 dB
Greater than -10 dB

Figure 13.1 CPICH E_c/I_0 plots for the network before (left) and after optimisation (right).

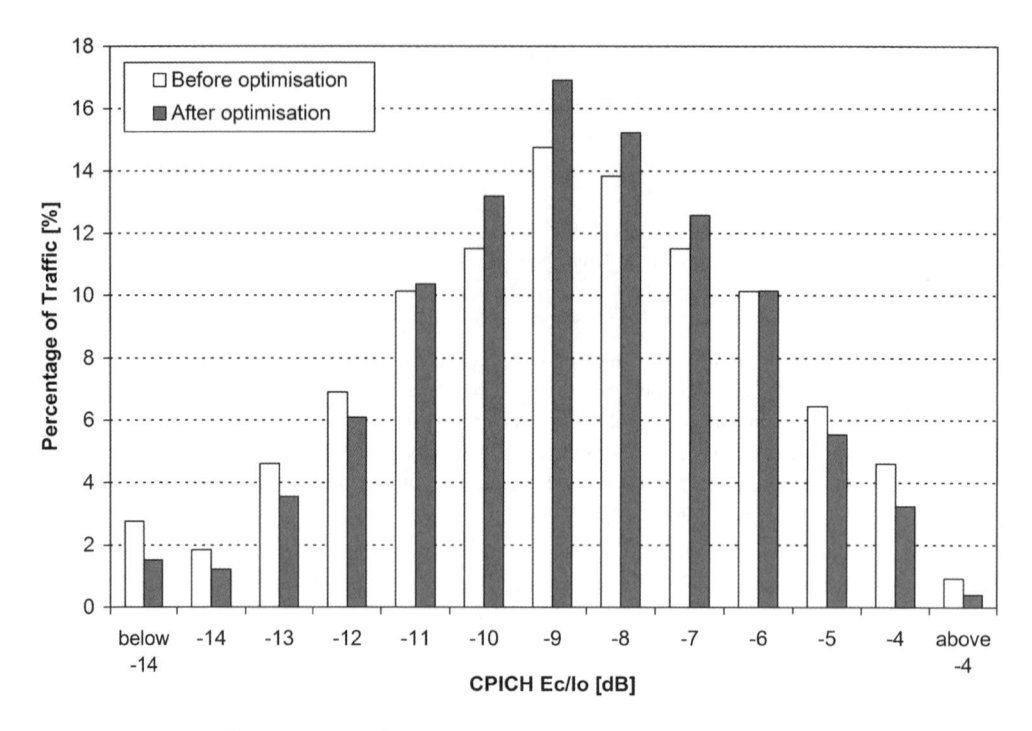

Figure 13.2 CPICH E_c/I_0 distribution before and after optimisation.

The same method can be applied to each of the above parameters. To do so, the optimisation engineer has to supply the following data:

1. traffic distribution data – for all types of traffic (RABs);
2. threshold values for each of the parameters (for the above example a threshold value of −11 dB was used for CPICH E_c/I_0);
3. A weighting factor for each of the above parameters, to get one number scoring for the network performance.

Histogram based analysis is a good way for analysing the performance of optimisation on the network level, but it obscures potential local degradation. Maps, as in Figure 13.1, that show differences between parameters before and after optimisation could then be used for analysing the geographical distribution of the improved and degraded local areas. These maps could be calculated according to the threshold values defined during the optimisation. From the example above, the map of differences between CPICH E_c/I_0 before and after optimisation could be calculated, taking into consideration the E_c/I_0 threshold of −11 dB. In this case, the positive values will represent improved areas and negative values will represent degraded bins.

13.3.2 SECTORS AND NETWORK STATISTICAL DATA

In order to complete the network analysis for the suggested optimisation changes, we also need to look at the sector performance data. The most important parameters are:

a. the total average transmit power per sector relative to the maximum available power or sector loads;
b. histogram of traffic channel power levels;

Introduction to Optimisation of the UMTS Radio Network 329

c. noise rise above thermal noise;
d. number of connections;
e. percentage of UEs with more than 1-way link (SHO state).

Unlike the previous section, the geographical distribution of the data above is less useful, and a tabular way and statistical distributions per sector are mainly used for optimising the results. An example of this data is shown in the Table 13.1 (the actual table usually has more columns, depending on the specific planning tool, and the availability of this information from the network statistical database). The last row supplies the average information per selected set of sectors. It may also be per RNC or the entire network.

Figure 13.3 is an example of such analysis done for the primary traffic per sector; results are presented before and after optimisation.

Table 13.1 Main statistical data for benchmarking optimisation results.

Sector ID	CPICH Power level (dBm)	Average Power level (dBm)	% of DCH above threshold	Reverse noise rise (dB)	Number of connections	Percentage of 1-way	Percentage of 2-way	Percentage of 3-way
0011	33	35.9	2	0.13	8	65	27	8
0012	33	37.2	15	0.45	17	59	33	8
0013	32	35.5	0	0.19	12	67	28	5
—	—	—	—	—	—	—	—	—
Total								

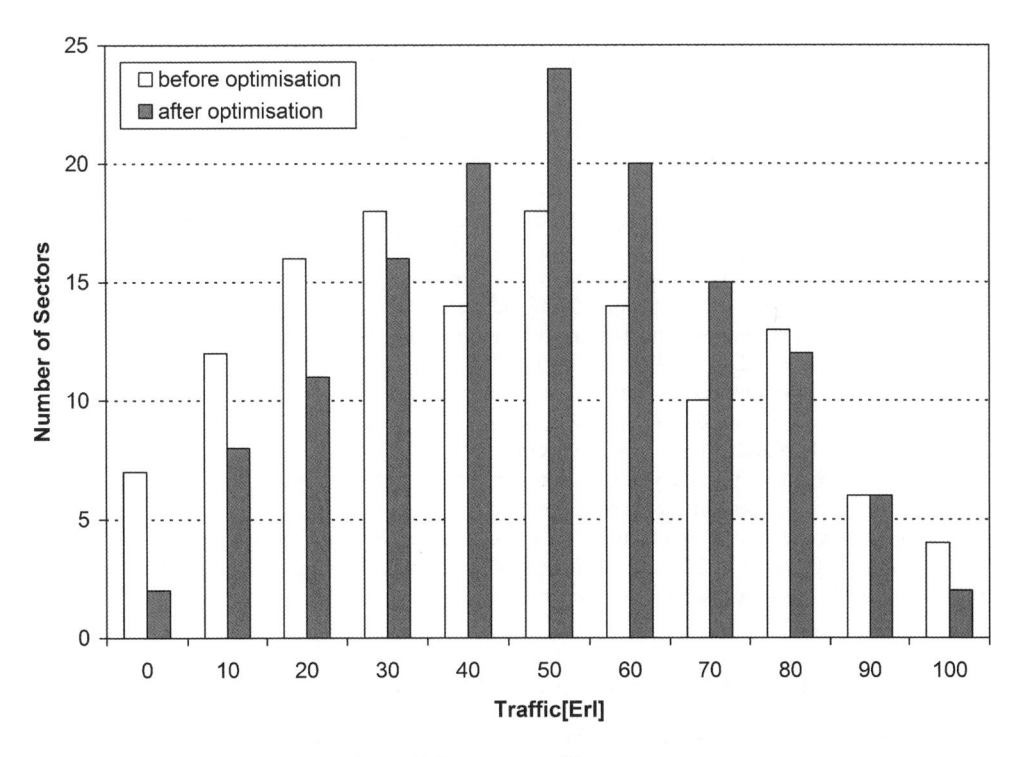

Figure 13.3 Primary traffic per sector.

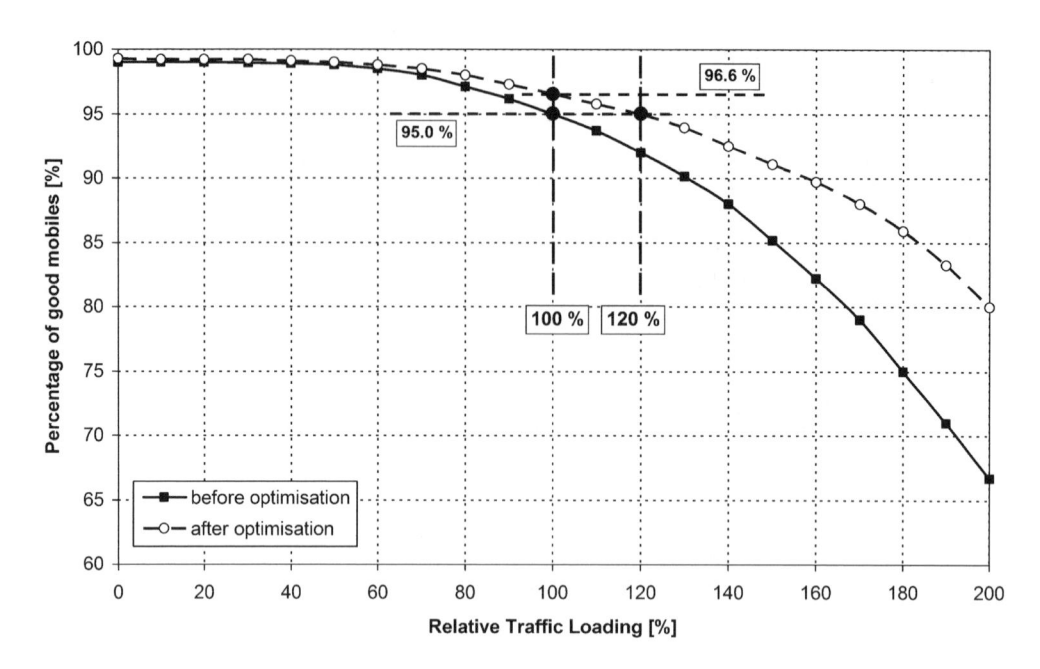

Figure 13.4 Traffic load analysis.

Again, this is an example for a successful optimisation as the carried traffic grew significantly, mostly by balancing traffic so that the number of either very low or very high traffic sectors went down, enabling support of additional traffic with the same performance.

Traffic Load Analysis: A powerful method of benchmarking the optimisation results, especially for loaded networks, is based on traffic load analysis. Here, instead of running network simulations only once using a single value for the traffic load, the simulations are repeated several times with varying traffic loads. Various parameters can be analysed using that method, including all of the above and more.

Figure 13.4 shows an example for such a Traffic Load Analysis, where the parameter shown is the percentage of 'good' mobiles, i.e. mobiles that were successfully connected to the network. Obviously, this percentage decreases as the network loading increases; therefore, analysing both absolute values and trends supplies a method of estimating the robustness of the network to increased traffic demands, be they temporary or fixed.

The reference traffic for this example was, for the original network configuration, set as the amount of traffic where the percentage of good mobiles crosses the 95 % line. After optimisation, it can be clearly seen that the relative traffic can be increased by 20 % maintaining the same quality level. Another way of looking at this analysis is to ask what the increase of percentage of connected mobiles is after optimisation: the result is an increase from 95 to 96.6 %. This increase seems small, but when we look what happened to the non-connected mobiles, we see that their number decreased from 5 to 3.4 %, yielding a decrease of 32 % which is quite significant. Further elaboration on this subject is found in the analyses of case studies in Chapter 15.

13.3.3 COST AND OPTIMISATION EFFORTS

Finally, it is required to estimate the cost and effort associated with the suggested optimisation plan. There are many cases where a similar improvement can be achieved by lower cost optimisation activities; for example, tweaking the CPICH power level is a very simple activity that is executed from the Network Operating Centre, in contrast to azimuth changes that require a field team and dedicated means, such as an expensive crane. Antenna changes induce additional costs of antenna purchase as well as field team costs for the antenna mounting. That means that the different costs can be defined for each type of optimisation activities. Implementation costs can be defined per cells and sites, taking into account antenna mounting type and simplicity of reaching the antenna. The engineer has also to minimise the number of visited sites to reduce costs and to enable implementing the optimisation changes over the network's maintenance window (usually late night until dawn).

The above briefly suggested proceedings and approaches provide an effective method of achieving a cost sensitive network optimisation.

REFERENCES

[1] T. Kürner, A. Eisenblätter, H.F. Geerdes, D. Junglas, T. Koch, A. Martin, *Final Report on Automatic Planning and Optimisation*, IST-2000-28088-MOMENTUM-D47-PUB; October 2003, http://momentum.zib.de/index.php.

[2] M. Vasquez, J.K. Hao, *A Heuristic Approach for Antenna Positioning in Cellular Networks*, J. of Heuristics, 7: 443–472, 2001, Kluwer Academic Publishers.

[3] R. Menolascino, M. Pizarroso: STORMS Project Final Report, ACTS 016 Project – Software Tools for the Optimisation of Resources in Mobile Systems, April 1999.

14

Theory of Automated Network Optimisation

Alexander Gerdenitsch, Andreas Eisenblätter, Hans-Florian Geerdes,
Roni Abiri, Michael Livschitz, Ziemowit Neyman and Maciej J. Nawrocki

Automated radio network planning and optimisation, as understood here, is the process of algorithmically arriving at a radio network design based on input data that is typically available within radio network planning tools. This data shall be adjusted to the real network conditions by fine-tuning the coverage prediction model, radio resource management (RRM) algorithm parameters, if dynamic simulators are used, and traffic figures by appropriate network measurements: drive test surveys, reports from mobile terminals and network statistics. Throughout the rest of this chapter, *optimisation* shall address network planning as well as network optimisation. The expected results of an optimisation process are suggestions on how to deploy, extend, or reconfigure a network.

We give an introduction to generic optimisation in Section 14.1. The basis for formulating optimisation problems associated to UMTS radio networks is then provided by describing the optimisation parameters (the degrees of freedom) in Section 14.2 and the criteria for which to optimise in Section 14.3. Optimisation methods are then differentiated according to whether or not they use simulation-based evaluation as a subroutine. Methodology and implementations of 'black box' approaches that iteratively determine the quality of intermediate solutions via simulation are described in Section 14.4. They have the advantage that they make decisions based on detailed performance analysis. However, the accuracy comes at the cost of large computation times. Optimisation methods that develop alternative evaluation approaches in order to estimate performance and avoid simulation are then presented in Section 14.5. Suitability of the presented algorithms for UMTS radio network optimisation is briefly comparison in Section 14.6.

14.1 INTRODUCTION

Optimisation problems occur implicitly or explicitly in many areas of pure and applied science, industry and economics. Consequently, many methods and algorithms are known for solving optimisation problems. This section briefly discusses how to translate a real-world problem into the domain of automated optimisation and which options are available in the optimisation toolbox.

14.1.1 FROM PRACTICE TO OPTIMISATION MODELS

Radio network optimisation is distinguished from network tuning, which is addressed in Chapter 16. In network tuning, a network's configuration is changed in reaction to performance deficiencies while the network is in operation (on-line). Network optimisation is assumed to be performed off-line; the optimisation software is not perceived as part of the operation and maintenance system. Life data is only used for purposes such as calibrating propagation predictions models (see Section 5.6), modifying specific path loss predictions, highlighting capacity shortages or selecting regions for optimisation.

14.1.1.1 Scope of Automatic Optimisation

Automatic radio network optimisation can serve various purposes. It is a means to improve network quality, to reduce the effort of running a network (operational costs, OPEX), to reduce infrastructure costs (capital expenditure, CAPEX), and to implement planning rules (please refer to Chapters 7 and 8 for more details on business aspects of network design).

The specific goals of employing automatic optimisation tools vary with the setting in which they are used. During initial roll-out, the emphasis is often on providing maximum coverage under budget constraints. The role of the optimisation tool is to produce a *pre-optimised* network design with good coverage, but not necessarily with good capacity. Once an initial network is set up, the continuing extension of the network aims at closing coverage holes and providing additional capacity where necessary. During this phase, the role of an optimisation tool can be to select new site locations for coverage or capacity enhancements, to propose revised configurations for antennas in the vicinity of new sites, or to propose antenna configuration changes to decrease interference coupling among cells. Automated optimisation is also used to assist classical manual network optimisation tasks, i.e., to reduce cell over-shooting or pilot pollution, to improve E_c/I_0 quality, to balance cell sizes and load among neighbouring cells, or to shape soft handover areas.

14.1.1.2 Distinguishing Goals and Constraints

The first step towards algorithmic network optimisation is to formalise the optimisation goal (or goals) and to clearly state the conditions under which the optimisation shall be performed. Basically, each of the network quality measures defined in Section 6.4 can be used as part of the optimisation goal. Each quantity may possibly occur within a constraint. Sector configurations may be constrained individually, independent from the others.

Some of the questions that need to be answered (implicitly or explicitly) are: Which quality indicator shall be optimised? Which indicators shall not drop or exceed a given threshold? Which changes of the network are allowed: tilt, azimuth, antenna type, antenna height, pilot power, etc.? To which extent may each of the parameters be changed?

Moreover, is the number of changes to be minimised that is necessary to achieve given quality targets? Alternatively, are quality measures to be optimised subject to a limited number of configuration changes? Budget constraints may be introduced together with a cost model for infrastructure and/or configuration changes. Once the goal and the constraints are precisely stated, the optimisation task is defined and optimisation may be started.

14.1.1.3 Combinatorial Optimisation

UMTS radio network optimisation can be stated as a combinatorial optimisation problem. The optimisation task is formally given through a set of solutions (solution set), an objective function f_{obj} assigning a real value $f_{obj}(x)$ to every solution (x), and the goal of either minimising or maximising this function over the underlying solution set. This is the simple case. In a more complex variant, where there are a number of objective functions, the objective function is multi-dimensional. Solutions are then no longer necessarily comparable with respect to their objective function value. This is the domain of Pareto optimisation or multi-objective optimisation. Models with a multi-dimensional objective are appealing from a practical point of view as there are several distinct quality measures for a UMTS network. From a computational complexity point of view, they are much harder to solve than ordinary models. In accordance with the majority of literature on UMTS optimisation, multi-dimensional objective functions are not addressed further.

The optimisation decisions are typically taken among discrete sets of options. An antenna type has to be chosen from a catalogue prescribed by the operator; antenna tilts are commonly set to integer values; and the azimuth is often adjusted in steps of 5° or even 10°. These are examples of discrete choices. If all choices are discrete, the problem is called a *combinatorial* optimisation problem. Combinatorial optimisation [1] is a comparatively young mathematical field, which has seen an impressive development over the second half of the 20th century. Combinatorial optimisation problems are quite different from continuous optimisation problems. There is, for example, no (standard) notion of a differentiable, smooth or even continuous function over a discrete domain. Hence, the methodologies for solving combinatorial optimisation problems differ greatly from those for solving classical continuous optimisation problems.

14.1.1.4 Complexity of Optimisation Problems

Although the configuration of a sector typically involves only a few hundred options (e.g. the number of available antenna types times the number of considered azimuth values times the number of considered tilts), the resulting number of possible network configurations is beyond imagination. If only 100 sectors are to be configured with 25 options each, then the optimisation has to consider 25^{100} configurations (if all are distinct). The number of particles in the observable universe is estimated to be around 10^{79}. This is an example of the phenomenon called *combinatorial explosion*: the number of solutions to a combinatorial optimisation problem grows extremely quickly with the problem size.

The large number of possible solutions to an optimisation problem is not necessarily an obstacle to finding the optimal solution with limited effort. For UMTS radio optimisation as discussed here, finding some (technically) feasible solution can usually be done quickly. Finding an optimal one, however, is typically hard. There is a formalised concept for measuring *hardness* of optimisation problems in complexity theory, see the classical text [2]. A central notion in complexity theory is *NP-hardness*. Most of the planning problems discussed later are NP-hard. It is commonly believed that *hard* optimisation problems cannot be solved to proven optimality *quickly* (in a number of steps bounded by a polynomial in the input size). This is a worst case statement on the infinitely many instances of an optimisation problem, not a statement on each individual instance. Fortunately, many practically relevant instances of NP-hard problems can be solved to proven optimality in acceptable running times.

14.1.2 OPTIMISATION TECHNIQUES

The spectrum of methods for solving hard optimisation problems is diverse, ranging from ad-hoc, easy-to-implement greedy methods to heavy-duty mathematical optimisation techniques. The choice of the method typically depends on the purpose, the time available to develop and implement specialised optimisation techniques, and the necessity to obtain solutions with guaranteed optimum objective value. The following paragraphs shall provide a brief overview of common techniques as an introduction to the detailed discussion of some optimisation techniques in Sections 14.4 and 14.5. The methods in Sections 14.1.2.1–14.1.2.4 are *heuristics*, they only strive to find 'good' solutions, without relating the solutions found to the global optimum. The methods in Section 14.1.2.5, in contrast, can give quality guarantees and are therefore also called *exact* methods.

A common classification of heuristic optimisation methods distinguishes between *construction* and *improving* methods. A construction method starts from nothing and builds up a solution in steps. While the final step has to produce a complete solution, intermediate steps may produce partial solutions. In the context of UMTS radio network design, such a partial solution may be a partially configured network. For example, only 50 out of a desired number of 150 sites are configured. Improvement methods, on the other hand, typically operate on complete solutions, iteratively changing some aspect in order to improve the result. Most of the neighbourhood search methods described next are of this kind.

A trivial solution strategy of the improvement type is *complete enumeration*. The idea is to enumerate the entire solution set, to memorise the solution with the best objective function value checked so far, and to return the best solution once all have been analysed. As the size of the solution set often grows at least exponentially with the size of the instance, complete enumeration is only practically feasible for very small problem instances.

14.1.2.1 Greedy Algorithms

A very popular type of construction method are greedy heuristics. The basic principle of greedy methods is to start with a trivial initialisation (e.g. no antenna is configured, or no base station is present). Then one parameter of the solution to the value that yields the largest improvement in the objective function is fixed. For example, in a base station selection problem, the base station which most improves coverage of yet uncovered points is selected. For problems with a very specific structure (called *matroids*), this algorithm yields optimal results. For all other problems, it is to be regarded as a heuristic that might well lead to suboptimal solutions. The advantage of greedy methods is that they can be executed very quickly. Greedy algorithms are often used as a starting heuristic.

14.1.2.2 Local Search Techniques

Search techniques use *neighbourhood* relations between solutions. The neighbourhood of a solution is then simply a subset of the solution set. For UMTS optimisation, for example, two networks may be considered neighbours if they only differ in the configuration of one sector. The idea of *search techniques* is to start at one solution and to explore the search space by iteratively moving from one solution to a neighbour.

A solution is a *local optimum* in the search space if none of its neighbours has a better objective function value. A solution is a *global optimum* if there is no better solution in the entire solution set. A locally optimal solution need not be globally optimal. In fact, a locally optimal solution can have an arbitrarily poor objective function value in comparison to the optimum value. As the optimum value is typically unknown, the question is when to terminate a search.

In *local search,* only moves to neighbours with a better objective function value are considered. The most straightforward search technique will always make a 'greedy' move to the neighbour with the best objective function value. This method is popular, however, it is obviously unable to escape from local minima. Numerous methods have been proposed in literature which address the problems with search methods in different ways: disconnected search spaces, widespread exploration of the search space, escape from local optima, and termination. Several popular search techniques for combinatorial optimisation problem and their theoretical underpinning are described in [3]. In the following, two local search methods that are particularly popular in the context of UMTS network optimisation are described.

14.1.2.3 Tabu Search

In the search method known as *tabu search*, the neighbourhood principle is extended by a so-called *tabu list* (TL). The tabu list is used to prevent cycling and to avoid getting stuck in a local optimum. After performing a valid *move* to the best alternative neighbour, the previous solution or representative properties of it are *tabu* for the next k iterations. With this mechanism, escape from a local minimum is promoted. The tabu list is an essential component of the algorithm, because it stores the history of the visited candidate solutions. There can also be aspiration criteria, which allow moves to tabu solutions if particular circumstances apply. A detailed description on tabu list management and aspiration criteria can be found in [4]. In Figure 14.1 a simple program structure for a Tabu Search algorithm is shown.

14.1.2.4 Simulated Annealing

Another popular search technique is *simulated annealing*. The name and inspiration come from annealing in metallurgy. By heating and controlled cooling of a material, the size of its crystals is increased and their defects are reduced [5]. The crucial point is the cooling time: If the annealing process is very fast, the atoms do not have enough time to find the correct position – the system freezes in a suboptimal local minimum. If the process is done sufficiently slowly, the atoms have time to find a state of minimum energy.

During a search process, the annealing paradigm is used to probabilistically control whether or not a step with decreasing objective function value $f(x)$ is taken. The chance of accepting a worse solution is controlled by a cooling temperature parameter T, which is decreased during the course of the algorithm. An improvement step is always admitted. A deteriorating step is only considered with a certain probability that decreases with the 'temperature'. Deteriorating steps are thus always possible, but their chances become ever smaller. For a detailed treatment of Simulated Annealing, see [6].

A simple program structure of a Simulated Annealing optimisation algorithm is shown in Figure 14.2. The variable Z is a uniformly distributed random value between zero and one, the chance for admitting

```
procedure Tabu Search
begin
        TL ← 0; // tabu list
        x ← initial solution;
        repeat
                X' ← subset of N(x) under consideration of TL;
                x' best solution of X';
                add x to TL;
                delete elements from TL, which are older than k iterations;
                x ← x';
                if x is better than best solution until now then
                        store x;
                end
        until termination condition is fulfilled;
end
```

Figure 14.1 Tabu Search algorithm.

```
procedure Simulated Annealing
begin
        i ← 0;
        T ← initial temperature;
        x ← initial solution;
        repeat
                x' ← N(x); // derive neighbourhood solution
                if x' is better than x then
                        x ← x';
                else
                        if Z < e^(-|f_obj(x')-f_obj(x)|/T) then
                                x ← x';
                        end
                T ← g(T, i);
                i ← i + 1;
        until termination condition is fulfilled;
end
```

Figure 14.2 Simulated Annealing algorithm.

a decreasing step is calculated according to the *Boltzmann* distribution that governs the temperature distributions in annealing systems. Implementations of the algorithm differ in their choice of the initial temperature and the cooling function g.

14.1.2.5 Advanced Optimisation Techniques

From a mathematical point of view, search techniques are unsatisfactory because they (normally) cannot give quality guarantees. The methods simply yield the best solutions examined so far, but they cannot state how far the solutions are from a global optimum. Quality guarantees can be achieved if some information on optimal solutions is known, even though no specific optimal solutions need to be known. Two methods that can give quality guarantees are linear programming and approximation algorithms.

Linear (Integer) Programming
Integer programming is today the optimisation tool for general purpose combinatorial optimisation, which is most developed mathematically and at the same time has a long track record of successful application to real-world problems. The important restriction of this method is that it can only be applied to optimisation problems with linear constraints and a linear objective function. A detailed theoretical treatment of this subject is [7].

For linear optimisation problems with continuous variables, very efficient algorithms exist, most notably the simplex algorithm and interior point methods. The linear programming problem can even formally be proven to be 'easy' in the polynomial time sense. In combinatorial optimisation, however, variables often only have a sensible interpretation if they assume integral values – for example, the number of antennas to be deployed on an antenna site cannot be 2.5, only 2 or 3 antennas make sense. In many cases, variables encode decisions – for example, whether or not a certain base station is used for a network. In this case, the variables are only allowed to take the value 0 or 1 and are called *binary* variables.

Most quality guarantees in this context are derived for linear integer programming problems by relaxing the integrality constraint. The problem is solved as a continuous one, which can be done quickly. While the optimal continuous solutions are in general no solutions to the original problem, they can be used to find good feasible solutions, and their objective function value bounds the optimum of the integral problem.

A detailed treatment on how to formalise optimisation problems as linear programs and finding a good formulation is given in [8]. If the application in question can be modelled adequately, even large problems with thousands of decision variables can often be solved quickly on standard computers with special software (e.g. [9]).

Approximation Algorithms

Approximation algorithms provide a general framework to efficiently find solutions with quality guarantees for combinatorial optimisation problems. No efficient exact algorithm for solving any NP-hard problem is known. In order to find solutions within polynomial time, one settles for non-optimal solutions. The difference to heuristics is that the solutions are required a priori to be provably close to the optimum after some polynomial running time. While the approximation algorithms and approximability are important for complexity theory [10], they have not yet played a major role in solving practical problems. One example of an approximation algorithm for a problem related to UMTS radio networks is [11].

14.2 OPTIMISATION PARAMETERS FOR STATIC MODELS

We now define the optimisation parameters, i.e. the dimensions in which the feasible solutions can be varied. The following list shows the most important parameters and subjects:

- Base station location and configuration
- Antenna settings

  - Antenna azimuth
  - Antenna tilt (electrical and mechanical)
  - Antenna height above ground or above rooftop
  - Antenna pattern (antenna type)

- Primary common pilot channel (P-CPICH) power level
- Soft handover parameters

  - Active set size
  - Active set window

- Neighbour relations

These parameters have been made subject to automatic optimisation, as they have a strong influence on the interference in the system and therefore on the performance, coverage and capacity of the network. There are many other relevant parameters that play an important role for the radio network. Those parameters are not considered here, as they have not been treated within the context of automatic optimisation on a larger scale.

Optimal base station location and its configuration lead to significant savings in the network cost of deployment and operation, thus it is an important factor in the business planning. The residual parameters – antenna configuration, power and SHO parameters – enable to increase the network coverage and capacity by:

- Reducing inter-cell interference and pilot pollution
- Optimising base station transmit power resources
- Traffic load sharing and balancing between cells
- Optimising SHO areas.

14.2.1 SITE LOCATION AND CONFIGURATION

14.2.1.1 Site Location

In the classical site location (or site selection) problem formulation, a set of potential site candidates is given. A subset of these potential sites has to be selected. The operation of each site is expensive, so cost plays a decisive role. Depending on the situation, the number of sites needs to be minimised under quality constraints (e.g. a certain amount of coverage), or a certain budget (cost constraint) is given and the task is to achieve maximum quality.

In practice, site location problems appear at different stages of the network design. If an operator already runs a 2G network and wants to deploy a 3G network in the same area, sites will be reused. In addition, the operator might need to deploy new site locations. This creates a set of potential locations which can be used for the 3G network. Often not all sites are actually needed, so a subset has to be selected that is to be equipped with 3G hardware (usually new locations for other 3G sites are needed). The goal then is to reach certain quality goals at minimum cost.

Another situation in which sites are selected occurs in the final stage of initial radio network planning. The radio network planning process ends up with geographical locations for base stations and their configurations. The final stage may vary considerably from the initial plan, since in practical network rollout difficulties in finding a suitable site in the desired area might arise (see Chapter 10). Possible reasons are community protests, legal restrictions (e.g. on electromagnetic emission), building constraints, or excessive cost. Since during the planning process only limited information is available, the responsible RF engineer uses 'rule-of-thumb' procedures to estimate the quality of a site candidate. Through an iterative process he produces a final cell plan. This plan is practicable, but it might be over-dimensioned because of the limited information. If a budgeted site location problem is solved for the final cell plan, significant savings in both CAPEX and OPEX might be achieved. This optimisation problem is often called 'minimum cost site reduction', when applied after a real cell plan has been created.

14.2.1.2 Base Station Configuration

The base station configuration problem applies to all hardware elements in a base station:

- Cabinets/racks
- Power amplifiers
- Antennas
- Cables
- Others.

In practice, it can be limited to the first two items from the list above (cabinet/racks and power amplifiers). Their purchase cost is often higher than the implementation and operational cost. For the remaining items, installation cost is dominant, so the hardware can be over-dimensioned. It is, for example, not efficient to implement only two cells in the first year and a third one in the following year at a site, since the cost of implementation would be almost doubled and could not be compensated by time shifted investment in antenna and its corresponding cabling. Instead, the full configuration should be installed at the beginning.

14.2.2 ANTENNA RELATED PARAMETER

The antenna parameters control the coverage and interference situation in the network since the antenna shapes the emitted energy. Besides the height of the antenna above the ground or rooftop level and the used pattern, the azimuth angle and the elevation angle (tilt) can be optimised. The height of the antenna can only be changed at high cost, if possible at all. Exchanging the antenna (hardware) itself is less complicated, but still requires significant handling and cost. However, changing antenna tilt (both electrical and mechanical) and azimuth is associated with less effort and yet has great leverage on network quality. In modern radio networks, Remote Electrical Tilt (RET) antennas are used to ease tilt changes.

The adjustment of antenna parameters will optimise first of all the path loss between base station and mobile. It will lead to minimising the required power for a connection; hence, there will be more power head room at the serving base station. This may be used for new connections or higher bandwidth applications. Therefore, less power required to establish a connection will cause less intra- and inter-cell interference. The inference decrease together with increased power head room will lead to an overall capacity increase.

14.2.2.1 Pattern (Antenna Type)

Antennas are characterised by a three-dimensional radiation pattern. For simplicity, usually the horizontal and vertical pattern is given (see Section 6.1). There are two means for influencing the antenna pattern:

1. Changing the *antenna type* of the base station.
2. Electronic reconfiguration of the antenna characteristics. This is supported through *electrical tilt*.

14.2.2.2 Azimuth Angle

Antenna azimuth is the direction in which the main beam of the horizontal pattern points. The cell coverage is determined among other things by the antenna's horizontal radiation pattern.

Base stations for UMTS are typically equipped with three-sectorised antennas to achieve an area-wide coverage. However, also base stations with one, two or more than three sectors are in use. For example, two-sectored base stations are often set up along highways, railways, tunnels, or to improve coverage in street canyons. More than three sectors may be used for cases where there is high traffic load or where there are unique RF conditions necessitating a tighter control on the radiation pattern. If, for example, a 120° sector covers a diverse topographic environment, where the optimal tilt angle is 3° for one side and 6° for the other side, the RF engineer may choose between using two separate antennas to the same sector and splitting this sector into two, thereby using more than three sectors at this site. The sector splitting leading to optimise the coverage shape is scarified by power reduction per each antenna sector, if the same number of power amplifiers needs to be used.

Figure 14.3, for example, shows the horizontal pattern of a KATHREIN 742264 antenna [12]. The pattern shows a difference in antenna gain of about 10 dB between the main direction of the antenna (0°) compared to an angle of ±60° (usually each cell of a three-sector base station occupies 120°). Due to that difference of 10 dB, the direction of the main beam of the antenna is quite significant and thus it is important to adjust the azimuth of the antennas in order to reach the highest antenna gain for the users in the own cell, as well as the highest attenuation for the other users located in neighbouring cells. This way, less power is needed for covering the area, and therefore less interference is generated.

In [13] and [14], algorithms are introduced, which are especially designed for optimising the azimuth angle for base stations equipped with three-sector sites.

14.2.2.3 Tilt Angle

The antenna tilt is defined as the angle of the main beam of the antenna relative to the horizontal plane. Since the tilt is usually set in the direction down to the ground, the term *downtilt* is often used.

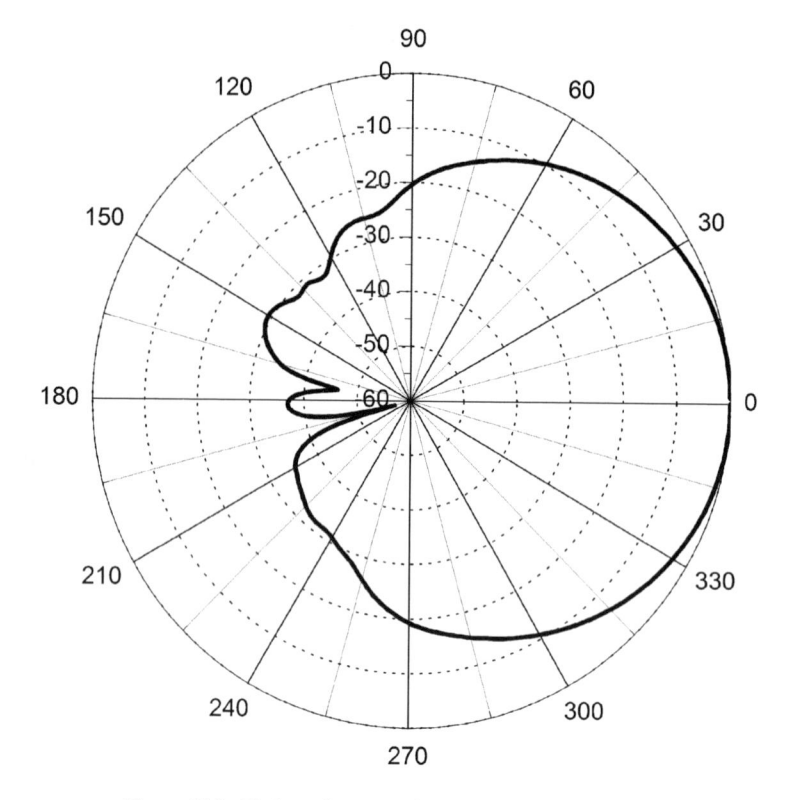

Figure 14.3 Horizontal pattern of base station antenna (in dB).

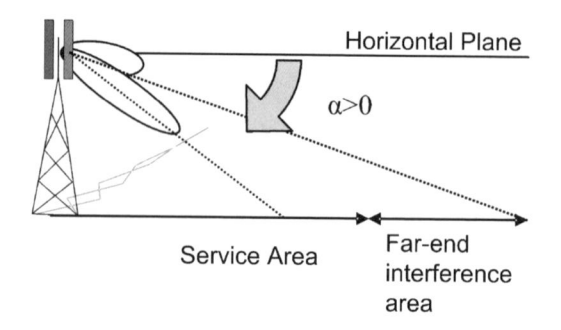

Figure 14.4 Adjustment of base station downtilt.

A *positive downtilt* is defined as the negative elevation angle of the main beam of the antenna relative to the horizontal plane. The service area in Figure 14.4 is the own cell and the far-end interference area is the area of the adjacent cells. Besides the terrain profile and the vertical antenna pattern, the downtilt influences the actual footprint of the sector and therefore the coverage and capacity of the network.

The antenna tilting can be implemented in *mechanical* and/or *electrical* way. These two tilting mechanisms have different effects: When using mechanical tilting, the antenna pattern itself stays constant and is only tilted in a mechanical way by a positioning motor or by manual fixation of the antenna on the mast (the antenna pattern is only moved). Electrical down- or up-tilting of the antenna changes the antenna pattern by electronic reconfiguration of the antenna characteristics. A very detailed examination on the effect of electrical and mechanical antenna tilting in UMTS networks can be found in [15] as well as in Section 10.4.5.

Detailed descriptions of the effect of antenna tilt on the system capacity are presented in [16–20]. The early experience gathered from optimising US CDMA networks shows that the average tilt angles used for CDMA, compared to those used for FDMA based technologies, were about 3° higher for networks operating in the 800 MHz band. As UMTS works in higher frequencies, the vertical beamwidth as well as cell range is smaller, so the difference in vertical tilt angles, for example, between GSM1800 and UMTS is estimated to be around 2°.

By down-tilting the antennas, the other-to-own-cell interference ratio i (see Sections 6.4.8.1 and 10.2.4.6) can be reduced: The antenna main beam delivers less power towards the neighbouring base stations, and therefore most of the radiated power goes to the area that is intended to be served by this particular base station [16]. Due to the fact that the interference in the system is decreasing, the capacity of the network increases. However, down-tilting the antenna will also reduce the sectorisation efficiency, which will influence the cell capacity in a negative way. It is worth mentioning that optimal tilt angles for UMTS networks are in general larger than for GSM networks because with WCDMA neighbouring cells interference plays a more decisive role. Multi-band GSM/UMTS antennas should hence be installed with special care.

In addition, antenna tilt adjustments affect the cells' coverage area. Too much down-tilting might cause that the service area becomes too small and coverage holes can occur. Furthermore, if the down-tilting reaches a certain value, the interference in the neighbouring cells increases again due to the side lobes of the vertical antenna pattern. In [20], it is shown that for smaller inter-cell site separation, higher downtilt is required to mitigate the inter-cell interference. As the inter-cell site separation increases, smaller downtilt is advantageous, offering higher gains to distant users. Hence, the impact on the cell coverage area limits the tilt to reasonable values.

Best network performance can be achieved when both tilts – electrical and mechanical – are set jointly. The mechanical one should be used for uptilt to reduce the back lobes influence on neighbouring cells. The electrical one should be used to optimise the cell's own coverage and minimise the interference to neighbours in the main and side directions. Therefore, it is recommended to use both possibilities at least for most problematic cells which are characterised by high traffic areas and significant number of neighbour relations.

14.2.2.4 Height above Ground

While azimuth and tilt might be modified in a running network, the antenna height has to be determined during the initial planning phase. This parameter is important, because the topology of the terrain as well as the height of the buildings have to be considered for the network deployment.

Two cases must be distinguished: Antenna height below or above the average clutter. As the initial UMTS deployment targets primarily coverage, most of the deployed antennas should be allocated above the clutter heights, that is above the rooftops in urban areas (in many GSM networks, antennas are placed below rooftop level). This calls for a more careful optimisation as the inter-cell interference may reach critical values. Furthermore, the engineer has to select one of the two throughout, as mixing them (when operating only one UMTS carrier) shrinks dramatically the service area of the antennas deployed under the roof tops due to high external to internal interference ratio.

Later modifications of antenna height are too costly or even impossible due to legal (e.g. EMC regulations) or constructional terms. It is consequently very important to set this up properly already in the planning phase.

14.2.3 CPICH POWER

The Common Pilot Channel (CPICH) is used as reference for handover, cell selection and cell reselection. It also shapes the effective cell size. When switched on, the mobile measures the received level of chip energy to interference plus noise density ratio (E_c/I_0) on the CPICH. While in the Radio Resource Control (RRC) Cell_DCH state (connected mode), the mobile measures and reports the level and quality of CPICH to the base station for handover procedures. E_c is the average energy per pseudo noise (PN) chip, and I_0 denotes the total received power density, including signal and interference. This E_c/I_0 ratio is given by Equation (14.1),

$$E_c/I_0 = \text{RSCP}_{\text{CPICH}}/\text{RSSI} \tag{14.1}$$

where $\text{RSCP}_{\text{CPICH}}$ is the received signal code power of the CPICH measured at the mobile station. This value can be used to estimate the path loss, since the transmission power of the CPICH can be read from the system information (broadcasted from the cell within System Information Blocks). The received signal strength indicator (*RSSI*) is the wideband received power within the relevant channel bandwidth in the downlink.

Usually, the primary pilot channel (P-CPICH) is the focus for coverage and capacity optimisation. The optional secondary pilot channel (S-CPICH) is usually set relative to the P-CPICH power level. The cell with the highest received CPICH level at the mobile station is selected as the *serving cell*.

As a consequence, by adjusting the CPICH power levels, the cell load can be balanced between neighbouring cells. Reducing the CPICH power of one cell causes part of the terminals to be served by adjacent cells, while increasing it invites more terminals to handover to the own cell, as well as to make their initial access to the network in that cell. Furthermore, inter-cell interference is reduced, network operation is stabilised and radio resource management is facilitated [21].

During the radio network planning process, the CPICH transmit power of the base stations should be set as low as possible, while ensuring that the serving cells and neighbouring cells can be measured and synchronised to and the CPICH can be used as a phase reference for all other downlink physical channels. Too high values of CPICH power will cause the cells to overlap too much and therefore create interference to the neighbouring cells, called *pilot pollution*, which will decrease the network capacity. However, a small amount of overlapping is necessary for soft handover (SHO), usually between 25 and 35%. These SHO areas can also be controlled by the strength of the CPICH power. By reducing the CPICH power, the SHO areas will decrease when keeping the Add/Remove window constant. A certain amount of overlapping is necessary for mobiles near the cell border to perform SHO and to counteract fluctuations of receiving signal power.

An interesting fact that may be noted is that lowering the CPICH power level of a given cell is usually followed by an increase of the inter-to-intra cell interference ratio to the mobiles served by that cell. This is because of two reasons: the first obvious one is that the received power of the serving cell decreases compared to the others and the second reason is due to the shrinkage of the service area of this cell. The surrounding cells are now serving users closer to it, causing an increase in both cell total transmitted power and UE transmitted power, thus increasing the total transmitted power in the network. This mechanism leads to a practical limitation of differences in CPICH power level settings between cells. The recommended range of CPICH power levels under normal conditions should fall under 5 dB from rail to rail. This limitation still leaves a decent room for optimisation as even 0.5 dB changes to CPICH power in urban area may influence the network's performance.

Most gain can be achieved when all cells within a given geographical area are undergoing CPICH power level optimisation. Here, the coverage holes due to decreasing CPICH power of one cell may be compensated by other serving cells. However, one should pay attention to the continuity of the cells' serving areas, as too many handovers are not recommended.

Furthermore, the CPICH power is part of the total transmit power of the base station, which is generally limited. Thus, less CPICH power would provide more power for the traffic channels, and therefore increase the capacity. On the other hand, the mobile stations are only able to receive the CPICH down to a certain threshold level of E_c/I_0, which determines the coverage area. Due to that fact, setting the CPICH power too low will cause uncovered areas between the cells. In an uncovered area, CPICH power is too weak for the mobile to decode the signal, and call setup is impossible. According to the specifications of the 3GPP, the mobile must be able to decode the pilot from a signal with E_c/I_0 of -20 dB [22]. The real service performance is limited by values of about -15 or -16 dB. Further information regarding the influence of the adjustment of CPICH power on the system capacity can be found in [17,19,21,23,24].

14.3 OPTIMISATION TARGETS AND OBJECTIVE FUNCTION

Besides defining the optimisation parameters (which define the search space), we also need to define the *targets* for optimisation. Therefore, an *objective function* f(x) is required, which maps the possible settings of the optimisation parameters to a (real-valued) number. In this way, each element of the solution space is assigned to a comparable numeric value $f(x)$. With this value it is possible to compare different solutions found, and make a statement on which one is better. In the field of UMTS optimisation, there is no definite objective function. It is possible to optimise for different targets such as capacity, coverage or a combination thereof.

The optimisation criteria that are related to the network's quality can be categorised into two groups: *mobile-based* criteria and *cell-based* criteria. Coverage is a typical mobile-based criterion, because it is defined per user. Capacity criteria are cell-based criteria, because the capacity calculations are done based on the cell's capacities.

A third factor that should be taken into consideration when optimising the configuration of existing networks is the cost of implementation. The implementation cost can be included as an additional target in the optimisation objective function or as additional constraint. In the first case, one tries to decrease the cost network improvement. In the second case the optimisation process will allow the antenna changes within the specific budget specified (budgeted *optimisation problem*).

14.3.1 COVERAGE

One optimisation target that has been requested by various network operators is *coverage*. The objective function for coverage optimisation $f_{\text{obj}}^{\text{cov}}(x)$ can be defined as

$$f_{\text{obj}}^{\text{cov}}(x) = \frac{A_{\text{cov}}}{A_{\text{tot}}} \qquad (14.2)$$

In Equation (14.2), A_{cov} represents the covered area in the considered terrain and A_{tot} is the total optimisation area. The term $f_{\text{obj}}^{\text{cov}}(x)$ thus represents the fraction of the planning area that is covered, ranging between 0 (no coverage) and 1 (all pixels are covered).

Calculation of A_{cov} is usually done on pixel basis. The side length of the pixels is defined by the resolution of the scenario; typical values are in the range from 10 to 100 m. A test function cov(x, y) can be used, which returns 1 if and only if the corresponding pixel is covered, and 0 otherwise.

According to the 3GPP standard [22], a pixel is defined as covered, if the signal to noise ratio (SNR) of the CPICH, $(E_c/I_0)_{\text{CPICH}}$ is higher than a certain threshold. This threshold is called $(E_c/I_0)_{\text{CPICH,tresh}}$. Equation (14.3) shows this definition.

$$(E_c/I_0)_{\text{CPICH}} = \frac{\text{RSCP}_{\text{CPICH}}}{\text{RSSI}} \geq (E_c/I_0)_{\text{CPICH,tresh}} \qquad (14.3)$$

In Equation (14.3) RSCP$_{\text{CPICH}}$ denotes the received signal code power of the CPICH as measured by the mobile and RSSI (received signal strength indicator) is the wideband received power within the relevant channel bandwidth in the downlink. Typical values for $(E_c/I_0)_{\text{CPICH,tresh}}$ are between -7 and -12 dB for unloaded networks and between -10 and -15 dB for loaded networks. This threshold depends on the required service E_b/N_0, allowed cell load and environmental conditions.

Since the coverage condition in a real network is 'fuzzy', some smooth or piecewise linear function could be used for the coverage criteria. The coverage criteria for the specific pixel (y) could be defined with a fuzzy threshold in the range

$$\text{cov}(x,y) = \begin{cases} 1, & (E_c/I_0)_{\text{CPICH}} \geq (E_c/I_0)_{\text{CPICH,tresh}} \\ 0, & (E_c/I_0)_{\text{CPICH}} < (E_c/I_0)_{\text{CPICH,tresh}} - \delta \\ \left((E_c/I_0)_{\text{CPICH,tresh}} - (E_c/I_0)_{\text{CPICH}}\right)/\delta, & \text{otherwise} \end{cases} \quad (14.4)$$

The coverage optimisation criterion is often weighted according to traffic density, as it is more important to cover areas with many users. A weighted definition of the test function is:

$$\text{cov}(x,y)_{\text{traf}} = \text{cov}(x,y) \cdot A_0(y) \quad (14.5)$$

where $A_0(x,y)$ is the traffic density in the pixel. In this case coverage is defined as a covered *traffic* rather than covered *area*. Other parameters that could also limit the coverage are maximum TCH power in downlink, minimum RSCP$_{\text{CPICH}}$ that defines the mobile sensitivity and maximum mobile transmitted power in uplink.

14.3.2 CAPACITY

Capacity is a key performance figure of a network. Capacity is often orthogonal to network cost: either the operator assumes a certain number of users that shall be served with minimum cost, or the number of served users shall be maximised with a given infrastructure budget. It has to be noted that capacity figures highly depend on the available service types and their quality requirements. In the context of 3G networks with their diverse service portfolio, this has far-reaching implications. Determining the capacity of standard voice services is fairly straight forward, but as soon as it comes to data services (specifically non real-time), some assumptions on radio resource management and quality of service need to be made. For running automatic optimisation, the evaluation methods and assumptions used have to be fixed beforehand. Only then one can speak of served users, throughput or bottleneck cells, and count or measure them.

14.3.2.1 Number of Served Users

The objective function for network capacity $f_{\text{obj}}^{\text{cap}}(x)$ can be defined as

$$f_{\text{obj}}^{\text{cap}}(x) = \sum_{k=0}^{N_{\text{cells}}} N_{k,\text{served}} \quad (14.6)$$

where N_{cells} is the total number of cells in the network, and $N_{k,\text{served}}$ is the number of served users in cell k. The number of users might be determined for a fixed reference snapshot or as an average over many snapshots. Thus, $f_{\text{obj}}^{\text{cap}}(x)$ calculates the total amount of served users N_{served}. Mobiles should not be counted twice, $N_{k,\text{served}}$ should thus only count best-server links (not soft or softer handover).

14.3.2.2 Throughput

Besides the maximisation of the number of served users, *throughput* is another possible optimisation goal. An optimisation of the throughput implicates a weighting of the users according to the service they use and favours users with higher data throughput. A possible objective function $f_{obj}^{cap,throughput}(x)$ can be defined as

$$f_{obj}^{cap,throughput}(x) = \sum_{k=0}^{N_{cells}} TP_k \tag{14.7}$$

In Equation (14.7), TP_k denotes the total throughput in cell k.

The objectives defined above are particularly important for a highly loaded network with many disconnected mobiles due to cell capacity limitations. If most mobiles can be served, optimisation of capacity criteria improves the network for supporting future traffic growth.

Setting an optimisation target for capacity only can, in extreme cases, cause the quality to drop. If the focus solely lies on connecting as many users as possible, users that are close to the base station might be preferred and more distant users neglected. The percentage of good connections may therefore be used to benchmark the optimised network. Similarly, for data services one may look at the percentage of re-transmissions and/or bit rate degradation where the network fails to supply the required bit rate for a specific service level. All these values are highly sensitive to the total absolute number of served users or throughput elaborated in the previous section. It is best to always use a combination of these targets, setting the flavour by the definition of the objective functions.

14.3.2.3 Bottleneck Cells

An alternative optimisation approach for capacity optimisation is to minimise the number of *bottleneck cells*. Bottleneck cells are the cells that first hit the capacity limitation and impede future traffic growth. Capacity limitations can be defined by RRM limits on code or power usage. Criteria for the power limitation will try to reduce the number of cells where the cell power load P_{tot} exceeds the predefined value $P_{MaxLoad}$:

$$f_{obj}^{cap,power}(x) = \sum_{k=0}^{N_{cells}} 1(P_{tot} - P_{MaxLoad}),$$

$$1(x) = \begin{cases} 1, & x > 0 \\ 0, & x \leq 0 \end{cases} \tag{14.8}$$

14.3.3 SOFT HANDOVER AREAS AND PILOT POLLUTION

Soft handover is a key feature of UMTS radio networks. For soft handover to work properly, cells have to overlap by a certain amount. If the overlap is too small, connections might be interrupted. On the other hand, if cells overlap too much, the mobiles receive too many strong pilot signals and cell selection and handover might fail – one speaks of *pilot pollution*. The soft handover properties of a network are commonly treated as a subordinate optimisation goal; they have to stay within certain acceptable bounds and are used as a tiebreaker.

The following expression can be used to evaluate the number of soft handover connections:

$$f_{obj}^{SHO}(x) = 1 - \frac{\sum_{k=0}^{N_{cells}} SHO(k)}{N_{served} \cdot SHO_{max}} \tag{14.9}$$

In Equation (14.9) SHO(k) is the number of soft handover connections (excluding the best server link) of mobile station k, and $\text{SHO}_{\max}$ is a network parameter defining the maximum possible number of SHO links per mobile. N_{served} is the total number of served users in the scenario. It is also possible to evaluate soft handover probabilities in the unloaded network on a pixel basis.

14.3.4 COST OF IMPLEMENTATION

An important optimisation target is network *cost*. This may refer to hardware cost (CAPEX), but also network operations (OPEX) and manpower needed for network configurations. For base station location or site reduction, usually the cost of a base station is fixed. A quality target (coverage and/or capacity) then usually has to be met at minimum cost.

If an existing network is reconfigured, cost values might be attached to specific configuration changes. For example, changing an electrical tilt is inexpensive, especially if it can be done remotely. Changes that require sending personnel to the site in order to adapt the configuration (mechanical tilt, azimuth) are usually considered more expensive. The cost of a configuration is only defined with respect to the initial, reference configuration. The objective function for cost is formed as the sum of all costs applicable in the current configuration.

14.3.5 COMBINATION AND FURTHER POSSIBILITIES

The optimisation targets described above shall usually all be increased at the same time, but they are partly contradictory: they usually cannot all be increased at the same time. The operator thus always has to specify a certain trade-off between the different optimisation goals. The usual approach in this situation – if multi-criteria optimisation is not considered a viable alternative – is to form a joint objective function. The joint objective is normally a simple linear combination of the different quality measures. Equation (14.10) shows a general formula for an extended objective function:

$$f_{\text{obj}}(x) = \alpha_1 f^1_{\text{obj}}(x) + \cdots + \alpha_n f^n_{\text{obj}}(x) \tag{14.10}$$

The scaling values α_i in Equation (14.10) are weights assigned to the individual factors. The values for α_i should be selected according to the priority of the optimisation targets and the specific situation. Different objectives are usually measured by different units. It is therefore important to normalise the targets to a comparable range (for example, 0–1).

It is often helpful to limit the objective function to very few, simple components that are easy to calculate. Especially in a pre-optimisation activity, this may allow fast computing performance. The criterion can, for example, be based on transmit power or received interference only, as most of the targets mentioned above are limited by finite power resources. A fine-granular evaluation and optimisation can then be used for fine-tuning.

14.3.6 ADDITIONAL PRACTICAL AND TECHNICAL CONSTRAINTS

In addition to the criteria described above, there are some constraints that are imposed either by operator-specific preferences or legislation. These constraints have to be considered for defining the set of feasible solutions. These are:

1. *Physical constraints* due to antenna deployment limitations, such as mechanical installation, obstructions, height, physical size, etc. These constraints are specific for each site.
2. *Environmental constraints*, mostly related to the permitted level of emission and defined by the state/municipality. This relates to the maximal permitted transmit power level.

3. *Electrical constraints* due to equipment limitations – maximal number of sectors, maximal transmitted power and power allocation for an individual channel.
4. Constraints on the *joint parameter settings* as to reflect dependencies between sectors. For example, a minimal angular separation between sectors may be defined, or transmit powers assigned to sectors belonging to the same site shall be balanced.
5. Constraints to prevent 'non-engineering' solutions. For example, within the optimisation process an area could be covered using the lower sidelobe of an antenna unless restrictions on tilts are imposed.
6. Constraints on the deviation of configurations from *existing network configuration*. For example, the operator may wish to change the azimuth of sectors by not more than ±30° relative to the current configuration.

14.3.7 EXAMPLE OF OBJECTIVE FUNCTION PROPERTIES

Closed characterisation of the objective function over the domain of feasible network configurations could considerably aid optimisation. Structural properties of the objective function revealed by a careful analysis could be exploited to improve the speed and quality of automatic network optimisation methods.

This section shows, on the basis of a simple case study [25], that such a closed, powerful characterisation is typically out of reach. The setting is as follows: two small network configurations are analysed under tilt and azimuth variation of all sectors. Each possible configuration with tilts between 3° and 17° and azimuth changes of up to ±28° (with a step size of 4°) are generated and evaluated using Monte-Carlo simulation. The objective function that is to be minimised is defined as the sum over the downlink cell loads.

Even in this simple setting, numerous local minima of comparable quality exist. These minima are not clustered in 'one corner' of the solution space; the area of attraction around the local minima varies greatly in size. Further analysis of these phenomena is certainly in demand. Some of the findings from [25] are sketched in the following to illustrate the above statements.

Network model for cost function definition. The objective function under consideration is a linear combination of the total cell transmit powers (energy minimisation). To compute these powers, the static model based on the system of linear equations is used (see Chapter 6.4). The objective function f_{obj} can be written as:

$$f_{obj} = \text{weight}_1 \cdot P^1_{sum} + \cdots + \text{weight}_k \cdot P^k_{sum}, \qquad (14.11)$$

where weight_k is the weighting factor for the *k-th* cell and P^k_{sum} represents the total transmit power in TCH channels for the *k-th* cell. Weighting factors can be used to prioritise cells of special importance for the operator. A possible simple setting is:

$$\text{weight}_k = \frac{1}{P^k_{max}}, \qquad (14.12)$$

where P^k_{max} stands for maximum transmit power of *k*-th cell. After substituting Equation (14.12) in (14.11), we obtain:

$$f_{obj} = \frac{P^1_{sum}}{P^1_{max}} + \cdots + \frac{P^k_{sum}}{P^k_{max}} = \text{load}_1 + \cdots + \text{load}_k, \qquad (14.13)$$

where load_k stands for the load of *k-th* cell. Consequently, the assumed objective function can be interpreted as the sum of total transmit powers in the network or as the sum of loads in all cells depending on the value and interpretation of the weighting factors.

Network scenarios for numerical tests. An analytical analysis of the objective function features seems to be extremely difficult. It is governed by antenna characteristics which are hard to parameterise as they are strongly nonlinear; hence, transmitted powers usually cannot be directly presented as analytical functions of antenna parameters. Furthermore, the need for terminal reassignment after a change of antenna orientation complicates the analytical analysis even more. Therefore, it was decided to analyse the objective function numerically for three selected network scenarios. The scenarios are summarised in Table 14.1 and presented in Figures 14.5 and 14.6.

The test networks consist of 3 or 6 cells with 140 or 180 active UMTS voice users in total. The propagation loss L is calculated as $L = -128.1 + 37.6 \cdot \log_{10}(d)$, where d is the terminal distance to the base station (expressed in km). Two base station antennas are used: Celwave APX 206513 (Scenario 1) and Kathrein 742 264 (Scenarios 2 and 3).

Base station antenna azimuth and/or antenna down-tilt angle were chosen to be optimised as shown in Table 14.1. Each parameter was assumed to be discrete. Antenna azimuth was adjustable in 15 steps between $-28°$ and $+28°$, with a resolution of $4°$ (relative to initial azimuths: $0°$, $120°$ and $240°$ for Scenario 1 and $20°$, $120°$, $250°$, $100°$, $230°$ and $120°$ for Scenarios 2 and 3). Antenna tilt was adjusted in 15 equally spaced values ranging from $3°$ to $17°$ (resolution $1°$). The whole solution space has thus

Table 14.1 Main parameters for analysed network scenarios.

	Number of cells	Number of active terminals	Optimised parameters	Antenna type
Scenario 1	3	140 (voice)	azimuth & tilt	Celwave APX 206513
Scenario 2	6	180 (voice)	tilt	Kathrein 742 264
Scenario 3	6	180 (voice)	azimuth	Kathrein 742 264

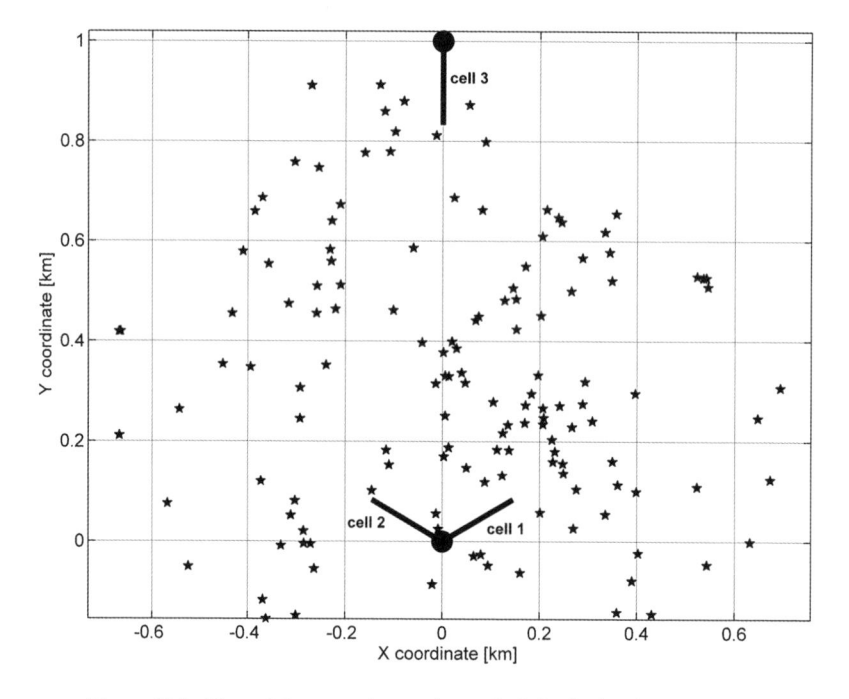

Figure 14.5 Network layout and example terminal distribution for Scenario 1.

Theory of Automated Network Optimisation

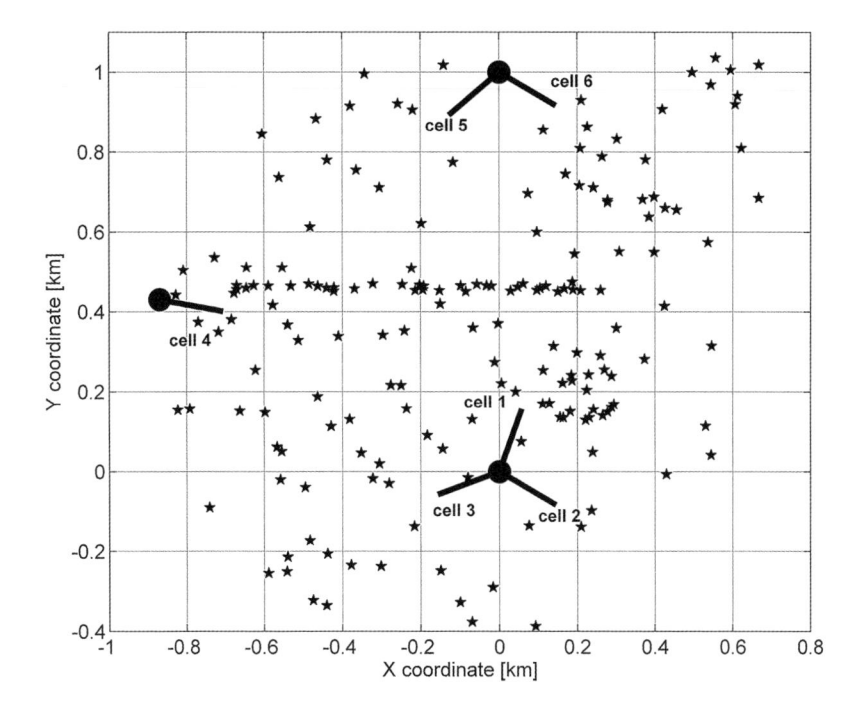

Figure 14.6 Network layout and example terminal distribution for Scenarios 2 and 3.

six dimensions and contains 11 390 625 discrete points. The objective function is the sum of the total cell transmit powers. Its shape is the same as the shape of the load based objective function with equal base station maximum total transmit powers. For each scenario, 1 and 25 Monte-Carlo (MC) snapshots are simulated.

Number and values of the objective function minima. For each scenario, all local minima were found. The number of local minima that differ no more than 10 % and 1 % from the global minimum value has also been determined. These results are summarised in Table 14.2.

There are 214, 649 and 215 local minima for the single snapshot case and 10, 104 and 23 local minima for 25 averaged snapshots for Scenarios 1, 2 and 3 respectively. It can be observed that increasing the number of Monte-Carlo snapshots acts as a low pass filter for the objective function providing significant reduction of the number of minima. This shows that the objective function becomes smoother when

Table 14.2 Computed minima statistics for assumed scenarios (based on [25]).

	Number of Monte-Carlo snapshots	Number of local minima	Number of local minima within 10 % range	Number of local minima within 1 % range
Scenario 1	1	214	13 (~6 %)	0
	25	10	5 (50 %)	0
Scenario 2	1	649	24 (~4 %)	3 (0.5 %)
	25	104	6 (~6 %)	0
Scenario 3	1	215	183 (86 %)	4 (~2 %)
	25	23	23 (100 %)	1 (~5 %)

an increased number of Monte-Carlo snapshots in static simulations is considered. However, this significantly increases the amount of computations. Optimisation algorithms could benefit from this fact offering faster convergence.

Figure 14.7 presents a histogram of all local minima values. There is a large amount of sub-optimum solutions which are not much bigger than the global optimum (please, also refer to Table 14.2). In particular, a large percentage of local minima values is close to the global optimum for antenna azimuth optimisation (Scenario 3). This can be indirectly interpreted as modest sensitivity of network performance on antenna azimuth changes (in the sense of power-based cost function).

Distance between global and local minima. Detailed information about the three best local minima for Scenarios 1 and 2 is presented in Table 14.3. The data in the table include minima values, distance to the global minimum and size of the minimum attraction region.

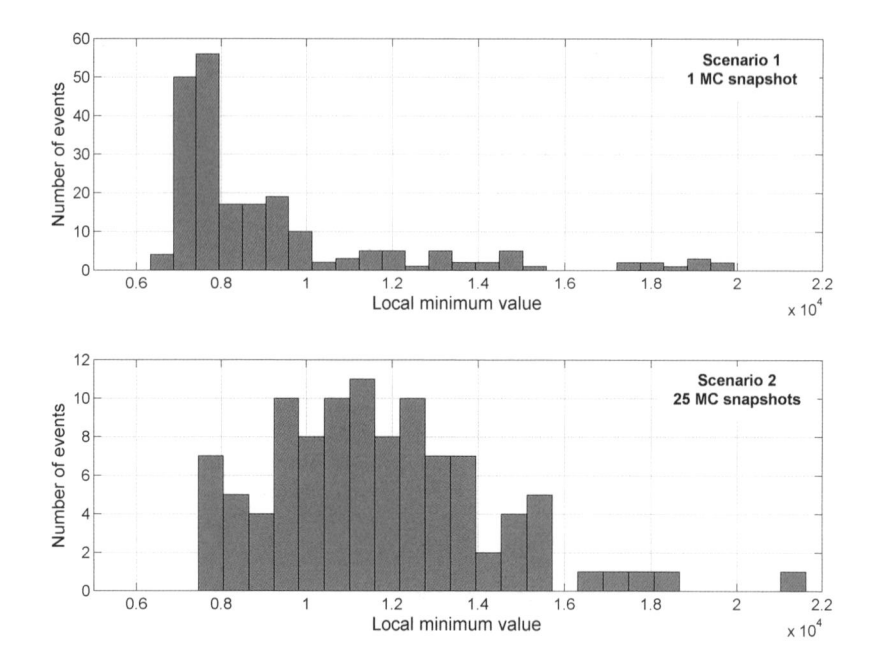

Figure 14.7 Histogram of local minimum values for Scenarios 1 and 2 for 1 and 25 Monte-Carlo snapshots respectively.

Table 14.3 Parameters of three best minima for two example scenarios (based on [25]).

Number of Monte-Carlo snapshots	Values of three best minima [W]	Distance between minimum and optimum	Size of minimum attraction region (%)
1 (Scenario 1)	6.31	0	29.3
	6.69	9.4	1.9
	6.81	13.6	2.5
25 (Scenario 2)	7.45	0	6.5
	7.59	9.0	4.4
	7.80	6.6	1.9

As an example, the global minimum was found for Scenario 1 with one Monte-Carlo snapshot at the point with coordinates $[(13° \ 5° \ 12°)_{tilt}(-16° \ 12° \ 28°)_{azimuth}]$ for tilt and azimuth angles respectively. The value of the objective function at this point is equal to 6313.3 mW. Detailed analysis of the computed data shows that *half* of the local minima represent values smaller then 8000 mW. This is around 25 % more than the value of the global minimum. The distance of these minimum points to the global minimum ranges from 8 to 19 units in 1-to-15 6-dimensional discrete argument space. The distance from the centre point of this hypercube to the most distant points is about 17 units. It must be noted that Scenario 3 yields an even more spread minima distribution.

The above analysis as well as further results in [25] clearly show that there are many local minima that have similar values and that are good enough to be considered reasonable (sub-optimum) configurations. The distance between them varies and very often is large (Figure 14.8).

Above considerations directly lead to the following remark: *Near-optimum solutions with comparable performance can represent very different network parameter configurations, i.e. being very distant from each other.*

In other words, two hypothetical networks can have similar performance but achieved through completely different network configurations. This remark is of significant importance. It means that an optimisation process can lead to very different network configurations but networks can have a similar performance. This remark can also bring an idea of the optimisation algorithm which reduces searching space because sub-optimum solutions are widely spread and one can be quite sure to have some of them in a reduced space.

Surroundings of the minimum and its attraction region. For optimisation results to be useful in practice, slight changes in the optimised parameter value (e.g. mechanical failure of antenna suspension, wind, installer mistake, etc.) or traffic changes must not lead to significant variations of network performance. Therefore, the attraction area of sub-optimum solutions needs to be analysed as well as the sensitivity of the objective functions to parameter value changes (sub-optimum must be 'stable' enough).

Table 14.3 lists some example attraction region sizes for the given minima as a percentage of the total argument space. Some of the minima have a very large attraction region but for others this region is really tiny. A number of examples was found during data analysis in which minima with a 'good' objective function value are close to attraction region of another minimum and where the border between these minima reaches large cost function values. A good example is the point with coordinates

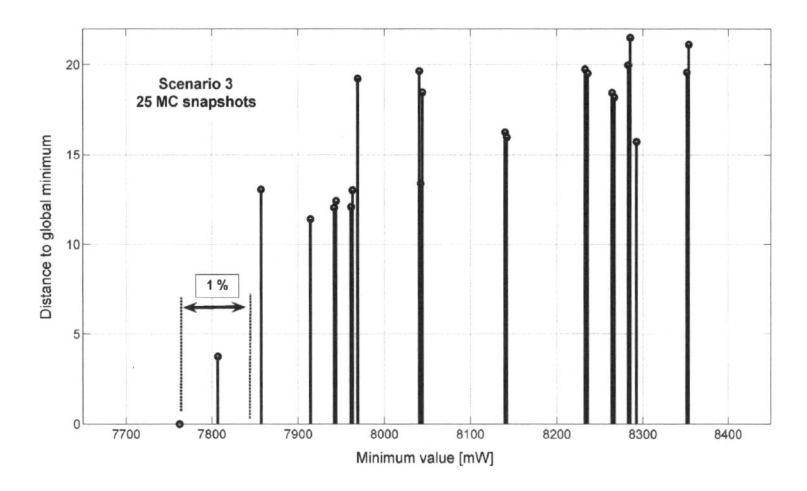

Figure 14.8 Distance between local minima and global one as a function of minimum value (Scenario 3 with 25 Monte-Carlo snapshots).

(6, 6, 6, 8, 8, 5) which represents the following tilt angles (8°, 8°, 8°, 10°, 10°, 7°) in Scenario 2 (25 MC snapshots). This is the third minimum in ranking and its objective function value equals 7.80 W. Surprisingly, the distance to other minimum attraction area equals 3 units while the fifth minimum, with value of 7.87 W has this distance equal to almost 10 units. Although the value is greater (but less than 1%) the second mentioned minimum is much better suited as a network stable working point. Consequently, it can be remarked that the quality of the optimum found is determined by optimality of its value *and* by the size of attraction region and values in its surroundings. Furthermore, the feature of the minimum of having a large attraction area can also rise potential for proper algorithm design.

The intention of the experiment presented in this section has been to analyse a simple example network layout and its performance. This is not sufficient to create general rules but it can reveal the specific behaviour of the network during an optimisation process. This can guide optimisation algorithm designers to create new, effective optimisation algorithms for which the knowledge of objective function properties is a prerequisite.

14.4 NETWORK OPTIMISATION WITH EVOLUTIONARY ALGORITHMS

Evolutionary Algorithms (EA) are a category of computer-based problem solving systems, which use calculable models of natural selection and evolution processes as their key elements.

A property of the search methods presented so far is the fact that they always have *exactly one* current solution x at a time. *Evolutionary algorithms*, in contrast, keep a whole collection (also called a *population*) of current solutions, and work on all of them in a single iteration. The method of organic evolution represents a useful strategy for the adaptation of living things to their environment, which motivated taking over some of its principles for the optimisation of technical systems [26]. Although the mechanisms that drive natural evolution are not fully understood, some generally accepted features have been identified [27]:

- Evolution takes place in the process of reproduction. *Mutations* may cause the *chromosomes* of biological children to be different from those of their biological parents, and *recombination* processes may create different chromosomes by combining material from the chromosomes of the parents.
- Natural *selection* is the link between chromosomes and the performance of the individuals represented by them. Processes of natural selection cause those chromosomes that represent successful structures to reproduce more often than those that do not.
- Biological evolution has no memory. Whatever it knows about producing individuals that will function well in their environment is contained in the gene pool – the set of chromosomes carried by the current individuals.

Evolutionary algorithms are considered a robust technique, which can deal successfully with a wide range of problem areas, including those that are difficult for other methods to solve. They are not guaranteed to find the global optimum solution, but they are generally good at finding acceptable solutions to problems in acceptable time [28]. In many areas, specialised techniques for solving particular problems are likely to outperform evolutionary algorithms in terms of speed and/or accuracy of the final result. The main applications for EAs are in difficult areas, where no such (sufficiently good) techniques exist. In this context, their main advantages are [29]:

- strength in global exploration of multimodal and rugged surfaces;
- general search principles, which makes them easy to adapt to the application problem by adding domain-specific knowledge;
- capability of self adaptation of the search parameters, which makes them capable of learning features of the search space and of gradually switching from global exploration to local optimisation;

- robustness to discontinuities (as caused for example by simulator failures); and
- the population-based search concept, which allows for flexible parallel computation.

The principal structure of an evolutionary optimisation algorithm is shown in Figure 14.9. In this figure, P denotes a population of solutions. In each iteration a new generation Q is produced, and from the union of Q and P the new population P is decided.

In the following two subsections, the most important and most popular representatives, *Genetic Algorithms* and *Evolution Strategies*, are described in more detail. Both are often used for optimising base station parameters.

14.4.1 GENETIC ALGORITHMS

The idea of using genetic approaches for optimisation originated from [30] and has been extended in [31]. Recent enhancements and research on theoretical properties can be found in [32] and [33]. Usually, a Genetic Algorithm (GA) operates on a coding of the function parameters called a *chromosome*. For the coding of the problem, a suitable structure should be used. For the optimisation of the parameters in a UMTS network, a numeric representation is most suitable [34]. Simple, stochastic operators (*selection, crossover* and *mutation*) are used to explore the solution domain in search of an optimal solution. The basic block diagram with the three operators is depicted in Figure 14.10.

This simple type of GA is known as *Canonical GA*. Successive *populations* of trial solutions are called *generations*. Subsequent generations are made up of *children*, produced through the selective reproduction of pairs of parents taken from the current generation. A list of some of the commonly encountered GA terms relating to the optimisation problem is presented below.

Population	Set of trial solutions.
Parent	Member of the current generation.
Child	Member of the next generation.
Generation	Successively created populations (GA iterations).
Chromosome	Coded form of a trial solution vector (string) consisting of genes made of alleles. A chromosome is also referred to as *individual*.
Gene	Each gene can have a value (allele) of a certain value set (e.g. one bit).
Allele	Concrete value of a gene (e.g. for a bit representation: zero or one).
Fitness	Positive number assigned to an individual representing a measure of quality.

The first population is usually initialised by a random setting. For the evaluation of all the individuals of the population, a *objective function* $f_{\text{obj}}(i)$ is needed (see Section 14.3). In the following the three

```
procedure EA
begin
        P ← set of initial solutions;
        Evaluate (P);
        repeat
                Q ← GenerateNewSolutionsByVariation (P);
                Evaluate (Q);
                P ← SelectBetterSolutions (P, Q);
        until termination condition is fulfilled;
end
```

Figure 14.9 The structure of an Evolutionary Algorithm.

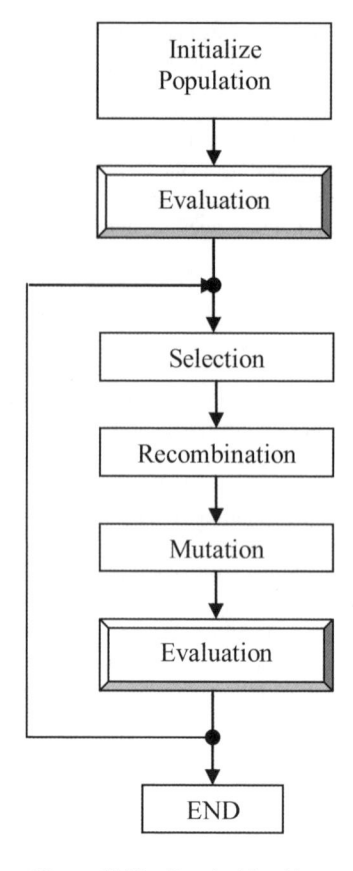

Figure 14.10 Genetic Algorithm.

operators are briefly explained. A practical implementation of a Genetic Algorithm for the optimisation of antenna tilt and CPICH power can be found in Section 14.4.3.

14.4.1.1 Selection

Parents for the next generation are selected at random, but according to natural selection: individuals with a higher objective value are selected more often than worse individuals. The selection process forces the population of the GA in the direction of better solutions. In the majority of cases, a *fitness proportional selection* scheme is used. This method is referred to as *roulette-wheel selection*, because it works like the roulette game with different probabilities. For each individual, the fitness value is scaled according to the sum of the fitness of all the individuals. The probability for an individual i to be selected in the selection process is shown in Equation (14.14).

$$p_s(i) = \frac{f_{obj}(i)}{\sum_{j=1}^{N_{pop}} f_{obj}(j)} \qquad (14.14)$$

$$\text{with} \quad f_{obj}(j) \geq 0 \quad \text{and} \quad \sum_{j=1}^{N_{pop}} f_{obj}(j) > 0$$

In Equation (14.14), N_{pop} denotes the size of the population.

| 0 | 0 | 0 | 0 | 0 | 0 | 0 | 0 | | 1 | 1 | 1 | 1 | 1 | 1 | 1 | 1 | parents

| 0 | 0 | 0 | 1 | 1 | 1 | 1 | 1 | | 1 | 1 | 1 | 0 | 0 | 0 | 0 | 0 | children

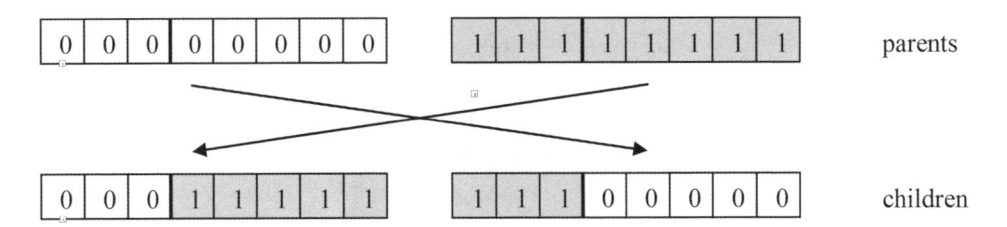

Figure 14.11 1-point crossover.

14.4.1.2 Recombination

Recombination is the primary operator for generating new individuals. In Figure 14.11, the *1-point crossover* is shown, as one possible example for a recombination operator. In this example eight parameters are *binary coded* on one chromosome.

The crossover is either performed for all individuals of the population, or for the majority of the population controlled by a certain randomness (e.g. crossover probability $p_c = 0.8$). The creation of couples as well as the choice of the *crossover points* is also performed randomly.

14.4.1.3 Mutation

The *mutation* operator is used as secondary operator in Genetic Algorithms. This operator makes small, random changes. Only one gene or very few genes of a chromosome obtains a new value. Mutation is responsible for introducing new and lost gene material (alleles) into the population. In Figure 14.12 an example of mutation for a chromosome with eight binary coded parameters is shown. Parameter no. 4 mutates from 1 to 0. Mutation is usually performed with a certain probability (smaller than one).

14.4.2 EVOLUTION STRATEGIES

Evolution Strategies (ES) are another popular manifestation of evolutionary algorithms. They typically use mutation and (optionally) recombination; the parameters are represented *as real-valued numbers*. First concepts of ES were investigated by H.-P. Schwefel in 1964. Based on this work, Rechenberg [26] developed a theory of convergence velocity for simple models and proposed the first population-based evolution strategy. This has laid the foundation for the present ES as introduced in [35], which is still considered a reference for ES today.

In Figure 14.13 a basic flow chart of an Evolutionary Strategy algorithm is shown. Before optimisation, a population of μ_p members is initialised. The optimisation loop starts with selecting a so-called

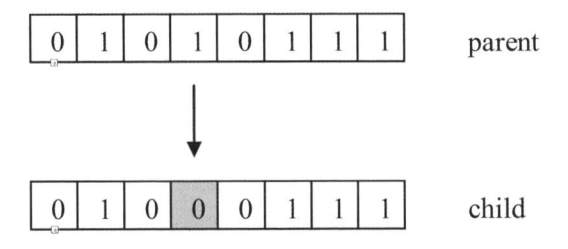

| 0 | 1 | 0 | 1 | 0 | 1 | 1 | 1 | parent

| 0 | 1 | 0 | 0 | 0 | 1 | 1 | 1 | child

Figure 14.12 Mutation.

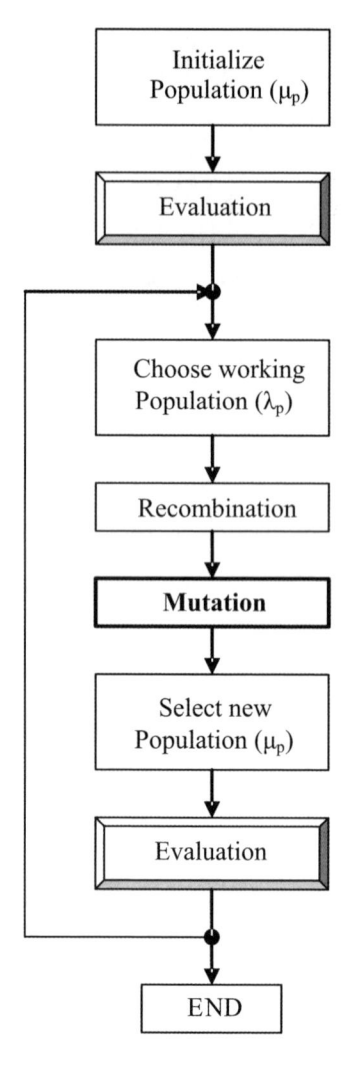

```
                    ┌─────────────────┐
                    │   Initialize    │
                    │ Population (μp) │
                    └────────┬────────┘
                             ▼
                    ┌─────────────────┐
                    │   Evaluation    │
                    └────────┬────────┘
                             │
            ┌────────────────┤
            │                ▼
            │       ┌─────────────────┐
            │       │ Choose working  │
            │       │ Population (λp) │
            │       └────────┬────────┘
            │                ▼
            │       ┌─────────────────┐
            │       │  Recombination  │
            │       └────────┬────────┘
            │                ▼
            │       ┌─────────────────┐
            │       │    Mutation     │
            │       └────────┬────────┘
            │                ▼
            │       ┌─────────────────┐
            │       │   Select new    │
            │       │ Population (μp) │
            │       └────────┬────────┘
            │                ▼
            │       ┌─────────────────┐
            │       │   Evaluation    │
            │       └────────┬────────┘
            └────────────────┤
                             ▼
                    ┌─────────────────┐
                    │       END       │
                    └─────────────────┘
```

Figure 14.13 Evolutionary Strategy algorithm.

working population of size $\lambda_p \geq \mu_p$ at random. Different operators such as recombination and mutation (main operator) are then invoked; each of them modifies some or all of the individuals. Subsequently, the changed working population is evaluated to obtain a new fitness value (according to the objective function $f_{obj}(x)$) for each individual. An iteration of the evolutionary algorithm is completed by the selection operator: according to its specific cost value, each individual can be selected between 0 and μ_p times into the new population of size μ_p.

14.4.2.1 Mutation

Unlike the Genetic Algorithms, where recombination is the driving force of evolution, mutation is the main operator in Evolution Strategies. In contrast to GA, the major part of a gene is modified in ES, but only by a small amount.

In many natural stochastic processes, small changes are more likely to occur than big changes. This principle is modelled by the Gaussian distribution $\mathbf{N}(\mu_g, \sigma_g)$ with a probability density function of

$$g(x) = \frac{1}{\sqrt{2\pi} \cdot \sigma_g(x)} e^{-\frac{(x-\mu_g)^2}{2\sigma_g^2(x)}} \tag{14.15}$$

where μ_g is the expectation value and σ_g is the standard deviation. The mutation operator perturbs the parameter vector of an individual according to a Gaussian distribution. It performs the following operation on the parameter vector $\mathbf{x} = (x_1, \ldots, x_n)$:

$$x_i' = \begin{cases} \max(\min(x_i + n_i, x_i^{\max}), x_i^{\min}) & : p_i \leq p_{\text{cell}} \\ x_i & : p_i > p_{\text{cell}} \end{cases} \tag{14.16}$$

where $n_i \sim \mathbf{N}(0, \sigma_g)$ is a vector of Gaussian deviates with expectation value 0 and standard deviation σ_g. The value $p_i \sim \mathbf{U}(0, 1)^*$ and $x_i^{\max}$ and $x_i^{\min}$ are the maximum and minimum values of the parameter x_i. The probability for a cell to be affected, p_{cell}, is an adjustable parameter of the operator.

The mutation operator is the main operator for Evolution Strategies. In typical algorithms a value between 0.8 and 1 is used for p_{cell}. The other adjustable parameter of Equation (14.15), σ_g, is usually controlled by the so-called *1/5-success-rule* or dynamically through *self-adaptation* (see section on self-adaptation).

14.4.2.2 Recombination

In contrast to mutation, a *recombination* operator produces one child from two (or more) parents. This corresponds to biological reproduction in natural evolution, where two parents exchange parts of their corresponding genetic material. This process provides the chance for diverse beneficial attributes found in different individuals to be combined into a single offspring. While some ES literature claims recombination to be of minor importance compared to (Gaussian) mutation, some other literature [13] found it to be helpful especially in speeding up the optimisation process.

Recombination can be used additionally before mutation in ES. One possibility is *Discrete Recombination*. This operator randomly selects each parameter of the child (i.e. each set of antenna tilt and CPICH values) from either the first or the second parent. This operation is described by the following equation:

$$x_i' = \begin{cases} x_{2,i} & : u_i < p_{\text{cell}} \\ x_{1,i} & : u_i \geq p_{\text{cell}} \end{cases} \tag{14.17}$$

In Equation (14.17), $x_{1,i}$ and $x_{2,i}$ are the values of parameter i in the first and second parent, respectively, $u_i \sim \mathbf{U}(0, 1)$, and p_{cell} is a parameter of this operator which can be used to bias the child in favour of either parent. Typical values for p_{cell} lie between 0.4 and 0.6.

14.4.2.3 Selection

After the operators have been applied and all newly generated individuals of the final working population have been evaluated and assigned a value indicating their fitness according to the cost function, a new population of size μ_p has to be selected from the available individuals in order to

---

* $U(v, w)$ denotes the uniform probability distribution with support (v, w).

Table 14.4 Significance of selection pressure.

Selection pressure too low	Selection pressure too high
• poor individuals remain in population • good individuals proliferate only weakly • slow algorithm convergence • in extreme cases, algorithm degenerates towards random search	• good individuals proliferate too quickly (so-called *superindividuals*) • population diversity is reduced • often premature convergence towards local optima

complete one full iteration of the ES. Selection is the process, which drives evolution in the right direction: individuals with a higher fitness shall have greater chances to make it into the new population, while individuals with poor fitness are less likely to do so – and in this case will not get a chance for producing further offspring (which is probably also poor). The ratio by which fitter individuals are preferred during the selection process is called *selection pressure*. The influence of this setting is illustrated in Table 14.4.

A typical selection scheme for Evolution Strategies is the so-called *greedy selection*. Let P denote the current population, and P' the new population. Greedy selection then chooses the μ_p best members for P'. Usually, the size of $P'(\mu_p)$ is smaller than the size of $P(\lambda_p)$; thus, the worst individuals will have no chance of being selected. In this approach, selection pressure is determined by the ratio μ_p/λ_p. Good values for the problem of UMTS optimisation have been found to be between 1/2 and 1/5 [13].

Note that with this $(\mu, \lambda) - ES$ selection scheme only individuals from the final working population are selected; in other words, the next generation is only selected from the offspring, excluding the μ_p parents from the old population. Thus, no individual may live longer than for exactly one generation. This also implies that an already found good solution will be lost again if it only produces offspring with lower fitness – maybe never to be rediscovered. On first impression, this looks like a disadvantage. However, according to several studies, this scheme has been found superior compared to a scheme which includes the parents in the selection process (so-called $(\mu + \lambda) - ES$); see [36] for a detailed rationale of this behaviour.

14.4.2.4 Self-adaptation

The course of an evolutionary algorithm is an intrinsically dynamic and adaptive process, and the use of fixed parameters runs contrary to this spirit. Obviously, different settings of these parameters might be appropriate in different stages of the evolutionary process. The main parameter in Evolution Strategies that can benefit from dynamic adaptation is the standard deviation (σ_g) of the Gaussian distribution in the mutation operator. In the initial phases of the algorithm, a big standard deviation might prove useful in order to enable a broad exploration of the search space; while in a later phase, smaller values could be preferable in order to concentrate on refining the most promising optima. This way the static parameter σ_g is replaced by a function $\sigma_g(t)$, with t being the generation counter.

In the *self-adaptation* approach, separate values are kept for each individual's adjustable parameter. The definition of an individual $(x_1, \ldots, x_l)$ is thus extended as follows:

$$I = (x_1, \ldots, x_n, \sigma_{g,1}, \ldots, \sigma_{g,n}) \tag{14.18}$$

Just like the x_i, we also subject the $\sigma_{g,i}$ to the evolutionary operators. The idea is that individuals with a better setting of their parameters, i.e. values that prove useful in the current position of the specific individual in the search space, will more often produce good offspring. These, in term, are more likely to survive and produce more children, hence propagating the better parameters. Thus, we now have a two-level evolution process, comprising evolution of the state parameters as well as evolution of the adjustable parameters. In detail, the adjustable parameters $\sigma_{g,i}$ are mutated according to Equation (14.19), while the state parameters x_i are mutated according to Equation (14.20):

$$\sigma'_{g,i} = \sigma_{g,i} \cdot e^{N(0,\tau_0)+N_i(0,\tau)} \qquad (14.19)$$

$$x'_i = x_i + N(0, \sigma_{g,i}) \qquad (14.20)$$

In Equation (14.19), $N(0, \tau_0)$ is calculated once per individual, while $N_i(0, \tau)$ is recalculated for each parameter. Some literature, e.g. [35], gives the following recommendations:

$$\tau_0 = \frac{1}{\sqrt{2\sqrt{N_{\text{pop}}}}} \qquad \tau = \frac{1}{\sqrt{2N_{\text{pop}}}}$$

14.4.3 PRACTICAL IMPLEMENTATION OF GA FOR TILT AND CPICH

(Fragments of Section 14.4.3 reproduced by permission of IEEE [34])

In this section, an example of a practical description of a Genetic Algorithm for the optimisation of antenna tilts and CPICH powers is given. The goal of the optimisation algorithm is to increase the capacity.[*]

For the implementation a deterministic fitness proportional selection scheme, a problem specific recombination operator and an improved mutation operator is used. In addition, a simple local optimisation based on [37] is introduced, which can be applied to the best individuals to improve their fitness.

14.4.3.1 Representation of the Individuals

For the Genetic Algorithm, a suitable representation of the parameters like CPICH power and antenna tilt is needed. In Figure 14.14 the used coding is shown.

Each *individual* of the population consists of $2N_{\text{cells}}$ *genes*, where N_{cells} is the number of cells. For one cell, one gene is used for the CPICH power and one for the antenna tilt.

14.4.3.2 Algorithm

In this section the optimisation process as well as the used genetic operators are described in more detail. The operators are used to incorporate knowledge about the quality of the cells in the network. In Figure 14.15, the flowchart of the implemented algorithm is shown.

The term *GoS* (*Grade of Service*) denotes the ratio of served users over all existing users[†]. During the optimisation process, GoS increases from its initial value of 95 % until it has reached 100 %. Then all users are served and the optimisation algorithm cannot proceed any further. However, the network could possibly accept more users. Thus, this practical algorithm applies the following approach: When

CPICH [dBm]	33	32	31	29	27	...	...	29
TILT [°]	2	0	5	0	6	...	...	3
cell	1	2	3	4	5	......		n

Figure 14.14 Representation of individuals for capacity optimisation [34]. (Reproduced by permission of © 2004 IEEE).

---

[*] This algorithm was developed during an optimisation project on the Institut für Nachrichtentechnik und Hochfrequenztechnik, Technische Universität Wien together with SYMENA, Software & Consulting GmbH [34,38].
[†] Note that some literature defines GoS as the probability of a cell being blocked or delayed for more than a specified interval.

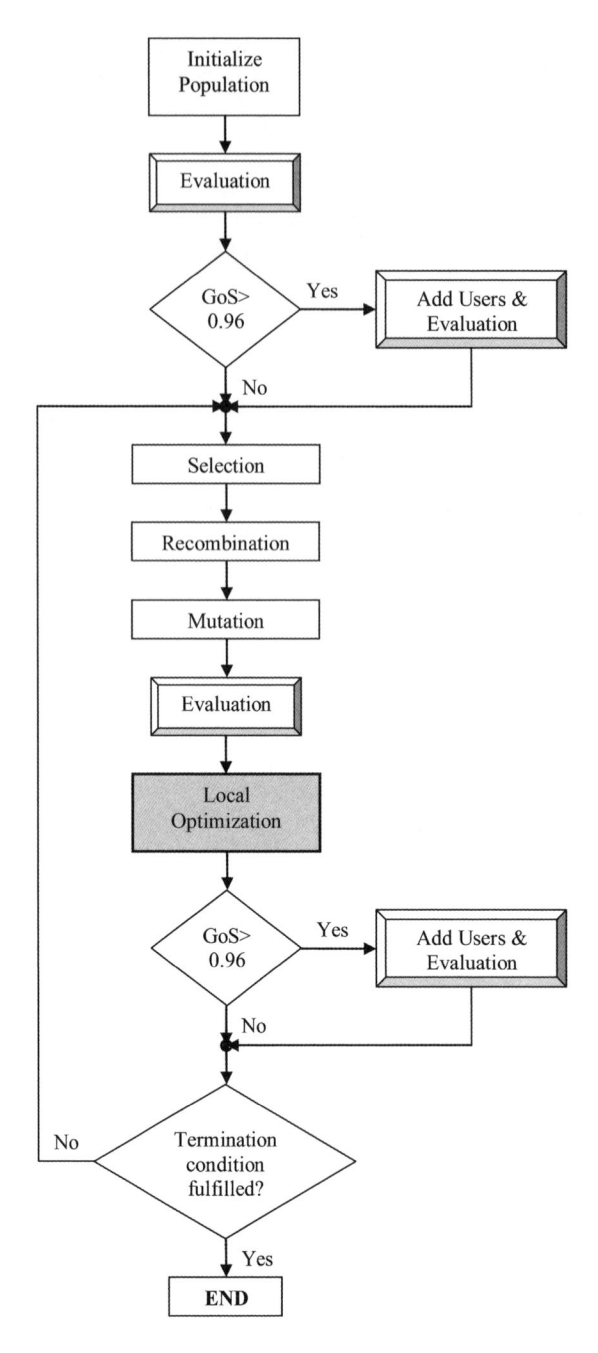

Figure 14.15 Flowchart of practical Genetic Algorithm implementation [34]. (Reproduced by permission of © 2004 IEEE).

GoS reaches a value of 96%, new users are added to the network until the initially defined GoS of 95% is reached again. In the following, this function is referred to as *Add Users* (see Figure 14.15).

The algorithm starts with the initialisation of all individuals of the population. For all individuals, the initial values for CPICH power and antenna tilt are chosen randomly within the defined search space, but with the same CPICH power and antenna tilt values for all cells. After the initial phase, the whole population is evaluated with a network simulator. The objective function $f_{\text{obj}}(i)$ as well the GoS are calculated for all the individuals ($i = 0, 1, \ldots, N_{\text{pop}}$). In the next step, the GoS of the best individual, i.e. the individual with the highest cost value, is compared to the limit of 96%. If the GoS is higher than this threshold, additional users are added to the network (*Add Users*). The whole population is then re-evaluated to obtain the new values for $f_{\text{obj}}(i)$ and *GoS*.

After this pre-processing of the population, the optimisation process is started. When the evolution process of one population is finished, the whole population has to be evaluated again to get the new values for $f_{\text{obj}}(i)$ and *GoS* of the individuals. On the best individuals of the population a local optimisation step is performed to improve the performance of the Genetic Algorithm. After the local step, the GoS of the best individual is compared to the threshold of 96%. If the GoS is higher, additional users are added and the population is re-evaluated. In the next iteration, the evolution process is repeated. The Genetic Algorithm stops, if a termination condition is fulfilled, e.g. after a certain number of iterations.

14.4.3.3 Selection

The selection of the individuals for the new population is implemented as a *deterministic fitness proportional selection*, because experiments with other methods (*roulette wheel selection, tournament selection,* etc.)* have indicated that this method fits best for the problem of UMTS base station parameter optimisation. The fitness values $f_{\text{obj}}(i)$ of the individuals are scaled to get normalised fitness values $f_{\text{obj}}^{\text{scal}}(i)$ for the selection process. A linear scaling function is used:

$$f_{\text{obj}}^{\text{scal}}(i) = a_1 \cdot f_{\text{obj}}(i) - a_2 \tag{14.21}$$

$$\text{with } a_1 = \frac{C_m \cdot \bar{f}_{\text{obj}} - \bar{f}_{\text{obj}}}{f_{\text{obj}}^{\text{max}} - \bar{f}_{\text{obj}}} \quad \text{and} \quad a_2 = a_1 \cdot \bar{f}_{\text{obj}} - \bar{f}_{\text{obj}} \tag{14.22}$$

In Equation (14.22), C_m denotes the *selection pressure*, which describes how much the algorithm favours good individuals compared to bad individuals. The mean value over all $f_{\text{obj}}(i)$ is denoted as $\bar{f}_{\text{obj}}$ in (14.22), and $f_{\text{obj}}^{\text{max}}$ is the highest fitness that occurs in the population.

The implementation of the fitness proportional selection works as follows: First, the best individual since the last function call of *Add User* is selected to guarantee that the best solution cannot get lost. This method is called *elitism*[†] in literature [33]. Next, the expected number of descendants for each individual is calculated with the following equation:

$$e(i) = \frac{f_{\text{obj}}^{\text{scal}}(i)}{\sum_{j=1}^{N_{\text{pop}}} f_{\text{obj}}^{\text{scal}}(j)} \tag{14.23}$$

In Equation (14.23), N_{pop} denotes the size of the population. From each individual $\lfloor e(i) \rfloor$[‡], descendants are produced. To complete the population, the best individual of the last population is repeatedly selected, until N_{pop} individuals are in the new population.

---

* In [33], all the selection methods are explained in detail.
† *Elitism*: Independent of the selection function, the best k individuals are taken into the new population to guarantee a monotonic increase of the fitness.
‡ The $\lfloor \cdot \rfloor$ operator denotes the floor function.

14.4.3.4 Recombination

The implemented recombination operator takes the quality of the cells into account. To describe the quality of a cell the *quality factor* (QF), introduced in [39], is used as performance indicator. The QF expresses whether a cell is heavily loaded or not. The range of the QF is between zero and one. A low value indicates a heavily loaded cell, and a high value indicates a weakly loaded cell.

The algorithm selects N_{pop} times randomly two individuals (referred to as parent 1 and parent 2) from the population and produces a new child with a recombination probability p_c. The recombination operator is shown in Figure 14.16.

For each cell in the network, the number of mobiles put to outage and the QF of the two individuals are compared. If the number of mobiles in outage of parent 2 is smaller than that of parent 1 and the QF of parent 2 is better for this cell, then the corresponding genes for this cell are taken from parent 2 for the new child, otherwise the two genes for CPICH power and antenna tilt are taken from parent 1. If the algorithm decides not to produce a child by recombination (with a probability of $(1 - p_c)$), then the first selected parent is taken unchanged. After recombination process, the population again has a size of N_{pop} individuals, consisting only of the children produced by the recombination.

14.4.3.5 Mutation

The mutation operator is performed with each individual of the population. The algorithm decides for each gene of the individual with a mutation probability p_m whether the value of the gene will be mutated (for CPICH power and antenna tilt separately). In Figure 14.17, the mutation for one cell is shown.

For both antenna tilt and CPICH power, the value is either left constant, increased by 0.5, or decreased by 0.5 (degree or dBm, respectively). The choice between those possibilities that still lie within the search space is taken at random (with uniform distribution).

14.4.3.6 Local Optimisation

After the evolution process, a local optimisation with the best *local_num* individuals is carried out. The flowchart of the local optimisation is shown in Figure 14.18.

First, the best *local_num* individuals are selected. For each of these individuals, *local_iter* local optimisation iterations are performed. Each iteration includes two steps. In the first step, the parameters are changed according to the quality of the cells in the network. The rules for changing the parameters are based on [37], with the extension that the parameters can be changed in both directions. In the second

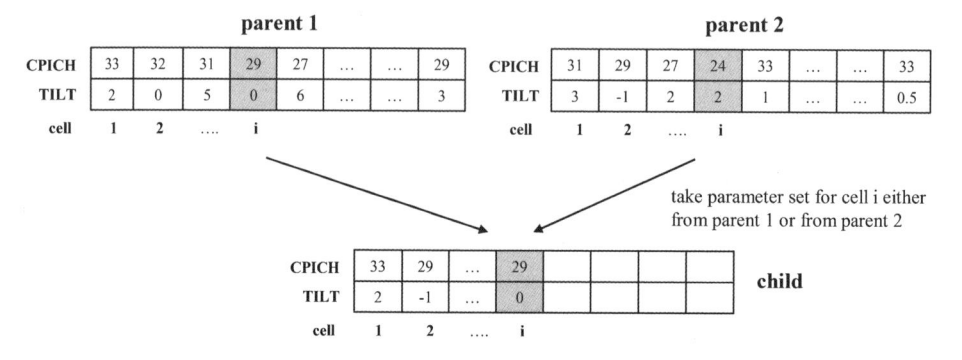

Figure 14.16 Recombination operator, practical implementation [34]. (Reproduced by permission of © 2004 IEEE).

Theory of Automated Network Optimisation 365

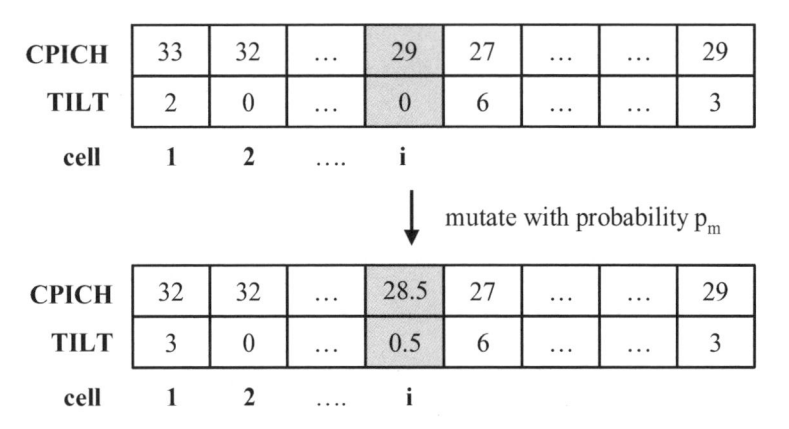

CPICH	33	32	...	29	27	...	...	29
TILT	2	0	...	0	6	...	...	3
cell	1	2	....	i				

↓ mutate with probability p_m

CPICH	32	32	...	28.5	27	...	...	29
TILT	3	0	...	0.5	6	...	...	3
cell	1	2	....	i				

Figure 14.17 Mutation operator, practical implementation [34]. (Reproduced by permission of © 2004 IEEE).

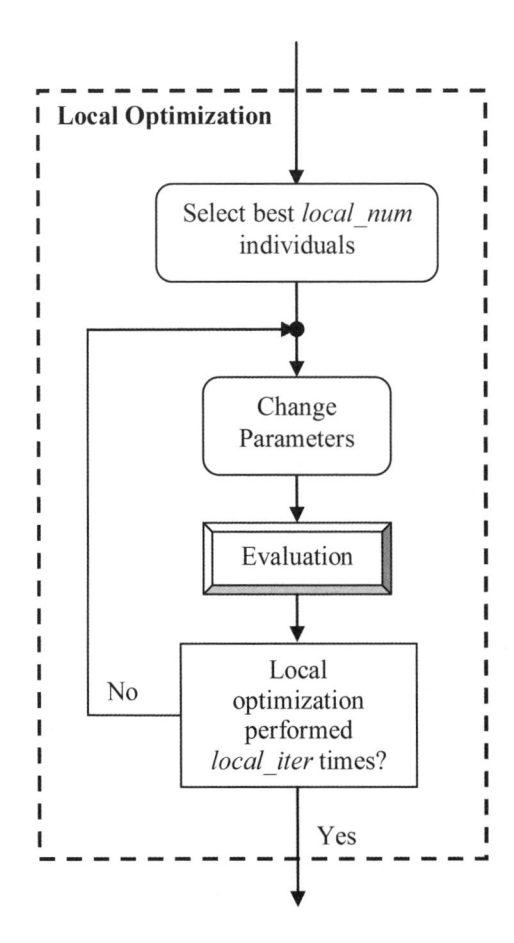

Figure 14.18 Detailed flowchart of local optimisation [34]. (Reproduced by permission of © 2004 IEEE).

step, the individuals are evaluated. If the fitness value for the new parameter setting of the individual is better than the old one, then this setting is taken, otherwise the old parameter setting is retained.

Two rules are used for the local optimisation. The first rule (*rule 1*) is to shrink a cell in the network and the second rule (*rule 2*) to enlarge a cell. In each local iteration for each individual, a random value decides, whether *rule 1* or *rule 2* is used. If *rule 1* is selected, then it is checked for each cell, if there are outaged mobiles and if the QF is bad (value for QF < 0.1). If this is the case, the CPICH power is decreased by 0.5 dB and the antenna downtilt is increased by 0.5° in this cell, both with a probability of 0.7. In the case of *rule 2*, in cells without outaged mobiles and with a good QF (value for QF > 0.1), the CPICH power is increased by 0.5 dB and the antenna downtilt is decreased by 0.5°, both with a probability of 0.5.

14.5 OPTIMISATION WITHOUT SIMULATION

Network optimisation using accurate performance evaluation, typically Monte-Carlo simulations, as a subroutine is a common practice. One of the advantages is that diverse and yet detailed statistics may be obtained. On the downside, however, is the large computational effort linked with such simulations. In many situations, it is unnecessary or ineffective to take this computational effort.

Developing *alternative system models* provide a means to target specific optimisation problems more efficiently by cutting short on the evaluation. The task in modelling is to sensibly simplify the system such that the remaining, abstract model is easier to handle, understand, or evaluate. This has to be done such that the system is still well represented by the resulting model – at least in the aspects that are of interest in the given situation.

This approach has several advantages for optimisation. Optimising abstract, simple models can often be done in significantly less time. Yet, good solutions can be found if the model is appropriate. A good modelling can also help to understand the system's behaviour and to conceive specialised optimisation algorithms. Finally, there is a large mathematical toolbox for certain types of optimisation problems (for example, convex ones), which can be accessed once the problem is modelled accordingly.

Already very simple methods such as those presented in Section 14.5.1 are able to find solutions that may serve as a good starting point for more sophisticated algorithms. Most optimisation approaches therefore use simple models for finding good starting solutions. This is, for example, the case in [38,40] and in the case study in Section 15.3.

Also when using alternative system models, the detailed assessment of a network's performance remains important to decide whether an optimisation solution meets the quality criteria. This can be done using Monte-Carlo simulation.

14.5.1 GEOMETRY-BASED CONFIGURATION METHODS

The most simple methods for determining appropriate antenna configurations are inspired by geometrical arguments. They are 'analytical' because they do not evaluate network performance at all, but rather allow to select a good configuration by easy calculations or a table lookup. Typically, analysis is made for a regular, simplistic scenario. Base station locations are distributed on a regular, triangular grid. Traffic is assumed to be uniform. In this setting, the optimal network configuration is easy to determine: cells are hexagon-shaped, tilts are adjusted for best coverage [41] (see Figure 14.19). In this setting, the influence of antenna tilt and configuration changes can be studied comparatively easily. This is done using simulation in [42] (for mechanical tilt) and [43] (for electrical tilt).

Most analytical approaches carry over the results of this model to real-world networks. Base stations are distributed and configured such that the resulting network resembles the idealistic configuration. Computational experiments [40] suggest that this can construct reasonably good starting points for refined optimisation from scratch.

14.5.1.1 Base Station Density Estimation

In the first steps of planning a UMTS radio network, the approximate number of base stations and their distribution has to be determined. This can be done analytically [44] in a simplified setting (not considering sectorised antennas) as described above. Depending on the user density, the required CIR target, and the maximum acceptable outage probability, an analytical expression for the optimum (maximum) base station distance can be derived. (In the cited paper, the optimum distance is inversely proportional to the square root of the user density.) For a given area to be covered and a forecast for the future traffic, an approximation to the minimum number of base stations needed to serve the area can be estimated.

14.5.1.2 Analytical Tilt Selection

The main idea in analytical tilt and azimuth finding heuristics is to 'interleave' the antennas' main lobes similar to the 'ideal' regular configuration. At the same time, cells should be created that are roughly comparable in size to balance the traffic load between cells. The tilt parameter is in this context basically viewed as a dependent one for which a canonical value can be found: the down-tilt has to be as large as possible for ensuring little interference. The limiting constraint is coverage. The 'idealised' tilt value is then calculated using simple trigonometrics. If the cell boundary (in the main lobe direction) is located at distance d, the tilt value for an antenna at height h having a vertical opening angle of α is $t = \arctan(h/d) - \alpha$.

14.5.1.3 Azimuth Selection

For azimuths, the regular hexagonal paradigm commonly is generalised as follows: antennas should point to an area in the base station's vicinity that no nearby antenna is yet servicing. Antennas are thus never directly opposed to each other. This is the main idea behind the analytical azimuth selection in [38,45]. In [46], the problem is formulated as a combinatorial optimisation problem. For each antenna, a small set of potential target areas is identified. The antenna directions are then chosen such as to minimise overlap, while irregular sectorisation is penalised.

Another example for azimuth (and tilt) selection based on similar geometrical considerations is given in [40]. (A two-stage process is used here, the second stage is a variant of the local search method described in Section 14.5.4.2.) The authors take it as an indication of validity of their method that for a scenario with three-sectorised sites on a regular triangular grid the 'idealised' configuration depicted in Figure 14.19(b) is reached when applying the method to random initial azimuth values.

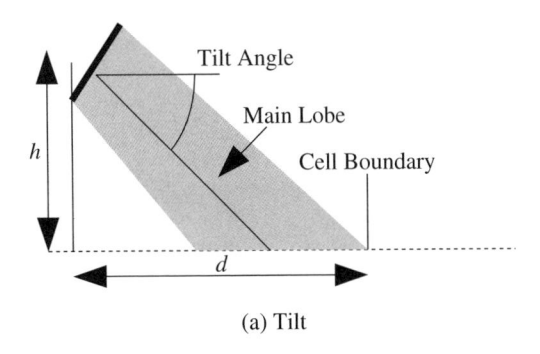

(a) Tilt

Figure 14.19 Geometrical approaches to radio network optimisation.

(b) Azimuth

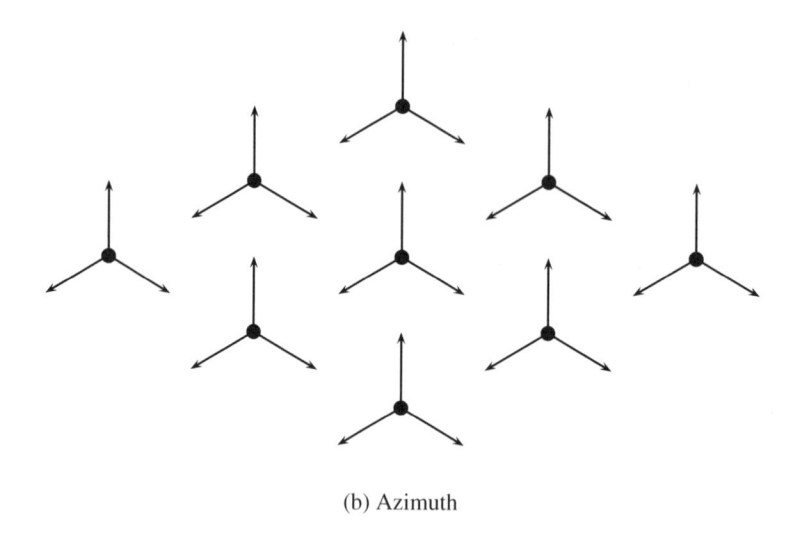

Figure 14.19 (*continued*)

14.5.2 COVERAGE-DRIVEN APPROACHES

Classical optimisation can be applied most easily to UMTS when simulations are dispensable for evaluating the optimisation target. This is the case with E_c coverage. For uplink and E_c/I_0 coverage, it only applies if interference is assumed at a fixed level, traffic independent. The models in this section can usually be solved using standard optimisation software, no specialised approaches are necessary. This allows to obtain optimal solutions or solution quality guarantees.

14.5.2.1 Set Covering Models

A simple question related to UMTS and other types of cellular networks is how to provide sufficient coverage for a planning area using as few base stations as possible. Usually, a set $\mathcal{B}$ of candidate base stations is prescribed. This is essentially a *set covering problem*: each base station $i \in \mathcal{B}$ serves a certain subarea $C_i \subset A_{planning}$ of the planning area and the goal is to cover the entire area with as few base station sets as possible.

Example. The crucial question is how to define the points covered by a base station. A popular way is to fix a certain threshold value $L^{(th)\uparrow}$, for the attenuation from pixel p to base station i. The threshold $L^{(th)\uparrow}$ is computed in radio link budget calculations. The coverage sets are then defined as

$$C_i := \left\{ p \in A_{planning} \mid L^{\uparrow}_{pi} \geq L^{(th)\uparrow} \right\}$$

As an MIP, the model reads

$$\min \sum_{i \in \mathcal{B}} x_i \qquad (14.24)$$

$$\text{s.t.} \sum_{C_i \ni p} x_i \geq 1 \quad \forall p \in A_{planning} \qquad (14.25)$$

$$x_i \in \{0, 1\} \quad \forall i \in \mathcal{B} \qquad (14.26)$$

The binary variables x_i determine whether a particular candidate i is selected ($x_i = 1$) or not ($x_i = 0$). The objective (14.24) aims at minimising the number of selected base stations. The only constraint (14.25) demands that each point p in the area be covered. This formulation can be found, e.g. in [47].

14.5.2.2 Facility Location Models

This pure set-covering formulation is normally not suited for practical purposes, especially if the resolution of the planning area is high. With realistic propagation data, there are always some pixels that cannot be covered by any base station, so the model has no feasible solution. After removing these pixels, there usually remains a large number of 'singletons', pixels that can only be covered by exactly one base station and thus force the opening of that base station, eventually leading to an exaggerated number of base stations in the final solution.

A variant of the above set-covering model thus states the problem differently: a fixed number of base stations may be chosen such as to maximise the covered area. One technical drawback of this formulation is that it requires another class of binary variables describing whether a pixel is covered. In mathematics, this is called the (uncapacitated) *facility location problem*. In engineering, it is referred to as *Maximum Covering Location Problem* [47]. (Both models are known to be $\mathcal{NP}$-hard.)

In [48, Chapter 6], a possibility to include a desired amount of cell overlap into this type of model is presented. Under that modification, however, the model is not solvable exactly anymore. A greedy algorithm is used instead.

14.5.3 ADVANCED MODELS

The models presented in this section so far are generic and can also be used for radio technologies other than WCDMA. Beyond these rather simple models, there are models that deal with the special features of WCDMA radio, notably signal quality measurements and soft capacity, and that aim at optimising base station configurations. These models can often be formulated as linear integer programs. Nonetheless, they are normally too complex for standard software and require specialised algorithms; meta-heuristics are often used to solve them.

14.5.3.1 Basic Principles

While advanced models are in general formulated for a specific optimisation task, they have common features.

Demand points. The area is covered with *demand points* (also called demand nodes, test points, service test points, traffic nodes). The subdivision often arises canonically from the format of the input data. In particular, propagation grids and geographic data are provided in pixel grids. Radio conditions are considered to be constant on one demand point. In consequence, demand points/pixels may be assigned as a whole to serving cells. The entire demand point is considered covered or not-covered.

Variables. In virtually all models, there are *configuration selection* decision variables. These variables denote whether a specific configuration is chosen for a base station or sector. There is only one decision per site (switching the site on or off) modelled by the variables x_i in the case of the simple coverage-based base station positioning models in Section 14.5.2. In more complex models, the decision between different configurations for one sector is usually broken into different variables for each potential configuration. For example, there might be variables $x_{i,t}$ for each potential down-tilt value t. The different decision variables are then linked by constraints ensuring that only one configuration can be chosen per sector, as in

$$\sum_{t} x_{i,t} \leq 1$$

Demand points are related to configuration decisions with additional, dependent *assignment variables* $y_{p,i}$. For a demand point p and a sector configuration i, the variable $y_{p,i}$ indicates whether the demand point is served (or can potentially be served) by the referring sector using the particular configuration. The decision is often made with respect to the level of received signal strength, e.g. of the CPICH power.

Power variables are also sometimes used. This is most often the case if pilot powers are to be adjusted. In this case, there are as many CPICH power variables as there are potential sector configurations. Several models in literature also contain link power variables. Usually there has to be a specific variable for the power on each potential link, that is as many link power variables as there are assignment variables. The large number of such variables leads to very large and hardly tractable models.

14.5.3.2 Modelling Traffic

Since in UMTS radio network planning, coverage and capacity cannot be separated, traffic has to be considered in network planning.

Weighted demand points. If inhomogeneous traffic distributions are considered, there is typically a non-negative number $A(p)$ specified per pixel/demand point p. These numbers can be used as a priority weighting over the set of pixels. This approach is popular for optimising coverage; see Section 14.3.1.

Multiple knapsack. For rough capacity estimations, the soft capacity feature of WCDMA is sometimes neglected. Instead, each cell is assumed to have a fixed capacity e. The capacity value can be considered an amount of Erlang that the cell can support. The set of demand points assigned to one cell can then easily be required to generate no more than e units of traffic.

$$\sum_{p} A(p) y_{p,i} \leq e \tag{14.27}$$

A single, linear constraint of the type in Equation (14.27) with positive coefficients is called a *knapsack* constraint. If several of these inequalities are taken into account (one per cell), this is called a *multiple knapsack*. Multiple knapsack constraints are used in [49].

Under the assumption of a fixed level of interference for all cells, the capacity of UMTS cells is modelled accurately with a multiple knapsack. In the uplink, this corresponds to an assumption on noise rise at the receiver; in the downlink, sometimes all cells are assumed to transmit at maximum power. While these assumptions greatly ease capacity calculations, they ignore the dynamic nature of WCDMA cells.

User snapshots. A technique that mimics Monte-Carlo evaluation is the inclusion of one or several user snapshots into the model. Demand points are not chosen regularly (e.g. according to the pixels) but according to a traffic distribution. An example for this kind of traffic modelling is given [50]. An advantage of this model is that service properties may be included. However, in order to predict the network's performance with some reliability, many snapshots have to be included, which increases model size and impedes solvability.

14.5.3.3 Modelling Interference

As WCDMA systems are often assumed to be interference limited, the modelling of interference plays a key role for UMTS radio network optimisation.

Overlap: counting interferers. A simple way of accounting for interference at a mobile station or test point is to count the number of signals received above a certain threshold (e.g. the receiver sensitivity). This can be done by summing up the values of assignment variables at a pixel. The interference measure is used in [51], where the overlap is to be minimised. In [48], the goal is to have a specified amount of overlap, and deviations from the desired overlap are penalised. However, the ratio of the serving signal and interferers is not determined with this measure.

Co-channel interference with fixed transmission powers. If the transmission powers of the interferers are known in advance, interference can be calculated exactly comparatively easily. This is a simple extension of the counting principle: interferers are counted weighted with their respective transmission power attenuated by the path loss. This model is used in [52,49].

Calculating transmit powers and CIR. None of the above interference models incorporates the reaction of mobiles to interference stemming from other users in the network. To duly consider UMTS power control, the transmit powers have to be adapted to the current network design. This leads to the inclusion of power variables. Models containing power variables are proposed in [53,50].

The basic principle is to include the system of linear inequalities that is used for determining transmit powers (see Section 6.4.9) as constraints into the model. Since the network design and thereby user assignment are not fixed, CIR inequalities have to be included for all *potential* links. They are switched on or off by the assignment variables. This leads to a large number of constraints in the model. The constraints are numerically difficult to handle and non-linear in their straightforward formulation. They can be linearised using standard reformulations, but this increases the model size further.

14.5.3.4 Solution Approaches

The advanced models presented in this section are usually too complex to be solved to optimality within reasonable time – at least for instances of interesting size. *Heuristics* – methods that strive for finding good solutions in reasonable time but without intrinsic quality guarantees or mere estimates of their quality – are therefore the most prominent approach.

Heuristics. The heuristic methods introduced before are used in many different flavours. How to best implement a given scheme depends on the specific problem formulation. Simulated annealing approaches are implemented for base station configuration in [51,38]. Greedy algorithms are specified in [48] for the problem of selecting base stations that have a desired degree of overlap under coverage constraints. In [52], greedy and simulated annealing methods are described for cell design problems considering interference. Examples of greedy-type base station configuration methods can be found in [51,53]. In the latter reference the greedy method is used for determining a good starting solution to a second stage that uses Tabu Search.

Integer programming. The exact mathematical optimisation methods from integer programming are in general not capable of producing provably optimal solutions to accurate models of UMTS radio network planning problems unless the instances are small. For small instances of base station configuration in networks without load, [49] reports on optimal solutions obtained with integer programming methods. Integer programming methods that are applied to larger models and are terminated before optimality is reached are used sometimes. If only selected parameters are optimised and interference modelling is kept simple, integer programming methods can be successful; an example for CPICH power adjustment is given in [54].

Problem reduction. As smaller instances of the planning problems are easier to solve, a common approach is to focus on problematic areas. For example, bottleneck cells can be identified together with a set of neighbours. Optimisation is then carried out in these special areas on a smaller scale. The contributions in [55] and [56] work along these lines.

14.5.4 EXPECTED COUPLING MATRICES

In Monte-Carlo methods, the evaluation of a radio networks is done by determining the *expected* values of cell powers and other performance measures (coverage, blocking, etc.) over a distribution of user snapshots. Occasionally, more information on the distributions is calculated, for example the variance. For each snapshot, the coupling matrix (see Section 6.4.9) determines the cell powers. The last optimisation approaches to be presented here are based on analysing the *expected coupling matrix*. They are used for the case study in Section 15.3 and different from the ones presented before as they do not consider test-points or individual users.

14.5.4.1 Calculating the Expected Coupling Matrix

Let A_{planning} denote the total planning area and $A_{\text{planning},i} \subset A_{\text{planning}}$ the best-server area of cell i. Let S be the set of services under consideration, let $A_s(\cdot)$ be the service-specific spatial traffic distributions, and let $A_s(p)$ denote the average traffic intensity of service s at location p (for some specific point in time, e.g. the busy hour). The traffic intensity counts simultaneous calls at location p.

Each service s is assumed to have fixed CIR targets $\text{CIR}^\uparrow_{\text{target},s}$, $\text{CIR}^\downarrow_{\text{target},s}$ and activity factors $\nu_s^\uparrow, \nu_s^\downarrow$; a location-specific noise N_p can be assumed for a mobile at position p. Since activity and CIR targets are fixed, the average load contribution in up and downlink (see the definitions in Section 6.4.9) by a point $p \in A_{\text{planning}}$ – for example, a pixel – is calculated as:

$$l_p^\uparrow := \sum_{s \in S} \frac{\nu_s^\uparrow \text{CIR}^\uparrow_{\text{target},s}}{1 + \nu_s^\uparrow \text{CIR}^\uparrow_{\text{target},s}} A_s(p), \quad l_p^\downarrow := \sum_{s \in S} \frac{\nu_s^\downarrow \text{CIR}^\downarrow_{\text{target},s}}{1 + \bar{\alpha}_p \nu_s^\downarrow \text{CIR}^\downarrow_{\text{target},s}} A_s(p) \quad (14.28)$$

This is straightforward due to linearity of expectation. The elements of the expected uplink coupling matrix are given by

$$M_{ii}^\uparrow := \int_{p \in A_{\text{planning},i}} l_p^\uparrow dp, \quad M_{ij}^\uparrow := \int_{p \in A_{\text{planning},j}} \frac{L_{ip}^\uparrow}{L_{jp}^\uparrow} l_p^\uparrow dp \quad (14.29)$$

The expected downlink coupling matrix and the expected traffic noise power are given by

$$M_{ii}^\downarrow := \int_{p \in A_{\text{planning},i}} \bar{\alpha}_p l_p^\downarrow dp, \quad M_{ij}^\downarrow := \int_{p \in A_{\text{planning},i}} \frac{L_{jp}^\downarrow}{L_{ip}^\downarrow} l_p^\downarrow dp, \quad P_i^{(N)} := \int_{p \in A_{\text{planning},i}} \frac{N_p}{L_{ip}^\uparrow} l_p^\downarrow dp \quad (14.30)$$

The notation is generic; note that in the common case where data is given in a discretised manner (pixel grids) the integrals become sums over pixels.

14.5.4.2 Local Search Using Average-coupling Power Estimates

The efficiency of a local search algorithm depends on the structure of the search space and on the speedy evaluation of candidate solutions. During execution, one has to be able to decide for the better one of two given configurations. The quicker this decision can be taken, the faster the local search algorithm can arrive at good solutions and eventually terminate. Performing a full-fledged Monte-Carlo evaluation for each improvement step is prohibitively expensive in terms of computational effort – it may also be considered an over-kill if the configurations differ only slightly.

An alternative to costly simulation for performing efficient local search is rough estimates on transmission powers and deduced performance measures. The expected coupling matrix can be used for this: instead of repeatedly solving coupling equation systems for different random realisations of the coupling matrix, the equation system is solved using the expected coupling matrix as described in Equations (14.29) and (14.30). The result is taken as an approximation to the average power vector.

With bookkeeping for quick calculation of the expected matrix entries in Equations (14.29) and (14.30), a rough estimate of transmission powers $\bar{P}^{\uparrow}$ and $\bar{P}^{\downarrow}$ can be computed within a fraction of a second on a standard computer for realistic problem sizes. These estimates can then be used for estimating cell load factors and E_c/I_0 coverage. The decisions on where to move in the search space are based on these performance measure estimates. A description of a complete local search based on these premises is given in Section 15.3. A similar implementation of a local search algorithm is described in [40].

14.5.4.3 Average-coupling Predicting MIP

The expected coupling matrix encodes information regarding traffic density, load distribution, and interference coupling in a UMTS radio network. This information gives hints on quality and performance of the network. Radio network design can thus be viewed as designing a networks coupling matrix. This is the motivation for the optimisation model outlined in this section. The model is described in detail in [57], it works by computing the contributions of the pixels in the area to the coupling matrix.

Contributions to the main diagonal and coverage. For each installation configuration, a configuration selection variable x_i is used. Again, the server of each point is determined using assignment variables y_{ip}. The assignment variables are held exactly: y_{ip} is only set to one, if i is the strongest server at p. This depends on the configurations of other sectors, so selection variables are linked in constraints on the assignment variables. As the coupling matrix's main diagonal only depends on the server of each point, it can be calculated from this information.

Off-diagonal elements. The off-diagonal elements depend on the relative strength of the signals at each point, so their calculation is more involved. Assignment variables of second order are therefore introduced for each pair of server and potential interferer. Since this can lead to very large models, only the strongest interferers are considered in practical applications.

Model usage. The resulting linear mixed integer programming model allows to calculate coverage and the expected average coupling matrix for any given set of candidate installations. A linear objective function formulated in the area coverage and the entries of the coupling matrix can then be optimised with integer programming software. This allows weighted combinations of the different optimisation goals: (a) maximising coverage, (b) balancing load between cells, and (c) reducing interference coupling.

14.6 COMPARISON AND SUITABILITY OF ALGORITHMS

A variety of algorithms for network optimisation problems have been introduced in this chapter. They differ in implementation effort, running time, and solution quality (see Table 14.5). The order of the methods in the table roughly reflects their complexity and running time. Which method is best suited for a given situation depends on the optimisation task and the desired results.

For picking the right strategy for a network planning problem, it is important to develop a feeling for the properties, advantages, and drawbacks of the respective methods. The most advanced optimisation methods towards the end of the table are not always the best choice. As time is usually a constraint, simpler and faster method may be preferable.

14.6.1 GENERAL STRATEGIES

Avoiding simulations. Performing detailed evaluation by simulation during optimisation is not always necessary. Methods that perform several Monte-Carlo simulations per iteration are computationally very costly. They can take hours or even days to terminate. All optimisation methods can, in principle, be implemented without Monte-Carlo simulation as a subroutine. (Geometric approaches and linear programming do not use it by definition.) Several key performance indicators – such as downlink coverage – do not depend on network simulation results. Simulation during optimisation can be sensible at advanced stages, but it should be avoided wherever possible. It is, however, recommendable to assess candidate solutions that have been produced by optimisation methods in detail in order to track down problems and also to more accurately compare different solutions.

Selective application of 'expensive' methods. If radio networks expose quality problems, these are usually limited to specific areas (hot spots, for example). For those areas, it might be useful to employ methods with longer running time. However, the area to treat with the specialised methods should be kept as small as possible. This will lead to good results in shorter time.

14.6.2 DISCUSSION OF METHODS

Simple approaches. The most prominent, simple algorithms are geometry-based and Greedy methods. For complex optimisation tasks in the context of UMTS radio network planning, these methods usually fail to come to competitive solutions. However, they are typically easy to implement and can serve as useful tools during optimisation. First, they can be used to generate reasonable starting solutions as an input to subsequent, more sophisticated methods. Second, they can be used as a reference for sophisticated methods. In addition to comparing the results of sophisticated methods to an arbitrary starting solution, it can be compared to the output of simple strategies. This sheds a more realistic light on the value added by sophisticated optimisation.

Search Methods. Search methods represent a compromise between running time and quality of results. They rely on evaluating a multitude of alternative network configurations. The level of detail of these evaluations directly scales their running time. Full simulation has to be exercised with care. Refined search methods have powerful mechanisms to escape local minima. Methods which can do so (such as Tabu Search or Simulated Annealing) need more evaluations.

Evolutionary Algorithms. The difference to the previous algorithms is that Genetic Algorithms and Evolutionary Strategies work on a population of solutions. This allows for a more widespread search for optimal solutions. The consequence of this is that in each iteration several simulations have to be performed and the running time is very high. The effort to implement these algorithms is only slightly higher than for greedy algorithms or search methods. The inherent structure of evolutionary algorithms facilitates a parallel implementation.

Linear Integer Programming. Linear optimisation problems are well understood in optimisation theory, and several good software packages exist to solve these problems. A linear formulation is hence desirable for any optimisation task. If a problem can be modelled as a continuous linear problem, there is usually no difficulty in solving it to optimality. If the formulation involves integer variables, the solvability and the approach's scalability depends on the specific formulation and typically requires some expertise in mathematical programming. In the context of UMTS network planning, integer programs have proven most useful for coverage optimisation.

Table 14.5 Comparison of optimisation methods.

Algorithm	Running Time	Suitable for	Comment
Geometry-based Approaches	Very short	Educated guess of starting solutions	Does not perform analysis of network
Greedy	Short	Reasonable starting solutions	Easy to implement
Local Search	Medium	Improvement of existing solutions	Performance depends largely on neighbourhood definition
Tabu Search	Medium	Improvement of existing solutions	Performance depends largely on neighbourhood definition, parameter adjustment and tabu list implementation
Simulated Annealing	Long	Optimisation from scratch	Probabilistic neighbourhood exploration, requires parameter tuning
Genetic Algorithm	Long	Optimisation from scratch, fine-tuning of existing solutions	Intrinsically diversified search, requires parameter tuning
Evolutionary Strategies	Long	Optimisation from scratch, fine-tuning of existing solutions	Intrinsically diversified search, as GA, complex to implement, requires parameter tuning
Linear (Integer) Programming	(depends on formulation)	Coverage planning, fine-tuning	Global perspective, powerful standard software packages, preferable whenever a linear formulation is at hand

14.6.3 COMBINATION OF METHODS

In practice, it often pays off to combine several approaches. The simplest optimisation methods find reasonable parameter settings for most networks. A more complex method should in all cases be applied afterwards. In any case, it can be useful to apply a simple search method at the end to find or exclude better solutions in the vicinity of the current solution. Table 14.5 compares the optimisation methods described above.

REFERENCES

[1] W.J. Cook, W.H. Cunningham, W.R. Pulleyblank and A. Schrijver. *Combinatorial optimization*. John Wiley & Sons, Ltd/Inc., New York, NY, USA, 1998. ISBN 0-471-55894-X.
[2] M.R. Garey and D.S. Johnson. *Computers and Intractability: A Guide to the Theory of NP-Completeness*. W. H. Freeman & Co., New York, NY, USA, 1979. ISBN 0716710447.
[3] E. Aarts and J.K. Lenstra, editors. *Local Search in Combinatorial Optimization*. John Wiley & Sons, Ltd/Inc., New York, NY, USA, 1997. ISBN 0471948225.
[4] F. Glover and M. Laguna, *Tabu Search*, Kluwer Academic Publishers, Dordrecht, 1997.
[5] W.H. Press, S.A. Teukolsky, W.T. Vetterling and B.P. Flannery, *Numerical Recipes in C++: The Art of Scientific Computing*, Cambridge University Press, 2nd ed., 2002.
[6] E. Aarts and J.H.M. Korst, *Simulated Annealing and Boltzmann Machines*, John Wiley & Sons, Ltd/Inc., 1989.
[7] A. Schrijver. *Theory of linear and integer programming*. John Wiley & Sons, Ltd/Inc., New York, NY, USA, 1986. ISBN 0-471-90854-1.
[8] G.L. Nemhauser and L.A. Wolsey. *Integer and combinatorial optimization*. Wiley-Interscience, New York, NY, USA, 1988. ISBN 0-471-82819-X.
[9] ILOG CPLEX Division. CPLEX 9.0 Reference Manual, 2003. www.cplex.com.

[10] G. Ausiello, M. Protasi, A. Marchetti-Spaccamela, G. Gambosi, P. Crescenzi and V. Kann. *Complexity and Approximation: Combinatorial Optimization Problems and Their Approximability Properties*. Springer-Verlag New York, Inc., Secaucus, NJ, USA, 1999. ISBN 3540654313.

[11] M. Galota, C. Glaßer, S. Reith and H. Vollmer. 'A polynomial-time approximation scheme for base station positioning in UMTS networks'. *In Proc. DIALM '01*, pages 52–59. ACM Press, New York, NY, USA, 2001. ISBN 1-58113-421-5.

[12] Kathrein, *790-2200 MHz Base Station Antennas for Mobile Communications*, 2001, Catalogue.

[13] S. Jakl, *Evolutionary Algorithms for UMTS Network Optimization*, PhD thesis, Technische Universität Wien, 2004.

[14] W. Karner, *Optimum Base Station Parameter Settings for UMTS Networks*, Master's Thesis, Technische Universität Wien, 2003.

[15] I. Forkel, A. Kemper, R. Pabst and R. Hermans, 'The Effect of Electrical and Mechanical Antenna Down-Tilting in UMTS Networks', *Proceedings of 3rd International Conference on 3G Mobile Communication Technologies*, pp. 86–90, London, Great Britain, 8–10 May 2002.

[16] S.C. Bundy, 'Antenna Downtilt Effects on CDMA Cell-Site Capacity', Proceedings of Radio and Wireless Conference, RAWCON 99, pp. 99–102, 1–4 August 1999.

[17] Jing Yang and Jinsong Lin, 'Optimization of power management in a CDMA radio network', *Proceedings of 52th IEEE Vehicular Technology Conference*, VTC 2000-Fall, vol.6, pp. 2642–2647, 24–28 September 2000.

[18] J. Laiho-Steffens, A. Wacker and P. Aikio, 'The impact of the radio network planning and site configuration on the WCDMA network capacity and quality of service', *Proceedings of 51th IEEE Vehicular Technology Conference*, VTC 2000-Spring, vol. 2, pp. 1006–1010, Tokyo, Japan, 15–18 May 2000.

[19] R.T. Love, K.A. Beshir, D. Schaeffer and R.S. Nikides, 'A Pilot Optimization Technique for CDMA Cellular System', *Proceedings of 50th IEEE Vehicular Technology Conference*, VTC 1999-Fall, vol. 4, pp. 2238–2242, 1999.

[20] S. Plimmer, M. Feenery, D. Barker and T. Normann, 'Adjusting antenna downtilt boosts UMTS optimization', *Wireless Europe*, pp. 34–35, November 2002.

[21] B.N. Vejlgaard, *Data Receiver for the Universal Mobile Telecommunications System (UMTS)*, PhD thesis, Aalborg University, Denmark, 2001.

[22] 3GPP, 'Requirements for support of radio resource management (FDD), TS25.133', v6.0.0, September 2002, http://www.3gpp.org.

[23] D. Kim, Y. Chang and J.W. Lee, 'Pilot Power Control and Service Coverage Support in CDMA Mobile Systems', *Proceedings of 49th IEEE Vehicular Technology Conference, VTC 1999-Spring*, vol. 4, pp. 2238–2242, Houston, TX, May, 1999.

[24] J.X. Qiu and J.W. Mark, 'A Dynamic Load Sharing Algorithm Through Power Control in Cellular CDMA', *Proceedings of 9th IEEE International Symposium on Personal, Indoor and Mobile Radio Communications*, vol. 3, pp. 1280–1284, 8–11 September 1998.

[25] Maciej J. Nawrocki, Mischa Dohler and A. Hamid Aghvami, 'On Cost Function Analysis for Antenna Optimisation in UMTS Networks', *Proceedings of 16th IEEE International Symposium on Personal, Indoor and Mobile Radio Communications* PIMRC 2005, 12–14 September 2005, Berlin.

[26] I. Rechenberg, Evolutionsstrategie: *Optimierung technischer Systeme nach Prinzipien der biologischen Evolution*, Fromann-Holzboog, Stuttgart, 1994 (in German).

[27] L. Davis (ed.), *Handbook of Genetic Algorithms*, International Thomson Computer Press, London, 1996.

[28] Beasley, Bull, Martin, 'An overview of genetic algorithms', University Computing, 15(2), Inter-University Committee on Computing, 1993.

[29] T. Bäck and M. Emmerich, 'Evolution strategies for optimization in engineering applications', *Proc. 5th World Congress on Computational Mechanics* (WCCM V), Wien, July 2002.

[30] J.H. Holland, 'Adaptation in Natural and Artificial Systems', *The University of Michigan Press*, Ann Arbor, 1975.

[31] K.A. Jong, *An Analysis of Behaviour of a Class of Genetic Adaptive Systems*, PhD thesis, University of Michigan, 1975.

[32] D.E. Goldberg, *Genetic Algorithms in Search, Optimization and Machine Learning*, Addison-Wesley, MA, 1989.

[33] Z. Michalewicz and D.B. Fogel, *How to Solve It: Modern Heuristics*, Springer-Verlag, Berlin, Heidelberg, New York, 2000.

[34] A. Gerdenitsch, S. Jakl, M. Toeltsch, 'The Use of genetic algorithms for capacity optimization in UMTS FDD networks', *Proc. 3rd International Conference on Networking* (ICN'04), Guadeloupe, French Caribbean, March 2004.
[35] H.P. Schwefel, *Numerical Optimization of Computer Models*, Wiley & Sons Ltd/Inc., Chichester, 1981.
[36] H.P. Schwefel, G. Rudolph, 'Contemporary evolution strategies', *Advances in Artificail Life – Proc. 3rd European Conf. Artificial Life* (ECAL'95), pp. 893–907, Springer, Berlin, 1995.
[37] A. Gerdenitsch, S. Jakl, M. Toeltsch and T. Neubauer, 'Intelligent Algorithms for System Capacity Optimization of UMTS FDD Networks', *Proc. 4th International Conference on 3G Mobile Communication Technologies*, pp. 222–226, London, Great Britain, 25–27 June 2003.
[38] A. Gerdenitsch, *System Capacity Optimization of UMTS FDD Networks*, PhD thesis, Technische Universität Wien, 2004.
[39] A. Gerdenitsch, S. Jakl, Y.Y. Chong and M. Toeltsch, 'An Adaptive Algorithm for CPICH Power and Antenna Tilt Optimization in UMTS FDD Networks', *Proc. 8th International Conference on CIC*, p. 378, Seoul, Korea, 28–31 October 2003.
[40] U. Türke and M. Koonert, 'Advanced site configuration techniques for automatic UMTS radio network design'. *In Proc. VTC-Spring 2005*. IEEE, Stockholm, Sweden, 2005.
[41] M.J. Nawrocki and T.W. Wieckowski, 'Optimal site and antenna location for umts – output results of 3G network simulation software'. *Journal of Telecommunications and Information Technology*, 2003.
[42] J. Niemelä and J. Lempiäinen, 'Impact of mechanical antenna downtilt on performance of WCDMA cellular network'. *In Proc. VTC-Spring 2004*, pp. 2091–2095. IEEE, Milan, Italy, May 2004.
[43] T. Isotalo, J. Niemelä and J. Lempiäinen, 'Electrical antenna downtilt in UMTS network'. *In Proc. 5th European Wireless Conference*, pp. 265–271. Barcelona, February 2004.
[44] S. Hanly and R. Mathar, 'On the optimal base station density for CDMA cellular networks'. *IEEE Trans. Comm.*, 50(8):1274–1281, 2002.
[45] S. Jakl, A. Gerdenitsch, W. Karner and M. Toeltsch, 'An approach for the initial adjustment of antenna azimuth and other parameters in UMTS networks'. *In Proc. 13th IST Mobile Summit*. Lyon, France, June 2004.
[46] T. Koch, *Rapid Mathematical Programming*. PhD thesis, Technische Universität Berlin, 2004.
[47] K. Tutschku, Models and Algorithms for Demand-oriented Planning of Telecommunication Systems. PhD thesis, University of Würzburg, 7 September 1999.
[48] K. Leibnitz, *Analytical Modeling of Power Control and its Impact on Wideband CDMA Capacity and Planning*. PhD thesis, University of Würzburg, 2 February 2003.
[49] R. Mathar, 'Mathematical modeling, design, and optimization of mobile communication networks'. *Jahresbericht Dt. Math.-Verein*, 2001.
[50] A. Eisenblätter, T. Koch, A. Martin, T. Achterberg, A. Fügenschuh, A. Koster, O. Wegel and R. Wessäly, *Modelling feasible network configurations for UMTS*. In G. Anandalingam and S. Raghavan, editors, Telecommunications Network Design and Management. Kluwer, 2002.
[51] S. Hurley, 'Planning effective cellular mobile radio networks'. *IEEE Trans. Vehicular Techn.*, 12(5):243–253, 2002.
[52] K. Tutschku, R. Mathar and T. Niessen, 'Interference minimization in wireless communication systems by optimal cell site selection'. *In 3rd European Persona Mobile Communication Conference (EPMCC'99)*, pp. 208–213. Paris, France, 3 1999.
[53] E. Amaldi, A. Capone and F. Malucelli, 'Planning UMTS base station location: Optimization models with power control and algorithms'. *IEEE Trans. Comm.*, 2(5), September 2003.
[54] I. Siomina and D. Yuan, 'Pilot power management in WCDMA networks: coverage control with respect to traffic distribution'. In *Proc. Of ACM MSWiM '04*, pp. 276–282. ACM Press, 2004. ISBN 1-58113-953-5.
[55] S.B. Jamaa, Z. Altman, J.-M. Picard and A. Ortega, 'Steered optimization strategy for automatic cell planning of UMTS networks'. *In Proc. VTC-Spring 2005*. IEEE, Stockholm, Sweden, 2005.
[56] A. Eisenblätter, A. Fügenschuh, H.-F. Geerdes, D. Junglas, T. Koch and A. Martin, 'Integer programming methods for UMTS radio network planning'. *In Proc. of WiOpt'04*. Cambridge, UK, 2004.
[57] A. Eisenblätter and H.-F. Geerdes, 'A novel view on cell coverage and coupling for UMTS radio network evaluation and design'. *In Proc. of INOC'05*. ENOG, Lisbon, Portugal, March 2005.

15

Automatic Network Design

Roni Abiri, Ziemowit Neyman, Andreas Eisenblätter and
Hans-Florian Geerdes

This chapter describes some of the optimisation challenges that result from the unique deployment of UMTS either on-top an existing GSM network, or competing with GSM networks. We will proceed by detailing some challenges related to the optimisation process, and will then deal with a few case studies which, albeit examples, serve to demonstrate tendencies in the automatic network design.

15.1 THE KEY CHALLENGES IN UMTS NETWORK OPTIMISATION

15.1.1 PROBLEM DEFINITION

UMTS was defined to meet the growing demands for wireless services, either voice calls or the more lucrative data services. Due to an intensive utilisation of the relevant frequency spectrum and the wish for harmonisation of the UMTS frequencies all over the world (which, in the end, was only partially achieved), the relatively high frequency band around 2000 MHz was selected. The propagation characteristics of this band (see Chapter 5) are worse compared to those used for GSM, posing new challenges to the wireless engineers. This chapter focuses on the following three challenges:

1. Matching UMTS coverage to GSM;
2. Supporting high bit rate data services;
3. Handling dual technology networks.

Understanding UMTS Radio Network Modelling, Planning and Automated Optimisation Edited by Maciej J. Nawrocki, Mischa Dohler and A. Hamid Aghvami © 2006 John Wiley & Sons, Ltd

15.1.2 MATCHING UMTS COVERAGE TO GSM

The basic plan of the network engineers was to re-use most GSM sites for the UMTS network, due to the following reasons:

1. Ease and speed of implementation;
2. Simpler maintenance and a lower cost of operations (OPEX);
3. Matching the demographic traffic requirements.

However, as mentioned above, the usage of a higher frequency band, together with the inherited differences between FDMA and CDMA technologies, makes this a very challenging task. To better understand the embedded differences, let us recapture some of the key methods used for GSM planning.

Frequency assignment. Each GSM cell is assigned frequencies not repeated in adjacent cells (for simplicity, the case of 1:1 synthesised frequency hopping is ignored here, it will be referred to later). The frequency assignment is done according to the following criteria:

- Number of frequencies is equal (or larger) than the number of transmitters (TXs) to satisfy the traffic demand;
- The C/I, as received by the Mobile Stations in the cell's service area and by the base station in the uplink, is above a pre-determined threshold, for example 9 dB.

Satisfying these conditions enable a high service quality. Failing to meet either of them impairs either traffic handling or call quality. As most of the cellular networks are quite heterogonous in terms of demography (as reflected by the traffic distribution), topography and cells effective coverage area, there is a large variance in the number of assigned frequencies and the cells which may be either interfering or interfered to. For example, it is quite common to find some high sites, especially in hilly areas that are required to cover large areas. It can be done by assigning them frequencies that are not repeated over a larger area. As long as there are enough frequencies to support both these 'boomer' sites and 'regular' sites – it is a feasible and commonly used solution. However, when it comes to CDMA-based technologies, the reuse of the same frequency makes it impossible, as these high sites create high interference all around them, impairing both coverage and capacity of the other regular sites. Hence, the planning engineer must avoid using these sites, or alternatively severely limit their propagation area. This usually means adding more sites to the network to compensate for the loss of coverage. It is not rare to find that the removal of one 'boomer' site necessitates adding three or more additional sites.

Another aspect related to 'boomer' sites is their traffic handling. In UMTS, as opposed to GSM, there is very little flexibility in shifting traffic from one cell to the other (as elaborated in the following sections). This results in high traffic loading in these 'boomer' sites – that generally cannot be handled. In the early days of US CDMA IS-95 deployment, this phenomenon was underestimated – a fact that led to a poor initial performance and that was rectified only after removing these 'boomer' sites.

Hierarchical Cell Structure (HCS). One of the key ways of GSM technology to handle non-homogonous traffic is by using a cell hierarchy. The servicing cell is not only determined by the 'best serve' algorithm, where the best serving cell is the one selected by the network to serve a specific mobile station, but also by a 'can serve' algorithm that may assign the mobile station to a weaker received base station that is still above service threshold. This mechanism enables addition of micro-site layers supporting local traffic hot-spots. As mentioned above, due to its nature, the UMTS network can only utilise the 'best serve' algorithm. Hierarchy can be applied only when more than one frequency is used, and the other frequency is dedicated for a Micro-site layer (an exception to this, however, has been introduced in Section 10.4.6). This is enabled by the compressed mode feature added to UMTS as a key learning from the deployment of US CDMA networks.

Handover planning. Generally speaking, handover planning consists of two sections: Neighbour list planning and Handover threshold planning. In a GSM network, the optimisation engineer may tweak these parameters, include or exclude neighbours to insure smooth transition along highways from one site to the other, move the cross-over point to shift a hot-spot (like a highway junction) from loaded cells to a less loaded one and so on. In UMTS, the degrees of freedom are significantly lower. For example, if the optimisation chooses to exclude a cell from the neighbour list, then this cell's received signal is converted to interference, lowering E_c/I_0, and potentially causing the call to drop, or the data rate to be lowered.

To conclude this section, poor RF conditions resulting from non-optimal site planning cannot be compensated by tweaking HCS or handover parameters. Also, the optimisation engineer obviously does not dispose over the most powerful tools for frequency optimisation to solve quality issues.

The design examples shown later in this chapter demonstrate how a careful site planning in terms of location and antenna parameters can significantly improve the planned network quality, compensating for the loss of the GSM specific tools utilising frequency, HCS and handover optimisation.

Addendum: GSM optimisation with 1:1 synthesised frequency hopping (SFH). There are quite a few GSM networks where frequency assignment is simplified because a 1:1 SFH is used. It is recommended to handle these networks similarly to the UMTS networks in terms of the RF footprint. Here, the existence of either 'boomer' sites or low covering micro-sites also inhibits globally applying 1:1 SFH for all sites. As a result, in most GSM networks, there are separated frequency domains for macro and micro sites, sometimes even using different frequency bands (900 and 1800 MHz), and the 'boomer' sites are planned and dealt with separately – even with a different frequency domain where feasible.

15.1.3 SUPPORTING HIGH BIT RATE DATA SERVICES

In order to better understand the limitations of a UMTS network to handle high speed data services, one has to understand the difference in handling voice and high rate data services.

One of the ingenious ideas applied in CDMA networks is the 'Soft Handover' concept. It was introduced to CDMA to compensate for the degradation of E_c/I_0 values at the cell boundaries that may otherwise have led to call drops. However, this feature has a cost attached to it, both in terms of additional equipment required for simultaneously serving the mobile station by two or more base stations, and also in terms of the total transmitted downlink power. Also, the handover in UMTS is a static state opposed to the dynamic nature of handover in GSM networks. A mobile station that is located in an area served with several cells, received approximately in the same level will be in *n*-way handover state as long it is not moving. Having a mobile station requiring high data rates at these *n*-way areas severely limits the network data throughput. To avoid this, the HSDPA works *without* the Soft handover feature. This causes degradation in the data transfer capability in all areas defined to be in *n*-way handover state for voice calls. The result is that the optimisation engineer, who has to optimise the network for two service types, may reach different schemes and parameters for both cases. Only careful optimisation that takes into account both requirements may lead to an optimised network for the pre-defined mixture of voice and data services.

Defining and maintaining the quality of data service. Capacity figures highly depend on the service types and their quality requirements. For example, standard voice services consume a relatively small bandwidth, but require a Frame Erasure Rate (FER) that is below 1 or 2 %. This is mapped to E_b/N_0 requirements that satisfy the FER over the varying radio conditions. However, looking at the other types of services, specifically the non real-time, the quality goals should be based on the Service Level Agreement (SLA) that is bindingly promised to that service. Setting different levels for the SLA can dramatically change the network's capacity handling, so a precise definition of the SLA is required prior to calculating the network's capacity.

Benchmarking the network's performance. The parameters used for benchmarking voice handling capacity are well defined: They are divided into hard blocking, when the network does not have physical resources to support the call, and soft blocking, when the limitation is due to the air-interface blocking, e.g. the percentage of FER is above the pre-defined threshold. Most simulation and optimisation tools only handle the second case, because this case cannot be calculated explicitly, requiring some form of Monte-Carlo analysis.

For the case of non-voice calls, other parameters should be used. Here, hard blocking is no more a go-no-go situation as the network can use its RRM mechanisms to decrease the data throughput and avoid blocking a new user. The network can also select to decrease other users' throughput in order to accommodate new users, and even give preferences to some users due to their SLAs. Therefore, in order to accurately predict the capacity limits of the network, it is required to have this mechanism modelled in the simulator, using the correct RRM settings.

The soft blocking characterisation of these services can be divided into two groups: real-time and non-real time services. The Real-time case is similar to the voice case as the requirement is to have a given FER or BER value, and if the information is lost, it is not re-transmitted. On the other hand, the quality of non-real time services is best modelled by the delivered throughput explained earlier together with the percentage of re-transmit requests. High values of re-transmit requests increases the service latency and also decreases the network's data handling efficiency; thus, the network operator has the incentive to minimise the number of users that ask repeatedly for re-transmissions, even allowing them to be dropped from the network.

15.1.4 HANDLING DUAL TECHNOLOGY NETWORKS

In most countries, excluding Japan and Korea, UMTS networks are built in co-existence with GSM networks. There may be two cases where the GSM network is either operated by the same carrier, or is maintained by a different one. In both cases, especially during first years of deployment, the GSM coverage is better compared to the one supplied by the UMTS network. Also, there are some cases, like in rural areas, where UMTS sites are not deployed to save on the networks costs. Hence, the optimisation engineer has to master both networks, and the relations between them to optimise the service to the end user. As GSM networks are going to be around for the foreseeable future, this is an ongoing task and the optimisation tools should support this task.

15.2 ENGINEERING CASE STUDIES FOR NETWORK OPTIMISATION

The following section details some case studies that demonstrate the usage and results of automatic optimisation for two scenarios. Both scenarios are based on the same network physical data, and the optimisation is carried out for different traffic loading conditions. As this is an already deployed network, the optimisation done here is for the cell's parameters only, i.e. not touching the site locations.

The first case is for an unloaded network where the traffic is not significant. This may describe the situation of an operator just before commercial launch. The major goal of optimisation would be to maximise the coverage reliability, described with CPICH levels and CPICH E_c/I_0 received by potential users in every location of the UTRAN network, since no real and reliable traffic distributions are known so far.

The second scenario contains some on-the-run network data and it can be used as an example of ongoing network optimisation. The major optimisation goal would be to maximise the number of connected mobiles or the data throughput to satisfy already existing users at their locations, since now real traffic data is known and could be scaled for network growth.

15.2.1 EXAMPLE NETWORK DESCRIPTION

The example network is based on an actually deployed network, located in a big city in Asia. The optimisation area with physical dimensions of 6.7 km × 6.2 km (41.54 sq km) consists of 43 sites (128 cells). It covers urban area, mostly with urban and suburban clutters and a river in the middle. Site-to-site distances range from 300 m to 1.2 km.

The optimisations have been performed using genetic algorithms (see Chapter 14) implemented in the optimisation tool OptiPlanner from Schema Ltd. The network snapshots for optimisation were achieved using a static Monte-Carlo simulator, based on standard RF predictions. As this is a real network, the sites are neither spaced in a perfectly triangular grid structure nor have standard sector orientations been used.

The first step in a new cellular network planning and/or optimisation is to check its performance under no traffic loading or light traffic loading conditions. This is done in order to estimate the best potential coverage and service quality that may be received, because a higher traffic load does not improve this achieved performance. Therefore, if the network in unloaded traffic conditions and after undergoing optimisation cannot meet the original performance goals, such as percentage of covered area or percentage of good links, the network deployment (like adding more cells or sites) needs to be reconsidered under assumed quality goals.

15.2.2 PRE-LAUNCHED (UNLOADED) NETWORK OPTIMISATION

The first case is for an unloaded network where the traffic is not significant. Traffic is considered insignificant if the following conditions are met:

1. The sector power amplifiers do not reach their saturation point; there is plenty of headroom for supporting additional traffic.
2. Uplink noise rise hardly exists.

The suggested way to achieve the above is to take e.g. only 10 % of the forecasted traffic loading. Also, for simplicity only, voice traffic can be assumed. The only requirement is to verify that the above conditions are met for all cells in the simulation. This approach is better demonstrated by taking very low, flat traffic values all over the network servicing area, as this optimisation stage is the precursor for the fully loaded network. This will lead to a better starting point for optimisation under real load conditions. The optimisation goal is to improve the quality of service as can be measured by the following direct indicators:

1. Minimise outage pathloss area: Minimise the area (weighted by traffic) that suffers from high pathloss. The best indication for such areas is the UE transmitted power level, especially for the unloaded case as the noise rise above thermal noise is practically zeroed, so the indicator may be the percentage of traffic (or area if we use even traffic distribution) that requires UE transmitted power level above a predefined value (usually lower than the maximal available power level to accommodate some degree of good engineering practice).
2. Maximise E_c/I_0: As already known for CDMA based air-interface technologies, having a low pathloss, or in other words a high receive power level (RSCP) does not necessarily guarantee an adequate level of E_c/I_0, mainly due to the other cells contribution to I_0; hence, this indicator (that may also be used as an optimisation goal) balances the first one to create an optimised network.

Additionally, there are more indicators that are less significant for the unloaded case, such as percentage of Soft Handover (SHO) users, headroom for the power amplifiers, etc. These indicators will be discussed in detail in the loaded network case study.

15.2.2.1 Initial Network Analysis

The initial network was planned manually by a leading engineering team working at a major UMTS operator. During this manual plan, cell parameters were tweaked to optimise performance. For the simulation, low and uniform distributed voice traffic has been assumed. All results here are based on averaging of 30 Monte-Carlo runs, consisting of total 1332 simulated mobiles. The low traffic (small number of mobiles) together with the high amount of simulation runs were necessary to weight evenly and significantly each service point in the network. This implies the area optimisation. The indicators used for assessing the quality of this network are as described below.

Percentage of non-connected mobiles and the reason for it. As one could have predicted, the initial performance are not so poor; however, ∼6.6% of mobiles are not able to access the network, mostly due to low values for E_c/I_0 (Table 15.1). It is also not a surprise to see that all mobiles were able to access the uplink, because the uplink noise rise for the unloaded case is negligible.

Histogram of CPICH E_c/I_0 distribution. Here, as the network is only lightly loaded, there is a relatively large percentage of traffic under high values of E_c/I_0; but already here, there are also mobiles with low values of E_c/I_0 that may not be connected to the network (Figure 15.1).

Histogram of cell transmitted power. The transmitted power in the initial states varies between 34 and 39 dBm (Figure 15.2). When compared with the maximum power of 43 dBm (20 W), there is still sufficient head room for further power increases due to increased network load.

Handover analysis. The initial network design causes a very high percentage of traffic being involved in the soft handover (Figure 15.3). Only 36% of all mobiles are in the single connection state. It clearly increases the probability that a mobile would be connected, but at the expense of consumed downlink capacity.

Figure 15.3 contains various types of handover connections which are described as follows:

- *single*: mobile connected to one cell;
- *soft*: 2-way soft handover, mobile connected to two cells, each one from different Node B;
- *soft soft*: 3-way soft handover, mobile connected to three cells, each one from different Node B;
- *soft softer*: 3-way soft handover, mobile connected to three cells, one belongs to the strongest Node B (*soft*), other two belongs to another Node B (*softer*), strongest two belong to two different Node Bs;
- *softer*: 2-way soft handover, mobile connected to two cells, each one from the same Node B;
- *softer soft*: 3-way soft handover, mobile connected to three cells, the strongest and one other belong to one Node B, the third belongs to a different Node B;
- *softer softer*: 3-way soft handover, mobile connected to three cells from the same Node B.

Later on, these data will be compared with the post-optimisation results to show the improvement.

Table 15.1 Percentage of non-connected mobiles – unloaded initial network.

Mobiles' state	Quantity	Percentage
Connected mobiles	1244	93.39
Not connected – Not enough DCH power	0	0
Not connected – Low E_c/I_0	81	6.08
Not connected – High FER	7	0.53
Not connected – Mobile Tx power above threshold	0	0
Not connected – Cell Power reached maximum	0	0
Total mobiles	1332	100

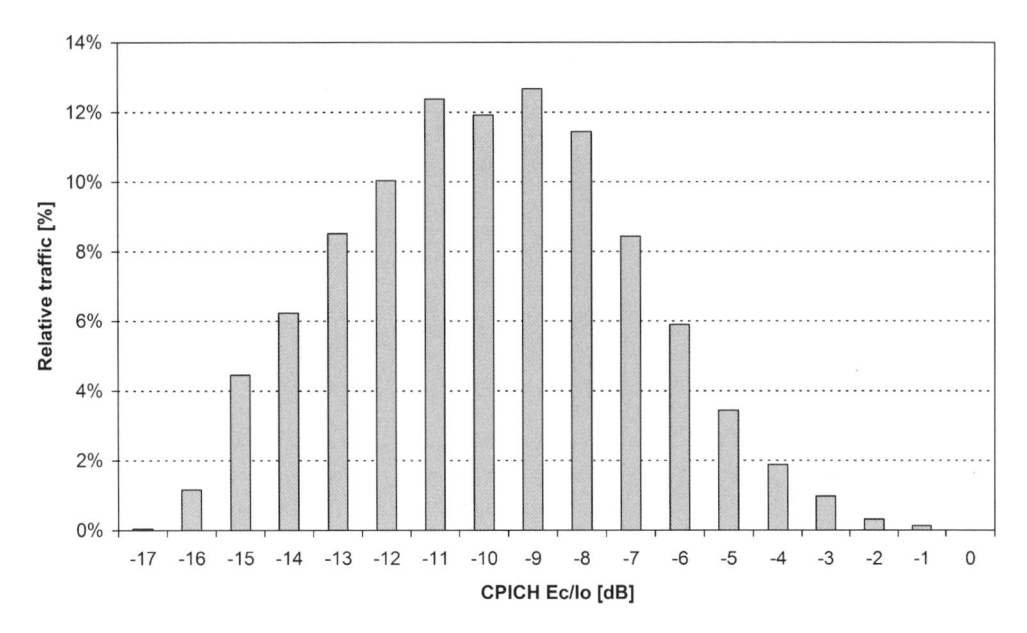

Figure 15.1 Histogram of CPICH E_c/I_0 distribution – unloaded initial network.

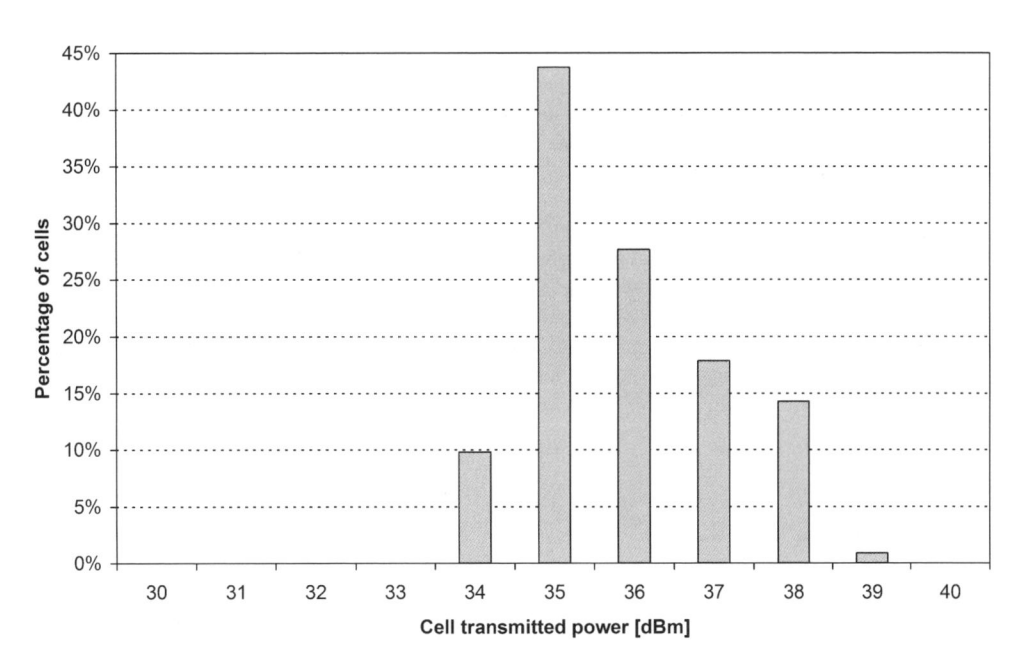

Figure 15.2 Histogram of cell transmitted power – unloaded initial network.

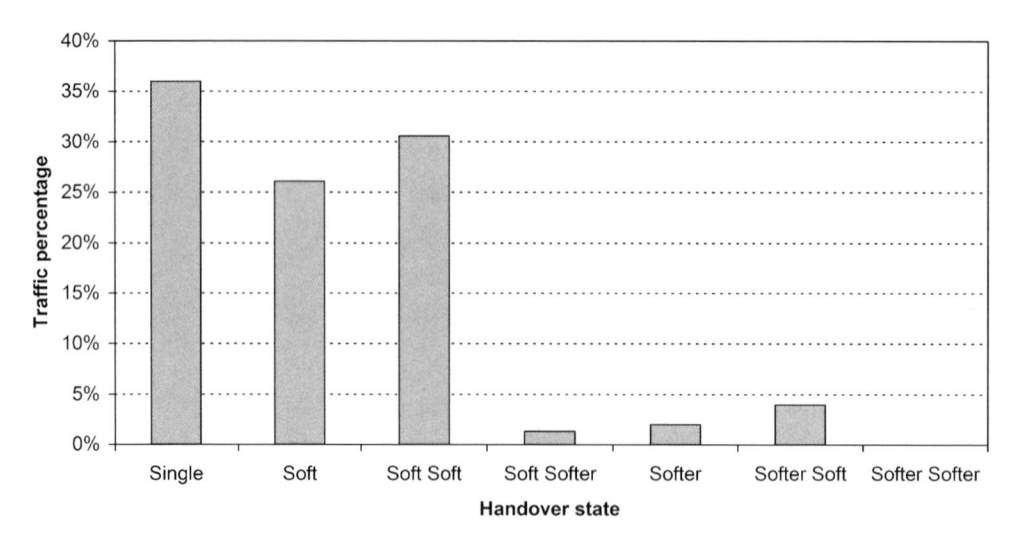

Figure 15.3 Handover analysis – unloaded initial network.

15.2.2.2 Optimisation Goals and Constraints

As mentioned before, the majority of non-connected mobiles are due to low values of E_c/I_0; thus, the primary goal should be to set E_c/I_0 above the coverage threshold. It is also required to set a goal of maximising the number of connected mobiles, to prevent deterioration in performance due to other reasons. To maximise future capacity, the soft handover areas have been optimised as well. The transmitted cell power can be anticipated as not the limiting factor, but for future traffic growth it should be maintained during the optimisation. The constraints were globally defined to all sites (the same limits) as follows:

- Antenna replacement – from the list of approved antennas;
- Change of both mechanical and electrical tilt – limited both relatively to initial state and absolutely [mechanical + electrical tilt absolute limits were set from 0° to 15°], with step of 1°;
- Change of azimuth to ±30° relative to initial azimuth, with step of 5°;
- Change of CPICH power level from 30 up to 33 dBm. It should be noted that CPICH power changes are limited within 3 dB margin from the initial state. Power is an efficient optimisation measure for loaded networks with traffic hot spots, as it is not a symmetrical parameter, like e.g. antenna patterns.

15.2.2.3 Optimisation Results

Most of the cell parameters were tweaked during the optimisation: for 41 out of the 43 sites, and 106 out of the 128 sectors, changes were suggested. A total of 13 site antennas were replaced by other types. The indicators used for assessing the quality of this network are shown below (Table 15.2, Figures 15.4–15.7).

Percentage of non-connected mobiles and the reason for it. As clearly shown, the initial performance was significantly improved and only less than 0.5 % of mobiles could not connect to the network, which is negligible (Table 15.2).

Table 15.2 Percentage of non-connected mobiles – unloaded optimised network.

Mobiles' state	Quantity	Percentage
Connected mobiles	1326	99.55
Not connected – Not enough DCH power	0	0
Not connected – Low E_c/I_0	6	0.45
Not connected – High FER	0	0
Not connected – Mobile Tx power above threshold	0	0
Not connected – Cell Power reached maximum	0	0
Total mobiles	1332	100

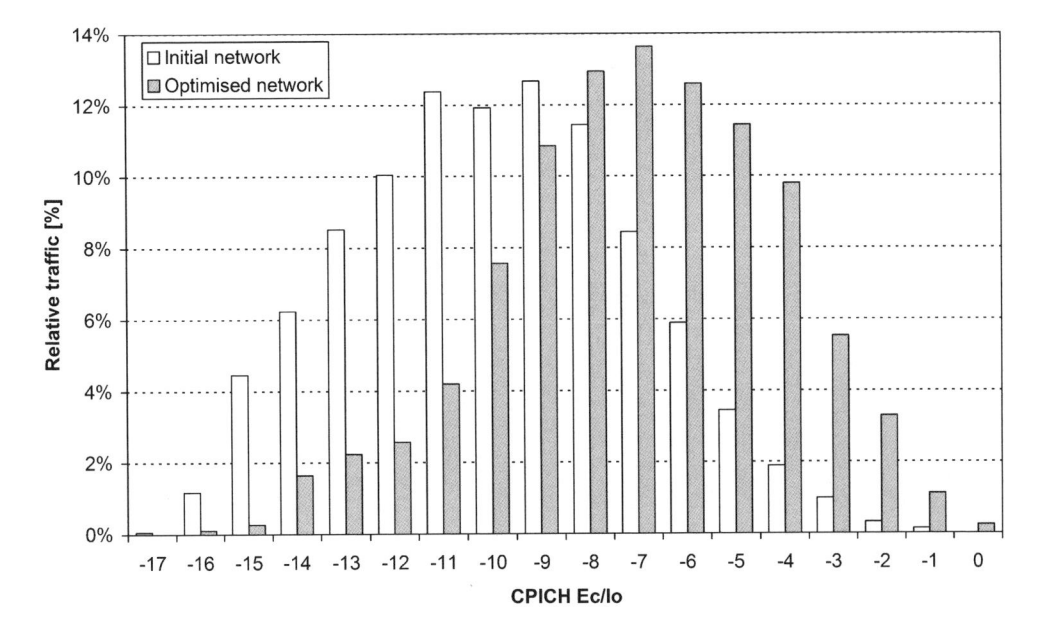

Figure 15.4 Histogram of E_c/I_0 distribution – unloaded optimised network.

Histogram of the E_c/I_0 distribution. Here, as the network is only lightly loaded, there is a relatively large percentage of traffic under high values of E_c/I_0; but already here, there are mobiles with low values of E_c/I_0 that may not be connected to the network (Figure 15.4). The histogram was considerably shifted to the right side, decreasing the percentage of traffic under low E_c/I_0.

Histogram of cell transmitted power. Due to the CPICH power level optimisation, there is a larger variance in the cell's total transmitted power, and one of the cells also reached its peak power level (Figure 15.5). Here, the fact that network was optimised using non-loaded traffic conditions led to results that could not be optimal for the loaded case, as some of the cells were already closer to their peak power. On the other hand, more than 20 % of the cells decreased the transmitted power; again, mostly due to the change of the CPICH power level.

Handover analysis. Changes in power together with tilt and antenna pattern optimisation led to a significant increase in single connection modes (Figure 15.6). Additionally, the contribution of soft

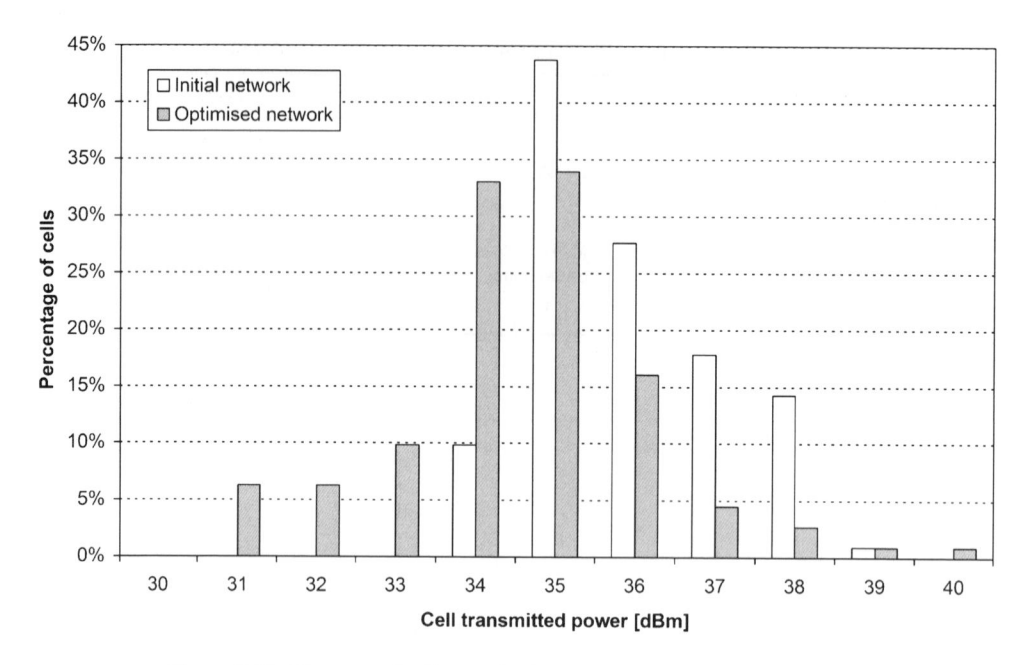

Figure 15.5 Histogram of cell transmitted power – unloaded optimised network.

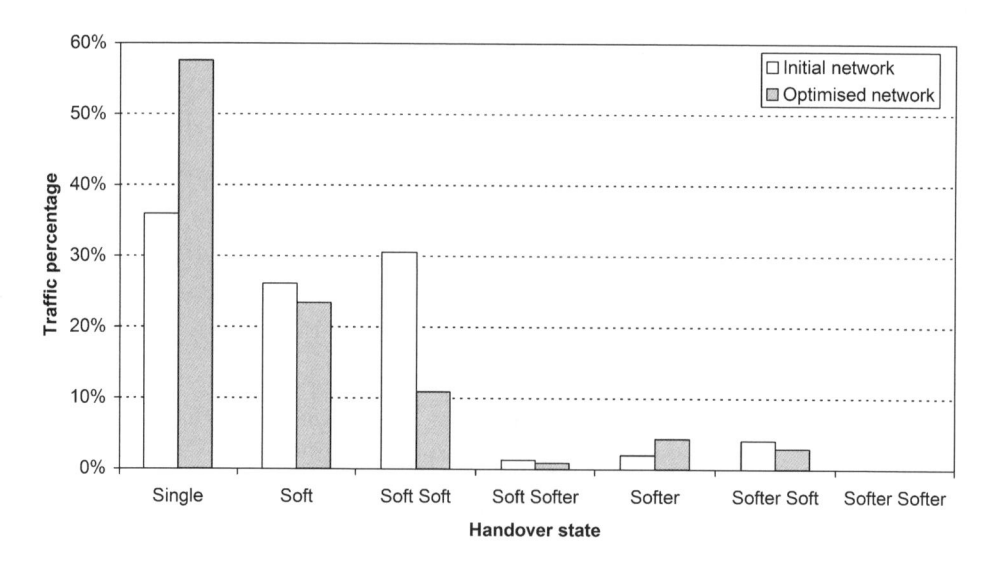

Figure 15.6 Handover analysis – unloaded initial and optimised networks.

and soft/soft modes has been minimised; thus, not only the downlink RF capacity were saved, but the transmission and signalling in fixed parts of the network as well. The increase in softer mode requires no more resources from the fixed network to be consumed, because the effect of antenna downtilt is not that crucial and the combining of the uplink is made in the Node B. It should be noted that softer/softer mode is not present at all; thus, strong overshooting from multiple cells does not exist to

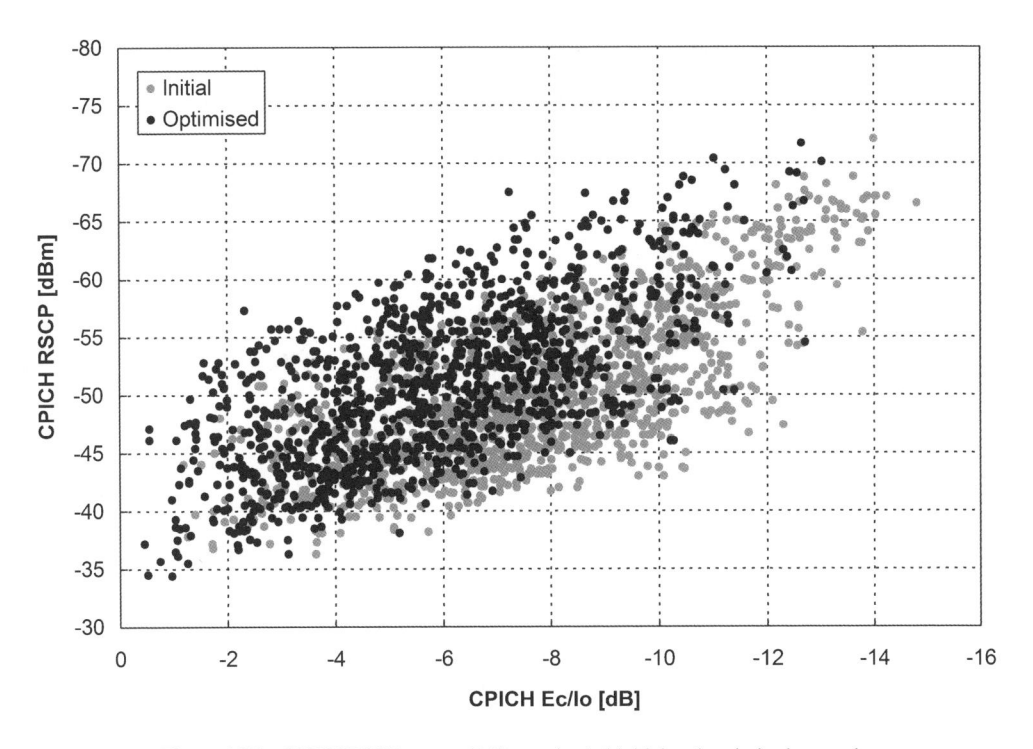

Figure 15.7 CPICH RSCP versus E_c/I_0 – unloaded initial and optimised networks.

a high extend. This leads to an improvement of the *i*-factor (inter-to-intra cell interference factor; see Chapter 10).

Quality of the RF network – distribution of CPICH RSCP versus E_c/I_0. The optimised network offers, at almost the same serving signal levels (RSCP of CPICH), an improved signal quality (CPICH E_c/I_0), indicated by a shifting of the points to the left (higher E_c/I_0 values). This means that at similar path loss the received interference is less, which leads directly to an improved system capacity.

15.2.3 LOADED NETWORK OPTIMISATION

After successfully demonstrating the unloaded case, the next step is to perform the loaded network optimisation. Here, as mentioned in previous sections, there is a need to include more goals and indicators in the optimisation process. As such, the optimisation results in terms of the suggested changes will differ from the previous case, as the balancing point is located in a different place. Several runs were done for the loaded case with increased load. The following chart shows the percentage of connected mobiles for each run, starting from a lightly loaded network and reaching the heavily loaded one. (*Remark*: These runs were carried out with a different mixture of voice/data users, compared to the case study in Section 15.2; the results for the non-loaded case are hence different.)

Setting the percentage of non-connected mobiles to 2.5%, enables the support of ∼4500 mobiles for the initial network. After optimisation, this value is increased to ∼13 400 mobiles under the same conditions, i.e. 3× the original capacity handling! The subsequent section describes in detail the optimisation scenario for approximately 10 000 mobiles that is shown in the centre of Figure 15.8.

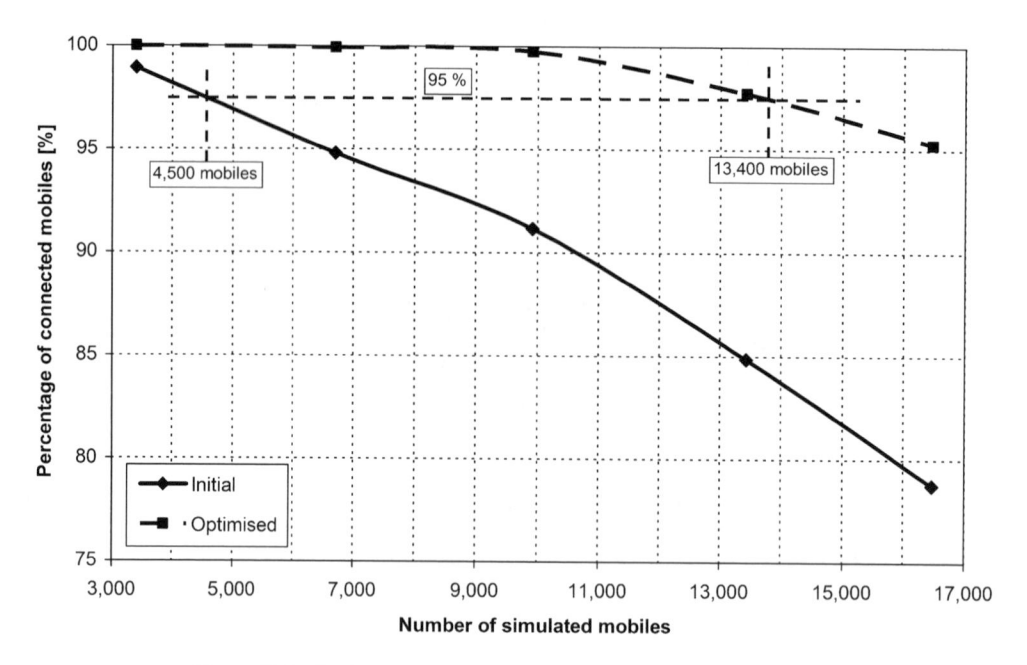

Figure 15.8 Traffic load analysis – loaded initial and optimised networks.

15.2.3.1 Initial Network Analysis

The same initial network as in Section 15.2.2.1 was simulated with 10 Monte-Carlo runs, consisting of a total of 9922 simulated mobiles (7.4 times more compared to the non-loaded case). The indicators used for assessing the quality of this network are the same as for the unloaded scenario.

The initial network quality presented in the Table 15.3, compared to the non-loaded case, is lower by 2.2%, mostly due to a higher percentage of mobiles with low values of E_c/I_0; this is in-line with the common understanding of CDMA dynamics.

15.2.3.2 Optimisation Goals

As the majority of non-connected mobiles in the initial network is due to low values of E_c/I_0, the main goal is to minimise the occurrence of low E_c/I_0 values. This goal is connected with the

Table 15.3 Percentage of non-connected mobiles – loaded initial network.

Mobiles' state	Quantity	Percentage
Connected mobiles	9046	91.17
Not connected – Not enough DCH power	2	0.02
Not connected – Low E_c/I_0	829	8.36
Not connected – High FER	45	0.45
Not connected – Mobile Tx power above threshold	0	0
Not connected – Cell Power reached maximum	0	0
Total mobiles	9922	100

general goal of increasing the percentage of connected mobiles. To distribute the load uniformly among cells, two measures have been taken into account: the maximum transmit power of a cell was kept 3 dB below the saturation point and the uplink noise rise was kept below the 3 dB margin. Similar to the unloaded case, the soft handover areas require to be optimised. Additionally the pilot pollution needs to be investigated, since it will lead at first to increase of network performance in a weak best server situation and minimise interference to dominant servers. All the goals achieved the results shown in the next section. The following constraints were globally defined to all sites (same limits):

- Antenna replacement – from the list of approved antennas;
- Change of both mechanical and electrical tilt – limited both relatively to initial state and absolutely [mechanical + electrical tilt absolute limits were set from 0° to 15°], with a step of 1°;
- Change of azimuth to ±30° relative to initial azimuth, with a step of 5°;
- Change of CPICH power level from 27 dBm up to 33 dBm.

15.2.3.3 Optimisation Results

Percentage of non-connected mobiles and the reason for it. Based on the results presented in Table 15.4, we can see a remarkable improvement in the number of non-connected mobiles compared to the initial network. The number of mobiles not connected due to low values of E_c/I_0 was almost zeroed after optimisation. Furthermore, even though the transmitted power of a cell was kept 3 dB below the maximum available value, there exist three mobiles that are not connected due to cell limiting transmit power, caused by the re-distribution of traffic between the cells; since the contribution of these mobiles to the disconnected category is low, the effect can be neglected.

Histogram of CPICH E_c/I_0 distribution. Here, as the network is nominally loaded, the initial network does not have many high values of E_c/I_0 (Figure 15.9), mostly due to high transmit levels of the DTCH; note that the transmit power level of DTCH is proportional to the CPICH level that has a fixed value for the initial network. After optimisation, the histogram was significantly shifted to the right, decreasing the percentage of low E_c/I_0 traffic. This effect is mostly due to a decrease in the total transmitted power in many cells, as will become evident in the next section.

Histogram of cell transmitted power. The cell distribution in the initial loaded network differs from the non-loaded case, because the additionally keyed channels create a wider distribution of the transmitted power levels. After optimisation of the CPICH power level, many cells transmit at much lower power, causing a decrease in the total I_0. This, in turn, leads to higher values for E_c/I_0, as was set to be the main optimisation goal (Figure 15.10).

Table 15.4 Percentage of non-connected mobiles – loaded optimised network.

Mobiles' state	Quantity	Percentage
Connected mobiles	9897	99.75
Not connected – Not enough DCH power	0	0
Not connected – Low E_c/I_0	10	0.10
Not connected – High FER	12	0.12
Not connected – Mobile Tx power above threshold	0	0
Not connected – Cell Power reached maximum	3	0.03
Total mobiles	9922	100

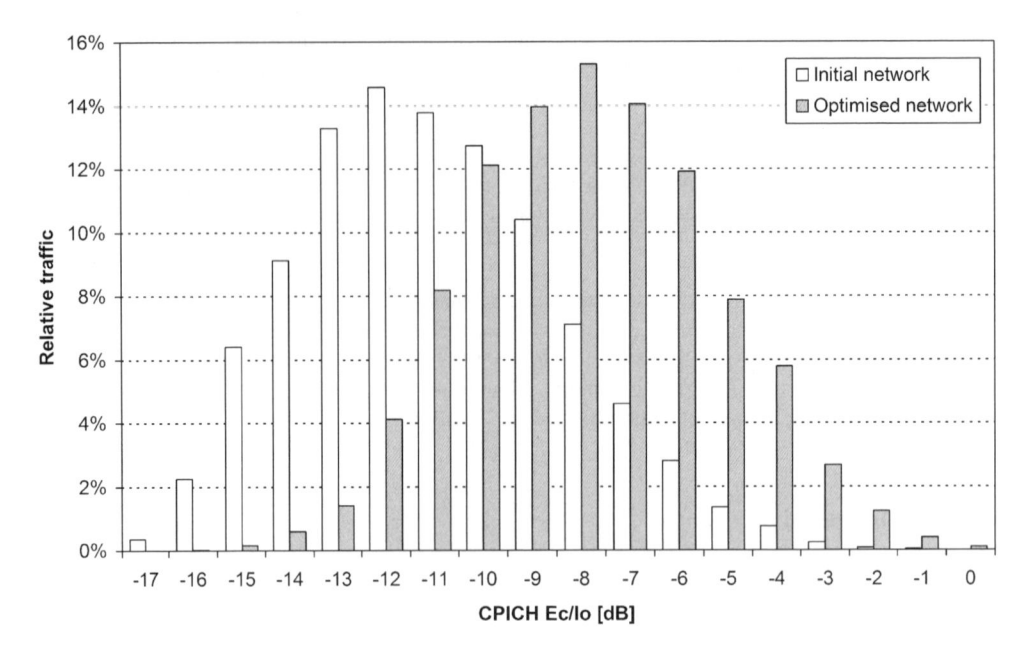

Figure 15.9 Histogram of CPICH E_c/I_0 – loaded initial and optimised networks.

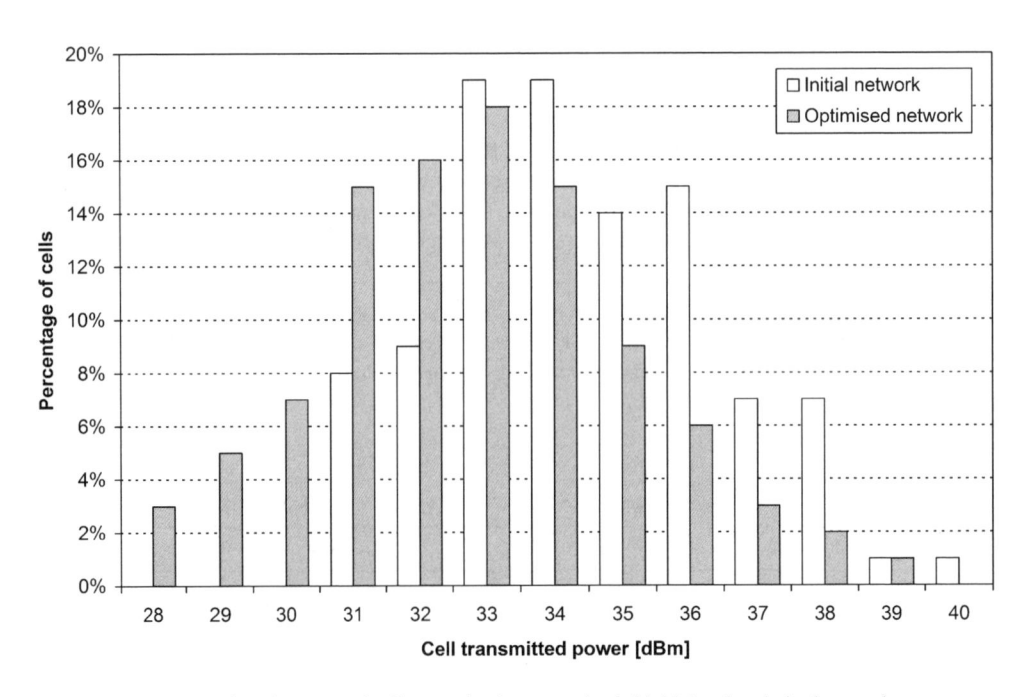

Figure 15.10 Histogram of cell transmitted power – loaded initial and optimised networks.

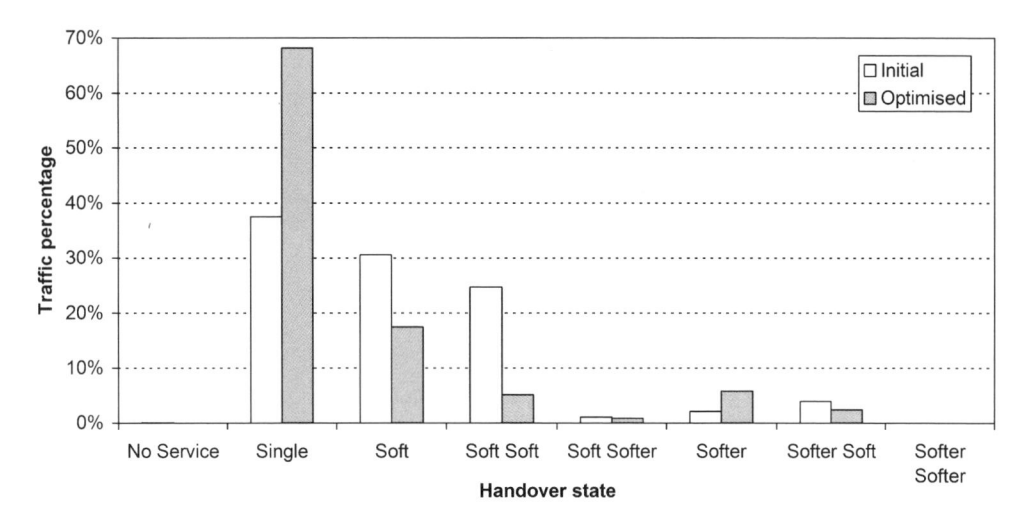

Figure 15.11 Handover analysis – loaded initial and optimised networks.

Handover analysis. After optimisation, the single connection mode was considerably improved (see Figure 15.11). The soft handover part dropped to an acceptable level. Similar to the non-loaded scenario, the contributions of soft and soft/soft states were minimised, leading to an improvement in the transmission and downlink capacity. More traffic has been captured by the softer connection state, albeit its influence on the capacity is kept low. This has again been achieved by downtilting of the antenna main beams.

Histogram of pilot pollution distribution. The optimisation process decreased the pilot pollution radically. The shift of the difference between serving cell and 4th pilot is obvious and varies around 7 dB (Figure 15.12). As a consequence, connections are established and maintained at lower interferences; thus, either the connection quality may rise (e.g. throughput) or more users may be accommodated, maintaining the same system level of quality.

Quality of the RF network – distribution of CPICH RSCP versus E_c/I_0. The optimised network offers, at the same serving signal levels (CPICH RSCP), a significantly improved signal quality (CPICH E_c/I_0). Consequently, at the same or lower (due to minimised cell output power) path loss between Node B and mobile, the received interference is minimised. This leads at both ends to a decrease of the required transmit power for a connection, and hence builds sufficient headroom for future capacity growth. This can be seen in Figure 15.13.

Distribution of non-connected mobiles. Finally, let us consider the location and number of non-connected mobiles at the centre of the optimised area. Figure 15.14 shows the randomly simulated mobiles in the initial state. The white squares represent mobiles that are connected to the network and serviced, while grey ones are representing mobiles with low values of E_c/I_0. The black squares represent non-connected mobiles due to high FER. Figure 15.15 shows the same network after optimisation. It is also possible to see some of the azimuth changes made to the antennas. A comparison between the two figures gives an indication that distant mobiles at cell edges were able to connect to the network. There are improvements in the mobiles being located close to the base station as well. Both were possible due to improvements in the best server areas and reduction of the total interference, illustrated by the improvement in the pilot pollution.

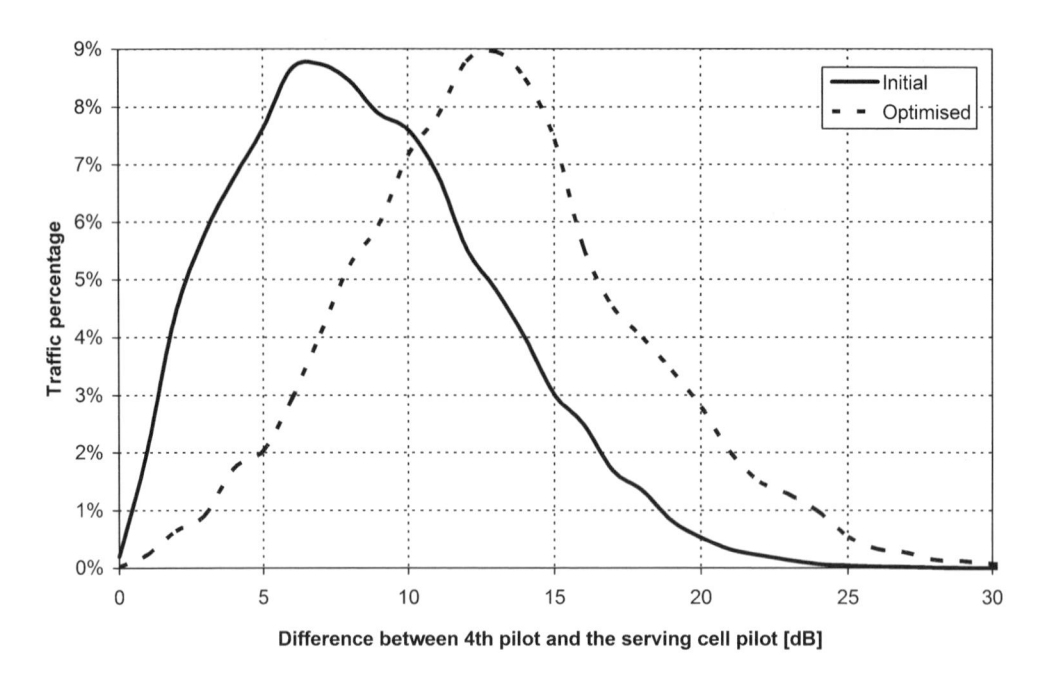

Figure 15.12 Pilot pollution analysis – loaded initial and optimised network.

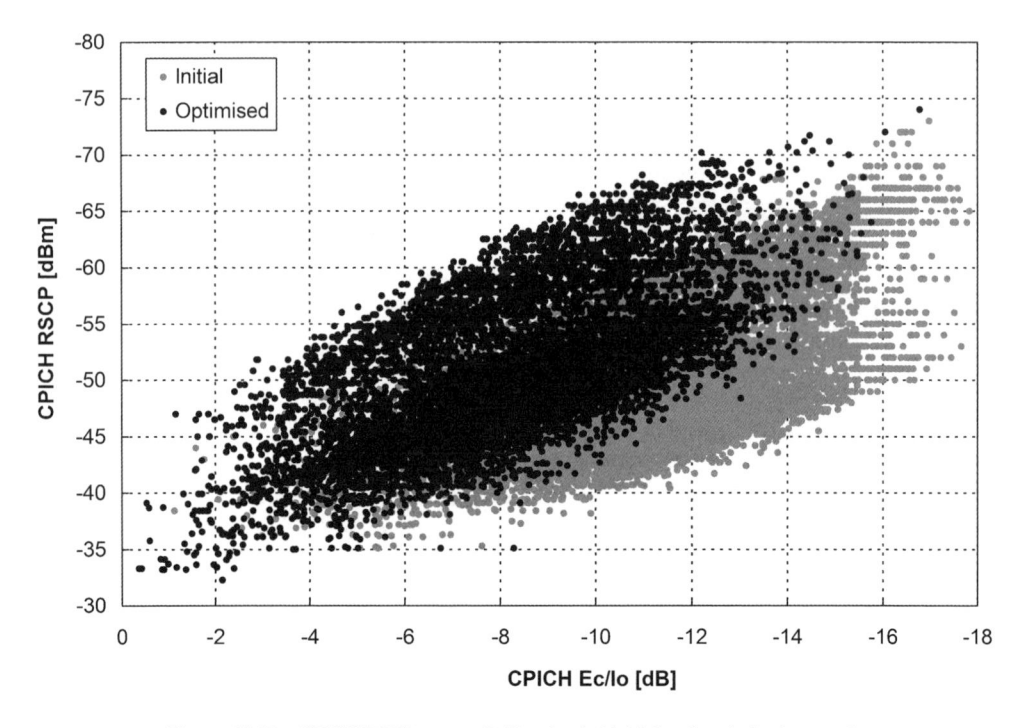

Figure 15.13 CPICH RSCP versus E_c/I_0 – loaded initial and optimised network.

Automatic Network Design

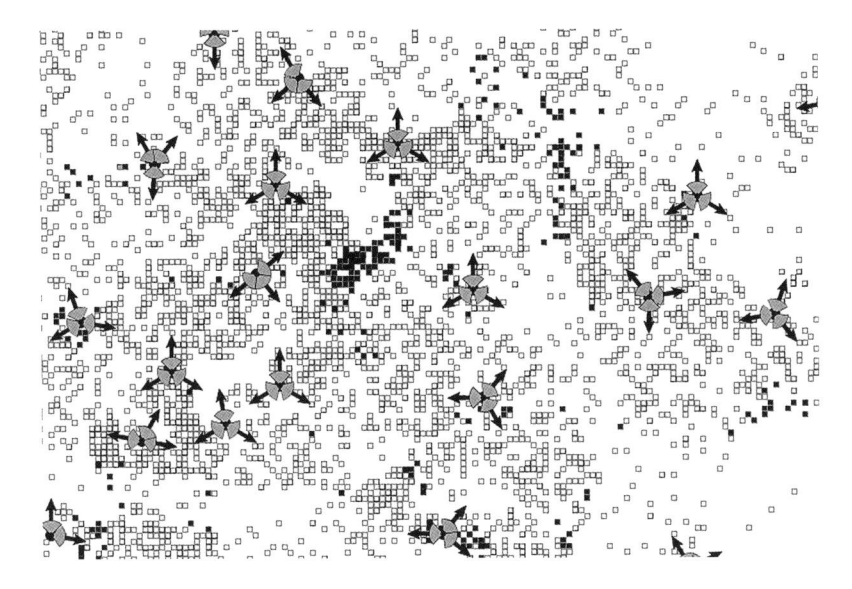

Figure 15.14 Distribution of simulated mobiles – loaded initial network.

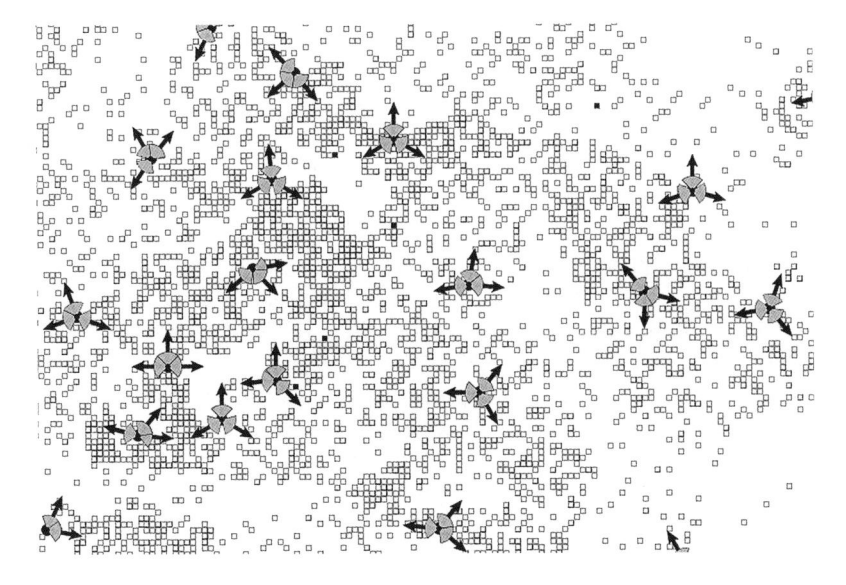

Figure 15.15 Distribution of simulated mobiles – loaded optimised network.

15.3 CASE STUDY: OPTIMISING BASE STATION LOCATION AND PARAMETERS

This section presents a case study of UMTS radio network optimisation. The study is conducted on publicly accessible data, so that the interested reader may take a close look at the underlying data and the outcomes.

We consider the following setting. A network operator is running a GSM network in an urban surrounding. The operator plans to introduce UMTS in addition to GSM. The take-up of UMTS is expected to be slow. The perception of the network quality of early adopters will be primarily based on network coverage. The goal is therefore to deploy an initial UMTS network that offers coverage similar to the existing GSM network. The capacity requirements are determined by traffic map derived from experience with GSM.

The objective of the optimisation is to design a UMTS network based on the existing site locations that retains GSM coverage, meets the capacity requirements, and saves on base station locations and base station sectors. The optimisation process thus has to

- select GSM base station locations for UMTS upgrade;
- decide the sectorisation for each UMTS site;
- decide the antenna configuration for each UMTS sector.

The CPICH power is set to a uniform value in all UMTS cells for this study.

15.3.1 DATA SETTING

15.3.1.1 Public Datasets

This case study is based on the *Berlin* scenario included in the publicly available datasets from the MOMENTUM project [1,2]. Besides Berlin (Germany), there are also datasets for Lisbon (Portugal) and The Hague (The Netherlands) available for download. The datasets comprise all kinds of data typically used in radio network simulation tools.

The Berlin scenario. The scenario covers an area of $7.5 \times 7.5 \, \text{km}^2$ in central Berlin. Like with the other public datasets from the MOMENTUM projects, all data available in state-of-the-art network planning tools and specifically all information for a detailed static system level model is included. The geographical database, elevation model and propagation grids have a resolution of 50 m. Further details can be found in [3].

Available sites and reference configuration. There are 69 potential site locations in the scenario. For this case study, 65 out of these sites are admitted. The scenario includes a reference configuration on the 65 sites shown in Figure 15.16a. The configuration uses three sectors at each site (with two exceptions), a regular sectorisation with sectors at 90°, 210° and 330° is most frequent. This leads to a total of 193 cells. No individual tilt values are present, a uniform value of 6° (electrical) tilt is used.

Traffic and services. In the Berlin scenario, a differentiated service mix comprising voice and video telephony, data streaming, web and background services (FTP, e-mail) is used. The significant presence of data streaming users makes the traffic mix downlink-biased. The traffic density varies inhomogeneously across the area. The downlink traffic distribution is depicted in Figure 15.16b. The figure show the pixels' average contribution to the downlink user load (see Section 6.4).

15.3.1.2 Parameters to be Optimised

For this case study, a uniform CPICH power value of 30 dBm is prescribed. Other common channels are fixed at sensible power values. The following degrees of freedom are considered.

Site position and sectors. A subset of the 65 candidate sites that will be equipped with UMTS hardware has to be selected. The other sites are not used for the UMTS network. It is also possible to remove individual sectors and use a site with one or two sectors only. More than three sectors are, however, not allowed because the underlying GSM network is assumed to have three sectors.

Automatic Network Design

(a) Available sites and cell footprints of reference configuration.

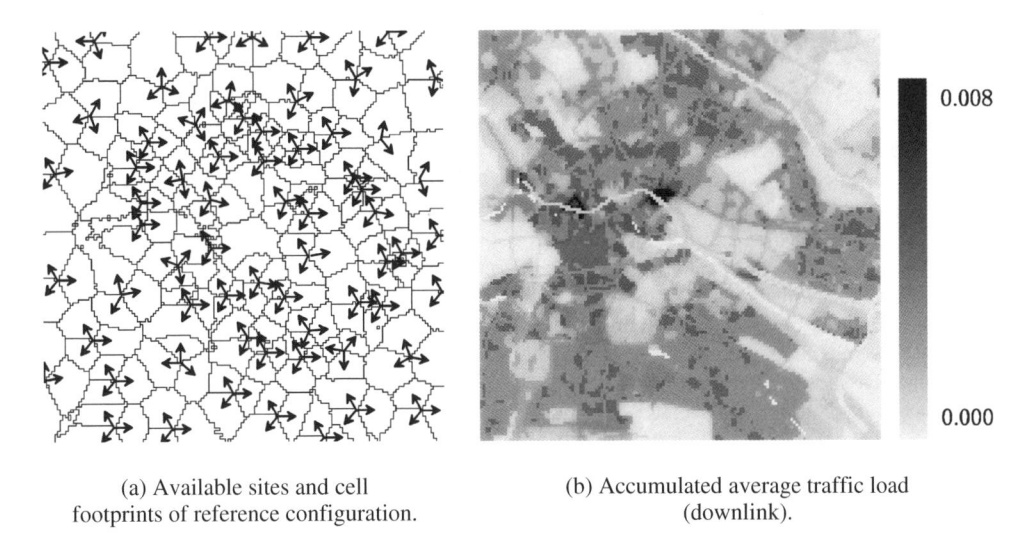

(b) Accumulated average traffic load (downlink).

Figure 15.16 Berlin planning scenario.

Tilt. Separate UMTS and GSM antennas are assumed to be used (no dual band equipment), so mechanical and electrical tilt can be varied independently. Tilts may be varied in the range from 0–14° with a step size of 2°. The antenna type in the scenario realises electrical tilts up to 8°. All higher tilt values have to be implemented with a combination of electrical and mechanical tilt. Electrical tilt is preferred, so the only possible combinations of electrical and mechanical tilt are 0° mechanical with 0–8° electrical tilt and 2–6° mechanical tilt with 8° electrical tilt.

Azimuth. The antennas are assumed to be installed at tilt brackets. Hence, the antenna azimuth is shared among the two systems. We assume that the joint azimuth can be changed up to ±30° with respect to the original direction in the reference configuration. The resolution for azimuth changes is 15°.

15.3.2 OPTIMISATION APPROACH

We adopt a two-phase optimisation approach in this study. Although the techniques applied in the two phases differ significantly, the underlying system model is the same. In both cases, the network load evaluation is based on the expected coupling matrices as described in Section 14.5. No Monte-Carlo simulation to assess the network performance is carried out during the optimisation process. Only the final evaluation of the network performance is conducted with a static Monte-Carlo network simulator.

In the first phase, an advanced local search technique (see Section 14.5) simultaneously optimises for network quality and tries to save on infrastructure. Due to the local scope of the technique employed in the first phase, there are typically areas where the outcome is not completely satisfactory. By chance, this may not be the case. But in general, it is a good idea to have some 'afterburner'. In our second phase, an MIP-based technique is used to optimise the sector configuration in confined areas.

15.3.2.1 First Phase: Local Search

The few introduced basic properties of the local search method are applied in this case study. These properties should help to appreciate the kind of optimisation performed during the first phase.

Throughout the entire run, one fixed objective function is used. This function measures the improvement of the current network configuration with respect to a reference network. The reference here is the network provided at the start. Improvement is determined independently in the dimensions E_c-coverage, E_c/I_0-coverage, traffic loss (due to cell overload), pilot pollution and interference coupling. Scores inferior to the reference value are heavily penalised, whereas improvements are rewarded. The scores per dimension are then combined linearly in order to obtain an overall network score (with respect to the reference network). This score is the objective function value for any given network.

Local search is based on configuration changes of one sector at a time. When a sector is chosen for optimisation, the objective function value for all alternative configurations of the sector are computed. The configuration with the best value is finally selected, and optimisation continues with another sector. The optimisation process is terminated in case no more (significant) improvement has been achieved for some number of steps.

Periodically, sites and sectors are checked for the eligibility for removal. An removal is executed if this does not (noteworthy) degrade the objective function value. Preference is given to removing entire sites. The objective function used here does not reward removals of sites and sectors explicitly.

15.3.2.2 Second Phase: Tuning MIP

An MIP, using the core model outlined in Section 14.5.4, is used to improve the solution found by local search in selected areas with *simultaneous* configuration changes in multiple cells. Neither the number of sites nor sectors are modified in this phase.

Objective function and constraints. The MIP model reflects the expected downlink interference coupling between cells. The objective function includes two components: traffic balancing between neighbours and interference coupling. Traffic balancing is determined by calculating the (absolute value of the) difference between the referring main diagonal elements of the coupling matrix. The measure for interference coupling is the sum of the off-diagonal elements of the average coupling matrix. Coverage is modelled as an additional constraint: if the reference configuration's coverage value is already satisfied, any feasible solution must achieve at least the same level of coverage. For details and a formal definition of the model, see [4].

Admitted configuration changes. The entire optimisation model including all possible configurations for all candidate cells is too large to be solved with reasonable effort. Optimisation is therefore focused on local improvements in selected areas consisting of neighbouring cells. The areas are selected on the base of (approximated) cell load: for each cell (or group of neighbouring cells) that has a high load compared to its neighbours and the scenario average, a group of 5–10 neighbours exposing strong coupling relations to the cell(s) with high load is selected.

For the set of cells to be jointly optimised, a 'neighbourhood' in the search space is included in the model. The allowed changes are the minimum allowed steps for both tilt and azimuth, that is (a) changing the tilt by 2° (up or down) and (b) changing azimuth by 15°. This process is repeated until no further improvement in the objective function is achieved.

15.3.2.3 Problem Sizes and Computation Time

Local search. The computations were carried out on a laptop computer equipped with a PentiumM 1700 MHz processor. The evaluation (average load approximates and coverage) of a candidate configuration took about 150 milliseconds in the example setting. About 7500 network configurations were evaluated within 20 minutes of computation time.

Tuning MIP. The tuning MIP was used in five subareas to locally improve the solution beyond the possibilities of local search. The corresponding areas are shaded in Figure 15.17(b). Per subarea,

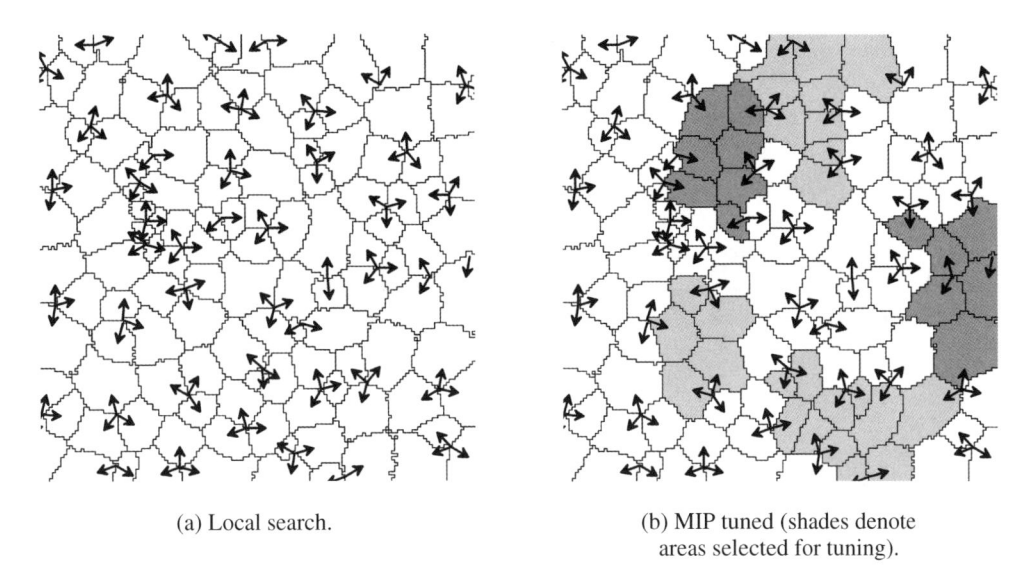

| (a) Local search. | (b) MIP tuned (shades denote areas selected for tuning). |

Figure 15.17 Antenna directions and cell shapes of optimised networks.

3–5 MIPs were solved using the optimisation software CPLEX [5]. The programs have a total of about 20 000–60 000 variables (mostly binary) and about 40 000–80 000 constraints (after CPLEX preprocessing). The solution process was interrupted after two hours computation time if the optimal solution had not been found yet. Not all MIPs were thus solved to optimality. However, the current incumbent solutions had a small optimality gap in all cases and were used for further optimisation.

15.3.3 RESULTS

15.3.3.1 Optimised Configuration

The results of the computation and the detailed optimised configurations are ready for download at this book's website [6]. This section highlights the most important features and changes of the optimisation process. After applying the local search procedure, a total of 122 sectors at 45 sites was selected out of the 65 available candidate sites with 193 sectors altogether. The configured cells are depicted in Figure 15.17(a).

Tilt values. The reference network had a uniform value of 6° (electrical) tilt. During optimisation, the tilt has been adapted according to the local situation. In the local search step, tilting down was used as a means for reducing interference coupling where possible, while coverage was increased by tilting up. A histogram of the resulting tilt values is given in Figure 15.18a. On average, the downtilt was slightly reduced in order to guarantee good coverage with a reduced number of sites. The fine-tuning MIP then reduced some of the (down-) tilt values again for load balancing in critical areas, see Figure 15.18b.

Azimuth values. The reference configuration's predominantly regular sectorisation with sectors at 90°, 210° and 330° is clearly visible in the azimuth histogram in Figure 15.19a (in some cases, the second sector has a slightly higher setting). In the optimisation process, the azimuth values, too, have been adapted to interact with their neighbours more favourably. In consequence, the histograms in

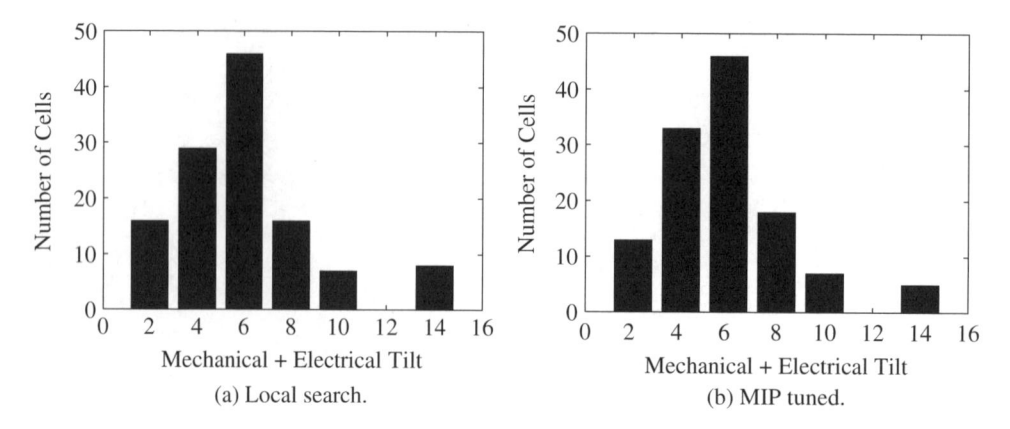

Figure 15.18 Histogram of tilt settings of optimised networks. The reference network has a uniform tilt of 6°.

(a) Reference
(193 sectors).

(b) Local search
(122 sectors).

(c) MIP tuned
(122 sectors).

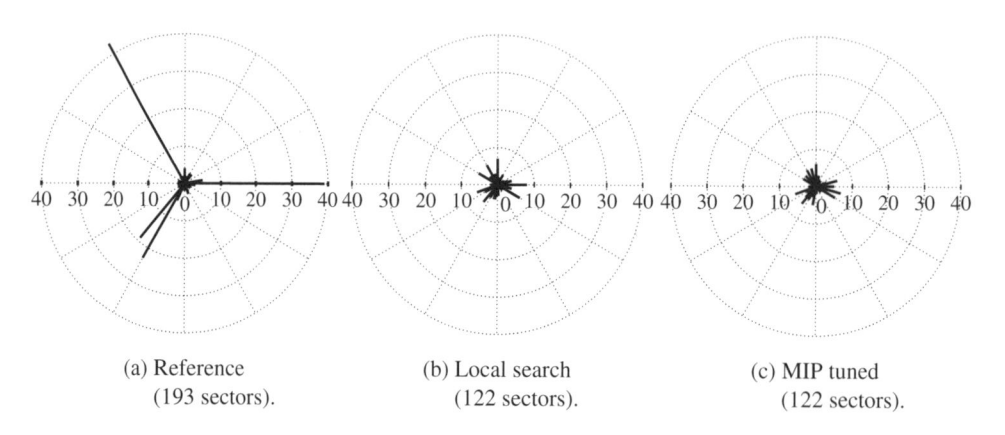

Figure 15.19 Histograms (polar) of azimuth settings in original and optimised networks.

Figures 15.19b and c show no uniform arrangement any more. (The previous sector settings are still discernible because only modifications of up to ±30° were admitted.)

15.3.3.2 Evaluation of Results

The reference configuration and the optimised configurations are evaluated with Monte-Carlo Simulation using 1000 snapshots; see Section 6.4 to determine average transmit and received powers as well as missed traffic.

Coverage. The E_c-coverage for a threshold of −85 dBm of the different configurations are listed in Table 15.5. Outdoor coverage was already excellent in the reference setting. However, the local search method still managed to increase the coverage, despite reducing the number of cells by more than a third. The increased coverage was retained in the tuning stage. For the assumed threshold value of −15 dB for E_c/I_0 coverage, E_c/I_0 coverage is given across the scenario for all configurations. However, the E_c/I_0 coverage plots in Figure 15.20 reveal that the E_c/I_0 situation has improved. The

Table 15.5 Global performance indicators.

Configuration	Cells (#)	E_c coverage (area %)	Average DL Load (%)	Average UL Load (%)
Reference	193	99.32	15.01	6.49
Local Search	122	99.99	17.55	8.99
MIP tuned	122	99.99	17.23	9.10

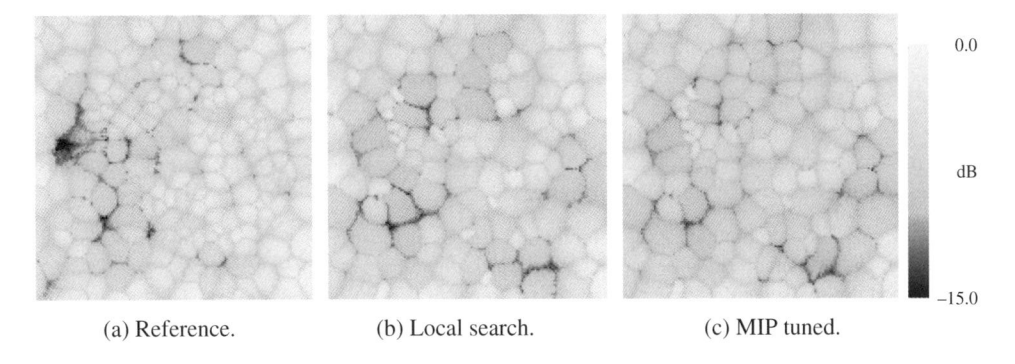

(a) Reference. (b) Local search. (c) MIP tuned.

Figure 15.20 E_c/I_0 coverage maps.

situation in the problematic area (E_c/I_0 close to -15 dB) in the left part of the scenario (dark areas in Figure 15.20a) is recognisably relieved by optimisation (Figures 15.20b,c).

Capacity/Load. The global average load figures for uplink and downlink are given in Table 15.5 for the three configurations. The global averages seem low, this is because cells are mostly deployed for coverage motives in the major part of the scenario. In the areas with high traffic, the scenario is downlink limited, uplink load is far lower than the downlink load. The reference configuration seems clearly over-dimensioned with respect to the expected traffic load. Many cells remain virtually empty, even in the area with comparably high traffic (see Figure 15.16b). The (uniform) common channels power $p^{(c)}$ makes for a base load of 10.66 % in each cell.

After eliminating 71 cells (see Figure 15.21b), the network is busier, but still at an acceptable level (about 35 % peak load). However, the load is locally not well-balanced in many cases (empty cells

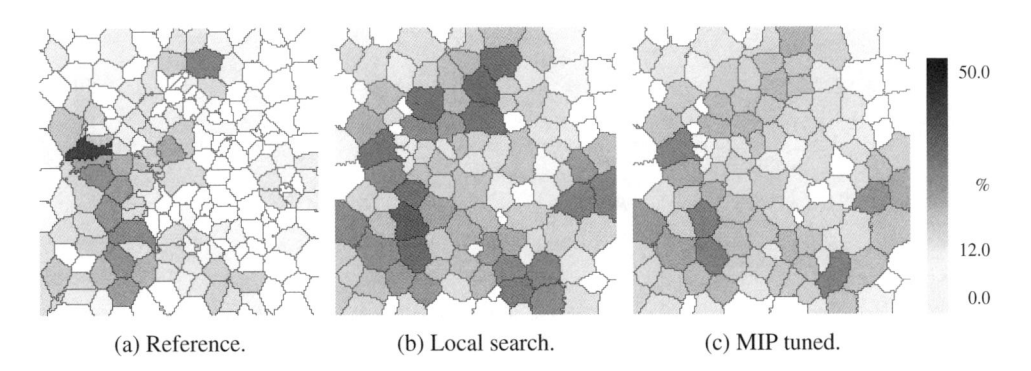

(a) Reference. (b) Local search. (c) MIP tuned.

Figure 15.21 DL load per cell.

close to the most crowded cells). The load imbalance is successfully remedied by the tuning MIP method, which can perform simultaneous configuration changes in several sectors. Figure 15.21(c) shows that the load is better balanced in the critical areas in the final configuration. This also makes for a reduction in the average load (from 17.55 to 17.23 %).

Why does the first phase end prior to reaching an optimum configuration?

First of all, the fact that this does happen should not come by surprise. Local search techniques (just as other neighbourhood search methods or genetic algorithms, etc.) have limited capabilities. Among others, they are not guaranteed to find optimal solutions to hard optimisation problems. But coming back to the question, our local search method proceeds from one network to the next by performing changes to one sector (leaving the site/sector removal aside). A change is performed only in cases where the objective function value improves (or more generally, does not deteriorate). This basic step is sometimes not powerful enough to achieve improvements that are still possible by changing several sectors at once. The local search is stuck at a local optimum. More powerful search steps or relaxed change criteria can reduce this problem (see the discussion in Section 14.5.1), but this typically leads to much higher computational efforts. The local search described here seems to be a good compromise between solution quality and computational effort.

Missed traffic. The predominant cause for missed traffic in the Berlin scenario is downlink power exhaustion. Uplink outage and user equipment power limits turned out to be negligible as all networks exhibit good coverage features and there is only comparatively little uplink traffic. The base stations are assumed to have a maximum nominal output power of 20 W; load and call admission are assumed to show effect in a cell when its average transmit power reaches 70 % of this value. In this case, the cells' blocking rates in a snapshot are estimated using the analytical methods from [7].

The blocking rates per cell are depicted in Figure 15.22 for the three network configurations. As expected, noteworthy amounts of blocking coincide with high average cell load. No cell features average blocking rates above 1 % in any case. As the average cell power is increased after turning off a large number of cells in the first optimisation step, the local search solution shows higher blocking rates in general. The improved load distribution in the second optimisation step's outcome then reduces the blocking rates noticeably; see Figure 15.22(c).

15.3.4 CONCLUSIONS

The automatic planning and optimisation process proved very powerful in this case study. The goal of providing a UMTS network that meets the capacity requirements, retains coverage, and saves on

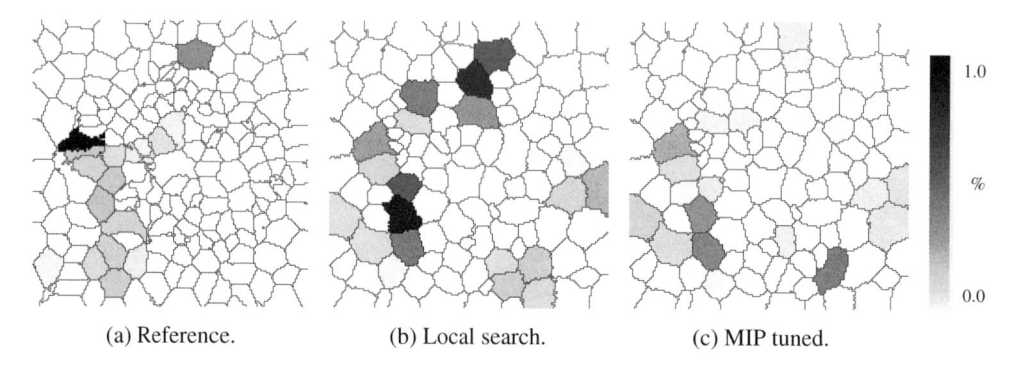

(a) Reference.   (b) Local search.   (c) MIP tuned.

Figure 15.22 Blocking (DL power limit) per cell.

base station locations and sectors was achieved with almost no manual intervention. The optimisation process comprised two phases. In the first phase, a local search method reduced the number of sites and sectors and simultaneously improved the overall network quality. A few regions were then manually selected and submitted to a second optimisation phase. This time, the full power of modern discrete mathematical optimisation was used. Each of the selected areas was optimised via a mixed integer program in order to improve the load share among neighbouring cells. This noticeably reduced the peak values for transmit powers and average blocking rates. In the end, a total of 20 sites and 71 sectors could be saved; the resulting network has excellent E_c- and E_c/I_0-coverage and does not show significant cell blocking or dropping.

In both phases, the optimisation relied on a novel system model for UMTS network performance that does not involve traffic snapshots. Monte-Carlo simulations were only used to determine the network performance of the final network configuration.

REFERENCES

[1] IST-2000-28088 MOMENTUM, 'Models and simulations for network planning and control of UMTS', 2001, http://momentum.zib.de.
[2] IST-2000-28088 MOMENTUM Scenarios, 'Momentum public UMTS planning scenarios', 2003, http://momentum.zib.de/data.php.
[3] A. Eisenblätter, H.-F. Geerdes, and U. Türke, 'Public UMTS Radio Network Evaluation and Planning', *Intern. J. on Mobile Network Design and Innovation*, 2005.
[4] A. Eisenblätter, H.-F. Geerdes, 'A novel view on cell coverage and coupling for UMTS radio network evaluation and design', *Proc. of INOC'05*, 2005, Lisbon, Portugal.
[5] CPLEX 9.0, *Reference Manual*, www.cplex.com.
[6] http://www.zrt.pwr.wroc.pl/umts-optimisation.
[7] A. Eisenblätter, H.-F. Geerdes, and N. Rochau, 'Analytical approximate load control in WCDMA radio networks', *Proc. IEEE VTC-2005 Fall*, 2005, Dallas, TX, USA.

16

Auto-tuning of RRM Parameters in UMTS Networks

Zwi Altman, Hervé Dubreil, Ridha Nasri, Ouassim Ben Amor, Jean-Marc Picard, Vincent Diascorn and Maurice Clerc

This chapter deals with issues related to an automated tuning of radio resource management parameters. As will be shown, automated tuning has a strong influence on the quality of the network performance. We will also describe the most important parameters of the tuning process. The remaining part of the chapter is dedicated to the description and performance analysis of chosen optimisation strategies.

16.1 INTRODUCTION

The variety of services and applications provided by UMTS networks with different traffic classes and quality of service (QoS) requirements imposes important challenges in network management. Radio Resource Management (RRM) functionalities such as mobility, admission control, load control or packet scheduling are central management functions that allow to control the network performance and QoS. These functionalities are defined in terms of algorithms and associated parameters and thresholds that can be set to control radio resources. By judiciously setting RRM parameters, the profitability of the network can be considerably enhanced.

An important effort has been invested by UMTS network operators in automating management tasks to improve network performance and to reduce operational expenditure. In this context, auto-tuning of RRM parameters has gained an increasing interest from vendors, operators, and academia, and has been the subject of intense research and development activity. Auto-tuning aims at dynamically controlling the network performance by adjusting certain RRM parameters (see Figure 16.1) such as load target thresholds for admission and QoS control, or add- and drop-windows for macrodiversity and mobility management.

Unlike automatic cell planning (see Chapters 14 and 15) that optimises the network for some average traffic distribution, auto-tuning is related to the dynamic nature of the network and seeks to adapt the network to traffic variations.

Understanding UMTS Radio Network Modelling, Planning and Automated Optimisation   Edited by Maciej J. Nawrocki, Mischa Dohler and A. Hamid Aghvami   © 2006 John Wiley & Sons, Ltd

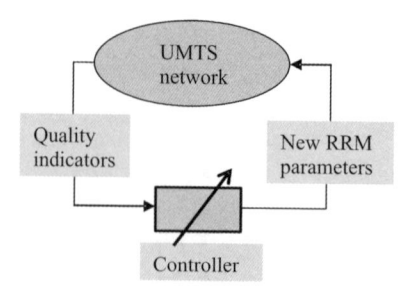

Figure 16.1 Auto-tuning scheme.

One of the first research experiments in auto-tuning has been reported for hierarchical cell structure (HCS) in GSM networks [1,2]. The objective has been to dynamically auto-tune the signal strength threshold to control traffic flux from the higher layer (macro-cells) towards lower layers (micro-cells), thus improving traffic balance between cells of the HCS network and increasing its capacity. Most of the reported contributions that followed focused on WCDMA networks, such as auto-tuning of the downlink load target and maximum link transmitted power [3], the transmitted pilot power [4], or the uplink (UL) and downlink (DL) E_b/N_0 for packet traffic [5]. Recently, the concept of auto-tuning has been extended to Joint RRM (J-RRM) or Common RRM (CRRM) of multi Radio Access Networks (RAN) to improve multi-system cooperation (see for example [6,7]).

The design of RRM controllers for auto-tuning tasks is a central topic discussed in this chapter. Fuzzy logic controllers (FLC) constitute an efficient framework for designing and performing control tasks [8–10]. They allow translating simple linguistic rules into mathematical form which are directly implemented by the controller. The design of high quality FLCs can be a difficult time-consuming task that calls for optimisation techniques to automate and improve the design process. Two approaches are considered here: off-line and on-line approach. In the former, the FLC is optimised on a computer prior to its introduction in the network. Three optimisation strategies based on the Particle Swarm (PS) [11] method are utilised to optimise the FLC parameters. The latter approach can be carried out in the operating network. The Reinforcement Learning (RL) with the fuzzy Q-learning implementation [12] is utilised in the on-line case [13].

The chapter is organised as follows: Section 16.2 briefly describes RRM for controlling network quality. Different aspects related to FLC design are presented in Section 16.3, including a case study of macrodiversity auto-tuning. Section 16.4 describes off-line and on-line optimisation strategies for the auto-tuning process.

16.2 RADIO RESOURCE MANAGEMENT FOR CONTROLLING NETWORK QUALITY

Radio resource management is responsible for allocating and controlling resources in the air interface to satisfy QoS requirements for different services. In UTRA FDD mode, RRM are divided into five functionalities: handover (including macrodiversity), admission control, power control, load control, and packet-scheduling functionalities [14–15]. In UTRA TDD mode, RRM includes the functionality of dynamic channel allocation. The RRM functionalities are distributed in the mobile station (power control), in the base station (power control and load control), and in the RNC for all the five functionalities. They are briefly summarised below, with relation to auto-tuning.

Macrodiversity. Macrodiversity (or soft handover) is responsible for seamless intra-frequency mobility in the UMTS network. The macrodiversity algorithm controls the creation and suppression of radio links between mobile and neighbouring base stations (BS). The BSs which are connected with the mobile form the mobile active set, and the list of stations that are continuously monitored by the mobile – the neighbour set. Three events are generated by the macrodiversity algorithm [14]: the addition of a new link or *Event 1A*, the removal of an existing link, or *Event 1B*, and the replacement of an existing link, denoted as *Event 1C*. The mobile continuously measures the pilot (CPICH) signal E_c/I_0, $(E_c/I_0)_{CPICH}$, which is the ratio between the received energy per chip of the pilot channel and the total power density in the bandwidth. A BS is added to the active set if the signal from that BS is higher than that of the best station minus a hysteresis window, *add_win* or *Hysteresis_event1A* during a *time-to-trigger* period ΔT:

$$\left(\frac{E_c}{I_0}\right)_{CPICH}^{\text{Best station}} - \left(\left(\frac{E_c}{I_0}\right)_{CPICH}^{\text{Station}} + CIO^{\text{Station}}\right) \leq \text{Add\_win} \quad (16.1)$$

where CIO^{Station} or *Cell Individual Offset* is an optional offset that can be added to the signal of a potential new link to favour the entry of this station to the active set.

A BS is removed from the active set if the corresponding signal is smaller than that of the best station minus a hysteresis window *drop_win* or *Hysteresis_event1B* during a period ΔT:

$$\left(\frac{E_c}{I_0}\right)_{CPICH}^{\text{Best station}} - \left(\left(\frac{E_c}{I_0}\right)_{CPICH}^{\text{Station}} + CIO^{\text{Station}}\right) \geq \text{Drop\_win} \quad (16.2)$$

A BS in the active set (superscript *In AS* in Equation (16.3)) is replaced by a new BS if the corresponding signal from the new BS is bigger than that of a BS in the active set plus a hysteresis window *rep_win* or *Hysteresis_event1C* during a period ΔT:

$$\left(\left(\frac{E_c}{I_0}\right)_{CPICH}^{\text{Station}} + CIO^{\text{Station}}\right) - \left(\frac{E_c}{I_0}\right)_{CPICH}^{\text{In AS}} \geq \text{Rep\_win} \quad (16.3)$$

By combining the received signals from the BSs of the active set using the maximum ratio combining mechanism, the radio link quality in DL is improved, the coverage in the cell border is increased, and the UL capacity increases.

Admission control. To ensure uninterrupted service provision and quality of service of ongoing communications, the admission control (AC) regulates the entering connections. Before admitting a new connection, the network estimates the contribution of this connection to the total BS load in uplink and downlink, using uplink interference and downlink transmitted power respectively [15]. The radio access bearer (RAB) of the new connection can be admitted if the new estimated load does not exceed a predetermined threshold in both uplink and downlink, otherwise the connection is denied. Certain implementations of AC algorithms consider also the neighbouring station loads. For packet switched service, the requested RAB can be modified, i.e. by reducing the corresponding bit rate. Exceeding the load thresholds could result in excessive interference, degraded QoS and loss of coverage. It is noted that the AC algorithms are manufacturer dependent.

Power control. WCDMA is an interference limited system, namely the capacity is limited by interference. Power control (PC) is the most important mechanism that allows to minimise interference in the air interface. It is responsible for transmitting the minimum power for each user that allows to decode the received information with a small and predetermined probability of error per bit of information. PC allows to adapt the transmitted signal to varying channel conditions due to fast fading and shadowing.

PC comprises three power loops: an *open-loop power control* which determines the initial power to be transmitted; an *inner power loop* that compares the signal-to-noise ratio (SIR) to a predetermined SIR target and adjusts the transmitted power to meet this target ratio; and an *outer-loop power control* that adjusts the SIR target in both uplink and downlink.

Load control. If an excess of BS load is detected, the load control brings back the load below the maximum load target to avoid power saturation that handicaps the power control, and degrades QoS of ongoing connections. Typically, system overloading is avoided by the AC and packet scheduling. However, if a congestion occurs, the load control can take different measures: deny power-up commands received by the mobile, reduce SIR target for the UL inner power loop, reduce throughput of packet data traffic or bit rate of real-time connections, hand over the user to another UMTS carrier or RAN, and if there are no other possibilities, drop certain connections.

Packet scheduling. The function of packet scheduling is to share the available capacity in the air interface between packet users, and is performed by the packet scheduler. The packet scheduler allocates packets to transport channels: dedicated, common and shared channels, and controls the allocated bit rate. The packet scheduler has also the function of monitoring the system load and the QoS of the packet users. Packet scheduling can be based on time division or code division scheduling. Different algorithms have been devised for the scheduling task, which are particularly important for real-time applications, such as *Round Robin*, and *Modified Round Robin*, or *C/I scheduling*. The latter two try to benefit from favourable channel conditions to transmit packets.

In the context of auto-tuning, the main effort has been invested in admission control and macrodiversity that will be the focus of the rest of this chapter. However, it is noted that some work and ideas have been explored for other RRM functionalities and parameters such as power control [5], maximum transmitted power of traffic channels [3], pilot channel transmitted power [4] or neighbouring cell list [15].

16.3 AUTO-TUNING OF RRM PARAMETERS

This section presents the design of fuzzy logic controllers for RRM auto-tuning. First, the parameters, indicators and targets for the auto-tuning process are described. Then, the mathematical framework for designing fuzzy logic controllers (FLC) is summarised. Finally, the design of two macrodiversity FLCs is described.

16.3.1 PARAMETER SELECTION FOR AUTO-TUNING

Macrodiversity. The most effective parameters for macrodiversity auto-tuning are:

- add_win (or *Hysteresis_event1A*);
- drop_win (or *Hysteresis_event1B*).

Auto-tuning of *rep_win* (or *Hysteresis_event1C*) could be of interest, especially in a dense urban environment, as well as the size of the active set. Other parameters defined by the macrodiversity such as neighbouring list, *Cell Individual Offset* or *time to trigger* period should be optimised; however, the gain of performing it dynamically is not obvious.

Admission control. Two important parameters for admission control auto-tuning are the load target thresholds in UL and DL (see Figure 16.2), sometime denoted also as *PrxTarget* and *PtxTarget* respectively [15]. Decreasing the threshold reduces the number of new entering calls, so that resources are shared among less connections and the QoS is enhanced. Conversely, increasing the load target threshold allows to serve more traffic with the risk of compromising QoS.

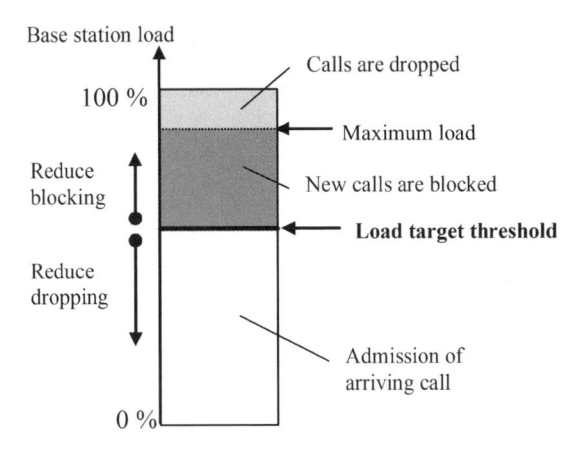

Figure 16.2 Load target threshold for auto-tuning.

The maximum load threshold is a congestion control parameter. This threshold is introduced to avoid saturation effects of the connections that render power control ineffective and deteriorate QoS. Typically, the maximum load threshold is fixed.

Auto-tuning can dynamically control resource allocation between real-time and non-real-time traffic, as part of the admission control process. In this strategy resources are allocated into two distinct bands separated by a *reservation target threshold* (see Figure 16.3). Two efficient strategies for auto-tuning can be considered:

1. Real-time reserved band and a mixed real-time and non-real-time band;
2. Non-real-time reserved band and a mixed real-time and non-real-time band.

The second strategy is of interest when pre-emption of real-time traffic is implemented, namely if the admission of a real-time connection can reduce resources (i.e. bit rate) allocated to non-real-time traffic.

Researches have been also conducted on auto-tuning of other parameters, such as the maximum transmitted power of traffic channels [3], pilot channel transmitted power [4] or neighbouring cell list [15].

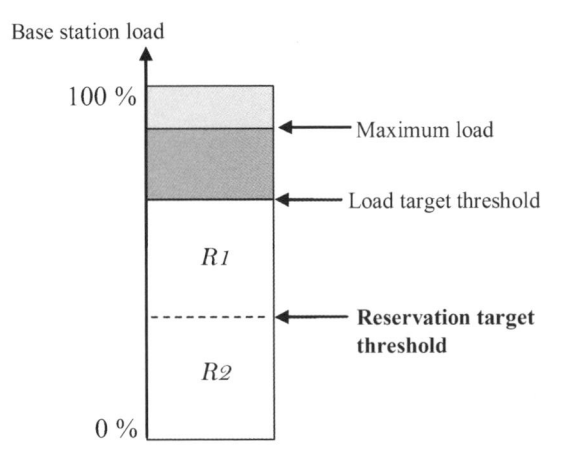

Figure 16.3 Reservation target threshold for auto-tuning.

16.3.2 TARGET SELECTION FOR AUTO-TUNING

Target selection refers to performance and quality indicators as well as to target thresholds for certain indicators that guide the auto-tuning process. The following quality indicators are relevant for macrodiversity auto-tuning:

- Blocking rate, defined as the ratio between blocked calls and all call-attempts;
- Dropping rates, defined as the ratio between dropped calls and admitted calls;
- Macrodiversity blocking rate, defined as the ratio between the number of blocked macrodiversity links and the total number of requests to establish those links;
- Downlink load, defined as the ratio between the transmitted and the maximum BS power; and
- Ping-pong effect, measured in terms of the frequency of active set updates or as the number of link establishments per time unit for each mobile station.

Quality indicators that could be considered for the admission control are:

- Uplink and downlink loads;
- Blocking rate;
- Dropping rate.

For non-real-time traffic, one could add:

- Buffer delay;
- Throughput per BS or throughput perceived by the user.

To render the control process efficient, it is recommended to use indicators that are not too correlated. For example, the last three indicators (buffer delay, BS throughput and user perceived throughput) are highly correlated and therefore only one of them should be selected.

16.3.3 FUZZY LOGIC CONTROLLERS (FLC)

The concept of a fuzzy set, utilised in fuzzy logic, has been introduced by L.A. Zadeh [16] as a generalisation of the ordinary or crisp set known in classical logic theory. A fuzzy set can be seen as a predicate whose truth value is drawn from the unit interval $I = [0,1]$ rather than the set $\{0,1\}$ in ordinary sets. If X is a set that serves as the universe of discourse, a fuzzy set A of X is associated with a characteristic function, or a member function μ_A:

$$\mu_A : X \to [0,1] \qquad (16.4)$$

where $\mu_A(x)$ indicates the degree to which x is a member of the set A. Fuzzy logic theory has been derived to integrate the concept of fuzzy sets. Figure 16.4 shows an example of four fuzzy sets, the universe of discourse for the variable *dropping rate* (often called the *linguistic variables* in fuzzy logic terminology). The fuzzy sets correspond to small, medium, high and very high linguistic variables, and are denoted by S, M, H and VH respectively. An input dropping rate value of 0.12 is mapped into two member function values of 0.8 and 0.2 for the medium and high fuzzy sets respectively. It is noted that the partition of the domain utilises overlapping sets allowing to describe different aspects of a phenomenon with different degrees of membership. Often, triangular sets are utilised although other functions such as splines or Gaussian functions can be used.

The fuzzy sets are used in a rule based model for controlling a system, and the set of rules constitutes a Fuzzy Inference Systems (FIS). The control process is denoted as fuzzy logic control. A rule has the

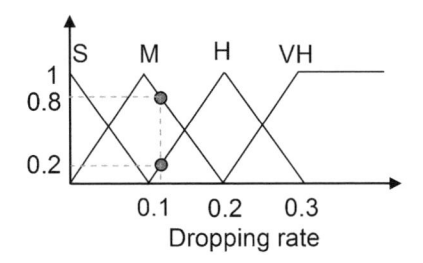

Auto-tuning of RRM Parameters in UMTS Networks 411

Figure 16.4 Fuzzy sets for the dropping rate indicator.

following form: in a given situation, perform this action (conclusion). An example of a rule of a FIS can be written as follows:

If (blocking_is_high) AND (dropping_is_low) THEN

(big_increase_AC_threshold) (16.5)

This formulation of rules simplifies the translation of human knowledge into the FIS, making this approach particularly attractive.

The rules of the form of Equation (16.5) can be aggregated into a matrix, denoted as a *decision matrix*, and the corresponding FIS is called the *Fuzzy Logic Controller* (FLC). The control process using an FLC is performed in three steps: *fuzzification, inference, defuzzification* [8]. The scheme of the FLC that could be used for RRM auto-tuning is presented in Figure 16.5.

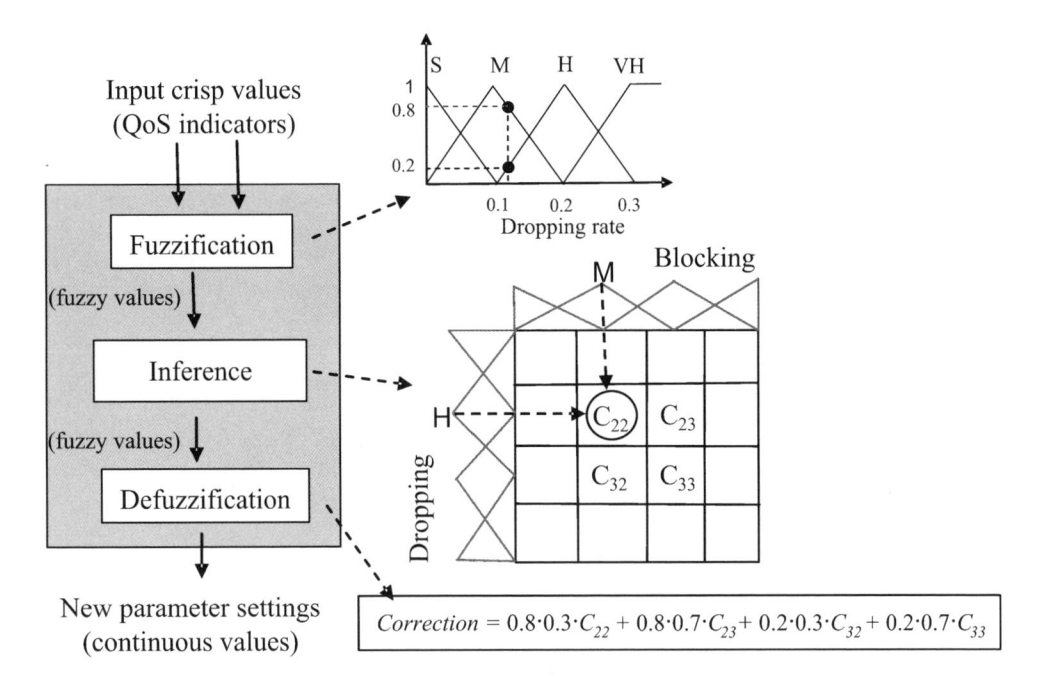

$$\text{Correction} = 0.8 \cdot 0.3 \cdot C_{22} + 0.8 \cdot 0.7 \cdot C_{23} + 0.2 \cdot 0.3 \cdot C_{32} + 0.2 \cdot 0.7 \cdot C_{33}$$

Figure 16.5 Fuzzy logic controller using a decision matrix.

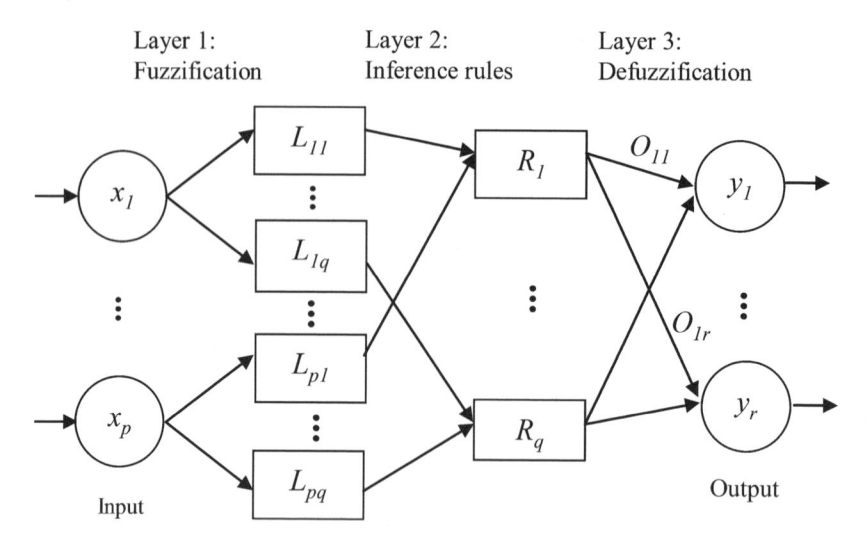

Figure 16.6 General structure of a fuzzy logic controller.

The first block of the FLC, *fuzzification*, receives as input continuous (*crisp*) values of quality indicators such as blocking and dropping rates from a station and its neighbours. For each input indicator, up to two member function values are computed (see Figure 16.4). The second block, *inference*, comprises one or several decision matrices. Each element of a decision matrix D corresponds to a distinct 'if – then' type of rule such as in Equation (16.5). The element D_{ij} represents the modification or correction to be applied to a given parameter. Let A^H – Dropping and A^M – Blocking be two fuzzy sets with a corresponding member function values (or degree of membership) of μ_i and μ_j respectively. Then the correction D_{ij} is triggered with a strength of $\mu_i \times \mu_j$ (see Figure 16.5). In the third block, *defuzzification*, a global crisp value correction is calculated by aggregating all the weighted corrections defined in the inference step. A weighted average (gravity centre) is often utilised.

The general scheme for an FLC [17] is presented in Figure 16.6. The three steps of the controller are presented in three layers. In the first (fuzzification) layer, the input indicators x_i are mapped into degrees of membership for the different fuzzy sets L_{ij}. The second layer defines the rules, R_i, via the connections between the input fuzzy sets and up to r output correction fuzzy sets O_{mn}. In the (deffuzification) layer 3, the output is computed using aggregated degrees of membership contributed by all rules. The rule R_i of the FLC is written as:

$$\text{if } (x_1 \text{ is } L_{1i} \text{ and } \ldots x_p \text{ is } L_{pi}) \quad \text{then} \quad (y_1 \text{ is } O_{i1} \text{ and } \ldots y_r \text{ is } O_{ir}) \qquad (16.6)$$

In the RRM application presented in this section the FLC has one output. Hence to each rule R_i there corresponds one fuzzy set O_i. It is noted however that more than one FLC can operate simultaneously, each of which controls one RRM parameter [10].

16.3.4 CASE STUDY: AUTO-TUNING OF MACRODIVERSITY

This section presents two models of FLCs for macrodiversity auto-tuning, which are characterised by the input quality indicators and the output correction. The first model utilises the blocking rate of macrodiversity requested links as input indicator, and the second model the downlink loads. The output corrections in both cases are the *add_win* (*Hysteresis_event1A*) and *drop_win* (*Hysteresis_event1B*) defined in Section 16.2.

To construct an FLC, a thorough understanding of the macrodiversity mechanism is required. When a BS increases the *add_win* parameters, more mobiles enter in macrodiversity and create extra links in the network. The improvement of link quality of mobiles in macrodiversity is accompanied by an average increase of downlink load. Sectors with high load will see their QoS reduced, and further loading of the network due to macrodiversity should be avoided. When two adjacent BS are considered, denoted as a central BS and a neighbouring BS, a highly loaded central BS can alleviate its load by increasing its *add_win* parameter. In this case, the neighbouring BS will create new links with the mobiles of the central BS and reduce its load. To a certain extent, the macrodiversity balances the traffic between the cells. When a mobile fails to create a new macrodiversity link, it may be dropped, for example when it is moving towards an overloaded BS. The corresponding quality indicator is the macrodiversity blocking rate, and is denoted here for the sake of brevity as *MD blocking*. Fluctuation of link quality due to fast fading, shadowing etc. may cause a *ping pong effect*, namely the repetitive creation and cancellation of macrodiversity links, which results in signalling overload.

To simplify the auto-tuning process, a constant difference of 2 dB between the two hysteresis parameters is assumed: *drop_win* − *add_win* = 2 dB. The FLC correction values for the *add_win* parameter (i.e. the rules) are determined automatically by the optimiser (see Section 16.4).

In WCDMA networks, BSs could be coupled due to the sharing of the frequency bandwidth between the users. Coupling effects can be seen when a change of a parameter in one BS modifies performance in neighbouring BSs. It is noted that neighbourhood is not necessarily expressed in terms of geographical distance, but rather in terms of interference [18] or traffic flux. To guarantee a robust auto-tuning process, and to avoid chaotic behaviour and oscillations in the network, the controller takes into account quality indicators from the (central) BS and its neighbours. Traffic flux between the BSs is computed and used to define neighbourhood relations (Figure 16.7).

The auto-tuning process is carried out using a semi-dynamic simulator. After each time step n, that typically varies between one and four seconds, the positions of the mobile users are updated and the powers and other quality indicators are calculated [10]. Every quality indicator, I, utilised by the FLC is first filtered using an averaging window,

$$I_{\text{filtered}}(n) = \frac{1}{M} \sum_{m=0}^{M-1} I(n-m) \tag{16.7}$$

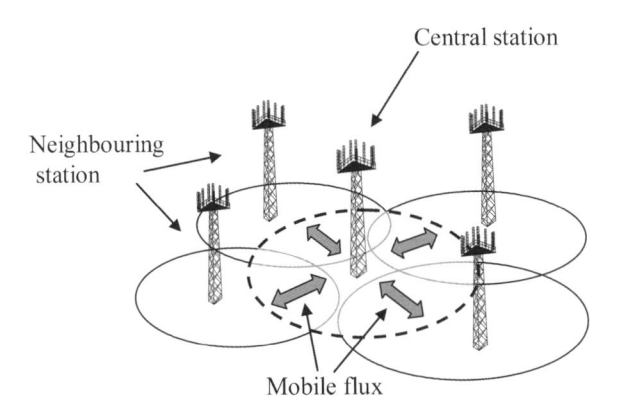

Figure 16.7 Traffic flux between base stations is used to define neighbourhood relations.

The averaging in Equation (16.7) can be seen as a low pass filter that stabilises the control process. Hence, the auto-tuning process is guided by the indicator trends and not by short time chaotic fluctuations. The filtering in Equation (16.7) is done using samples from the last 100 seconds.

MD blocking based model. The first FLC model, denoted herein as *MD blocking based model*, utilises *MD blocking* (i.e. macrodiversity blocking rate) as input to the FLC and will be described here using the decision matrix formulation. Three decision matrices in two levels are used: the first level comprises two matrices, and their corresponding corrections are aggregated by a third (second level) matrix (see Figure 16.8). The zeros in the squares correspond to no correction, and the '+++' and '– – –' to strong increase and decrease of the *add_win* of the central BS respectively.

The first matrix based on 'local reasoning' utilises the maximum MD blocking of the neighbouring BSs and MD blocking of the central BS as input indicators. The correction of this matrix should account for the neighbouring BS in the worst situation. For example, when the central BS suffers from high MD blocking while all neighbouring BSs experience low MD blocking, then a strong increase of

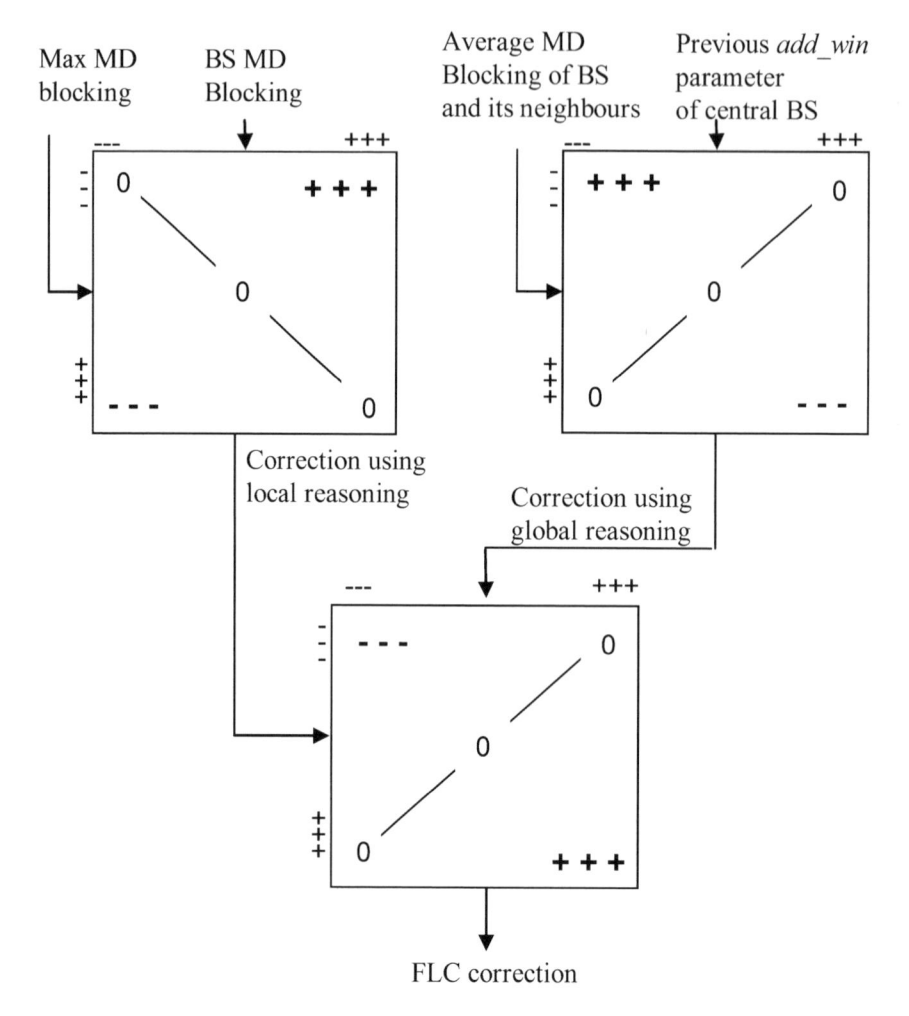

Figure 16.8 Two level fuzzy logic controller for macrodiversity auto-tuning.

the *add_win* of the central BS is applied, causing the neighbouring BSs to alleviate the central BS. On the contrary, when a high maximum MD blocking occurs for one of the neighbouring BSs while the central BS remains with low MD blocking, then a strong reduction of the *add_win* parameter of the central BS will alleviate the corresponding neighbour BS by reducing the number of macrodiversity links.

The second decision matrix based on 'global reasoning' utilises the average MD blocking of the central BS and its neighbours as input indicator together with the previous value of the *add_win* parameter. When the average MD blocking is low and the previous *add_win* is low, a strong increase of the *add_win* value is applied. The idea is to prevent a drift of the system towards low values of *add-* and *drop_win* that would handicap mobility. The other extreme is when both the average MD blocking and the previous *add_win* are high (i.e. important macrodiversity resources are allocated). In this case a strong decrease of the *add_win* is applied.

The third (second level) matrix aggregates the corrections from both first level matrices. When a strong *add_win* decrease/increase is demanded by both (first level) matrices, the same correction strength is used for the aggregated correction.

Load based model. The second model, denoted as *load based model*, auto-tunes the macrodiversity parameters as a function of downlink (filtered) loads of the central BS, $load_s$, and its neighbours. The equivalent load of the neighbouring BSs, $load_{nbr,s}$ is computed by weighting the load $load_i$ of each neighbour station of BSs with the normalised traffic flux, $w_{i,s}$, between BS i and s:

$$load_{nbr,s}(n) = \sum_{i \in NS(s)} w_{i,s} load_i(n) \qquad (16.8)$$

where NS(s) is the cell-neighbouring set. The total traffic flux is normalised to one:

$$\sum_{i \in NS(s)} w_{i,s} = 1 \qquad (16.9)$$

To design the FLC, triangular fuzzy sets are defined for the input indicators, $load_s$ and $load_{nbr,s}$, that span the domain [0,1]. The determination of rules, i.e. the correction to apply to each couple of input fuzzy sets for $load_s$ and $load_{nbr,s}$ is carried out automatically using the Q-learning, and is described in detail in Section 16.4.2. The resulting FLC can be seen as a function, f_{FLC}, that maps the input indicators ($load_s(n)$, $load_{nbr}(n)$) into a correction $C(n)$:

$$C(n) = f_{FLC}\left(load_s(n), load_{nbr,s}(n)\right) \qquad (16.10)$$

The load based model has been found to be both simple and effective.

16.4 OPTIMISATION STRATEGIES OF THE AUTO-TUNING PROCESS

Optimisation of the FLC is essential to obtain a high performance of the auto-tuning process. It allows to adapt the controller to new conditions of utilisation such as a new environment or traffic composition. Each FLC solution produced by the optimiser is evaluated using around 1000 correlated snapshots of the semi-dynamic simulator to assess its performance. A snapshot corresponds to a full static network evaluation. Hence, the number of iterations of the optimisation algorithm is limited.

Two approaches to FLC optimisation are considered here: Off-line and on-line optimisation. Off-line optimisation is carried out on a computer, and can utilise robust optimisation techniques such as combinatorial methods, in conjunction with a dynamic simulator for the network evaluation. On-line optimisation on the other hand aims at optimising the FLC on a real network. The two approaches are discussed presently.

16.4.1 OFF-LINE OPTIMISATION USING PARTICLE SWARM APPROACH

Off-line optimisation gives the designer the flexibility of choosing robust optimisation techniques that are too complex to be implemented on a real network. In the present case, the Particle Swarm (PS) optimisation technique has been adapted to adjust the FLC parameters. PS optimisation is a robust technique belonging to the category of *Swarm Intelligence* methods which is inspired by the social behaviour of flocking organisms [11,19]. Other optimisation techniques such as the Genetic Algorithm (see Chapter 14) could also be candidates for the optimisation task. The PS method has been chosen due to its convergence properties, namely a relative rapid improvement can be achieved in the first tens of iterations of the algorithm. The block diagram of the optimisation process is depicted in Figure 16.9.

The PS method utilises a population (or *swarm* in the PS terminology) of individuals, called *particles*, each of which represents an FLC, to probe promising regions in the optimisation space. Each particle is assigned a fixed number of neighbours. The exploration of a particle of the solution (search) space is described metaphorically in terms of a *velocity*, which is added to the current position to bring the particle to its next position (Figure 16.10). The velocity of a particle comprises three components which depend on its own cumulated best position, p_d, the best cumulated position of its neighbours, g_d, and its own (current) velocity [10]. The position of a particle, x_d, represents a vector of K parameters of an FLC, $x_d = (x_1, x_2, \ldots, x_K)$. The velocity v_d, which is added to the particle position, is a vector of K elements, $v_d = (v_1, v_2, \ldots, v_K)$. The particle evolution is given as:

$$\begin{cases} v_d \leftarrow c_1 v_d + c_{max}\text{rand}(0,1)(p_d - x_d) + c_{max}\text{rand}(0,1)(g_d - x_d) \\ x_d \leftarrow x_d + v_d \end{cases} \quad (16.11)$$

where *rand* denotes the random function. It has been shown that c_1 and c_{max} can be analytically derived from a single parameter φ [11], which has been set here to 4.14.

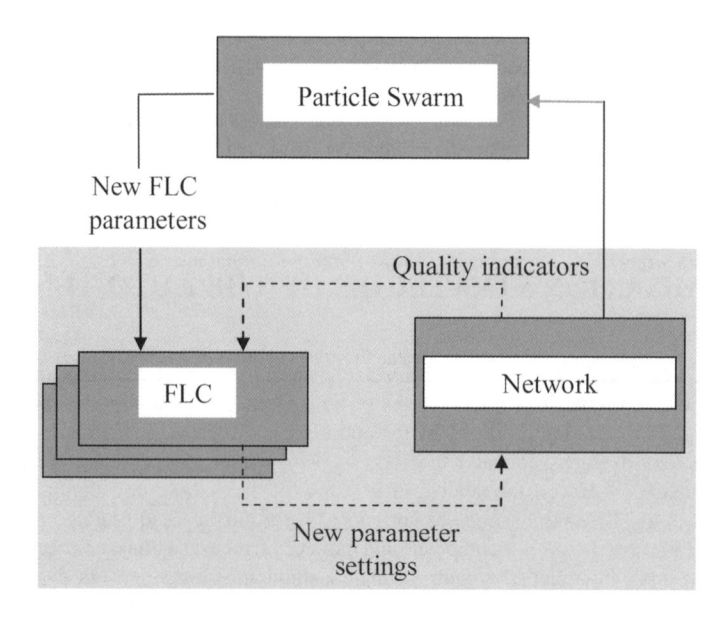

Figure 16.9 Block diagram of the FLC optimisation.

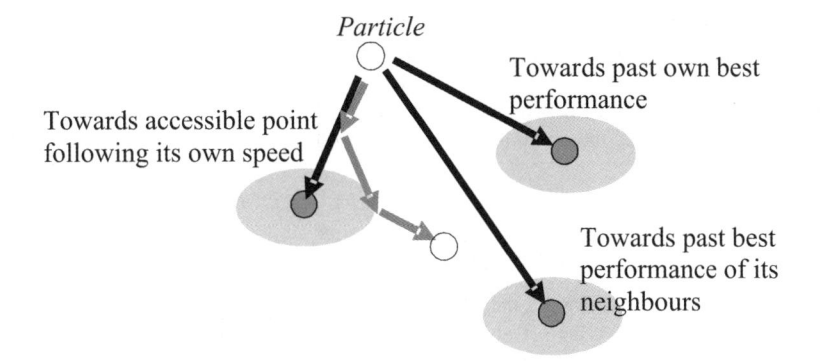

Figure 16.10 Update of a particle position in the solution space.

The PS optimises the following parameters of the FLC:

- The width of the member functions of the input indicators;
- The width of the correction member functions;
- The values of the decision matrix elements.

To reduce the optimisation complexity and time considerably, the parameters for optimisation are determined as follows: A linear variation of the member function width is assumed, namely each set of member functions (for the input indicators and output corrections) is defined by the width and position of the first and the last member functions and their number. The decision matrices are constructed as follows: each matrix element represents an index (pointer) to a correction member function. The matrix is then defined by its extreme four elements using linear interpolation. For example, for an N by M decision matrix D, the element d_{ij} is given by the closest integer value of the linear interpolation, $Interp(d_{11}, d_{1M}, d_{N1}, d_{NM})$. Typically, around 20 parameters are set for optimisation of an FLC.

Two objectives have been chosen to guide the PS optimisation of the network performance: the blocking and dropping rates. Three optimisation strategies are now described for the macrodiversity FLC optimisation.

16.4.1.1 Mono-objective Optimisation

In mono-objective optimisation, the objective (cost) function, f_{obj}, aggregates all the objectives defined by the optimisation problem. In the present case, the following cost function is used:

$$f_{obj} = \text{blocking} + \beta \cdot \text{dropping} \tag{16.12}$$

where

$$\text{blocking} = \sum_{i=1}^{N_{stations}} w_i \text{BR}_i \tag{16.13}$$

$$\text{dropping} = \sum_{i=1}^{N_{stations}} u_i \text{DR}_i \tag{16.14}$$

where *blocking* and *dropping* are the weighted sum of the filtered blocking and dropping rates, BR_i and DR_i, respectively, of all the BS in the network. The weighting coefficients w_i and u_i allow to assign more importance to problematic sectors in the optimisation process, but are taken here equal to 1. The coefficient β expresses the relative importance assigned to one objective with respect to the other. When β varies, it guides the optimisation towards different points on the *dropping–blocking* Pareto front (see Section 16.4.1.2). Typically one needs to perform several optimisations with different values of β and then choose the solution that represents the most suitable compromise between the objectives.

16.4.1.2 Multi-objective Optimisation

In multi-objective optimisation (MOP), the objectives are not aggregated into a single cost function, but instead, they are considered as distinct objectives in the optimisation process. Consider n objectives, $f_{\text{obj},i}, i = 1, \ldots, n$, each of which is a function of a set of parameters given by the vector $\mathbf{x}$. The MOP is formulated as follows:

$$\text{minimise } \mathbf{y} = \mathbf{f}_{\text{obj}}(\mathbf{x}) = \left(\mathbf{f}_{\text{obj},1}(\mathbf{x}), \mathbf{f}_{\text{obj},2}(\mathbf{x}), \ldots, \mathbf{f}_{\text{obj},n}(\mathbf{x})\right) \quad (16.15)$$

subject to

$$\begin{cases} \mathbf{x} = (x_1, x_2, \ldots, x_m) \in X \\ \mathbf{y} = (y_1, y_2, \ldots, y_n) \in Y \end{cases} \quad (16.16)$$

The vector $\mathbf{x}$ denotes the parameters which are set for possible optimisation (i.e. *add_win* or admission control threshold); X stands for the parameter space; $\mathbf{y}$ denotes the objective vector (i.e. blocking and dropping rate) and Y the objective space.

The set of optimal solutions of an MOP problem is the set of solutions for which the objective vectors cannot be improved in any dimension without degradation in another dimension. These solutions are known as non-dominated or Pareto optimal solutions. For any two vectors $\mathbf{x}$ and $\mathbf{x}'$, the Pareto dominance is defined: A parameter vector $\mathbf{x}$ dominates $\mathbf{x}'$ if and only if

$$\begin{cases} \forall i \in \{1, \ldots, n\}, f_{\text{obj},i}(\mathbf{x}) \leq f_{\text{obj},i}(\mathbf{x}') \\ \text{and} \\ \exists j \in \{1, \ldots, n\}, f_{\text{obj},j}(\mathbf{x}) < f_{\text{obj},j}(\mathbf{x}') \end{cases} \quad (16.17)$$

A solution is said to be non-dominated if there exists no solution that dominates it. The set of parameter vectors that are non-dominated within the entire search space constitute the Pareto optimal front.

In a single execution of the PS optimisation, the entire Pareto front is obtained, from which a solution is chosen according to the operator strategy. To transform a mono-objective PS algorithm into a multi-objective algorithm, one has to modify only the comparison criterion between particles, namely the relation of dominance (Equation (16.17)).

As a first example, a 21-sector (BS) network in a dense urban environment is considered with an arrival rate of 3.4 mobiles/sec, and average call duration of 100 seconds of voice service mobiles. Each solution is evaluated using 1000 correlated snapshots with a time interval of 4 seconds between adjacent snapshots. The quality indicators are filtered using a 100 second averaging filter. Every 12 seconds the FLC controls the *add_win* and *drop_win* parameters of the macrodiversity algorithm. A swarm of 20 particles is chosen with three neighbours allocated to each particle. 50 iterations of the PS algorithm are performed corresponding to 1000 solution evaluations which require several hours of computation on a workstation. Figure 16.11 compares results for the mono-objective with $\beta = 5$ and multi-objective optimisation of the FLC. Each solution corresponds to a distinct macrodiversity FLC and represents the network quality in the average dropping–blocking plane. As expected, the solutions so obtained by the MOP are better spread in the dropping–blocking plane.

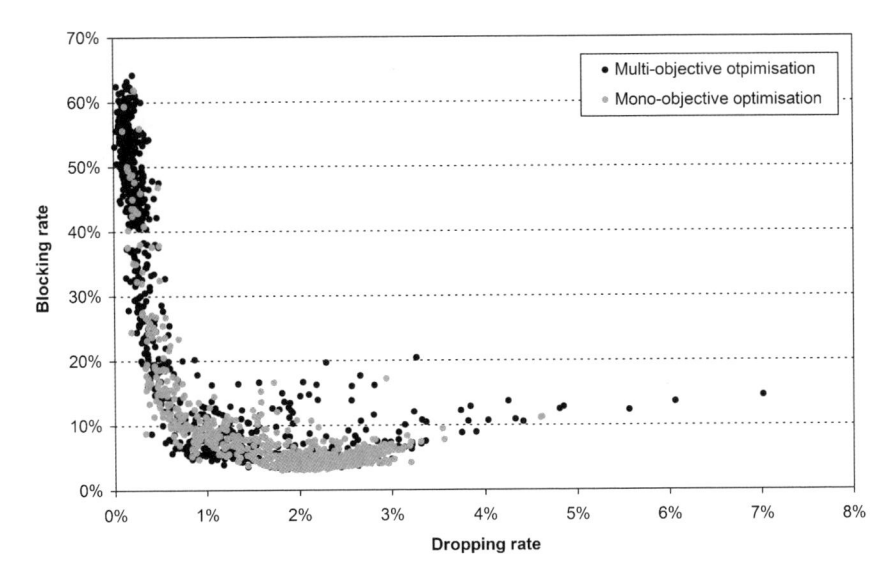

Figure 16.11 Solutions visited by the PS algorithm during the mono- and multi-objective optimisation of the FLC in the blocking–dropping plane.

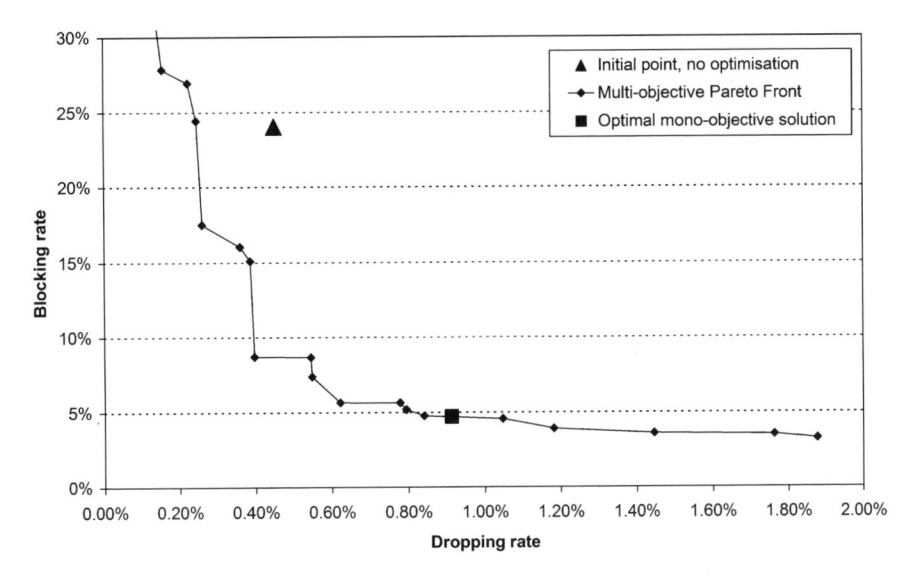

Figure 16.12 Initial (large triangle) and optimised solutions using the mono (large square)- and multi-objective optimisations. The MOP produces the entire Pareto front (line with diamonds) of optimal solutions.

Figure 16.12 presents the best results for the mono- and multi-objective optimisations. The network without auto-tuning, used as the initial point in the two optimisations, is presented by the black triangle. The MOP produces the Pareto front of optimal solutions (black diamond) that dominates the initial solution. The mono-objective optimisation solution (black square) is obtained using $\beta = 5$ in Equation (16.12) and is situated on the MOP Pareto front. This solution does not dominate the initial

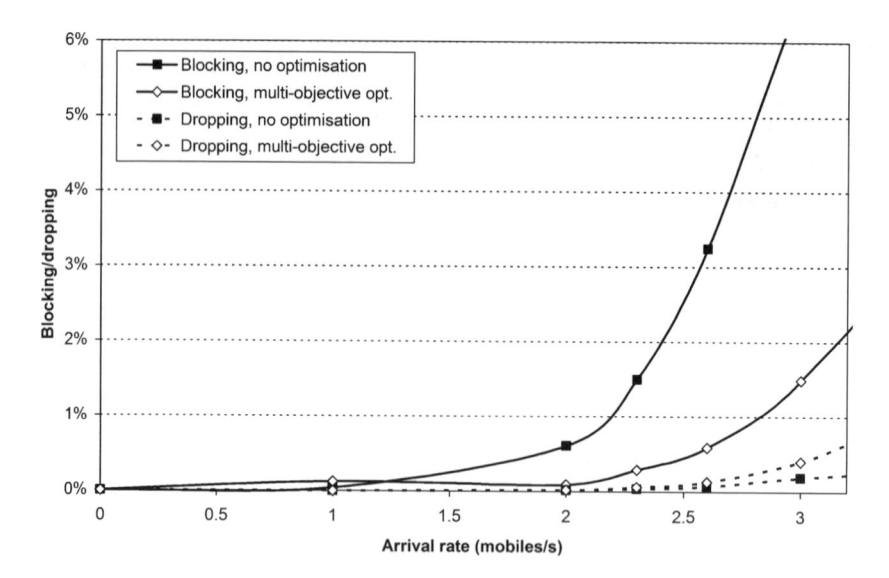

Figure 16.13 Blocking (continuous lines) and dropping (dashed lines) rates as a function of mobile arrival rate for the auto-tuned network with MOP (diamonds) and for the network without auto-tuning (squares).

solution but presents a better compromise between blocking and dropping rates, namely it gives a smaller cost function value.

To test the robustness of the FLC solution, the auto-tuning is performed for different traffic levels, by varying the arrival rate of the mobiles. Figure 16.13 presents the results for dropping and blocking rates for the initial and the optimised solutions using the MOP PS algorithm. For 2% of blocking rate for example (which is a typical target for QoS), an increase of the arrival rate from 2.4 to 3.2 mobiles/sec is brought about by the MOP optimisation which represents a capacity gain of 33%. The capacity increase is accompanied by a negligible degradation in the dropping rate.

16.4.1.3 Adaptive Optimisation

The PS algorithm is characterised by three parameters that should be fixed, namely the swarm size, the number of neighbours for each particle and the parameters c_1 and c_{max} in Equation (16.11) (or equivalently the parameter c). For each new problem one has to determine the PS parameters for which the algorithm is efficient. An adaptive variant of the PS algorithm based on the TRIBES program has been developed [20,21] that does not require any parameters in the algorithm, and adapts itself during the optimisation problem. The basic ideas of the adaptive PS algorithm are summarised presently.

In the TRIBES program the particles are gathered into tribes which constitute groups of particles. In each tribe, all the particles are connected, namely each particle is a neighbour of all the other particles. The tribes can find local minima and should be able to communicate with each other to pass on information. Hence, loose connection remains between the tribes by means of neighbourhood connections between certain particles. The quality of the tribes is defined and continuously evaluated, and according to it a tribe can generate or remove particles. The strategies used for moving in the parameter space (*the velocity*) are based on different probability distributions that are chosen as a function of the recent particle performance [21].

As in the standard PS algorithm, the adaptive PS can be used in both mono- and multi-objective versions. Figure 16.14 compares the solutions constituting the Pareto front using the standard (square)

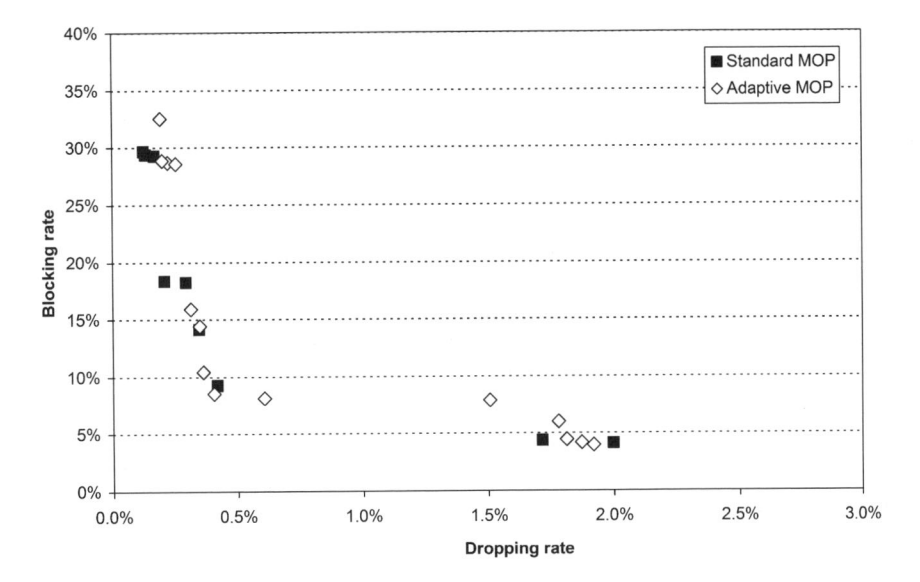

Figure 16.14 Solutions comprising the Pareto fronts for the standard (diamonds) and adaptive (squares) MOP PS.

and the adaptive (diamond) MOP algorithms. The results obtained using the two approaches are equivalent.

The FLC has been optimised for a small size network of 21 BS. Scalability tests have shown that the FLC remains effective when used with a large network size and for similar environment [17]. Finally it is noted that the auto-tuning process can be carried out simultaneously for more than one RRM parameter. In [10], the joint auto-tuning of admission control and macrodiversity RRM parameters has been studied.

16.4.2 ON-LINE OPTIMISATION USING REINFORCEMENT LEARNING

On-line optimisation aims at optimising and adapting the FLC while it is operating in the network, namely during the process of auto-tuning. This section presents the Reinforcement Learning (RL) [12] and its fuzzy Q-learning (FQL) implementation [22] to perform on-line optimisation. RL is often used in robotics and in control applications in engineering. It is applied here to the load based model for macrodiversity auto-tuning described in Section 16.3.4. In RL, an agent (an FLC in the present context) evolves while analysing consequences of its actions, thanks to a scalar signal (function), the reinforcement, given by the environment (i.e. UMTS network). The reinforcement signal that is generally perceived in terms of reward or punishment allows the agent to modify its behaviour. The agent has the tendency to replicate actions that, in the same circumstances, led to success. The agent uses the reinforcement signal to determine a policy that maximises future rewards. The policy defines the action to be taken for each state of the system.

Figure 16.15 describes the FLC optimisation scheme using the Q-learning algorithm. At each time step n, the agent receives as input the system state vector $\mathbf{x}(n)$, $\mathbf{x}(n) = \left(\text{load}_s(n), \text{load}_{nbr,s}(n)\right)$, i.e. the BS load and the equivalent load of its neighbours, and a reinforcement value $r(n)$, and performs the action $a(n)$ (modification of the *add-* and *drop_win* parameters). The network moves to a new state $\mathbf{x}(n+1)$ and produces a new reinforcement value $r(n+1)$ that are both introduced to the agent in the next time step.

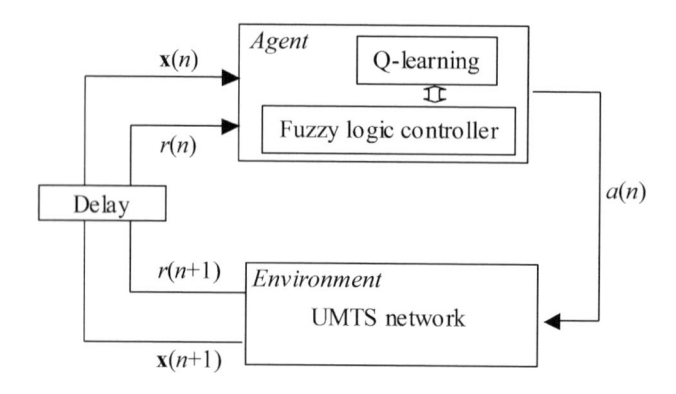

Figure 16.15 Learning by reinforcement scheme.

A policy π is defined by a mapping between the states of the system and the actions, $\pi : \mathbf{x}(n) \mapsto a(n) = \pi(\mathbf{x}(n))$. An optimal policy is derived by the Q-learning algorithm by maximising a discounted returned function $R(n)$ utilising the present and future rewards:

$$R(n) = \sum_{i=0}^{\infty} \gamma^i r(n+i) \qquad (16.18)$$

where γ is the discount factor and is taken here as 0.95. The Q-learning algorithm is based on a temporal difference equation of order 0. It is constructed using two functions that are initialised to zero and are updated using the reinforcement values: the Q-function that gives the value of a state and action, $Q_n(\mathbf{x}(n), a(n))$, and the *value* function $V_n(\mathbf{x}(n+1))$ that gives the value of a state, and is defined as:

$$V_n(\mathbf{x}(n+1)) = \max_{a \in A_{n+1}} Q_n(\mathbf{x}(n+1), a) \qquad (16.19)$$

where the maximum is calculated over all possible actions and the corresponding states that can be reached from the current state $\mathbf{x}(n)$. The update equation for the Q-function is given by:

$$Q_{n+1}(\mathbf{x}(n), a(n)) = Q_n(\mathbf{x}(n), a(n)) + \eta \{r(n) + \gamma V_n(\mathbf{x}(n+1)) - Q_n(\mathbf{x}(n), a(n))\} \qquad (16.20)$$

where η is the learning rate. The Q-learning algorithm is carried out in two phases. The first is an exploration phase which optimises the controller by exploring new region of space: at each step a new action is chosen at random with a small probability ε; and an action that maximises the corresponding *value* function (Equation (16.19)) is chosen with the probability $1-\varepsilon$. It is noted that in practical problems, one can control the choice of random actions to minimise possible deterioration of the network quality.

The second phase is the exploitation phase which utilises the optimised FLC. In each state, the controller chooses the action that maximises the *value* function (Equation (16.19)). The full Q-learning algorithm is presented in [12,13]. The reward function is defined as:

$$r(n) = (T_{block} - \text{blocking}(n)) + \beta \cdot (T_{drop} - \text{dropping}(n)) \qquad (16.21)$$

where T_{block} and T_{drop} are the target values for the blocking and dropping rates which are set to 5 % and 1 % respectively, and β is chosen here as 4.

In the Q-learning algorithm, the interaction between the agent and the environment is modelled as a *Markovian Decision Problem* which assumes a discrete state space. To handle continuous input indicators such as the BS loads, a simple interpolation procedure is introduced in the Q-learning algorithm that is known as the fuzzy Q-learning algorithm [13,22]. In this algorithm, a quality value q is assigned to each rule, namely to each set L_j of input fuzzy sets (for quality indicators) and an output fuzzy set (for the correction). The Q-value is calculated as a linear interpolation of the q-values:

$$Q(\mathbf{x}, a) = \sum_{j=1}^{m} \alpha_j(\mathbf{x}) \times q_j(L_j, O_j) \quad (16.22)$$

The sum is performed over all the m rules of the FIS. O_j is the selected output correction of the j-th rule (in the exploration or exploitation phases), and $\alpha_j(\mathbf{x})$ gives the truth value of j-th rule defined as the product of the degrees of membership of the different input fuzzy sets μ_{ij}:

$$\alpha_j(\mathbf{x}) = \prod_{i=1}^{n} \mu_{ij}(x_i) \quad (16.23)$$

It is assumed that $\sum_{j=1}^{m} \alpha_j(\mathbf{x}) = 1$. The action a is given as the linear interpolation of the output corrections weighted by their truth values:

$$a = \sum_{j=1}^{m} \alpha_j(\mathbf{x}) \times O_j \quad (16.24)$$

Next, some results for the macrodiversity auto-tuning are presented for the similar example of Section 16.4.1. A 32 sector (BS) network is simulated in a dense urban environment with voice service. 80 % of the mobiles are indoor with speed set to zero, and 20 % move at 60 km/h. Figure 16.16 presents the evolution of the blocking rate as a function of arrival rate. The optimised controller using the FQL algorithm considerably decreases the blocking rate. For 5 % blocking rate, for example, a capacity increase of 30 % is obtained. The dropping rate in this example remains negligible for all input traffic.

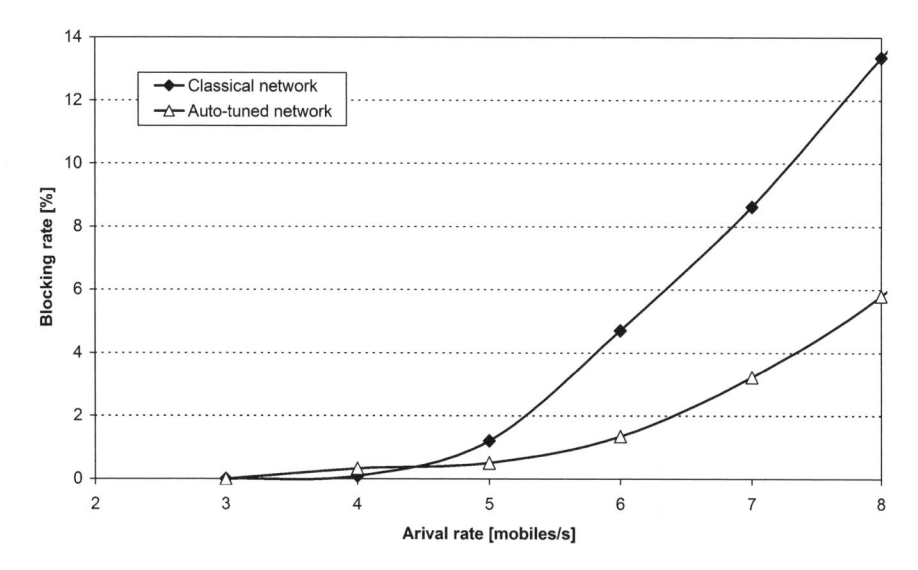

Figure 16.16 Blocking rate as a function of mobile arrival rate for the network with (triangles) and without macrodiversity auto-tuning (diamonds).

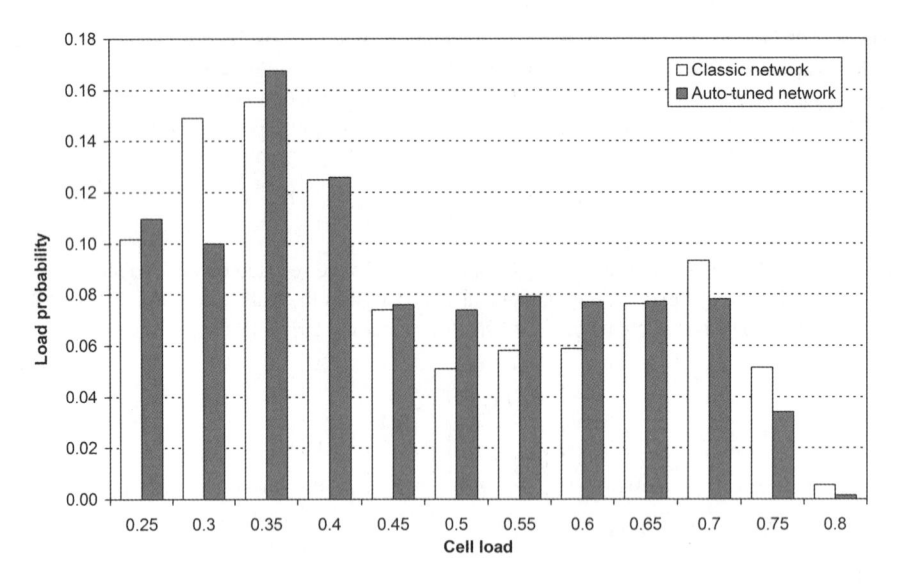

Figure 16.17 Downlink load distribution for the network without and with auto-tuning optimised using the fuzzy Q-learning algorithm.

The downlink load probability distribution for network without and with auto-tuning optimised by the FQL is depicted in Figure 16.17. One can see that for the auto-tuned network there are less BSs with very low and very high loads, and more BSs with medium load, namely the load histogram is more centred. Hence, to a certain extent, the macrodiversity auto-tuning performs traffic balancing.

Figure 16.18 shows that for high traffic level, the auto-tuning process decreases faster the percentage of mobiles in macrodiversity with respect to the network without auto-tuning with fixed *add-* and

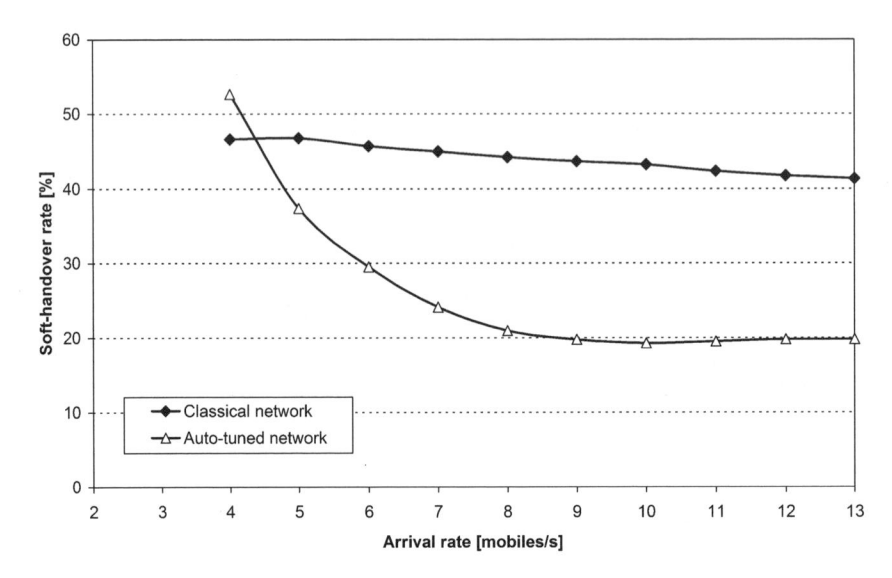

Figure 16.18 Percentage of mobiles in macrodiversity as a function of arrival rate for the network without (diamonds) and with auto-tuning (triangles).

drop_win parameters equal to 4 and 6 respectively. The auto-tuning process tries to alleviate overloaded BSs which suffer from poor QoS and allows the network to provide better capacity. For low level traffic, this tendency is reversed, namely the auto-tuned network allows more mobiles to be in macrodiversity.

16.5 CONCLUSIONS

This chapter has presented auto-tuning of RRM parameters of UMTS network based on fuzzy logic controllers, with a special focus on design and optimisation. The fuzzy logic controller (FLC) is shown to be a simple and effective framework for designing a controller that orchestrates the auto-tuning process. In the design of an FLC, engineering rules are presented in terms of a linguistic form, and are directly translated into mathematical form. The optimisation of the controllers is shown to be essential to guarantee a high quality auto-tuning process, and may be required when the conditions of utilisation of the FLC change, i.e. network environment or traffic composition. Two optimisation approaches are presented: off-line approach, performed on a computer using the Particle Swarm optimisation; and on-line approach using the Reinforcement Learning, that can directly be implemented on the network. The two optimisation approaches fully automate the FLC design process. The main task remaining for the user is to identify the best input indicators that are best suited to the parameter to be auto-tuned. The design and optimisation of the macrodiversity FLC has been described in detail. Two auto-tuning models have been presented and analysed: *MD blocking based model* and *load based model*. The latter utilises downlink load of the BS and its neighbours as input indicators and has shown to be both simple and effective. The two models bring about a capacity increase of around 30 % in a network in a dense urban environment. This example illustrates the importance of auto-tuning in UMTS engineering. Studies have shown that auto-tuning of macrodiversity and admission control can be carried out simultaneously using two controllers. The extension to auto-tuning of several RRM functionalities needs further investigation. Finally, the extension of auto-tuning in a multi-system context opens important perspectives for mobile network engineering. The challenge is to auto-tune Joint-RRM parameters that could improve cooperation between different radio access technologies.

ACKNOWLEDGEMENT

The authors of this chapter would like to thank Mr Christophe Gay for his help in implementing the optimisation algorithms in this chapter and Professor Pierre Yves Glorennec and Mr. Xavier Le Guillou for helpful discussions on Fuzzy Q-Learning.

REFERENCES

[1] P. Magnusson and J. Oom, 'An architecture for self-tuning cellular systems', *2001 IEEE/IFIP International Symposium on Integrated Network Management*, pp. 231–245, 2001.
[2] P. Gustas, P. Magnusson, J.Oom and N. Storm, 'Real-time performance monitoring and optimisation of cellular systems', *Ericsson Review*, n. 1, pp. 4–13, 2001.
[3] A. Höglund and K. Valkealahti, 'Quality-based tuning of cell downlink load target and link power maxima in WCDMA', 56th *IEEE Vehicular Technology Conference 2002-Fall*, 24–28 Sept. 2002.
[4] K. Valkealahti, A. Höglund, J. Parkkinen and A. Flanagan, 'WCDMA common pilot power control with cost function minimization', 56th *IEEE Vehicular Technology Conference 2002-Fall*, 24–28 Sept. 2002.
[5] A. Hämäläinen, K. Valkealahti, A. Höglund and J. Laakso, 'Auto-tuning of service-specific requirement of received $E_b N_0$ in WCDMA', 56th *IEEE Vehicular Technology Conference 2002-Fall*, 24–28 Sept. 2002.

[6] P. Stuckmann, Z. Altman, H. Dubreil, A. Ortega, R. Barco, M. Toril, M. Fernandez, M. Bary, S. McGrath, G. Blyth, P. Saidha and L. M. Nielsen, 'The EUREKA GANDALF project: monitoring and self-tuning techniques for heterogeneous radio access networks', *IEEE Vehicular Technology Conference 2005*, Stockholm, Sweden, 29 May–1 June 2005.

[7] W. Zhang, 'Handover decision using fuzzy MADM in heterogeneous networks', *IEEE Wireless Communications and Networking Conference 2004 (WCNC 2004)*, 21–25 March 2004, Atlanta, USA.

[8] D. Diankov, H. Hellendoorn and M. Reinfrank, *'An Introduction to Fuzzy Control'*, Springer-Verlag, 2nd ed., 1996.

[9] J. Ye, X. Shen, J.W. Mark, 'Call admission control in wideband CDMA cellular networks by using fuzzy logic', *IEEE Transaction on Mobile Computing*, vol. 4, pp. 129–141, March–April 2005.

[10] H. Dubreil, Z. Altman, V. Diascorn, J.M. Picard, and M. Clerc, 'Particle Swarm optimisation of fuzzy logic controller for high quality RRM auto-tuning of UMTS networks', *IEEE International Symposium VTC 2005*, Stockholm, Sweden, 29 May–1 June, 2005.

[11] M. Clerc and J. Kennedy, 'The particle swarm: Explosion, stability, and convergence in a multi-dimensional complex space', *IEEE Tran. Evol. Comput.*, vol. 6, pp. 58–73, Feb. 2002.

[12] P.Y. Glorennec, 'Reinforcement learning: an overview', Proceedings of the ESIT 2000 conference, Aachen, Germany, Sept. 2000.

[13] R. Nasri, Z. Altman and H. Dubreil, 'Fuzzy Q-learning based automatic management of macro-diversity algorithm in UMTS networks', to be published in Annals of Telecommunications.

[14] 3GPP TR 25.922, 'Radio resource management strategies', Release 6, V6.0.1, May 2005.

[15] J. Laiho, A. Wacker and T. Novosad, *Radio Network Planning and Optimisation for UMTS*, John Wiley & Sons, Ltd/Inc., England, 2002.

[16] L.A. Zadeh, 'Fuzzy sets', *Information and Control*, vol. 8, pp. 338–353, 1965.

[17] L. Jouffe, 'Fuzzy inference system learning by reinforcement methods', *IEEE Transactions on Systems, Man, and Cybernetics*, vol. 28, pp. 338–355, Aug. 1998.

[18] S. Ben Jamaa, H. Dubreil, Z. Altman and A. Ortega, 'Quality indicator matrices and their contribution to WCDMA network design', *IEEE Trans. on Vehicular Technology*, pp. 1114–1121, May 2005.

[19] R. C. Eberhart and J. Kennedy, 'A new optimiser using partical swarm theory', 6th *Symp. Micro Machine and Human Science*, Nagoya, Japan, pp. 34–44, 1995.

[20] G. C. Onwubolu, 'TRIBES application to the flow shop scheduling problem', *New Optimisation Techniques in Engineering*. Heidelberg, Germany: Springer, pp. 517–536, 2004.

[21] M. Clerc, *L'optimisation par essaims particulaires. Versions paramétriques et adaptatives*. Hermès Science, 2005 (in French).

[22] P.Y. Glorennec and L. Jouffe, 'Fuzzy Q-learning', 6th *IEEE International Conference on Fuzzy Systems*, 1–5 July, 1997.

17

UTRAN Transmission Infrastructure Planning and Optimisation

Karsten Erlebach, Zbigniew Jóskiewicz and Marcin Ney

17.1 INTRODUCTION

The fundamental changes between 2G to 3G transmission networks and their resulting risks and chances are often overlooked in the first phase of a 3G network deployment. Most of the current mobile network operators originate from a classic telecom background and had only little experience and affinity to implement direct IP-vendors during the initial 3G rollout, which was often before 2004. On the other side, nearly all IP- and ADSL-operators do not comply with the same level of quality, which classic telecoms were used to. The reason is based on the different type of services. While data services, such as download, e-mails or surfing, can easily tolerate end-to-end outages of ten or more seconds, this is impossible for the performance of voice services, which has for a long time been the core product of mobile operators. 3G experiences a merge of both main service categories into one access network, which implies two main requirements for an access transmission network:

1. Quality to transport conversational services like voice and video telephony with the same availability standards such as a 2G telecom provider.
2. Cost efficiency to compete with IP-companies in offering high bandwidth services to customers.

Currently, most of the revenues are still being generated by voice services, which means that high availability standards still have to be ensured. With data services emerging, this strategy will change, particularly if a significant portion of them was not conversational or streaming. Furthermore, the option of throughput asymmetries in forward and reverse traffic lead to new challenges, since Plesiochronous Digital Hierarchy (PDH) and Synchronous Digital Hierarchy (SDH) transmission systems show here significant constraints.

Understanding UMTS Radio Network Modelling, Planning and Automated Optimisation Edited by Maciej J. Nawrocki, Mischa Dohler and A. Hamid Aghvami © 2006 John Wiley & Sons, Ltd

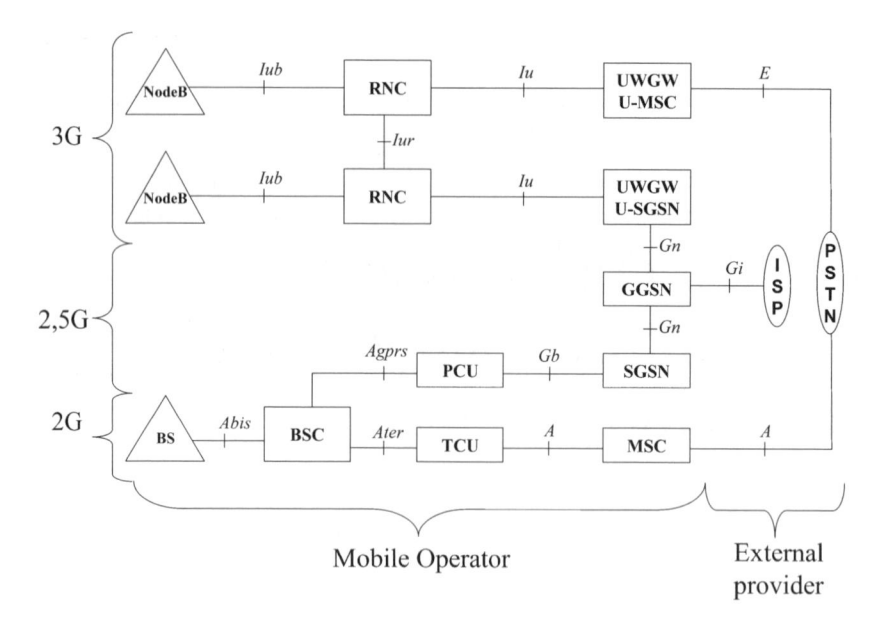

Figure 17.1 UTRAN overview.

17.1.1 SHORT UTRAN OVERVIEW

The UTRAN transport network has been standardised by the UMTS forum, 3GPP and 3GPPiP. Figure 17.1, derived from [1], [2] and [3], shows the UTRAN network nodes and interfaces and gives an example of the integration of 2G and 2.5G network nodes into the UTRAN access transmission structure.

The abbreviations can occasionally differ throughout the vendors, but the principles are similar: While 2G transmission networks are designed to transport all mobile-originated payload back to the MSC or SGSN this does not necessarily apply to 3G transport networks. Compared to a 2G BSC, a 3G RNC can execute the mobility management as well as the routing of traffic within an access network without routing it via an UMTS Wireless Gateway (UWGW), and UMTS Main Switch controller (U-MSC) or an UMTS Serving GPRS Server Node (U-SGSN). If call control features like authentication, user registration (HLR, VLR) and billing are implemented on RNC-level, then this setup allows a complete separation between the connectivity plane and the control plane.

17.1.2 REQUIREMENTS FOR UTRAN TRANSMISSION INFRASTRUCTURE

All mobile operators currently experience the paradigm change of a complete circuit switched network topology – mostly based on PDH and SDH-physicals – to a packet-oriented topology like Ethernet/IP routing. Since the very most of the mobile operators have implemented their transmission infrastructure based on circuit switched topologies, most of them use packet emulation technologies to transport packet data, in particular Frame Relay for 2.5G and ATM AAL2 or AAL5 for 3G data traffic. Compared to direct IP router technologies like Ethernet these emulators produce a significant transport overhead [4].

This has economical consequences, when high user data rates have to be transported. On the other hand, packet technology offer a great variety of routing options allowing a more flexible and thus economical utilisation and scalability of physical infrastructure. A variety of wireless transport systems are available in several markets. Beside conventional PDH and SDH microwave links as well as fibres, new wireless systems like

- PMP-systems based on ATM or IP cross connects (see Section 17.4.3),
- WiMAX systems (see Section 17.4.4), and
- Wireless Gateways from various vendors

exist now on the market. If allowed by the national regulator, this offers new, better performance and cost-efficient opportunities to cope with the requirements.

17.1.2.1 Technical Requirements

Compared to circuit switched traffic, a guarantee of service quality for dedicated sites or services is difficult, because neither ATM nor IP evaluates its payload on transport level (layer 2). The ATM Adaptation layers (AAL1, 2, 3/4 and 5) and service classes (CBR, VBR, UBR, ABR) can prioritise services on the session level (layer 3), provided that the transport of such services was enabled end-to-end on dedicated Virtual Channels (VC). If not, the only option is the encapsulation of session bits on OSI-layer 4 or higher to ensure service quality [5]. Beside the service quality and network availability, further requirements have to be met:

- Accessibility and interoperability with third-party carriers, particularly ISPs. Particularly the establishment of virtual home environments incorporating high data rate services via two or more carriers requires optimised UNIs (User Network Interfaces) and quality enablers.
- Insurance of traffic contracts and policing, if required by third-party carriers, corporate customers or companies.
- Interoperability with 2G and 2.5G user equipment as well as with all 3G user equipment types defined by IMT-2000.
- Guarantee of maximum delay standards for interactive and streaming classes. Most of the vendors tolerate a net transmission delay of 10 ms between Node B and RNC depending on the delay budgets of its components. Since ATM introduces significant packetisation and transport delays, and IP v4 can produce remarkable defragmentation delays, the limitations can be severe.
- Assurance of ITU synchronisation standards, particularly [6–8]. While the synchronisation of all 2G and UTRAN network elements via SDH and PDH can usually be established, there are limitations to a number of IP routers.
- Insurance of minimised transmission overheads as well as retransmission on packet transport. This includes the establishment of channels that enable bandwidth-on-demand. This aspect has become a critical issue for the initial 3G transmission network products. Some UTRAN-vendors keep code capacity allocated during a session even when no user traffic is present. This leads to a significant transmission overhead, particularly during web page changes or interactive gaming. Other vendors re-establish instead a session when new data is transmitted. This leads to significant delay or even drops on a congested network, when for example the CAC (Call Admission Control) delays a PVC (Permanent Virtual Circuit) or SPVC (Semi-Permanent Virtual Circuit) re-establishment.

With the growing complexity of services, user equipment and data throughput the requirements to meet technical KPIs will rise drastically. However, the economical objectives will become increasingly important in the future as well.

17.1.2.2 Economical Requirements

There are many requirements for 3G networks that are not driven by technical but by economical needs. Mobile operators will enter a direct and stiff competition with Internet service providers. Presumably 50 % of the mobile operators' revenues within the next 5 years will not be generated by airtime but by portal content. ASPs will be in the position to choose and prioritise between mobile and fixed carriers. However, capacity roadmaps and price erosion of IP-vendors occur significantly faster than that of telecom vendors, putting IP-carriers in an economical advantage. The recent growth of WLAN hotspot technology is just one example for the potential of wireless IP carriers.

Particularly the economic circumstances will be the main driver for the backhaul network strategy. Unlike the radio access network, these strategies cannot be standardised but will be unique to each operator. This has the following reasons:

- The access transmission network has no direct customer visibility. For this reason it does not have to be designed at all in a plug-and-play manner like air interface user equipments. In fact, each of the ATM vendors differ in the detailed form in which they have configured VCs, Interfaces or PNNI (Private Network-to-Network Interface).
- Each country has a different leased line carrier structure offering various products and prices. With 3G nodes being able to operate layer-1 products like PDH and SDH as well as higher layer-products like Frame Relay, ATM or Ethernet, the options are tremendous.

Beside the economical constraints and opportunities defined by the market and the equipment vendors, some more requirements defined by the operator's product marketing have to be met:

- An entirely different billing and tariff structure. Billing can be applied on time, volume, number of sessions or interactions (like number of web pages) as well as on various type of content. This differs significantly from 2G applications where one call data record retrieved from the MSC is sufficient for most service applications.
- Enhanced options for resilience and dynamic routing. Conventional static $1+0$ protection of circuit switched connections is cost intensive compared to a dynamic routed meshed network.
- The ability to route sessions via the core network edges (like the Iur) instead via the switch. Since all UTRAN elements including Node B are equipped with ATM or IP functionalities the packetisation, transport and routing of user payload of and to virtually each network node is a scenario. And if traffic does not have to be routed to the switch, then the backhaul network can be relieved.

The development of decentralised routing and billing in the mobile access networks is comparable to the migration of IT-mainframe systems to Client/Server architectures. The result is a better scalability of the required network infrastructure and cost.

17.2 PROTOCOL SOLUTIONS FOR UTRAN TRANSMISSION INFRASTRUCTURE

Currently, the 3G mobile transmission protocol solutions are almost all based on ATM. End-to-End-IP, if configured, is very often operated on ATM AAL5 or on Frame Relay. Section 17.2.3 will describe the way to IP-network protocol solutions, which is seen as the path to the future mobile network architectures.

17.2.1 MAIN CONSIDERATIONS FOR ATM LAYER PROTOCOLS IN CURRENT 3G NETWORKS

ATM was developed by the ITU Study Group XVIII Standardization Sector as a Wideband-ISDN-standard in the late 1980s [9]. The fact that it became by far the most used layer-2 technology by 3G-operators was based on its ability to establish stringent quality standards for circuit switched as well as for interactive and background packet services. ATM Packets have been standardised as 53-byte packets, with 7 bytes required for the header [10] as shown in Figure 17.2.

GFC/VPI	VPI
VPI	VCI
VCI	
VCI	PTI	C
HEC	

Header 5 bytes

Information
48 bytes

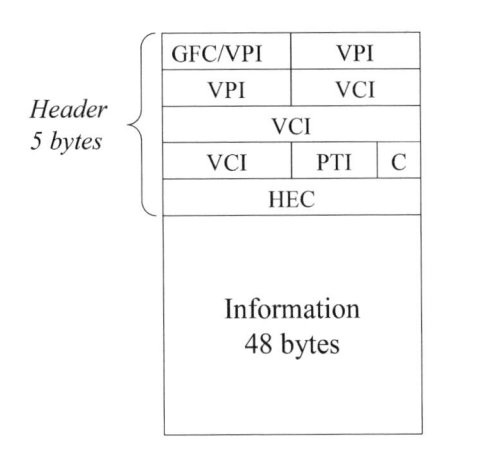

GFC: Generic Flow Control
VPI: Virtual Path Identifier
VCI: Virtual Channel Identifier
PTI: Payload Type Identifier
C: Cell Loss Priority
HEC: Head Error Control

Figure 17.2 ATM cell structure.

Generic Flow Control (GFC) is only used in UNI, and supports the user configuration. The Virtual Path Identifier (VPI) field contains the address of the virtual path and incorporates usually several virtual channels. The channels are addressed in the Virtual Channel Identifier (VCI). The Payload Type Identifier (PTI) describes the type of payload. Payload can be distinguished between user and network data. The content of the CLP-field (Cell Loss Priority) triggers whether a cell will be dropped in case of congestion. Cells with CLP = 1 will be dropped earlier than cells with CLP = 0. The Head Error Control (HEC) field is required for error control and correction of the header data. With HEC, an ATM receive node can synchronise to the ATM cell offset. For error detection, a CRC-mechanism is used, which is based on the division of the header field by the polynomial $x^8 + x^2 + x + 1$.

Similar to all other interfaces, Iub traffic can be divided into traffic dedicated to user services (like Circuit Switched and Packed Switched Traffic), traffic dedicated to Node B signalling like C-NBAP (Common Node B Application Protocol) and D-NBAP (Dedicated Node B Application Protocol) and traffic dedicated for operation and maintenance (OAM). In ATM, each traffic type can be assigned to a VC (Virtual Channel) on a dedicated VCI. The VC is then mapped on a VP (Virtual Path). This assignment is done in the ATM-CC-Unit of the Node B or the RNC. VCs can be configured as permanent connections on a fixed route (PVC) or as switched connection (SPVC). SPVCs appear as permanent connections if seen from the ingress and egress port of establishment. Between the ingress and egress port, the route can be established dynamically depending on traffic load and congestion within the network. In general, it is desirable to have the option of a complete dynamic routing between the Node B and the RNC. However, if the ATM vendor differs from the 3G-vendor, the establishment of an end-to-end dynamic network is often impossible due to missing vendor interoperability to establish a PNNI. In such cases, some sections have to be designed as PVC originating on UNI.

To handle load and overbooking on an ATM trunk during connection establishment, the first measure is the configuration of a suitable CAC to each PVC or SPVC. During a session, the efficient utilisation of the ATM trunk can be ensured by the design of a suitable 'transmit-corridor' by applying a peak, minimum or sustainable cell rate. Example: An E1-Trunk consists out of 30 transmission channels (64 kbps) for payload plus 2 for transmission management. The peak information rate (PIR) is then 1920 kbps, which corresponds to a peak cell rate of PIR (in Byte/s) /53 byte, which means 4.528 cells/s. When the E1 resource shall be used for up to five Node Bs (overbooking), the minimum cell rate – which has to be guaranteed by the ATM-CC – is then 9.05 cells/s.

The prioritisation of PVC towards other PVC can be controlled by the application of a traffic class [11]. The traffic classes defined by ATM forum [10] are:

- Constant Bit Rate (CBR), which is real circuit emulation. In this category, the ATM network receives a continuous stream of bits. It usually implies a very low delay and very low delay variation.
- Real-Time Variable Bit Rate (RT-VBR). This service class has very tight bounds on delay but might not have very tight bounds on cell loss. There are certain kinds of traffic such that if the delay gets too large it might not deliver it at all.
- Non-Real-Time Variable Bit Rate (NRT-VBR) is the complement of RT-VBR. This class puts a low priority in the delay but focuses into not losing cells instead. E-mail service is an example of this type of traffic.
- Unspecified Bit Rate (UBR) is kind of "best effort" method, since UBR has no guarantees.
- Available Bit Rate (ABR) involves flow control. The goal here is to have a very low cell loss within the network.

With this set of options, the dimensioning of a well-performing and cost-efficient transport network is large. However, the first UTRAN vendor releases show severe limitations with regard to the described ATM functionalities, particularly a limited number of traffic classes, which cannot be translated from one ATM vendor to another. Additionally, the separate end-to-end transmission of different services (conversational, interactive, etc.) has not yet been established by several vendors or vendor interoperability, which means that most of the traffic has to be transported in one or two VCs (like one for circuit switched and one for packet switched traffic), and not in VCs dedicated to a specific access bearer service.

Beside the design limitations, several other constraints with regard to performance retrieval and optimisation exist. This will be described in the next sections.

17.2.1.1 ATM Adaptation Layers for Different 3G Bearer Services

The task for the ATM Adaptation Layer (AAL) is the adaptation of the payload data from upper layers to the format of the ATM cell. The adaptation happens in relation to the required services. Furthermore, it reassembles the payload stream at the ATM egress node and equalises cell delay variation. To satisfy the various requirements of the different services, four layer types have been created, namely AAL1, AAL2, AAL3/4 and AAL5.

All AAL are divided into the CS (Convergence Sub-layer) and SAR (Segmentation and Reassembly Sub-layer) [9]. The SAR assembles the upper layer data into segments optimised for the layer type size. The CS performs error correction, re-synchronisation as well as error checks.

AAL1. This standardised protocol is used for the transport of time critical applications and conversational services with a constant bit rate such as voice or video telephony. In addition, it is used for the emulation of circuits like E1, T1 or DS0.

The AAL1-SAR requires one of the 48 payload bytes therefore reducing the user payload to 47. The SAR header byte consists of a 4-bit sequence number and a 4-bit SNP, which generates a CRC-3 checksum. The AAL1-SAR adds error detection by calculating 4 bytes out of 124 data bytes and adding them into the cell stream. The entire 128-byte load is then reassembled as depicted in the Figure 17.3 [12]. This means that at least 128 cells have to be buffered before reassembling and transmission can begin [10].

AAL2. This type has a particular use in mobile applications since it is able to transport time critical services that incorporate a variable bit rate. AAL2, as defined in ITU-T-I366, supports CBR, RT-VBR and NRT-VBR. The principle of the AAL2 payload is illustrated in the Figure 17.4 [13]:

Each VC is assigned a VCI, which is stored in the CID-byte (Channel-ID-Byte, which indicates the Virtual Channel) after the ATM header. With VCI-Nr 0-7 reserved, it means that 248

Figure 17.3 AAL1 structure.

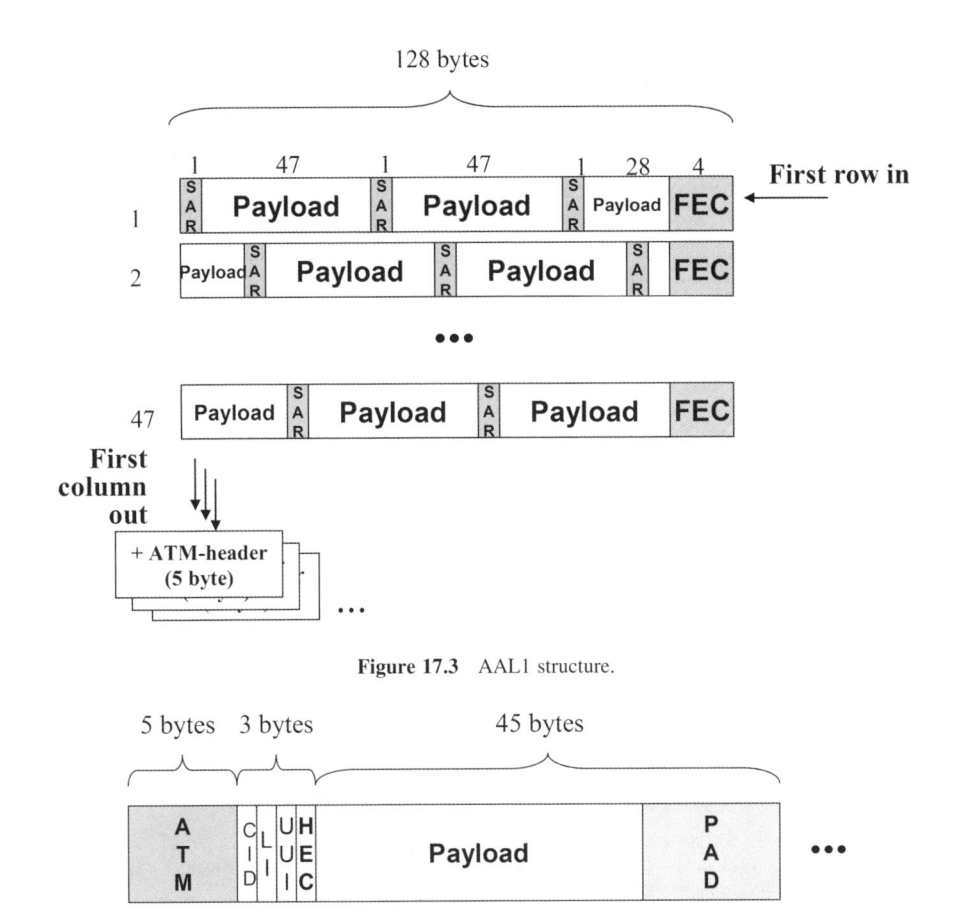

Figure 17.4 AAL2 structure.

VCI can be mapped to one VP. Two more bytes of the ATM cell are needed for further AAL2 functions:

- LI (Length Indicator): indicates the length of the payload (0–45 bytes). If the payload is smaller than 45 bytes, the rest of the cell will be filled with padding octets;
- UUI: (User-to-User-Indication). This indicator provides an information link between CPS (Common Part Sublayer) and SSCS (Service Specific Convergence Sublayer);
- HEC: Header Error Control (see FER in AAL1 description).

As a difference to AAL1, which supports only CBR, the maximum delay and minimum throughput of a channel can be configured by setting a MCR and a SCR as well as limits for the cell transfer delay and variation.

AAL3/4. The main function of the AAL3/4 type is the adaptation of connection-oriented and connectionless data transfer to the ATM cell format. Its main area of application is the connection of LANs and ATM transmission.

In AAL 3/4, the protocol first inserts error-checking functions before and after the original data. Then the information is segmented into 44-byte packets. The cell payload includes two bytes of header and two bytes of trailer so this whole construct is exactly 48 bytes. There is a CRC check on each cell to

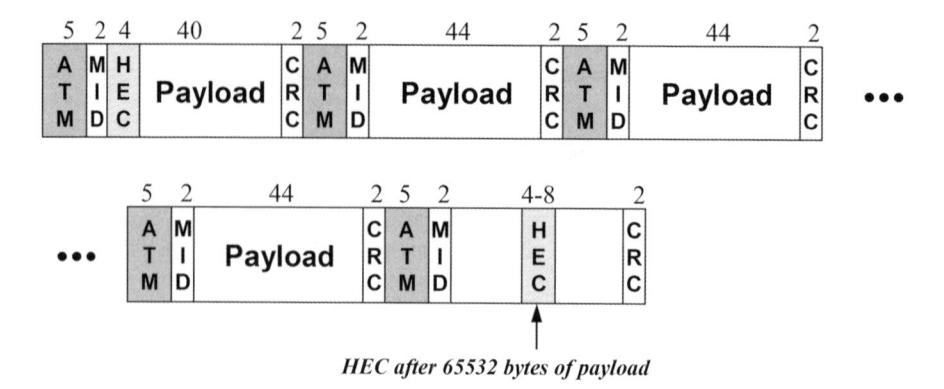

Figure 17.5 AAL3/4 structure.

check for bit errors as well as Message Identifier (MID). The MID allows multiplexing and interleaving of large packets into a single virtual channel. This is useful in a context where the cost of a connection is very expensive since it would help to guarantee high utilisation of that connection [14] (Figure 17.5).

Cyclic redundancy check is performed in each ATM cell, while HEC is performed after 65532 bytes of payload.

AAL5. The AAL5 type has been created for special requirements of packet-oriented applications. AAL5 is a downgraded version of AAL3/4 but with a lower overhead. Multiple conversations may not be interleaved in a given connection. Here the CRC is appended to the end and the padding is such that this whole construct is exactly an integral number of 48-byte chunks. This fits exactly into an integral number of cells, so the construct is broken up into 48-byte packets and put into cells [15]. Figure 17.6 shows AAL5 structure.

Table 17.1 gives a summary of all AAL and traffic quality options available, and assigns them to the applications and the bearer classes defined by the ATM Forum [16].

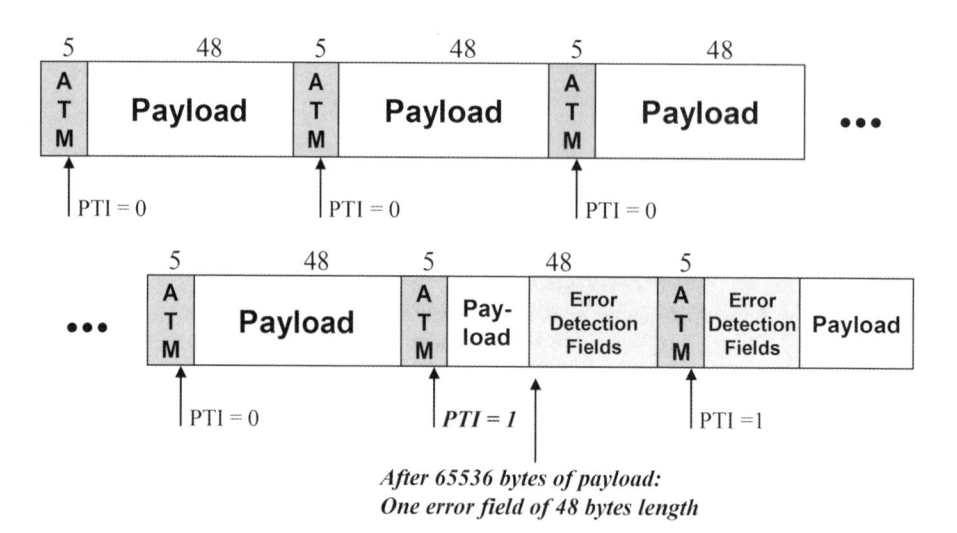

Figure 17.6 AAL5 structure.

Table 17.1 Summary of AAL types and qualities.

Service	CBR	rt-CBR	Nrt-VBR Connection oriented	Nrt-VBR Non-connection oriented	UBR	ABR
Bearer Class	Class A	Class B	Class C	Class D	Class X	Class Y
Applications	Voice and Clear Channel	Packet, video and voice	DATA			
Connection Mode	Connection oriented			Connection less	Connection oriented	
Bit rate	Constant	Variable				
Timing	Required		Not required			
Services	Private Line	None	Frame Relay	SMDS	Raw Cell	
AAL type	1	2	3/4 and 5	3/4	Any	3/4 and 5

17.2.1.2 ATM Key Performance Indicators

ITU has defined a set of standards for OSI layer 1, 2 and 3-performance retrieval on ATM and IP-systems, which are illustrated in Figure 17.7.

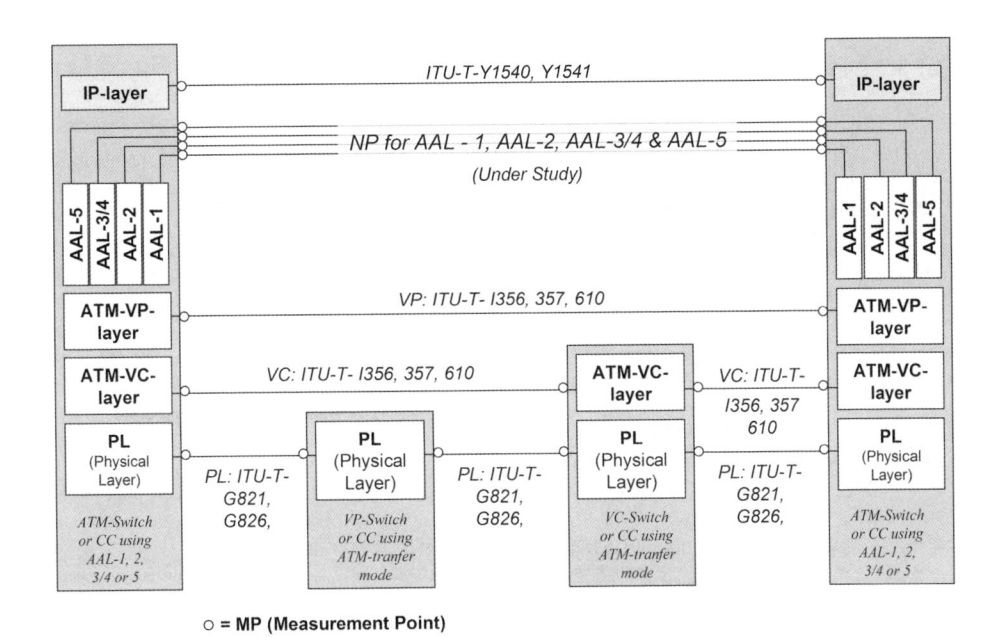

Figure 17.7 ITU Performance Measurement Regulations for physical layer, ATM and IP.

The main specifications for ATM performance are defined in [17–19]. The principle is comparable to the definition of unavailable and errored seconds, which are for example defined for SDH in [20]. For ATM systems ITU substitutes seconds by 53-byte ATM cells. Between two MPs (Measurement Points) six cell events are defined as in Table 17.2.

The second main criterion beside Cell Error is the transmission delay as well as the delay variation of the ATM cell stream [21]. Cell Transfer Delay (CTD) and Cell Delay Variation (CDV) are illustrated in Figure 17.8.

The mean Cell Transfer Delay (CTD_{mean}), 1-Point Cell Delay Variation at MP2 (CDV_{Cell}^{1Point}), and 2-point Cell Delay Variation between MP1-MP2 (CDV_{Cell}^{2Point}) are defined as follows:

$$CTD_{mean} = \frac{1}{n} \cdot \sum_{i=1}^{n} \Delta t_i,$$

$$CDV_{Cell}^{1Point} = t_{cell} - D_{ref},$$

$$CDV_{Cell}^{2Point} = \Delta t_{cell} - D_{12},$$

where:
D_{ref} – reference point of time that an ATM cell should reach the point MP2,
t_{cell} – point of time that the specific ATM cell reaches Time reaches MP2,
D_{12} – reference delay for an ATM cell between MP1 and MP2,
Δt_{cell} – delay that the specific ATM cell has required between MP1 and MP2.

ITU gives recommendations for KPI thresholds, but each operator will set its own values depending on its hardware, network or Leased Line infrastructure, products and Service Level Agreements (SLA).

17.2.1.3 New Challenges and Opportunities Using ATM Transmission

A real challenge for ATM transmission is the correct and quick retrieval of its end-to-end performance. Also, the detection of traffic congestion spots in meshed networks, particularly when overbooking, IMA (Inverse Multiplexing on ATM) and PNNI are configured, can become a real challenge. Additionally, the KPIs for ATM performance have only been ITU-specified for PVC, but not yet for SPVC. The tools for optimisation are illustrated in Section 17.2.1.5.

In the classic 2G-world, fault detection and optimisation of e.g. a GSM-circuit is commonly handled on a section-by-section basis. This is not applicable for a PNNI-'cloud', because a SPVC-route cannot be mapped at all to a physical transmission system. For this reason, it is required to ensure performance and optimisation on both the physical systems and the logical network connections [9]. We shall proceed with some examples.

Usage of IMA. IMA usually groups several physical ports (e.g. E1, T1, E3) to a virtual (logical) port. This accelerates ATM packetisation and reduces the complexity to route a VC, because it is then mapped exclusively to this logical port. Likewise, all retrieval of utilisation or errored cell data is based on this virtual port. A malfunction of a port in an IMA-group is then less easy detectable, particularly when the malfunction is not generating an alarm [10].

Usage of dynamic routing in a PNNI-cloud. Most of the operators configure PNNI, because it increases the bandwidth efficiency, which reduces cost. This, however, means that an SPVC can be routed via any physical connection and route in this cloud. This requires that certain KPIs and components to be constantly monitored and optimised:

Table 17.2 Events defined by ITU-T I356 and I357.

Event	Description	KPI
STCO (Succesful Transferred Cell Outcome)	The cell has been transferred without error	
ECO (Errored Cell Outcome)	The cell header or payload has been transmitted in error	$CER = \dfrac{EC \text{ (Errored Cells)}}{STC \text{ (Succesful Transferred Cells)}}$
TCO (Tagged Cell Outcome)	The cell has been tagged due to a transmit delay	CLR (Cell Loss Ratio)
LCO (Lost Cell Outcome)	The cell has been lost	$\begin{cases} \text{High Priority cells } CLR_0 = \dfrac{LC \text{ (Lost Cells)} + TC \text{ (Tagged Cells)}}{TTC_0 \text{(Total Transferred Cells w.CLP} = 0)} \\ \text{All cells } \quad CLR_{0+1} = \dfrac{LC \text{ (Lost Cells)}}{STC_{0+1} \text{(Total Transferred Cells)}} \\ \text{Low Priority cells } CLR_1 = \dfrac{LC \text{ (Lost Cells)}}{TTC_1 \text{(Total Transferred Cells w. CLP} = 1)} \end{cases}$
MCO (Misinserted Cell Outcome)	The cell has been misinserted in the received ATM cellstream	$CMR = \dfrac{MC \text{ (Misinserted Cells)}}{\Delta T \text{(Time Interval)}}$
SECBO (Severely Errored Cell Block Outcome)	An entire block of ATM cells has been received in error	$SECBR = \dfrac{ECB \text{ (Errored Cell Blocks)}}{TTCB \text{ (Total Transferred Cell Blocks)}}$

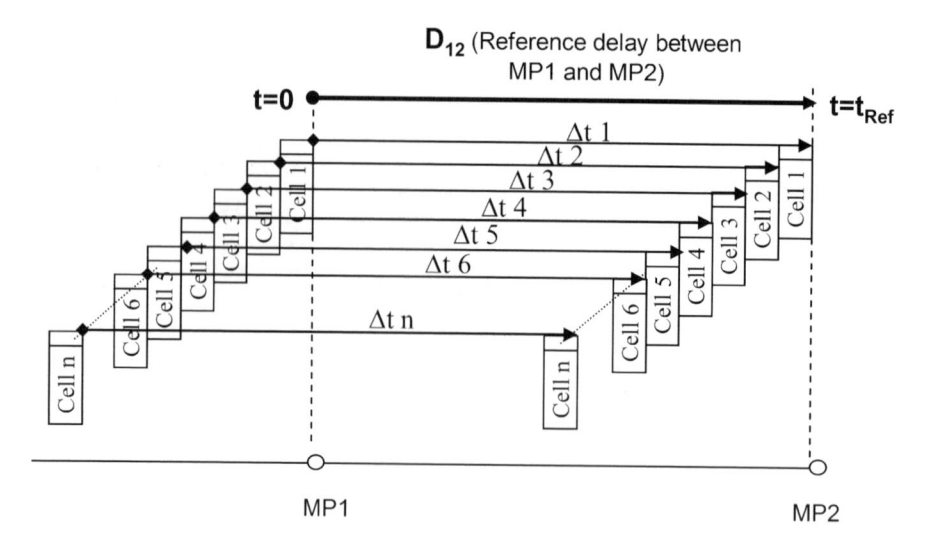

Figure 17.8 ATM transfer delay.

1. The load and cell performance of all physical ports and interfaces in the cloud.
2. The load and cell performance of the monitored connection.
3. The average length of the established SPVC-route, which determines the average cell delay.
4. The average length of an SPVC-reestablishment.

Particularly (3) and (4) can drastically increase with the load and complexity of the ATM cloud. This can have fatal consequences. If a SPVC is re-established due to congestion or physical unavailability, it does not commonly generate an alarm to the UNI. Depending on the service transported on the SPVC, this leads to call or session drops or even service resets. This is of particular concern in conversational services [9].

Overbooking. ATM allows overbooking of physical trunks with PVC or SPVC peak cell rates. This means that if a PVC has been established but the traffic volume decreases, this volume can be filled by traffic of other PVC. As a consequence, PVC can transmit a higher volume than its configured Peak Cell Rate (PCR) if the physical environment allows it, and if traffic policing and shaping are disabled. It should always be remembered that PCR is the criteria for the Connection Admission Control during PVC establishment, but not during its entire transmission [9].

Another new challenge and opportunity is the emulation of PDH or SDH circuits on OSI-Layer 3 (Circuit Emulation Services, CES). The design for CES transport will be described in Section 17.2.1.6.

17.2.1.4 Dimensioning Link Load and Occupancy in ATM Networks Using PNNI

Dimensioning the load of ATM networks carrying mobile application traffic depends on the service mix. A safe approach to create a rule of thumb is to split all services into two main services:

1. Conversational services like voice or video telephony as well as data services that require CBR quality.
2. All remaining services, which are more background based or interactive.

For both type of services, 2G and 2.5G have developed dimensioning methods that also can be applied on a service mixed port or connection: conversational services have been successfully dimensioned by the Erlang B formula for a long time. If several service bit rates were applied, the usage of its multirate derivation is applicable [22].

For background and interactive services, the dimensioning experience of mobile data networks, particularly WAP and GPRS, can be applied. Most vendors here recommend a maximum GPRS-Bearer (Gb) load (see Figure 17.9) of not more then 80 % of the accommodating Gb-capacity, because of the burstiness of data traffic. Taking into account the session and link layer overhead (DLCI-framing, PDP-context, etc.), this means that the user payload should not exceed more than 55 % of the Gb-capacity. However, it has to be considered that the real overhead for interactive services is significantly higher because the radio channel bearer set-up is not immediate therefore decreasing its throughput after a user interaction. For this reason, a safe design threshold is the dimensioning of no more than 70 % of the entire connection or interface capacity, if the services are entirely non-conversational [23].

With the further consideration that conversational services have to be prioritised and the ATM overhead for CBR and AAL1 can be considered as a stable 17.5 % (see Section 17.2.1.1), the dimensioning method for PVC without overbooking can be performed as depicted in Figure 17.9; the dimensioning of required bandwidth for data services by 120 % / 70 % is based on the minimum ATM AAL2 overhead plus the 70 % threshold described earlier.

If overbooking and dynamic routing is introduced, the additional risk of congestion has to be considered, the dimensioning threshold depends heavily on the overbooking factor. It is recommended that the engineering limit for a SPVC five times overbooked ATM trunk or ring should be reduced to 50 % instead of 70 %.

The described dimensioning approach is of particular interest if the load and the ATM performance cannot be retrieved on an end-to-end basis or the load needs to be simulated by software tools. For more in-depth evaluation and optimisation, live network tools are required, which are described in Section 17.2.1.5.

17.2.1.5 Optimisation of ATM Network Parameters

Similar to the RF-Interface, the optimisation of ATM networks is performed in two steps [22,24]:

1. Optimisation of nodes.
2. Optimisation of clusters.

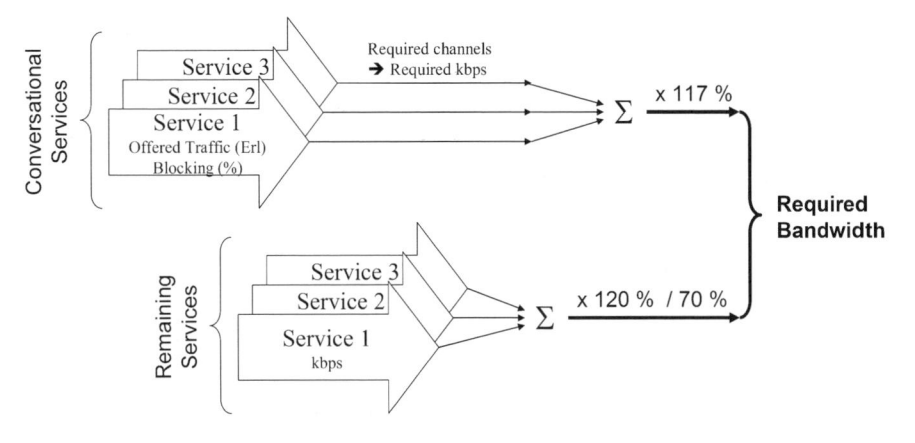

Figure 17.9 ATM dimensioning method.

In step 1, all physical and logical connections have to be checked. The checklist contains issues like:

- Synchronisation: Are all nodes properly synchronised? When did the transmission systems undergo a 24-h stability check for its last time? How is its CRC-4, LSS (Loss of Synchronisation Signal) or AIS-record (Alarm Indication Signal)? Has it been considered that only AAL1 and AAL2 connections require and supply timing? Etc.
- Service classes: Are PVC-connections within an end-to-end-route configured with the same service class (CBR, VBR, ABR, UBR). Do all ATM vendors and components in the network support and interoperate these classes? Have they been applied with the desired AAL types? Etc.
- Physicals: Are the physicals (e.g. the Node Bs) evenly distributed to the ATM ports? Are the ports, the cards and the connections properly protected? Etc.
- Throughput: Are the PCR, MCR and SCR configured correctly? Are they plausible considering the expected service mix and throughput? Does the combination leave sufficient throughput for UBR and ABR-traffic? Etc.
- Cell loss: Are the buffer sizes and the cell delay tolerances implemented as recommended by ATM Forum or the component vendor? How is traffic policing and shaping implemented on this component, this port or this VC? Etc.

In step 2, the end-to-end performance has to be established by an entire ring or mesh optimisation [25]. The questions for this session can be quite different:

- What average Cell Transfer Delay (CTD) and Cell Delay Variation (CDV) can be retrieved from the system? What or how many ATM nodes are in the route on average? Etc.
- How high is the utilisation of the ports? Can clusters or nodes be identified that are over- or under-utilised? Etc.
- Are Peer Groups configured in a way that meshes or rings are divided? How much load is on the Peer Group lead component? Etc.
- Do the SPVC or PVC contain significant amounts of tagged, errored and lost cells? At which component do they occur on the way? Are they retrieved properly? Etc.

It can generally be stated that, the more thoroughly step 1 is performed, step 2 takes a much shorter time. Step 2 requires the implementation of specific tools that are specified e.g. in ITU T-I610. Some ATM vendors, however, have not yet implemented tools like this and almost all UTRAN-vendors do not offer I610 Performance retrieval at all. Instead, they retrieve performance like CLR as depicted in the scheme in Figure 17.10.

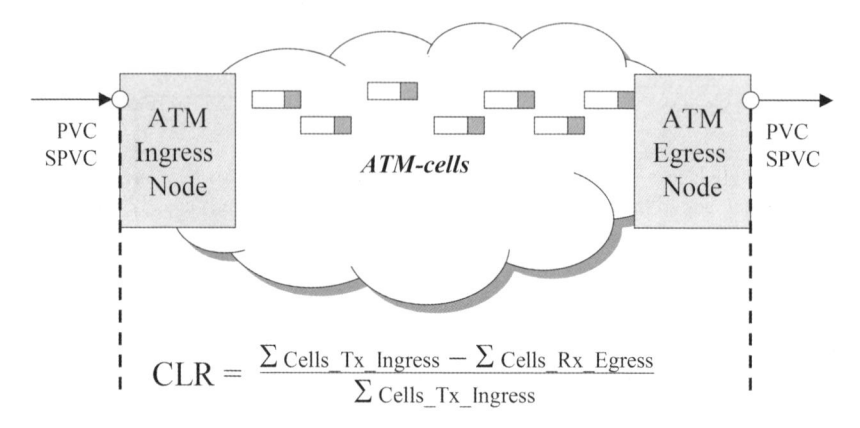

$$CLR = \frac{\Sigma\ Cells\_Tx\_Ingress - \Sigma\ Cells\_Rx\_Egress}{\Sigma\ Cells\_Tx\_Ingress}$$

Figure 17.10 ATM Performance retrieval without I610 tools.

This scheme, however, does only work for utilisation, but not at all for error or loss performance.

Example: Take an E1-connection between Node B and RNC with an average CTD of 10 ms. Assume further that a management system is able to retrieve the valid cells transmitted and received every hour exactly synchronised with μs-accuracy. Even in this case, 55 of 19,660,800 ATM cells in an hour would be within the 10 ms route giving an inaccuracy of already $2.7 \cdot 10^{-6}$, making it impossible to establish KPI of $1 \cdot 10^{-6}$ resolution as defined by ITU.

In real life operations, it will not be able to synchronise retrieval even in one second accuracy, making any result retrieved on a two node basis completely unusable.

For this reason ITU-T-I610 defines the establishment of tools that add ATM cells containing accurate information on the amount of cells and a time stamp. The process is depicted in Figure 17.11.

The implementation of these tools allows an accurate retrieval of CLR, CMR and CER. In general, [19] defines four optimisation methods and tools:

1. Alarm Indication Signal/Remote Defect Indication (AIS/RDI): This tool monitors any indication of an upstream interruption at the ATM (or physical) layer and generates an alarm.
2. Loopbacks: This method works similar to an IP 'ping'. It is useful for example when provisioning an ATM connection. Note that these ATM layer loop-backs are not to be confused with physical layer loop-backs such as may be created by manually inserting bridging cables at the remote end of a physical layer path.
3. Continuity Check (CC): This tool detects an interruption in an ATM connection (e.g. in an ATM switching matrix).
4. Performance Monitoring (PM): This tool detects and measures errored cells or lost/misinserted cells in an ATM connection.

The implementation of ITU-I610 is not cheap, making it not likely that all UTRAN vendors will implement it soon and thoroughly. For this reason, it is important that it is at least available at the core sections of the transmission network.

17.2.1.6 Exploitation of 3G-ATM Networks for 2G Voice and Data Networks

Several operators are in the position of having implemented separate backhaul networks for 2G and 3G services. This situation is an economical driver in the coming years to shift 2G and 2.5G traffic onto the ATM networks that have actually been implemented for the transport of 3G. One main reason

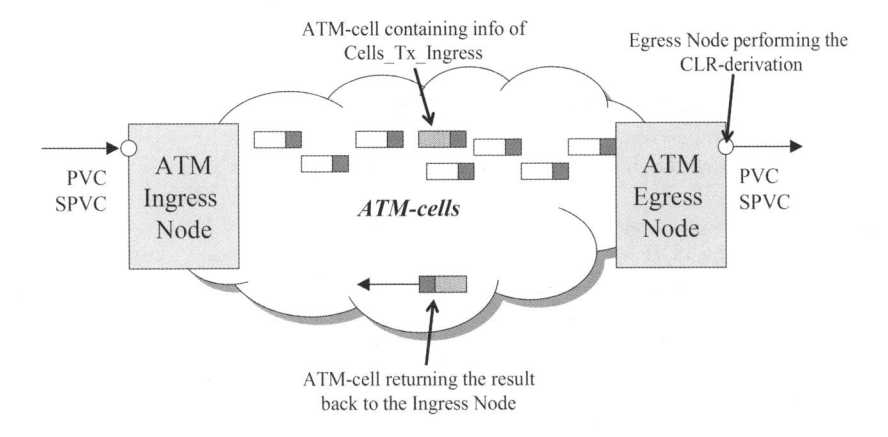

Figure 17.11 ATM performance retrieval using ITU-T-I610 tools.

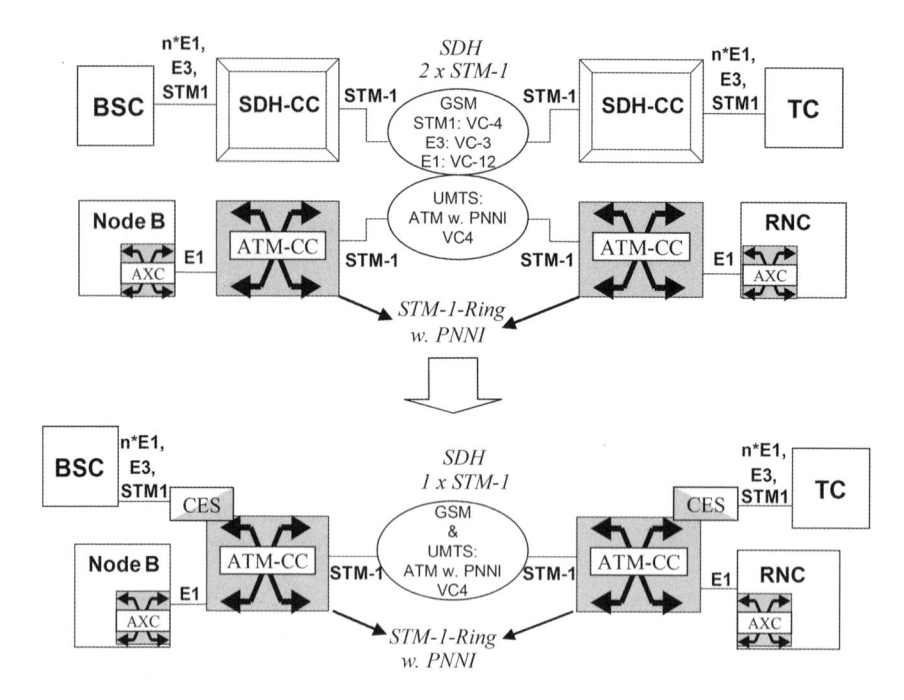

Figure 17.12 Circuit Emulation on ATM.

is a feature that has been standardised by the ATM Forum, namely CES (Circuit Emulated Services) as shown in Figure 17.12. CES emulates a physical E1, E3 or STM-1, converts the emulated payload to ATM, and then transports the signal via the ATM Network [24].

This has several economical advantages:

- Only one physical trunk has to be used for the transport of 2G and 3G traffic, which means no or later upgrades of microwaves, Leased Lines, fibres, switches or cross connects on growing traffic.
- The implementation of resilience for several 2G interfaces is easier and thus more cost-efficient using dynamic routing features like SPVC-reestablishment and PNNI instead of a 1+0 protection.
- The complexity of the backhaul network is lower, which has a positive impact for maintenance, because ATM allows several QoS classes; the prioritisation of 2G, 2.5G or 3G traffic can be configured on demand.

According to the number of ATM specifications, a variety of configurations can be implemented on port level. Table 17.3 shows some options used for an E1 port as well as the impact on the packetisation delay (the delay to pack the bits on the PCM frames into one ATM VC) as well as the bandwidth saving [24].

In general, ATM generates an overhead of at least 17.5 % when CBR and AAL1 is used. Considering that an ATM frame on PDH takes 125 μs, this results in a minimum delay of 367 μs to pack and unpack 2048 bit into ATM cells. When reducing the payload by half – for example by transmitting 16 of 32 DS0 or compressing the 64 k DS0 to 32 k – the delay doubles, while the required bandwidth is reduced to 58.3 %. In the field some microseconds have to be added to the equipment. Secondly, every additional ATM hop generates an additional delay of 200 μs, if STM-1 transport is used. IMA will add 1.5 ms to the delay.

Actually, the main delay limit according to GSM-spec 3.3 is defined to be 180 ms round trip between the mobile and the transcoder unit. Considering the actual delay budgets of 2G network elements (for

Table 17.3 E1 Trunk types using CES.

CES type	Description	Packetisation delay (Ingress + output port)	Bandwidth compared to E1
Unstructured	Emulation and transmittal of the entire E1-trunk	$\geq 367\,\mu s$	$\geq 113\%$
Structured and unchannelised	Emulation and transmittal of all 31 DS0 of an E1-Trunk, except Time Slot 0	$\geq 378\,\mu s$	$\geq 109\%$
Structured and channelised	Emulation and transmittal of between 1 and 31 DS0 of an E1 Trunk	$378\,\mu s$–$11.75\,ms$	109%–3.52%
Structured and channelised and compressed	Emulation of 1 to 31 DS0, compression from 64 kbps to 32, 16 or 8 kbps, conversion and transmission over ATM	$378\,\mu s$–$46.0\,ms$	109%–0.88%

example 30 ms for the TCU), the allowed BS-TCU delay commonly shrinks to 10 ms. This corresponds in practice to an ATM route with no more than 30 ATM hops.

Beside the delay other technical constraints have to be considered, particularly resilience, alarm transmission or synchronisation issues. The migration of 2G traffic to ATM or IP-networks offers various economical opportunities as it contains various technical traps and risks depending on the complexity of equipment manufacturers, services and configurations. For this reason, the implementation strategy cannot be standardised but needs to be tailored exactly to the operator's circumstances.

17.2.2 MPLS-ARCHITECTURE FOR FUTURE 3G TRANSMISSIONS

The rapid growth of data volumes and services in the World Wide Web led to new QoS requirements for IP. Particularly the establishment of specific user groups or services with different requirements for availability, throughput and performance in one common IP-network are of a particular interest for mobile operators, since they apply commonly more complex tariff structures to their customers than IP-carriers. Most of the current IP technologies have limited potential for traffic engineering and aggregation. MPLS (Multi Protocol Layer Switching) addresses these requirements. Actually developed for IP networks, MPLS integrates the label-swapping paradigm with network layer routing. It further supports the delivery of services with QoS-guarantees, which can improve the price per performance of network-layer routing [26]. The main quality improvement of MPLS is depicted in Figure 17.13, which shows in particular the MPLS-ATM interworking (MPOA) as defined in [27].

MPLS encapsulates ATM or IP-Packets by adding a label in the Inter Working Function (IWF), which contains an LSP-Information (Label Switched Path) that is used by the LSR (Label Switched Router). ATM Packets of the same VC can be switched to different LSPs. Likewise, several VCs can be mapped to one LSP, which is a significant improvement for the efficient transport of broadcast information to multicast services. For IP-networks, MPLS is a significant step forward to multicasting as well as prioritised IP-routing, which minimises the end-to-end delay. If the IWF is implemented in the service unit (e.g. the Node B), then services with the same priority (like 2 voice services) can be routed or prioritised differently, which increases the options to establish QoS for instance to premium users.

One great strategic advantage to current 3G-operators, who deployed a significant amount of classic ATM switches, is the interoperability of MPLS over IP and MPLS over ATM, which allows the usage of the same routing protocols for IP, LAN or WAN sessions. The most common protocols are [9]:

- Transmission Control Protocol/Internet Protocol (TCP/IP);
- Routing Information Protocol (RIP);

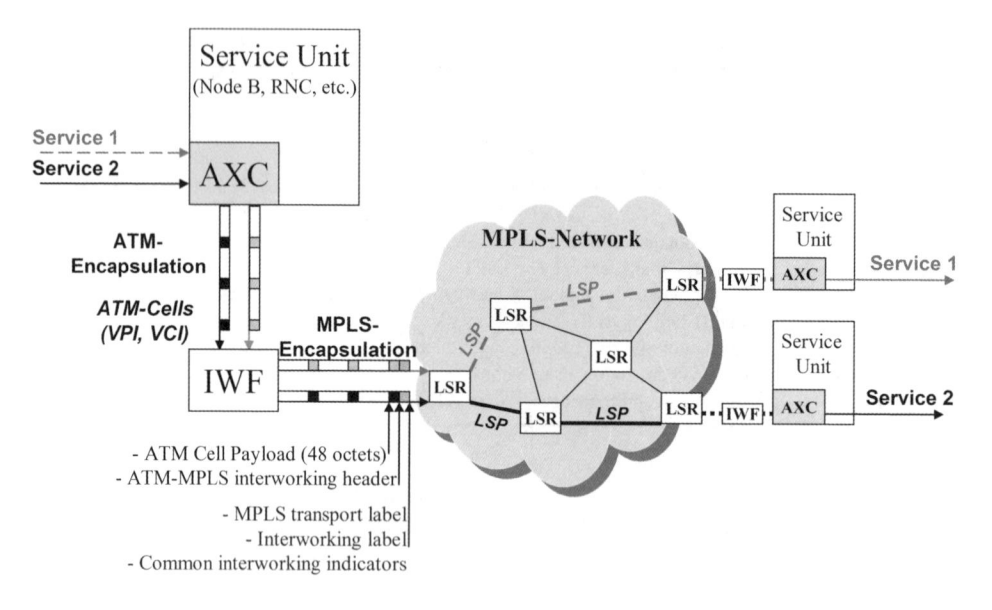

Figure 17.13 MPLS network structure.

- Open Shortest Path First (OSPF) or Multicast OSPF (MOSPF);
- Intermediate System to Intermediate System (IS-IS);
- Border Gateway Protocol-4 (BGP-4);
- Protocol Independent Multicast (PIM);
- Distance Vector Multicast Routing Protocol (DVMRP).

This also allows the implementation of end-to-end QoS-tools based on these protocols, particularly of Differentiated Services (DiffServ) or Integrated Services (IntServ). The interoperability of IP and ATM via MPLS could finally allow a smooth and successive migration from ATM/PDH/SDH towards IP on either PPP/SDL or LLC/SNAP on Ethernet. This will be described in the next section.

17.2.3 THE PATH TO DIRECT IP TRANSMISSION NETWORKING

A key learning of 2G deployment was the fact that building up a separate transmission infrastructure for GSM-voice and data (GPRS) was costly. Furthermore, it achieved a low performance given the invested infrastructure since backhaul capacities had to be kept separate on a shared physical link. For this reason, the main idea of 3G Release 04 was the convergence of voice and data networks, as shown in Figure 17.14 from IPv6-forum [28].

As described earlier, many operators implemented UTRAN R99 using classic ATM switches. MPLS networks can use traditional ATM equipment as a migration step in introducing MPLS to an existing ATM network. Traditional ATM switches can be used in three ways:

1. Backhauling, when the access device is remote from the edge LSR. The access device is connected to the edge LSR by PVCs switched through an ATM network.
2. Tunnelling through ATM switches between an edge LSR and an ATM LSR. In this case, the edge LSR does not need to be adjacent to an ATM LSR, but can be connected through an ATM network.
3. Tunnelling through ATM switches between ATM LSRs. In this case, the core network uses traditional ATM switches as well as ATM switches.

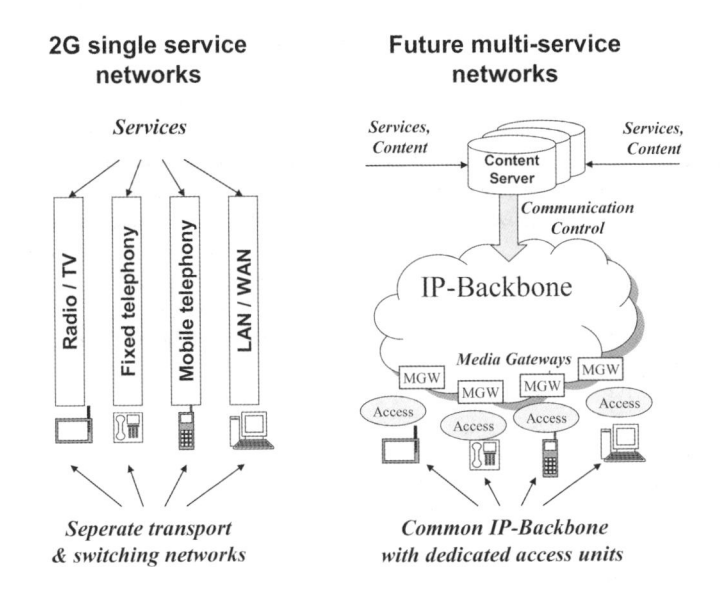

Figure 17.14 IP evolution scenario.

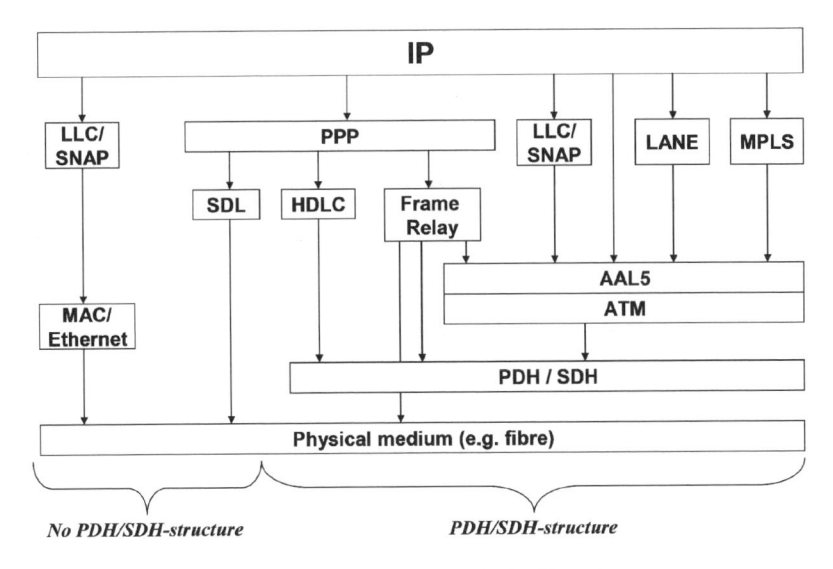

Figure 17.15 Transport protocols for IP-services.

The use of traditional ATM equipment for IP-services has the disadvantage of excessive overhead and higher delay [4,23,22]. The reason can be found in the protocol structure, which is depicted in Figure 17.15.

Each protocol layer adds overhead as well as end-to-end delay. As can be seen in Figure 17.15, the use of PPP/IP on FR on AAL5 on PDH incorporates more protocol layers and thus more overhead and delay compared to IP on Ethernet or SDL. Secondly, PDH based on ITU G703 is defined as symmetric for uplink and downlink. This is efficient for conversational services, but not at all for data streaming, online-TV or downloads.

On the other hand, many operators have implemented PDH or SDH-systems, and not MAC/Ethernet. For this reason, with the growth of IP-based services, a trigger for a migration to MAC Ethernet or SDL needs to be defined. Recent scenarios see such a trigger when at least 80% of the transport traffic is IP-based or when the uplink/downlink asymmetry is more than 5:1. A feasible scenario is the successive migration of clusters from PDH to Ethernet by interoperating them via IP and MPLS.

A key driver for the time frame is still the development of voice and video telephony. Particularly, the recent success of VoIP services could decrease the time frame to a migration away from PDH/SDH towards MAC/Ethernet.

17.3 END-TO-END TRANSMISSION DIMENSIONING APPROACH

In classic 2G networks, the dimensioning of transmission networks has been a straightforward process based on the Erlang B formula. Since the backhaul systems were almost all circuit switched, the dimensioning had only to consider the air interface and the interface between the base station controller and the switch [23].

In 3G this changes considerably, due to the following reasons:

- In the Node B, not only the throughput can be limited by the air-interface due to cell breathing and interference, but also by the OSVF-codes (amount of Node B baseband units), which is a viable scenario when HSDPA is introduced.
- Secondly, the Node B ATM Cross Connect (AXC) allows overbooking of VCs in the ATM network.
- Thirdly, dynamic traffic routing like PNNI introduces new challenges on delay and transmission availability triggering new dimensioning requirements.
- Finally, new end-to-end services introduce additional overheads, retransmissions and delays that have to be considered on a dimensioning approach.

As long as no retrievals of utilisation and blockings of errors are available, an end-to-end simulation approach would start, section by section, with the Node B first, then with the ATM network, and finally with the services on top of it. Section 17.3.3 will therefore use IP-services as an example.

17.3.1 DIMENSIONING OF NODE B THROUGHPUT

As already described in Section 17.2.1.4, the most practical approach to dimension a Node B capacity requirement is the grouping of all services either into circuit switched and packet switched traffic or delay sensitive and non-delay sensitive traffic. To model the different services and data rates, an adequate multirate model has to be applied to the Iu-interface. To begin with the multirate model, we assume a mix of circuit switched services with rates at a granularity $n \cdot 16$ kbps; commonly, a voice service requires $1 \cdot 16$ kbps on a PDH/SDH frame. This is illustrated in Figure 17.16.

17.3.1.1 Dimensioning Methods

Generally, there are two methodologies to evaluate the required bandwidth [23]:

1. *Statistical approach* The services during a busy hour are modelled as calls with an individual data rate and data activity as shown in Figure 17.17. This approach can be executed with several resources (like Excel). On the other hand, it cannot model the time dependencies of protocols, and the effects of buffer sizes, latency and cell delay variation.
2. *Analytical approach* This approach models the protocols of the network elements involved. The effort of this method is quite high, since an in-depth knowledge of the vendor equipment is required.

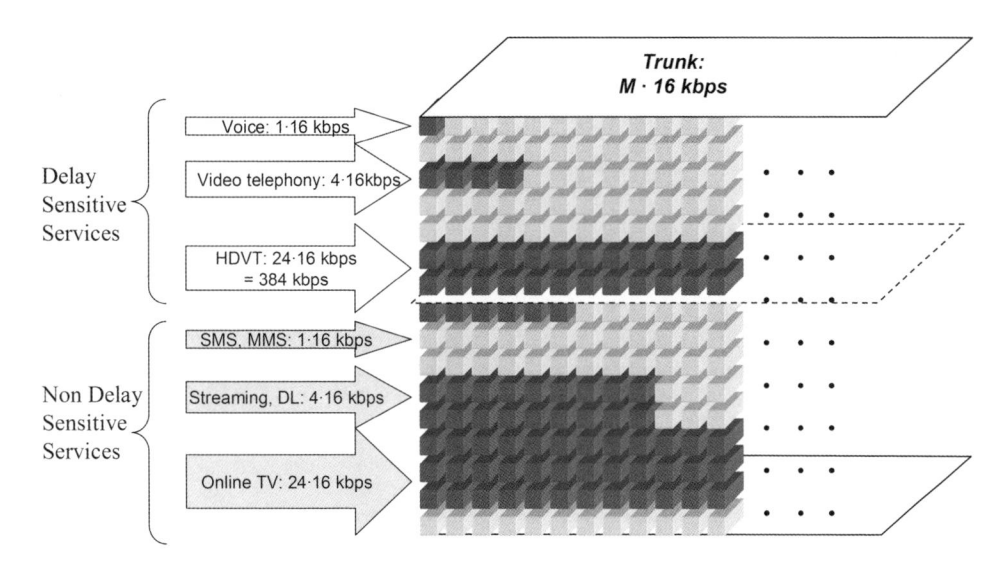

Figure 17.16 Dimensioning of Node B transmission resources.

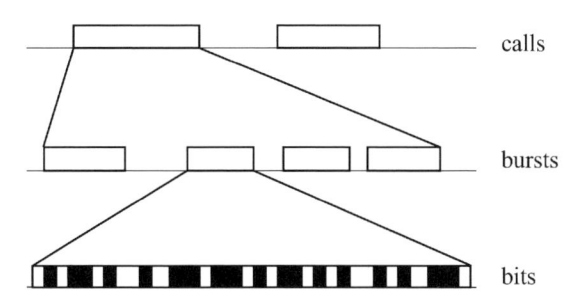

Figure 17.17 Statistic model for transmission dimensioning.

The suitable approach is then:

1. Prioritise delay sensitive traffic by calculating the required bandwidth on a busy hour first.
2. Calculate the required bandwidth of the non-delay sensitive traffic. Here, it should be distinguished whether the service requires a stringent delivery time or is a background service.
3. Calculate the required bandwidth for the Node B control channels. These are usually one channel for OAM, one for the Baseband Cards (D-NBAP) and one for the Node B radio modules (C-NBAP).
4. Add the ATM overhead. This will be done in Section 17.3.2.1.

The main factors for the traffic calculation are:

- User Data rate.
- Service activity factor: As a difference to the air interface the activity for circuit switched services has to be assumed as 100 %, except silence detection and suppression has been implemented in the Node B transmission unit.
- Burst efficiency: Most vendors support a sustained transmission of a FACH for 5–20 seconds on bursty services. The intention is to keep alive higher layer protocol sessions like telnet or TCP/IP.

This can be severe, if the service involves mainly the transmission of small packets like on web browsing or interactive gaming. In the worst case, 384 kbps FACH resources could keep allocated up to 20 seconds after the transmission of a 16k packet making the service extremely burst inefficient.
- Number of users using the service in a busy hour.
- Offered traffic (in Erlang) and required blocking (in %) for delay sensitive services.
- Retransmission percentage and maximum delay of 95 or 99 % packet quantile.
- Soft handoff overhead, if soft handoff is applied to the service.

All user channels as well as the Node B signalling channels (OAM, CNBAP, DNBAP) will then be ATM encapsulated. The required ATM overhead will be shown in Section 17.3.2.1.

The analytical approach for a Node B requires a higher effort. The example in the next section shows a FDD and TDD simulation of a number of users having the same user profile. As an example, the computational tool give the option to configure the number and length of sessions during the busy hour (BH), the data volume as well as the maximum FDD or TDD burst size. The principle is as follows:

1. Generate an offered traffic for each of the users for every second of the BH.
2. Simulate the absorption of the offered traffic given the configured FDD and TDD burst sizes for every second of the BH.
3. Optional: rerun the calculation with a varying number of users.

The example of the main configuring screen of the tool can be seen in the Figure 17.18.

In the tool, the limit for maximum throughput can be set to the limit at the RF side of the Node B. As will be seen, one of the most sensitive parameters when configuring the throughput for a service mix mainly consisting of data services, is the maximum burst size. This will be illustrated in the next section.

Offered Internet traffic

Number of users during a busy hour per cell	10
Internet user profile	
Number of sessions per user during a BH	2
Number of Web Pages per session	20
Data volume per Web site (kByte)	15
Number of Downloads per session	3
Size per Download (Mbyte)	6
Average online seconds per user during busy hour excluding data downloads	600
Number of Web Page changes per user per BH	40
Number of Downloads per user	6
Number of CS-sessions per user	2
Time per CS-second (s)	100
Data rate of CS-second (kbps)	16
Average download size per user during busy hour (MByte)	0.30

Simulate ALL

Simulate Offered Traffic

| Simulate TDD | Simulate FDD |

Variate LOAD

From	10	user
Until	10	user
Intervals	5	

	TDD	FDD
Number of 5 MHz-carriers	2	1
Chiprate (Mbit/s)	7.68	3.84
Maximum burst rate per user (Kbit/s)	1536	768
Eb/No (dB)	5	5
Code Orthogonality	85%	85%
Frequency reuse efficiency	50%	50%
Downlink Pole capacity during a Downlink time slot (maximum number of users per cell using the maximum burst rate)	5.270462767	5.270462767
Load	60%	60%
Maximum throughput rate during a Downlink time slot	4857.258486	2428.629243
Number of Time slots per frame usable for downlink traffic (maximum 15)	9	15
Maximum Downlink throughput per second (Kbit)	2914.355092	2428.629243

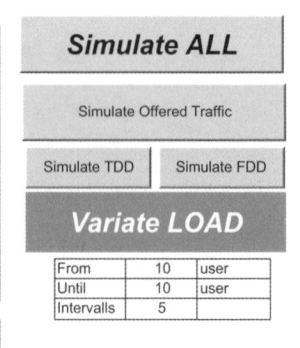

Figure 17.18 Main screen of VBA dimensioning tool.

17.3.1.2 Example Calculations

Table 17.4 shows the result of the statistical approach calculations. Adding an additional Layer 2 overhead, which is commonly ATM & AAL1-5, the required bandwidth would increase at least another 30 %; this means that at least four E1 links would be required to transport the load in this scenario.

The results of the analytical approach show the burstiness of internet browsing and download compared to circuit switched sessions. Commonly, downloads contain file sizes of several Mbytes, which are offered (in theory) in a split of a second. The offered traffic graph in Figure 17.19 shows that the highest peaks appear whenever a download is requested.

It then depends on the maximum burst size of the transport unit how to absorb the sudden DL traffic load. From Figures 17.20 and 17.21, it can be seen that a maximum bearer burst size of 128 kbps requires a significantly longer time compared to a 384 kbps bearer.

The results show also that for Internet services a low maximum bearer size leads to a long download time, even when the total Node B resources are not congested. The higher bearer size can absorb bursty traffic much better, which leads to higher internet user satisfaction. Many classical telecom operators have configured bearer sizes of not more than 384 kbps to ensure congestion rates smaller then 5 %. This is a classical circuit switched approach, which dissatisfies Internet users. If an internet user could download a file on 512 kbps with an assumed 20 % of blocking, the total download time would still be faster compared to a download on 384 kbps and 1 or 0 % blocking. Surfing and download Internet users can tolerate blocking of 20 % during download time, which would, of course, be disastrous for circuit switched services.

This shows that Node B dimensioning heavily depends on the type of service transported, and that traffic cannot be regarded as traffic without the context to the applied service.

Table 17.4 Results of statistical Node B dimensioning.

User Spectrum Required

Delay sensitive-Service	1	2	3	4	5	6
Name	Voice	Video telephony	HDVT	Online Gaming		
Data rate (kbps)	12.2	64	384	12	0	0
Activity factor (%)	100%	100%	100%	5%	0	0
Burst efficiency	100%	100%	100%	10%	0	0
Number of users	35	12	2	200	0	0
Blocking (for DS-services)	0.20%	0.10%	0.10%	0.10%	0.10%	0.10%
Offered traffic (mErl) per user	23	15	10	20	0	0
SHO-overhead	35%	35%	35%	35%	0%	0%
Total traffic	0.81	0.18	0.02	2.00	0.00	0.00
Required Channels	5	4	2	8	0	0
Required Spectrum (kbps)	82.35	345.6	1036.8	129.6	0	0

Non Delay Sensitive-Service	1	2	3	4	5	6
Name	SMS, MMS	Internet surfing	Online TV	Download		
Data rate (kbps)	12	12	384	384	0	0
Activity factor (%)	10%	2%	80%	5%	0	0
Burst efficiency	50%	15%	100%	30%	0	0
Number of users	200	150	2	30	0	0
Retransmission percentage	10%	0%	0%	24%	0	0
SHO-overhead	0%	35%	35%	0%	0	0
Required Spectrum (kbps)	528.0	324.0	829.4	2380.8	0.0	0.0

User Spectrum required for Node B				5656.59	kbps

Spectrum for Signalling

Type of node B-signalling	CNBAP	DNBAP	OAM
Required Spectrum (kbps)	50.0	50.0	50.0

Complete Spectrum required for Node B (except ATM)	5806.59	kbps

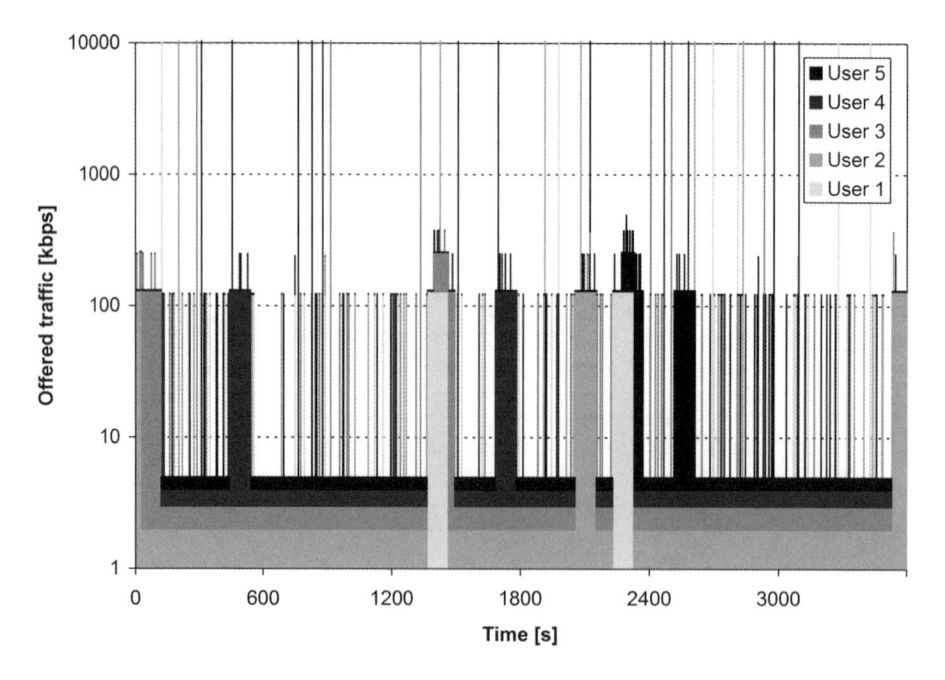

Figure 17.19 Offered traffic for data services during a busy hour.

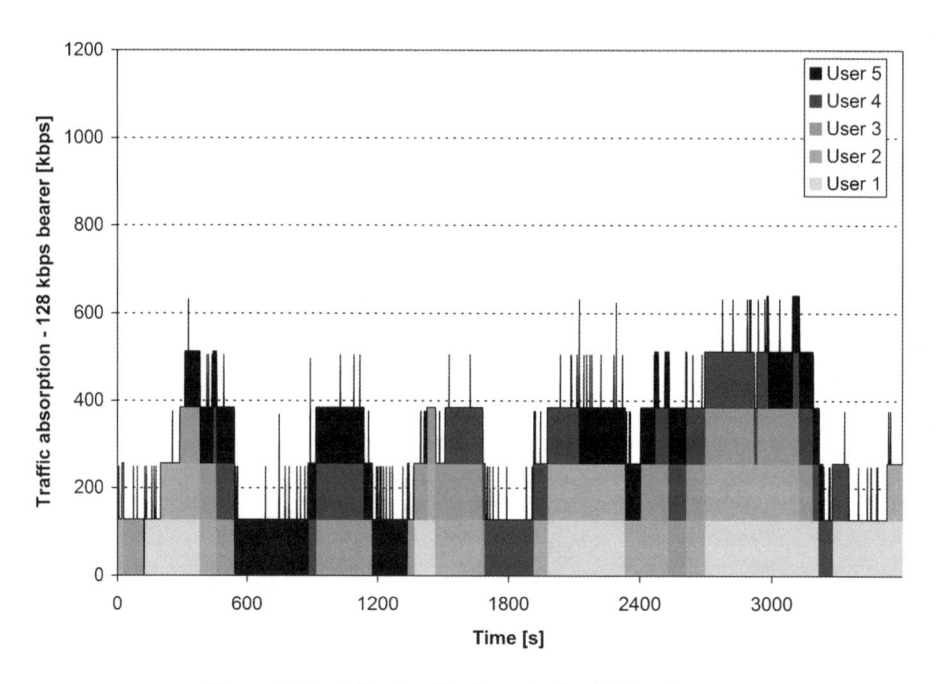

Figure 17.20 Offered traffic absorption for 128 kbps bearer.

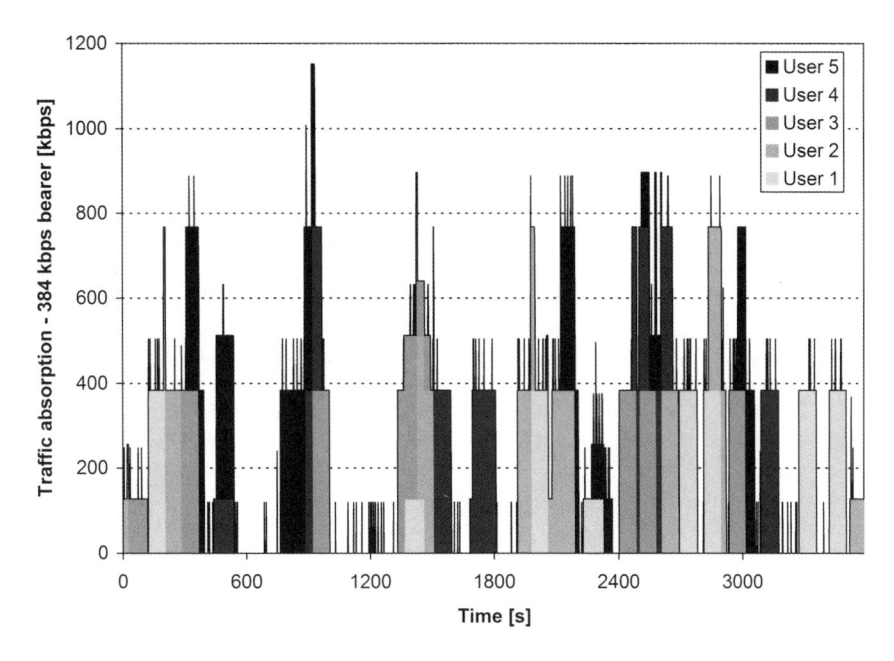

Figure 17.21 Offered traffic absorption for 384 kbps bearer.

17.3.2 TRAFFIC DIMENSIONING OF THE ATM NETWORK

As already mentioned, the Node B AXC allows overbooking of VCs in the ATM network. Secondly, dynamic traffic routing like PNNI introduces new challenges on delay and transmission availability. Section 17.3.2.1 illustrates the ATM overhead of all ATM adaptation layers.

While overbooking of an ATM network can be handled by linear multiplication of the Node B traffic, the handling of PNNI on a meshed network requires more complex formulas. Even more complex is the handling of buffer sizes and transmission delays on ATM networks.

17.3.2.1 Dimensioning Methods

The principle dimensioning methods for Node Bs, described in Section 17.3.1, apply also for the dimensioning of the ATM network. Table 17.5 shows the overhead to be considered for the ATM adaptation layers.

Except for AAL1m the maximum overhead can deviate significantly depending on the burstiness of the traffic. On AAL2, empty bytes of an ATM container are padded, which does not apply to AAL5. Additionally to this overhead, the AAL Signalling VC has to be added.

While the handling of ATM overhead can be considered straightforward, the consideration of PNNI in a ring or even a meshed network is rather complex. Section 17.3.2.2 shows an example of how to calculate this traffic.

Table 17.5 Minimum overhead of different ATM adaptation layers.

Adaptation type	AAL1	AAL2	AAL3/4	AAL5
Minimum overhead	16.40 %	17.78 %	20.47 %	10.50 %

Even more complex is an analytical approach, which considers a ring or meshed network topology as well as the ATM protocol structure and the configured buffer sizes of the network elements. Current simulation software is capable of simulating meshed networks, including 200 network elements with most common transport protocol structures. In comparison to air interface simulation, this shows that the analytical simulation of a transport network requires significantly higher resources.

17.3.2.2 Example Calculations

When applying the same example for the statistical approach of Section 17.3.1.2, ATM adds a significant overhead, particularly when the payload packages are small, e.g. during on-line gaming or web surfing. Compared to the initial payload, the requirement for the transport capacity would then increase from four E1 links to more than six E1 links. To ease complexity of the calculation of the network presented in Figure 17.22, equal link costing, transmission capacity and one peer group was assumed.

As already mentioned, the analytical approach is very complex, and requires enhanced IT capabilities. However, to give a hands-on approach for a fixed network planner, the mesh just illustrated has been used for an analytic link failure scenario, which is described below.

The example presented in Figure 17.23 shows the scenario of the described ATM cluster with a link failure on link number 4 after 1800 seconds during a busy hour. As can be seen, the traffic on link 4 has to be re-routed after 1800 seconds. As a result on the remaining links, the traffic increases by 25 % on average the moment the link number 4 fails. The exact amount depends on the link capacity, the costing and the existing traffic.

17.3.3 TRAFFIC DIMENSIONING OF THE IP-NETWORK

Like in ATM, it is important to know the type of service and its session protocols when dimensioning the overhead generated by IP. The most suitable adaptation layer for IP traffic in mobile networks is commonly AAL5 [29]. However, the most common adaptation layer in the mobile application world

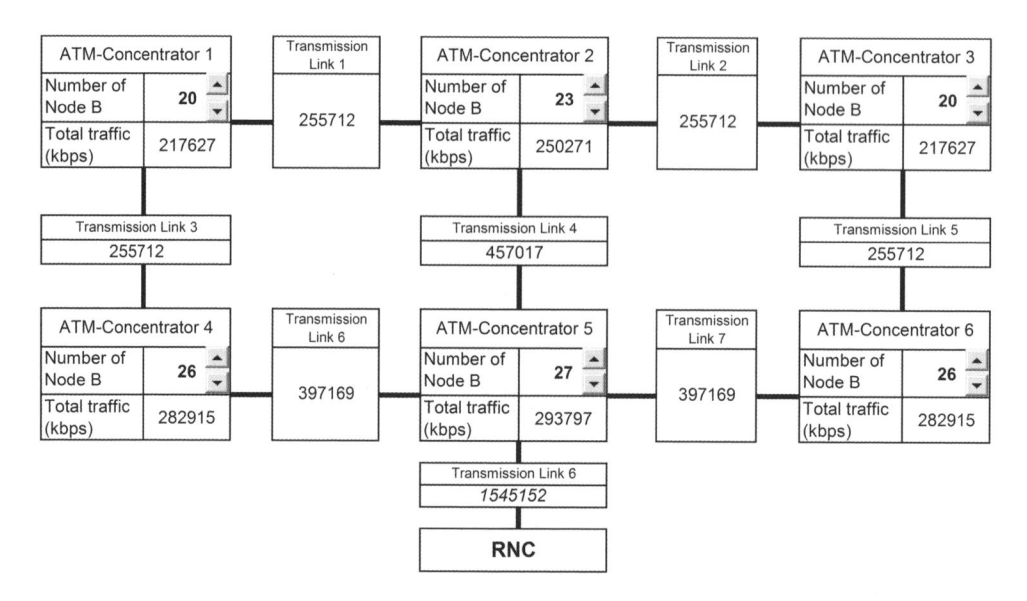

Figure 17.22 ATM dimensioning of a meshed network.

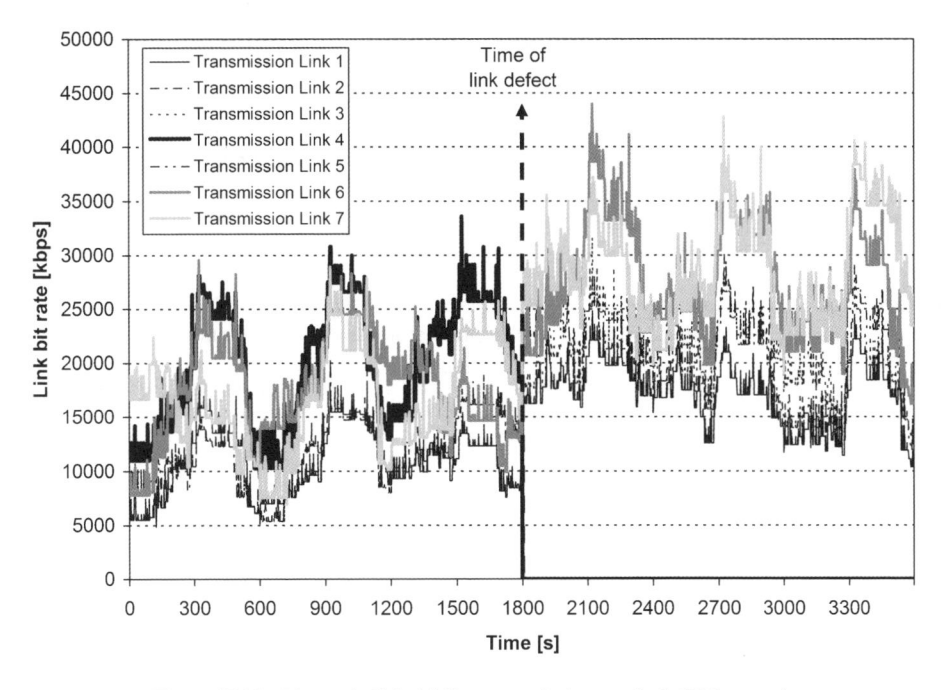

Figure 17.23 Bit rate in link 4 failure scenario in a meshed ATM network.

is AAL2, which increases the overhead. As will be seen in the next section, the main drivers for the overhead are the payload size in the IP-packet as well as the size of the Maximum Transfer Unit (MTU).

When dimensioning the IP overhead, the easiest approach is to work from the OSI application layers to the session layer, to the link layer, and finally to the physical layer. Table 17.6 describes the minimum overheads of the most common protocols used for the application and session layer [30].

Table 17.6 Overhead of IP protocols.

Type of protocol	Description	Minimum overhead
TCP	TCP adds a header of 20 bytes to the payload. TCP enables packet retransmission, multiplexing and data integrity	2.13 % on Ethernet G802, 0.36 % on Gigabit Ethernet
UDP, RTP	UDP or RTP adds a header of 8 bytes to the payload used for multiplexing and data integrity	0.53 % on Ethernet G802, 0.09 % on Gigabit Ethernet
IPv4	IPv4 encapsulates the payload into a 20 byte header. The IPv4 address has only 4 byte. The IPv4 allows refragmentation in a transmission router.	1.33 % on Ethernet G802, 0.22 % on Gigabit Ethernet
IPv6	Ipv6 encapsulates the payload into a 40 byte header. The IPv4 address has 8 byte. The IPv6 does not allow refragmentation in a transmission router	2.61 % on Ethernet G802, 0.44 % on Gigabit Ethernet

In contrast to IPv4, IPv6 does not provide for fragmentation and reassembly. If an IPv6 packet received by a router is too large to be forwarded over the outgoing link, the router simply drops the packet and sends a 'Packet Too Big' ICMP error message back to the sender. The intention is to optimise the end-to-end delay of the IP network. The sender can resend the data, using a smaller IP packet size. Fragmentation and reassembly is a time-consuming operation; removing this functionality from the routers and placing it squarely in the end systems considerably speeds up IP forwarding within the network. If the packet has been transmitted by UDP, retransmission is not executed. This configuration ensures an acceptable small overhead, and is optimal for streaming, but not for downloads.

The protocol encapsulation on the link layer leads to the minimum overhead, as displayed in Table 17.7. Finally the overhead imposed by the physical transmission medium has to be considered, as shown in Table 17.8.

Tables 17.9 and 17.10 present an example tool with input information as provided in the previous Tables 17.6 and 17.8.

Figure 17.24 illustrates the main driver for the overhead to the IP payload. It can be clearly seen that the overhead generated by IP depends not only on the protocols used, but particularly on the

Table 17.7 Overhead of Layer-2 protocols.

Type of protocol	Description	Minimum overhead
LLC/SNAP	LLC adds an additional 8 byte header to each packet. The packet does not have to exceed the MTU-size	0.09–17.7 %, depending on MTU
ATM	The protocol structure has been explained in Section 17.2.1.1. If AAL5 is applied the ATM cell will eventually be padded, if there are still free octets after the 8 byte AAL5 trailer	11–40 %, depending on AAL type and payload size

Table 17.8 Overhead of fibre transmission systems.

Type of protocol	Description	Minimum overhead
Ethernet G802	Ethernet adds 38 byte of overhead plus 4 byte, if the optional 4 byte VAN Tag is used. The maximum Ethernet payload is 1500 bytes	Ca. 2.27 %
Gigabit Ethernet	Gigabit Ethernet has the same overhead structure like Ethernet G802, but a payload size of 9000 bytes	Ca 0.46 %
Sonet OC-3 or OC-12	Sonet OC-3 has a 2430 byte frameout of which 90 bytes are overhead. Sonet OC-12 has a 9720 byte frame with a 360 bytes overhead	Ca. 3.77 %
Multimode Fibre 100 Mbps	A multi mode fibre adds a 2 byte overhead between a 53 byte ATM cell	Ca. 3.95 %
DS3	A DS3 has a 44736 byte frame, out of which 40704 bytes are payload	Ca. 9.97 %

Table 17.9 Example of dimensioning sheet for IP Ethernet.

	Gigabit Ethernet with Jumbo Framing	Ethernet G.802
User Payload-Size per Data-packet	9000	1500
Application Protocol	TCP	TCP
Overhead due to application protocol	0.36%	2.13%
IP-Version	v6	v6
Overhead due to IP	0.44%	2.61%
MTU-Size	6134	6709
LLC/SNAP-Overhead	0.18%	0.12%
Ethernet G802-Tagging	☑ VLAN-Tag (Y/N)	☑ VLAN-Tag (Y/N)
Overhead due to Ethernet	0.46%	2.66%
Total overhead	1.44%	8.13%
Physical line Rate (Mbps)	1000	100
Maximum User line Rate (Mbps)	985.76	92.48
Transmission Rate after 4B/5B-encodding acc. To FDDI	1250	125

Table 17.10 Dimensioning sheet for IP on SDH and PDH.

	STM-1 or OC-3	STM-4 or OC-12	MM-Fibre 100 Mbps	DS3
User Payload-Size per Data-packet	1000	600	400	200
Application Protocol	TCP	TCP	TCP	TCP
Overhead due to application protocol	3.20%	5.33%	8.00%	16.00%
IP-Version	v4	v4	v6	v6
Overhead due to IP	1.98%	3.29%	9.26%	17.24%
MTU-Size	1500	1500	1500	1500
LLC/SNAP-Overhead	0.78%	1.27%	1.69%	2.94%
AAL-Mode	5	5	2	2
Overhead due to ATM	12.55%	16.67%	21.46%	32.50%
Overhead due to physical medium	3.77%	3.77%	3.95%	9.97%
Total overhead	21.00%	28.33%	51.50%	104.00%
Physical line Rate (Mbps)	155.52	622.08	100.00	44.74
Maximum User line Rate (Mbps)	128.53	484.74	66.01	21.93

payload of the datagram of the IP-packet payload. For payload sizes less than 50 bytes, the overhead introduced by the protocols increases to more than 250%. This is severe for interactive gaming, ping, SMS or any small band transport traffic.

The notches in the graph for all ATM protocols are due to the padding imposed by the ATM adaptation layers. For packet sizes bigger then 200 kB, the limits of ATM-based IP compared to IP on Ethernet are very transparent. The highest overhead on high packet sizes is on DS3 making it the medium of last choice. The graph is based on an MTU size of 576 bytes, which is the Internet network default. For higher MTU sizes, Gigabit Ethernet can utilise its higher packet size.

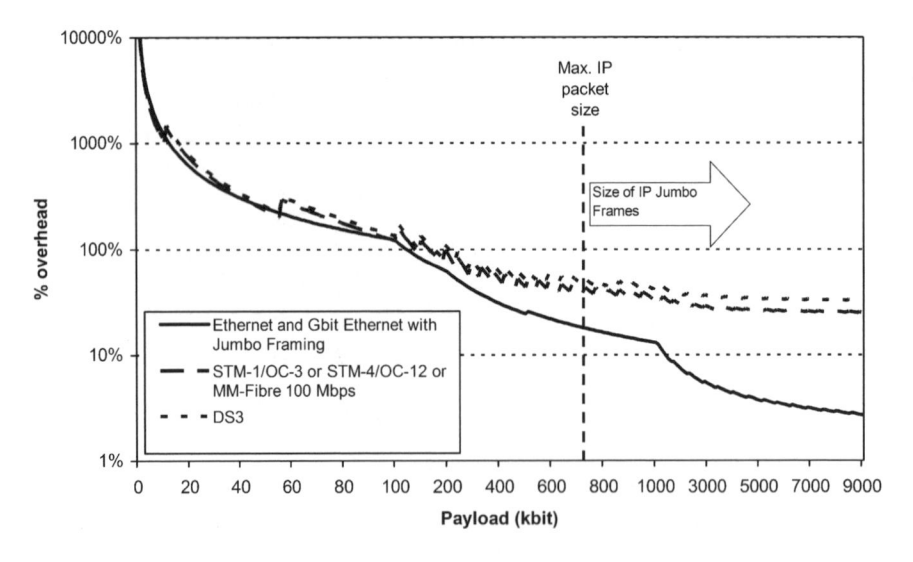

Figure 17.24 Overhead dependency to IP packet payload size.

17.4 NETWORK SOLUTIONS FOR UTRAN TRANSMISSION INFRASTRUCTURE

17.4.1 LEASED LINES

Leased lines (LL) are a main type of circuit-switched WANs, which allow permanent connection between two points set up by a telecommunications common carrier, therefore sometimes referred to as private lines or dedicated lines. Businesses, service provider network operator and even individuals typically use leased lines to connect geographically distant facilities (i.e. offices, network devices) or for Internet access. There are two types of the leased lines [31]:

1. Point-to-point leased lines;
2. Multi-point or multi-drop leased lines.

The point-to-point leased line is used to connect two separate facilities directly one to another, providing full-time and full-capacity communications. Leased line point-to-point channels are often terminated into multiplexers that connect to other telecommunication devices.

The multi-point or multi-drop leased lines are used to join several different facilities to a single central facility with common transmission channels. The devices attached to those channels share the capacity of the channel. Using a polling protocol, the central facility controls each attached device for occurrence of data traffic. When the pooled device has data to send, it transmits the data. Otherwise, it sends a short response signal to the central facility to enable the polling of the next device onto the multi-point channel. The central facility receives data inputs from all remote devices, prepares responses to the data inputs, and transmits the responses to the appropriate multi-point device. Both types of leased lines can be accomplished by using a number of approaches with throughputs from 64 kbps to 622 Mbps and more:

- Digital Data Services (DDS), which essentially use 56 or 64 kbps channels;
- ADSL which is currently offered using 768, 1536 or 3072 kbps;
- T-1 (DS-1) with bit rate 1.544 Mbps or T-3 (DS-3) with 44.736 Mbps carrier backbone;

- E-1 (CEPT-1) with bit rate 2.048 Mbps, E-3 (CEPT-3) with bit rate 34.368 Mbps or E-4 (CEPT-4) with bit rate 139.246 Mbps carrier backbone;
- J-1 with bit rate 1.544 Mbps or J-3 with bit rate 32.064 Mbps carrier backbone;
- SDH/SONET (Synchronous Optical Network Technologies) backbone:
  - STM-0/STS-1 (OC-1) with bit rate 51.48 Mbps
  - STM-1/STS-3 (OC-3) with bit rate 155.52 Mbps
  - STM-3/STS-9 (OC-9) with bit rate 466.56 Mbps
  - STS-12/STM-4 (OC-12) with bit rate 622.08 Mbps;
- *Dark fibre* backbone (customer leases the fibre itself), using commonly Ethernet like G802 or Gigabit Ethernet.

To set up the leased line, the transmission media and Channel Service Unit (CSU) are needed. The transmission media might be copper twisted-pairs or fibre with higher capacities. The CSU terminates each end of the carrier facility (e.g. E-1, T-1). The CSU equalises the received signal, filters the transmitted and received waveforms, and interacts with another customer's devices to perform diagnostics. Essentially, the CSU is used to set up the T-1/E-1/J-1 line with a customer-owned PBX, channel banks as stand-alone devices, intelligent multiplexers (for example, T-, E-, or J-carrier multiplexers), and any other DS-*x*/CEPT-*x*–compliant DTE, such as digital cross-connects. Using intelligent multiplexers, it is possible to manage transmission resources and to aggregate onto a higher-speed transmission line.

Leased lines provide high throughput and can come with SLAs (Service Level Agreement) and SLGs (Service Level Guarantee) that govern the 'uptime' of the circuit and technical support of the telecommunications provider.

On the benefit side, leased lines are required to support large networks, high speed private circuits, host web servers, transfer large amounts of data and files and run multimedia applications. Especially new 2G and 3G mobile operators need leased lines to complete their own backhaul networks.

The main disadvantages of this solution are the highest cumulative costs (CAPEX plus OPEX) as well as limited implementation. The CAPEX depends on installation and activation expenses. The OPEX is mainly connected with leasing expenses, which depend on the country, the operator, the geographical location, the transmission rates and the length of the leased line. Annual rentals for 2 Mbps at the distance of 2 km national leased line amounts to about a few thousands Euro a year. For instance, in Poland, the rentals for LL from the national telecomm operator are regulated by the decision of the president of the national telecommunications regulator (UKE). In other European markets, like UK or Germany, several city carriers have started to offer 3G backhauling using Ethernet G802, which is marketed as 'Metro Ethernet'.

Furthermore, the technical means of an LL customer to ensure SLA-guarantees are very limited, because most of the LL carriers disclose neither the route or used medium, nor the methods of how unavailable seconds are retrieved. This problem will increase with the rapid growth of Ethernet-products, which tolerate significantly lower availability standards for circuit emulated leased lines (see Section 17.2.1.6).

The limited implementation of leased lines to UTRAN transmission infrastructure concerns limited access to leased lines at many base station locations, especially out of urban areas. In many cases it requires additional investments that increase the CAPEX. Additionally, mobile operators using leased lines become dependent on other telecommunication operators; however, this is not always avoidable.

17.4.2 POINT-TO-POINT SYSTEMS

One of the main and the most popular elements of all backhaul and backbone networks are point-to-point (PTP) systems. They provide high capacity and high availability connections between two separate fixed facilities. For mobile cellular networks, the most flexible and cost-effective solutions are fixed-wireless broadband systems called radio relay systems or radio links. They operate with microwave frequencies (from 2 GHz up to 60 GHz) and match the capabilities of cable-based transmission systems as well as provide the same type of protocols and interfaces as leased lines (from E1 to STM-1). Radio relay systems enable outstanding generic benefits in planning and installing new backhaul networks. Fixed wireless relay systems are suitable solutions over difficult terrain, in rural areas or in old city centres where costs of wireline infrastructure or building restrictions may prevent from new construction [32]. For these reasons, the radio relay systems are being used as links between RNCs and Node Bs, between RNC and MSC or between RNCs. Their range and capacity depends on the used frequency band, the transmitter power, the transmit and receive antenna gains, the receiver sensitivity and the radio channel bandwidth with capacity ranging from a single E1 to two times STM-1 (311 Mbit/s).

Typical radio relay systems consist of two terminal stations (end points), which are very close to the source or destination of traffic. The terminal station is equipped with a set of identical radio transmitters, a set of identical radio receivers, filters, a directional antenna and multiplexer. Transmitters and receivers are connected to the shared antenna by coaxial cable or waveguide. The terminal stations can be divided into two parts: an indoor unit (IDU) and an outdoor unit (ODU). The indoor unit is connected with the traffic source or the traffic destination and supports some ODUs. The outdoor unit is integrated with antenna and mounted on a mast or a high tower. Parabolic reflectors, shell antennas and horn radiators are typical antenna types used in radio relay systems. Their diameter can reach up to 4 m. High antenna directivity permits efficient use of the available spectrum and long transmission distances, even when the transmit power equals to only 1 W.

Backhaul networks involve a very high availability and reliability, which typically ranges from 99.9 to 99.999 % of the time. The radio relay system availability is strongly connected with propagation conditions, deep fading and device malfunctions. To obtain the best propagation conditions, the terminal stations are located on hill-tops, mountain-tops, very high buildings and towers to achieve line of site (LOS) conditions with the first Fresnel zone free of obstacles (for more details, see Chapter 5). For the LOS propagation in the lower atmosphere strata (troposphere), the radio wave propagation is only affected by changes in barometric pressure, temperature, snowfall, rain, water vapour, turbulence and stratification (but of course the problem with multipath propagation still exists). For the frequencies above 10 GHz, scattering on hydrometeors and the molecular absorption are of great importance. For these reasons, the service range of radio relay systems depends on the climate zone where devices are to be installed. Figure 17.25 shows an example of the range of a radio relay system installed in regions with different rain intensity.

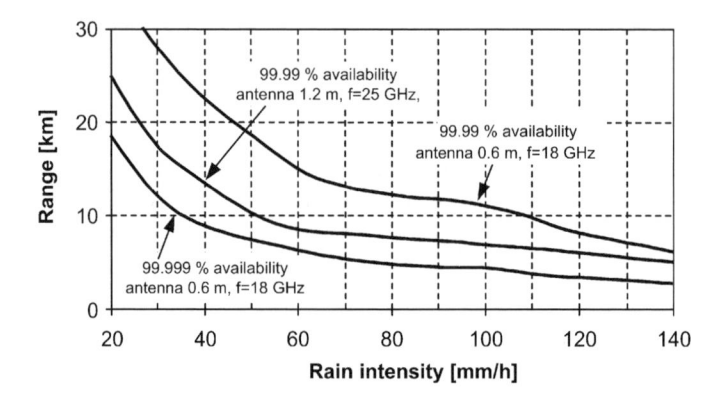

Figure 17.25 Example of the radio relay link ranges.

In the frequency range below 10 GHz, multipath propagation is considered as a dominant factor, which limits the range of the radio relay systems. The obstacles in the first Fresnel zone as well as atmospheric effects lead to problems of signal interferences and distortions (amplitude, frequency and phase dispersion). Only adaptive channel equalisation and diversity techniques are the best way to reduce multipath propagation effect. The multipath fading is overcome by using frequency diversity or space diversity. In frequency diversity the same information is transmitted by two separate links, which operate simultaneously on different frequencies (radio channels). In space diversity, additional antennas and receivers are required. These antennas have to be mounted at the same terminal point, but separated by a distance equivalent to some wavelengths.

Usually, the radio relay system capacity is not concentrated in a single channel but divided into some channels. Each channel uses an independent set of transmitter and receiver. In configurations without redundancy, all the sets are being used for data transmission between terminal points, providing maximum link capacity. In configurations with redundancy, only a part of the sets is used for transmission. The remaining sets are capable to take over any of the other sets in case of its malfunction. A typical configuration for radio relay systems is know as an '$m+1$', where there is only one reserve radio set and m active sets [33].

Obstacles, the curvature of the Earth, the area configuration and reception issues have to be taken into consideration before radio relay links can be established. A single link can be established if the terminal stations are close enough so that the Earth's curvature can be neglected and if there is no obstruction on the radio path. Long distance relay links may require intermediate stations (repeater or relay station) to receive and retransmit the signals. Links between terminal station and relay station or between relay stations are called hops.

Large-scale mobile networks demand branching stations to concentrate the traffic from base station to the controller. Branching stations allow to comprise individual low capacity links into the one higher and multiple capacity link and arrange star-topology networks with limited antenna numbers in the RNC. An example of a UMTS radio infrastructure with radio relay links is shown in Figure 17.26.

Microwave radio relay systems are used to connect backhaul traffic at the local level or at the access level with capacity ranging from a single $n \times E1$ to double STM-1 (311 Mbit/s). Utilised frequencies allow a connection over a link longer than 30 km. The access distances are typically short and capacity ranges from a $2 \times E1$ up to $16 \times E1$. At the local level, distances might be longer and the capacity requirement is larger, ranging from 16 E1 to STM-1. The fixed-wireless broadband systems deployed in 3G systems are a mix of SDH (Synchronous Digital Hierarchy) and PDH (Plesiochronous Digital Hierarchy).

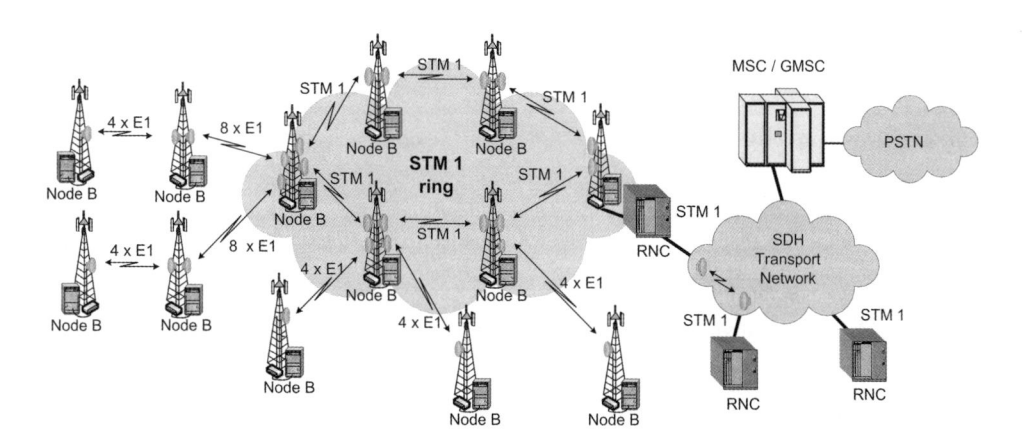

Figure 17.26 An example of UMTS backhaul network.

17.4.3 POINT-TO-MULTIPOINT SYSTEMS – LMDS

In metropolitan areas, high capacity cellular networks require a large number of base stations. All the base stations are usually located within a few hundred meters from each other. In order to serve the high densities of UMTS base stations, a large number of links to the RNC has to be established. In this case, the range of point-to-point microwave radio lines is not a serious problem; however, the number of links and required antennas on rooftops as well as costs and lose of flexibility constitute a problem. The Point-to-MultiPoint (PMP) system seems to be a better solution for the backhaul network to reduce costs and simplify network installations and future extensions. A single PMP base station can serve many remote Node Bs. Within the range of PMP entities, an operator has only to equip Node Bs with appropriate terminal stations and configure transmission parameters at the base station [32].

The radio transmission in PMP systems is not symmetrical as in PTP, and also line of sight (LOS) conditions are required. In the downlink (from PMP base station to remote terminals), the PMP base station broadcasts signals to all served entities in the cell or sector. The capacity of the broadcast channel is split and dynamically allocated between all the remote terminals. Each terminal is able to receive only information from broadcasted signals, which is dedicated to it. In the uplink, radio resources have to be shared between all the active terminals.

Due to the asymmetric transmission in uplink and downlink, the PMP systems are called Local Multipoint Distribution System (LMDS). The acronym LMDS is derived from the following:

- **L** (local) denotes that propagation characteristics limit the potential coverage area of a single cell site; ongoing field trials conducted in metropolitan centres place the range of an LMDS transmitter at up to a few kilometres;
- **M** (multipoint) indicates that signals are transmitted in a point-to-multipoint or broadcast method; the wireless return path, from subscriber to the base station, is a point-to-point transmission;
- **D** (distribution) refers to the distribution of signals, which may consist of simultaneous voice, data, Internet and video traffic;
- **S** (service) implies the subscriber nature of the relationship between the operator and the customer; the services offered through an LMDS network are entirely dependent on the operator's choice of business.

At the beginning, LMDS systems provided simple services, e.g. POTS data transmission, with low throughput as an alternative for point-to-point systems. At present, dependent on the configuration, LMDS systems provide a full range of broadband services with ATM or IP transmission for the last mile. The LMDS system architectures, access techniques, modulation methods as well as solutions are discussed in following sections.

17.4.3.1 System Architecture

LMDS systems are being designed for transport of multimedia services. This requires detailed technical specification for the system architecture to provide simultaneous access of many users to different services with defined quality of service and system availability. Special demands apply to an access and transmission network to allow dynamic transmission resource assignment on demand. A standard LMDS setup has a central facility connected to the operator's network hub via point-to-point microwave links. Basically, the point-to-multipoint wireless system architecture consists of four parts:

1. Network operations centre (NOC);
2. Fibre-based infrastructure;
3. Base station;
4. Customer Premise Equipment.

The network operations centre includes the network management equipment for managing regions of customer networks and can be interconnected with other NOCs. The fibre-based infrastructure basically consists of SDH or SONET links, the ATM and IP switching systems, interconnections with the network and the central office equipment.

The conversion from fibered infrastructure to wireless infrastructure happens at the base stations. The base station provides interface to fibre termination, modulation and demodulation functions, microwave transmission and reception equipment, as well as optionally local switching, which allows communications for customers without entering the fibre infrastructure. This function implies that billing, channel access management, registration and authentication have to be implemented locally within the base station.

The customer premise equipment varies widely from vendor to vendor. All configurations include indoor digital equipment and outdoor-mounted microwave equipment.

The system architecture depends on the type of delivery media to transport rendered services. The typical broadcast LMDS network architecture for Digital Video Broadcasting (DVB) networks with an interactive channel is presented in Figure 17.27. The interactive system is composed of Forward Interaction path (downstream) and Return Interaction path (upstream) [34]. The customer premise equipment, which is called Set Top Box (STB), consists of Network Interface Unit (NIU) and Set Top Unit (STU). The NIU comprises two network-dependent elements: Broadcast Interface Module (BIM) and Interactive Interface Module (IIM). The general concept is to use downstream transmission from the Interactive Network Adapter (INA) to the NIUs to provide synchronisation and information to all NIUs. This allows the NIUs to adapt to the network and send synchronised information upstream. The STB is equipped with interfaces to the broadcast and interactive channels, but only module IIM enables STB communication with the network.

Upstream transmission is divided into time slots, which can be used by different users. One downstream channel is used to synchronise upstream channels. A counter at the INA is sent periodically to the NIUs, so that all NIUs work with the same clock. This gives the opportunity to the INA to assign network resources to different users.

Three major access modes are provided with this system. The first one is based on contention access, which lets users send information at any time with the risk to have a collision with other users' transmissions. The second and third modes are contention-less based, where the INA either provides a finite amount of radio resources to a specific NIU, or a given bit rate requested by a NIU until the INA stops the connection. These access modes are dynamically shared with collision avoidance for the contention-less based access modes.

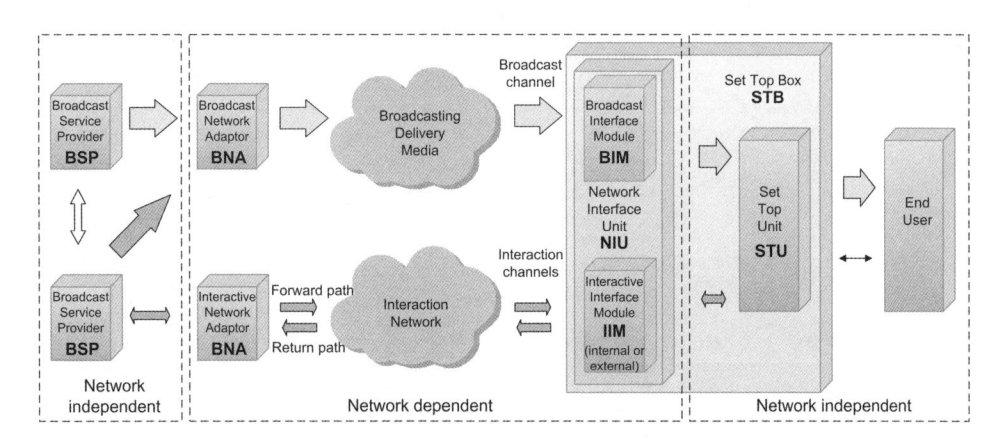

Figure 17.27 A generic LMDS reference model for interactive systems (based on [34]).

By periodically leaving a large time interval, the INA indicates to new terminals the possibility to execute the sign-on procedure, in order to synchronise their clock to the network clock without collisions with already active STBs.

All STBs have to support at least one of those solutions. Each of them differs in the overall system architecture, but provides the same quality of service. Both can be implemented in the same system under the condition that different frequencies are assigned to each system.

Nowadays, the complete implementation of the generic model presented in Figure 17.27 in interactive terrestrial broadband radio systems is too expensive and there is not any reason to do that in one system. Terrestrial broadband radio access systems can be split into two categories. The first category fulfils demands of telecommunication operators and the second one meets the requirements of cable television operators, who consider provision of telecommunication services. However, the interactive terrestrial broadband radio systems have immense competitors in various satellite systems. For this reason, as well as operators' demands and some technical aspects, mainly the terrestrial data transmission networks are being developed and implemented. The architecture of these systems corresponds with point-to-multipoint systems, the configuration of which depends on the used radio access methods. The radio is often the ideal way of obtaining communications at low cost, large distances and almost independent of the difficult topography.

According to ETSI EN 301 213-1 [35], a PMP system is comprised of a base station, which is called Central Station (CS), and a number of terminals (Figure 17.28). The CS may be subdivided into two Central Controller Station (CCS) and Central Radio Station (CRS) or more units. The CCS enables connection to the telecommunication networks via SNI interface and controls at least one or more number of CRS. The CRS provides the air interface to the terminal station or to repeater stations. Each CRS's radio transceiver is connected to a separate directional sector antenna to increase the capacity of the PMP system and to connect to other repeater stations or terminal stations within the CRS in MultiPoint–MultiPoint (MP–MP) systems.

The performance of the LMDS system transmission can only be achieved for an unobstructed line of sight between radio stations or terminals, e.g. CRS and Terminal Stations (TS). Repeater Stations (RS) are deployed to expand the range of the PMP network or to establish radio communication to some TSs without LOS in the direct radio path. The number of served TSs and RSs depends on the assigned channel bandwidth and provisioned services. The terminal station provides the interface to the terminal equipment, the subscriber premise equipment. A RS can serve one or more TSs or other

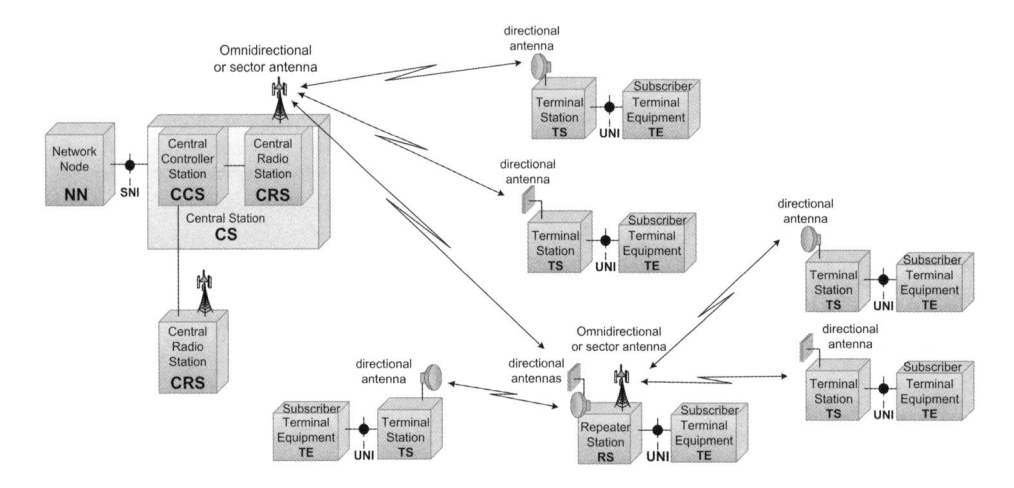

Figure 17.28 General PMP system architecture (based on [35]).

RSs. The RS may also provide the fixed interfaces to the Terminal subscriber Equipments (TE). The UNI is the point of connection to the subscriber equipment TS. All system elements are not necessarily deployed in a particular network, but the single CS is indispensable. In PMP systems, radio terminals from different manufacturers are not intended to interwork at radio frequency. Network operators have to choose solutions according to performance and availability requirements in order to extend the possible area of application, thus fitting to their network needs.

17.4.3.2 Access Techniques and Modulation Methods

PMP wireless broadband access systems use different access methods, taking into account the basic physical parameters of frequency, code and time. This leads to the four access methods of:

1. Frequency Division Multiple Access (FDMA);
2. Time Division Multiple Access (TDMA);
3. Code Division Multiple Access (CDMA);
4. Multi-Carrier Time Division Multiple Access (MC-TDMA).

The FDMA PMP system transmits a RF-signal from the Terminal Stations or Repeater Stations to the Central Radio Station only utilising a spectral bandwidth corresponding to that capacity which is requested from and assigned to the customer by Pre-Assigned Multiple Access (PAMA) or by Demand Assigned Multiple Access (DAMA). The Central Radio Station receives from each TS a single modulated carrier being processed independently within the CRS [36]. Thus, the CRS is receiving an FDMA PMP signal. For the FDMA links, the terminal station is allocated a fixed bandwidth, or a bandwidth varying slowly over time. FDMA access links fit in well if the user requirement is constant bandwidth and continuous availability.

In TDMA PMP systems, a central station broadcasts information to terminal stations in a continuous Time Division Multiplex (TDM) or in a burst TDMA mode [37]. The Terminal stations transmit in TDMA mode. The users may have access to the spectrum by sharing it through time multiplexing. The TDMA mode makes sense for customers who do not have a very heavy upstream traffic and just needs a 10 BaseT port for access to the Internet.

The DS-CDMA Central Radio Station transmits simultaneously and continuously to all active Terminal Stations within its coverage area, utilising a specific set of codes allocated to each active Terminal Station [38]. The terminal stations use the same, or a different, set of codes when transmitting to the CRS. Transmissions from CRS to TS are distinguished from transmissions on the other directions by using different frequency channels (FDD mode) or different time slots (TDD mode). Repeater stations may be placed for cell coverage enhancing. But CDMA supports a significantly smaller number of users than TDMA.

It is also possible to use Multi-Carrier Time Division Multiple Access (MC-TDMA) as a more flexible alternative to TDMA, FDMA and CDMA. The CRS and/or TS may transmit one or more sub-carriers at various frequencies, bandwidths, modulation and power levels. In these multiplexing access modes, the normative requirements relating to channel bandwidths and spectrum masks must be met. The terminal stations transmit in TDMA mode. The users may have access to the spectrum by sharing it through time multiplexing.

All of the abovementioned multiple access methods are implemented in the uplink channels (from TS or RS to CRS). In the downlink channels, most companies supply Time Division Multiplexing (TDM) or a burst TDMA mode.

The data rate capacity and the bandwidth spectrum efficiency of LMDS systems depend on the used modulation schemes. Modulation methods for broadband wireless LMDS systems are generally separated into Phase Shift Keying (PSK) and amplitude modulation (AM). Broadband transmission is only possible using higher order modulation schemes; for example, in LMDS systems with FDMA,

BPSK, QPSK, DQPSK, 8PSK, 4QAM, 16QAM or 64QAM modulation methods can be used. Typical modulation schemes for LMDS systems in the downlink channels are QPSK, 16QAM or 64QAM. In the uplink channels, most applied modulations are QPSK and 16QAM. The 64QAM scheme is identified as optional due to the cost impact of supporting this format, resulting from stringent constraints on power amplifier linearity and radio phase noise requirements.

The Reed-Solomon forward error coding (FEC) and convolution coding methods protect radio transmission. Coding methods are combined with modulation schemes and are changed simultaneously to maximise the capacity of the available bandwidth for each radio channel condition. In particular, terminal stations located near to the base station could take advantage of 64QAM even during a rainstorm, while subscribers at the edge of a cell would be limited to QPSK.

The selection and implementation of an appropriate modulation scheme is driven from two primary perspectives, the first being the desired efficiency of the system and the second being the reliability of the network.

17.4.3.3 Capacity

The capacity of LMDS networks can be measured in terms of data transmission rate and maximum number of associated terminal stations (customer premises). For data rate calculations, the capacity of each LMDS base station has to be known. The LMDS base station capacity depends on its configuration, e.g. the number of sectors, types of used modulation schemes, sectors' overlapping areas, available channel number in the sector and channel bandwidth. The maximum number of terminal stations (customer premises) is related to the LMDS base station capacity, as well as the data transmission rate required by each customer. In this case, the maximum range of the LMDS system is very important.

Modulation and channel bandwidth have a main impact onto the capacity of a LMDS and its range. The capacity depends on the spectral efficiency of used modulation schemes and the system range depends on the required E_b/N_0 at the receiver input for signal demodulation at a given bit error rate (BER). Some examples are presented in Table 17.11.

Spectrum efficiency not only depends on the used modulation method, but also on the channel filter mask roll-off factor, which is important to determine the correct relationship between channel bandwidth and data rates. For these reasons, real values of spectrum efficiency for LMDS systems differ from the theoretical and maximal ones, as presented in Table 17.11.

The system capacity increases for wider channel bandwidths, but the range decreases because of the higher noise level. The channel spacing in Europe is usually obtained by successive division of 112 MHz by 2. Table 17.12 contains detailed information about frequency bands preferred for LMDS systems, regulations and channel spacings. The channel arrangements and minimum PMP base station transmission capacity have been presented in Table 17.13.

The capacity in LMDS uplink and downlink channels usually differs because of physical layer function differences of both channels, even if the bandwidth is the same. In PMP systems physical

Table 17.11 Spectral efficiency and required E_b/N_0 for the modulations used in LMDS systems.

Modulation	QPSK	4QAM	16QAM	64QAM
Theoretical maximum spectral efficiency [bps/Hz]	2	4	6	8
Maximum spectral efficiency of LMDS systems [bps/Hz]	1.5	1.5	3.5	5
E_b/N_0 @ BER = 10^{-6}	10.5	15	18.5	24

Table 17.12 Frequency bands available to PP and PMP systems.

Band [GHz]	Frequency range [GHz]	Regulations and guidelines for channels arrangements	Channel spacing [MHz]
3.5	3.4–3.6	ERC/REC 14-03	1.75; 3.5; 7; 14
3.7	3.6–3.8	ERC/REC 12-08	1.75; 3.5; 7; 14
10.5	10.15–10.30 10.50–10.65	ERC/REC 12-05	3.5; 7; 14; 28
26	24.25–26.6	ITU-R F.748, ERC/REC 13-04; ERC/REC 13-02; ERC/REC 00-05	3.5; 7; 14; 28; 56; 112
28	27.5–29.5	ITU-R F.748, CEPT REC 13-04; 13-02; DEC (00)09	3.5; 7; 14; 28; 56; 112
32	31.8–33.4	ITU-R F.1571, ERC/REC 01-02	3.5; 7; 14; 28; 56
38	37–39.5	ERC/REC T/R 12-01	3.5; 7; 14; 28; 56; 140

Table 17.13 Channel arrangements and minimum PMP base station transmission capacity (based on [36,37]).

Minimum CRS transmission capacity [kbps] for modulation	\multicolumn{7}{c}{Channel spacing [MHz]}
	1.75	3.5	7.0	14	28	56	128
4QAM	21×64	42×64	84×64 4×2048	8×2048	16×2048	32×2048	64×2048 or STM1
16QAM	42×64	84×64 4×2048	8×2048	16×2048	32×2048	64×2048 or STM1	128×2048 or $2 \times$ STM1
64QAM	3×2048	6×2048	12×2048	24×2048	48×2048	96×2048 or STM1	192×2048 or $2 \times$ STM1

layer and medium access control layer issues (i.e. channel coding, filtering, access and multiplexing methods) have to be taken into consideration during network planning. Each sector of PMP base station (CRS) shall provide minimum transmission capacity within the channel spacing. Requirements for each frequency band are defined in the appropriate standards (i.e. EN 301 213-2, EN 301 213-3 for systems in the frequency range 24.25 to 29.5 GHz which use FDMA and TDMA respectively).

The PMP system capacity can be considered as the maximum number of terminal stations simultaneously connected to the CRS sector and transporting their maximum payload bit rate. But the minimum payload (expressed either as the number of 64 kbit/s signals or an aggregate bit rate for the used traffic type, e.g. ATM cells) shall be taken into account as well. The maximum number of terminal stations simultaneously connected to the LMDS base station shall be declared by the manufacturer. That number can be calculated for each access method. PMP systems are limited by capacity and range. The network capacity can be increased through the use of sectorisation and frequency reuse mechanisms that allow the simultaneous use of the same frequencies in base station and the network. Typically, the LMDS base station provides service with 4, 8, 14, 16 or 24 sectors using directional antennas with 90, 45, 30, 22.5 or 15-degree beam-width respectively. Capacity can be increased by using different modulation schemes for different sectors. Some configuration examples of an LMDS base station are presented in Figure 17.29. The available configurations depend highly on the manufacturer's solutions.

The LMDS system coverage depends on several factors: frequency band, bandwidth, modulation type, type of the terminal being used, system availability and rainfall. The system availability is guaranteed by appropriate fading margins depending on frequency band and installation place, i.e. rain zone.

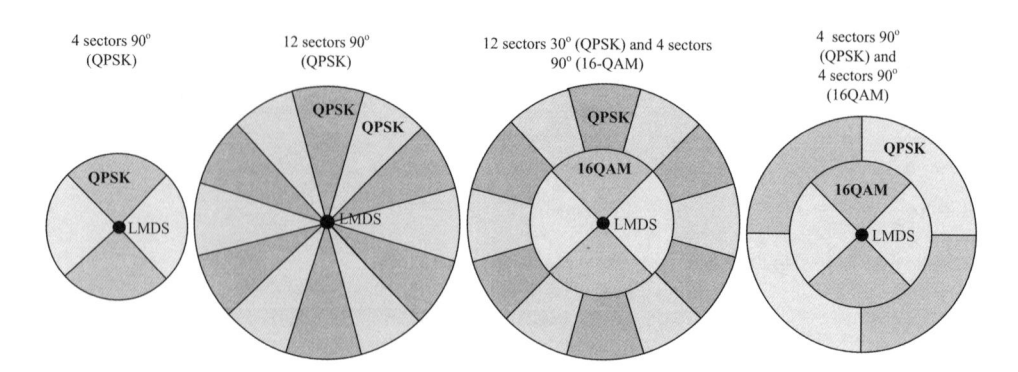

Figure 17.29 Examples of LMDS base station configurations.

The transmitter output power for CRS, RS and TS is limited to +35 dBm. For the same system capacity, the system coverage depends on transmitting and receiving antenna gains. Increasing the antenna gain of the terminal station, the service range can be expanded (Table 17.14). In this case, antennas with higher gains can be applied. But high gain antennas have to comply the standards (e.g. ETS 300 833). It must be noted that the LMDS station range can also be noticeably dependent on the used polarisation. The range difference for horizontal and vertical polarisations rises with frequency and depends on the ITU rain zone. Usually, the terminal station antennas are designed to be highly directional with high gain and narrow beamwidth (e.g. 18 dBi and 18° @ 3.5 GHz, 35 dBi and 2.5° @ 26 GHz).

Co-channel interference is caused by frequency reuse at terminal station receivers by remote LMDS base station transmitters and at LMDS base station receivers by remote terminal station transmitters. The interference imposed by TS transmitters on remote LMDS BS receivers is more detrimental than the interference imposed by LMDS BS transmitters onto remote TS receivers, because the former hinders reliable reception for all terminal stations served from the remote BS, while the latter only affects a few TSs.

The co-channel and adjacent-channel interferences can be reduced by physically separating potential interferers from each other and maximising the isolation between adjacent sectors in the network through orthogonal polarisation (horizontal and vertical) [39]. The antennas with narrow beamwidths located as high above ground as possible allow reducing multipath propagation effects, cross-polarisation and co-channel interferences in the network.

The lower the interference levels, the higher the frequency reuse factor and the network capacity for the same radio resources. Higher reuse factors reduce interference more than lower reuse factors, but not without cost; for a fixed total spectral allocation, higher reuse factors reduce the amount of usable spectrum per LMDS base station. Conversely, to maintain the same amount of usable spectrum per LMDS BS, higher reuse factors require larger total allocations. A smaller allocation requires a denser grid and more LMDS base stations, increasing costs (also backbone and switching). Conversely, a larger allocation allows a sparser grid with fewer LMDS base stations and lower costs. Some frequency allocation models for LMDS networks, which also include polarisation separation, are presented in Figure 17.30.

17.4.3.4 Local Multipoint Distribution Systems (LMDS) Solutions

The LMDS can support a variety of services over the same infrastructure and a wide range of customers can be served from the same base station. For these reasons, it is the perfect *last mile* solution for many communication and information technologies: IP, ATM, Ethernet, Frame Relay, Leased Line,

Table 17.14 Examples of LMDS range in kilometres.

Frequency band [GHz]	Modulation schemes	Base station sector [°]	Terminal antenna	Rain zone E 7 MHz channel 99.999%	Rain zone E 7 MHz channel 99.99%	Rain zone E 14 MHz channel 99.999%	Rain zone E 14 MHz channel 99.99%	Rain zone H 7 MHz channel 99.999%	Rain zone H 7 MHz channel 99.99%	Rain zone H 14 MHz channel 99.999%	Rain zone H 14 MHz channel 99.99%
10.5	QPSK	90	integrated	13.7	21.4	10.8	16.7	9.9	16.3	8.2	13.3
	16-QAM	90	integrated	9.6	15.0	7.6	11.7	6.9	11.4	5.7	9.3
26	QPSK	30	integrated	3.1	4.9	2.8	4.3	2.5	4.2	2.2	3.6
			390 mm	4.1	6.6	3.7	5.8	3.3	5.7	2.9	5.0
			650 mm	5.1	8.4	4.6	7.5	4.0	7.3	3.6	6.5
		90	integrated	2.7	4.3	2.4	3.7	2.2	3.7	1.9	3.2
			390 mm	3.6	5.7	3.2	5.1	2.8	5.0	2.5	4.4
			650 mm	4.4	7.3	4.0	6.5	3.5	6.4	3.1	5.6
	16-QAM	30	integrated	2.2	3.4	1.9	3.0	1.8	4.2	1.6	2.5
			390 mm	2.9	4.6	2.6	4.1	2.3	2.9	2.1	3.5
			650 mm	3.6	5.9	3.2	5.2	2.8	5.1	2.5	4.5
		90	integrated	1.9	3.0	1.7	2.6	1.5	3.7	1.4	2.2
			390 mm	2.5	4.0	2.2	3.5	2.0	2.6	1.8	3.1
			650 mm	3.1	5.1	2.8	4.5	2.4	4.5	2.2	4.0

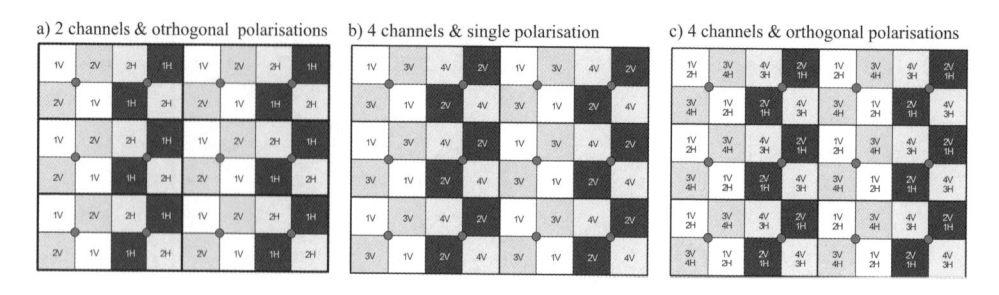

Figure 17.30 Frequency allocation models for LMDS network.

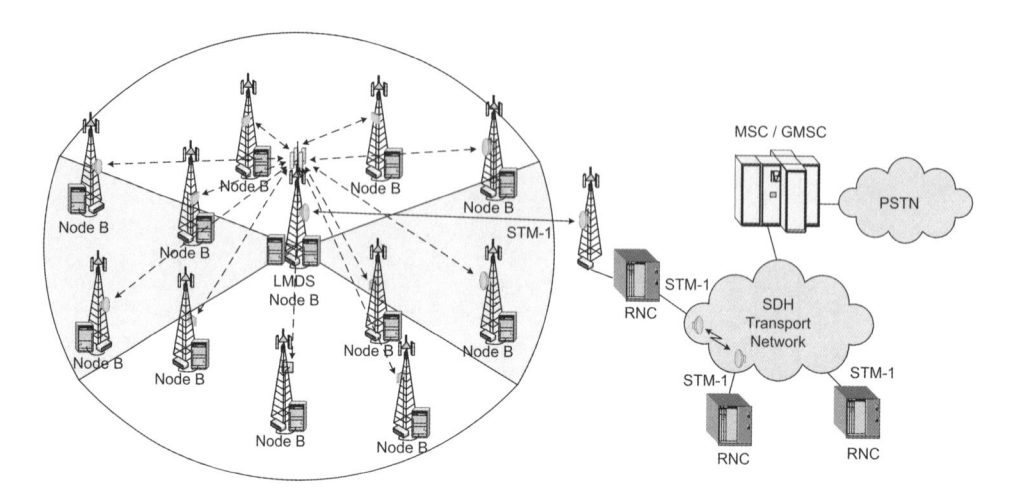

Figure 17.31 An example of LMDS backhaul network for UMTS.

POTS and ISDN, which provide end users different services: VoIP, data transmission and access to the Internet. It is a flexible solution for interconnecting 2G, 3G and collocated 2G/3G base stations to the transmission network (Figure 17.31). The LMDS system provides telecommunication services to small and medium enterprises (SME), small offices/home offices (SOHO) and residential customers, and ensures wireless local loop backhaul. The modular and scalable system architecture allows the operator start deploying small networks with low CAPEX and low risk and expands the network if necessary. Additionally, LMDS and PMP systems enable efficient and fast installation of new elements (terminal stations), re-deployment and network modifications according to new operator demands. Dynamic bandwidth allocation and control allow an efficient use of system resources. Efficient backhaul connectivity provides coverage for mobile and fixed narrowband wireless system base stations with cost effective transmission infrastructures. In the last several months, operators have become interested in the new WMAN systems, which offer much higher transmission rates then LMDS systems without LOS requirement. In the future WMAN systems will replace LMDS systems.

17.4.4 *WiMAX AS A POTENTIAL UTRAN BACKHAUL SOLUTION*

The system, which recently gained big market popularity, is WiMAX. Because its main application deals with point-to-multipoint broadband wireless access, it was natural to assess its UTRAN backhauling possibilities. The following section gives a WiMAX system overview, whereas more practical guidelines can be found in Section 17.5.

17.4.4.1 WiMAX Standard Overview

As already elaborated on in Section 4.5.2, the WiMAX system concept originated from the IEEE 802.16 standard family; this relation is very similar to the one between WiFi and the IEEE 802.11 standards. The entire IEEE 802.16 group is designed for carrier class metropolitan access. The family consists of the following standards:

- IEEE 802.16 – main Fixed Wireless Access standard;
- IEEE 802.16a – enhancement for 2–11 GHz band;
- IEEE 802.16b – Quality of Service;
- IEEE 802.16c – enhancement for 10–66 GHz profiles (interoperability);
- IEEE 802.16d – adapts 802.16a to mobility requirements;
- IEEE 802.16e – completes support for mobility (including handover);
- IEEE 802.16f/g – network management extensions;
- IEEE 802.16h – license exempt operation.

The main standard in the family is IEEE 802.16 [40]. It specifies the radio interface for point-to-multipoint broadband wireless access systems working on frequencies from 10 to 66 GHz. Medium Access Control (MAC) layer and three physical layers (OFDM, OFDMA and single carrier) are specified. The most popular standard currently is called 802.16 Revision D (or 802.16-2004). It was approved in July 2004 and focuses on fixed applications and smart antenna enhancements for indoor applications. It also extends the system band to frequencies below 11 GHz. It consolidates the base standard and all the amendments ('a' and 'c'). Therefore, most of the WiMAX equipment currently available on the market is compliant with the Revision D standard.

The future standard which specifies Mobile WiMAX is 802.16e. It incorporates features and protocols needed for portability and mobility. It also includes some new modes, such as SOFDMA, to enhance the portability and mobility performance. Commercial applications of this standard are expected in 2007–2008.

In parallel to the IEEE effort to standardise WiMAX, there are also related works in ETSI, dealing with the HiperMAN standard. The European standard is in fact identical to the IEEE 802.16 Revision D, except that the only supported physical layer is OFDM.

The third important body for the system standardisation is the WiMAX Forum. It is an industry organisation appointed for a worldwide promotion of IEEE 802.16 and ETSI's HiperMAN standards. The main goal was to have one choice for physical layer and optional parameters, to enable interoperability and industry usage of the common technology. The WiMAX Forum also drives the effort for test specifications and performs test lab equipment certification processes.

It is also worth mentioning that WiMAX is under evaluation as a future potential UMTS access method (802.16e standard) in the framework of 3GPP standardisation. More information on WIMAX standards can be found on [41,42].

17.4.4.2 System architecture

The typical WiMAX system architecture as depicted in Figure 17.32 consists of:

- Base Station (BS);
- Customer Premises Equipment (CPE);
- Repeater;
- Network Management System.

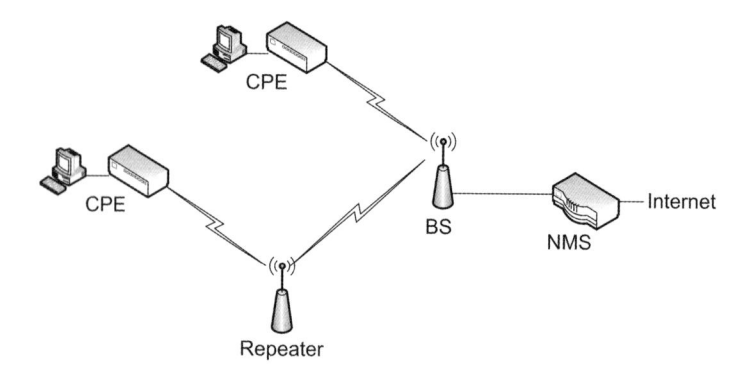

Figure 17.32 WiMAX System architecture.

The typical base station configuration is from one (omnidirectional) up to six sectors. The repeater is usually similar to the base station, but without backhaul connectivity; this enables range extension. There can be two main types of CPE: outdoor and indoor. While using outdoor ones, the same system coverage can be achieved with a less number of base stations (only outdoor coverage service is needed). The role of NMS is WiMAX network management. Additionally, the system should have some routers and switches for traffic connectivity and aggregation.

17.4.4.3 System Characteristics

The WiMAX system, compatible with IEEE 802.16 Revision D standard, can yield up to 70 Mbps in a 20 MHz channel in the frequency range from 2–11 GHz. However, the equipment available now uses mainly 3.5 MHz channels that can achieve 12.7 Mbps throughput in one channel. The second currently available channel size is 1.75 MHz. The other possible channel sizes are: 5, 7, 10 and 20 MHz. The system can use both FDD and TDD duplexing methods, which depends on the frequency band used. The usual FDD duplex separation is 100 MHz. The bands available for WiMAX are:

- 3.4–3.6 GHz (currently the most popular band; 1.75 MHz, 3.5 MHz channels in FDD mode available now; 7 MHz channel in TDD mode available in future);
- 2.5–2.7 GHz (5 MHz channels in TDD mode);
- 3.6–3.8 GHz (possible extension band for 3.4–3.6 GHz);
- 5.1–5.3 GHz (possible future extension band);
- 5.4–5.8 GHz (5 and 10 MHz channel in TDD mode; license exempt band);
- 698–746 MHz (possible future extension band).

The only available physical layer is OFDM with 256 sub-carriers (FFT size). The modulations and code rates supported are:

- BPSK (1/2, 3/4);
- QPSK (1/2, 2/3, 3/4);
- 16QAM (1/2, 3/4);
- 64QAM (2/3, 3/4, 5/6).

In fact, the system is using an adaptive modulation scheme. Therefore, the optimal modulation is chosen in dependency of the particular environment scenario. Table 17.15 presents example throughputs that can be achieved on a single channel.

Table 17.15 Example WiMAX single channel throughput.

Modulation	Channel bandwidth	
	3.5 MHz	1.75 MHz
BPSK 1/2	1.41 Mbps	0.71 Mbps
BPSK 3/4	2.12 Mbps	1.06 Mbps
QPSK 1/2	2.82 Mbps	1.41 Mbps
QPSK 3/4	4.23 Mbps	2.12 Mbps
QAM16 1/2	5.64 Mbps	2.82 Mbps
QAM16 3/4	8.47 Mbps	4.24 Mbps
QAM64 2/3	11.29 Mbps	5.56 Mbps
QAM64 3/4	12.71 Mbps	6.35 Mbps

The other important system features are protocol independent core and QoS mechanisms, as well as security. From the core perspective, it is important that WiMAX can transport IPv4, IPv6, Ethernet, ATM or others, supporting multiple services simultaneously. The standard which deals with QoS is 802.16b. It enables non line-of-sight (NLOS) operation without severe distortion of the signal from buildings, weather and vehicles. It also supports intelligent prioritisation of different forms of traffic according to its urgency. The security mechanism provided in WiMAX includes measures for privacy and encryption, such as authentication with x.509 certificates and data encryption using DES in cipher block chaining (CBC) mode with hooks defined for stronger algorithms like Advance Encryption Standard (AES).

17.4.4.4 Examples of System Applications

According to the WiMAX Forum vision, the goal for the system is to create a global mass market for deployment of broadband wireless networks, which will enable fix, portable and mobile users to maintain high speed connectivity and to lead the 'access everywhere' revolution supporting delivery of data, voice and video applications at home, in the office and on the go. Thus, and because of wide features and functionalities of WiMAX, the system can address different market segments:

- Business access market;
- Residential broadband access (Internet + other services);
- Cellular network backhaul;
- WiFi hotspots connectivity.

For business users (enterprises), WiMAX provides an equivalent of N times E1/T1, ATM, Frame Relay or Fast Ethernet connectivity and fractions of them. The system can be used instead of fibre networks or xDSL connections with greater cost effectiveness for subscribers. Furthermore, for residential access the system is either equivalent of xDSL and cable TV access. Especially, while designing the network for outdoor CPE usage, the cost effectiveness seems to be far better than for classical copper line systems.

The main WiMAX usage, from this chapter's perspective, are cellular networks (including UMTS) backhauling. The details on backhauling possibilities and limitations of PMP system are presented in Section 17.6. The subset of this, but with slightly different demands, is WiFi hotspot connectivity; particularly for cities with a large density of WiFi hotspots it provides more cost effective solutions for WiFi hotspot transport.

17.4.4.5 Possible Evolution Paths

The main WiMAX evolution path paves the way to mobility dimensioning. As current WiMAX market implementations may be named fixed outdoor systems, the fixed indoor and nomadic ones are only one step ahead. All of these are connected with the 802.16 Revision D standard, and the main differentiating factor for them is the CPE type (outdoor or indoor). Furthermore, fixed outdoor requires fixed location, installation outside the subscriber's house and provides applications of E1/T1 or fractional E1/T1 level services for enterprises, backhauling solutions and limited broadband access (early adopters, rural and developing countries). Whereas fixed indoor enables consumer self installing and auto provisioning, gives possibility of nomadism (subscriber can move CPE to another location in service area) and provides broader range of last mile broadband access for residential consumers.

Further evolutionary steps are related to the 802.16e standard; they are hence about portability and mobility. The portable version is dependant on WiMAX enabled chipsets incorporated into laptop PCs (plans have already been announced by Intel). It will provide some handover mechanisms, but with 1–2 second interruption type. The seamless handover for real-time services will be available in the full mobile version only. The devices will be PDA or mobile phone types, rather than mobile PCs. As mentioned before, this version is expected to be incorporated into the 3GPP standards framework, as a complementary UMTS access method.

17.5 EFFICIENT USE OF WiMAX IN UTRAN

WiMAX systems seem to be a big opportunity for usage as UTRAN complementary backhaul solutions. For the UMTS system planner, every possibility that can optimise investments should be evaluated. Therefore, it is very important to assess CAPEX and OPEX savings that can incur because of a WiMAX implementation and to estimate all the technical drawbacks and limitations of such solution. This section deals with both of them.

17.5.1 DIMENSIONING OF WiMAX FOR UTRAN INFRASTRUCTURE

To assess the reasonability of using WiMAX for UTRAN backhauling, a similar method as described in Chapter 8 should be used. All the system implementation and maintenance costs should be calculated and compared with alternative solutions (e.g. LMDS, point-to-point microwave links, leased lines). Since the results can be different per clutter type, the case that suits a particular environment best should be used for comparison purposes. Probably the best method is to compare live transmission network costs in a particular area with the costs of hypothetical WiMAX rollout. To enable such a comparison, some initial WiMAX network assumptions should be made, e.g. Node B collocated with 2G BTS, WiMAX BS not collocated with 2G BSC, connection between Node B and RNC via WiMAX and ATM network, WiMAX system equipped with E1 interfaces, etc.

Given the assumptions, both coverage and capacity dimensioning phases should be performed. First of all, the WiMAX BS configuration should be decided upon. An example BS configuration could be:

- 4 sectors;
- 7 MHz of bandwidth per sector;
- resulting average capacity of 14 Mbps per sector.

All the other BS physical parameters should be properly set as well, in order to enable power budget construction and pathloss (cell range) calculation. For ease of comparison, it is useful to convert the total site throughput to E1s. By dividing the particular network area by the WiMAX coverage (cell) area, the number of WiMAX sites is obtained. Furthermore, having the sum of all 3G sites' transmission needs (in E1s) divided by the WiMAX BS capacity, the number of WiMAX capacity sites is calculated. The bigger of these values is the number of WiMAX sites needed.

The WiMAX CAPEX consists of the number of sites multiplied by the single site cost (equipment, civil work, etc.) plus the WiMAX license cost (it is important to mention that, across most countries, the WiMAX license can be nationwide or assigned for a particular administrative area only). The WiMAX OPEX consists mainly of equipment maintenance that can be easily calculated with drivers related to CAPEX (as presented in Chapter 8).

While considering a completely new network and CAPEX related solutions (microwaves or LMDS), a credible financial comparison with other backhauling methods can be done by simply comparing related CAPEXs. When comparing WiMAX versus leased lines, the sums of cumulative CAPEX plus OPEX should be compared. In this case, the advantage of WiMAX is only the matter of time (thus, also the PayBack Time can be calculated). Furthermore, to evaluate the savings because of a possible migration of existing backhauling technologies to WiMAX, similar methods can be used, but with serious attention paid to PayBack Time.

The results of such business case scenarios performed by the authors clearly showed WiMAX to be applicable in relatively dense and clustered towns. While comparing to leased lines (according to current pricings), a WiMAX solution is always cheaper (different PayBack Times). And finally, a WiMAX solution is also cheaper than microwave links when one WiMAX BS can support more than 10–12 Node Bs.

17.5.2 CURRENT WiMAX LIMITATIONS

From the previous section, it can be incurred that WiMAX could be a very good and cost effective UTRAN backhauling technology, at least in particular scenarios. However, looking into deep technical details, some serious limitations may also be noticed. One of them is the total WiMAX network capacity limit, which is not sufficient for future UTRAN capacity needs. The other one is the network synchronisation issue.

The mobile network radio interface requires a synchronisation stability of $5 \cdot 10^{-8}$ s [43–45]. Such stability cannot currently be provided while connecting Node Bs via a packet switched network (like WiMAX). It hence seems that synchronisation can be a serious blocking factor for using WiMAX as a UTRAN backhauling system. Fortunately, there are three candidate solutions that can overcome this issue. These are:

1. Network Time Protocol (NTP);
2. IEEE 1588;
3. SYNCoIP.

The already established protocol and readily available solution at the server side is a local reference steered by the NTP. However, the possibility to achieve the required accuracy is uncertain. Furthermore, because of the long time constant, the cost of a local reference is relatively high.

The next possible solution for synchronisation over packet is IEEE 1588. With this standard, the reference is maintained and distributed through the Ethernet network. It is an already standardised protocol, which has a more than required accuracy. It could also easily be integrated into WiMAX terminals. But it requires an Ethernet based network only and must be supported at each network element; this is clearly more difficult, because the protocol is relatively new, with not so many commercial implementations.

The last way to overcome the synchronisation issue is SYNCoIP: a reference is transported over IP in 'CE style'. The drawback of that method is that the development is not finalised and IP QoS support may be needed for that method to work. On the other hand, it provides the required accuracy, relies on IP only and is developed from existing techniques.

Concluding, the synchronisation issue seems to be a hard blocking factor for WiMAX implementation as a UTRAN backhauling solution now. Current methods to overcome this issue are either not widely available on the market or not even mature enough for implementation. However, there are real chances for having synchronisation over WiMAX in the very near future.

17.6 COST-EFFECTIVE RADIO SOLUTION FOR UTRAN INFRASTRUCTURE

In the previous sections, it has been shown that the UMTS network imposes some strict demands on the UTRAN transmission infrastructure. These demands, in comparison to other systems as e.g. GSM, arise from the higher capacity of UMTS networks, especially in dense urban areas where a significant increase in the number of Node Bs and a growth of the bit rate are being observed. Additionally, the number and type of served services as well as transport format (ATM or IP) effect the capacity of UTRAN transmission links. In many cases, the transmission infrastructure has to serve base stations of two networks (GSM and UMTS), and requirements for both systems have to be considered during backhaul network planning.

The UTRAN backhaul network has to fulfil high transmission requirements of Node Bs and provide a low purchase and operating cost, good transmission quality and reliability, flexibility and scalability. Some possible solutions for UTRAN transmission infrastructure have been presented in Sections 17.4 and 17.5. Due to an easy network installation and extension, the fixed radio systems (e.g. PTP, PMP, LMDS and WiMAX) seem to be the most suitable. For the PTP systems, the spectrum efficiency is the highest, but the costs, large number of required antennas and inflexibility of this approach rise serious problems. The PMP or WiMAX systems are much more flexible, but their cost-effectiveness depends on the application. In the case of supporting a UMTS backhaul network, dedicated methods of finding a cost-effective solution are required. These methods and some comparative results for applications of PTP and LMDS in UTRAN will be presented in following sections.

17.6.1 RF PLANNING ASPECTS

Radio backhaul network planning for UMTS covers a series of problems related to both UMTS and transmission system planning. Before commencing with the design of the backhaul network, which is to serve the UMTS system, it is necessary to define the key parameters of UMTS. These parameters affect the architecture and parameters of the UTRAN transmission network. The most important for the analysis are UMTS base station locations and required throughput according to service types and transmission overhead. All this information can be derived from the real UMTS network or any suitable planning tool. Having the UMTS environment defined, the radio backhaul network can be introduced.

Correct RF engineering and planning of the backhaul network requires ensuring optimal deployment of the network. The output of these tightly coupled engineering exercises results in a multi-year plan for the deployment and installation of radio systems (PMP and PTP) and its backbone transport and switching infrastructure. The goal in all RF engineering and backhaul network planning aspects is to provide high-quality services over high-availability wireless links to as many Node Bs as possible within each localised region in a resource-efficient manner. This means maximising the RF range, taking into consideration the statistical nature of the radio performance. The radio system specifications (gains, transmit powers, receiver sensitivities, etc.), when matched to worst-case rainfall and atmospheric attenuations and interference-induced degradations, yield the radio system range of the fixed radio system; the latter is defined by an operator at a minimum required BER and a percentage of the time the network can provide services. For backhauling applications, the operator shall demand a BER of at least 10^{-6} and a system availability of 99.99 or 99.999 %; poor service quality or system availability of the radio backhaul network can damage business viability of the UMTS network.

It has to be noted that parameters and functionality of fixed radio transmission systems differ from vendor to vendor; therefore, ranges shall be considered for each system installation independently. In particular, it concerns the rain region, used frequency band, channel spacing, modulation scheme, terminal types and antenna gains. The maximum service radius is an important design parameter of the backhaul network. It mainly concerns systems with a central base station (PMP, LMDS and WiMAX), because the number of base stations required for a complete coverage is inversely proportional to the maximum cell radius of each of them. Usually, manufacturers publish only maximum idealised interference-free service ranges for some typical configurations. The service range examples of LMDS for different configurations, consisting of 90- and 30-degree sectors and three different subscriber type antennas, have been presented in Table 17.14.

Frequency reuse is necessary in the real backhaul network with many base stations; however, this causes interferences at some locations within the service area. These interferences can degrade radio transmission performance (range, throughput, availability and quality) of some radio links and for this reason shall be taken into account for planning complex and wide networks to reach required transmission performance to all Node Bs in the UMTS network.

For all radio systems (PTP, PMP, LMDS, WiMAX), the best transmission performance is achieved for LOS conditions with the first Fresnel zone devoid of any obstacles. Only WiMAX systems can operate without direct line of sight, but in this case their ranges and availability decrease dramatically. For this reason, during UMTS radio planning, installation of additional radio equipment for a backhaul network requiring LOS conditions to other devices shall be considered. The only practical solution is the use of sophisticated planning procedures, implemented in planning tool. These procedures should study inter-PMP interferences and PTP-on-PMP interferences; they should also be able to find the best Node Bs locations according to UMTS *and* the backhaul network performance. The choice of a solution for the UTRAN transmission infrastructure depends on the radio planning results and the cost analysis.

17.6.2 THROUGHPUT DIMENSIONING

A very important part of designing the UTRAN transmission network is the dimensioning of the required throughput for each Node B in the UMTS network. For this reason, the output of radio network planning is taken as input for the radio access network design. All the data can be divided into a number of parts, which represent clusters in RNC service areas. For each cluster, the number of base stations, their location and the maximum traffic demand (including all system procedures and services) have to be known. The calculation of maximum traffic demands shall consider not only traffic in the UMTS radio interface, but also control information indispensable for managing Node Bs. The transmission formats used in the interface between Node B and RNC in conjunction with served services are of great importance to the calculations to the Node B's throughput. It is mainly related to the transmission overhead, which varies from service to service. Detailed requirements for UTRAN transmission infrastructure and throughput dimensioning have been presented in Sections 17.1–17.3.

The accurate analysis of maximum Node B throughput requirements, information about distance and propagation conditions in the radio path to each other Node B in the cluster are sufficient to design the UTRAN transmission infrastructure using Point-to-Point radio systems. Some types of radio lines used in backhaul networks and their parameters have been presented in Table 17.16.

For the backhaul radio network solution, which makes use of the central station, the designer has to know the total capacity of all Node Bs in the cluster and the cluster size, which is defined as the maximum distance from between remote Node Bs in the cluster. To design the radio backhaul network, the operator has only to know the radio resources, which are being assigned to the operator by the

Table 17.16 Parameters of radio relay systems used in UTRAN backhaul networks.

Type	No. of E1 interfaces	Total throughput (kbps)	Channel spacing (MHz)
PDH 2E1	2	4096	3.5
PDH 4E1	4	8192	7
PDH 8E1	8	16384	14
PDH 16E1	16	32768	28

national spectrum regulator. The amount and bandwidth in assigned channels, as well as frequency band and Node B layout, determine the central station configuration, service area range and number of base stations in the cluster. Of course, during the designing process, the operator has to take into consideration all the aspects mentioned in this chapter.

Because analysis and design are quite similar for PMP, LMDS and WiMAX systems, only the LMDS solutions will be considered here. The LMDS base station parameters and configuration have a vital impact onto the system capacity. On the other hand, they highly depend on the actual system being used. For the purpose of analysis, let us consider the potential capacity of modern commercial LMDS systems from a major manufacturer. The LMDS central station (CS) capacities for some configurations and assigned radio resources with 7 MHz channel spacing have been presented in Table 17.17.

All considered configurations provide the co-channel and adjacent-channel interference reduction by using other radio channels in adjacent sectors for QPSK and 16QAM modulation. For this reason, LMDS base station configurations with QPSK and 16QAM require at least four radio channels. This assumption does not allow configuring LMDS base stations if the spectrum regulator has assigned the operator only two radio channels. It is theoretically possible to assign the same channels to all sectors (QPSK or 16QAM), but the orthogonal polarisations in the adjacent sectors have to be used. The transmission with 16QAM modulation is more susceptible to interference than QPSK, given its higher carrier-to-noise requirements; therefore, one-frequency solutions with 16QAM LMDS base stations are not recommended.

Regardless of the frequency band, the sector capacity only depends on the modulation scheme, channel number and channel spacing. For this system, the useful net bit rate in the 7 MHz channel with QPSK modulation is 8.192 Mbps; however, due to the required overhead (forward error correction, encryption, MAC messages) the raw bit rate is about 12 Mbps. The sector capacity doubles if 16QAM modulation is implemented instead of QPSK, or if 14 MHz channels are used instead of 7 MHz channel.

Denser sectorisations provide higher spectrum efficiency, higher LMDS base station capacity and larger ranges. However, this should only be done for a high Node B density in the LMDS coverage area, otherwise it is possible that some too narrow sectors will not serve any Node B.

17.6.3 METHODS OF FINDING OPTIMAL LMDS NETWORK CONFIGURATIONS

Optimal network configuration means that the UTRAN transmission infrastructure will provide required capacity and will operate with a minimum number of equipment and radio resources. To solve this problem, a suitable optimisation method is necessary; however, there is no single method available, which allows finding the optimal radio backhaul network solution for UTRAN. Due to the many possible solutions, we limit our considerations only to some methods of finding the optimal LMDS network configuration. All these methods can be adapted to any other backhaul network solution, which makes use of radio systems with a central base station, e.g. PMP or WiMAX.

The definition of optimal network configuration has to be modified for LMDS used for the UTRAN transmission infrastructure. In this case, the LMDS base station shall serve all Node Bs in the coverage

Table 17.17 LMDS central station capacity for 7 MHz channel spacing.

Configuration	Radio resources	14 MHz	28 MHz	56 MHz	112 MHz
	Modulation	QPSK	QPSK	QPSK	QPSK
	Channels in sector	1	2	4	8
	Sector capacity Mbps	8	16	32	64
	Total CS capacity Mbps	32	64	128	256
	Channels in CS	4	8	16	32
	Modulation	QPSK	QPSK	QPSK	QPSK
	Channels in sector	1	2	4	8
	Sector capacity Mbps	8	16	32	64
	Total CS capacity Mbps	96	192	384	768
	Channels in CS	12	24	48	96
	Modulation	—	QPSK+16QAM	QPSK+16QAM	QPSK+16QAM
	Channels in sector		1 1	2 2	4 4
	Sector capacity Mbps		8 16	16 32	32 64
	Total CS capacity Mbps		192	384	768
	Channels in CS		8	16	32
	Modulation	—	QPSK+16QAM	QPSK+16QAM	QPSK+16QAM
	Channels in sector		1 1	2 2	4 4
	Sector capacity Mbps		8 16	16 32	32 64
	Total CS capacity Mbps		160	320	640
	Channels in CS		16	32	64
	Modulation	—	QPSK+16QAM	QPSK+16QAM	QPSK+16QAM
	Channels in sector		1 1	2 2	4 4
	Sector capacity Mbps		8 16	16 32	32 64
	Total CS capacity Mbps		288	576	1152
	Channels in CS		24	48	96

area with a transmission quality and availability required by the operator, or better. It shall limit all implementation and operational costs of the LMDS network and provide an efficient utilisation of all the equipment as well as allocated spectrum.

All the methods can be divided into two groups. In the first one, the analysis is performed only for a theoretical deployment of UMTS base stations according to the basic cellular system design methods with regular and triangular grid and hexagonal cells. The range and capacity of each Node B are the same for the entire network. The UMTS base station range depends only on the link budget calculated for the given transmitter and receiver parameters, environment, service types, number of frequency channels being used, loading, sector types and cell loading. The capacity is calculated for a given service or mix of services, required transmission overhead and control data. In the second type of methods, the analysis is preformed for a real UMTS network or for data obtained from UMTS radio network-planning tools. In this case, information about the Node B locations as well as their transmission link capacity is only required.

17.6.3.1 Approximation Method

To check the optimal LMDS network configuration, the total capacity of all Node Bs located in the LMDS serving area has to be compared to the total LMDS base station capacity. The single UMTS base station capacity and a number of such base stations in the LMDS coverage are solely necessary for this kind of calculation, given that a uniform UMTS network with regular cell range is considered. Due to many available LMDS configurations, the most convenient way is to do the analysis for a single LMDS sector and then multiply the obtained result by the number of sectors. For uniform UMTS networks, the function which defines the number of Node Bs in a single sector of an LMDS base station can be approximated by a polynomial given in [46]. The method is universal given that the approximation is applied to the relation between LMDS range and Node B radius. The number of Node Bs in an LMDS sector depends on the mutual location between LMDS and UMTS base stations; therefore, the approximation has to be prepared separately for each LMDS sector. There are two main LMDS base station locations with reference to the UMTS Node B. The first one is called *collocation* (Figure 17.33a), because the LMDS base station is placed together with one of the Node Bs. In the second one, which is called *no collocation*, the operator has to find an independent location for the LMDS base station (Figure 17.33b).

Knowing both system ranges and the polynomial function, it is possible to calculate the number of Node Bs in the LMDS sector and the total transmission capacity for a given service profile. The comparison of the maximum capacity offered in the LMDS sector and the required one by the UMTS base station located in it yields some information about efficiency. If the transmission capacity offered in LMDS sector is much higher than required by UMTS, this means that the LMDS range is limited and the LMDS base station serves too few UMTS Node Bs. Such a situation occurs usually in rural areas, where PTP solutions are much cheaper than LMDS ones. If the transmission capacity offered in the LMDS sector is not sufficient for UMTS requirements, this means that the LMDS capacity is limited and only some but not all Node Bs in the sector are served by the LMDS base station.

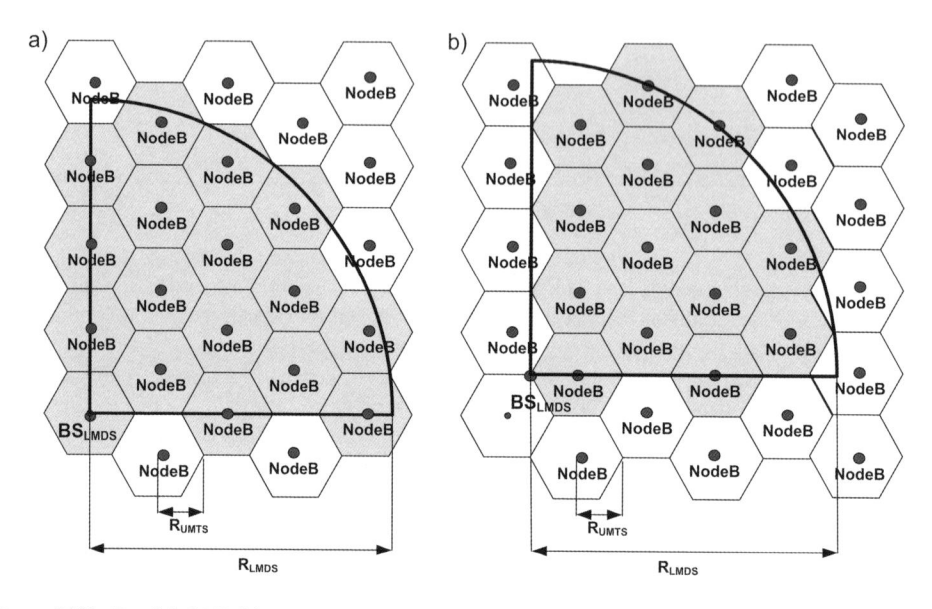

Figure 17.33 Possible LMDS base station location in reference to UMTS Node B: (a) collocation with Node B and (b) no collocation.

Thus, some efficiency factor of applying LMDS for UTRAN's backhaul network has to be known to find the optimal solution. The knowledge of the maximum LMDS sector range, which allows as many as possible Node Bs located in its coverage area to be served, is very important for planning purposes. This factor is referred to as the Maximum Sector Capacity Limited Range (M-SCLR). It has to be noted that the M-SCLR depends on the LMDS sector capacity, as well as on the Node B density and transmission requirements.

When comparing the range of a real LMDS sector with the calculated SCLR, it is possible to estimate its usefulness for the UTRAN backhaul network. If the range of a real LMDS system is the same or imperceptibly smaller then the maximum sector capacity range, this means that the LMDS configuration is (close to) optimum for the UTRAN transmission infrastructure. The LMDS sector capacity has to decrease, if the SCLR is greater than the real LMDS, because the LMDS configuration has likely been chosen wrongly (for example: too many sectors, or too many channels used in each sector, or channel bandwidth too broad). The LMDS capacity ought to be increased or the LMDS range decreased, if the sector capacity range is smaller than the real range.

In Table 17.18, some examples of the maximum sector capacity limited range for a 90-degree LMDS system and a uniform (theoretical) UMTS network with a loading of 25 % are presented. The sign '>>' in Table 17.18 means that the maximum sector capacity limited range of LMDS is much bigger than that for the longest LMDS range in the 10 GHz frequency band. The UMTS network design has been developed using the usual power budget and the COST-Hata propagation model. The Node B's maximum indoor range and required capacity to the RNC have only been determined for voice services (12.2 kbps) to UMTS low transmit power terminals (21 dBm). The Node B capacity was calculated for three sectors, taking into consideration the transmission overhead in ATM. Comparing this data to real LMDS ranges, it is possible to estimate whether the LMDS sector ranges are capacity limited or not. Such an analysis has been carried out for some of the (theoretical) UMTS networks serving only one type of service.

Calculation results for 90-degree LMDS sectors using a single channel of 14 MHz bandwidth for data transmission to and from UMTS base stations are presented in Table 17.19. In the frequency band of 26 GHz, the LMDS ranges do not exceed 7 km; thus, only for a few cases, an LMDS capacity limitation has occurred. This is usually observed for high bit rate services in dense urban environments, where QPSK is used; italic/bold fonts indicate all these cases. In other cases, the Maximum Sector Capacity Limited Range has exceeded the real range, and only the real range is written down in the table. Comparing these results to Table 17.18, it is clear that for a 16QAM scheme, the range is too short and capacity too high.

17.6.3.2 System Transmission Density

There is another way than presented in Section 17.6.3.1 to find an optimal LMDS base station configuration for UTRAN transmission networks. To determine the maximum LMDS sector capacity limited range for UTRAN backhaul networks, a comparison of the transmission density ratio for Node Bs and LMDS sectors versus the LMDS sector range has to be done [47]. The transmission density ratio is defined between the radio system transmission capacity (in bps) and the related surface of the served area (in m^2). The transmission density is hence expressed in $b/(s \cdot m^2)$ and combines two of the most important parameters. The Node B's transmission density function is calculated as the quotient of the Node B's required link throughput to the RNC and the surface of the coverage area. The LMDS sector transmission density is determined from dividing the LMDS sector throughput by its serving area.

The Node B transmission density can be assumed constant for specified environments and services. Additionally, it is independent from the LMDS sector range and distinct from the LMDS sector transmission density, which decreases with an increase of the LMDS sector range, achieving minimum value for maximum range.

Table 17.18 Maximum Sector Capacity Limited Range for LMDS 90° sector.

Environment				Maximum Sector Capacity Limited Range in km							
				Dense urban				Suburban			
Modulation scheme	LMDS channel bandwidth	Single LMDS channel throughput	Number of channels assigned to single sector	Voice 12.2 kbps	LCD64 kbps	LCD144 kbps	LC384 kbps	Voice 12.2 kbps	LCD64 kbps	LCD144 kbps	LC384 kbps
QPSK	7 MHz	8 Mbps	1	3.00	3.40	2.75	2.10	6.95	7.95	2.75	4.90
			2	4.45	5.10	4.15	3.05	10.35	11.85	4.15	7.10
			3	5.80	6.40	5.20	3.80	13.45	14.75	5.20	8.80
			4	6.70	7.45	6.05	4.40	15.50	17.20	6.05	10.20
	14 MHz	16 Mbps	1	4.45	5.10	4.15	3.05	10.35	11.85	4.15	7.10
			2	6.70	7.45	6.05	4.40	15.50	≫	6.05	10.20
16QAM	7 MHz	16 Mbps	1	4.45	5.10	4.15	3.05	10.35	11.85	4.15	7.10
			2	6.70	7.45	6.05	4.40	≫	≫	6.05	10.20
			3	8.35	9.10	7.40	5.50	≫	≫	7.40	12.75
			4	9.75	10.60	8.65	6.35	≫	≫	8.65	14.70
	14 MHz	32 Mbps	1	6.70	7.45	6.05	4.40	≫	≫	6.05	10.20
			2	9.75	10.60	8.65	6.35	≫	≫	8.65	≫

Table 17.19 Examples of 90° LMDS sector capacity limited range analysis results for some UMTS network configuration.

Environment					Sector Capacity Limited Range in km							
					Dense urban				Suburban			
UMTS loading	LMDS type	LMDS terminal type	LMDS availability %	LMDS range km	Voice 12.2 kbps	LCD 64 kbps	LCD 144 kbps	LCD 384 kbps	Voice 12.2 kbps	LCD 64 kbps	LCD 144 kbps	LCD 384 kbps
			Node B range in km		0.929	0.669	0.576	0.440	2.145	1.545	1.330	1.017
25 %	QPSK @ 26 GHz	Integrated antenna	99.999	2.40	2.40	2.40	2.40	2.40	2.40	2.40	2.40	2.40
		Antenna 390 mm	99.99	3.71	3.71	3.71	3.71	*3.08*	3.71	3.71	3.71	3.71
			99.999	3.19	3.19	3.19	3.19	*3.08*	3.19	3.19	3.19	3.19
		Antenna 650 mm	99.99	5.06	5.06	5.06	*4.16*	*3.08*	5.06	5.06	5.06	5.06
			99.999	3.99	3.99	3.99	3.99	*3.08*	3.99	3.99	3.99	3.99
			99.99	6.48	6.48	*5.13*	*4.16*	*3.08*	6.48	6.48	6.48	6.48
	16QAM @ 26 GHz	Integrated antenna	99.999	1.68	1.68	1.68	1.68	1.68	1.68	1.68	1.68	1.68
			99.99	2.60	2.60	2.60	2.60	2.60	2.60	2.60	2.60	2.60
		Antenna 390 mm	99.999	2.23	2.23	2.23	2.23	2.23	2.23	2.23	2.23	2.23
			99.99	3.54	3.54	3.54	3.54	3.54	3.54	3.54	3.54	3.54
		Antenna 650 mm	99.999	2.79	2.79	2.79	2.79	2.79	2.79	2.79	2.79	2.79
			99.99	4.54	4.54	4.54	4.54	*4.43*	4.54	4.54	4.54	4.54
			Node B range in km		0.829	0.597	0.514	0.393	1.915	1.379	1.188	0.908
50 %	QPSK @ 26 GHz	Integrated antenna	99.999	2.40	2.40	2.40	2.40	2.40	2.40	2.40	2.40	2.40
			99.99	3.71	3.71	3.71	3.71	*2.75*	3.71	3.71	3.71	3.71
		Antenna 390 mm	99.999	3.19	3.19	3.19	3.19	*2.75*	3.19	3.19	3.19	3.19
			99.99	5.06	5.06	*4.58*	*3.72*	*2.75*	5.06	5.06	5.06	5.06
		Antenna 650 mm	99.999	3.99	3.99	3.99	*3.72*	*2.75*	3.99	3.99	3.99	3.99
			99.99	6.48	6.48	*4.58*	*3.72*	*2.75*	6.48	6.48	6.48	**6.37**
	16QAM @ 26 GHz	Integrated antenna	99.999	1.68	1.68	1.68	1.68	1.68	1.68	1.68	1.68	1.68
			99.99	2.60	2.60	2.60	2.60	2.60	2.60	2.60	2.60	2.60
		Antenna 390 mm	99.999	2.23	2.23	2.23	2.23	2.23	2.23	2.23	2.23	2.23
			99.99	3.54	3.54	3.54	3.54	3.54	3.54	3.54	3.54	3.54
		Antenna 650 mm	99.999	2.79	2.79	2.79	2.79	2.79	2.79	2.79	2.79	2.79
			99.99	4.54	4.54	4.54	4.54	*3.96*	4.54	4.54	4.54	4.54

The intersection of the straight line defining the transmission density for UMTS with the curve characterising the transmission density of the LMDS sector gives the maximum sector capacity limited range (M-SCLR) for the LMDS system. This method is the most suitable for a (theoretical) deployment of UMTS base stations according to basic cellular system design methods with regular and triangular grid and hexagonal cells. For real UMTS networks, the cells with mean or maximum capacity should be considered. If only the LMDS sector transmission density is higher than the Node B transmission density, then the LMDS base station is able to serve all real UMTS base stations located within the LMDS coverage area.

The Node B transmission density has been calculated for basic UMTS propagation environments. The UMTS network design is the same as described in Section 17.6.3.1; respective calculation results are presented in Table 17.20.

The results of the Node B transmission density have been compared with some sector transmission densities for various LMDS base station configurations. The analysis was conducted for different numbers of radio channels in a single LMDS sector, QPSK and 16QAM modulation, as well as for two channel bandwidths (7 and 14 MHz). Figures 17.34 and 17.35 present these results for UMTS base stations deployed in dense urban and suburban environments, respectively, assuming a transmission density factor for a 90-degree sector and LMDS channels of 14 MHz.

It must be noted that the calculation results of the LMDS sector capacity limited range are the same for the approximation method and the system transmission density method; however, the method presented in the current section is much easier to implement, because it does not require any information about the Node B deployment and additional approximations.

17.6.3.3 Real Network Analysis Method

Both methods presented in Sections 17.6.3.1 and 17.6.3.2 allow to estimate the usefulness of various solutions that utilise radio systems with a central base station, e.g. PMP or WiMAX, for UTRAN backhaul networks; however, they presumed a UMTS network with a regular and uniform triangular grid. These methods are not suitable for real networks with irregular Node B deployment, different ranges and capacity, because they solely allow estimating the initial technical and financial effectiveness of LMDS-like systems as a transmission network. In this case, a method is needed, which allows analysing real UMTS network parameters and which find the best solution for the UTRAN transmission infrastructure. The detailed analysis of the real network has to show where an operator can expect profits using various configurations of the LMDS base stations (number of sectors, type of modulation, overlapping areas, available bandwidth and range with respect to ITU Regions, etc.). For these reasons, the comparison of implementation and operating costs of alternative solutions (i.e. PTP, PMP, LMDS, WiMAX) have to be taken into account by this method.

Typical configurations and deployments of LMDS base stations with respect to Node B locations have been depicted in Figure 17.36. In the presented cases, the number of Node Bs in neighbouring

Table 17.20 Node B transmission density in Mbps/km^2 (UMTS loading 25%).

Service	Environment		
	Dense urban	Suburban	Rural
Voice 12.2 kbps	0.8271	0.1535	0.0197
LCD 64	0.7158	0.1338	0.0173
LCD 144	1.0713	0.2009	0.0260
LCD 384	1.9636	0.3705	0.0481

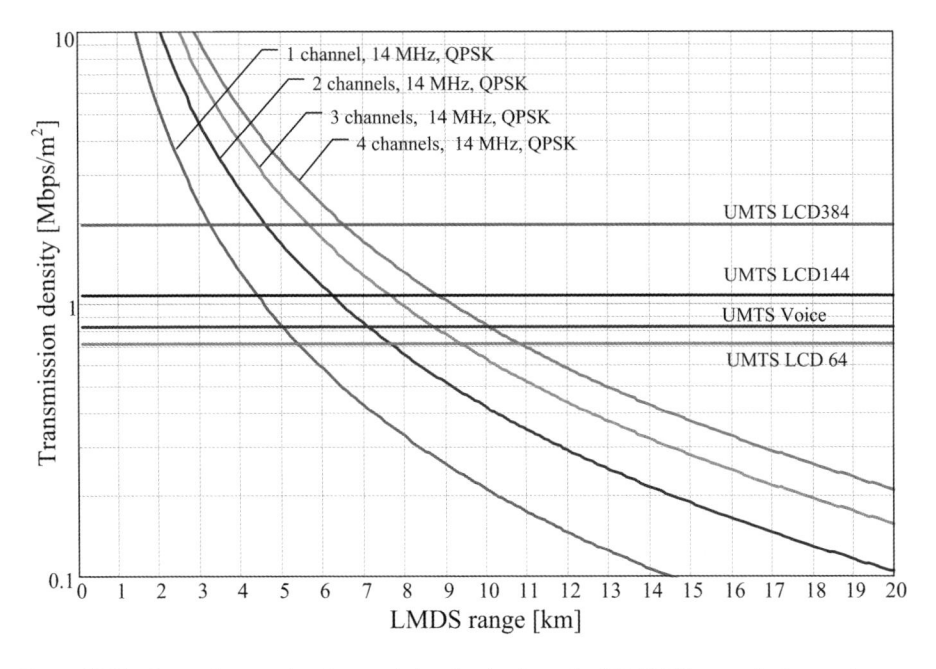

Figure 17.34 Comparison results of transmission density factor for 90° LMDS sector in dense urban area.

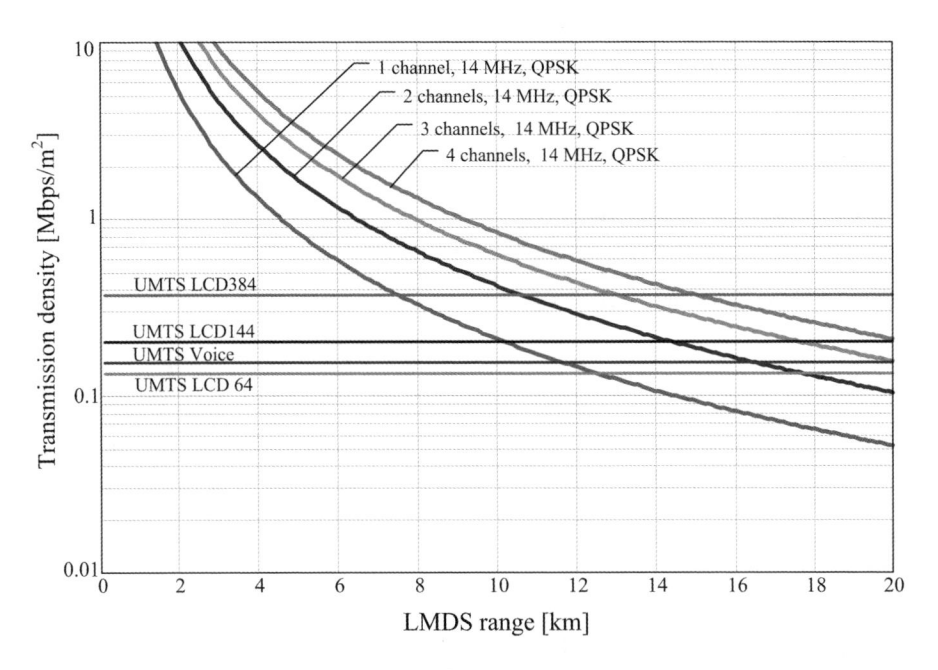

Figure 17.35 Comparison results of transmission density factor for 90° LMDS sector in suburban area.

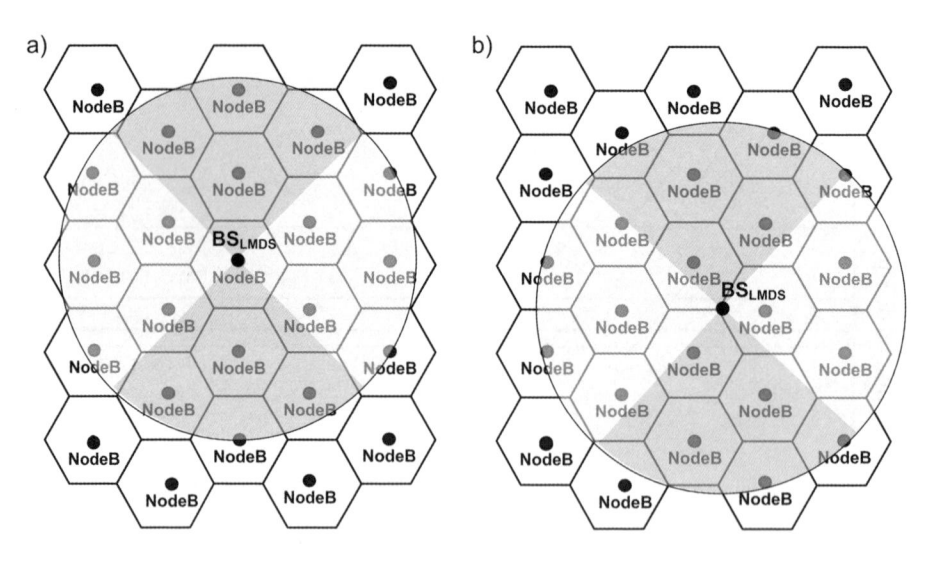

Figure 17.36 Influence of LMDS base station location on the number of served Node B: (a) collocation with Node B and (b) no collocation.

LMDS sectors can be different (from 3 up to 6 in the presented example). The sector orientations and range are of main significance, as well as the allocation procedure of Node Bs to the LMDS sectors. Thus, a generalisation of correct analysis results obtained for a single sector to the entire LMDS network can lead to incorrect results. This problem is much more complicated for real UMTS networks with non-uniform a Node B distribution.

The only method to obtain correct results is an analysis of the real LMDS base station and Node B locations, so as to get correct information about the number of Node Bs situated in each LMDS sector coverage area. Some additional parameters have to be taken into account: These concern LMDS base station configurations (sectors, modulation schemes, number of channels), coverage area and range of each LMDS sector according to the used modulation, frequency band, assumed availability and applied LMDS terminal stations. The function, which allows balancing the transmission load in adjacent LMDS sectors, should be applied as well. It should be done for Node Bs located at borders of the LMDS sector or the sector overlap area, if they are defined. It must be noted that LMDS systems vary greatly in their parameters, architecture, configuration, functionality and costs, depending on the manufacturer. Therefore, analyses have to be carried out independently for every of the manufacturer's solutions.

Some tailored computer applications are clearly necessary to study all these aspects and find some cost-effective solutions. An example of such a software [48] and some obtained cost analysis results are presented in Sections 17.6.4 and 17.6.5, respectively.

The description of other applications for automatic planning of PMP solutions for UTRAN backhaul networks as well as some calculation results for real network scenarios are presented in [49]. The authors focused on the optimisation of access networks to reduce purchase and operating costs. In the developed four-phase optimisation method, a fast heuristic algorithm has been implemented to determine the location of the PMP equipment. The algorithm uses a four-step procedure and hence solves the Weighted Independent Set (WIS) Problem for the PMP locations. Due to the underlying assumptions, the analysis was limited to Node Bs located in the same clusters and PMP base stations collocated with a Node B. PMP planning aspects or the overlapping of sectors have not been incorporated in the optimisation method.

17.6.4 COSTS EVALUATION OF UTRAN INFRASTRUCTURE – SOFTWARE EXAMPLE

For the case studies presented in the subsequent Section 17.6.5, the authors of this chapter have developed a UTRAN transmission network cost evaluation software for LMDS and PTP solutions. The software has been developed as an Excel Visual Basic for Application script, which allows configuring all required parameters of the LMDS and UMTS systems (Figure 17.37).

The software analyses the costs of real networks, hence requiring information about the real geographical Node B positions and transmission capacities. The application gives the user two possibilities to perform the calculations. The first one provides calculations for user defined data (for example coming from UMTS planning tools or real UMTS networks). The second one incorporates a simple network planning tool, which automatically distributes identical Node Bs using a regular grid. In the latter case, the UMTS base station locations depend on their range, which is defined based on the link budget for uniform services (voice transmissions at 12.2 kbps, data transmissions LCD64, LCD144, LCD384) and the Cost-Hata propagation model (only for urban areas). The user can define the service, system loading, building attenuations and the number of channels in each of three Node B's sectors. The required transmission capacity is computed as described in Section 17.2. The user can also define the percentage of the total bit rate, which is to be reserved for delivering user services. The remaining part can then be used for transmission of control information and for packet transmissions for UMTS users. If this percentage is declared as 70 %, this means that the required data bit rate, calculated from the link budget to carry out the given service by the UMTS base station, will comprise 70 % of the total bit rate. The remaining 30 % of the total bit rate can be used for other purposes, e.g. packet transmissions or control data for GSM base stations.

As a result of the above procedure, the program creates a grid of uniformly distributed UMTS base stations, forming regular hexagons with a given side length. The network size depends on the serving area of the chosen LMDS base station configuration.

In the automatic mode, the calculations can be carried out for LMDS base stations collocated with Node Bs or for the LMDS base station placed at the edge of three UMTS cells (Figure 17.36). In the

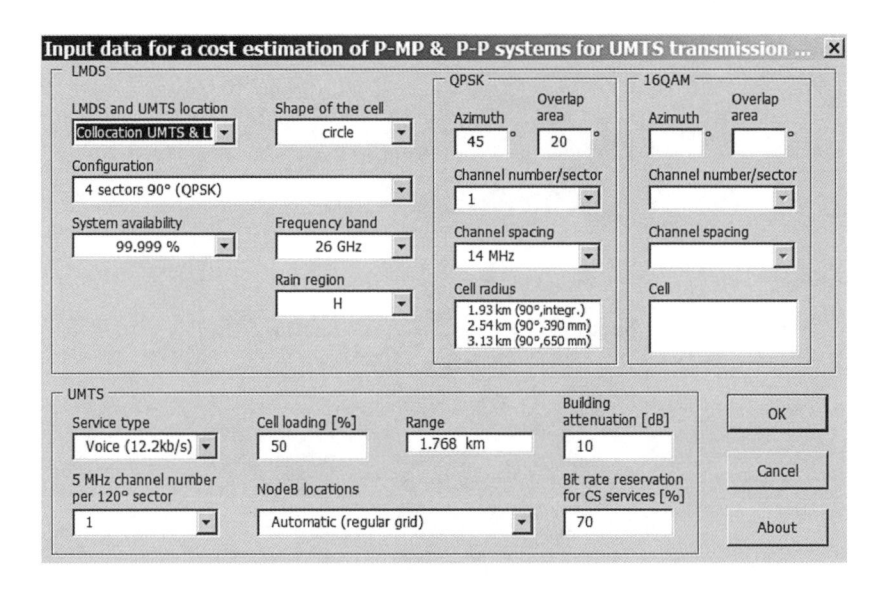

Figure 17.37 Configuration data for cost analysis.

first case, it is not necessary to use a radio data link to/from the UMTS base station, which is already at the same location as the LMDS.

The application allows also to consider the following LMDS base station configurations of a modern commercial LMDS system from one of the major manufacturers:

- base station with four sectors, all using QPSK modulation;
- base station with 12 sectors, all using QPSK modulation;
- base station with 12 sectors using QPSK modulation and four sectors using 16QAM modulation;
- base station with four sectors using QPSK modulation and four sectors using 16QAM modulation.

Choosing the configuration which uses both QPSK and 16QAM requires first to analyse the UMTS stations in the area of LMDS terminals using 16QAM (having a shorter range) and only afterwards in the area of QPSK (having a longer range). If the UMTS stations cannot be serviced by the LMDS system using 16QAM (because of the limited capacity or shorter range), then they are analysed for the possibility to be serviced by an LMDS system using QPSK.

The software application facilitates transmission loads balancing when the serviced area is divided into sectors and overlap areas are defined for adjacent sectors of LMDS. By stating appropriate angles, it is possible to independently define common areas and the angle of the first sector for each of the modulation types. Node Bs located at the sector borderline are assigned to the sector to which this borderline is to the right of the azimuth. Often, the loads of individual sectors are not equal, because of the uneven distribution of UMTS base stations within the sectors of the LMDS system. In order to overcome this, a procedure was developed allowing for sector load balancing if common areas larger then 0° are present. In the above case, stations located within predefined azimuths, excluding the common areas, are assigned to appropriate sectors. The Node Bs located within the common areas are assigned to the sector which, at the time of analysis, is less loaded. The graphical representation of the common area is shown in Figure 17.38.

The analysis can be conducted for any arbitrary number of radio channels assigned to the sectors of the LMDS system. Nevertheless, only a uniform assignment of channels is possible within a LMDS site, i.e. the same number of radio channels for each of the sectors. The selection of the number of

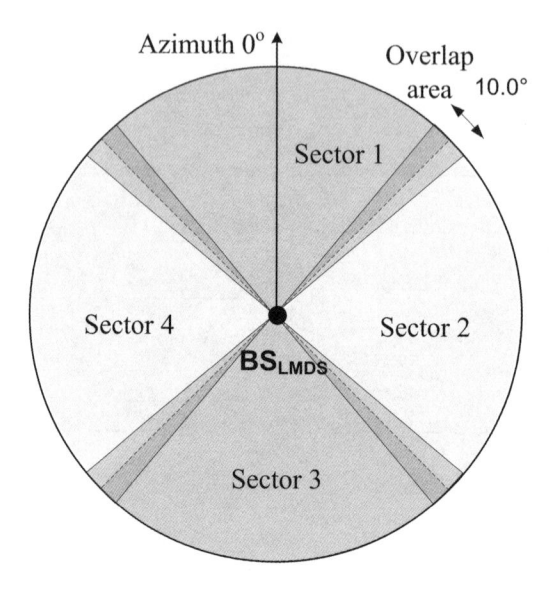

Figure 17.38 Definition of the common area and first sector azimuth.

channels is independent for systems using QPSK and 16QAM, but it must relate to the frequency band available to the operator and should rely on the guidelines for optimal use of the electromagnetic spectrum. It should be noted that, e.g., in order to have one channel in the sector of an LMDS system, the operator needs to have at least two radio channels. This rule applies when determining the extent and costs of electromagnetic spectrum utilisation.

In order for the comparison of radio resources and costs to make sense, it is also required to define the frequency requirements and the purchase costs of radio lines (PTP) required to setup links for UMTS base stations in the same configuration as defined for the LMDS system. In order to simplify the analysis, a star topology has been assumed and that each link between Node B and the UMTS RNC system is one radio link. During the analysis of the data transmission capabilities through the radio links to and from the Node B, their type is also defined for each of the links.

17.6.5 EXAMPLE CALCULATIONS AND COMPARISON OF RESULTS

The subsequent sections present some detailed case studies for cost calculations of implementing PTP and LMDS into a UTRAN backhaul network. The calculation results only take into account the real cost of equipment, excluding spectrum costs. Spectrum costs have been omitted because of differences in rentals, which depend on the country, the geographical location, service range, frequency band and assigned number of channels and their bandwidth. The calculations have been carried out for two kinds of UMTS networks. The first one is a theoretical network, which was generated by a uniform Node B deployment according to calculation results obtained for voice services. The second one is the real UMTS network, which was described and optimised in Section 15.2.

17.6.5.1 Regular Node B Deployment

All comparison results presented in this section have been obtained for UMTS network with regular and uniform triangular grid and the same size of hexagonal cells. Each Node B consists of three sectors. The Node B's requirements for the backhaul network have been calculated only for voice 12.2 kbps transmissions with a cell loading of 50%. Additionally, it was assumed that the voice service's data rate comprises 70% of the total bit rate. The remaining 30% of the total bit rate are used for packet transmissions or control data. The UMTS cell size has been varied by an additional attenuation (building attenuation) in the link budget. In all cases, only LMDS base stations working in the 26 GHz frequency band and ITU rain zone E with an availability of 99.99% have been applied.

A first analysis has been carried out for the UMTS network in non-dense urban areas, where the main problem is to have an insufficient LMDS range. In all of these three cases, the Node B's range amounts to 2.043 km. The network configurations are presented in Figure 17.39 and summarised in Table 17.21. To increase the LMDS range, the narrowest channel bandwidth has been chosen (Figure 17.39a); however, the LMDS base station capacity is poor. Thus, only less than half of all Node Bs in the LMDS range are served. These Node Bs, which are served by the LMDS base station, are drawn as hexagonals; otherwise, they are drawn as a dot. The achieved LMDS sector capacity allows to serve only two Node Bs. An increase of the LMDS capacity with additional high capacity sectors (16QAM) does not offer the required effect (case 'b'). The LMDS solution is hence more expensive than PTP radio lines, because additional equipment for LMDS base stations is needed. The limited range of the 16QAM sectors allows to serve only one or two Node Bs, and the LMDS base station capacity factor for 16QAM sectors does not exceed 30%. The cost of the LMDS solution is a little bit lower when two out of 7 MHz radio channels are assigned to a single QPSK sector. In this case (Figure 17.39c), more LMDS equipment increases the total purchase cost.

Assigning one of the 14 MHz channels to each LMDS sector, instead of 2×7 MHz, seems to be a much better solution (Figure 17.40). It allows achieving the same LMDS base station capacity as in

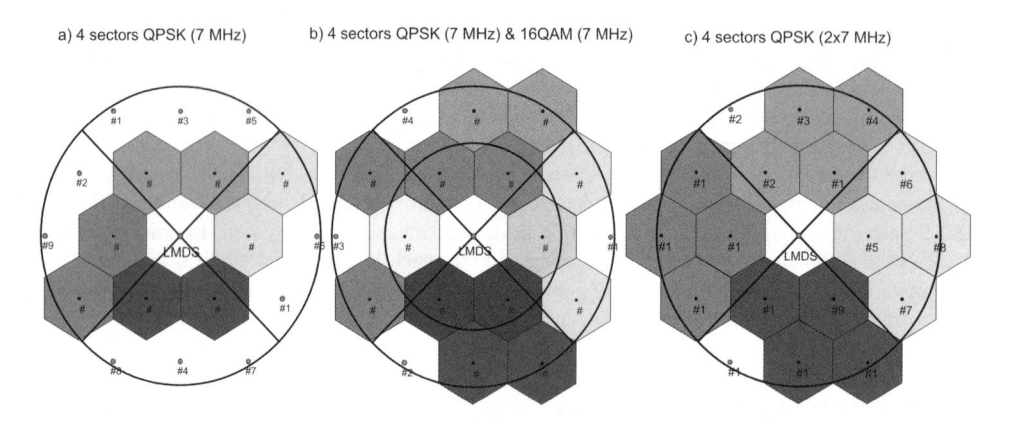

Figure 17.39 Examples of LMDS applications in UTRAN backhaul network (non-dense urban area case).

Table 17.21 Cost calculation results for LMDS and PTP applications in UTRAN backhaul network (non-dense urban area case).

Case	a		b		c	
Maximum Node B range in km	2.043		2.043		2.043	
UMTS service type	Voice (12.2 kbps)		Voice (12.2 kbps)		Voice (12.2 kbps)	
Configuration of LMDS base station	4 sect. 90° (QPSK)		4 sect. 90° (QPSK) 4 sect. 90° (16QAM)		4 sect. 90° (QPSK)	
LMDS configuration type	QPSK	16QAM	QPSK	16QAM	QPSK	16QAM
Maximum range of LMDS BS in km	7.3	—	7.3	5.10	7.300	—
Distance to the furthest serving Node B in km	6.128	—	7.076	3.538	7.076	—
Number of channels per LMDS sector	1	—	1	1	2	—
Radio channel spacing in MHz	7	—	7	7	7	—
Throughput of air interface for single radio channel in kbps	8192	—	8192	16384	8192	—
Total capacity of LMDS BS in kbps	32768	—	32768	65536	65536	—
Number of Node Bs served by LMDS	8	—	8	6	16	—
Total bit rate of served Node Bs	25894	—	25894	19421	51789	—
LMDS base station capacity factor	79.02 %	—	79.02 %	29.63 %	79.2 %	—

Table 17.21 (*continued*)

Number of not served Node Bs	10	—	4	0	2	—
Total bit rate of not served Node Bs	32368	—	12947	0	0	—
Cost of LMDS Base station	€72 853		€160 471		€138 423	
Cost of LMDS terminals	€28 752	—	€28 752	€30 222	€43 128	—
Total cost of LMDS equipment	€101 605		€219 445		€195 927	
Total cost of PTP equipment	€99 217		€173 629		€198 433	

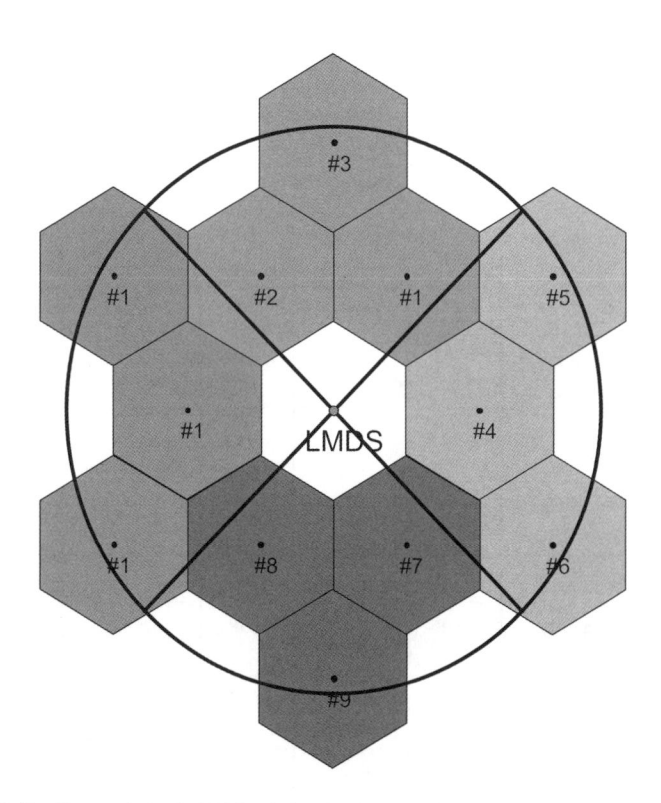

Figure 17.40 Cost-optimised LMDS solution for UTRAN backhaul network in non-dense area.

presented case 'c' with only a slightly decreased range. This solution does not require additional base station modules (compared to case 'a') and hence limits the LMDS system cost to 115 981€. This is competitive to the 148 825€ for the respective PTP radio line costs. A higher capacity allows serving 12 Node Bs; however, because of the limited range, the LMDS base station capacity factor does not exceed 60%. For this reason, it is recommended to use higher range LMDS systems working in the 3.5 and 10 GHz bands in the case of suburban and rural areas.

Above analysis shows that if there is no more than two Node Bs served with one radio channel in each LMDS sector, the LMDS based solution is not competitive to the PTP one (for an assumed LMDS equipment vendor). On the other hand, the LMDS base station capacity factor is low; thus, if remaining LMDS resources can be used for other purposes, the cost-effectiveness of this solution can be increased.

Let us also consider an LMDS system used in a UMTS dense urban environment. With reference to the results presented in Section 17.6.3, the cases with 25 % of UMTS loading and a Node B range of 929 m assuming voice services will be analysed here. The results are presented in Table 17.22 and Figure 17.41.

Usage of the 14 MHz channel does not assure the required LMDS system capacity, so as to serve all Node Bs within its range (the same situation was for non-dense urban areas). However, this solution

Table 17.22 Cost calculation results for LMDS and PTP applications in UTRAN backhaul network (dense urban area case).

Case	a		b		c	
Maximum Node B range in km	0.929		0.929		0.929	
UMTS service type	Voice (12.2 kbps)		Voice (12.2 kbps)		Voice (12.2 kbps)	
Configuration of LMDS base station	4 sect. 90° (QPSK)		4 sect. 90° (QPSK) 4 sect. 90° (16QAM)		4 sect. 90° (QPSK)	
LMDS configuration type	QPSK	16QAM	QPSK	16QAM	QPSK	16QAM
Maximum range of LMDS BS in km	6.480	—	6.480	5.110	6.480	—
Distance to the furthest serving Node B in km	5.571	—	6.433	4.825	6.433	—
Number of channels per LMDS sector	1	—	1	1	2	—
Radio channel spacing in MHz	14	—	14	7	14	—
Throughput of air interface for single radio channel in kbps	16384	—	16384	16384	16384	—
Total capacity of LMDS BS in kbps	65536	—	65536	65536	131072	—
Number of Node Bs served by LMDS	40	—	24	36	60	—
Total bit rate of served Node Bs	64736	—	38841	58262	97104	—
LMDS base station capacity factor in %	98.78	—	59.27	88.90	74.08	—
Number of not served Node Bs	20	—	0	0	0	—
Total bit rate of not served Node Bs	32368	—	0	0	0	—

Table 17.22 (*continued*)

Cost of LMDS Base station	€72 853		€160 471		€138 423	
Cost of LMDS terminals	€143 759	—	€86 255	€181 334	€215 638	—
Total cost of LMDS equipment	€216 612		€428 060		€354 061	
Total cost of PTP equipment	€496 083		€744 124		€744 124	

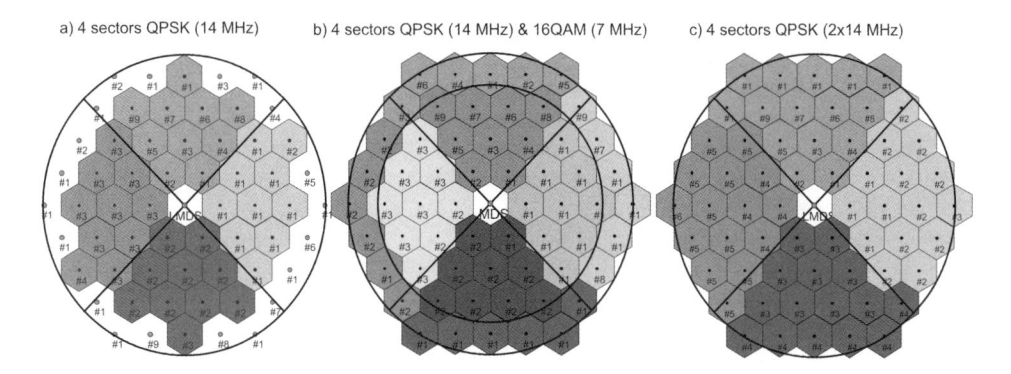

Figure 17.41 Examples of LMDS and PTP applications in UTRAN backhaul network (non-dense urban area case).

is about twice as cheap as PTP radio lines used in UTRAN. Usage of two channels per LMDS sector seems to be reasonable in the case of a uniform Node B distribution, because of the significantly lower costs of equipment purchase. Usage of 16QAM guarantees sufficient transmission reserves for potential high speed services or for utilisation by other systems belonging to the same operator (e.g. GSM, WiFi access points, etc.).

It should be noted that in analysed case (case 'b' in Table 17.22), 7 MHz channels for 16QAM have been used. As for QPSK, at least 14 MHz channels should be used in dense urban areas; transmission capacity otherwise deteriorates.

17.6.5.2 Real Network Case

To present the real case scenario, an analysis for a network based on the case study from Section 15.2 was done. Figure 17.42 presents a UMTS network layout with LMDS sectors on top of it (QPSK and 16QAM sectors). The LMDS base stations were collocated with Node Bs. The results are presented in Table 17.23. The total cost of equipment for LMDS compared to PTP microwave radio lines is about 30% lower, which is a significant gain.

In summary, use of Point-to-MultiPoint systems can give significant savings in the costs of UTRAN transmission infrastructures, particularly in urban areas. As every wireless system, also PMP must be planned carefully to achieve the best results and a high spectrum utilisation. Even greater savings can be expected when WiMAX will overcome synchronisation issues (see Section 17.5.2) and will become mature enough to be used in the same way as LMDS systems. Finally, it is clearly an interesting idea to have one wireless system (UMTS) served by another wireless system (PMP), where UMTS base

492 Optimisation

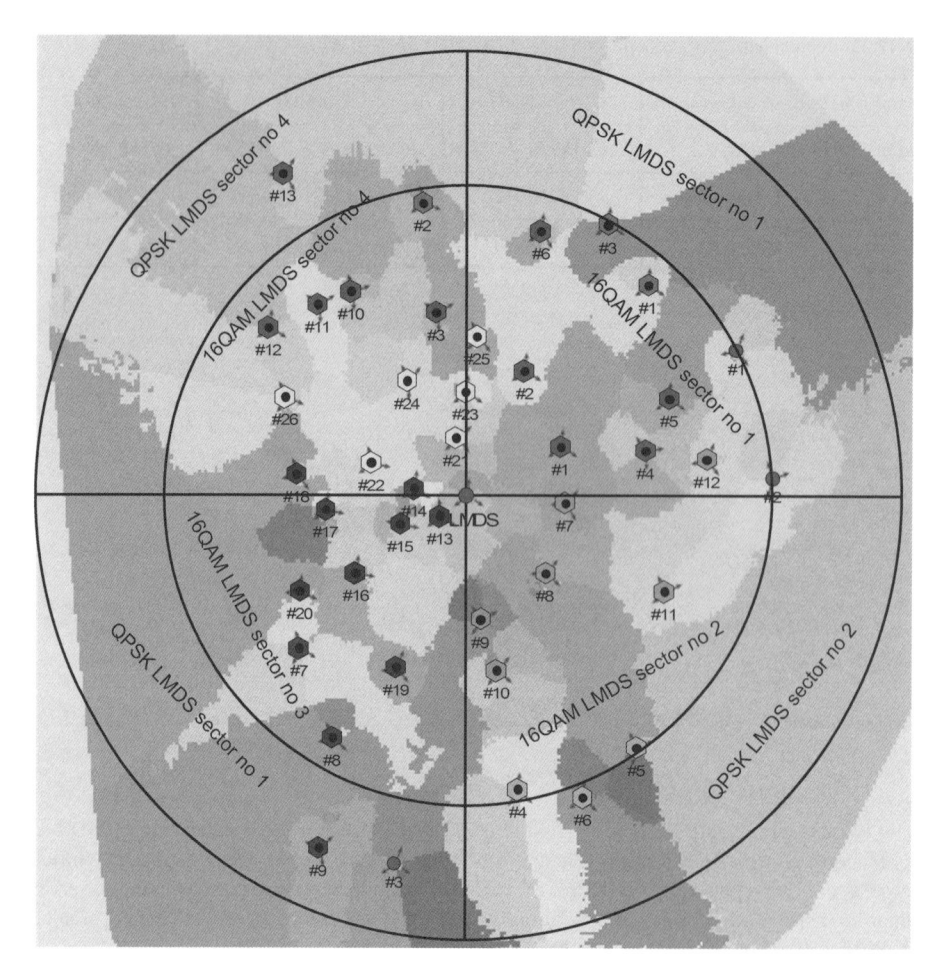

Figure 17.42 LMDS applications in real UTRAN backhaul network (collocation with Node B).

Table 17.23 Cost calculation results for LMDS and PTP applications in UTRAN backhaul network (real network case).

Configuration of LMDS base station	4 sectors 90° (QPSK) and 4 sect. 90° (16QAM)	
LMDS sector configuration	QPSK	16QAM
Maximum range of LMDS BS in km	3.99	2.79
Distance to the furthest serving Node B in km	3.446	2.474
Number of channels per LMDS sector	1	1
Radio channel spacing in MHz	14	14

Table 17.23 (continued)

Throughput of air interface for single radio channel in kbps	16384	32768
Total capacity of LMDS BS in kbps	65536	131072
Number of Node Bs served by LMDS	13	26
Total bit rate of served Node Bs	58054	122016
LMDS base station capacity factor in %	88.58	93.09
Number of not served Node Bs	3	0
Total bit rate of not served Node Bs	21983	51470
Cost of LMDS Base station	€160 471	
Cost of LMDS terminals	€46 722	€130 964
Total cost of LMDS equipment	**€338 156**	
Total cost of PTP equipment	**€490 704**	

stations become terminals for PMP. Let this idea guide the operators' transmission departments to the development of a more efficient and cheaper network infrastructure.

REFERENCES

[1] 3GPP, Technical Specification TS 25.401 'UTRAN overall description', version 3.1.0 Release 1999 by Pierre Lescuyer.
[2] 3GPP, Technical Specification TS 23.060 'General Packet Radio Service (GPRS); Service description'; Stage 2 (version 3.2.1), by Hans-Petter Naper.
[3] 3GPP Technical Specification TS 22.001 'Principles of circuit telecommunication services supported by a GSM Public Land Mobile Network (PLMN)', version 3.1.1 Release 1999.
[4] Trillium Digital Systems, Inc. 'Comparison of IP-over-SONET and IP-over-ATM Technologies', Web-version 1072006.11, November 26, 1997, 12100 Wilshire Blvd., Suite 1800 Los Angeles, CA 90025-7118, http://www.trillium.com.
[5] Brian Williams, Ericsson Australia: 'Quality of service, Differentiated services and Multiprotocol label switching', March 2000, Ericsson Inc. Datacom Networks, 77 South Bedford Street, Burlington, MA 01803 USA, www.ericsson.com/iptelephony.
[6] ITU-T G.811, Recommendation 'Timing requirements at the outputs of primary reference clocks suitable for plesiochronous operation of international digital links', Revision 1 Sept 1997 Circ 45/72 COM 13-R12.
[7] ITU-T G.812, Recommendation 'Timing requirements at the output of slave clocks suitable for plesiochronous operation of international digital links', Revision June 1998 Circ 94/126 COM 13-R26 & Amendment 1 June 2004 Circ AAP 77/79 SG15.
[8] ITU-T G.813, Recommendation 'Timing characteristics of SDH equipment slave clocks (SEC)', Revision Aug 1996 Circ 201/221/238 COM 13-R57 & Corrigendum 1 June 2005 Circ AAP 14/16 SG 15.
[9] Cisco Systems Internetwork Design Guide, ©1992–2005 Cisco Systems, Inc, http://www.cisco.com/univercd/cc/td/doc/cisintwk/idg4/index.htm.
[10] Victoria Wright, Jill Kaufmann, ATM Forum computer based training course, 1996, ATM forums Education Working Group of the Market Awareness Committee, http://www.mfaforum.org/education/atm_presentations.shtml.

[11] ICT Electronics 'ATM introduction', Application Note Nr NAATMBAS10, 1999 www.ict.es.
[12] ITU-T I.363.1, 'B-ISDN ATM Adaptation Layer (AAL), types 1 and 2 specification' New Aug 1996 Circ 201/221/238 COM 13-R51.
[13] ITU-T I.363.2, 'B-ISDN ATM Adaptation Layer (AAL), type 2 specification', New Sept 1997 Circ 45/72 COM 13-R6, Revision Nov 2000 Circ 283/15 COM 13-R73.
[14] ITU-T I.363.3, B-ISDN ATM Adaptation Layer (AAL), types 3/4 specification, New Aug 1996 Circ 201/221/238 COM 13-R51.
[15] ITU-T I.363.5, B-ISDN Adaptation Layer (AAL), type 5 specification, New Aug 1996 Circ 201/221/238 COM 13-R51.
[16] ITU-T I.375.1, 'Network capabilities to support multimedia services – General aspects', June 1998 Circ 94/126 COM 13-R17.
[17] ITU-T I.356, 'B-ISDN ATM layer cell transfer performance', Revision 2 Mar 2000 Circ 232/265 COM13-R64.
[18] ITU-T I.357, 'B-ISDN semi-permanent connection availability', Revision Nov 2000 Circ 283/15 COM 13-R78.
[19] ITU-T I.610, 'B-ISDN operation and maintenance principles and functions', Revision 3 Feb 1999 Circ 145/184 COM 13-40 & Addendum 1 Mar 2000 Circ 232/265 COM 13-R64.
[20] ITU-T G.826, 'Error performance parameters and objectives for international, constant bit-rate digital paths at or above the primary rate', Revision 3 Dec 2002 Circ A43/A45 SG11.
[21] ITU-T G.821, 'Performance objectives for international, constant bit-rate digital paths below the primary rate', Revision 1 Nov 2002 Circ B43/B45 SG11.
[22] ITU-D Study Group 2, Question 16/2 'Handbook of TELETRAFFIC ENGINEERING' Geneva, December 2003.
[23] Martin Pohl, 'Verkehrstheoretische Untersuchungen im Mobilfunknetz von O_2 Germany', Abschlussbericht (März 2005), in German.
[24] Peter McShane, Ian Horsley, 'PSAX NETWORK DESIGN AND PLANNING GUIDELINE or O2' Issue A.3, 28 May 2003, Reference LWS-O2-NDP-001.
[25] Nortel Wireless Service Provider Solutions: 'UMTS Performance Management for Access Network' UMT/DCL/DD/0021 03.08c/EN Standard September 2004. 411-8111-534.
[26] Göran Hågård and Mikael Wolf Multiprotocol label switching in ATM networks, Ericsson Review No. 1, 1998, http://www.ericsson.com/about/publications/review/1998_01/files/1998015.pdf.
[27] ITU-T Y.1411 'ATM-MPLS network interworking – Cell mode user plane interworking', Prepared by ITU-T Study Group 13 (2001–2004) and approved under the WTSA Resolution 1 procedure on 22 February 2003.
[28] Alcatel Technical Paper 'The role of MPLS technology in next generation networks', October 2000, Alcatel Compagnie Financiýre Alcatel, Paris, France.
[29] IETF: Heinanen, Juha: 'Multiprotocol encapsulation over ATM adapation layer 5', RFC 1483, July 1993.
[30] IETF: Atkinson, Randall J. Default ITU MTU for use over ATM RFC 1626, May 1994.
[31] L. Golen, Telecommunications essentials: the complete global source for communications fundamentals, data networking and the Internet, and next-generation networks, Lillian Goleniewski, Addison Wesley Publishing Company, 2002, ISBN 0-201-76032-0.
[32] Wireless Access Networks: Fixed Wireless Access and WLL Networks – Design and Operation, Martin P. Clark, John Wiley & Sons, Ltd/Inc. 2000, ISBN 0-470-84151-6.
[33] Microwave Radio Links, from Theory to Design, Carlos Salema, John Wiley & Sons, Ltd/Inc. 2003, ISBN 0-471-42026-3.
[34] EN 301 199 Digital Video Broadcasting (DVB), Interaction channel for Local Multi-point Distribution Systems (LMDS).
[35] ETSI EN 301 213-1 V1.1.2 (2002–02), Fixed Radio Systems, Point-to-multipoint equipment, Point-to-multipoint digital radio systems in frequency bands in the range 24.25 GHz to 29.5 GHz using different access methods, Part 1: Basic parameters.
[36] ETSI EN 301 213-2 V1.3.1 (2001–06), Fixed Radio Systems, Point-to-multipoint equipment, Point-to-multipoint digital radio systems in frequency bands in the range 24.25 GHz to 29.5 GHz using different access methods, Part 2: Frequency Division Multiple Access (FDMA) methods.
[37] ETSI EN 301 213-3 V1.4.1 (2002–02), Fixed Radio Systems, Point-to-multipoint equipment, Point-to-multipoint digital radio systems in frequency bands in the range 24.25 GHz to 29.5 GHz using different access methods, Part 3: Time Division Multiple Access (TDMA) methods.
[38] ETSI EN 301 213-4 V1.1.1 (2001–08), Fixed Radio Systems, Point-to-multipoint equipment, Point-to-multipoint digital radio systems in frequency bands in the range 24.25 GHz to 29.5 GHz using different access methods, Part 4: Direct Sequence Code Division Multiple Access (DS-CDMA) methods.

[39] ETSI TR 101 205 V1.1.2 (2001–07), Technical Report, Digital Video Broadcasting (DVB), LMDS Base Station and User Terminal Implementation Guidelines for ETSI EN 301 199.
[40] IEEE 802.16-2004 IEEE Standard for Local and metropolitan area networks Part 16: Air Interface for Fixed Broadband Wireless Access Systems, IEEE.
[41] Ohrtman, Frank, WiMAX Handbook, ISBN: 0071454012, 2005 McGraw-Hill.
[42] Clint Smith, John Meyer, 3G Wireless with 802.16 and 802.11, ISBN: 0071440828, 2005 McGraw-Hill.
[43] Daniel Sweeney, WiMax Operator's Manual: Building 802.16 Wireless Networks, ISBN: 1-59059-357-X, 256 pp., Published: June 2004, Apress.
[44] Ron Olexa, Implementing 802.11, 802.16, and 802.20 Wireless Network: Planning, Troubleshooting, and Operations (Communications Engineering), 2005, Elsevier.
[45] IEEE 1588-2002 Standard for a Precision Clock Synchronisation Protocol for Networked Measurement and Control Systems, IEEE.
[46] M.J. Nawrocki, T.W. Więckowski, K. Zięcina, *Systemy LMDS jako element infrastruktury UMTS*, KKRRiT, Poznań 2001 r. ISBN 83-907067-09 (in Polish).
[47] Z.M. Jóskiewicz, T.W. Więckowski, R.J. Zieliński, K. Zięcina, *Analiza możliwości zastosowania systemu LMDS do budowy infrastruktury sieci UMTS*, Krajowa Konferencja Radiokomunikacji, Radiofonii i Telewizji, Gdańsk, czerwiec 2002, pp. 151–154. ISBN (in Polish).
[48] R.J. Zieliński, Z.M. Jóskiewicz, T.W. Więckowski, *Efficient Use of LMDS in UMTS Network*, 16th International Wrocław Symposium and Exhibition on Electromagnetic Compatibility, Wrocław, czerwiec 2002, pp. 275–280.
[49] M. Gebala, M. Kutyłowski, B. Rozanski, M. Zawada, J. Vossnacker, T. Winter, *Optimised UTRAN Topology Planning Including Point-to-multipoint Equipment*, 12th GI/ITG Conference On Measuring, Modeling and Evaluation of Computer and Communication Systems, 3rd Polish-German Teletraffic Symposium.

Concluding Remarks

Was it yet another book on UMTS? Hopefully not!

We hope that you, as a reader, have gained novel insights into the behaviour of CDMA-based networks. We have tried to explain in great detail the parameter and performance dependencies within UMTS Radio Access Networks, underpinned by in-depth analyses and numerous examples. With an understanding of these analytical interdependencies, we hope you will be able to characterise and synthesise current and future CDMA-based radio communication systems, leading to better, if not optimum, wireless networks.

It was important for us to expose latest techniques related to radio access optimisation, be they automated or not. With this in mind, we reached the topic of optimisation via modelling and planning. Indeed, a typical roll-out procedure is to model part of the network first, then plan it and finally optimise it; some solutions provide planning and optimisation in one step, others do not. We also included a fairly 'rare' chapter on UTRAN backhaul planning and optimisation, which is usually neglected in books related to 3G radio design.

We hope that the exposure of both theory and examples have made and will make this book a viable and complete compendium to professional and academic network designers. The diverse background of the people having co-authored this book ensures that all herein exposed issues have been treated in a fair and comprehensive manner; their expertise guarantees that the topics are up to date and dealt with knowledgeably.

We have arranged the contributions by all co-authors in four main parts, i.e. an introductory part followed by the modelling, planning and optimisation parts. In the introductory part, the main emphasis was to acquaint the reader with basic issues related to UMTS in general. The modelling part highlighted the importance of choosing the correct model and modelling approach; we have also described a diverse range of models needed for the planning and optimisation stage(s). The planning part, being the third part in the book, has been dedicated to procedures and principles related to various planning aspects of the UMTS radio network. Finally, the fourth part is entirely devoted to numerous issues in optimisation, both general and automatic.

As for the first introductory part, Chapter 1 has exposed historical developments of radio network modelling and planning. We have endeavoured to emphasise the need for a more modern approach to

---

Understanding UMTS Radio Network Modelling, Planning and Automated Optimisation   Edited by Maciej J. Nawrocki, Mischa Dohler and A. Hamid Aghvami   © 2006 John Wiley & Sons, Ltd

the subject. We also exposed limitations of modelling tools in general and in the UMTS context in particular. In this chapter, we hence discussed the advantages and disadvantages of both manual and automated optimisation. In Chapter 2, we have introduced the principles and fundamentals of the UTRA FDD radio interface. This has been facilitated by means of a description of some general CDMA-based principles and some UTRA FDD key mechanisms. This chapter also included a first list of parameters which we deemed important for optimisation. Chapter 3 then dealt with 3G spectrum allocation and service provision, both of which need to be understood in order to comprehend the drive behind UMTS network optimisation. Finally, Chapter 4 shed some light onto the historical developments in the past and likely future related to the UMTS radio access network. We have described new developments within the 3G standardisation community, as well as their relations to the topics of this book.

As for the second part on modelling, in Chapter 5, we have given an overview of existing deterministic and site-specific propagation models. An understanding of the models has been underpinned by a description of physical phenomena, such as free-space propagation, reflection, diffraction and scattering. Equipped with these models, Chapter 6 then provided an in-depth analysis of the theoretical modelling approaches in UMTS radio network planning. It included analysis of antenna and link level modelling, as well as static and dynamic system level modelling. Chapter 7 has been dedicated to business modelling and its goals. It incorporated the 'rare' topic of how to prepare a proper business plan, how to project infrastructure developments and how to estimate associated budgets.

As for the third part on planning, Chapter 8 has concentrated on some more detailed planning issues related to the business side of mobile 3G networks. It incorporated a detailed description of the various planning processes, such as market analysis and forecasting, calculation of CAPEX and OPEX, calculation of revenue and non-technical related investments, etc. As for Chapter 9, it shed some light onto the behaviour of power limited WCDMA networks. We attempted to expose various aspects related to power, including the power dependency on the distance between mobile and base station, the load, any irregularities of cell layout, as well as the size of the actual UMTS network. Chapter 10 then dealt with the fundamentals of practical RAN design. We have hence discussed important network dimensioning metrics, such as coverage, capacity, their trade-off, etc. Chapter 11 was dedicated to the planning exercises required to properly manage compatibility issues of UMTS with other wireless communication systems, as well as within its own system. Finally, Chapter 12 has explored some specialised aspects in the radio network design, which are often neglected in the 3G planning process; in particular, it dealt with issues related to network infrastructure sharing, adjacent channel interference control and ultra high sites.

As for the fourth and final part on optimisation, in Chapter 13, we have given an introduction to problems arising in optimisation, be it automated or not and also motivations behind network optimisation. This has been further elaborated in Chapter 14, where we have dealt with the theory of automated network optimisation in great detail. It included methodologies and implementations of 'black box' approaches that iteratively determine the quality of intermediate solutions via simulation, and also alternative evaluation approaches in order to estimate performance and avoid simulations. This has been further elaborated upon in Chapter 15, where we have described some of the optimisation challenges that result from the unique deployment of UMTS either on-top an existing GSM network, or competing with GSM networks. Chapter 16 dealt with a very 'hot' topic related to an automated tuning of radio resource management parameters, which have a profound influence on the performance of the UMTS network. Finally, Chapter 17 has concluded the optimisation part with a fairly 'rare' theoretical and practical treatment on the UTRAN backhaul design by means of, e.g., PMP or WiMAX solutions.

Some parts, mainly to corroborate the understanding of background and state-of-the-art, have clearly been published before in books; however, the majority of the content of this book has never been published before in current form. We hence hope that you have benefited from an understanding and comprehension of, e.g., the following topics:

- spectrum aspects of UMTS;
- detailed propagation modelling including interference propagation modelling;
- comprehensive models of UMTS system components, including some fresh approaches to antennas, link level and static/dynamic system level simulators;
- business models and planning of network roll-out and operation;
- compatibility of UMTS systems, including cross-border coordination;
- network infrastructure sharing and the Ultra High Site concept;
- detailed theory for automated optimisation of the UMTS radio access network, including an in-depth presentation of the state-of-the-art optimisation algorithms;
- real network case studies presenting optimisation results obtained with use of exposed algorithms;
- auto-tuning aspects, which move us closer to the new world of *dynamic optimisation*;
- use, planning and optimisation of the UTRAN transmission network, including a cost comparison of selected solutions; and
- many other topics.

The main conclusion we would like you to take away after having read this book is that there is no alternative to automated network planning and optimisation solutions. Next generation networks are becoming increasingly complicated, complex and interdependent, prohibiting a human engineer to cope with all aspects of such a dynamic system. We therefore hope that this book will contribute in one form or another to the development of automated optimisation software, as well as increase the confidence of operators in automated solutions.

Writing this book was an enormous effort, since it required over 25 top-world specialists from universities, operators, vendors and consulting companies to be involved. The aim was not to write a book about all aspects of UMTS optimisation, since this is virtually impossible, but rather to have our efforts of several man-years concentrated onto dealing with important and involving issues in this area. We hope you will be in touch with us, be it in terms of feedback on the book, your own optimisation experiences or even any achieved breakthroughs in automated optimisation.

See you at **http://www.zrt.pwr.wroc.pl/umts-optimisation!**

Index

3rd Generation Partnership Project (3GPP) 37, 286, 302
802, IEEE family (.11x, .16x, .20x) 57, 64–5, 469–72

Absorption
 gaseous 101–2
 tropospheric 101
Acquisition 126–7
Activity factor 139, 146, 148, 235, 242
Adjacent channel 273, 275, 279, 282, 284, 289, 291, 293–6
 channel protection (ACP) 144, 315
 interference 143–4
 interference ratio (ACIR) 282
 radio leakage ratio (ACLR) 143–4, 282, 289
 selectivity (ACS) 143–4, 282, 289, 293
Admission control 31, 35, 50–1, 150–2, 158, 311, 405–8, 410, 418, 421, 425, 429, 438
 see also Radio Resource Management (RRM)
Algorithms
 approximation 338–9
 evolutionary 354–5
 genetic 355–75
 greedy 336, 360
 linear programming 338
 local search 336–7
 simulated annealing 337–8, 371
 tabu 337, 371
 see also Optimisation
Angle 95, 98–9, 103
 clearance 95, 98–9
 elevation 98–9, 103
 reference 99

Antenna
 base station, modelling 118
 definition 115
 electrically small 116, 117
 mobile terminal, modelling 117
 modelling 115–22
 parameters, see Antenna parameters
Antenna parameters
 azimuth 34, 212–14, 256, 260–1, 277, 279, 334–5, 339, 341, 348–53, 367, 397–400, 486
 bandwidth 117
 directivity 117, 142
 gain 117, 277, 279–80, 282, 287
 height 334, 339, 343
 polarisation 117
 radiation pattern 116
 realised gain 117
 tilt 121–2, 334–5, 339, 341–3, 348–50, 353–4, 359, 361, 363–4, 366, 369
 see also Electrical tilt; Mechanical tilt
Artificial neural network 77–80
Asynchronous Transfer Mode (ATM)
 adaptation layers (AAL) 429, 432–3, 451–2, 455
 cell outcome 437
 cell transfer delay 433, 436, 440
 circuit emulation (CES) 438, 442–3
 CS and SAR 432
 dimensioning methods 446–8, 451–2
 overbooking 431, 436, 438–9, 451
 overhead 451
 performance acc. ITU T-I610 440–1
 performance and KPI's 435–6
 PVC or SPVC 429, 431–2, 436, 438–40, 442
 service classes 429, 440

Understanding UMTS Radio Network Modelling, Planning and Automated Optimisation   Edited by Maciej J. Nawrocki, Mischa Dohler and A. Hamid Aghvami   © 2006 John Wiley & Sons, Ltd

Attenuation 141, 142, 144, 146, 147, 153, 156
 building 82, 91
 diffraction 87, 102, 105, 113
 ducting/layer reflection 102
 floor 81
 line-of-sight (LOS) 101
 path 15, 81, 83, 88, 94, 95, 101, 104–6, 136, 138, 167
 penetration 83
 propagation 81, 85, 91
 transmission 15, 101–3
 wall 82–3
Auto-tuning 405–6
 hierarchical cell structure 406
 macrodiversity 412–15
 optimisation 415–25
 parameter selection 408–9
 target selection 410
Autocorrelation function (ACF) 19, 22
 see also Function
Automatic Frequency Planning (AFP) 4
Availability, system, see System, availability
Average Revenue Per User (ARPU) 180, 189, 199, 201, 324
Azimuth, see Antenna parameters

Bandwidth conversion factor 282
Base station
 classes 222, 232–3
 location 396
Bearer 37, 46–52
 physical 140–1, 146, 150–2
 radio 43, 50
 radio access 47
 signalling radio 50
Benchmarking 181
BER 46, 48–50, 75, 162–3, 223–4, 235, 325, 464, 474
Berg model 84, 88, 89, 94
Best-server assignment 156, 210, 346, 372
Binary variable 338, 369
Blocking 275, 279–82, 284, 289, 291–2, 402–3
Body loss 147
 see also Propagation
Bottleneck cells 346–7, 371
Branching station, see Station
Break point 84, 85, 88, 108–9, 110
Broadband global access network (BGAN) 59–60
Broadcasting Satellite Service (BSS) 273, 289, 292
Budgeting process 177, 179
Business analysis 178
Business modelling 177–83
Business plan 177–9, 181, 190, 196–201

Cable
 loss 226
 system 223, 226
Capacity 210–17, 334, 339, 341, 343, 344, 345, 346, 370, 396, 401–2, 452, 456–60, 462–6, 472–82, 486–91, 493
 hard 249–50
 loss 210, 212
 prediction error 215–17
 soft 31, 250–1
 transmission 452, 465, 478–9, 485, 491
Capital expenditure (CAPEX) 179–81, 187, 189–200, 310, 314–15, 334, 340, 348, 457, 468, 472–3
Carrier spacing 41
Carrier-to-interference ratio (CIR) 123, 135, 138, 139–42, 145, 147–57, 160, 266, 367, 371–2
CDMA2000 55–6, 58
Cell 272, 275, 284, 286–7, 289, 295–6, 298–300
 breathing 29
 hexagonal 4–5
 hierarchical cell structure (HCS) 263, 272
 irregular 203
 macro 263, 287, 295–6, 298
 micro 263, 287, 293, 296, 298
 pico 287, 298
 triangular 4
Cell breathing 29, 60, 239, 242, 262, 325, 446
 see also Cell
Cell coupling 214
Cell load 149, 154, 158, 215, 237, 239, 245, 344, 346, 373, 398, 402
Channel 272–3, 275, 279, 282, 284, 286, 289, 291, 293–6, 298, 300–1
 access indicator (AICH) 26
 bandwidth 458, 462, 464, 471, 479, 480, 487
 broadcast (BCH) 25
 common control physical (CCPCH) 25
 common pilot (CPICH) 23, 25, 126–8, 140, 145, 149, 210, 245–8, 327–9, 339, 344–6, 364, 370, 389–94, 407
 control 144–6, 208–10
 dedicated physical (DPCH) 27
 dedicated physical control (DPCCH) 28
 dedicated physical data (DPDCH) 28
 forward access (FACH) 25
 high-speed dedicated physical control (HS-DPCCH) 29
 high-speed physical downlink shared (HS-PDSCH) 28
 high-speed shared common control (HS-SCCH) 27
 paging (PCH) 25
 paging indicator (PICH) 26
 physical 15, 20, 27
 physical downlink shared (PDSCH) 28

physical random access (PRACH)   27
  shared   144–5, 152
  spacing   464–5, 475–7, 488, 490, 492
  synchronisation (SCH)   22
Channel bandwidth   135
Channel element   147, 150
Channel estimation   127–8
Channel estimation error   172
Channel raster   41
Channelisation code   15
  see also Code
China Wireless Telecommunications Standard
    (CWTS)   57
Chip rate   275, 287–9, 300
  high (HCR)   275, 287–8
  low (LCR)   275, 287–8
Circuit switched   22, 48, 146, 194, 428–32,
    446–9, 456
Clutter   94–5, 98, 101, 105
  local   101, 105
  nominal   101
Co-channel interference   286, 295–6, 371, 466
Co-location   291
Code   279, 297–8, 300–4
  channelisation   15
  coordination   297, 300–1, 304
  Gold   19
  group   301–4
  orthogonal variable spreading factor (OVSF)   15
  preferential   298, 300–4
  primary synchronisation   22
  scrambling   15, 26, 297, 300, 302
  spreading   11
  Walsh   15
Code Division Multiple Access (CDMA)   4–5, 11–12,
    56, 59, 75, 134, 205, 214, 287, 343, 463
  see also Multiple access
Coding   464–5
  convolution   464
  forward error   464
Combinatorial optimisation, see Optimisation
Common Pilot Channel (CPICH), see Channel,
    common pilot (CPICH)
Compressed mode   33
Conductivity   72
Conference Preparatory Meeting (CPM)   38
Connection Admission Control   438
Controller, see Fuzzy Logic Controller
Conversational speech traffic   166
  see also Traffic
Coordination   275, 279, 284–6, 289, 293–304
  distance   293–4
  time slot   297
  zone   294–5, 303
Correlation length   142

Cost, network   339, 346, 348
COST 231 Indoor   81
COST 231 Walfisch–Ikegami   84, 85–7,
    88, 91, 92
Coupling matrix   154–5, 157, 372–4
Coverage   142, 149–50, 152, 334, 339, 341, 343–5,
    366, 368, 373, 380–2, 383, 386, 396, 398–403
  E_c/I_0 398, 400–1
  holes   214
CPICH, see Channel, common pilot (CPICH)
Cross correlation function (CCF)   16, 126–7, 169
  see also Function
Crossover   355, 357
Cyclic redundancy check (CRC)   123

Data source model   146
Deadzone   316
Delay spread   74–5, 89, 141, 234, 262
Delivery Duty Paid (DDP)   196
Delivery Duty Unpaid (DDU)   196
Despreading   11, 32, 127, 171, 315
Diffraction, theory of   71–3, 76, 85, 87–8, 90
  geometrical, GTD   72
  uniform, UTD   72
  see also Loss
Digital Elevation Model (DEM)   6
Digital Enhanced Cordless Telecommunications
    (DECT)   39, 41, 273, 284, 292–3
  Cordless Terminal Adapter (CTA)   292–3
  Fixed Wireless Access (FWA)   293
  portable profile (PP)   292–3
  Radio Fixed Profile (RFP)   292–3
  Wireless Local Loop (WLL)   292–3
Directional antenna, see Antenna parameters,
    directivity
Distribution, see Probability distribution
Diversity, see Macrodiversity
Dominant path   80, 81, 82, 83
Driver   181, 187, 190, 191, 194–8
Dual technology networks   382
Duplex   40–4
Dynamic channel selection   293
Dynamic model   161–72
  see also Model

Earth radius
  effective   101–3
  true   102
EBIDTA   200
Efficiency, spectrum   463–4, 474, 476
Electrical tilt   341, 348, 366, 396–7, 399
Electronic Communications Committee (ECC)   37–8,
    41, 43, 272–3, 294–5, 299

Emission 274–5, 279–84, 289, 291–2, 294–5
  fundamental 279
  mask 284
  out-of-band 279, 291, 294–5
  parasitic 279
  spurious 275, 279, 289, 291–2
  unwanted 280–4
Enumeration 336
Environment
  adaptation 78, 91
  indoor 73, 80, 81, 93
    see also Model, picocell
  Manhattan 87–8, 94
    see also Berg
  microcellular 84, 85, 88, 89
    see also Model, microcell
  MSE 80
  open 91, 100, 108
    see also Modified Hata
  outdoor 81, 82–3, 93, 94, 104
    see also Model, microcell, macrocell
  rural 74, 79, 90, 98, 105
    see also Modified Hata
  SOHO 80
  suburban 87, 90, 91, 95, 98, 106
    see also Modified Hata
  urban 73, 84, 85, 91, 95, 100
    see also Model, microcell
European Common Proposals (ECPs) 38
European Conference of Postal and
  Telecommunications Administrations (CEPT)
  37, 41, 43, 94, 104, 271, 279, 284–5, 292, 295–6,
  299, 301–3
European Radiocommunication Committee (ERC)
  37–8, 41, 272–3, 286, 292–3, 299
European Radiocommunication Office (ERO)
  272, 284
  Frequency Information System (EFIS) 272
European Space Agency (ESA) 59
European Telecommunication Standard Institute
  (ETSI) 37, 271, 279, 284–5
Evolution Strategies 355, 357, 359
Evolutionary Algorithms 354, 357, 374
Exchange rate 190, 201

Facility location 369
Factor
  capacity 487–90, 493
  efficiency 479
  reuse 466
  transmission density 482–3
Fading 94, 100, 104
  fast 141, 145, 147, 236
  fast, margin 236
  slow 240
  slow, margin 240
  slow, standard deviation 240
FDD 448, 463, 470
FDMA 463, 465
FEC, see coding, forward error
Fixed Service (FS) 272–4, 295
Frame Erasure Rate (FER) 32, 381–2, 384, 387, 390,
  391, 393
Frequency 94–5, 99–102, 271–80, 282, 284–7,
  290–301, 305–6
  adhering 275–6, 290–1
  coordination 297–9, 301
  nominal 95, 99
  offset 279, 282
  preferential 296–301
  range 94, 100
Frequency band 271–5, 278, 280, 286, 289–97, 300
  additional 38–9, 46
  adjacent 273–5, 282, 286
  core (base) 39–41, 271–3, 275, 290, 292, 297, 300
  extended 39, 42–3, 271–5, 286, 291, 295–6
  future 38
  guard 275, 278, 280, 290–2, 296
  satellite component 39
  terrestrial component 39–43
Frequency block 41, 43
Frequency Division Multiple Access (FDMA) 4–5, 11
  see also Multiple Access
Fresnel 70, 72, 85, 101
Friis propagation formula 70, 228
Function
  autocorrelation (ACF), see Autocorrelation function
  cross correlation (CCF), see Cross correlation
    function
Fuzzy Inference System 410
Fuzzy logic 410
  fuzzy set 410
  member function 410
Fuzzy Logic Controller 411–21
  decision matrix 411–12
  defuzzification 411–12
  fuzzification 411–12
  inference 411–12
Fuzzy Q-learning 421, 423–4

Gain
  antenna 79, 101, 105
    see also Antenna
  floor height 83
  height 83
Generation 355–6, 360
Genetic algorithms 355, 357, 374
Geographical Information System (GIS) 7, 9
Geosynchronous Orbit (GEO) 59

Index

Global optimum 336, 338, 352–4, 375
Global Positioning System (GPS) 58, 105
Globalstar 58
Gold code 19
   see also Code
Grade of service 361
Greedy algorithms, see Optimisation
Greenfield operator 189, 190, 194–9
Grid
  hexagonal 4–5
  irregular 210–12
  triangular 4
  see also Cell
GSM
  frequency assignment 5, 380–1
  handover planning 381
  Hierarchical Cell Structure (HCS) 380–1
  planning 3, 5, 343, 380

Handover
  planning 381, 383–8, 393
  SHO probability 212
  soft handover 32, 147–8, 152, 234, 339, 344, 347
  soft handover gain 147, 234
  softer handover 143, 147–8, 234
Hard blocking 382
Height of antenna 98–101, 103, 105
  effective 98
  reference 98
Heterogeneous network 64
Heuristic 336, 339, 367, 369, 371
Hierarchical Cell Structures (HCS) 263–8
High Altitude Platform System (HAPS) 286
  see also System
High Earth Orbit (HEO) 59
High-Speed Downlink Packet Access (HSDPA) 27, 28, 29, 62–3, 65, 162, 178, 288, 381, 446
High Speed Uplink Packet Access (HSUPA) 63
Hockey stick curves 180–1
Horizons 59
Hot spot 25, 55–7, 255, 263–4, 380, 381, 386
HSDPA, see High-Speed Downlink Packet Access
HSUPA, see High Speed Uplink Packet Access

IMA 436, 442
Image method 75–6, 79
Impedance 71, 76
IMT-2000 37–46, 55–9, 61, 62, 94, 271–4, 288–9, 294–6
Incumbent operator 178, 189, 190, 194, 195, 197–9
Individual 354–5, 359, 361, 363
Inmarsat 59–60
Inter-cell interference 205, 207, 215, 217
  see also Interference

Interference 271, 273–80, 282–7, 289, 291–302, 446, 466, 474, 476
  adjacent band 275
  adjacent channel 275, 279, 282, 284, 291, 476
  aggregated 278–80
  co-channel 279, 284, 286, 295–6, 466, 476
  inter-cell 205, 207, 215, 217, 236
  intra-cell 205, 207, 215, 217, 236
  margin 237
  mutual 271, 274, 291–2, 295, 301
  other-to-own cell 236
  scenario 275–80, 283–5, 290–2, 295
  signal 277–8, 283–4, 293
  situation 278–80, 284–5, 299–300, 302
  threshold 280, 282, 289
Interference raise 146, 204
Interference Rejection Combining 131
Intermodulation 275, 279, 284, 291
Internal Rate of Return (IRR) 200
International Telecommunication Union (ITU)
    37–45, 271–3, 278–9, 283, 285–6, 292, 294, 298
  Radio Regulations (RR) 272, 279, 294, 296
  Radiocommunication Sector (ITU-R) 37–8, 41–5, 94, 100–2, 271, 273, 279, 285–6, 294, 298
  Telecommunication Standardisation Sector (ITU-T) 37
Intersymbol interference (ISI) 74, 75
Intra-cell interference 205, 207, 215, 217
  see also Interference
IP
  IPv6 453–4
  migration to IP 444
  transport protocol types 445–6
Iridium 59
Irregular base station distribution 210–12, 214
Isolation 277, 280, 282–3, 294–5
  adjacent band 282
  loss 282–3

Knapsack 370
  see also Multiple knapsack

Label switched router 443
Leased line 430, 436, 456–7, 466
Line-of-sight (LOS) 74, 83, 86, 94, 101–2, 458, 460, 462, 468, 475
Linear (integer) programming 338–9, 371, 373–5
Linear minimum mean square error (LMMSE) 131, 133
Linear solvers 159–61
  Krylov subspace projection method 159
  preconditioning 159, 161
  stationary iterative method 159

Link budget 6, 117, 147, 239, 245–8, 299, 368, 477, 485
Link level model 7, 122–34
  see also Model
LL, see Leased line
LMDS, see Local Multipoint Distribution System (LMDS)
LMMSE equaliser 64
Load 396–403
Load control 149, 152, 157, 158, 408
Loading
  downlink 242–4
  uplink 242
Local Multipoint Distribution System (LMDS) 4, 460–8, 474–93
  see also System
Local optimum 336, 337, 352
Local search 336–7, 367, 372–5, 397–403
Long-term plan (LTP) 179
LOS, see Line-of-sight
Loss, see Attenuation
Low Earth Orbit (LEO) 59

Macrodiversity 33, 407
  add_win 407, 413–15, 418, 421
  drop_win 407, 413, 415, 418, 421
  ping pong effect 410, 413
  rep_win 407
Market forecasting 187–9, 198
Market share 189, 199
Markovian Decision Problem 422
Matched filter 126–8, 222
Maximum likelihood 128
Maximum ratio combining (MRC) 128–30, 143, 147
MC-CDMA 463
Mechanical tilt 342, 348, 366, 397
  see also Antenna
Medium Earth Orbit (MEO) 59
Mid-term plan (MTP) 179
Minimum
  attraction region 349, 352–4
  distance to global 352–4
  global 351–3
  local 349, 351–3
  number of 351
  see also Global optimum; Local optimum
Minimum Coupling Loss (MCL) 280–1, 283–5, 291–2, 294
Missed traffic 150, 152, 400, 402
Mixed integer program (MIP) 397–403
Mobile Satellite Services (MSS) 39, 58–9, 193, 271–5
Model
  deterministic 70, 75, 79, 80, 84, 90
    see also Image method; Ray launching; Ray tracing

empirical 78, 80, 81, 85, 90
link level 124–6
macrocell 90, 105
microcell 84, 88, 93
mobility 164
picocell 80
system level, dynamic 161–72
system level, static 139–61
traffic 165
see also Propagation model
Modified Hata model 90–1, 104, 105
Modulation 460–1, 463–5, 467, 470, 471, 475–7, 480, 482, 484, 486
  PSK 463
  BPSK 464, 470, 471
  QPSK 464, 467, 470, 471, 476–7, 479–82, 486–8, 490–2
  DQPSK 464
  8PSK 464
  4QAM 464, 465
  16QAM 464, 465, 467, 470, 476–7, 479–82, 486–8, 490–2
  64QAM 464, 465, 470
Monte Carlo (MC) 105, 139, 141, 149, 151–2, 169, 280, 283–4, 291–2, 349, 351–3, 366, 370, 372, 374
Motley–Keenan 80, 94, 105
  see also Model, picocell
MPLS 443–4
Multi-Wall Model (MWM) 81, 105
Multipath component (MPC) 123–4, 127
  see also Propagation
Multipath interference canceller (MPIC) 134
Multiple Access
  Code Division Multiple Access (CDMA) 4–5, 9, 11
  Frequency Division Multiple Access (FDMA) 4–5, 11
  Time Division Multiple Access (TDMA) 4–5, 11
Multiple-input, multiple-output (MIMO) 62, 79, 128–34
Multiple knapsack 370
  see also Knapsack
Multipoint Multimedia Distribution System (MMDS) 272–4, 295–6
Mutation 354–5, 357–8, 360, 364
MVNO 178, 313

National roaming 309, 311–15
Neighbourhood 336–7, 375
Net Present Value (NPV) 200–1
NETCO 313
Network
  backbone 457
  backhaul 430, 442, 457, 458, 459, 460, 468, 474–6, 479, 487–92
  dimensioning 178–9, 189–94

Index 507

roll-out 178, 189–91, 193
size, minimum required 214–18
statistical data 326, 328–30
synthesis/analysis 323–4
Network dimensioning, *see* Network
Network roll-out, *see* Network
Network sharing, *see* Sharing
Network size, *see* Network
Noise
　floor 142
　thermal 142, 205–7, 210
Noise figure 223, 227–9, 250
　base station 224
　terminal 227
Noise rise 30, 150, 237
Non Line of Sight (NLOS) 74, 83–4, 86, 94
Non-real-time traffic 265
NP-hardness 335, 339

Objective function 335, 345–54, 358, 363, 373
　properties 349–54
Operational expenditure (OPEX) 179–81, 187, 192, 196–9, 310, 315, 334, 340, 348, 457, 472, 473, 498
OPEX 179–81, 187, 192, 196–9, 310, 315
OPEX, *see* Operational expenditure (OPEX)
Optimisation
　adaptive 420–1
　automated 323, 334, 339
　benchmark 326–31, 347, 382
　case studies 382–403
　combinatorial 335–6, 338–9, 367
　constraints 334, 338, 345, 367, 369–71, 373, 374, 386
　goals 381, 383, 386, 390–1
　greedy algorithms 335–6, 360, 369, 371, 374–5
　loaded network 383, 386, 389–95
　local 354, 361, 363–4
　manual 323–4
　mono-objective 417–18
　multi-objective 418–20
　off-line 416–21
　on-line 421–5
　parameters 323, 333, 339–45
　sub-optimum solutions 352–3
　targets 345–54, 368
　techniques 335–9, 354–75
　TRIBBES program 420
　unloaded network 382, 383–9
Orthogonal Frequency Division Modulation (OFDM) 64, 469, 470
Orthogonal variable spreading factor (OVSF) 15, 18–20, 232, 242

Orthogonality 135, 140, 141, 149
　factor 20, 141, 170, 232–4, 238, 243, 250
Other-to-own cell received power ratio 150, 152, 215, 217

Packet scheduling 51, 63, 408
　opportunistic 263–8
Packet switched 146
Particle swarm 416–21
Pathloss
　average 136
　worst case 138
　see also Loss
Payback period 200
Peak financing need 200
Penetration, building 82, 84
Permittivity, dielectric 82
Personal Communication System (PCS) 41
　see also System
Personal Handyphone System (PHS) 41
　see also System
Pilot
　pollution 150, 334, 339, 344, 347, 393–4
　power 334, 344–6, 366, 370
　see also Channel
Planning
　automated 7–9
　manual 7–9
Planning tool 3, 7, 9
PMP, *see* Point-to-MultiPoint
PNNI 430–1, 436, 438–9, 442, 446, 451
Point-to-MultiPoint 4, 429, 460–8, 471, 474–6, 482, 484, 491, 493
　see also System
Point-to-Point 457–9, 460, 474, 475, 478, 482, 485, 487–93
　see also System
Polarisation 70, 72
Pole capacity 136, 208, 210, 211, 214–17
Population 354–5, 357, 374
　working 358–9, 360
Power
　allocated to users 136
　of interfering signals 135
　terminal location dependant 203–6
　total 208–10, 212–13, 216
　traffic load dependant 204–5, 207–10
　transmit available 136
　transmitted by the base station 204–6
　transmitted by the mobile station 203–4, 211
Power control 31, 128, 140, 142, 144–5, 152, 158
　at cell level 152, 158
　headroom 236
　perfect 145
Power control range 204–6

Power delay profile (PDP)   141
Power spectral density (PSD)   276–80, 289
  aggregated   278–9
Primary synchronisation code   22
  see also Code
Probability distribution
  Chi-squared   266
  Gaussian (normal)   94, 100, 102, 104–5
  lognormal   94, 104–5
  Nakagami   94
  Rayleigh   94
  Rice   94
Processing gain   14
Propagation   69–111, 276–7, 280, 282–6, 291–2, 294, 296–8, 302
  body loss   147
  diffraction sub-path   101–2
  free-space   70, 73, 85, 88, 91, 95
  line-of-sight (LOS)   94, 101–2
  model (method)   277, 280, 282, 284–6, 291–2, 294, 297
  multipath   73, 75, 81, 93, 94, 100, 143, 169
  non line-of-sight (NLOS)   94
  path   276, 280, 284
  pathloss   136, 138
  phenomena   70, 75, 94
  shadowing   94, 104, 141–2, 151
  trans-horizon   101–3
  urban/suburban   95, 99
  see also Loss
Propagation curve   95–8
  nominal   95
Propagation model
  dual slope   88, 108–10
  ITU-R 1546   94–100
  ITU-R 452   100–4
  modified Hata   104–5
  Multi-Wall   105
  single slope   92, 106–10
  tuning algorithm   106–8
Propagation scenario
  indoor–indoor   104–5
  indoor–outdoor   104
  outdoor–outdoor   94, 104
Protection ratio   282
PTP, see Point-to-Point

Q-learning   421–3
Quality factor   364
Quantitative business models   181

Rack sharing, see Sharing
Radiation pattern
  modelling   117–22

multiplication   119
role of separation   119, 120
Radio Astronomy Service (RAS)   273–5
Radio channel   69
  coherence bandwidth   74
  delay spread   74–5, 89
  mean excess delay   74
  narrowband   75
  Power Delay Profile (PDP)   74, 89
  wideband   69, 73–5, 81, 89
Radio Resource Management (RRM)   50, 405–9
  Admission Control   50, 51, 54
    see also Admission Control
  handover   46, 51, 54
  joint-RRM   425
  load/congestion control   51, 54
  packet scheduling   51, 54
  power control   51
Radio resources   146–7, 149–50, 158
Rake receiver   143, 231–2
RAN sharing, see Sharing
Range
  LMDS   460, 464, 466–7, 478–82, 487, 488, 490, 492
  Node B   474, 487–8, 490
  PMP   462, 465
  radio relay   458–9
Ray launching   76–7, 79, 80, 81
Ray tracing   70, 75–6, 77, 79, 80
Rayleigh criterion   71
  see also Variations
Receiver   275–85, 288–9, 292, 294–8
  affected   276–8, 282, 284, 297
  noise   280, 288–9
  selectivity   277, 279
  victim (interfered with)   280–3, 284–5, 292
Recombination   357, 359, 361, 364
Regional Radiocommunication Conference (RRC)   272
Reinforcement learning   421–2
Reliability   458, 464, 474
Return of Investment (ROI)   200
Revenue   187–8, 198–9, 201
RNC   428, 429, 431, 458–60, 472, 475, 479, 487
Root-raised cosine, see Matched filter
RRM, see Radio Resource Management (RRM)

Sanity checking   197–8
Satellite
  component   38–9, 45–6, 273, 274, 275, 294
  Digital Multimedia Broadcasting (S-DMB)   294
  Radio Interface (SRI)   294
  segment of IMT-2000   58–60
  see also Mobile Satellite Services, Spectrum

Index

Scrambling code 15, 26
  see also Code
SDH 436, 438, 455, 457
Sea-level refractivity 102
Sectorisation 396, 399
Selection 354–6, 358–9, 363
  pressure 360, 363
Self-adaptation 359–60
Sensitivity analysis 190, 199, 201
Separation 271, 277–80, 282–6, 291, 293–6, 298
  channel 293, 298
  distance (geographical) 277–8, 280, 282, 286, 291, 294–6, 298
  frequency 278–9, 282, 284, 286, 291, 296
Service Level Agreement 381, 457
Service Level Guarantee 457
Serviced area 486
  shape definition 120–2
Services 46–52, 140, 146, 149, 151, 158
Serving cell 344, 369
Set covering 368–9
Shadowing 94, 104, 141–2, 151, 168
  see also Propagation
Sharing 271–5, 280, 286–7, 289, 292, 295
  matrix 272–4, 292
  network infrastructure 309–15
  rack 310
  RAN 309–11, 314–15
  site 309–10, 314–17
Signal-to-interference ratio (SIR) 14, 170
Simulated annealing 337–8, 371, 374–5
Simulator 415
  dynamic system level 161–72
    see also Model
  semi-dynamic 413, 415
  snapshot 415
  static system level 139–41
    see also Model
Site 191–2, 194–8
  location 254–6, 334, 340
  sharing, *see* Sharing
SLA, *see* Service Level Agreement
Slant polarisation
  X-pol 117
SLG, *see* Service Level Guarantee
Smart antennas 141, 158, 160–1
Snapshot 140, 145–6, 149–53, 159–60, 346, 351–3, 370, 372
Soft capacity 31
  see also Capacity
Soft(er) Handover, *see* Handover
Space Service (SS) 273, 292
Spectrum 271–2, 278–80, 282, 284, 296, 301
  additional 38, 46
  amount 38

demand 45–6
requirement 37–9
unpaired 44
Spreading 11
Spreading code 11
Standard deviation
  received interference power 208–9
  transmit power 208–9
Static system level model 139–61
  see also Model
Station
  branching 459
  central 462, 463, 476
  intermediate 459
  terminal 458, 459, 460, 462–6
Stochastic reliability 152
Subsidies 190, 198–9
SWOT analysis 182
Synchronous uplink 58
System
  availability 457–8, 460, 463, 465, 467, 474–5, 477, 481, 484, 487
  coverage 465–6, 470
  fixed-wireless 458
  High Altitude Platform System (HAPS) 61, 286
  Local Multipoint Distribution (LMDS) 4, 460–8
  Personal Communication (PCS) 41
  Personal Handyphone (PHS) 41
  Point-to-Multipoint 4, 460–8
  point-to-point 5, 457–60
  radio relay 458
System capacity
  power-limited 134–9
  single cell, downlink 136
  single cell, uplink 138
  see also Capacity
System modelling 187, 189–90

Tabu list 337, 375
Tabu search 337, 371, 374–5
TDD 448, 463, 470
Terminal power class 227
Throughput 346–7, 457, 460, 475
  dimensioning 475
Tilt, *see* Antenna
Time division multiple access (TDMA) 4–5, 11, 463, 465
  see also Multiple access
Traffic
  conversational speech 166
  load analysis 330, 390
  video streaming 166
  web browsing 166

Traffic classes   48–50
  Background Class   49
  Conversational Class   48
  Interactive Class   48, 49
  Streaming Class   48
Traffic distribution
  non-uniform   212–13, 215
Transmission dimensioning
  of ATM-layer   438–9, 451–2
  of Ethernet and Sonet   455–6
  including IP-Network protocols   452–56
  at Node B Iu-interface   446–51
Transmit power limits   142, 149–50, 158
Transmitter   275–85, 287, 289, 291–2, 294–5, 297
  interfering   277, 280–5
  mask   279, 282

Ultra High Sites (UHS)   61, 318–20
UMTS, see Universal Mobile Telecommunications System (UMTS)
UMTS Forum   37–8, 45–6
UMTS Terrestrial Radio Access (UTRA)   41, 272–5, 286, 296–301
Universal Mobile Telecommunications System (UMTS)
  coverage   379, 380–1
  fee   190, 195, 196
  handover planning   381
  license   177–8
UTRAN   428–30

Value chain   182
Variability   95, 100, 106
  location   95, 100
  time   95
Variations   94, 100, 104
  Gaussian   83, 90, 100
  Rayleigh   71, 83, 94
Vertical polarisation, see Antenna parameters
Video streaming traffic   166
  see also Traffic
Voice over IP (VoIP)   48, 63, 446

Walsh code   15
Wave, electromagnetic
  equation   69
  reflection, specular   70–1, 75, 76, 84, 102
  scattering   71, 75, 89, 100, 101
  transmission   70, 75, 76, 101, 103
  see also Attenuation
Web browsing traffic   166
  see also Traffic
Weighted Average Cost of Capital (WACC)   190, 201
WiMAX   468–74, 475, 476, 482, 491
  synchronisation   473–74
WLAN   56, 57, 64, 65
WLL   56, 292–3, 468
Working population, see Population
World Administrative Radio Conference (WARC)   38–40
World Radiocommunication Conference (WRC)   38–40, 43, 46, 272